Solar Thermal Energy Systems

Werner Platzer · Robert Stieglitz

Solar Thermal Energy Systems

Fundamentals, Technology, Applications

Volume 1

Springer

Werner Platzer
Umwelt und Natürliche Resourcen
Albert-Ludwigs-Universität Freiburg
Freiburg im Breisgau, Baden-Württemberg,
Germany

Robert Stieglitz
Institut für Neutronenphysik und
Reaktortechnik (INR)
KIT
Eggenstein-Leopoldshafen
Germany

ISBN 978-3-031-43175-3 ISBN 978-3-031-43173-9 (eBook)
https://doi.org/10.1007/978-3-031-43173-9

This Springer imprint is published by the registered company Springer Nature Switzerland AG
The registered company address is: Gewerbestrasse 11, 6330 Cham, Switzerland

Paper in this product is recyclable.

Before the final publication of this textbook, my highly esteemed co-author Prof. Robert Stieglitz, who was indispensable and tireless in the production of the book, passed away unexpectedly. He was a great person and an internationally recognized scientist who excelled in his broad knowledge of materials science and renewable energy. His loss affects not only his family, but also his colleagues and the scientific community as a whole.

Preface

Against the background of mankind's fight against climate change and its consequences, technologies and systems to provide heat, cold and electricity from the sun play a key role to support the necessary transformation of our energy systems towards a truly renewable and sustainable one. In solar thermal technologies, solar electromagnetic radiation is primarily transformed into heat, which may serve as a second step to provide us with heat and cold for thermal applications, but also electricity, chemical reactions, and fuels. The aspect of conserving energy and resources is gaining particular importance when introducing new technologies. Solar thermal systems allow the provision of a wide spectrum of temperature ranges for different purposes by means of suitable collectors. The scope of heat applications exceeds by far the use for room heating and hot water preparation in buildings. The use of solar heat for cooling and refrigeration systems, the integration into heat networks, the customized provision of industrial process heat for all kinds of industries as well as the conversion into electricity in thermal power plants are equally important applications in our current energy matrix. Although the technologies and systems are mature, future optimization using digitalization, new sustainable materials and integrated system designs will revolutionize current products on the market. Considering the technological development, the situation could be compared with that of the car industry before World War II and now. The available solutions already show great potential, but missing market capacity has not yet allowed to develop designs with minimum material content and based on real mass production. On the other hand, raw materials exploitation status and energy and climate-friendly production will force us to improve our designs soon.

This script is addressed to technically interested readers, as well as students in the Bachelor's and Master's area up to interested technically educated career changers, and is modularly structured in two parts. In the first part, the basics are worked out and the conversion of solar radiation into heat is presented, while the second part shows and discusses the applications in building technology, heat process technology, solar thermal power plants as well as solar thermal air conditioning.

Part I

Heat is intrinsically linked to energy, so at the beginning of the first part the concept of energy is highlighted, the temporal consumption profile of heat and electricity is addressed using the example of a household or the heat levels of different industrial processes. The role of heat, electricity, and fuels, which must be generated CO2-neutrally in the future as far as possible, is shown for our economies worldwide. A key lesson is that demand and generation do not necessarily coincide, and therefore, storage and load-shifting have to be developed if renewables have to cover all requirements. The vision is to develop a future decentralized and sector-coupled energy system, at a reasonable cost and with reliable and resilient supply at all times. Chapter 2 we devote to one primary renewable source of energy, our sun with its virtually infinite radiation resource. Solar irradiation on our earth has to consider the sun's extra-terrestrial radiation intensity, atmospheric attenuation, the influence of geographical location and temporal changes. The quantifying of irradiation needs suitable measuring devices for its detection. In Chap. 3 the physical basics for the optical conversion of the short-wave solar radiation into heat, as well as methods for concentrating the radiation, are presented. Measurement methods are used to quantify the optical quality of materials and devices, and mathematical description and modelling are presented for designing and optimizing the optical components. The physical and engineering description of the heat transfer by thermal conduction, thermal radiation and natural convection is included in Chap. 4, where practical problems with coupled heat transfer mechanisms are also highlighted. In general, the generation of heat and its demand occurs at different locations, so the heat must be transported by heat transfer fluids. The flow of these fluids can be either single-phase or multiphase, laminar, or turbulent. Chapter 5 examines these energy and momentum transport in solar thermal systems and discusses fundamentals, approaches to solutions, and points out engineering correlations. In Chap. 6 stationary low-temperature collectors and in Chap. 7 concentrating and tracking high-temperature collectors, solar thermal collector types with their different characteristic properties, operating modes and design alternatives are presented and discussed.

Part II

The second part of the book contains the descriptions, concepts and system designs for several application fields of the thermal energy generated by the collectors. In most solar thermal systems, the ability of storage is already conceptually included, contrary to other renewable energies like grid-integrated photovoltaics and wind, which rely on the flexibility to accept additional power of the grid. Several concepts and designs of thermal energy storage are described first in Chap. 8, which allow decoupling of the generation of heat from the consumption profiles of heat. The focus of Chap. 9 is the targeted use of solar generated low-temperature heat well below 100 °C for domestic applications, namely the production of hot water and the use in the building heating system. The chapter describes not only the essential components and basic algorithms for dimensioning and planning, but also describes the concepts for integration into heat networks. The subject of Chap. 10 is the optimal design and layout of line and point concentrating collectors in a large field consisting of mirrors, absorbers and tracking devices with the aim of efficiently generating heat. Control and regulation systems as well as the operation and maintenance of the solar field are shown. In Chap. 11, solar thermal power plants are addressed, where the heat generation is linked to a heat engine. In order to have a high conversion efficiency of the heat engine, high temperatures above 400 °C are preferable. In addition to the various concepts of solar only thermal power generation, hybrid concepts are also shown and illustrated. Industry as a sector with a huge heat demand is able to use solar thermal heat from low temperatures below 100 °C to a medium range between 100 °C and 400 °C up to more than 1000 °C. Solar heat can be integrated using very different integration concepts into supply for industrial processes. In Chap. 12, starting from the determination of the sectoral heat demand, methods for integration are shown as well as procedures for the classification of solar integration. In the final Chap. 13, methods, procedures and concepts are presented on how cooling and air conditioning can be realized with the help of thermal technologies. Finally, in two appendices, methods for the design of hydraulic components such as pumps and heat exchangers and measuring devices for the detection of thermal variables are presented, since these are often required for the design of solar thermal systems but are not specific to solar thermal energy.

The script is designed in such a way that it can also be used as a reference work. In particular, the individual chapters are written in such a way that they are largely self-contained and require little back-and-forth flipping. Nevertheless, a basic education in physics is useful for understanding the chains of argumentation and trains of thought.

Freiburg im Breisgau, Germany — Werner Platzer
Eggenstein-Leopoldshafen, Germany — Robert Stieglitz

Acknowledgements

We would like to thank our families for bearing with us, and our institutions for giving us the freedom to dedicate the necessary amount of time to the creation of this book.

Contents

About the Authors

Prof. Dr. rer.nat. Werner Platzer is a physicist and received his Ph.D. from the Albert-Ludwigs-University Freiburg in 1988. He has been working for more than 35 years for the Fraunhofer Institute for Solar Energy Systems ISE in Germany, performing research and development of solar thermal energy material, components and systems, facade technology and energy efficiency in buildings. His focus in research and teaching is on concentrating solar thermal technology for process heat and power. Until 2015, he served as Director of the Division Solar Thermal and Optics. He cofounded the Fraunhofer Centre for Solar Energy Technology in Santiago de Chile and was Executive Director there in the years 2016 and 2017. Since 2018, he coordinated the international business development of Fraunhofer ISE. He is teaching as an honorary Professor at the Faculty for Environment and Sustainability, University of Freiburg. He was appointed chairman of the IEC TC 117 on the standardization for solar thermal electricity (CSP) and has been actively involved in many European and international projects on CSP and Solar Process Heat. He has authored more than 200 articles and conference papers.

Prof. Dr.-Ing. Robert Stieglitz obtained a doctoral degree in engineering in 1994 at the University of Karlsruhe. In his professional career, he has spent some time in the power industry and joined the research centre Karlsruhe (today Karlsruhe Institute of Technology-KIT) in 1995 leading functions in several science fields associated with heat and mass transfer, instrumentation, material qualification and nuclear as well as accelerator technologies. In 2009, he was appointed as director of the Institute for Neutron Physics and Reactor Technology (INR) and he simultaneously holds a chair at the university campus as professor in Applied Thermo-Fluidics. Since 2009, he has been the Director of the CEA KIT Frederic Joliot-Otto Hahn School and served many associations as an advisor and expert such as German Nuclear Reactor Safety Commission, the Alexander von Humboldt Foundation and others. He continued the INR's more than 30-year tradition in solar thermal research through a wide range of contributions to national and international solar thermal R&D projects and industrial collaborations. His scientific record is characterized by more than 240 scientific articles/conference papers, 2 books and some book contributions in several science fields.

Part I

Fundamentals

1 Energy

Abstract

Renewable energies meet an energy market that is characterized by an energy demand or rather an exergy demand, a supply by energy carriers and disposal of end products such as carbon dioxide (CO_2). Energy supply must satisfy the worldwide population growth as well as the increasing per capita energy consumption and this simultaneously without excessive consumption of resources. The potential of regenerative energies thus depends essentially on their properties, with which they can meet the market characteristics. The profile of the energy demand, its follow-up and production costs as well as the boundary conditions determine the energy market and thus define the framework conditions for the development of regenerative energy sources, which in the future must increasingly satisfy the aspects of sustainability and resource efficiency. This chapter sheds light on these aspects, introduces key terms and discusses in particular the energy market environment in which thermal solar energy operates and finally introduces the concept of sector coupling.

Keywords

Energy demand • Energy resources • Impact on climate change • Energy forms and carriers

1.1 Energy Demand and Resources

Energy is a multifaceted term and a number of energy forms and energy carriers exist, the meaning of which will be explained later in this section. One of these forms of energy, which forms the core of this book, is heat produced by thermal solar energy systems, which can be classified as regenerative or renewable energy. Here, the term renewable or regenerative means that the energy carrier is integrally and infinitely available or can renew itself within a human generation time scale. Thus, regenerative energies differ fundamentally from fossil chemical energy

R. Stieglitz and W. Platzer, *Solar Thermal Energy Systems*,
https://doi.org/10.1007/978-3-031-43173-9_1

sources, whose availability is finite. The basically unlimited availability of the carrier does not imply, however, that a temporal and spatial availability is guaranteed at any time. Furthermore, the development of renewable sources requires the use of technical tools and raw materials, which must be available in terms of both time and resources. In a competitive market, therefore, renewable energies face other energy sources, where price is oriented on the demand-based supply of energy. This market price principle based on the match of energy production and consumption at the lowest possible price to be paid by customers, needs to modified and possibly overruled by additional considerations, public directives and regulations, in the light of climate change. These considerations include environmental damage caused by the operation, e. g. the emission of carbon dioxide (CO_2) from the use of fossil fuels, but also the consumption of resources to be used for the production of the generating facilities and for their disposal or recycling in order to ensure the long term sustainability.

A key factor in the integration of renewable energy generation technologies, especially solar sources, into a regional energy service system is therefore its geographical location. Large temporal differences between generation and consumption require either the provision of energy storage technologies or supra-regional energy balancing infrastructures where energy or energy carriers can be transported with low losses, for example electricity grids or chemical carrier networks.

Since the beginning of the industrial revolution, energy demand has been met primarily by the combustion of fossil fuels and the associated CO_2 emissions. Against the background of the climate change, a future supply system must have a footprint that is as CO_2-neutral as possible. It should be emphasized that a holistic assessment of any energy generation technology may not only consider the CO_2 emissions from the operation, but also includes corresponding CO_2 equivalents originating from production and follow-up costs, life cycle duration, resource consumption or recyclability. The environmental footprint, e. g. in terms of land use or consumption, shall also be taken into account in an assessment. Energy supply concepts, which are essentially based on volatile renewable generation sources, also require energy storage concepts as well as distribution grids and thus infrastructures that provide the energy when generation from renewable sources is not able to cover the demand. This so-called residual load, which results from the difference between consumption and demand, can be generated either fossil or largely CO_2-neutral, for example from pumped storage power plants, accumulators, or other storage technologies, but includes not only the generation facility but also corresponding distribution networks, which in turn contain a CO_2 footprint. The design of a future resilient energy supply system therefore includes the integration of energy storage and distribution in addition to a large number of generation technologies to be considered.

Due to variable geographic, meteorological and economic conditions, as well as social and cultural differences in the various regions of the world, there can be no single "*correct*" energy supply concept. Socio-political assessment variables play an dominant role and set different priorities, because they have to bear the investment burden and the consequences resulting from the implementation of the respective energy supply concept.

This chapter highlights the issues raised, introduces key concepts and discusses the energy market environment in which thermal solar energy wants to find its future role.

1.2 Energy Consumption, Economic Performance and Resource Availability

If one analyses the global per capita energy consumption of a person, it quickly becomes apparent that, in addition to historic developments, it depends in particular on the economic performance and the degree of industrialization of the respective country. If gross domestic product per person (GDP) is used as a parameter, a high value is to a significant degree associated with the provision of a large amount of energy per capita. Conversely, economic prosperity (expressed in terms of gross domestic product) allows corresponding amounts of capital to be mobilized via targeted incentive measures to enable effective energy use and thus a reduction in per capita consumption in the first place. Figure 1.1 shows the energy consumption in kilogram oil equivalents (OE; $1kg\,OE = 41.868\,MJ/kg$) per capita of different countries of the world in relation to the gross national product per capita.

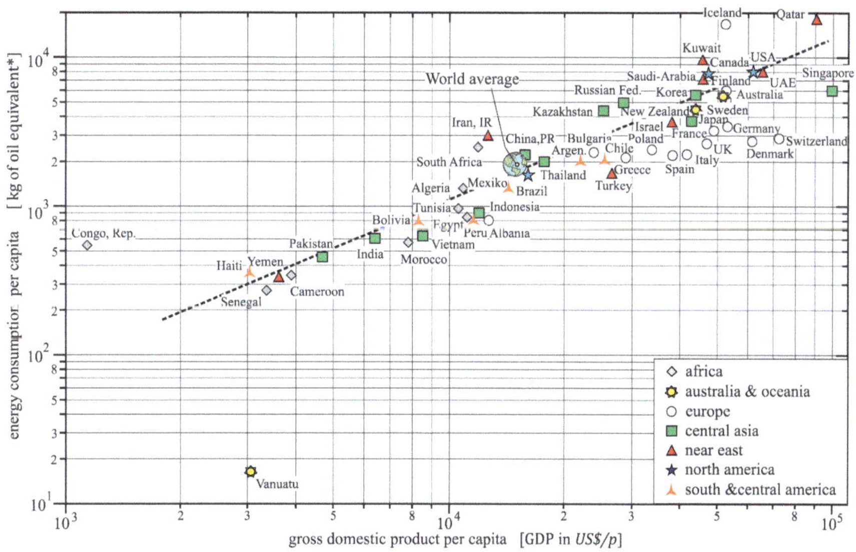

Fig. 1.1 Annual per capita energy consumption in kilograms of oil equivalents as a function of gross national product per capita per year in U.S. Dollars [US$] for various countries (World Bank, 2020). The further to the right of the centre line a country is drawn, the more effectively the energy is used

The dashed line in the graph represents an estimate for which there is no compelling functional relationship. However, it can be deduced from the figure that there are countries whose economies use energy more effectively than others that use more primary energy to achieve the same gross domestic product. For a direct comparison, the composition of the gross domestic product, the geographic structure of the countries, the weather conditions and, most importantly, the degree of industrialization or the importance of industrial production would have to be considered, which are not the focus of this book. Emerging economies often are very inefficient with their energy consumption. Most of them lack the necessary capital to improve their infrastructure. The capital required for a rational energy use is therefore in competition with the capital required to satisfy basic consumption, stabilize political power or expand production. It must be emphasized here that the size of energy consumption per capita, is not necessarily related to CO_2 emissions; for example, although Iceland has the second highest per capita energy consumption in the world, it has a renewable energy share in electricity generation of more than 99%, most of which comes from hydropower and geothermal energy.

The importance and relevance of efficient and at the same time resource-saving energy production becomes particularly obvious if one looks at the development of the world population over time as well as the average global energy consumption per capita, which is shown in Fig. 1.2. Both variables show a monotonous growth over the last twenty years. A further continuation of this trend would require the provision and generation of ever greater amounts of energy, so that measures to reverse it are imperative.

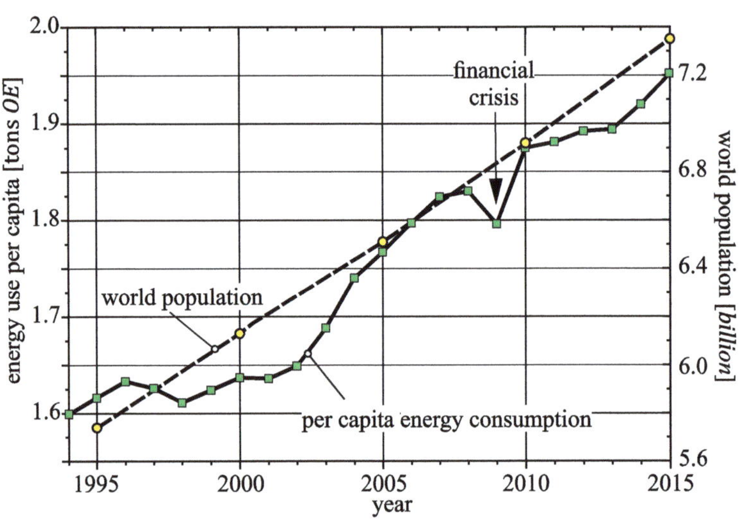

Fig. 1.2 Development of global per capita energy consumption and world population from 1994 to 2015 (World Bank, 2020), (United Nations Population Division, 2020)

Even the slowdown of one of the two trends means a steady further growth of the total energy demand. However, availability in time and place is the anchor and pivot of any competitive market economy, as it is part of all manufacturing processes of economic goods, the provision of services, communication and information as well as mobility. Until today, most of the energy production and supply in the industrialized countries is based on chemical carriers of mostly fossil nature such as oil and gas and/or coal and their established transport chains. If this annual resource consumption figure is covered solely by the available known resource reserves, then the time period in which these resources will be spent can be determined.

The global annual consumption, the secured reserves and their static range are compiled in Table 1.1. The "static range" is a hypothetical quantity and represents the ratio between the reserves and the annual world production currently determined in the survey year. The figure thus indicates how many more years the raw material will be available if consumption remains constant and reserves are fixed. The term reserve differs from that of resource. To be classified as a reserve, an energy resource must meet three conditions: It must have been confirmed by drilling/mining, it must have been accurately measured, and it must be economically recoverable with current technology at current prices. Resources, which are also listed in Table 1.1., on the other hand, describe the geologically proven reserves of an energy carrier that are not yet considered economically recoverable because the extraction requires new and/or expensive technologies. The table shows that the fossil carriers will be consumed within a few generations even at today's consumption rates, not considering their climate-damaging consequences during exploration or combustion here. At a first glance, this seems to apply to nuclear "fuels" as well; however, this is relativized by a factor of at least 60–100 if multiple recycling of the fuel in breeder reactors is considered, and by further orders of magnitude if thorium is used in nuclear reactors. If nuclear fusion is included in the consideration, range time scales result which are in comparable dimensions of renewable energies, however, they are also subject to a raw material demand as well as considerations of nuclear waste and risks..

1.3 Impact of Energy Demand and Population Growth

This continuously growing energy demand, mainly due to the thermal use of fossil fuels, is reflected in a monotonous increase of the CO_2 concentration in the atmosphere. To the natural CO_2 present in the atmosphere the anthropogenic amount is added, i.e. the amount produced by human activities. A large part of this comes from the combustion of fossil carriers, as Fig. 1.2 indicates (Fig. 1.3).

Besides the limitation of fossil fuel reserves, the release of CO_2 into the atmosphere has other serious consequences. The content of CO_2 in the atmosphere has increased from about $280 ppm$ to more than $410 ppm$ since the beginning of industrialization. The CO_2 present in the atmosphere absorbs the thermal radiation emitted from the earth's surface, especially in the far infrared spectral range, as

Table 1.1 Annual global consumption of different energy sources, "static" range of secured reserves and resources at today's consumption, data extracted from (British Petroleum BP p.l.c., 2021), (ENI SpA, 2021), (OECD/NEA and IAEA, 2020), reference year 2020

Energy carrier	Worldwide consumption in $[EJ = 10^{18} J]$	Static reserve $[a]$	Static resource $[a]$
Oil	173.73	95	160
Gas	152.9	85	300 – 600
Coal	156.56		
Hard coal		115	2600
Lignite		320	5500
Nuclear	23.82		
One through cycle		145	315
Multi-recycling		>8000	≫10000
Renewable energy production	31.72	∞	∞
Electricity production	96.96	–	–

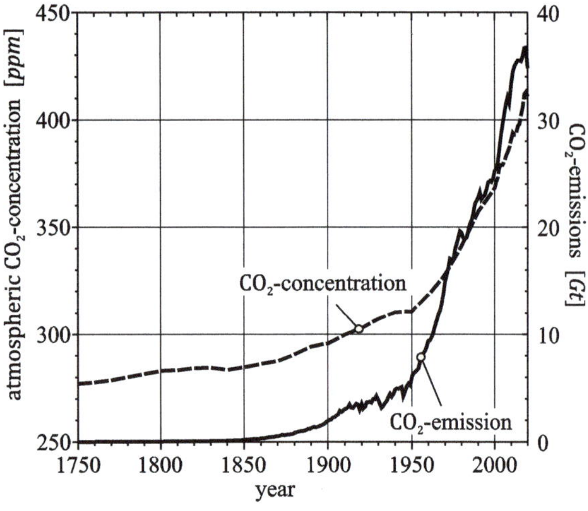

Fig. 1.3 Increase in atmospheric CO_2 concentration and CO_2 emissions from fossil fuel combustion as a function of time (Ritchie and Roser, 2020)

shown in Fig. 1.4. The Earth's atmosphere consists of about 78% nitrogen, about 21% oxygen and about 1% argon. The CO_2 portion amounts to approx. 0.04%. Due to the increased absorption caused by the increased CO_2 concentration in a previously relatively transparent waveband, the heat radiation from the earth into space is reduced and global warming occurs. Other greenhouse gas emissions like the one of methane (see Fig. 1.4) from agriculture and perma frost soil adds to the problem.

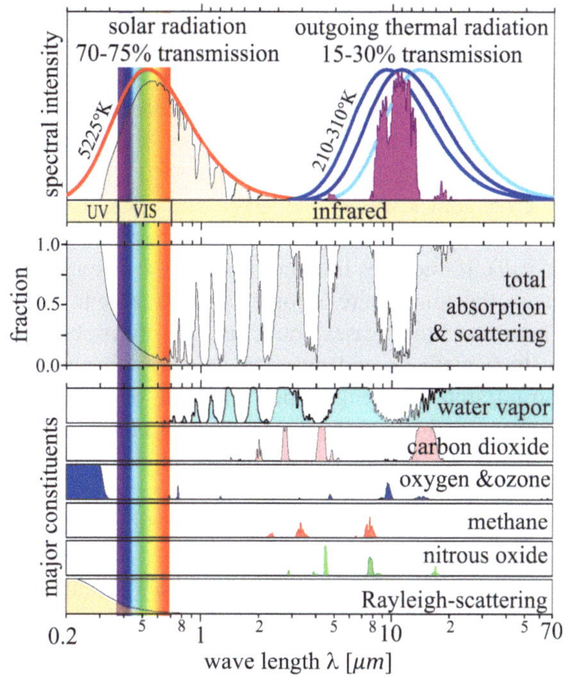

Fig. 1.4 Spectral dependence of radiation incident on and radiated from the Earth, and of the major species or mechanisms contributing to absorption/ scattering

As Fig. 1.4 shows, water vapor is the most important greenhouse gas, and its contribution to the natural greenhouse effect under clear skies is about 60%. . The origin of water vapor is mostly the natural water cycle (ocean - evaporation - precipitation - storage in the ground) and a smaller fraction from volcanism. Water vapor is thus the main source of atmospheric counter radiation and carrier of the "*natural*" greenhouse effect. The natural greenhouse effect is an important effect for the radiation budget of the earth because it leads to a global average temperature of $15°C$, which makes life on earth possible in the first place. Human activities indirectly increase the water vapor content in the atmosphere because global warming increases the air temperature and thus the evaporation rate. This is the most important feedback factor amplifying global warming. In the stratosphere, between $12 - 50km$ above the earth's surface, the atmosphere is always stably stratified and very dry. There, water vapor occurs only in traces, for example from air traffic and from the oxidation of methane to CO_2 and H_2O, which also contributes to the greenhouse effect. If the natural water vapor content is essential for human life, traces of water vapor present in the stratosphere are considered particularly relevant to the climate.

The earth re-radiates the absorbed heat of the sun in a wavelength range between 3 and $100\,\mu m$ (IR radiation, the blue curve, Fig. 1.4 top right). Only in a narrow range of a radiation window between $8 - 13\,\mu m$ an almost unhindered radiation is possible (the violet marked area, top right of the picture). The

IR weather satellites also look through the so-called radiation window in the wavelength range between $8-13\,\mu m$. Here the thermal radiation of the earth can escape almost unhindered. This allows a direct view down to the lower cloud cover and Earth's surface. Above all, the water vapor in the atmosphere is responsible for the confinement to this radiation window. It absorbs most of the radiant energy and limits the permeability of the window to $8-17\,\mu m$.

CO_2 comes second as a greenhouse gas. It limits the radiation window mainly in the range of wavelengths between 13 and $17\,\mu m$. (The maximum absorption is at $15\,\mu m$). If one goes into the range of the absorption lines/bands of the greenhouse gases, the atmosphere becomes impermeable to thermal radiation and thus opaque. There are other gases found in the atmosphere due to anthropogenic effects, such as methane and nitrogen oxides with climate-relevant absorption bands in the infrared wavelength spectrum. The increase in atmospheric concentrations of methane and nitrogen oxides also correlates with the increase in world population and the accompanying industrialization and agricultural use of the Earth's surface, as Fig. 1.5 illustrates. The climatic effectiveness of methane is more than 80 times higher than that of carbon dioxide (CO_2), because the absorption of methane has not yet reached saturation. Anthropogenic methane (CH_4) release is caused, among other things, by the use of natural gas, excessive agricultural use, and livestock farming.

The individual contributions of the trace gases to the greenhouse effect can be seen in Table 1.2. The table also shows the average residence time of the greenhouse gases in the atmosphere. Especially carbon dioxide (CO_2), nitrogen oxides (N_2O) as well as methane (CH_4) show a considerable residence time in the atmosphere, so that their effect on the climate is of a long-term nature. Their importance in terms of climatic consequences is quite of the order of chlorinated fluorochlorohydrocarbons (*CFC*'s), the use of which has been banned in many

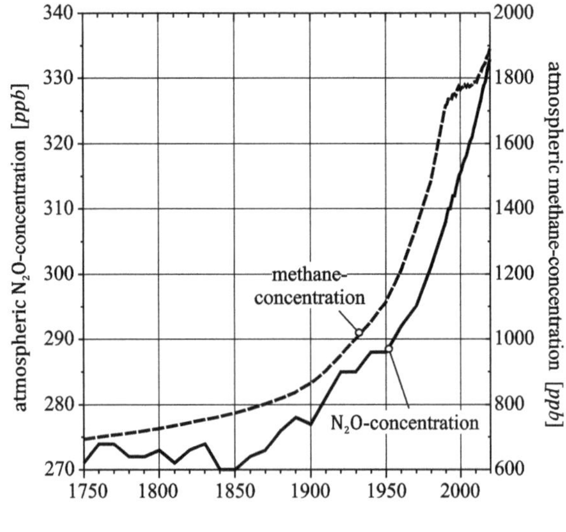

Fig. 1.5 Increase in methane and nitrogen oxide concentrations in [ppb] in the atmosphere since the beginning of industrialization from 1750, (Butler and Montzka, 2021)

Table 1.2 Contributions of climate-relevant trace gases to the greenhouse effect, their residence time in the atmosphere and the temperature effect on thermal radiation (*RF*) and the global warming potential (*GWP* reference value 20 years)

Gas	Residence time in [*a*]	Regional impact scale	Radiative forcing *RF* in $[W/m^2]$	Global warming potential*1 GWP_{20} [/]
H_2O	10^{-2}	Continental	10^{-3} – 10^{-2}	0
H_2	4 – 7	Local	*2	33
CO_2	~30	Global	2.16	1
CH_4	10 – 12	Global	0.54	84
N_2O	110	Global	0.21	264
CFC (chloro-fluoro-carbons)	640	Global	0.18	>5000
O_3 Stratosphere troposphere	0.1	Continental-global	–0.05 0.65	62 – 69

*1 from (Myhre et al., 2013); *2 under discussion; from (World Meteorological Organization, 2021) and the literature listed there

technical applications under international agreements. The extreme challenge is to quantify the influence of anthropogenically emitted greenhouse gases, since it is not known how each individual type of human activity affects ground-level air temperature and how the retarding effects of huge sinks like the oceans have to be considered. Since many processes influence each other, i.e. create feedbacks, the individual effects cannot simply be added together. Two numbers are often given in the assessment, and they must be clearly distinguished from each other.

One figure is a dimensionless quantity the global warming potential (*GWP*) or CO_2 equivalent of a chemical compound, which is a measure of its relative contribution to the greenhouse effect, i.e. its average warming effect on the earth's atmosphere over a certain period of time usually 100 years. It thus indicates how much a given mass of a greenhouse gas contributes to global warming compared to the same mass of CO_2. However, the global warming potential cannot be equated with the actual contribution to global warming, since the emission quantities of the various gases differ greatly. With this concept, the different contributions of individual greenhouse gases to global warming can be compared given known emission quantities.

In the Intergovernmental Panel on Climate Change (IPCC) report, another quantity used for quantification is the "stratospheric adjusted " radiative forcing (*RF*), which is the net radiative flux density at the tropopause due to the changing concentration of greenhouse gases. The term radiative flux density (*RF*) indicates the radiant energy passing through the tropopause per second and per square meter in (W/m^2). The stratospheric adjustment assumes that the temperature in the stratosphere has already adjusted to the new equilibrium temperature, but the temperature profile in the troposphere still corresponds to the old equilibrium, which is given by the stable stratification of the stratosphere. If greenhouse gas concentrations are increased, additional energy goes into the lower atmosphere and the radiative forcing and thus the *RF* factor is positive. In contrast, an example of a negative radiative forcing would be if aerosols reflect more sunlight back into space. Then less energy reaches the lower atmosphere and it cools down. The advantage of this concept is that the *RF* does not depend on complicated and poorly understood feedbacks in the climate system. The major disadvantage is that the *RF* says nothing about the changes in measurable and sensible quantities at the ground, which are the quantities of real interest for assessing climate change. Another disadvantage of this concept is that only influences that affect the temperature via radiation can be considered, so that effects such as evaporative cooling or diffusion are not taken into account, although they also affect the temperature.

The release of CO_2 into the atmosphere due to the combustion of fossil fuels can be reduced by some extent if the exhaust gas from the power plants is captured, separated and stored in natural gas reservoirs or coal mines (sequestration by carbon capture and storage technology-*CCS*). The conversion of mobility requires high energy densities, especially for heavy-duty transport by water and land, or for flight operations, which are difficult or costly to realize in terms of battery technology; here, research is still in its infancy. In the short term, the only alternatives are synthetically produced hydrocarbons or a generation and distribution network based on hydrogen as a carrier. However, the development and conversion of both electrochemical and synthetic carriers also requires a stringent analysis of the resource vector, recyclability, and an evaluation of efficiency to ensure a sustainable long-term transformation. It is obvious that any energy transformation in the sectors of households, traffic, industry as well as agriculture, trade, commerce and services consume enormous efforts in infrastructure changes, and thus initially leads to a further shortage of energy and resources. Moreover, the lower degree of availability of all intermittent renewable generation sources, such as solar and wind, forces the construction of generation facilities with installed capacities significantly higher than those of fossil power plants, in addition to the provision of new infrastructures for energy storage and distribution. However, waiting is no option, and a substantial transformation in all sectors is needed, otherwise the limitations of fossil carriers and the onset of a phase of general shortage will come sooner than expected. As the recent past shows, competition for essential resources may give rise to armed conflicts. It is therefore necessary to develop alternatives even further and, above all, to diversify all forms of renewable and far-reaching energy generation technologies to the point where a self-sustaining renewable industry can be built on the basis of renewable energies, resource efficiency and recyclability of required materials in a circular economy.

At present, using all forms of renewable energy (wind, solar, geothermal, hydro) seem to be necessary for developing a long-term solution to the energy supply. Whether a potentially sustainable use of nuclear technology, including nuclear fusion in the long term, is capable to contribute to a low risk supply of the needs of a growing global population, is an open question for the future.

1.4 Energy Forms, Carriers and Conversion Processes, Exergy-Energy

For the practical human use of energy, four primary physical forms of energy can be distinguished,

- Nuclear energy,
- Chemical energy,
- Thermal and radiant energy, and
- Mechanical (potential and kinetic) energy.

The particular form of energy requires an energy carrier either in the way of a nuclear or chemical fuel, heat, solar radiation, or mechanical energy. The type of energy carrier, in turn, determines the conversion process(es) to produce the generated form of energy (electricity, heat, or fuels). Figure 1.6 illustrates the chain from the primary energy form with the primary energy carrier used to the final energy produced, including the conversion options potentially available.

Renewable energy forms and carriers are those whose availability is either permanent or renewable up to a period of one human generation (approx. 30 – 40 years). The term renewable energy also covers a whole range of energy carriers. They also have different local and temporary availability and exergy content. A

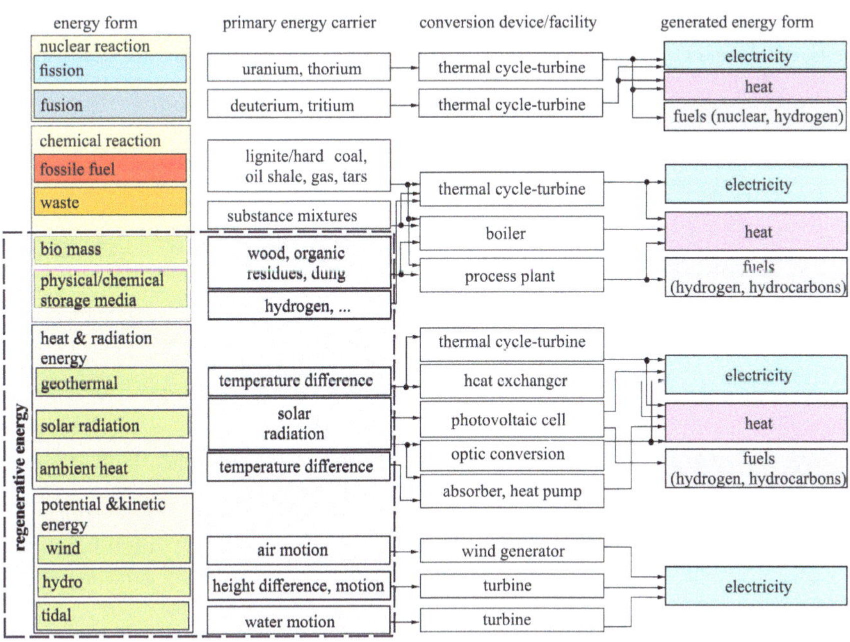

Fig. 1.6 Schematic overview of the energy conversion chain for different energy carriers by means of a conversion technology to different usable forms of energy

list of initial supply, conversion processes and forms of application is also shown in Fig. 1.6.

The conversion of the primary energy carrier into a form that can be used by the consumer requires one or more subsequent conversion process steps. In the first step, secondary energy is obtained from primary energy. In a solar thermal system this refers to the optic conversion of solar radiation in heat. In the subsequent steps, final energy or useful energy is obtained, as illustrated by the sequence of the conversion chain in Fig. 1.7. Each of these conversion processes in the chain is associated with a scalar quantity, the efficiency η_i of the respective conversion process, e.g. the absorber efficiency (η_A) or that of the thermodynamic cycle (η_{th}), where for each efficiency in the chain $\eta_i < 1$. Along an efficiency chain, the individual conversion efficiencies multiply by those of the previous step. As a result, only a fraction of the potential energy content of the primary carrier is converted into final energy directly applicable to the user, which is he described by the total efficiency along the entire chain (η_{tot}).

The general and so far, also in this book used terminology to speak of an energy problem is strictly speaking not correct. Energy is available in the world in practically unlimited quantities, if one thinks for example only of the heat content of the oceans. However, this does not initially have anything to do with local energy that can be used by the end consumer.

This approach requires a clear definition of the concept of energy. What is generally needed and partly scarce is rather exergy E. Exergy (E) is energy which is convertible into mechanical work. In the case of heat, this is the heat content that is exceeding the ambient temperature level T_{amb}. Figure 1.8 illustrates the division of heat into exergy and the fraction B that cannot be used to generate mechanical work, which is also referred to as anergy.

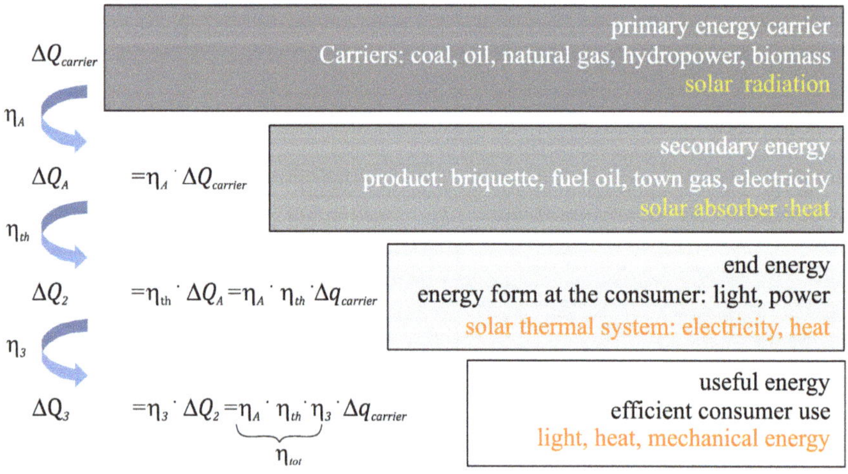

Fig. 1.7 Conversion chain of energy from primary energy source to useful energy; solar thermal accessible indicated on right in lower line

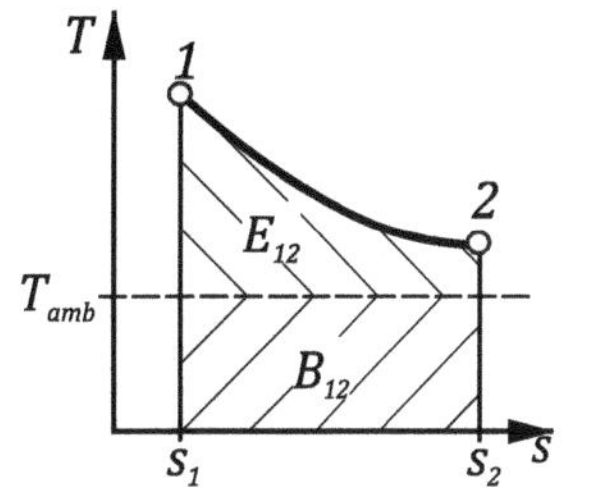

Fig. 1.8 Splitting of heat into exergy (E_{12}) and anergy (B_{12}) in the temperature-entropy diagram ($T - s$) related to the ambient temperature (T_{amb})

If we consider the heat Q_{12} crossing the system boundary in a process leading from point 1 to point 2, we must distinguish the energy E_{12} lying above the ambient temperature T_{amb} termed as exergy from the energy B_{12} situated below the ambient temperature called anergy.

For the use and for considering the value of energy the ambient temperature plays an important role. We value heat only, if it is at a different temperature level than the ambient. In physics this is reflected by the concepts of exergy and anergy. In the temperature-entropy diagram ($T - s$) of process, a change of the physical state, the corresponding energy quantities are proportional to the areas under the respective curves describing the physical process. When we split up the energies into two parts, the one above the ambient temperature, and the one between the absolute zero and the ambient, the following relationships result for anergy B_{12} and exergy E_{12} (Fig. 1.8):

$$Q_{12} = \int_1^2 T\,ds, \tag{1.1}$$

$$Q_{12} = E_{12} + B_{12}, \tag{1.2}$$

$$B_{12} = T_{amb} \int_1^2 ds \quad \text{with} \quad ds = \frac{dQ}{T}, \tag{1.3}$$

$$E_{12} = \int_1^2 (T - T_{amb})\,ds = \int_1^2 \left(1 - \frac{T_{amb}}{T}\right) dQ, \tag{1.4}$$

$$E_{12} = Q_{12} - T_{amb} \int_2^1 \frac{dQ}{T}, \tag{1.5}$$

where s describes the specific entropy. It can be seen from Eq. (1.3) that the fraction of the total heat which lies as exergy above the level of the ambient temperature T_{amb} is determined by a factor corresponding to Carnot's efficiency. In the terminology used here, the first law of thermodynamics has the wording:

In a closed system, the sum of exergy and anergy remains constant.

Accordingly, the second law of thermodynamics can be formulated:

1. in irreversible processes of a closed system, exergy is converted into anergy,
2. in reversible processes the exergy remains constant,
3. anergy cannot be converted into exergy.

As soon as energy is to be converted into mechanical work, it is immediately clear that only exergy can be used. Exergy is also required for heat demand, such as for heating purposes. Of course, the required exergy input can be minimized. Figure 1.9 illustrates the problem using the example of a heat flux from a room through a wall into the environment with the temperature T_{amb}.

When a room is heated above the level of the outside temperature (ambient temperature) as indicated in Fig. 1.9, heat is added to provide temperatures above T_{amb}. The energy contained in the room is split in two fractions as exergy (E_Q) and as anergy (B_Q). Due to the heat conduction in the outer wall, as long as exergy is in the room, heat is flowing outside and is added to the infinite reservoir of ambient, thus all exergy is converted into anergy. If electric resistance heating were used for the room as illustrated in Fig. 1.10, the electric energy, which is pure exergy (E_{el}), would be converted in the resistor partly to exergy (E_Q) and partly to anergy (B_Q). In this case, the heat pumps, which will be discussed later, are more favourable, as they use only not only exergy, but also anergy for heating.

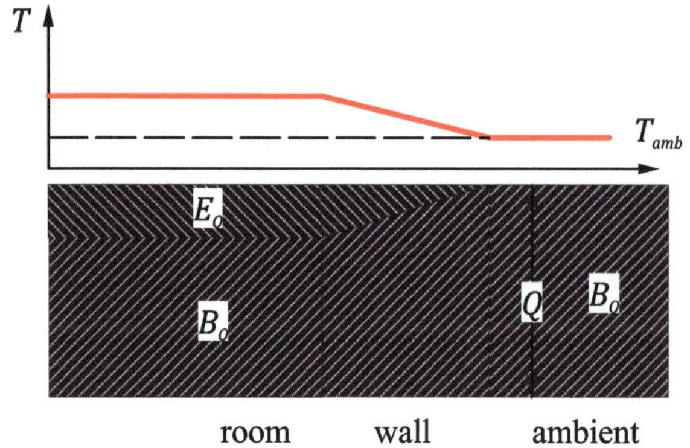

Fig. 1.9 Exergy (E) and anergy (B) using the example of a heat flow from a room through a wall into the environment with T_{amb}.

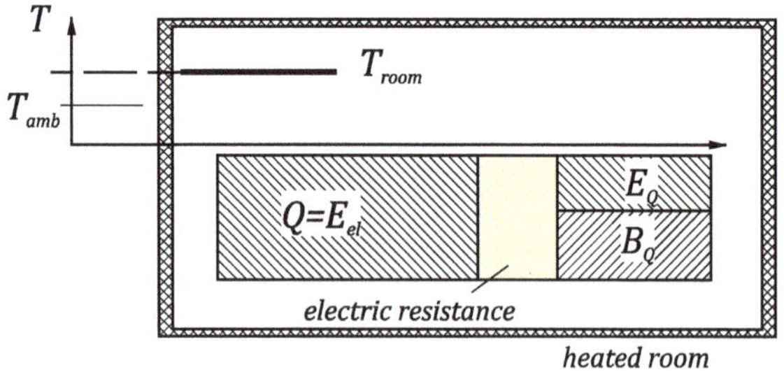

Fig. 1.10 Share of exergy and anergy for electric resistance heating of a room

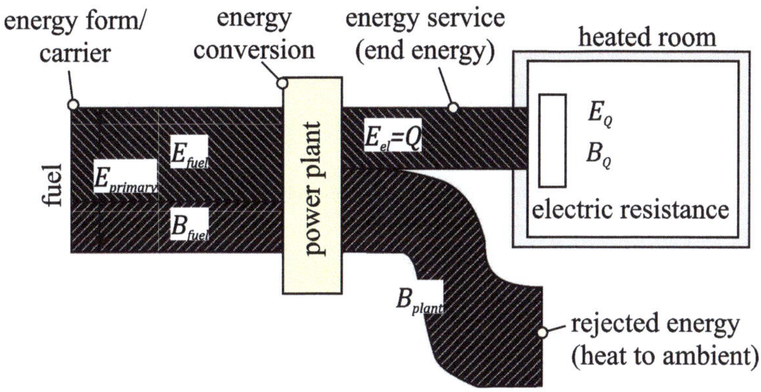

Fig. 1.11 Energy flow diagram of a coupled fuel-power plant-electric resistance heating system with division into anergy (B) and exergy (E)

This consideration can be extended to include the primary energy carrier, in the example of the heated room the fuel for the electrical energy (E_{el}). Figure 1.11 shows the heating process in a closed heated room, taking into account the power plant generating electricity. Already in the thermodynamic cycle of the power plant, part of the energy supplied by the fuel ($E_{primary}$) is lost to the environment as anergy (divided into B_{fuel} and B_{plant}). The largest fraction is the rejected heat of the thermodynamic cycle to the environment, which can be shown most clearly in the $T - s$ diagram.

The consideration of the energy flow from the fuel as energy carrier via the power plant as conversion technology to the final energy use of a single heated room can be extended to an entire national economy. The energy supply is based on different energy carriers and the final energy is broken down into different consumer sectors, such as transport, mobility, trade and residential. The result is a Sankey diagram showing an integrated national energy or energy carrier flow representation. This, however, is not an exact representation in the sense of the previously introduced anergy-exergy definition but at best a simplification. For example, waste heat flows are not identical to anergy flows, because the latter may still contain usable heat. Figure 1.12 shows such an economic Sankey diagram using the example of the United States of America.

This expresses electricity generation from renewable energy sources such as hydropower, wind power, geothermal power, and solar energy based on the heat of fossil-fuel power plants to more easily account for the differences between the conversion efficiency of renewables and the fuel utilization of combustion- and nuclear-powered systems. In this way, hydro, wind, and solar energy can be included in the energy flow on a similar basis as coal, natural gas, and oil. For this reason, the sum of the input variables for electricity differs slightly from the total electricity production shown. Distributed electricity includes only retail electricity sales and does not include self-generation. Electricity generation efficiency is calculated as the total retail electricity supplied divided by the primary energy input

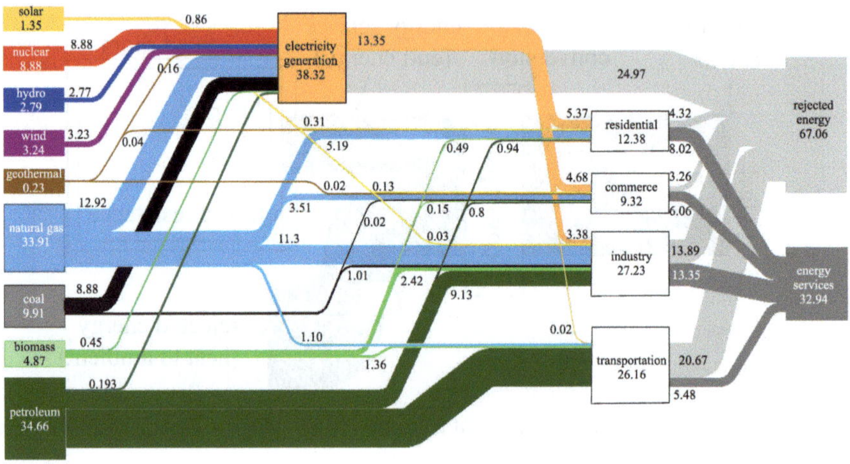

Fig. 1.12 Sankey diagram of United States energy flow in [%], where 100% $= 98EJ$, modified from (Lawrence Livermore National Laboratory, 2021)

used in electricity generation. End-use efficiency for the residential, commercial, and industrial sectors is usually estimated to be about 80%; the mobility sector is assumed to be 25% efficient.

The numbers in such a Sankey diagram indicate either the absolute amount of energy in joules, in the case of the USA $98 \cdot 10^{18} J = 98EJ$, or as a percentage of a total amount of energy to be indicated. In addition to the main energy sources, the chart also shows the breakdown of consumption in the individual sectors of the economy. The diagram also provides information on the efficiency with which primary and secondary energy is utilized. In the USA, more than 60% of the secondary energy is released into the environment in the form of excess heat. If one analyses exactly the transition from primary to secondary energy, especially for fuel carriers, one recognizes that apparently a part of the energy disappears; which of course is not the case. Rather, it is a so-called non-energy consumption, i.e. the conversion of fuels into products, such as paints, varnishes, or the like. Sankey diagrams thus provide information about the integral energy organization of a national economy and the energy-related interconnection of the sectors. However, they are only suitable to a limited extent for comparisons between countries, since these either have a different industrial development or a different geographical location. For example, the share of heat demand in Canada or the Scandinavian countries is different from that of the USA.

Likewise, Sankey as a year-integral representation does not provide information about the temporal and local availability of energy but only about the integral flow. In other words, the control and distribution structure of energy, which are the basis of the resulting integral energy flows as well as their accessibility and utilization, are not part of a Sankey diagram.

1.5 Energy Demand and Economic Considerations

The aim of this book is to show the use of thermal solar energy in the whole range from heat for domestic applications, process heat, desalination to electricity generation. The technological challenge is to analyse the solar energy supply and to transform it by means of components, processes, and systems in such a way that the solar energy is converted with high efficiency into a usable form of exergy while at the same time making rational use of resources at the end user.

Fluctuation in solar supply must be addressed in detail. If the solar supply does not match the temporal energy demand, storage must be provided in the system or substitute energy sources must be kept available. In the second case, the energy costing for renewable energy generation cannot be based on the savings of the original energy source alone (e.g., fossil carriers). Rather, a cost approach must then be considered that is based on the fuel-saving mode of use through the use of renewable energy. This is mandatory because a substitute energy carrier is still required, the provision and use of which is associated with a CO_2 footprint in the form of constructional expenditure, energy distribution or even the direct use of fossil carriers.

If the solar energy is used to generate a product such as drinking water, the storage can also be in the form of the end product. However, the cost of providing the energy production facility then shifts to the expense of a production facility that can only be partially utilized and is therefore may be oversized. In the field of industrial production, for economic reasons, the product value stream must be considerably higher than that of the energy used. To the pure energy cost calculation from expenditure (for the solar plant and maintenance) as well as energy (exergy) yield comes a further aspect. The competitiveness of an energy generation technology for service and production requires at the same time the maintenance and guarantee of a supply security.

In the light of these considerations, the statement that solar energy is free of charge, is not true, since a user-oriented supply of heat, electricity or fuel must be made available through a technical equipment, based on material resources and work. In a business sense, the use of technology turns the "free good" of solar radiation into an "economic good" that has a price. Moreover, in addition to the direct costs of operating and maintaining the system, there are also indirect costs for equipment, facilities and distribution systems that must be operated to maintain security of supply when the renewable source is not available. Currently, these additional costs in the electricity market are largely borne by consumers in the form of levies, while market deployment is supported by legal regulations such as tax depreciation or operational market regulation interventions such as the feed-in priority.

As well the electricity, the mobility and the heat market are served by different energy carriers worldwide. As we need renewable energy sources to penetrate this market, they must be either equivalent to them in terms of a 100% supply, or have advantages in at least some aspect of.

- Economic efficiency,
- Security of supply and,
- Environmental protection.

Environmental protection requirements, such as landscape consumption, can shift the market situation in the future. This aspect, which is based on social consensus, is not the subject here, but is of central importance for market penetration. If renewable energy systems do not meet the first two conditions of the general market requirements, an economic and secure supply of energy, market niches are nevertheless existing, which make a further development not only worthwhile but mandatory. Subsidies for the market introduction of renewable technologies are justified, because they offer the unique chance to reduce the long-term dependence on fossil fuels, and it is a major asset in the fight against climate change.

The treatment of solar radiation and the provision of adequate technical solar conversion systems also involves an economic analysis in terms of cost optimization. Figure 1.13 shows the diurnal variation of the energy demand broken down into electricity, heating, and hot water of an average 2.5-person household on an average cloudy working day in the transitional period of April in the course of a day in Germany in comparison with the solar area-specific irradiation. First of all, it is noticeable that the daily energy demand is subject to considerable temporal fluctuations and also does not correlate with the diurnal variation of the solar irradiation.

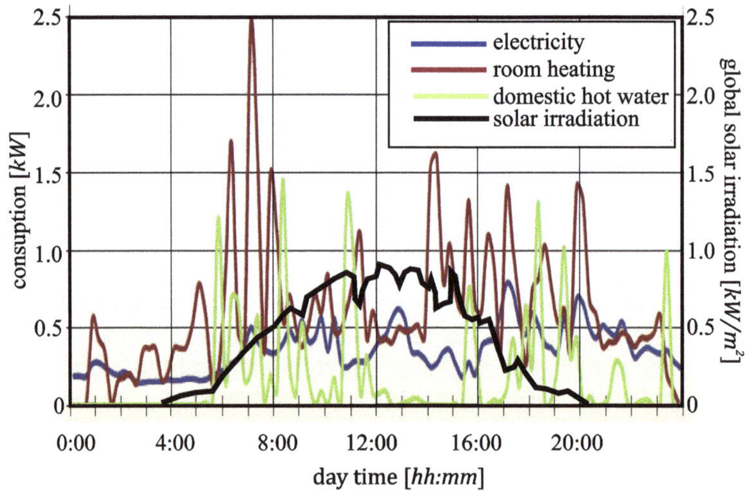

Fig. 1.13 Power demand for electricity, heating and domestic hot water of a standard household (2.5 persons) in the course of a day during the transition season (April) in comparison with the mean area-specific gross solar irradiance in Germany from (Beck,2007)

Therefore, in order to ensure the security of supply of the overall system at all times of the year and day, the integration of storage solutions into renewable energy system concepts is indispensable.

Whereas in the residential environment the temporal match of demand and generation has been discussed, for the industrial utilization of thermal energy, the level of the temperature plays a decisive role as well. With the help of solar thermal energy, a wide range of temperatures is technically accessible that would otherwise have to be provided via fossil carriers for process heat in the low-temperature range ($< 100°C$) or medium process heat range (up to approx. $400°C$). The amounts of energy used in the industrial sector are enormous, as shown by the breakdown according to process temperatures used and consumption for the production of industrial goods using Germany as an example, cf. Figure 1.14. To ensure meaningful competition, the energy to be thermally generated must correlate with the demand profile of the different industries and their time requirements for process heat. For solar thermal energy supply, the ranges up to about $400°C$ are interesting. In the upper temperature level, this simultaneously corresponds to the requirements for thermal power generation. For this purpose, systems other than those used in the domestic sector must be used, both in terms of the achievable temperature and the size of the system. In anticipation of the following chapters, concentrating collectors are required for this purpose. Leaving climate change aside, from the individual project it can be more favourable, however, to operate the low-temperature preheating with the solar plant and to reach the peak temperature with a fossil auxiliary firing in a hybrid or dual-source installation. The second source can then also compensate solar power fluctuations. Solar technology itself offers a number of possibilities for optimization, which will be examined in the course of the following chapters.

1.6 Sector Coupling—Power, Heat and Chemical Carriers

The term sector coupling does not refer to economic sectors, such as private households, commerce and services, industrial production and transport, but refers in general to the linkage of "energy sectors" electricity, heat and chemical carriers. The basic idea of the concept of sector coupling is that the sectors of the energy economy as well as the industry are to be interconnected and optimized in a common holistic approach to the energy sector, thus enabling a better and more favourable overall system. In perspective, sector coupling offers three main advantages:

- It allows decarbonizing all sectors of the economy with the help of a large/predominant share of renewable energies.
- A significant reduction in primary energy consumption through intelligent coupling of sectors using energy-efficient technologies.

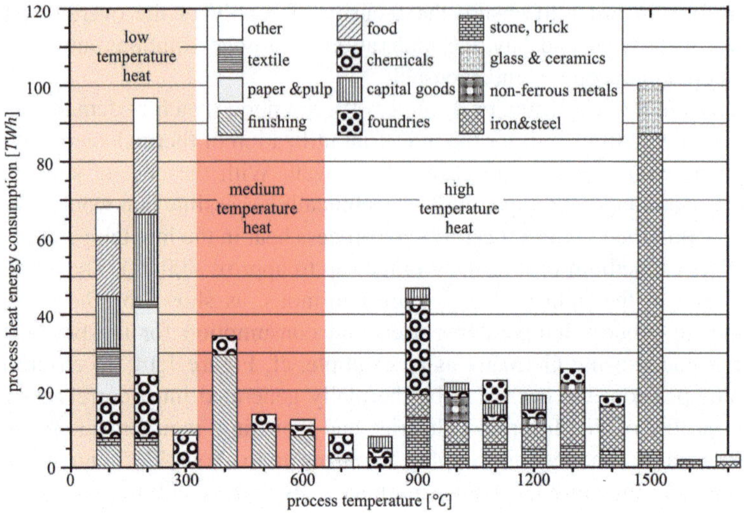

Fig. 1.14 Process heat energy consumption in [*TWh*] of Germany classified by temperature levels and different industrial goods of the manufacturing industry, modified from (Deutsche Energie-Agentur (DENA), 2016)

- Use of large and cheap energy storage outside the electricity sector through functional electricity storage (e.g. synthetic chem. carriers, heat) to satisfy demand flexibility for electrical energy during fluctuations in supply-dependent renewables.

The last point aims to reduce investments in resource-intensive, expensive electrical energy storage and to increase energy security by means of sector coupling.

But, it is also clear from the previously formulated perspective goals of sector coupling that it requires considerable challenges and material efforts that include the following aspects:

- Construction and operation of efficient conversion infrastructures (Power to Gas-$P2G$, Power to Peat-$P2H$).
- Planning, construction and maintenance of flexible transmission and distribution networks.
- Intelligent and efficient load management to meet demand-driven fluctuations while ensuring security of supply.

The concept of sector coupling is based on a diversified interaction of the different energy sources and there is no singular master solution for the different regions of the world due to different infrastructural prerequisites (degree of industrialization, orientation of the economy, etc.) but also different needs (e.g. heat demand due to geographical location or availability of renewable resources).

Figure 1.15 top outlines the current energy supply system with the generation sources participating in the energy market, the supply forms, the distribution system and the energy consumer side. Similar to the Sankey diagram, the energy transfer to the end consumer is mainly carried out via fossil energy sources in the horizontal level by means of corresponding distribution systems, for example heat or power grids, pipelines or gas stations. Vertical coupling of the consumer sectors occurs only to a very small extent on the distribution side. The functionality of this traditional energy concept is based on the fact that energy production and consumption are *balanced at all times.* Whereas this balancing of generation and consumption for electricity consumption is achieved through dedicated load management of the power grid at regional and supra-regional levels, where differences in the absolute magnitude of energy between generation and consumption at any given time t are regulated by market mechanisms (electricity trading arbitrage) and security of supply is ensured by a system of primary, secondary and ternary reserves, for example reserve power plants, hydroelectric power plants and the like. This ensures that the following equation is fulfilled within a control cycle dt.

$$\int_{production} \dot{P}(t)\,dt = E_{loss,distribution} + \int_{consumption} \dot{P}(t)\,dt, \qquad (1.6)$$

The majority of the difference between generation and consumption is ensured by the use of dispatchable fossil-based power plants or quasi-loss-free storage in the form of fossil carriers. Depending on the geographic characteristics of each country, pumped storage power plants can provide base load or residual load when electricity generation is insufficient. A production characteristic on the consumer side is desired but not required, nor is an intersectoral exchange of energy with an integrated storage function. As Eq. (1.6) illustrates, there is a direct coupling of generation (left side) and consumption (right side), so that the energy flow (electricity, heat, fuels) is almost exclusively from left to right. Buffering in the form of storage is not required either on the generation side, since it is provided there by the loss-free storage of fuels, or hydropower, but especially not on the demand side. An advantage of the classical system is that the absolute of the generation capacities compared to the consumption side are small compared to a sector concept and also the hierarchically organized distribution networks of the supply side are small. However, this is bought by the fact that large amounts of energy are either released unused into the environment as thermal waste energy and large amounts of fossil carriers on the consumption side are transported to the consumer in the transport network with low efficiencies, as shown in the Sankey diagram using the example of the USA, cf. Figure 1.12.

An integral concept based on sector coupling must not only consider the volatile character of renewable sources such as wind and solar technologies, but also their conflicting goals of temporal and local availability, as well as sometimes considerable differences between generation and consumption. In order to realize this concept, which is simplified illustrated in Fig. 1.15b a massive expansion of infrastructures and systems is required on the energy generation side as well as on the

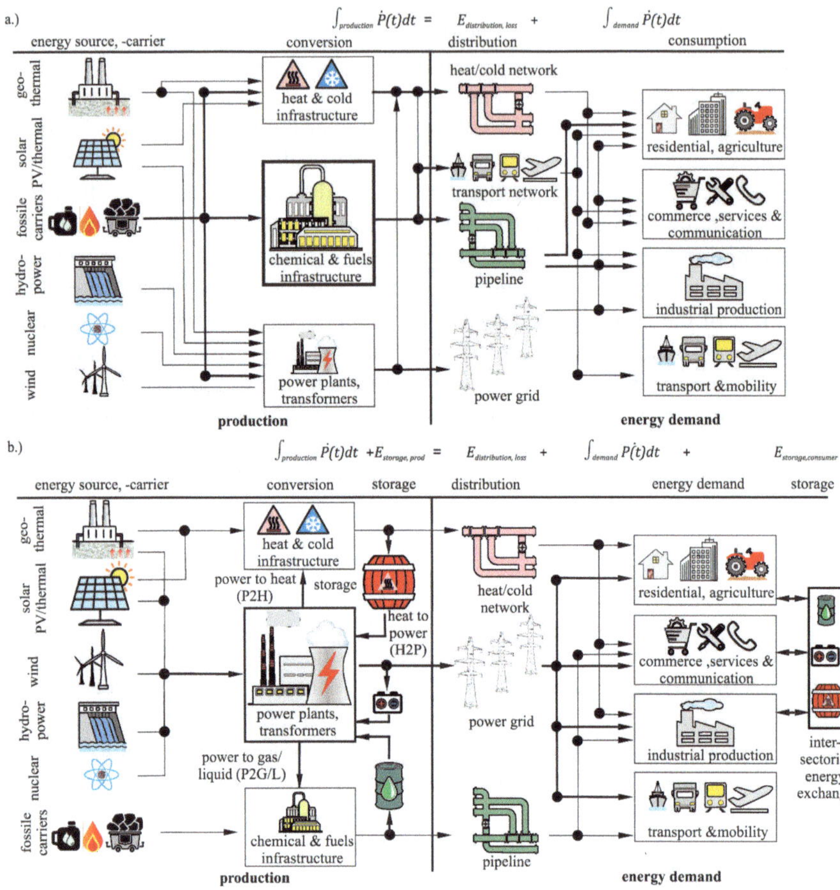

Fig. 1.15 Schematic of the generation, distribution and consumption structures of a conventional fossil based energy system **a** and a sector-coupled energy system based mainly on electricity **b** essentially fed by renewable (and nuclear) sources

distribution and consumer side. Electricity is at the centre of the generation side. In order to generate energy as CO_2-free as possible, the generation capacity of renewable sources and the associated distribution network must be massively expanded and is thus its generation capacity exceeds by far that of a conventional system. To compensate for the differences between generation and consumption at different times of the day and year, additional conversion technologies such as Power to Heat ($P2H$) and Power to Liquid/Gas ($P2G/L$) and their corresponding inversion technologies as well as additional storage facilities are required. This vertical linkage ensures not only the provision of a balancing function on the generation side, which the conventional system lacks, but also the feeding of storage to compensate for environmentally induced fluctuations in generation, thus contributing to the stabilization of the generation side, (left side of Eq. (1.7)). The distribution grid itself hardly changes its distribution function, but the preferred use of electricity as the hub of energy transport and distribution means a large-scale expansion of the power grid, since it must additionally provide energy requirements for mobility, industrial production, trade, commerce and services, as well as the private sector. Ideally, the distribution role of transport for energy products can be largely eliminated. On the consumer side, a sector-coupled concept means a mobility shift to electromobility, fuel cells or hydrogen, but also a shift to CO_2-free production technologies organized close to the consumer and thus relieve the distribution grids. Another aspect of a sector-coupled concept is that the consumer side is no longer passive, but is also assigned a producer or storage role, so that an intersectoral energy exchange is guaranteed. As a consequence of a sector-coupled energy system, both sides of Eq. (1.7) change dynamically, whereby the energy balancing quantities ($E_{storage,prod}$, $E_{storage,consumer}$) are of different dynamic sizes and can have both positive and negative signs.

$$\int_{production} \dot{P}(t)\,dt + E_{storage,prod} = E_{distribution,loss} + \int_{consumption} \dot{P}(t)\,dt + E_{storage,consumer} \tag{1.7}$$

The networking of all sectors, i.e. sector coupling in the vertical plane, should therefore contribute to the achievement of climate policy goals. This requires technical facilities, infrastructures and also market models of the various sectors to be more closely coordinated with one another in order to establish a comprehensive, intelligent energy system.

In order to drive forward the energy transition in a targeted and consistent manner, it is of great importance that the different areas and sectors of the energy system work together. So far, the focus has mostly been on the "electricity turnaround." However, in order to achieve the international climate targets, a heat and mobility transition is also imperative. Whether heat can be realized exclusively by electricity, for example via heat pumps, can be questioned, but the necessary establishment of a decentralized structure of heat networks, as well as regeneratively produced fuels with high energy density, is less so. Thus, a second important pillar of a smart energy system is a consistent expansion of generation and storage systems (especially electricity and heat) at the regional or neighbourhood level. This not only enables a consistent use of all types of waste heat and, as a consequence, a saving of fossil resources but also the relief of superregional distribution networks.

The integration of additional interconnected technologies also means a massive data flow based on modern information technologies as well as regulation and control strategies. The structure may look complex, but it also opens up the possibility of participation of new market players and the establishment of innovative business models.

What are the advantages and disadvantages of sector coupling?
Sector coupling brings with it various advantages. On the one hand, it promotes the decarbonization of all sectors through the use of renewable energies. This means that fossil fuels are gradually replaced by renewable energy sources. Sector coupling promotes flexibility in the demand for electrical energy, so that fluctuations associated with the use of renewable energies can be balanced out without having to make too large investments in energy storage. This circumstance, in turn, has a positive impact on energy security. Overall, the clear advantage of sector coupling is that it acts as a catalyst for the energy transition. By linking different sectors, synergy effects are brought about, allowing the use of renewable energy sources to be expanded further and more efficiently.

Unfortunately, compared to liquid or gaseous fuels, electricity cannot easily be stored in a vessel. Therefore, electricity must be stored electrochemically in the form of accumulators or in a modified form, such as pumped storage. However, this storage or conversion always involves energy losses. However, accumulator (often termed battery) storage is rather expensive overall and still exhibits a considerable consumption of scarce raw materials. Therefore, in some applications fossil fuels, which are expensive but can be stored at lower costs, are cheaper in business terms than technically complex electricity to heat (power to heat-$P2H$) or electricity to liquid/gas technologies (power to liquid/gas-$P2G/L$). Despite these challenges, it is essential that the energy transition is driven forward across sectors in a holistic approach with the help of intelligently deployed flexibilization and linking of power generation, power consumers and storage. It must also be clear that the realization of a sector coupling concept means a correspondingly high level of fragility with respect to attacks on the data infrastructure, since the data volumes are growing significantly and the number of active market participants is much greater. Another aspect is that the development towards a sector-coupled energy supply system must be designed to be barrier-free and technology-neutral on both the generation and consumer sides so that the most resource-conserving and energy-efficient solutions are applied and introduced into the market. This calls for research and technological development, but also for socially accepted frameworks and rules.

In this context, it is of central importance that the evaluation of a technology must be carried out on a holistic level. A simple approach to assessing a technical product or energy system in terms of CO_2 value is to allocate a CO_2 equivalent to all consumption of raw materials and inputs (i. e. self-consumption, infrastructure costs, provision of distribution networks) during the tentative projected operating life, in addition to manufacturing and recycling. This lifetime accumulated CO_2- equivalent can then be correlated with the benefit, for example in the case of an energy generation plant in CO_2 per kWh generated. If this kind of consideration is applied, no technology is really CO_2-free, including renewable energy production methods or processes, which is in accordance with the second law of thermodynamics, but it is possible to consistently search for Pareto front-optimal solutions, which represent an optimal solution in a multi-objective parameter problem.

References

British Petroleum BP p.l.c. (2021). *Statistical Review of World Energy 2021–70th edition.* London SW1Y 4PD, 1 St James's Square, UK: Whitehouse Associates, London. Retrieved from https://www.connaissancedesenergies.org/sites/default/files/pdf-actualites/bp-stats-review-2021-full-report.pdf

Butler, J. H., & Montzka, S. A. (2021). The Noaa Annual Greenhouse Gas Index (AGGI). Boulder, CO 80305–3328, Colorado, USA. Retrieved 2 28, 2022, from https://gml.noaa.gov/aggi/aggi.html

Deutsche Energie-Agentur (DENA). (2016). *Process heat in industry and commerce-Technology solutions for waste heat utilisation and renewable provision.* Chausseestraße 128 a, 10115 Berlin, Germany: Deutsche Energie-Agentur GmbH (dena). Retrieved from www.dena.de

ENI SpA. (2021). *O&GR-Volume 2: World Gas and Renewables Review 2020.* 00144 Roma, Piazzale Enrico Mattei, 1, Italy . Retrieved from https://www.eni.com/assets/documents/eng/scenari-energetici/WORLD-GAS-AND-RENEWABLES-REVIEW-2020-vol2.pdf

Lawrence Livermore National Laboratory. (2021). *Energy Flow Charts: Charting the Complex Relationships among Energy, Water, and Carbon.* 7000 East Avenue Livermore, CA 94550. Retrieved 3 1, 2022, from https://flowcharts.llnl.gov/

Myhre, G., Shindell, D., Bréon, F.-M., Collins, W., Fuglestvedt, J., Huang, J., Zhang, H. (2013). Chapter 8: Anthropogenic and Natural Radiative Forcing. In *Climate Change 2013: The Physical Science Basis. Contribution of Working Group I to the Fifth Assessment Report of the Intergovernmental Panel on Climate Change.* Cambridge University Press.

OECD/NEA &IAEA. (2020). *Uranium 2020-Resources, Production and Demand.* Organisation for Economic Co-Operation and Development. 75775 Paris Cedex 16,2 rue André-Pascal,. France: OECD/NEA Publishing.

Ritchie, H., & Roser, M. (2020). CO_2 and Greenhouse Gas Emissions. (O. W. Data, Ed.) Retrieved from https://ourworldindata.org/co2-and-other-greenhouse-gas-emissions

United Nations Population Division. (2020). *World Population Prospects.* Population Division, United Nations,, 2 United Nations Plaza, Rm. DC2–1950, New York, NY 10017 USA. Retrieved from https://population.un.org/wpp2019/Download/Files/1_Indicators%20(Standard)/EXCEL_FILES/1_Population/WPP2019_POP_F01_1_TOTAL_POPULATION_BOTH_SEXES.xlsx

World Bank. (2020). *Annual Report.* The World Bank, 1818 H Street, NW Washington, DC 20433, United States of America. Retrieved from https://data.worldbank.org/indicator

World Meteorological Organization. (2021). *WMO Greenhouse Gas Bulletin: The State of Greenhouse Gases in the Atmosphere Based on Global Observations through 2020.* Atmospheric Environment Research Division. Geneva, Switzerland: World Meteorological Organization. Retrieved from https://community.wmo.int/activity-areas/gaw

Solar Radiation

2

Abstract

In this chapter the physical processes and the composition of the sun leading to the sun as radiation source will be described. From that the properties of extra-terrestrial irradiation close to the earth are derived. Depending on the orientation of a surface different amounts of irradiation is received (horizontal, tilted, tracked). The atmosphere is interacting with this radiation through absorption and scattering depending on a number of parameters (aerosols, clouds) and modifies the incoming terrestrial radiation on the surface of the earth. For collectors not only the beam radiation directly from the sun, but also diffuse components from sky and ground reflected albedo are important. Short-wave diffuse sky radiation depends on the clearness of the sky and sun position and is in general anisotropic. Models for the angular distribution of diffuse irradiation describe this behaviour. Statistical methods exist to derive the diffuse fraction of global irradiation based on the clearness index. The basic methods for measuring solar radiation for different locations are presented, with terrestrial methods and satellite measurements. Using long-term time series representative meteorological data sets for planning purposes can be constructed.

Keywords

Sun physics • Sun geometry • Directional irradiation • Spectral solar radiation • Diffuse and global irradiation • Atmospheric attenuation • Terrestrial and satellite measurements • Statistical solar data

R. Stieglitz and W. Platzer, *Solar Thermal Energy Systems*,
https://doi.org/10.1007/978-3-031-43173-9_2

2.1 The Sun as Energy Source

2.1.1 Structure and Composition of the Sun

The sun is a star which provides the energy for life on our planet for billions of years. It draws its energy from the fusion of hydrogen nuclei, which are present as plasma in the inner core of the sun. As a result of the hot fusion process the sun has a temperature of about 10^7 °K in its interior. It is not a rigid body but is consisting of several zones as shown schematically in Fig. 2.1.

The age of our sun is estimated at about 4.5 *billion* years. During this time the hydrogen concentration has decreased by about 8% and the helium abundance therefore has reached this value 8%. However, the hydrogen supply is still sufficient to ensure that the current output can be delivered for more than additional 10 *billion* years. The sun radiates into space 4×10^{26} W of electromagnetic radiation power, of which only 2×10^{17} W reach the earth. For comparison, the current energy consumption worldwide is at about 1.6×10^{13} W. Thus, although the energy of the sun is not renewable, it may provide energy to our planet for an incredible long time and is not exhausted within a few generations of man as conventional energy sources might be.

90% of the energy is generated in the central *core* within $0.23R$, with R being the radius of the sun. The core is solid due to the high pressure and has an average density of 10^5 kg/m^3, which is about eight times the density of lead. The temperature rises to 40 *million* °K in the centre. Although the core is only 15% of the volume, it contains about 35–40% of the solar mass. From the core outwards, the energy transport takes place through radiation. However, the absorption is so strong at this high density that the radiation diffusion takes a long time, and therefore the radiation which we see now was generated near the centre of the sun about 1–2 *million* years ago.

Faster transport is achieved in the next zone, the *convection zone* extending from about $0.7R$ nearly to the edge. Convection cells exist there, which can be recognized from the outside as granules. The lifetime of such cells is a few minutes. The

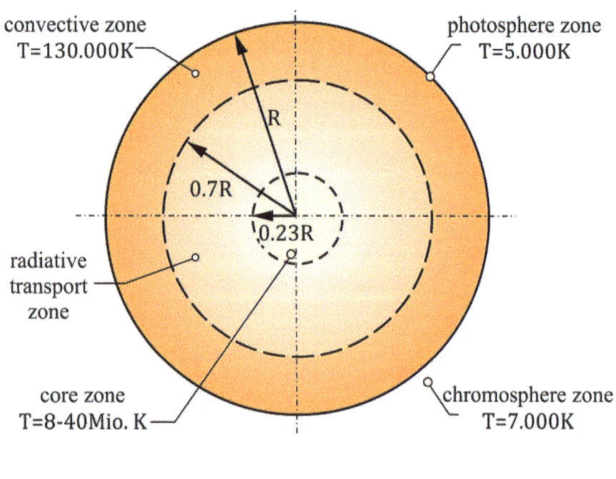

Fig. 2.1 Schematic structure of the sun

Table 2.1 Some key data related to radiation emission of the sun from McCracken and Scott (2005)

	Phys. unit	Value
Diameter	m	1.392×10^9
Surface	m^2	6.088×10^{18}
Mass distribution	H_2	0.75
	He	0.23
	Other	0.02
Total mass	kg	1.99×10^{30}
Mean density	kg/m^3	1.41
Core temperature	K	1.6×10^7
Surface temperature	K	5776
Core pressure	MPa	4×10^{10}

Table 2.2 Some general data of the sun from McCracken and Scott (2005)

	Phys. unit	Value
Rotation of the poles	days	34
Rotation of the equator	days	25.4
Age	years	4.6×10^9
Lifetime (expected 1/2 burning time)	years	5×10^{10}
Fuel consumption H_2	Mt/s	600
Helium production	Mt/s	596
Mass difference	Mt/s	4.3

density there is about $70\,kg/m^3$ and the average temperature about 1.3×10^5 K. Some characteristic sun data are condensed in Tables 2.1 and 2.2.

The thin *photosphere* adjoins the convection zone. The temperature at the edge is about 5000 °K. In the low density photosphere again radiation takes over the heat transport. Here is the origin of the radiation energy emitted by the sun. In the photosphere, which is also called the reverse layer, the temperatures rise to about 7,000 K towards the outside. Its thickness is estimated to be 400–500 km and the density varies from 10^{-4} to $10^{-5}\,kg/m^3$.

Above the photosphere lies the *chromosphere*, which is about 8,000–10,000 km thick and has an even lower density of 10^{-8} to $10^{-12}\,kg/m^3$. The temperature in the chromosphere increases from 7,000 K inside to about 3×10^5 K outside. Taking into account the different densities and temperatures of the photo- and the chromosphere, which constitute the thin surface layer of the sun, an equivalent "*surface*" temperature of 5776 K has been derived (McCracken & Scott, 2005). Even further outside is the *corona*, where the temperature rises to 10^6 K due to magnetic heating.

2.1.2 Fusion Process Within the Sun

In order to understand the nature of the radiation generation by the sun one has to have a look at the fusion process. The sun draws its energy from the fusion of hydrogen nuclei, which due to the high temperatures are present as plasma in the inner core of the sun. In the plasma state the hydrogen atoms are present as protons (p). According to the enormous pressure and gravitational force of the sun, a multi-stage fusion process of hydrogen into helium takes place. The first step is the proton collision process ($p = {}^1H$), which leads to the generation of a deuteron ($D = {}^2H$).

$$p + p \rightarrow D + e^+ + \nu + 0.42\,\text{MeV}. \tag{2.1}$$

Here, e^+ is a positron and ν an electron neutrino. Like all neutron physical processes, the collision of protons is a statistical process in which the effective cross section, the density of the reaction partners and their relative velocities play a role. The reaction cross section of a proton is comparatively small and amounts to 10^{-55} barn (1 barn = 10^{-24} cm^2). The resulting positron e^+ annihilates with an electron e^- from the plasma, i.e. both are completely converted into energy. The mass of the electron–positron pair is released as kinetic energy of two gamma quanta γ

$$e^+ + e^- \rightarrow 2\gamma + 1.022\,\text{MeV}. \tag{2.2}$$

The resulting deuterium (D) at some time reacts with another proton to form the light helium isotope ^{3}He:

$$D + p \rightarrow {}^3\text{He} + \gamma + 5.49\,\text{MeV}. \tag{2.3}$$

Two light helium nuclei ^{3}He then fuse into a single ^{4}He-nucleus (α-particle), releasing two protons. These are available for new fusion processes:

$${}^3\text{He} + {}^3\text{He} \rightarrow {}^4\text{He} + p + p + 12.86\,\text{MeV}. \tag{2.4}$$

Figure 2.2 illustrates the complete sequence of the fusion process in its individual steps. In the mass balance, the following evaluation results for hydrogen–helium fusion:

$$\begin{array}{l} 4p \equiv 4 \times 1.00727647\ u \\ {}^4\text{He} - 4.0015060883\ u \\ \underline{2e^+ - 2 \times 0.00054858\ u} \\ difference\ \ 0.02650263\ u \\ => \Delta m = 4.4 \times 10^{-29}\,\text{kg}, \end{array} \tag{2.5}$$

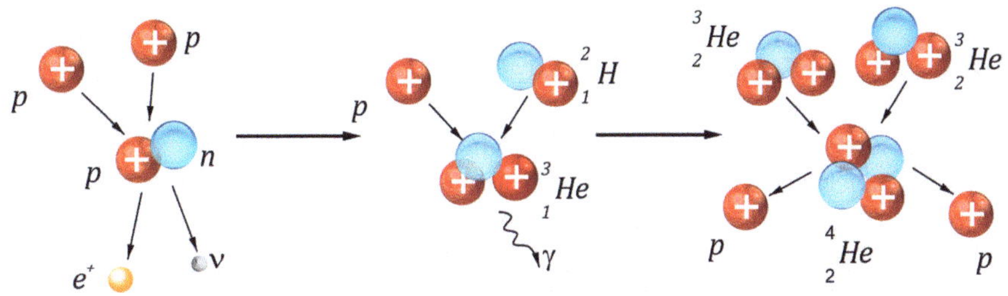

Fig. 2.2 Single steps in hydrogen fusion to Helium (Gerthsen et al., 1995)

where u is the atomic mass unit with 1.66054×10^{-27} kg. Using the Einstein formula (2.6) with c being the speed of light ($=2.9979 \times 10^8$ m/s), the mass defect Δm results in an energy production per fusion process of 3.955×10^{-12} J = 24.68 MeV.

$$E = m \cdot c^2. \tag{2.6}$$

The energy from the positron–electron reaction has to be added with 1.022 MeV because the positron e^+ is annihilated thereby creating kinetic energy. This yields a value of 26.7 MeV for hydrogen–helium (H–He) nuclear fusion, equivalent to the sum of the reaction energies from Eqs. (2.2) to (2.5). However, the kinetic energy of 0.26 MeV carried away by each neutrino escaping the sun does not contribute to the sun's internal energy. Therefore, only 26.2 MeV are used for the heating process of the sun.

When use with the total mass loss rate of Table 2.2 then using Einstein's relation we can estimate solar power emitted by the sun by (2.7) and dividing it by the mass defect of a single fusion reaction we arrive at the number of fusion processes occurring per second.

$$P_{fusion,\,sun} = \Delta\dot{m} \cdot c^2. \tag{2.7}$$

The results are 3.86×10^{26} W emitted power and 9.8×10^{37} fusion reactions per second. Now, using the average distance sun-earth as the radius of a sphere around the sun we can calculate the average extra-terrestrial irradiance at this position and obtain 1374 W/m^2. This value represents the so-called solar constant (i_{SC}) being introduced in the next paragraph.

2.2 Extra-terrestrial Solar Radiation

2.2.1 Solar Constant

The solar constant i_{SC} describes a specific, long-term averaged extraterrestrial solar irradiance integrated over all wavelengths, which hits the earth perpendicularly per square meter at an average distance between earth and sun without the influence of the atmosphere. It can be calculated using the average orbit radius of the earth around the sun. From the geometry between sun and earth Fig. 2.3 the view angle θ_s of the sun disk as seen from the earth can be calculated as $\pm 0.267°$ or ± 4.65 mrad. Although the luminosity decreases from the centre of the solar disk to the outside, for most terrestrial applications, it is sufficient to assume a uniform radiation density over the entire solar disk.

The determination of the solar constant has been the subject of several investigations in the past. A measurement of the extra-terrestrial irradiance yields only a momentary picture, of course, and is dependent on solar activity. Already in the early twentieth century Abbot and colleagues from the Smithsonian institute did pioneering work using terrestrial measurements on high mountains to extrapolate to the solar constant. Abbot's value of $1322\,\text{W/m}^2$ was later corrected upwards by Johnson to $1395\,\text{W/m}^2$ (Johnson, 1954), showing the large error inherent in these methods. The availability of ballone and spacecraft permitted the determination outside of the atmosphere with a variety of instruments. This resulted in a value of $1353\,\text{W/m}^2$ with an estimated error of $\pm 1.5\%$ (Thekaekara, 1976). The measurements and accuracy have been re-examined by Fröhlich who published a new value of $1373\,\text{W/m}^2$ with an uncertainty of 1–2% (Fröhlich, 1977). This is about 1% above the previously used value of $i_{SC} = 1353\,\text{W/m}^2$. Based on additional new spacecraft measurements (Willson, 1978) in 1982 by the World Meteorological Organization in Geneva and the World Radiation Centre (WRC) have proposed

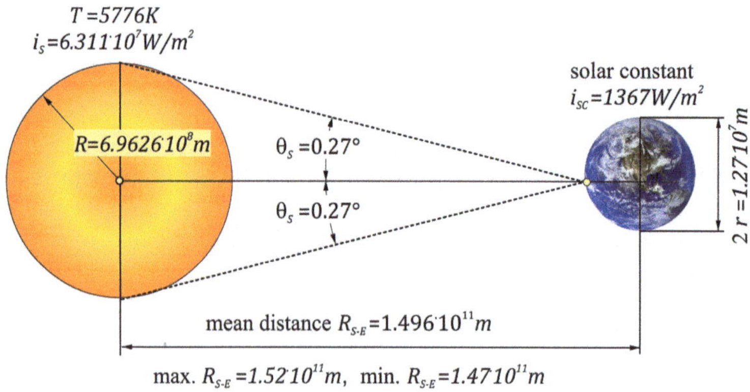

Fig. 2.3 Derivation of the intensity i_S of solar radiation as a function of the distance sun-earth R_{S-E}

$i_{SC} = 1367\,\mathrm{W/m^2}$ as solar constant based on measurements by various authors. This will be the value which will be used in this book:

$$i_{SC} = 1367\,\mathrm{W/m^2} \pm 1\%. \tag{2.8}$$

A more detailed discussion on the history of irradiation measurement and the determination of the solar constant can be found in Duffie and Beckmann (2020). They also show how to derive the magnitude of the solar constant using the Stefan-Boltzmann law and the sun's equivalent surface temperature.

More about the periodicity and the relation of the fluctuations with the activity of the sunspots can be found at Pecker (1994) and Fröhlich et al. (1992). These reports show that in periods of about one year the solar intensity varies by less than $\pm 0.1\%$. In the UV range with wavelengths below 170 nm, the radiation can vary by a factor of 2. In the X-ray range at 0.2–3 nm the radiation power changes by up to 2 orders of magnitude. The current most accurate total spectral irradiance (TSI) values from the Total Irradiance Monitor (TIM) on NASA's Solar Radiation and Climate Experiment (SORCE) is $1360.8 \pm 0.5\,\mathrm{W/m^2}$ during the 2008 solar minimum as compared to previous estimates of $1365.4 \pm 1.3\,\mathrm{W/m^2}$ established in the 1990s (NASA, 2022).

Of greater importance are the fluctuations of extra-terrestrial radiation as a result of the elliptical orbit of the Earth around the Sun. As a result of the orbital eccentricity, the distance between the sun and the earth fluctuates between 1.471×10^{11} m and 1.521×10^{11} m yearly. The resulting periodically changing distance causes the irradiance on a unit area of one square meter perpendicular to the direction of irradiation to fluctuate by $\pm 3.3\%$; this results in a day-dependent extraterrestrial solar constant $i_{dir,extra}$. It can be calculated for the nth day of the year according to Eq. (2.9)

$$i_{dir,extra} = i_{sc} \cdot \left(1 + 0.033\,\cos\left(\frac{360}{365} \cdot n\right)\right), \tag{2.9}$$

in which $i_{dir,extra}$ the extraterrestrial intensity of solar radiation in watts on a surface of $1\,\mathrm{m^2}$ is perpendicular to the direction of irradiation or to the direction of the sun/earth. Figure 2.4 shows the annual variation in extra-terrestrial solar radiation intensity per square meter as a function of the season.

As mentioned already, from the solar constant the radiant power of the Sun can be calculated by multiplying it by the surface A of that envelope around the Sun, which has the radius of the mean distance from Earth to the Sun with $R_{S-E} = 1.496 \times 10^{11}$ m. With the help of Eq. (2.10) the solar radiation power Φ is obtained:

$$\Phi = i_{SC} \cdot A = i_{SC} \cdot 4\pi \cdot R_{S-E}^2 = 3.845 \times 10^{26}\,\mathrm{W}. \tag{2.10}$$

If that radiation power is equated with the radiation emitted by a blackbody sphere with radius of the sun, an equivalent emitting temperature of about 5800 K results—this is approximately the temperature of the sun at the inner part of the

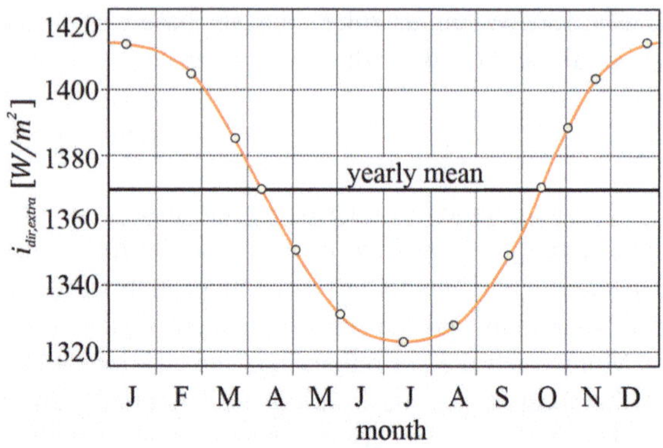

Fig. 2.4 Derivation of the extra-terrestrial intensity $i_{dir,extra}$ of solar radiation as a function of the distance sun-earth R_{S-E} along a year

relatively transparent chromosphere. With an average distance of the earth from the sun with $R_{S-E} = 1.496 \times 10^{11}$ m it results in a mean radiation density of the sun of $i_S = \Phi/(4 \cdot \pi \cdot R^2) = 6.311 \times 10^7$ W/m^2, where R is the solar radius with $R = 6.9626 \times 10^9$ m.

Using the Stefan-Boltzmann law formulated in (2.11), one can indicate the radiant power Φ emitted by a "black body" of area A and absolute temperature T. Therefore, if the radiant power Φ of the sun is known from Eq. (2.11) one can assume that the sun is a "*black radiator*" and indicate its surface temperature.

$$\Phi = \sigma \cdot A \cdot T^4, \tag{2.11}$$

where σ is the Stefan Boltzmann constant with $\sigma = 5.67.10^{-8}\,\mathrm{W}/\left(\mathrm{m}^2\mathrm{K}^4\right)$. This results in a value of the solar surface temperature $T_S = 5776\,\mathrm{K}$.

2.2.2 Spectrum of Extra-terrestrial Irradiation

The simplified picture of the structure of the sun shown above indicates that the radiation emitted by the sun into space is composed of several components and stemming from different zones close to the surface of the sun. This implies that the extra-terrestrial solar radiation has a complex spectrum that can only be approximated with a "*black body spectrum*" with a temperature of $T = 5776$K. The real extraterrestrial spectrum of solar radiation hitting the earth is shown in the Fig. 2.5 and compared to the 5776K blackbody spectrum. It covers the following wavelength ranges:

- Ultraviolet (UV) $\lambda < 380$ nm with an energy share of 6.4%,
- Visible $380 < \lambda < 780$ nm with an energy share of 48%,
- Infrared (IR) $\lambda > 780$ nm with an energy share of 45.6%.

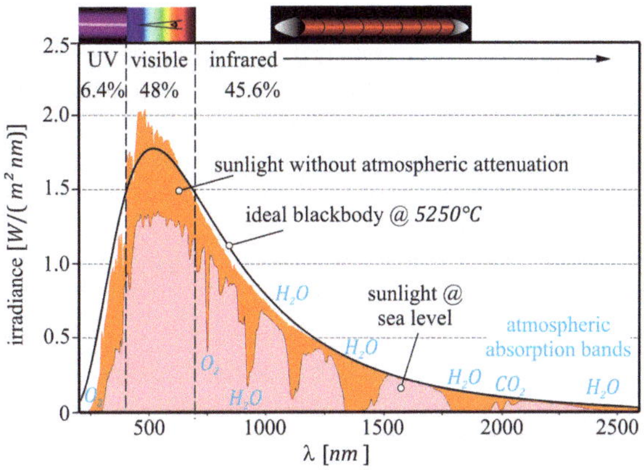

Fig. 2.5 Spectrum of solar radiation Φ as a function of wavelength λ extraterrestrial from measurements and in comparison the spectrum of a black body with 5800 °K. The intensity of the black body reduced by $((R/R_{S-E})^2 = 2.165 \times 10^{-5}$. The visible range is $380 < \lambda < 780$ nm

The visible part is determined by the spectral sensitivity of the human eye. The integration of the specific spectral irradiance over the individual wavelengths provides the energy content of the individual components. As a result, *the infrared portion of solar radiation is of the same energetic importance for solar thermal use as the visible light range.*

A real body cannot emit more thermal radiation at any wavelength than a "*blackbody*", which is therefore an ideal source of thermal radiation. Moreover, since the spectrum of the blackbody does not depend on any parameters other than temperature, and in particular on any material properties, it is a useful reference for many purposes. Its spectrum is described by Planck's radiation law (2.12).

Planck assumed in his empirically found formula that light is not absorbed and emitted continuously, but only discretely in *quanta* (today we speak of photons). According to this, the electromagnetic radiation of frequency ν can only be emitted and absorbed in discrete energy packets of the size $E = h \cdot \nu$. The Planck constant h is the quantum of electromagnetic action that relates a photon's energy to its frequency. Its magnitude was derived from the adaptation the formula law (2.12) to the experimentally determined values of the cavity radiation and assumes the value $h = 6.6261 \times 10^{-34}$ Js.

The spectral specific radiant intensity $\phi(\lambda, T)$ is dependent on wavelength (λ) and temperature (T). $\phi(\lambda, T)$ is the radiant power radiated into the entire hemisphere by the area element dA in the wavelength range between λ and $\lambda + d\lambda$. Planck's law, however, can also be formulated as a distribution in the frequency range between ν and $\nu + d\nu$. It is calculated according to Eq. (2.12):

$$\phi(\lambda, T) = \frac{2\pi\, h\, c^2}{\lambda^5} \frac{1}{exp\left(\frac{hc}{\lambda k_B T}\right) - 1}, \tag{2.12}$$

where c is the speed of light with $c = 2.9979 \times 10^8$ m/s and k_B the Boltzmann constant with $k_B = 1.36 \times 10^{-23}$ J/K. The relationship between wavelength (λ) and frequency (ν) is given by $c = \lambda \cdot \nu$ and the unit of the spectral specific radiant intensity $\phi(\lambda, T)$ is $\mathrm{Wm^{-2}\,\mu m^{-1}}$ in the SI-system.

According to Planck's law, the thermal radiation emitted by a black body consists of a mixture of electromagnetic waves of a wide range of wavelengths. The maximum of the distribution depends on the temperature and is given by Wien's displacement law

$$\lambda_{max} = \frac{2897.8\,\mu\mathrm{m\,K}}{T}, \tag{2.13}$$

where the constant 2897.8 (μm K) is called Wien's displacement or Lummer's constant.

Figure 2.6 shows Planck's radiation spectra for different temperatures as a function of the wavelength between 300 and 1000 K in linear representation. The result is a bell shape with a distinct maximum, a steep decrease to short wavelengths and a longer decrease to long wavelengths. As required by Wien's displacement law (2.13) the position of the radiation maximum shifts to shorter wavelengths with increasing temperature. At the same time, according to the Stefan-Boltzmann law (2.11) the total specific radiation increases with the fourth power of the absolute temperature.

This disproportionate increase in radiation intensity with increasing temperature makes it difficult to display curves over a wider temperature range in a diagram. Therefore, a double-logarithmic division is chosen in the Fig. 2.7. At 300 K (≈ambient temperature) the maximum of this curve is $\lambda = 10\,\mu\mathrm{m}$. This means that the exchange of thermal radiation from objects at room temperature takes place in long-wave infrared. Typical infrared thermometers and thermographic cameras work in this area. The curve for 3000 K corresponds to the typical radiation spectrum of an incandescent lamp. Now already a part of the emitted radiation according to Fig. 2.7 is emitted in the visible spectral range. At the effective temperature of the sun of $T = 5776\,\mathrm{K}$, the radiation maximum lies in the centre of the visible spectral range.

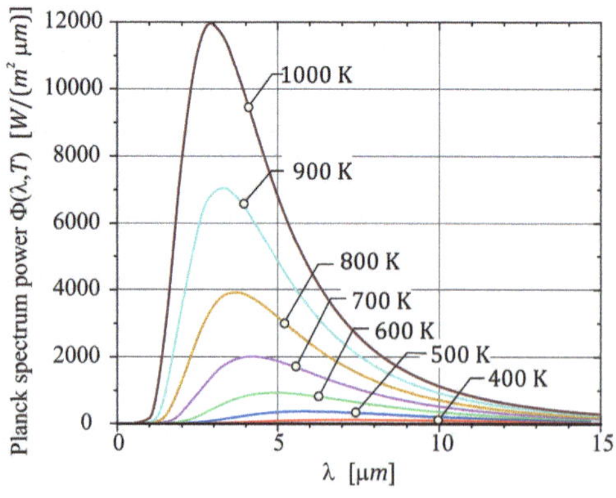

Fig. 2.6 Planckian radiation spectrum of a "*black*" radiator at different temperatures as a function of wavelength λ (μm)

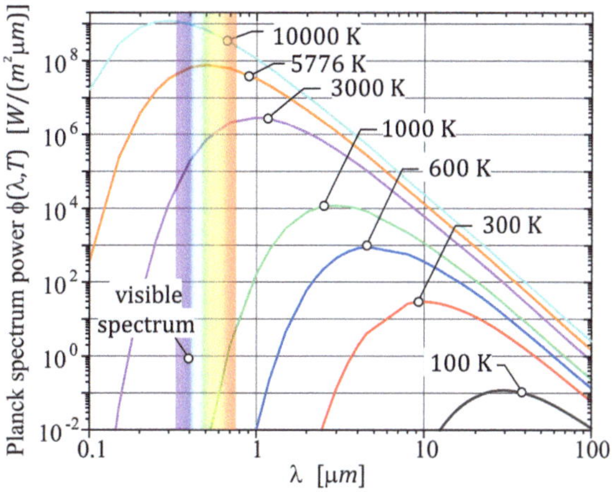

Fig. 2.7 Planck's radiation spectra as a function of wavelength λ for different temperatures in a double-logarithmic plot

2.2.3 Directional and Temporal Variation of Extra-terrestrial Irradiation

2.2.3.1 Direction of Direct Radiation

The intensity of the direct solar radiation on a surface and hence the energy that can be obtained depends mainly on the position of the earth in its orbit, the direction of the incident solar beam, and the orientation of the solar absorber (collector). Many factors play a role here, which are explained in the following section and can be expressed algebraically with the help of geometric relations. It's these:

- the elliptical orbit of the earth around the sun, the so-called ecliptic,
- the declination of the Earth, i.e. the inclination of the Earth's north–south polar axis to the ecliptic (seasonal influence),
- the rotation of the earth (influence of the time of day),
- the dependence of the position of the sun on the geographical latitude Φ and
- the influence of short term climatic aspects (weather).

The irradiation direction of the solar direct radiation onto an arbitrarily oriented, flat surface A is determined by angular relationships. Figure 2.8 shows the different angles of such a configuration. In a first step we assume that the area of the terrestrial site is horizontal.

- θ is the incident angle of the beam radiation between the direction of irradiation and the normal of an arbitrarily oriented and tilted surface A.
- θ_z is the solar zenith angle. This is the angle between the sun and the normal on the horizontal plane where area A is located.
- Φ is the angle of the geographical latitude (e.g. Freiburg $\Phi = 48°N$) and is $\Phi = 0°$ at the equator and $\Phi = 90°$ at the poles. Latitudes on the Southern hemisphere are negative.

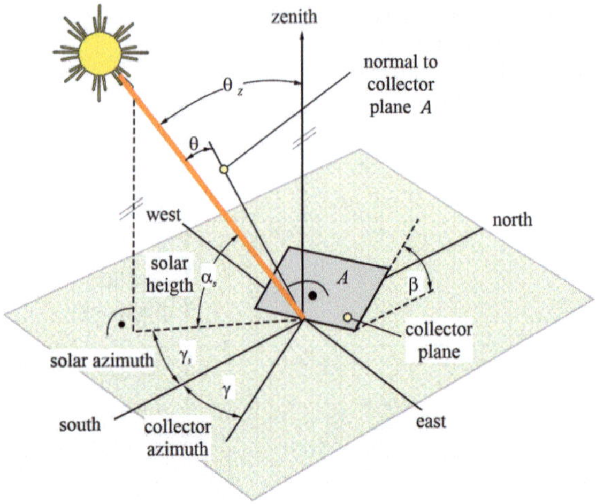

Fig. 2.8 Definition of the angles for calculating the direction of incidence of direct radiation on a surface A, which is arbitrarily oriented towards the Earth

- α_s is the sun's elevation angle or solar altitude angle. The following applies: $\alpha_s + \theta_z = 90°$.
- β is the tilt angle of the area A against the horizontal plane.
- δ is the declination of the earth and describes the angle between the sun's highest point and the equator and varies between $-23.5°$ and $+23.5°$.
- γ is the azimuth angle of the surface normal of A. It describes the angular difference of the horizontal projection of the surface normal from the south direction (West is $+90°$, East $-90°$). Observe that different systems exist to characterize the angles, e.g. in some literature South is indicated 180°.
- γ_s denotes the azimuth angle of the irradiation direction. It indicates the deviation of the horizontal projection of the irradiation direction from the south direction.
- ω is the hour angle in the plane perpendicular to the earth's rotation axis: (1 h = 15°; morning $\omega < 0$ and afternoon $\omega > 0$).

Looking into the geometrical relation between different planes of earth's movement and the position of a horizontal area on the earth's surface, the declination angle, and the latitude angle, as well as the hour angle affect the angle of incidence. If we describe the direction of the solar beam by a unit vector $\vec{s}$ and the normal of the tilted plane with the vector $\vec{n}$, then the cosine of the incidence angle is given by the scalar product of the normal vector and the solar vector:

$$\theta = \cos^{-1}(\vec{s} \cdot \vec{n}) = \cos^{-1}(\cos\,\alpha_s\ \sin\,\beta\ \cos\,(\gamma_s - \gamma) + \sin\,\alpha_s\ \cos\,\beta). \quad (2.14)$$

When the expressions for the solar altitude and azimuth angles are used, this expands to the following angular relationship:

$$\cos\theta = \sin\delta\sin\Phi\cos\beta - \sin\delta\sin\Phi\cos\beta\cos\gamma + \cos\delta\cos\Phi\cos\beta\cos\omega + \cos\delta\sin\Phi\sin\beta\cos\gamma\cos\omega + \cos\delta\sin\beta\sin\gamma\sin\omega, \quad (2.15)$$

in which the hour angle ω as well as all other angles have to be used in radian. The hour angle ω here corresponds to the true solar time in degree (at 12 noon $\omega = 0°$; during the morning is $\omega < 0$ and negative, while during the afternoon is ω positive). Within a day, the hour angle can be given with an accuracy of ≈10 s with a daily correction around noon with $\omega = 0$ using Eq. (2.16)

$$\omega = \frac{360\,2\,\pi}{24}(12 - t_{loc}). \quad (2.16)$$

t_{loc} in Eq. (2.16) refers to the true local solar time, which must be corrected by the legal time difference in case of summertime. t_{loc} is counted from 0 to 24 h.

In the relation for the true solar time t_{loc} a correction Δt_{loc} for the geographical longitude L_{loc} must also be included. The longitude-dependent correction Δt_{loc} results from the difference between the longitude of the local time zone of the location L_{loc} and the standard longitude L_{st} on which the legal time is based. For Δt_{loc}, the following relationship results:

$$\Delta t_{loc} = \frac{24}{360}(L_{loc} - L_{st})\,(\text{h}). \quad (2.17)$$

The daily time differences mentioned above add up and result in deviations of up to 15 min between noon of true solar time and noon of mean solar time. The causes are the elliptical orbit around the sun and the inclination of the earth's axis to the ecliptic. After one or more rotations, the sun is seen from a different angle because of the rotation around it. As a result, the Earth needs more or less than 24 h to turn the same side to the sun again. In order to obtain the local true solar time t_{loc}, therefore, a so-called equation of time is additionally required, which describes the difference between the true and mean solar time depending on the season. This equation of time (ET) is described by Eq. (2.18).

$$ET = \sum_{k=1}^{3}[a_k\cos(k\cdot J) + b_k\sin(k\cdot J)]\ \ (\text{h});\ \ J = \frac{2\pi}{365.25}\cdot n, \quad (2.18)$$

where n is the number of the day in the year starting with 1, January 1st and a_k and b_k are constants with the values $a_1 = 0.0072$, $a_2 = -0.0528$, $a_3 = -0.0012$, $b_1 = -0.1229$, $b_2 = -0.1565$ and $b_3 = -0.0041$. The maximum error of the Eq. (2.18) is 0.8 s. The result of the ET equation is shown in the Fig. 2.9 as a curve over one year.

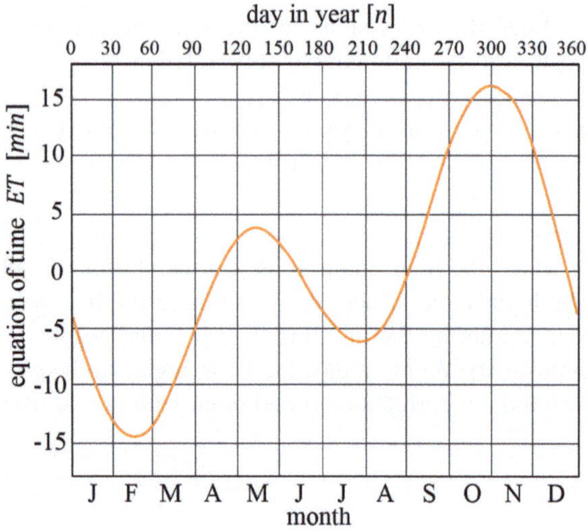

Fig. 2.9 Time correction ET from Eq. (2.18) of the true solar time due to the variation of the Earth's rotation over the year

The Eq. (2.18) ET is important if a concentrating mirror with a small acceptance angle θ_a of a few degrees is tracked with the computer. Here too, 1 h corresponds to about 15°. After all corrections over season, legal time, and longitude the local true solar time t_{loc} is given by Eq. (2.19).

$$t_{loc} = t_{st} + \frac{24}{360}(L_{loc} - L_{st}) + ET \quad \text{(h)}. \tag{2.19}$$

where t_{st} denotes the time of the local time zone (from 0 to 24 *o'clock*), L_{loc} is the longitude of the location considered and ET is the equation of time in hours according to Eq. (2.18).

The declination δ fluctuates due to the inclination of the earth's axis to the ecliptic. The declination of the Earth relative to the Sun can be determined for each day n of the year using Eq. (2.20).

$$\begin{aligned} &\delta = \arcsin\left[0.398 \cdot \sin\left(2\pi\,\tfrac{n-Eq}{365{,}25}\right)\right] + \Delta \quad (°) \\ &\Delta = 0.026 \cdot \sin\left(2\pi\,\tfrac{n-173}{365{,}25}\right) \quad \text{for} \quad 173 < n < 365; \\ &\Delta = 0 \qquad\qquad\qquad\qquad\qquad\quad \text{for} \quad n \le 173, \end{aligned} \tag{2.20}$$

where Eq indicates the spring equinox (lat. equinox). The value of Eq fluctuates between 79 and 79.75 (leap year: 79, leap years +1 : 79.25, leap years +2 : 79.5, leap years +3 : 79.75). The declination is zero at the equinox, +23.45° at the summer solstice and −23.45° at the winter solstice, as Fig. 2.10 shows.

The approximation of Eq. (2.21) gives the declination with a maximum error of 0.2°. A simplified relationship is represented by the following equation:

$$\delta = 23.45 \cdot \sin\left(360\frac{284 + n}{365}\right). \tag{2.21}$$

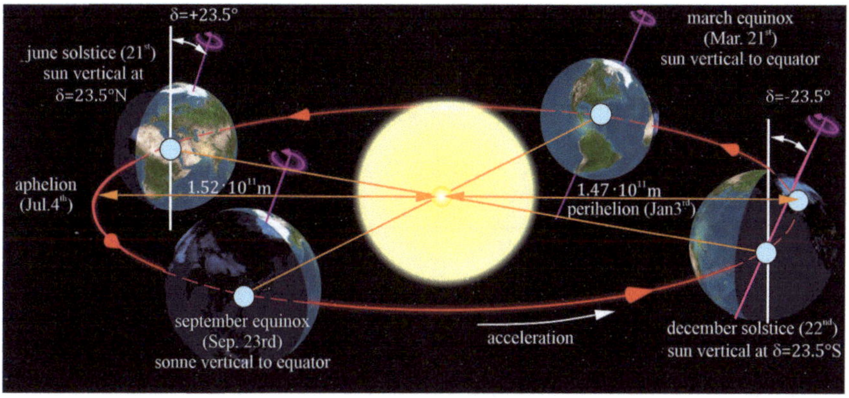

Fig. 2.10 Declination as a function of the season modified from https://allgeographynow.wordpress.com/tag/suns-declination/

The leading and trailing time compared to the mean time is called the equation of time (ET). If they are plotted against the angular distance of the Sun from the celestial equator, they take the form of an eight, a so-called analemma, which is illustrated in Fig. 2.11. In order not to have to commit to a particular location, however, the equation of time is usually shown as a function of the declination of the sun.

2.2.3.2 Direction of Irradiation on a Horizontal Surface

When irradiating a horizontal surface, the two angles are $\beta = 0°$ and $\gamma = 0°$. The angle to the surface normal changes to the zenith angle. This simplifies the Eq. (2.15):

$$\cos\theta_z = \sin\Phi \cos\delta \cos\omega. \tag{2.22}$$

For the derivation of Eq. (2.22) the cosine theorem has been used, for which the following applies: $\cos(\alpha \pm \beta) = \cos s\alpha \cos\beta \mp \sin\alpha \sin\beta$. The sun elevation angle α_s is linked to the zenith angle θ_z in the following way $\alpha_s + \theta_z = 90°$; as also shown in the Fig. 2.12. So it follows:

$$\cos\theta_z = \sin\alpha_s. \tag{2.23}$$

If the equation of time (2.18) for ET is included in the calculation of the angle of the sun's elevation, the sun elevation is obtained as a function of time for each day of the year. The graphical representation shows that for each day there is a loop-shaped form of the solar elevation angle. Figure 2.12 shows the result with the zenith angle θ_z or the solar elevation angle α_s as a function of the azimuth angle γ_s, the declination δ or season and the time of day. An analysis of Eq. (2.15) shows that in summer in the northern hemisphere the sun rises north of the

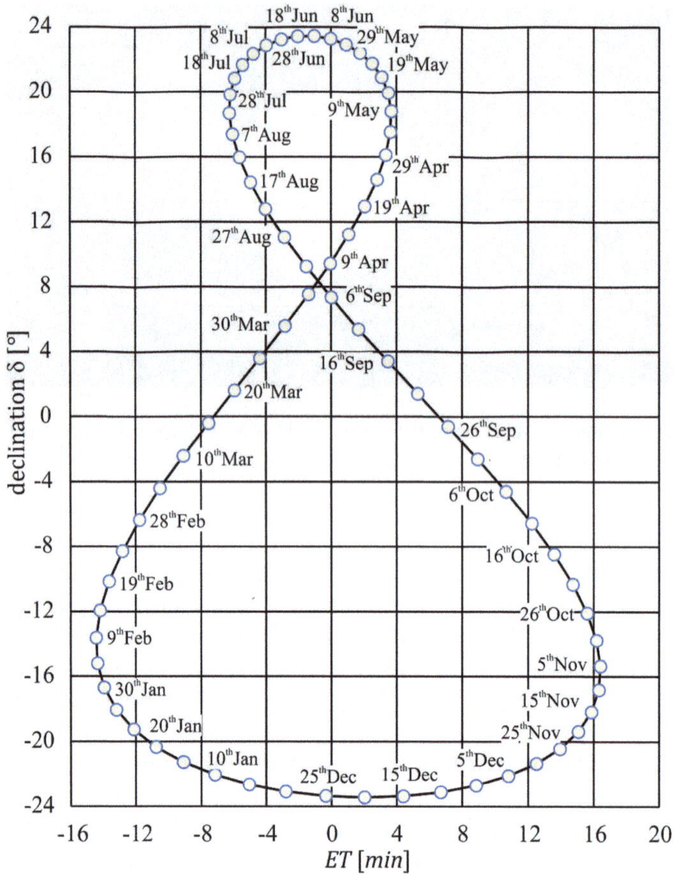

Fig. 2.11 The analemma shows the declination and equation of time (ET) for each day of the year

east–west axis. North of the polar circle depending on the latitude the sun doesn't even set at all. This means that a vertical collector at latitude 48° that is exactly oriented to the south without any tracking sees the sun on the front only between 8 and 16 o'clock (Fig. 2.12) and thus only 8 h (!!).

Figure 2.13 shows for the same location the solar elevation angle α_s as a function of the azimuth angle γ, the time of day and the declination δ (here in the form of the day in the year). In simpler representations the hourly values are entered as simple lines without consideration of the equation of time.

2.2.3.3 Direction of Irradiation on a Vertical Surface

For a vertical surface, with $\beta = 90°$ and $\gamma = 0°$ the Eq. (2.15) becomes:

$$\cos\theta_z = \sin\ \delta\ \cos\Phi\ \cos\gamma + \cos\ \delta\ \sin\Phi\cos\gamma\cos\omega + \cos\delta\sin\gamma\sin\ \omega. \quad (2.24)$$

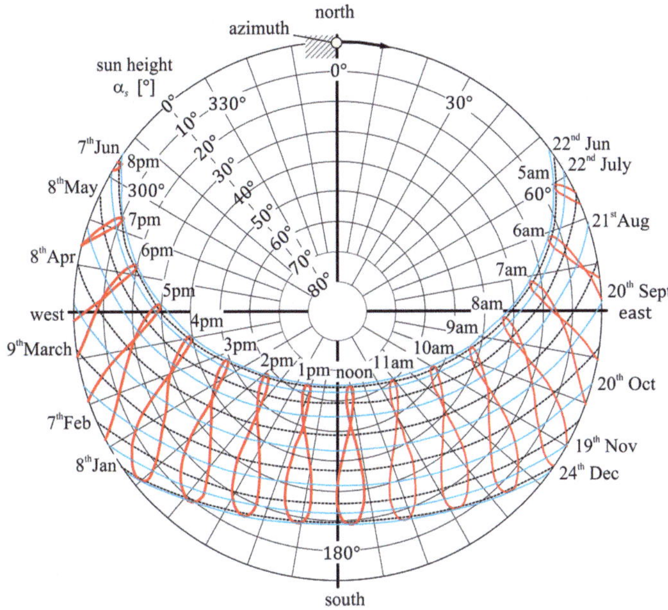

Fig. 2.12 Sun elevation angle α_s and θ_z as function of the azimuth angle γ, the time of day and the declination δ for the location Munich, latitude $\Phi = 48°\ 10'$ and $11°\ 30'$ eastern longitude; the curves to the full hours contain the time shifts according to Eq. (2.20)

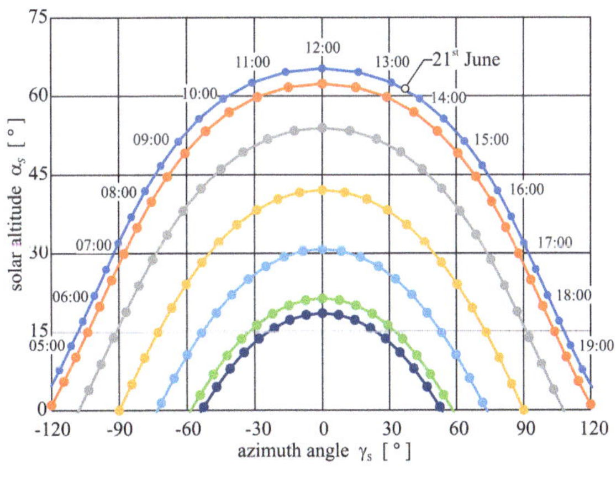

Fig. 2.13 Sun elevation angle α_s as function of the azimuth angle γ, the time of day and the declination δ (here in the form of the day in the year)

The angles of incidence of surfaces oriented to the south or north can be displayed with the difference $(\Phi - \beta)$ at an azimuth angle of $\gamma = 0°$. Figure 2.14 shows the situation schematically.

Equation (2.15) thus changes for the northern hemisphere:

$$\cos\theta_z = \sin(\Phi - \beta)\sin\delta + \cos(\Phi - \beta)\cos\omega \text{ (northern hemisphere).} \quad (2.25)$$

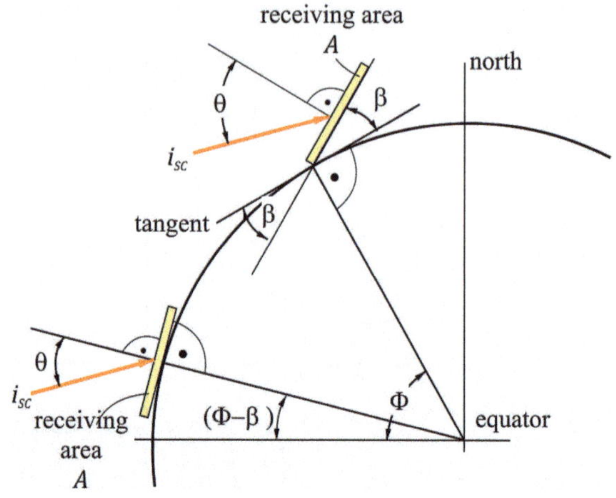

Fig. 2.14 Schematic sketch for the calculation of the radiation flux on an inclined surface A, which is aligned exactly in south-north direction ($\gamma = 0°$)

For the southern hemisphere ($\Phi + (\Phi - \beta)$) if convention of + and − is followed for Φ and δ the following relationship is obtained:

$$\cos\theta_z = \sin(\Phi + \beta)\sin\delta + \cos(\Phi + \beta)\cos\delta\ \cos\omega \text{ (southern hemisphere)} \tag{2.26}$$

2.2.3.4 Sunrise and Sunset

Equation (2.22) can also be solved according to the hour angle. At sunrise and sunset the zenith angle is $\theta_z = 90°$. If you enter the hour angle of the sunrise with ω_R and the hour angle of the sunset with ω_S and set for the hour angle of one hour $\omega = 15°$ ($\triangleq$ 1 h), you get the exact sunrise and sunset times $\omega_{R/S}$, resulting in Eq. (2.27):

$$\cos\omega_{R/S} = -\frac{\sin\Phi\ \sin\delta}{\cos\Phi\ \cos\delta} = -\tan\Phi\ \tan\delta. \tag{2.27}$$

The astronomical sunshine duration S_0 is the time difference between the rise and set times:

$$S_0 = 2 \cdot \frac{24}{360}\left|\omega_{R/S}\right|. \tag{2.28}$$

The daily solar maximum α_s is given by the latitude Φ and the declination δ according to the relation in equation

$$\theta_z = \Phi \pm \delta;\ \text{respectively}\ \ \alpha_{s,max} = 90° - (\Phi \pm \delta). \tag{2.29}$$

2.2.3.5 Direct Extra-terrestrial Radiation on Horizontal or Inclined Surfaces

The larger the incidence angle of the solar irradiance on a surface, the lower the surface-related radiation density, as the Fig. 2.15 schematically shows. This is considered by the angle $\cos\theta_z$ from Eq. (2.22) for a horizontal surface A_{hor}. The radiation intensity $i_{dir,h}$ results from the product of the solar constant i_{sc} and the solid angle of irradiation θ_z

$$i_{dir,h} = i_{SC} \cdot \cos\theta_z, \tag{2.30}$$

in which the index h indicates the horizontal component.

For a surface A arbitrarily oriented towards the sun, however, the cosine of the angle from the direction of irradiation θ and the normal of an arbitrarily oriented surface A from Eq. (2.15) must be used. This results in

$$i_{dir,t} = i_{SC} \cdot \cos\theta, \tag{2.31}$$

where the subscript t is for tilted.

Often irradiation data is only available for horizontal surfaces. The conversion of irradiance on a horizontal surface A_{hor} to irradiance on an arbitrarily oriented surface is done with the help of the Eq. (2.32). The value Ψ represents the ratio of the radiation on an inclined surface compared to a horizontal one.

$$\Psi = \frac{i_{dir,t}}{i_{dir,h}} = \frac{\cos\,\theta}{\cos\,\theta_z}. \tag{2.32}$$

This will be illustrated by two examples:

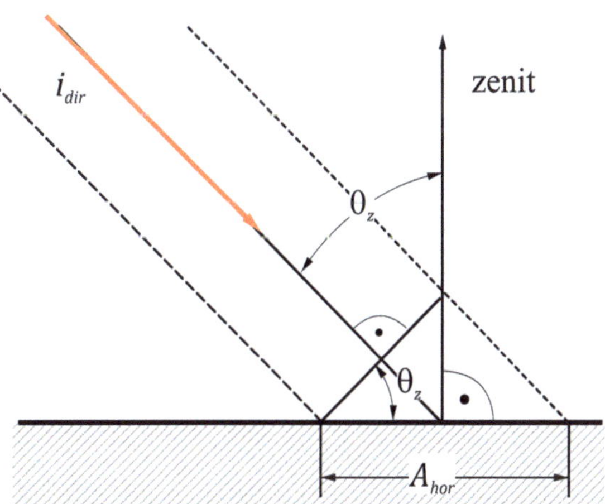

Fig. 2.15 Schematic diagram of the direct radiation of a radiation density i_{dir} on a horizontal surface A_{hor}

- On a summer day, the sun strikes a flat surface ($\beta = 0°$) at an elevation angle $\alpha_s = 60° \rightarrow \theta_z = 30°$. Direct radiation on the horizontal surface $i_{dir,h}$ with a value of $i_{dir,h} = 650\,\text{W/m}^2$ is measured. If at the same location the irradiation surface is inclined by $\beta = 30°$, the angle of incidence would be $\theta = 0°$. From Eq. (2.32) a Ψ-value of $\Psi = \cos 0° / \cos 30° = 1.15$. Thus the direct radiation component for the inclined surface $i_{dir,t} = \Psi \cdot i_{dir,h} \approx 750\,\text{W/m}^2$.
- On a winter day the elevation angle is $\alpha_s = 30°$ and an $i_{dir,h}$ of $150\,\text{W/m}^2$ is measured. If the surface is inclined at the same angle by $\beta = 30°$, the result is $\theta = 30°$ and $\theta_z = 60°$. The Ψ-value then assumes a value of $\Psi = 1.732$ and the direct radiation component on the inclined surface is then $i_{dir,t} \approx 260\,\text{W/m}^2$.

This ratio is still important for the calculation of irradiance on arbitrary surfaces from horizontal radiation values. With the help of these geometric relationships, the diurnal and annual courses of solar radiation outside the atmosphere can be calculated relatively quickly from the solar constant. They show the influence on the power density of the radiation when varying different parameters, such as orientation or location of the collector surface.

2.2.4 Daily and Monthly Sums of Extra-terrestrial Irradiation on Arbitrarily Inclined Surfaces

The sunrise and sunset angles and hence the times and the duration of sunlight were derived in Eq. (2.27) when we remember that the hour angle of 15° is equivalent to 1 h and the following relationship is assumed for the sunrise and sunset angles $\omega_R = -\omega_S$.

Under these conditions, the daily sum of the extra-terrestrial radiation E can be calculated from the integral from sunrise to sunset over time. Here it must be considered that the radiation hits the surface at an angle. The power density is thus reduced according to the Eq. (2.33) by cos θ. The corresponding functional correlation is

$$E = i_s \cdot \frac{24}{2\pi} \int_{\omega_R}^{\omega_S} \cos\,\theta\; dt. \qquad (2.33)$$

For a horizontal surface with $\beta = 0°$ and $\gamma = 0°$ the Eq. (2.22) can be used as angle of incidence ($\cos\,\theta_z = \sin\,\Phi \sin\,\delta + \cos\,\Phi \cos\,\delta\, \cos\,\omega$). Knowing the latitude and declination, the following simplification can be applied

$$\cos\,\theta_z = a + b \cos\,\omega. \qquad (2.34)$$

Using $\omega_R = -\omega_S$ the integration results in the daily sum of the extra-terrestrial irradiation E on a horizontal surface:

$$E = \frac{24 \cdot i_S}{\pi} \cdot a\omega_S + \frac{24 \cdot i_S}{\pi} b \sin \omega_S = \frac{24 \cdot i_S}{\pi}(a\omega_S + b \sin \omega_S), \qquad (2.35)$$

Table 2.3 Hour angle for sunrise and sunset, the mean monthly duration of the day as well as the mean declination $\overline{\delta}$ for the respective months in the last column for a location with $\Phi = 48°$ northern latitude (Freiburg)

Month	ω_S (°)	Duration (h)	$\overline{\delta}$
January	64.2	8.6	21.4
February	74.4	9.9	13.4
March	86.8	11.6	2.9
April	93.2	12.4	9.3
May	112.2	15.0	18.8
June	118.7	15.8	23.4
July	116.1	15.5	21.6
August	105.9	14.1	13.4
September	92.6	12.3	2.3
October	79.3	10.6	9.5
November	67.1	8.9	19.3
December	61.6	8.2	23.4

where for ω the value to be entered here is in radians (1 rad $= 180°/\pi$).

Table 2.3 lists the values of the Eq. (2.35) for a location with $\Phi = 48°$ northern latitude. The maximum intensity of the extra-terrestrial irradiation is obtained around noon when $\omega = 0°$.

The following figures illustrate the effects of the angular relationships on the maximum intensity or the maximum daily sums of the extra-terrestrial radiation depending on the inclination of a surface to the horizontal (β) or the installation site (geographical latitude Φ).

Figure 2.16 shows the diurnal variation of the extra-terrestrial irradiance incident on a flat horizontal surface ($\beta = 0°$) at a latitude of $\Phi = 48° N$. Naturally, in summer the days may become almost twice as long as in winter and, additionally, due to the larger solar elevation angle α_s (Fig. 2.13) one observes also a higher direct irradiation intensity. In total the time-integrated daily irradiation (the area below the curve) in June is a factor of 8 larger than in December.

Inclined receiver surfaces oriented towards South (in the Northern hemisphere) resp. North (in the Southern hemisphere), are more favourable in winter, if they are tilted with a much larger angle of inclination β. In summer, on the contrary, less inclined surfaces (small β) are more advantageous, as the Fig. 2.17 shows. This is a combined effect of latitude and solar declination which both affect the prevailing incidence angles. Moderate inclination angles in between lead to higher intensities and daily totals, especially during the interseason periods due to their "hump curve" shape. As a rule of thumb an inclination angle of Φ should be chosen to maximise the annual irradiation sum on a surface.

Special requirements determined from the load curve of a consumer producing uniform daily totals over a certain time period or maximisation of output in a specific season may lead to an optimal collector tilt, differing from the standard orientation. Figure 2.18 shows this for the example of Cologne. With a flat horizontal orientation ($\beta = 0°$), the ratio of integrated energy per square metre from

Fig. 2.16 Daily course of the intensity of extraterrestrial solar radiation i_{extra} in kW/m^2 as a function of the time of day t in hours (h) on different days of the year for the latitude of $\Phi = 48° N$ (Freiburg); angle of inclination: $\beta = 0°$ (horizontal)

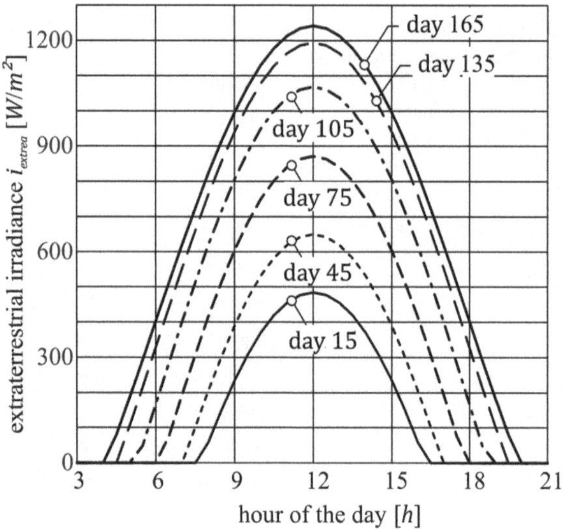

Fig. 2.17 Annual course of the daily maximum intensities of the extraterrestrial solar radiation i_{extra} in W/m^2 for the geographical latitude of $\Phi = 48° N$ (Freiburg) for different angles of inclination β; installation direction: south

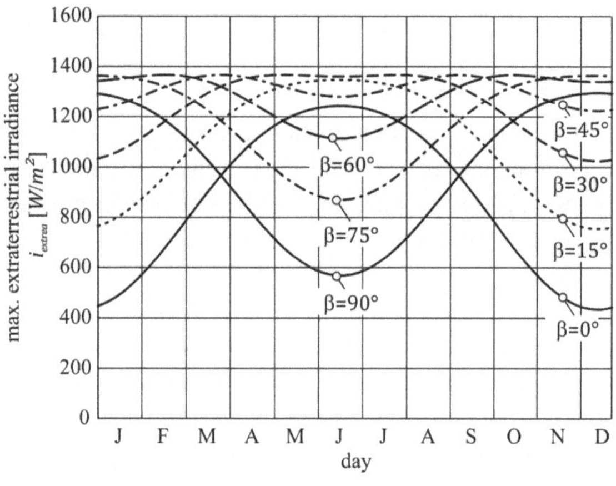

summer to winter is almost a factor of four, whereas with a surface inclined at 45°, the ratio summer to winter is only 1.5. Morevover, the solar irradiation is relatively constant during the summer half year. On the other hand a tilt of 75° yields the highest values in the heating season.

Figure 2.19 shows the extra-terrestrial irradiance of a receiver surface inclined by $\beta = 45°$), for different latitudes of the northern hemisphere. At the equator ($\Phi = 0°$), maximum values of irradiation intensity occur in winter, whereas at the polar caps, solar irradiation is limited to summer only. In the polar winter of the northern hemisphere there is no irradiation due to the polar night and the irradiation power is zero.

This is also reflected in the achievable daily totals for the year as shown in the Fig. 2.20. It is noteworthy that the daily sum at an inclination of $\beta = 45°$ without

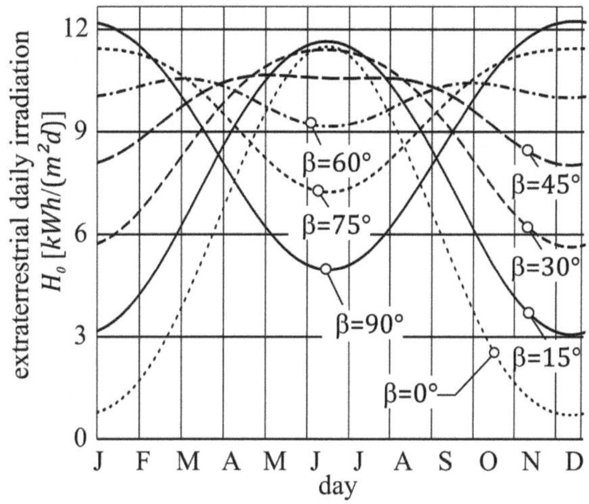

Fig. 2.18 Annual course of the daily sums of the extraterrestrial solar radiation i_{extra} in $\mathrm{kWh/(m^2d)}$ for the geographical latitude of $\Phi = 51°N$ (Cologne) for different angles of inclination β; installation direction: south

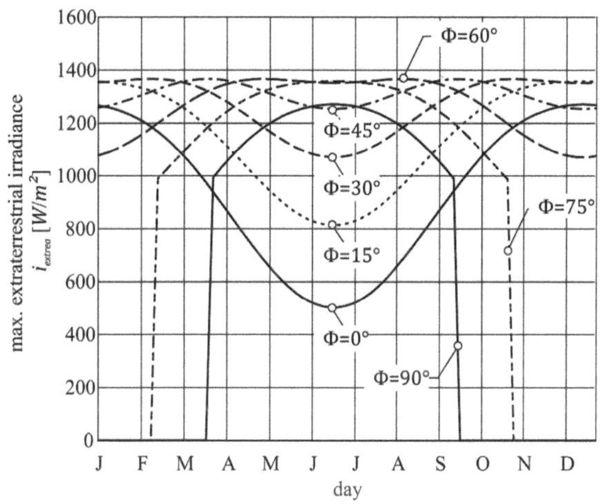

Fig. 2.19 Annual course of the daily maximum intensity of the extraterrestrial solar radiation i_{extra} in $\mathrm{W/m^2}$ for different geographical latitudes Φ; installation direction: south; inclination angle: $\beta = 45°$

tracking varies only slightly at a latitude of $\Phi = 45°$. In addition, the seasonal fluctuations of the daily totals are much smaller there than, for example, near the equator.

Figure 2.21 the annual course of the daily sums of the intensity of the extra-terrestrial solar radiation for the geographical latitudes $\Phi = 34°$ (Rabat) and $\Phi = 59°$ (Stockholm) for the case that the collector surface always follows the achievable maximum intensity value both in the installation direction and the angle of inclination. A comparison of the daily sum from the Fig. 2.21 with that of a rigid collector (cf. Fig. 2.18) shows that a considerable improvement can be achieved, especially in winter. Here up to a factor of four (!) higher harvest values can be achieved. In summer the difference is smaller but still considerable, with a factor of two compared to a rigid collector. However, the Fig. 2.21 also shows the

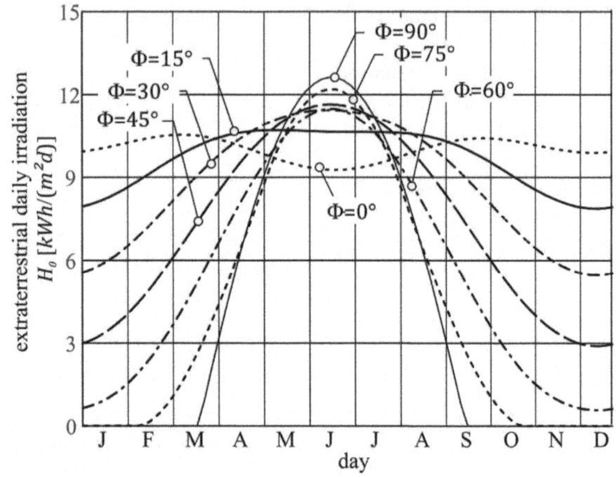

Fig. 2.20 Annual course of the daily sums (kWh/(m²d)) of the extraterrestrial solar radiation as a function of the geographical latitude Φ, installation direction: south; angle of inclination $\beta = 45°$

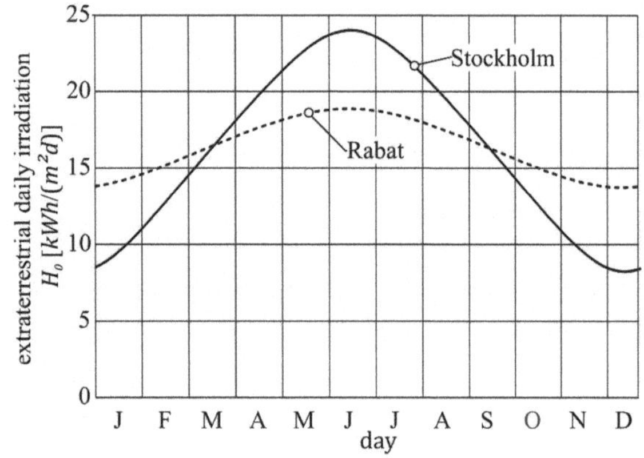

Fig. 2.21 Annual course of the daily sums (kWh/(m² d)) of the intensity of the extraterrestrial solar radiation for the geographical latitudes $\Phi = 34°$ (Rabat/Morocco) and $\Phi = 59°$ (Stockholm/Sweden); biaxial tracking with respect to the maximum achievable irradiation power in the direction of installation

influence of the length of day. The daily sums of the irradiation on a biaxially tracked surface are calculated. In Eq. (2.33) cos = 1 is set and integrated over the length of the day. The result shows that although Stockholm is further north than Rabat, the daily sum of irradiation in summer exceeds that of Rabat. With regard to the energy yield of a collector, however, this only has an effect with a two-axis tracking system.

Due to the polar summer comparatively high maximum daily totals also occur at the North Pole, Fig. 2.22 shows. As expected, an integration of daily totals over the year shows the highest yields near the equator. The reason for this is the inclination of the earth's axis of 23.5° to the ecliptic and thus the irradiation occurring throughout the year.

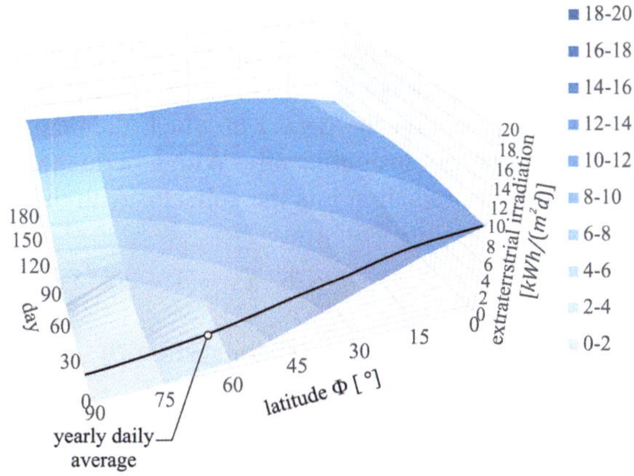

Fig. 2.22 Intensity of daily sum and time-integrated annual radiation on a flat horizontal surface as a function of latitude Φ, ecliptic skew 23° (Gerthsen et al., 1995)

2.2.5 Tracking Surfaces

According to the Fig. 2.23 the irradiation power on a surface decreases with increasing angle of incidence θ_z. For the technical use of solar energy it can therefore be advantageous to track the collector surface to the changing position of the sun. With ideal biaxial tracking, the radiation always hits the collector perpendicularly and the angle of incidence θ_z is always the same:

$$\cos\theta_z = 1. \tag{2.36}$$

However, a two-axis tracking system is technically very complex. For this reason, solar collectors are usually only tracked on one axis. The preferred tracking axes run in north/south or east/west direction. A third variant is the polar tracking with an axis parallel to the earth axis. A detailed analysis of the Fig. 2.12 allows first estimates of a reasonable control of the tracking angle.

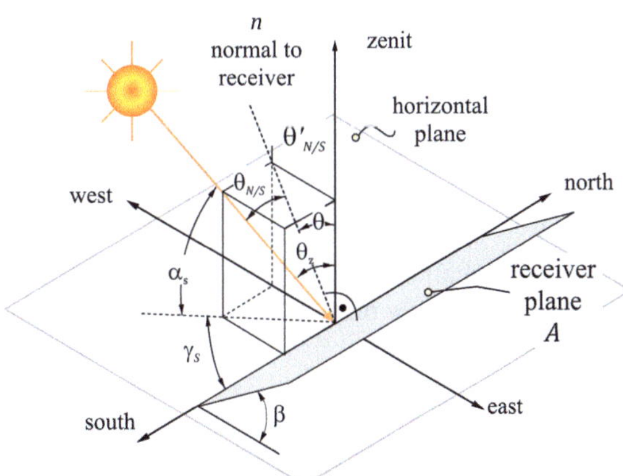

Fig. 2.23 Angular arrangement for converting the irradiation intensity to arbitrarily oriented surfaces

In this context, for the design of a solar thermal system, the demand for exergy (e.g. process heat) must be linked to the demand of the connected work process in relation to its usage time, for example the daily operating time. This requires tracking only within the period in which the demand is requested and limits the tracking effort not insignificantly. With a permanent tracking around the east/west axis, the direct radiation hits the collector with the angle $\theta_{E/W}$:

$$\cos\theta_{E/W} = \left(1 - \cos^2\delta\sin^2\omega\right)^{1/2}. \quad (2.37)$$

The inclination of the collector surface $\beta_{E/W}$ to be set is thus given by:

$$\tan\beta_{E/W} = \tan\theta_z \cdot |\cos\gamma_s|, \quad (2.38)$$

which contains the azimuth angle of the sun γ_s. For the azimuth angle γ_s applies

$$\cot\ \gamma_s = \frac{\sin\Phi\cos\omega - \cos\Phi\tan\delta}{\sin\omega} \quad \text{or} \quad \sin\gamma_s = \frac{\cos\ \delta\ \sin\ \omega}{\sin\theta_z}. \quad (2.39)$$

For an East/West (E/W) oriented plane, which is adjusted once a day at noon in such a way that the irradiation hits vertically, the angle of incidence depends on the hour angle. The following relationship results.

$$\cos\theta = \sin^2\delta + \cos^2\delta\cos\omega. \quad (2.40)$$

The inclination angle of the plane β corresponds to the zenith angle θ_z at noon:

$$\beta = |\theta_z| = |\Phi - \delta|, \quad (2.41)$$

where Φ is the latitude and δ the declination angle. In the case of a plane rotating around a north/south axis (N/S), the angle of incidence $\theta_{N/S}$ results from the equation

$$\cos\theta_{N/S} = \left(\cos^2\theta_z + \cos^2\delta\sin^2\omega\right)^{1/2}, \quad (2.42)$$

and the angle of inclination $\beta_{N/S}$ is calculated from the relationship

$$\tan\beta_{N/S} = \tan\theta_z|\cos(\gamma - \gamma_S)|. \quad (2.43)$$

In a constellation in which the plane is continuously polar tracked, the angle of incidence θ_{pol} varies only in the range of the declination δ:

$$\cos\theta_{pol} = \cos\delta. \quad (2.44)$$

According to the Eq. (2.31) maximum $\cos\delta = \cos 23.45° = 0.92$, i.e. only approx. 8%, the losses that occur in uniaxial polar tracking due to the deviation

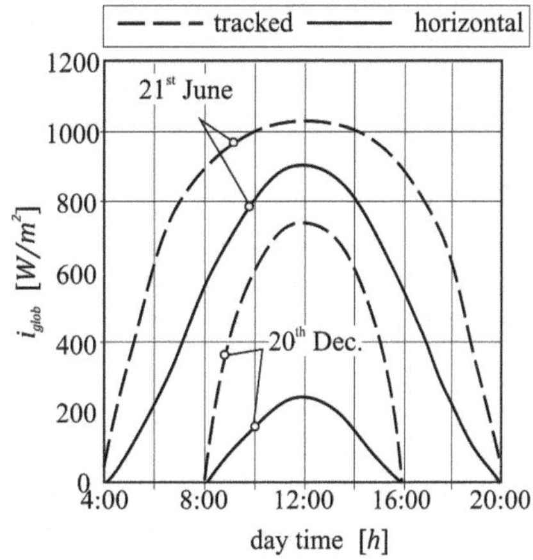

Fig. 2.24 Global radiation intensity i_{glob} (W/m^2) as a function of the time of day on a horizontal surface A_{hor} compared with a tracked surface at a geographical latitude $\Phi = 50°$ on a cloudless day

from the vertical direction of irradiation are the rotation around the polar axis corresponds to the hour angle ω. Further observations on tracking angle are described in Duffie and Beckmann (2020). However, they can also be easily derived using the trigonometric functions themselves by placing a cuboid above the angular distribution as shown in the Fig. 2.23. With the help of the projections on the surfaces of the cube, the figure allows the calculation of tracking angles using the cosine and sine theorems. For the angle of incidence of the north–south correction θ_{NSC} results:

$$\cos\theta_{NSC} = \left(\cos^2\theta_z + \cos^2\beta\sin^2\theta_z\right)^{1/2}, \tag{2.45}$$

and respectively for the tracking angle follows:

$$\tan\theta'_{NSC} = \tan\theta_z \sin\gamma_s. \tag{2.46}$$

There are large differences in the irradiation intensity between a horizontal surface and a tracked surface. On cloudless days at a geographical latitude of $\Phi = 49°$ (Karlsruhe), up to 30% higher radiant power densities in relation to the collector surface can be achieved by tracking even in summer, as the Fig. 2.24 shows. In winter this can even vary by factors. Integrated over one year, about 30% more radiant energy per square meter and year can be achieved in Germany through a targeted tracking of the collector surface.

2.2.6 Shading of Surfaces

Shading of solar surfaces occurs frequently, sometimes being a problem and in other cases a wanted regulation of the solar gain. For example the shading of collectors by neighbouring buildings restricts the positioning of these collectors in the building envelope. On the other hand, a shading of a window in summertime might avoid overheating of rooms. It is especially important, of course, for the direct beam irradiation. Therefore we treat this blocking of direct beam radiation in the chapter of extra-terrestrial radiation. The reduction of diffuse sky radiation for collector apertures is less important and can be quantified by using the view factor calculation for diffuse surfaces.

For the determination of shading we distinguish between irregular shapes like trees, mountains or the skyline of buildings, and regular geometries like collector rows or overhangs or wingwalls. In modern raytracing methods and object shapes in principle can be added and thus shading simulations for different sun positions may be performed. However, this is a very time-consuming approach and sometimes a quick assessment is needed.

For regular shapes the shading is much more amenable to analytical calculation and at least good approximations can be developed for these geometries. A second approach, however, is the qualitative use of sun path diagrams. Plots of solar altitude α_s versus solar azimuth γ_s for different days of the year may be prepared, on which shapes of obstructions (shading profiles) can be superimposed. From these diagrams one may determine when the path from the sun to the point in question is blocked.

At any point in time and at a particular latitude, φ, δ, and ω are fixed. From the equations of this see Sect. 2.2.3, the zenith angle θ_z or solar altitude angle α_s and the solar azimuth angle γ_s can be calculated. A solar position plot of θ_z and α_s versus γ_s for latitudes of $\Phi = \pm 45°$ is shown in Fig. 2.25. Lines for mean days of the months are labelled with their dates and lines of constant hour angles are labelled by hours. The angular positions of buildings, collectors, overhangs, or also natural obstructions like trees and mountains can be entered on the same plot.

The angular positions of outlines of an obstructing element may be calculated by trigonometric calculations. To illustrate the principle we give an example here.

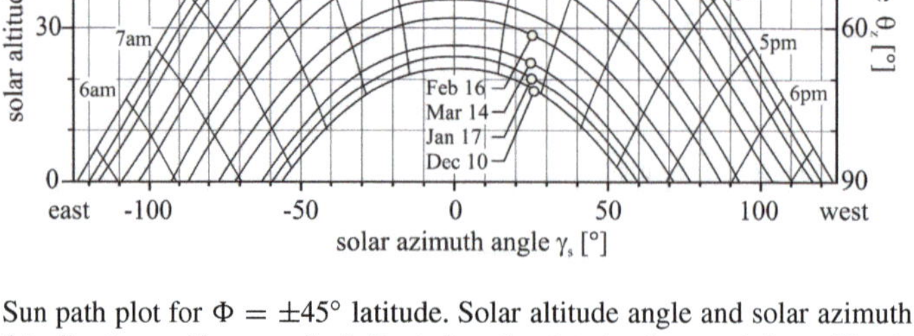

Fig. 2.25 Sun path plot for $\Phi = \pm 45°$ latitude. Solar altitude angle and solar azimuth angle are functions of declination and hour angle, indicated on the plots by dates and times. The dates shown are for northern hemisphere (Duffie & Beckmann, 2020)

Table 2.4 Calculation of angular coordinates for a building according to Eq. (2.47)

Point	D_i (m)	Δh_i (m)	Δw_i (m)	α_i (°)	γ_i (°)
1-East	45.71	26.00	−8.00	29.63	−9.93
2-South	45.00	26.00	0.00	30.02	0.00
3-West	68.77	26.00	52.00	20.71	37.10

Let us assume a collector installed on the roof of a two storey house at a height of 8 m above ground. In front of the collector at a distance of $L = 45\,\text{m}$ is a rectangular building wide $w = 60\,\text{m}$, 34 m high parallel to the collector. The Eastern corner of the building is $\Delta w = 8\,\text{m}$ to the left, the Western façade $\Delta w = 52\,\text{m}$ to the right of the point being exactly in Southern direction. First, we calculate the angular coordinates of the 2 top corners plus the roof position right in front in direction South. Thus, we arrive using the formulae for point i

$$\alpha_i = \tan^{-1}\left(\frac{\Delta h_i}{\sqrt{L^2 + \Delta h_i^2}}\right) \text{ and } \gamma_i = \tan^{-1}\left(\frac{\Delta w_i}{\sqrt{L^2 + \Delta h_i^2}}\right). \tag{2.47}$$

Table 2.4 illustrates the computational results according to Eq. (2.47).

Vertical lines are vertical in the diagram, but horizontal lines are curved. We may calculate more points along the roof line, plot these points in the sun path plot, and shade the area of the obstructing element. The result of that example can be seen in Fig. 2.26. The precedent discussion would be exactly true only if the collecting area is an infinitesimal element. That means it is shaded or not. In reality the area is finite, and one edge of the collector might be shaded while the other is not. As a consequence a more precise treatment of shading has to take into account partial shading and the different relative angular positions of the shading objects. Then an integration over the whole collector area has to be performed. This is especially important if the shading object is close, smaller or of the same magnitude as the shaded collector aperture.

In order to illustrate this we want to consider the case of an overhang over a window. For simplicity we start with the approximation of a very long overhang so that the solar altitude angle projected onto the vertical incidence plane determines the position of the horizontal border between shadow and irradiated area. This projected angle is called the profile angle α_p. Assume a window of height H below a projection/overhang of extension L, which is a distance H_2 above the window. The plane defined by the overhand protrusion and the upper edge of the window hits the vertical plane at an angle of α_2, the plane through the lower edge at an angle $\alpha_1 > \alpha_2$. The angles are calculated as $\arctan(H_2/L)$ and $\arctan((H_1 + H_2)/L)$, respectively. For solar profile angles larger than $\alpha_p > \alpha_1$ the window is completely shaded, for profile angles $\alpha_p < \alpha_2$ it is completely irradiated. For angles in between a partial shading is observed.

A similar problem is encountered if we consider the shading of collector rows by identical rows in front of them, as illustrated in Fig. 2.27. If the solar profile

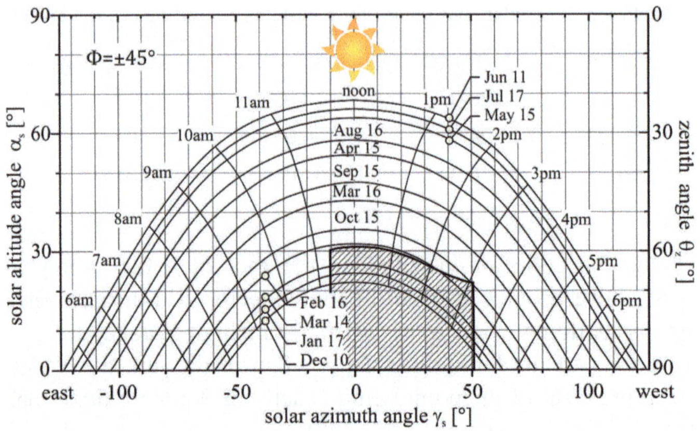

Fig. 2.26 Example of obstructing building within the sun path plot of Fig. 2.27 modified from Duffie and Beckmann (2020)

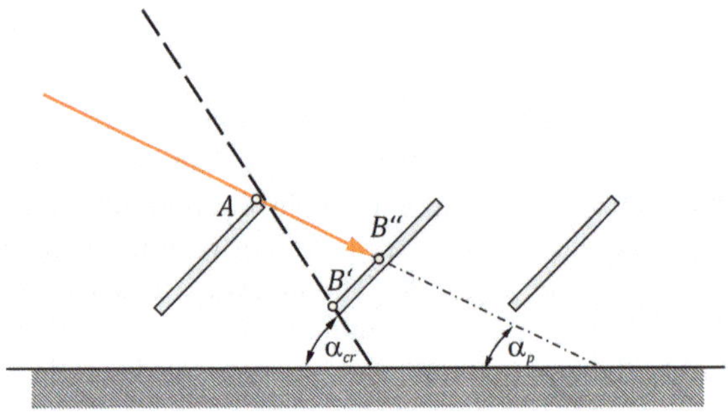

Fig. 2.27 Cross section of two collector rows

angle α_p is larger than the angle defined by the plane AB' and the horizontal plane, the collector is fully irradiated. For smaller positive angles a shading edge will appear on the second collector row in position B'' on the collector, α_p being the angle between plane AB'' and the horizontal plane. So if the collector rows are very close to each other, the times of low altitude sun with significant shading may be substantial. The optimization of collector row distance and loss of energy is a cost–benefit optimization and depends amongst other factors on the availability of ground area, natural horizon, and collector row orientation.

2.3 Terrestrial Solar Radiation

Due to the atmosphere and its interaction with the solar radiation the latter one changes intensity and spectral composition substantially before reaching the earth's surface. The direct radiation is weakened on its way through the atmosphere by

- absorption and
- scattering.

There are basically three different types of scattering mechanisms in the atmosphere, Rayleigh scattering, Mie scattering and geometrical scattering. The latter one occurs when we have large rain drops in sun light, and we can observe a colourful rainbow. The other two mechanisms are omnipresent. Rayleigh scattering occurs even in very clean atmospheres and is produced by molecular particles in the atmosphere. It causes the blue coloration of the sky and occurs on particles that are much smaller than the wavelength of light. The scattering intensity is wavelength dependent and scaled with:

$$\approx \frac{1}{\lambda^4}. \tag{2.48}$$

This means that the smaller the wavelength, the greater the scattering of the light. The direct radiation suffers a small redshift, which is also visible to the eye when travelling long distances through the atmosphere.

Mie scattering in the atmosphere is done by aerosols, dust, or salt particles with sizes of the scattering agglomerates or particles larger or approximately equal to the wavelength of the light λ. The scattering intensity of Mie scattering scales approximately with $1/\lambda$. In most cases, however, due to a distribution of particle sizes, the resulting scattered intensity is independent of the wavelength λ. Diffuse light after Mie scattering is therefore white or grey.

Several constituents of the atmosphere are responsible for absorption. CO_2, O_2 and ozone (O_3) as well as water vapour are the dominant absorbing gases in the atmosphere. The absorption of ozone in the ultraviolet range is of no energetic significance, but is of great importance for life on earth, protecting cells against high energy ultraviolet radiation. Significant losses occur in the infrared range due to absorption of water vapour at wavelengths of 930, 1120, 1400 and 1850 nm. There are also absorptions by oxygen and ozone in the infrared range. While the extra-terrestrial spectrum contains still about 5% of power at wavelengths longer than 2.5 μm, it drops to very small values due to the absorption of water vapor and CO_2. Of course, the absorption and scattering processes are dependent on the density of the atmosphere, the prevalent particles and aerosols, and the overall integrated extinction is dependent on the length of the light path through the atmosphere. Therefore, the solar spectrum on high mountains is very different from a spectrum a sea level. Industrial atmospheres have higher aerosol content. The light path through the atmosphere is changing during the course of a day. For the latter effect, assuming that other atmospheric parameters are identical, the concept of the air mass factor AM has been developed. Imagine that the atmosphere is a plane layer above the earth. Then by pure geometry depending on the solar altitude angle as, the light ray passes a different path length. The shortest path is when the sun is right above in the zenith. This is called $AM1$ for the observer's position at sea level. The length of the path increases with zenith angle θ_z. The zenith angle θ_z and the elevation angle α_s are complementary angles for which ($\theta_z = 90° - \alpha_s$)

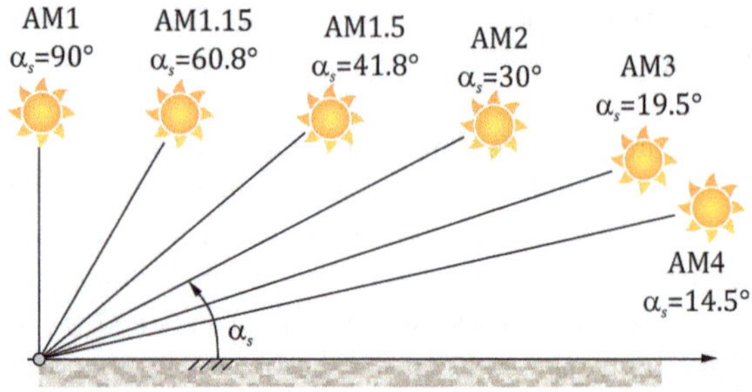

Fig. 2.28 Air mass factors AM according to Eq. (2.49)

applies. For a solar altitude of $\alpha_s = 30°$ we have twice the length of the imaginary light path through the layer. This is called $AM2$. The extraterrestrial radiation consequently is characterized as $AM0$. Generally, the optical air mass factor AM is approximated as

$$\mathrm{AM} \cong \frac{1}{\cos\theta_z} = \frac{1}{\sin\alpha_s} \to \begin{matrix} \mathrm{AM} = 1 & \text{for} & \theta_z = 0° \\ \mathrm{AM} = 2 & \text{for} & \theta_z = 60° \end{matrix}, \tag{2.49}$$

and is schematically shown in Fig. 2.28. More exactly one should consider a path length L of incident solar radiation through the atmosphere, where L_0 is the path length at zenith (i.e., normal to the Earth's surface) at sea level. The optical air mass coefficient AM is defined as:

$$\mathrm{AM} = L/L_0. \tag{2.50}$$

The approximation of Eq. (2.49) $L = L_0 \cdot \cos\theta_z$ applies to a flat atmosphere and overlooks the atmosphere's curvature and the refraction of the light beam entering the atmosphere. It predicts an infinite air mass at the horizon. It uses the direction to an apparent position of the sun, which is not the exact true position. However, it is reasonably accurate for values of θ_z up to around 70° where the error is less than 1%. A number of refinements have been proposed to model the path thickness more accurately towards the horizon, a very popular one in solar energy is that proposed by Kasten and Young (1989)

$$AM = \frac{1}{\sin(\alpha_s) + 0.50572\,(6.07995 + \alpha_s)^{-1.6364}}. \tag{2.51}$$

A more comprehensive list of such models is provided in Rapp-Arrarás and Domingo-Santos (2011) for various atmospheric models and experimental data sets. At sea level the air mass towards the horizon is approximately 38.

Figure 2.29 shows a terrestrial spectrum $AM1$ and the extra-terrestrial spectrum as a function of wavelength in (nm). For comparison, the spectrum after passing through the atmosphere is shown as a line. It can be clearly seen that a large part of the extra-terrestrial radiation does not penetrate the atmosphere in the near infrared (IR) and in the ultraviolet (UV) range. The radiation transmission is

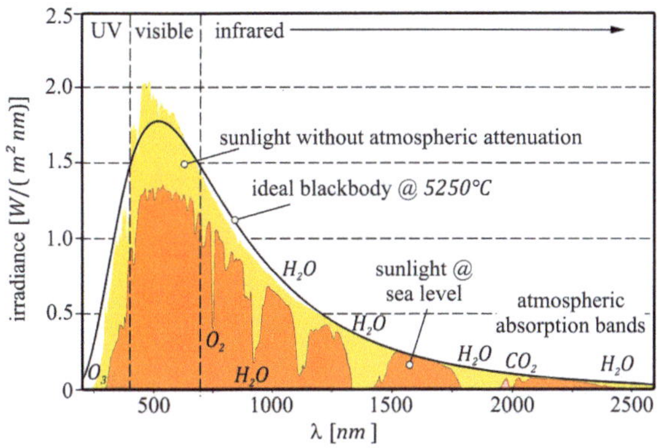

Fig. 2.29 Example of spectrum change due to Rayleigh scattering and gas absorption when passing through the atmosphere compared to extraterrestrial radiation. In particular, the absorption of water vapor in the infrared range ($\lambda > 780$ nm) should be noted. The influence of the wavelength dependence is shown in the adopted representation of Iqbal (1983). The relative air mass factor (AM) is $AM = 1$, i.e. vertical passage. The water vapour in the atmosphere was assumed to be 2 cm condensable

significantly influenced by absorption and scattering processes. Qualitatively, lower solar altitudes (larger AM) leads to a red shift of the direct solar spectrum, because of the large scattering of blue light in the long light path. This effect can be observed in an extreme case when the sun sets as a red disk when the horizon is very low.

As consequence of the atmospheric processes which affect the direct solar light beam, we have not only direct irradiance on earth. The scattering processes in the clear sky, but even more for cloudy atmospheres lead not only to a reflection of solar intensity to outer space, but also to a transmitted diffuse light coming from all over the sky hemisphere to the observer. The terrestrial irradiance coming from the sun consists of a direct beam irradiance part G_b and a diffuse irradiance part G_d. The sum is called global radiation G

$$G = G_b + G_d. \tag{2.52}$$

For the usual case that the irradiance is measured on a horizontal plane the index h is used. When a tilted plane is taken as reference (e.g. the collector plane) the index t is used. Figure 2.30 shows schematically the composition of the global irradiance.

2.3.1 Attenuation Mechanisms in the Atmosphere

In the following section we want to describe in detail and more quantitatively the attenuation mechanisms in the atmosphere.

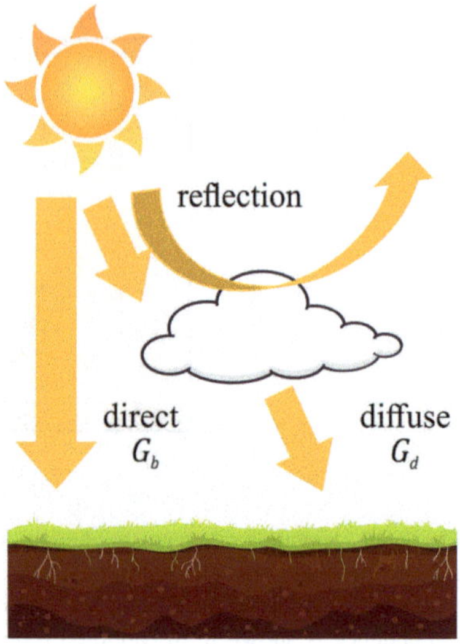

Fig. 2.30 The influence of the atmospheric layer on the solar irradiation

2.3.1.1 Absorption in the Atmosphere

Absorption is a dissipative process of a wave in a body, which means a loss of intensity. The absorption law for photon radiation has an exponential form as it is a probabilistic process. The intensity of light absorbed in a thin layer of thickness dx at penetration depth x is proportional to the intensity $I(x)$ of the radiation and proportional to an absorption coefficient (μ) dependent on the medium. It is therefore:

$$\frac{dI}{dx} = -\mu \cdot I(x) \rightarrow I(x) = I_0 \cdot \exp(-\mu \cdot x). \tag{2.53}$$

The absorption coefficient depends on the density of the material (and thus the number of particles N) and the absorption cross section (σ). If the material density is constant, this is summarised in the Lambert–Beer law, which reads to

$$I(x) = I_0 \cdot \exp(-N \cdot \sigma \cdot x) = I_0 \cdot \exp(-\delta(x)), \tag{2.54}$$

where the product $(N \cdot \sigma \cdot x)$ referred to as optical thickness $\delta(x)$. In the atmosphere, however, with increasing altitude, the density of the atmosphere and thus the number of particles in a control volume decreases. The particle density is a negative exponential function with altitude x according to the barometric height formula, as the following equation shows.

$$N(x) = N_0 \cdot \exp(-x/x_0), \tag{2.55}$$

where N is the number of molecules in the atmosphere as a function of altitude x relative to a reference altitude x_0. Inserting Eq. (2.55) into Eq. (2.54) gives a relationship for the height-dependent intensity $I(x)$, known as the Chapman function:

$$I(x) = I_0 \cdot \exp(-\delta(x)) = I_0 \cdot \exp(-N_0 \cdot \sigma \cdot x_0 \cdot \exp(-x/x_0)). \tag{2.56}$$

If one uses selected substances such as ozone or water vapour, one can see from Fig. 2.31 that an even small ozone concentration at low atmospheric heights yields a considerable reduction of the total intensity, while for small concentrations at larger heights x almost all intensity passes through the atmosphere. Not only ozone but also water vapour can lead to large losses in intensity even in small quantities depending on the atmospheric altitude. For constant steam content the intensity attenuation effect decreases with increasing atmospheric height.

The absorption increases significantly with increasing density of the atmosphere, as illustrated in Fig. 2.32. Note that there is a dramatic attenuation for wavelengths smaller than 300 nm, which is mainly due to ozone absorption.

The light absorption on surfaces or when passing through matter is dependent on material and frequency. Depending on the possible band structure of the

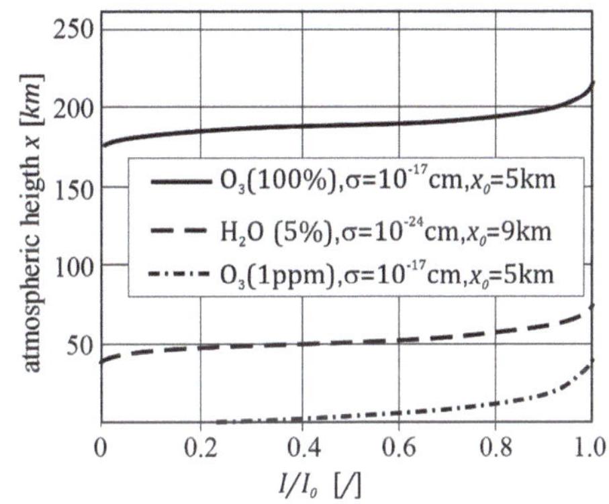

Fig. 2.31 Relative intensity change I/I_0 as function of height x for different gas concentrations and species from Hippler (2005)

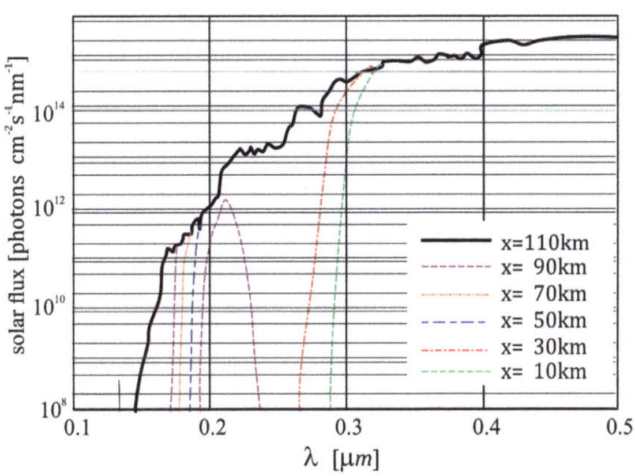

Fig. 2.32 Change in solar flux density as a function of wavelength λ at different heights from Hippler (2005)

molecules, different frequency ranges of light can be absorbed differently. This means that the absorption is different depending on the colour. Adjacent frequency components are either reflected or transmitted, depending on the material and the angle of incidence. If, for example, a surface that appears yellow is irradiated with white light, the green and red light is reflected/transmitted and blue light is absorbed. When light is absorbed, the absorbed energy is not only converted into heat, but can also be lost through other mechanisms such as fluorescence or through scattering at aerosols.

2.3.1.2 Scattering in the Atmosphere

Scatterers in the atmosphere are air molecules, aerosol particles, cloud, and rain droplets. These scatterers affect all types of electromagnetic radiation: solar radiation, thermal radiation, microwaves, and radar. Contrary to absorption scattering leads to a directional redistribution of the radiation, i.e. the radiation is deflected and usually (for elastic scattering) no energy conversion takes place as in absorption. The beam attenuation of light due to scattering can also be formulated using the Lambert–Beer's law (Eq. (2.54)) with a scattering coefficient, but additionally scattering theory yields a scattering function (scattering indicator matrix, phase function). The latter describes the distribution of the scattered radiation in the individual directions of space. As a result, the energy of the primary radiation beam decreases in the direction of the beam similarly to the absorption, but there is a gain in radiance in other directions.

All extinction due to scattering processes is described by the Lambert–Beer law. The change in wavelength-dependent intensity density $i(\lambda)$ as the radiation passes through an infinitesimal segment of thickness dx results in:

$$di(\lambda) = -\beta_\lambda \cdot i_\lambda \cdot dx, \tag{2.57}$$

where β_λ is the volume scattering coefficient $[1/m]$, which is calculated as

$$\beta_\lambda = \sigma_\lambda \cdot N. \tag{2.58}$$

In Eq. (2.58) σ_λ is the wavelength dependent scattering cross section and N the number density of particles. The total volume scattering coefficient β_λ results from the integral of all particles involved in the scattering process including multiple scattering and is evaluated by integration across the control volume:

$$\beta_\lambda = \int_{r=r_1}^{r=r_2} \sigma_\lambda(r) \cdot \frac{dN}{dr} dr. \tag{2.59}$$

Now the scattering cross section of an individual particle can be very well described with theories depending mainly on the size of the particle. For the interaction with light not only the geometric size is important. This can be understood when looking in the physics of the different scattering processes. Rayleigh scattering is an interaction of electromagnetic radiation with a dipole. When the particle reaches the size of the wavelength, the Mie scattering theory deals with the particle as a continuous medium with geometric shape and permissivity. When the particles become even bigger, the wave character of the light is less important and there is no coherent interaction of electro-magnetic (EM)-waves anymore. Then the realm of geometric optics describing the medium with a refractive index starts. Figure 2.33 shows the scattering cross section σ_λ as a function of the wave number $\kappa = (2 \cdot \pi \cdot r/\lambda)$. The range of Rayleigh scattering therefore occurs at wavenumbers $\kappa < 1$ while the laws of geometric optics apply only to wavenumbers $\kappa \gg 1$. The scattering cross-sections of Mie scattering vary considerably due to interference of waves.

In Table 2.5 the wave numbers κ, particle sizes (r) and wavelengths for some atmospheric particles are given. The particle size in particular determines the type of scattering process.

Figure 2.34 shows in double logarithmic order the areas of geometric optics, Mie- and Rayleigh scattering as a function of the mean particle radius (r) and the wavelength (λ).

Rayleigh Scattering

Rayleigh scattering describes the scattering of light in a sphere containing only scattering particles with a diameter d, whose size is much smaller than the wavelength λ. The typical particle size condition for elastic Rayleigh scattering is;

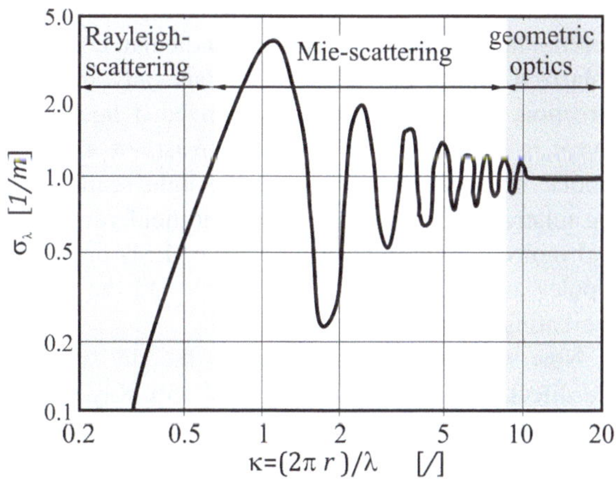

Fig. 2.33 Scattering cross section σ_λ as function of the wave number $\kappa = (2\pi r/\lambda)$ and limits of the different scattering processes

Table 2.5 Wave number κ and order of magnitude of typical air component sizes

	Typical size r	Wavelength λ (nm)	Wave number $\kappa = (2\pi r/\lambda)$
Molecule	0.1–1 nm	550	0.001–0.01
Aerosol	100 nm	550	1.2
Raindrop	1 mm	550	0.06–6

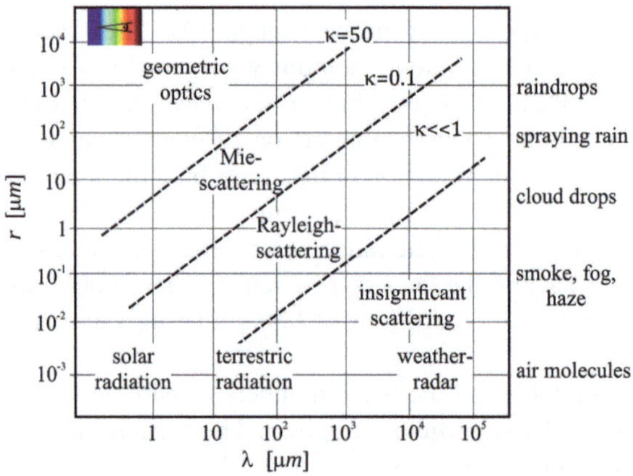

Fig. 2.34 Areas of geometric optics, Mie and Rayleigh scattering as a function of particle radius r and wavelength λ, and some typical particle representatives

$d \ll 100 \cdot \lambda$. In order to understand the physics, we first have to introduce the concept of polarization of light, i.e. of electromagnetic waves.

Light can be considered as a transverse electromagnetic wave, as shown in Fig. 2.35a, in which the wave vector $\vec{k}$ points in the direction of propagation. Perpendicular to the propagation direction are the field vectors of the electric and of the magnetic field $\vec{E}$ and $\vec{B}$, which in turn are perpendicular to each other ($\vec{E} \perp \vec{B}$). This allows a rotational degree of freedom in three-dimensional space, that of rotation around the wave vector (with time). Both vectors can be decomposed into two independent perpendicular components. Light is called linearly polarized if the field vectors oscillate only in one plane along the propagation direction. It is called circular polarized if this field amplitude rotates with time, and it is called unpolarized if it consists of a statistical mixture of both perpendicular components. Generally, the light scattering of particles is dependent on the relative orientation of the electric field vector to a particle orientation. Atoms and molecules in an electromagnetic (EM) field can be considered as oscillating dipoles as the negative electron shell and the positive nucleus interacts with the oscillating field.

Now we consider a molecule in the sky which, for the observer B, floats in the direction φ shown in the Fig. 2.35b. Rayleigh scattering describes the interaction of the EM-wave with the molecule. Two dipole moments oscillate excited by unpolarized light perpendicular to the propagation vector $\vec{k}$ with equal frequency in the plane defined by the points SAB (dipole 1, $I_{||}$) and perpendicular thereto (dipole 2, $I_{\perp}$). Any dipole radiates maximum perpendicular to its direction of oscillation, but it emits no intensity into the direction of oscillation. The angular distribution of the emitted field amplitude E_{emit} is proportional to $\sin\theta$, where θ is the angle between dipole axis $\vec{p}$ and propagation direction $\vec{k}_{emit}$. Therefore dipole 2 radiates towards B with maximum strength, while dipole 1 radiates only with the intensity $\sin^2\theta$ or $\cos^2\varphi$. The total brightness of the sky therefore changes accordingly with $\left(1 + \cos^2\varphi\right)$ as shown in Fig. 2.36b. It is darkest at 90° from the

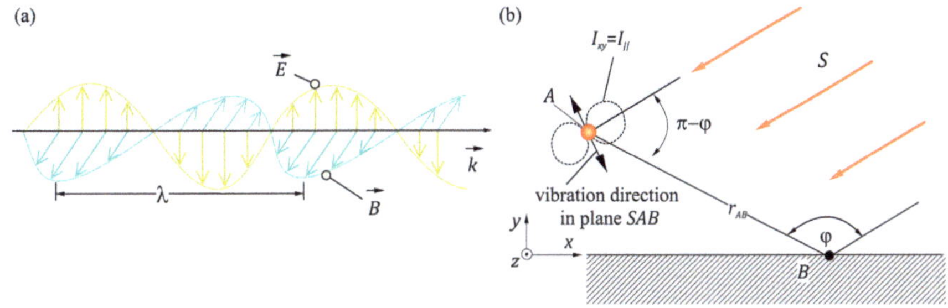

Fig. 2.35 **a** Schematic diagram of the propagation of an electromagnetic wave, wave vector $\vec{k}$, electric field $\vec{E}$ and magnetic field $\vec{B}$; **b** in an air molecule A, the portion of sunlight polarized in the plane SAB excites a dipole oscillation. The amplitude radiated to the observer B is proportional to $\cos\varphi/r_{AB}$, the intensity $I_{||} \sim \cos^2\varphi/r_{AB}^2$. For the polarization direction perpendicular to $SAB(x-y\text{-plane})$ $I_\perp \sim 1/r_{AB}^2$ is valid, because here the dipole oscillates perpendicular to AB into the surface

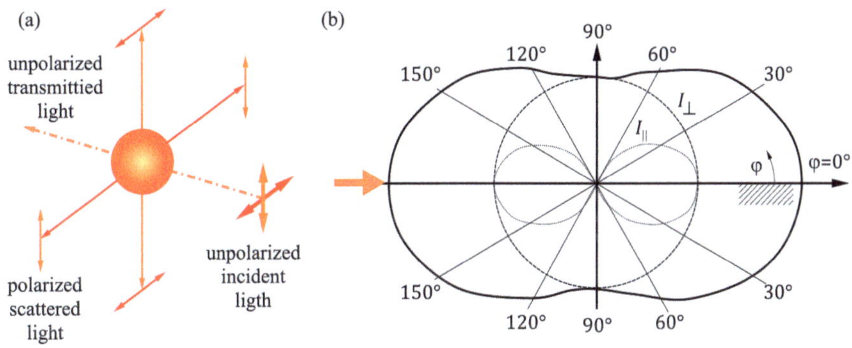

Fig. 2.36 **a** Scattering processes polarize the light. Here, dipoles emit polarized light along their axis. **b** Intensity distribution of the component $I_\perp$ perpendicular to the plane of vision and the plane $I_{||}$ parallel to the plane of vision of the incident unpolarized light as a function of the angle of incidence φ

sun. The eye hardly notices this slight difference, especially because the glare of the direct sun and the pale blue towards the horizon originating from mist obscure it. However, it easily can be detected using a polarization filter such as sunglasses.

As a summary the scattering molecule causes a polarization of the light, as shown in 2.36b. It emits polarized light along its axis. Thus, for Rayleigh scattering the intensity of the component perpendicular to the viewing plane $I_\perp$ remains constant with angle, whereas the intensity parallel to the viewing plane radiates only with $\cos^2\varphi$.

As a summary the scattering molecule causes a polarization of the light, as shown in Fig. 2.36a. It emits polarized light along its axis. Thus, for Rayleigh scattering the intensity of the component perpendicular to the viewing plane $I_\perp$

remains constant with angle, whereas the intensity parallel to the viewing plane radiates only with $\cos^2\varphi$.

Now let us now introduce the complete scattering function. The intensity of unpolarized light I of wavelength λ scattered by a molecule with polarizability α in Rayleigh scattering is calculated as:

$$I = I_0 \cdot \frac{1 + \cos^2\theta}{2R^2} \cdot \left(\frac{2\pi}{\lambda}\right)^4 \cdot |\alpha|^2, \tag{2.60}$$

wherein I_0 represents the intensity of the incident wave, θ the scattering angle and R the distance to the particle. The intensity loss due to Rayleigh scattering thus scales inversely proportional to the fourth power of the light wavelength. Let us assume that the particle is a small dielectric sphere with radius a and the refractive index n. Here the polarizability can be calculated as (Van de Hulst, 1981)

$$\alpha = \frac{n^2 - 1}{n^2 + 1}. \tag{2.61}$$

The scattering cross-section of such a particle is hence

$$\sigma_{\lambda R} = \frac{8\pi}{3} k^4 \cdot |\alpha|^2. \tag{2.62}$$

The Rayleigh scattering explains why the clear sky appears blue. The wavelength of blue light (450 nm) is smaller than the wavelength of red light with about 650 nm. Thus, the ratio of the effective cross-sections $\sigma_{blue}/\sigma_{red}$ is approximately 4–5.

Blue light—depending on the exact wavelength—is thus scattered about 4–5 times more strongly in the atmosphere than red light. According to Planck's radiation law, the maximum of the power radiated by the sun is green (approx. 550 nm). Rayleigh scattering shifts the power distribution of the scattered light, according to the Eq. (2.60) with λ^{-4}. As a result, the maximum moves into the UV range, as the curve of the intensity distribution of the sunlight scattered in the air shows in Fig. 2.37.

During the day, when the sun is high in the sky, the light only travels a short distance through the atmosphere. Only little blue light is scattered in other directions. Therefore, the sun appears yellow. From high-flying airplanes the sun appears "whiter" because less blue light is missing. The sum of all scattered light makes the sky appear blue from all other directions. On the moon, where a dense atmosphere is missing, the sky appears black even during the day. At low sun elevation the distance of sunlight through the earth's atmosphere is much longer. This causes a large proportion of the high frequency light components (blue) to be removed from the direct beam, leaving mainly light with long wavelengths which shifts the colour impression of the sun towards red. This effect is further may be enhanced by additional particles in the air (e.g. dust, sand).

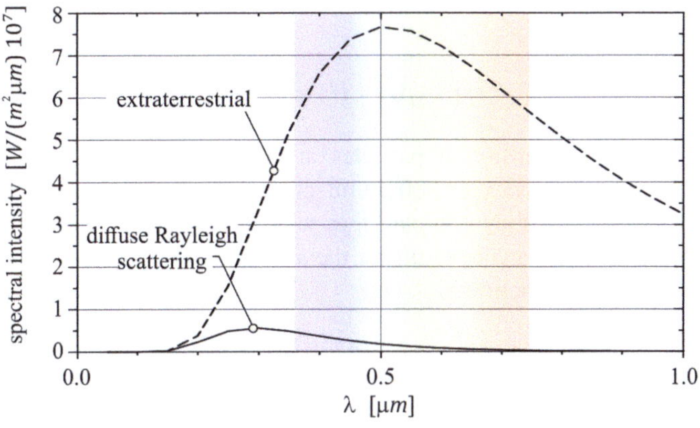

Fig. 2.37 Spectral power distribution distribution of solar radiation by Rayleigh scattering

In order to calculate how much light passes through the Earth's atmosphere due to Rayleigh scattering, the intensity formula (Eq. (2.60)) must first be integrated over the scattering angle θ and the distance from the scattering particle. The total density of scattering molecules in a volume is then considered. For an ideal gas the Avogadro number is contained for normal conditions of atmospheric pressure (1 atm) in a molar Volume of $22.4\,\mathrm{m}^3/\mathrm{kmol}$. From this the Loschmidt number N_0 gives the density of particles as $\approx 2.7 \times 10^{25}\,\mathrm{m}^{-3}$. In scattering theory the extinction of a gas containing many scatters can be calculated from the cross section of the particles and the particle density. This relates also to the effective index of refraction (the real part) of the gas n_0 which is (Van de Hulst, 1981),

$$n_0 = 1 + 2\pi \cdot N_0 \cdot \alpha. \tag{2.63}$$

The optical extinction coefficient β_λ of the atmosphere has to be integrated over all particles in the vertical atmospheric profile. Since the particle density of the atmosphere decreases approximately exponentially with height (see barometric height formula, Eq. (2.55)), the true air envelope can be replaced by a homogeneous layer whose thickness H is equal to the scale height and whose particle density is equal to the particle density at the ground with an optical depth δ_λ:

$$\delta_{\lambda R} = \beta_{\lambda R} \cdot H = N_0 \cdot \sigma_{\lambda R} \cdot H = N_0 \cdot \frac{2k^4}{3\pi} \cdot \frac{(n_0 - 1)^2}{N_0^2} \cdot H, \tag{2.64}$$

which results in

$$\beta_{\lambda R} = \frac{32\pi^3}{3} \cdot \frac{(n_0 - 1)^2}{\lambda^4} \cdot \frac{1}{N_0}. \tag{2.65}$$

The equivalent layer of thickness H now yields the total transmission T_λ for the earth's atmosphere

$$T_{\lambda R} = \exp(-\beta_{\lambda R} \cdot H) = \exp(-\delta\tau_{\lambda R}). \tag{2.66}$$

N_0 here gives the density of particles at ground level $2.7 \times 10^{25}\,\text{m}^{-3}$ and $n_0 = 1.000292$ is the index of refraction of air. The extinction coefficient has the dimension m^{-1}. We may compare that with the historical results of Paetzold (1952) who gives the following result for the attenuation of light in the atmosphere for the astronomical extinction k_{ex}

$$k_{ex} = \frac{160}{3} \cdot \pi^3 \cdot \frac{(n_0 - 1)^2}{\ln 100} \cdot \frac{H}{N_0} \cdot \frac{1}{\lambda^4}, \tag{2.67}$$

where in astronomy the relation between atmospheric extinction and transmittance of an atmosphere usually is not given as an exponential law but defined as

$$T_\lambda = 10^{-0.4 \cdot k_{ex}}. \tag{2.68}$$

The scale height in this formula has been taken as $H = 7990$ m.

The optical thickness describes the attenuation of light at vertical incidence. With oblique incidence at an angle θ against the vertical, the following effective layer thickness is approximately obtained as $H(\theta)$:

$$H(\theta) = \frac{H(\theta = 0)}{\cos\theta}, \tag{2.69}$$

if plane-parallel layering is assumed. The optical thickness of the Rayleigh atmosphere must then be corrected accordingly. Stöcker (1994) specifies the value $n_0 = 1.000292$. An insertion yields:

$$\delta_{\lambda R} = \frac{0.084\ \mu\text{m}^4}{\cos\theta \cdot \lambda^4}, \tag{2.70}$$

where λ represents the wavelength to be used in micrometres. Figure 2.38 shows the transmission as a function of the angle of incidence compared to the vertical. The graph clearly shows that, for example, already at a wavelength of $0.4\,\mu$m the transmission through the atmosphere decreases significantly with a change of the incidence angle from $\theta = 0°$ to $\theta = 60°$. With flat solar radiation, i.e. large angles of incidence ($\theta \gg 0°$), the intensity drops rapidly.

Table 2.6 shows the scattering coefficients of a Rayleigh atmosphere as a function of wavelength. As can be seen they decrease considerably with increasing wavelength.

Due to the large number of small molecules in the atmosphere, multiple scattering occurs, and the resulting diffuse irradiance from the sky which behaves similarly wavelength-selectively as the extinction on the passage of light through the atmosphere, resulting in a wavelength and angle-dependent diffuse Rayleigh irradiance as shown in the Fig. 2.39.

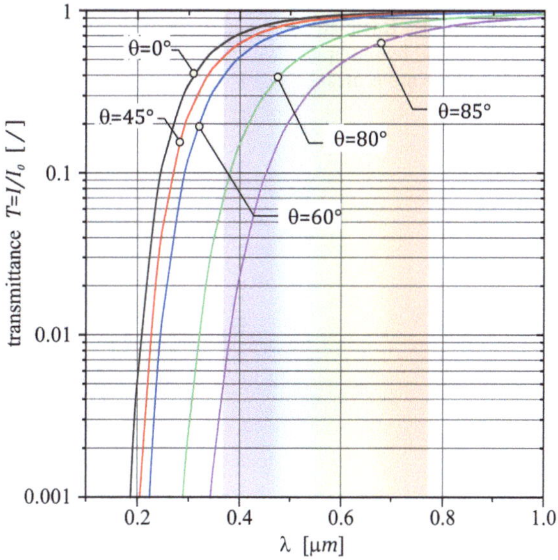

Fig. 2.38 Transmission T of Rayleigh scattering as a function of the angle of incidence θ versus the vertical incidence ($\theta = 0°$) according to Eq. (2.70)

Table 2.6 Dependence of the scattering coefficient $\beta_{\lambda R}$ on the wavelength λ according to Eq. (2.65)

Wavelength λ (nm)	Scattering coefficient $\beta_{\lambda R}$ (m^{-1})
300	147.9×10^{-6}
400	43.6×10^{-6}
800	2.6×10^{-6}
1000	1.1×10^{-6}
2000	0.07×10^{-6}

2.3.1.3 Properties of Mie-Scattering

Mie scattering occurs in particles whose dimensions d are approximately equal to the wavelength of the radiation. Strictly speaking, the historical Mie theory only applies to spherical particles, but solutions have been extended also to cylinders and ellipsoidal particles. While hygroscopic aerosols can be considered spherical, others are often non spherical (e.g. desert aerosol). The calculation of the scattering function requires the solution of the electromagnetic wave equation for a homogeneous body, using Maxwell's equations. The solution is given by an infinite series of angular functions. On the Internet one can find freely available optical simulation tools (e.g. http://omlc.ogi.edu/software/mie/).

For Mie scattering, with increasing particle size as larger part of the radiation is scattered forward. In Mie theory, the scattering coefficient is calculated as a function of wavelength λ divided by particle diameter d. The wavelength dependence of the scattering coefficient $\beta_{\lambda M}$ is less than in the case of Rayleigh scattering; the scattering coefficient can be scaled approximately as $\beta_{\lambda M} \sim 1/\lambda$.

In principle, Mie-scattering in the atmosphere behaves similar to Rayleigh-scattering. It also shows a spectral (i.e. wavelength-dependent) behaviour and is

Fig. 2.39 Diffuse Rayleigh scattering caused by multiple scattering on air molecules

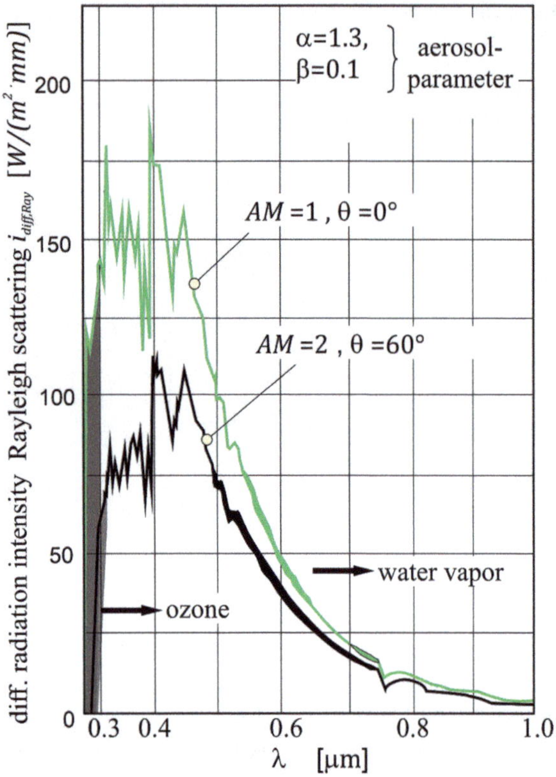

often referred to in the technical literature as aerosol scattering. However there are two important differences: The first is that for Rayleigh scattering backward and forward scattering effect are of the same size, whereas for Mie scattering the forward scattering dominates. Therefore, Mie scattering contributes more to the diffuse irradiance from the sky, as a comparison of the Fig. 2.39 with the Fig. 2.40 shows. The second difference is that the scattering cross section strongly depends on the parameter d/λ. As aerosols have a wide distribution of particle sizes, the resulting diffuse irradiance from Mie scattering is a mixture of different colours, as for each particle diameter a different wavelength scatters best (see Fig. 2.33). This mixture results in a white/grey diffuse irradiance $i_{diff,Mie}$ without any colour preference.

2.3.1.4 Multiple Reflections and Polarization of the Diffuse Sky Radiation

Both scattering mechanisms never occur singularly, but inevitably result in interactions in the form of multiple scattering processes of many particles and molecules, where the individual scattering mechanisms can be described by Rayleigh and Mie scattering theory. A further complication in the calculation of diffuse irradiance is the presence of the ground reflection (albedo). There are multiple reflections

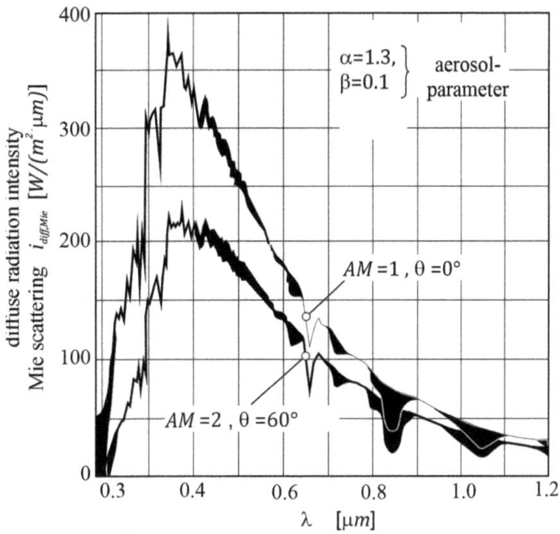

Fig. 2.40 Diffuse radiation intensity due to Mie-scattering $i_{diff,Mie}$ as a function of wavelength λ at different sun elevations (air mass factors = AM)

between ground and atmosphere, especially when clouds are present. This multiple reflection intensity is a third component of the spectral diffuse radiation intensity from the sky. Overall, the diffuse radiation intensity from the sky is created by many particle interactions consisting of Rayleigh, Aerosol (Mie) scattering plus diffuse light from multiple reflections between atmosphere and ground. The Fig. 2.41 shows the order of magnitude of these individual components as a function of wavelength λ. The addition of all scattered components can take up a considerable amount of the total radiation intensity and plays an important role especially when considering non-concentrating solar thermal collectors.

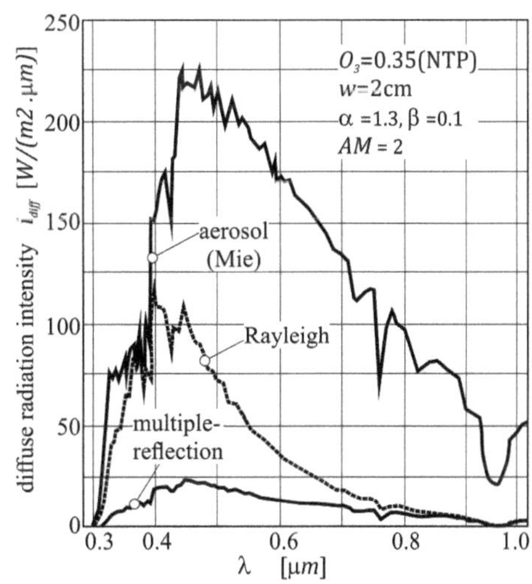

Fig. 2.41 Diffuse radiation intensity i_{diff} of Aerosol, Rayleigh scattering as a function of wavelength λ and by multiple reflection processes at an $AM = 2$

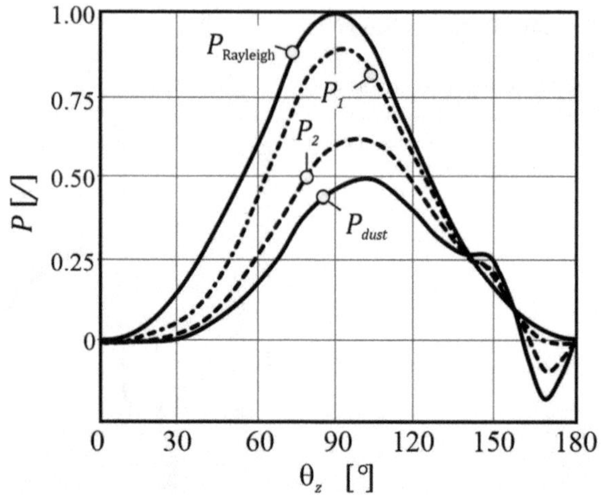

Fig. 2.42 Degree of polarization as a function of the zenith angle θ_z in a Rayleigh atmosphere and a dust atmosphere. Here is the mist component in $P_2 \gg P_1$ (Factor 60)

The polarization degree of diffuse sky light is an indicator of the accumulation of aerosols in the atmosphere. For this purpose, the degree of polarization P of the light is determined, especially in twilight. Whereas pure Rayleigh scattering exhibits full polarization P, aerosol scattering is unpolarized. The deviation from the values of the pure Rayleigh atmosphere indicates the degree of haze in the atmosphere, as shown in the Fig. 2.42. At a zenith angle of $\theta_z = 90°$ the greatest polarization occurs in the Rayleigh atmosphere. Atmospheres with a higher mist content are more turbid and exhibit a lower polarization. A high dust content initially causes a significantly reduced polarization of the light. Moreover, it leads to a shift in polarization towards larger zenith angles.

2.3.2 Technical Calculation of the Attenuation in the Atmosphere

Now we have discussed the different attenuation mechanisms of solar radiation in the atmosphere. All individual absorption and scattering processes are different and independent of each other. They therefore contribute to the overall attenuation with different spectral transmission coefficients. The direct spectral radiant power results from the product of the transmission coefficients of all components of the atmosphere $T_{i,\lambda}$ in the following way following a split-up of Iqbal (1983).

$$i_{dir,\lambda} = i_{SC,\lambda} \cdot \left(T_{Rayleigh} T_{ozone} T_{gases} T_{water\ vapour} T_{aerosol} T_{clouds} \right), \tag{2.71}$$

which is equivalent to the addition of all individual optical thicknesses:

$$\delta_{tot,\lambda} = \delta_{Rayleigh} + \delta_{ozone} + \delta_{gases} + \delta_{water\ vapour} + \delta_{aerosol} + \delta_{clouds}. \tag{2.72}$$

The individual transmission coefficients are again defined with an exponential function, which is described by Eq. (2.73)

$$T_{i\lambda} = \exp(-\delta_{i\lambda} \cdot AM). \tag{2.73}$$

Nowadays the spectral aerosol optical depths are becoming available and are being measured, which allows the determination of clear sky spectral irradiance.

Atmospheric turbidity characterizes the extinction effect of aerosols on solar radiation. It is well known that turbidity is the most important extinction affecting the direct and diffuse components of solar radiation under clear skies and that, as turbidity increases, the former decreases and the latter increases in about the same proportions so that the total effect of turbidity on global radiation is considerably smaller. However in many solar applications the effects have to be distinguished, for example in concentrating collectors which only accept beam radiation. Turbidity is used to characterize atmosphere in comparison to an ideal clean atmosphere without clouds.

The spectral turbidity can be defined using Ångström's law in practice, because of the complexity of Mie scattering calculation. It expresses the spectral aerosol optical depth ($SAOD$), $\delta_{a\lambda}$, at wavelength λ (in μm) through

$$\delta_{a\lambda} = \beta \cdot \lambda^{-\alpha}, \tag{2.74}$$

where the wavelength exponent α is generally observed to have values between 0 and 2.6. In this expression Ångström's turbidity coefficient, β, varies from 0 for an ideally aerosol-free atmosphere to more than 1 under extreme hazy conditions caused by sandstorms or heavy smoke plumes. The spectral aerosol transmittance is obtained from

$$T_{a\lambda} = \exp(-\delta_{a\lambda} \cdot AM). \tag{2.75}$$

where AM is the optical aerosol mass.

Instead, historically the concept of turbidity has been also used for broadband calculations. The Linke turbidity factor (T_L, defined for an air mass $AM = 2$) is a very convenient approximation to model the atmospheric absorption and scattering of the solar radiation under clear skies. It describes the optical thickness of the atmosphere due to both the absorption by the water vapor and the absorption and scattering by the aerosol particles relative to a dry and clean atmosphere. It summarizes the turbidity of the atmosphere, and hence the attenuation of the direct beam solar radiation. The larger T_L, the larger the attenuation of the radiation by the clear sky atmosphere.

The T_L denotes the transparency of the cloudless atmosphere. If the sky were perfectly dry and clean, T_L would be equal to unity (1). When the sky is deep blue, the T_L is just above 1 and still very small. In summer, in Europe, the water vapour is often large, and the blue sky is close to white. The T_L is larger than 3. In turbid atmosphere, e.g. in polluted cities, the T_L is close to 6–7.

The Rayleigh transmission factor T_{RS} is multiplied by an exponential function containing T_L and direct solar irradiance may be estimated using the solar constant i_{sc}:

$$i_{dir,h} = i_{sc} \cdot \varepsilon \cdot \cos\theta_z \exp\left(-0.8862 \cdot AM \cdot \frac{p}{p_0} \cdot \overline{\delta_R} \cdot T_L(AM2)\right), \quad (2.76)$$

where ε is the intensity correction of the solar constant due to the seasonal position of the sun in the elliptic orbit. The altitude correction with respect to sea level is done by the relative density of the atmospheric pressure p/p_0. The exponential function, which is the total broadband atmospheric transmittance T_{atm} can be split up into a transmittance of the $AM1$ Rayleigh atmosphere T_{RS} and a Linke transmittance factor T_{turb}: with an exponent of air mass factor AM:

$$T_{atm} = T_R \cdot T_{turb}^{AM} = \exp(-\overline{\delta_R}) \cdot \exp\left(-0.8862 \cdot T_L(AM2) \cdot \frac{p}{p_0} \cdot AM\right). \quad (2.77)$$

An alternative formulation can be the multiplication of broadband transmittance factors for the different attenuation processes (for the sun in the zenith $AM = 1$) absorption T_{ab}, Rayleigh scattering T_{RS} and aerosol scattering T_{ae}:

$$T_{atm}(AM) = T_{AB}^{AM} \cdot T_{RS}^{AM} \cdot T_{AE}^{AM}. \quad (2.78)$$

Tables 2.7 and 2.8 give transmission factors and turbidity values for different terrain situations. The tables show that there are considerable differences between urban areas and the high mountains, especially at low sun elevations. The seasons also lead to a variation in turbidity; for example, the higher temperatures in summer lead to a higher water vapour content in the air and consequently to greater turbidity. Both aspects must be considered when designing solar thermal systems.

Figure 2.43, however, shows that in a clear atmosphere the attenuation is mainly caused by water vapour in the near infrared wavelength range. The attenuation of the radiation intensity at an air mass factor of $AM = 1$ corresponds to the intensity

Table 2.7 Transmission factors and annual averages of turbidity factors as a function of solar altitude α_s

		Solar elevation angle α_s (°)				
	Turbidity	90	60	30	10	5
Ideal atmosphere T_{RS}^m	1.0	0.906	0.895	0.837	0.681	0.563
Gas absorption T_{ab}^m	1.9	0.91	0.91	0.89	0.88	0.81
Aerosols in high mountains T_{ae}^m	2.75	1.0	0.99	0.96	0.85	0.75
Lowlands T_{ae}^m	3.75	0.93	0.91	0.82	0.61	0.67
Cities T_{ae}^m	5	0.84	0.81	0.69	0.42	0.27
Industrial areas T_{ae}^m	5					

Table 2.8 Monthly averages of turbidity factors in different settlement types as a function of the month

	Jan	Feb	Mar	Apr	May	Jun	Jul	Aug	Sep	Oct	Nov	Dec
City	3.1	3.2	3.5	3.9	4.1	4.2	4.3	4.2	3.9	3.6	3.3	3.1
Countryside	2.1	2.2	2.5	2.9	3.2	3.4	3.5	3.3	2.9	2.6	2.3	2.3
Mountains	1.8	1.9	2.1	2.2	2.4	2.7	2.7	2.7	2.5	2.1	1.9	1.8

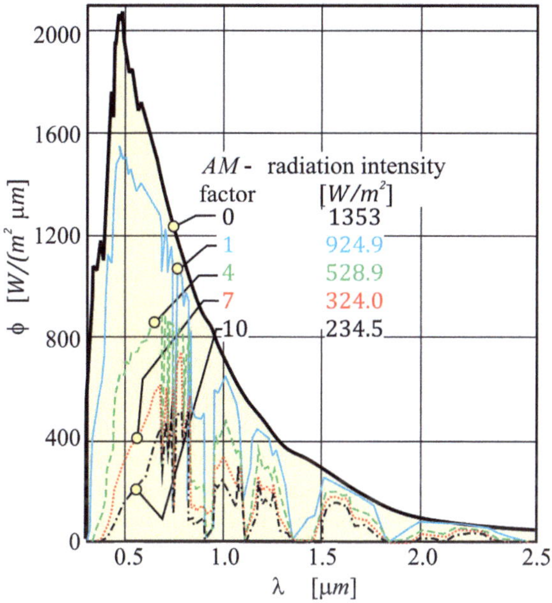

Fig. 2.43 Spectrum of direct solar radiation for different air mass bodies AM of $AM = 0\text{–}10$ as a function of wavelength λ in μm, assuming a clear atmosphere with water vapour corresponding to a precipitation of 20 and 3.4 mm ozone

attenuation of light when a layer of water of approximately 20 mm passes through. The effect of increasing air mass factors as a function of wavelength λ on the spectral radiant power is illustrated in Fig. 2.43. At the transition from an $AM = 0$ to $AM = 10$, the average radiant intensity already decreases by a factor of five (!).

2.3.3 Global, Direct, and Diffuse Irradiation

The irradiance on Earth is typically assessed on the horizontal surface. The instruments used for determining the irradiance value are able to distinguish between the direct beam radiation from the sun and the scattered diffuse irradiation from the sky. The latter one also has an angular distribution over the sky dome which in some circumstances can be quite complex. Whereas the clear sky diffuse intensity is dominated by the Rayleigh scattering and the intensity distribution due to polarization as explained already, the presence of clouds may complicate the situation.

In some cases a geometric arrangement of white clouds may even serve as a kind of concentrating element as beam radiation from the sun may be directly reflected towards the collector aperture for certain places. Other parts of the sky due to dark clouds might be contribute little to the overall diffuse irradiance. When the atmosphere is hazy with a lot of humidity, the whitish haze is especially bright around the horizon. And around the position of the sun we can detect an enhanced brightness which is called circumsolar radiation. For horizontal surfaces we look at the sum of diffuse sky and direct beam radiation. This quantity is called global irradiance. This is why in the context of solar collectors often G is used as the symbol for terrestrial irradiance instead of the more general i.

$$G_h = G_{b,h} + G_{d,h}. \tag{2.79}$$

When we have tilted surface, also reflected radiation from the ground can be accepted. But of course a part of the sky irradiance is not seen any longer by the aperture. The total irradiance on the tilted aperture is denoted by G_t:

$$G_t = G_{b,t} + G_{d,sky,t} + G_{d,refl,t}. \tag{2.80}$$

Even with clear skies, the proportion of diffuse irradiation depends on the length of the path through the atmosphere. Figure 2.44 shows this dependency in the form of the proportions of total radiation $i_{glob}(G_h)$, direct radiation $i_{dir}(G_{b,\,h})$ and diffuse radiation $i_{diff}(G_{d,\,h})$ as a function of the solar elevation angle α_s. The diffuse component hardly changes at solar elevations $\alpha_s > 40°$, while the direct radiation component increases significantly. At small solar elevation angles ($\alpha_s < 20°$) the intensity is determined by the diffuse component. Even with clear skies and vertical irradiation ($\alpha_s = 90°$), the diffuse part of the radiation is about 25% of the total irradiation intensity. Lower values are more likely to be the exception in Central Europe. They occur, for example, in mountainous areas with little aerosol concentration, environmental pollution and reduced density of air, in hours of cleaned air after heavy rains or at low temperatures where humidity is low.

Another aspect is that the total radiation i_{glob} is wavelength dependent due to scattering and absorption processes in the atmosphere. The blue sky radiation of clear skies without clouds is produced by wavelength dependent Rayleigh scattering. Therefore in this case, the largest part of the diffuse intensity i_{diff} is limited to the visible wavelength range, while at long wavelengths it has a negligible influence on the total radiation intensity i_{glob}, as the Fig. 2.45 depicts. The direct radiation intensity i_{dir}, on the other hand, is represented in the visible as well as in the near infrared wavelength range. On the other hand, heavy clouds with water droplets produce a different spectral pattern as here the larger particles result in Mie scattering. The resulting spectrum is more colour neutral (grey clouds).

The sum of both components diffuse radiation (i_{diff}) and direct radiation (i_{dir}) is called global radiation (i_{glob}). Since the distribution between diffuse and direct radiation intensity depends on the altitude, it also depends on the annual cycle. In Fig. 2.46 the typical annual cycle of global radiation in the form of the average daily values for a location in the Federal Republic of Germany. In Central

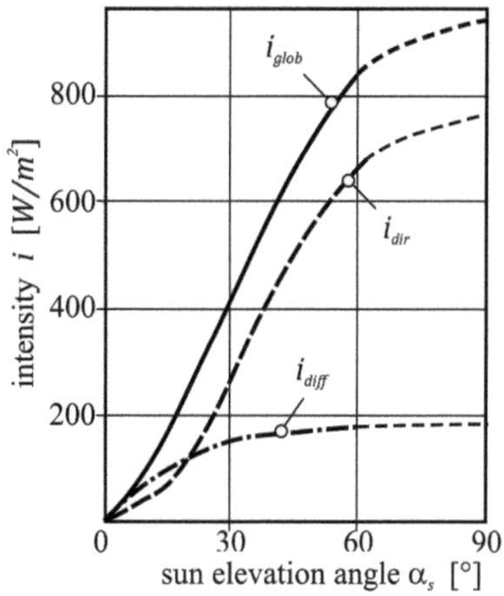

Fig. 2.44 Solar irradiance on a cloudless day (16.7.1969) at Sutton Bonington $\Phi = 53° N, 1° W$); i_{glob} = total irradiance; i_{dir} = direct irradiance on a horizontal surface; i_{diff} = diffuse irradiance; measured values are extracted curves, dashed lines are extrapolated from Monteith and Unsworth (2013)

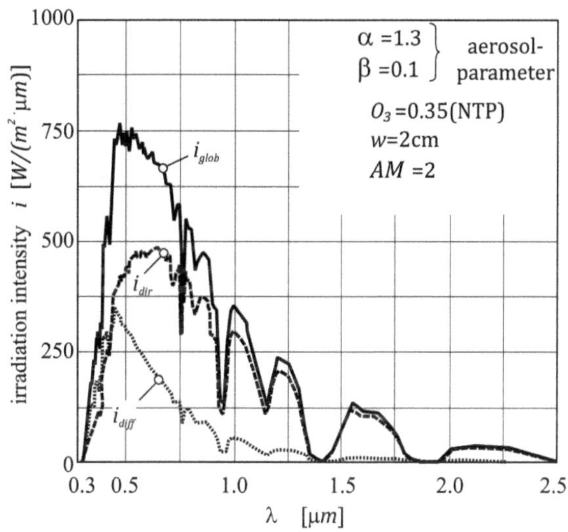

Fig. 2.45 Global, direct and diffuse radiation as a function of wavelength at the specified atmospheric parameters for a clear sky ($\Phi = 53° N, 1° W$) (Iqbal, 1983)

Europe, the diffuse part predominates. The annual average ratio of diffuse to direct radiation is around 60:40. Only in the summer months do the two components roughly balance each other out in terms of magnitude. The possible uses of both components are discussed in more detail in the following chapter.

Figure 2.47 shows the influence of different cloud types on the scattering in the atmosphere. Naturally, cumulus clouds with their high moisture content lead to a high optical thickness and reduce the proportion of radiation reaching the earth. Again there is an increased scattering with decreasing sun elevation angles and, as a consequence, the radiation intensity arriving at the ground is significantly reduced.

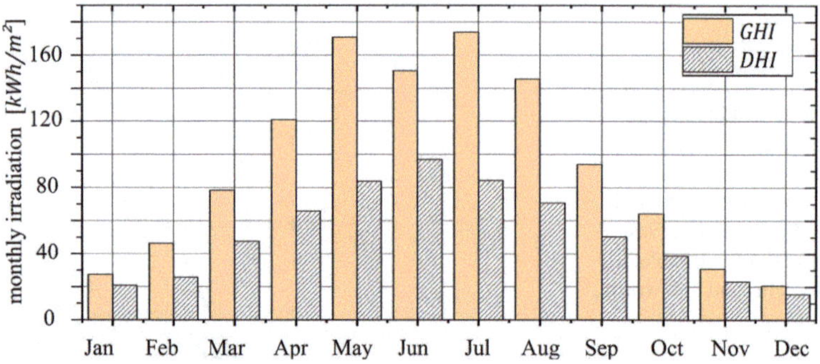

Fig. 2.46 Annual distribution of monthly global and diffuse irradiation sums on a horizontal plane for Munich/Germany

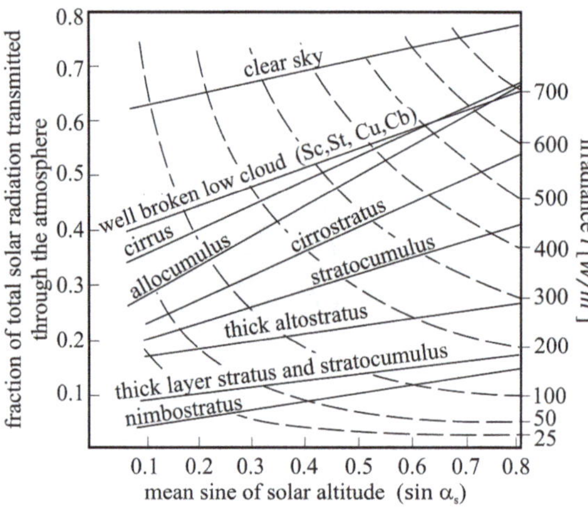

Fig. 2.47 Influence of different cloud types. The straight lines represent empirical relations between solar radiation and solar altitude for different cloud types according to measurements in the North Atlantic ($\Phi = 52° N, 20° W$). The curves represent isoplets of the illumination densities (according to Lumb). Sc = Stratocumulus; St = Stratus; Cu = Cumulus; Cb = Cumulonimbus modified from Monteith and Unsworth (2013)

2.4 Calculation of Irradiation on Arbitrarily Inclined Surfaces

Irradiation on a collector aperture consists of three main parts: direct beam radiation, diffuse radiation from the sky and the reflected radiation from the ground reaching the collector. Although the latter part also may contain directional components (e.g. reflection from a neighbouring glazed building or from a smooth water surface) very often it is assumed as diffuse. The diffuse component from the sky is produced by the scattering processes discussed previously. These lead to a generally variable distribution of radiance or sky brightness over the whole sky dome. One dominant and important part here is an enhanced brightness of the circumsolar part around the sun, the second one a typical horizon brightening at the horizon. Both effects can be observed already qualitatively with the bare eye.

Let us look onto the qualitative distribution of normalized intensity as a function of the zenith angle θ_z for clear and cloudy weather shown in the Fig. 2.48. It can be observed that within a few degrees to the sun position a strong decrease of diffuse intensity occurs. This is the so-called circumsolar radiation, which is solar beam radiation scattered in single events in the forward direction. A second part for a clear sky is the rather isotropic background intensity received rather uniformly from most of the sky hemisphere. The decrease of diffuse radiation in an angular range about 90° from the sun's position on a clear day, which is shown in the logarithmic scale of the Fig. 2.48 is explained by the dipole behaviour of the molecules at which Rayleigh scattering of matter occurs. In clear weather the Rayleigh scattering on air molecules predominates. The third part is horizon brightening from large zenith angles which originates from the fact that the number of scatterers in horizontal direction contributing to a diffuse intensity is high when we have good visibility. These three parts will be treated in an idealized way in the different diffuse sky models.

When we measure diffuse and direct components of irradiance on the horizontal plane, all these effects in summary lead to the measured value. However, when the measured irradiance shall be used to calculate the irradiance on a tilted aperture plane of a solar thermal collector, the angular view factors of the different parts

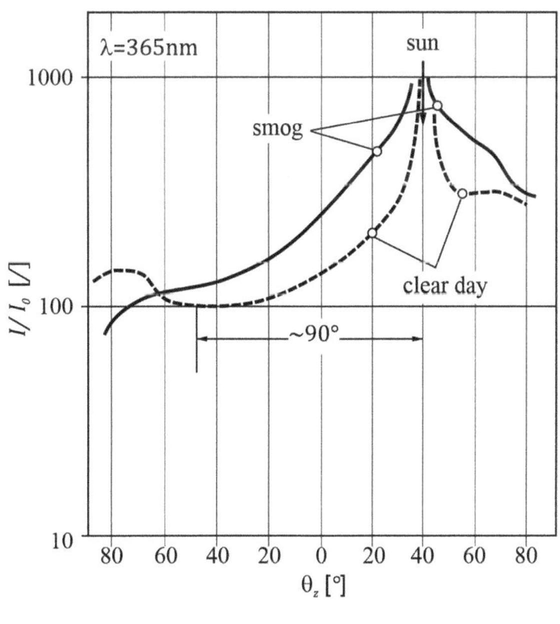

Fig. 2.48 Relative intensity I/I_0 of solar radiation (at $\lambda = 0.365\ \mu m$) as a function of the zenith angle θ_z in a plane where the sun is located, for Los Angeles on a clear day and on a day with smog (adapted from Coulson (1975))

change, and therefore a model is needed for the consideration of the different components.

2.4.1 Circumsolar Radiation

Depending on the atmospheric conditions, different directional distributions of the radiation intensity I result. When the sky is clear, about 75% of the diffuse sky radiation comes with approximately constant intensity from all directions of the firmament and about 25% comes from the direction of the sun. For simulation calculations of solar systems, this proportion is often referred to as the circumsolar part of direct radiation. This is permissible because direct radiation meters have an aperture angle of about 5° anyway.

The following Fig. 2.49 shows measured intensities of direct beam and circumsolar intensity. The measurement is taken in the angular range $[0, 30]$ mrad around the sun. The intensity drops exponentially fast outside the range of the sun disk rim at 4.65 mrad, where the intensity is nearly constant. If one integrates the circumsolar intensity Φ_{cs} from the $\delta_s = 4.65$ to $\delta_\Delta = 30$ mrad, the outer limit of the circumsolar region, θ indicates the radial angular displacement in the solar disk and $\phi(\theta)$ expresses the solar intensity profile assuming radial symmetry and divides it by the total intensity Φ_{tot} in this angular region one gets the circum-solar ratio CSR:

$$\Phi_{tot} = 2\pi \int_0^{\delta_\Delta} \phi(\theta) \sin\theta d\theta \approx 2\pi \int_0^{\delta_\Delta} \phi(\theta)\theta d\theta; \tag{2.81}$$

$$\Phi_{cs} = 2\pi \int_{\delta_s}^{\delta_\Delta} \phi(\theta) \sin\theta d\theta \approx 2\pi \int_{\delta_s}^{\delta_\Delta} \phi(\theta)\theta d\theta; \tag{2.82}$$

$$CSR = \frac{\Phi_{cs}}{\Phi_{tot}}. \tag{2.83}$$

The circumsolar profiles now can be classified by their CSR value. $CSR =$ 4–5% is a typical value for a clear sky. $CSR = 17\%$ is often taken as an average for slightly overcast sky.

Depending on the optical concentrator system, a large part of the CSR radiation moving with the sun's position can also be used in tracking collectors. As the acceptance angle (θ_a) of typical pyrheliometers anyway is about 4°–5° the CSR is recorded in meteorological measurements as beam radiation from the sun.

High circumsolar ratios (CSR) can significantly reduce the capture efficiency of high concentration collectors, due to a larger fraction of flux spillage. A changing CSR yields to a modification of the sun-shape distribution (intensity versus angle), see Fig. 2.50b. The actual sun shape is most strongly influenced by prevailing atmospheric conditions, particularly the level of particulate matter or moisture in

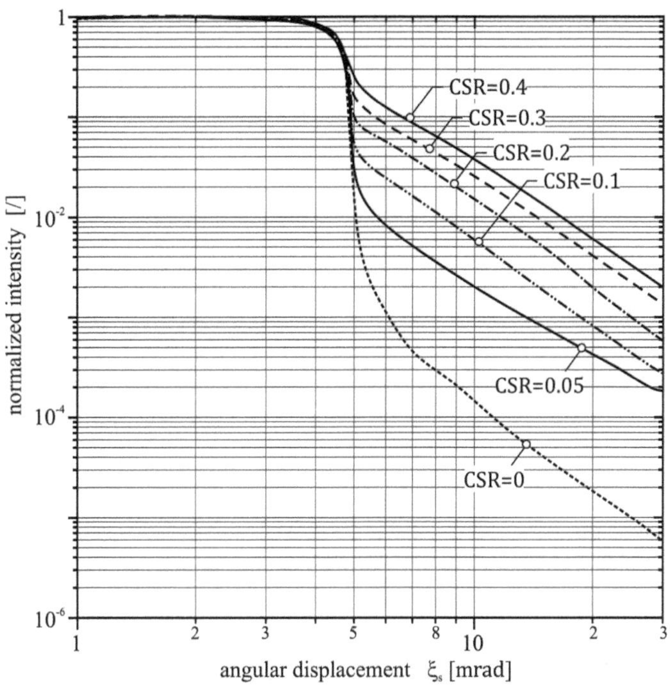

Fig. 2.49 Measurement data of different circumsolar conditions from Buie et al. (2003)

the air. A number of different sun-shape distributions can be found in literature however, for computational purposes the most commonly used is the one proposed by Buie et al. (2003). In Buie's model, the brightness at any point is normalized against the central intensity, which yields the following expressions:

$$\phi(\theta_s) - \begin{cases} \frac{\cos(0.326\,\theta)}{\cos(0.308\,\theta)} & \text{for } 0° \leq \theta \leq 4.65\,\text{mrad} \\ \theta^{\gamma}\ \exp\kappa & \text{for } \theta > 4.65\,\text{mrad} \end{cases}, \tag{2.84}$$

with

$$\kappa = 0.9\ln(13.5\,CSR)\,CSR^{-0.3};\ \gamma = 2.2\ln(0.52\,CSR)\,CSR^{0.43} - 0.1. \tag{2.85}$$

The exact solar shape distribution for yield calculation only plays a relevant role for concentrators with very high optical concentration ratios ($C > 10^3$), if the optical surface is also very good there. For collectors with lower concentrations, a flat intensity distribution, often called pill-box sunshape, is sufficient and provides only a few percent deviation.

The directionality of diffuse sky radiation is complex and can be in principle detected by sky cameras using lenses taking pictures of the holė hemisphere. When calibrated correctly, the radiance L of the different angular regions may be derived.

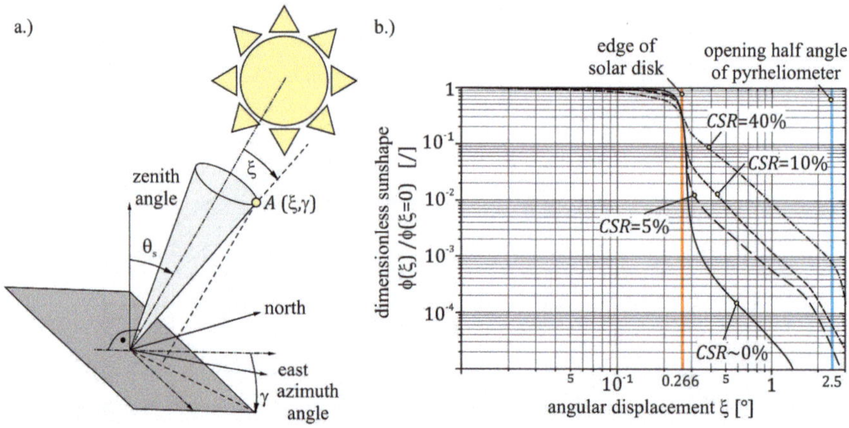

Fig. 2.50 **a** Coordinate system describing sky radiance starting from point $A(\xi, \gamma)$; **b** dimensionless solar radiance distribution as a function of angular distance ξ from the solar center ($\xi = 0$) at different atmospheric densities expressed as CSR. With sun disc limit (red line) and acceptance angle of a typical pyrheliometer indicated (blue line)

In practice however, simplified sky models serve to calculate the acceptance of solar irradiation on collector apertures. These are certainly simplifications, but for a large aperture say of a flat-plate collector it seems less important to model the exact angular distribution of the sky intensity. Also in thermal systems modelling we are mainly interested in time-averaged energy densities. And even if the time period is as short as 5 min, the changing radiances of moving clouds average out quite a lot.

The circumsolar part is also included in the typical measurement of the direct normal irradiance DNI. We consider an element of the sky whose angular position is defined by the angular distance ξ from the center of the Sun and the associated azimuth angle γ, as shown in Fig. 2.50, and from which the broadband celestial radiance $L(\xi, \gamma)$ emanates. The ISO definition of DNI is given by,

$$DNI = \int_0^{2\pi} \int_0^{\alpha_l} P(\xi, \gamma) L(\xi, \gamma) \cos\xi \, \sin\xi d\xi d\gamma, \tag{2.86}$$

in which $P(\xi, \gamma)$ is the penumbra function or often termed acceptance function. $P(\xi, \gamma)$ vanishes if ξ excceds a limit angle α_l. The aperture solid angle precisely defined with the help of the penumbra function can be roughly described by means of an equivalent angle half-width, which is called the acceptance, aperture half-width angle. The value of the penumbra function $P(\xi, \gamma)$ results from the fraction of parallel light rays falling on the aperture from the angles (ξ, γ) and reaching the sensor element of the pyrheliometer. The penumbra function can thus be calculated from the geometric dimensions of the pyrheliometer. For angles ξ_l smaller than 5° the deviation of the $\cos\xi$ term from 1 is negligible and assuming a radial symmetry of the celestial ray density in the vicinity of the sun's position in a clear sky we get:

$$DNI = \int_{0}^{2\pi} P(\xi)L(\xi)\sin\xi d\xi, \tag{2.87}$$

in which $P(\xi)$ and $L(\xi)$ are the azimuthal averages of $P(\xi,\gamma)$, $L(\xi,\gamma)$. The radiance $L(\xi)$ is a function of the optical density of the atmosphere and thus the aerosol fraction as well as the Rayleigh spectral and aerosol optical depth, the aerosol phase function, and other variables. The DNI is thus calculated from the sky radiance and the penumbra function, of the sun disk and the circumsolar region. In the definition of the DNI according to Eq. (2.86) lies the actual dilemma since it contains neither a prescription of limiting angle α_l nor a penumbra function. If different values are used for both variables, different amounts of the integrated radiance near the sun are obtained. At the outer edge of the atmosphere the angular extent of the sun is equal to the extent of the solar disk, i.e. δ_s, but the scattering of particles and molecules in the atmosphere results in an extension around the circumsolar region, which can be up to 5° and from which the radiation can also fall on a collector. Figure 2.50b shows measured data of the normalized radiation distribution as a function of the angular distance ξ from the solar center at different atmospheric states.

The direct radiance thus results from the radiance emanating from the Sun and the circumsolar region and is a function of the angular position relative to the centre of the Sun. The so-called sunshape, or $s_s(\xi)$, is then a dimensionless quantity in which the azimuthal mean radiance profile is normalized by the radiance at the centre of the sun including the circumsolar region.

$$s_s(\xi) = \frac{S_s(\xi)}{S_s(\xi = 0)} = \int_{0}^{2\pi} L(\xi,\gamma)d\gamma. \tag{2.88}$$

Even without the influence of the Earth's atmosphere, the radiance of the solar disk decreases with increasing angular distance from the centre of the sun. This effect is called limb darkening and is a function of wavelength. For example, the radiance at the edge of the solar disk at 1000 nm is about 55% of the radiance at the solar centre, while the radiance at 370 nm drops to about 20% (Blanc et al., 2014).

2.4.2 Isotropic Sky Model

The simplest model of the sky (and ground reflected) diffuse radiation is the isotropic model. Here the sky intensity is assumed to be constant over the hemisphere. Similarly the ground reflected radiation is taken as isotropically diffuse.

Table 2.9 Ground reflection coefficients (ρ_g) of various soil types for solar radiation

Soil type	ρ_g (/)
New snow	0.75–0.95
Old snow	0.40–0.70
Sandy soil	0.15–0.40
Asphalt	0.15
Meadows and fields	0.12–0.30
Streets and houses	0.15–0.25
Forests	0.05–0.20
Dark floor	0.07–0.10
Water (depending on incidence angle of radiation, pollution, algae content, ...)	0.03–0.22

This means that all diffuse radiation falling on a tilted collector aperture with tilt angle β can be quantified by the isotropic intensity and the view factor for the surface to the sky and the ground respectively. The direct irradiance has to consider the incidence angle with the cosine-factor:

$$G_t = G_{b,n} \cdot \cos\theta + G_{d,h} \cdot \frac{1+\cos\beta}{2} + \rho \cdot G_h \cdot \frac{1-\cos\beta}{2}, \tag{2.89}$$

where ρ is the ground reflectivity or albedo. $G_{b,n}$ is the direct normal beam irradiance and θ the incidence angle on the aperture area.

The isotropic sky model is easy to apply and to understand. It tends to be conservative in the sense that it underestimates G_t in typical situations. However, improved models have been developed which consider the horizon brightening or the circumsolar radiation.

The reflectivity of different ground surfaces (ρ_g) is given in the Table 2.9. This reflectivity is often referred to as albedo (Latin albus = "white") and is a measure of the reflectivity of diffusely reflecting surfaces. The albedo is the ratio of reflected light to incident light and is therefore between 0 and 1. It is particularly important in meteorology, as it allows estimations to be made about how much air warms up over different surfaces. The albedo of ice and snow in the polar regions influences considerably the radiation balance of the Earth. Even different types of snow show a considerably different but high degree of reflection, while that of water is almost negligible except for low solar altitudes. It is important that the diffuse albedo is nearly independent on the incidence angle of the solar radiation, in contrary to the direct reflectance of a smooth water surface.

2.4.3 Anisotropic Sky Models

In the situation of Fig. 2.51 let us consider the circumsolar part which is measured—depending on the sky conditions—partly as diffuse fraction (for large CSR

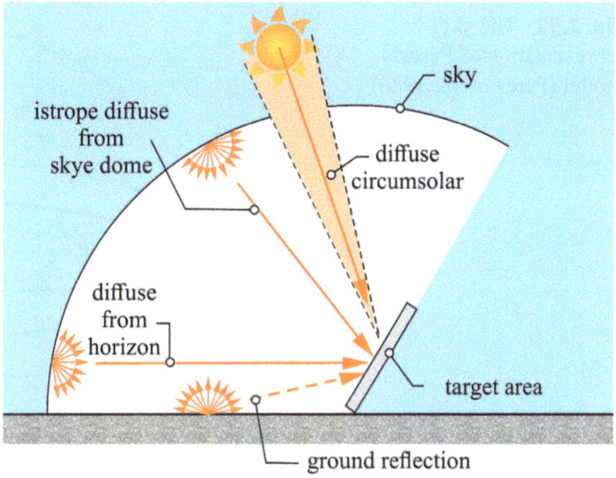

Fig. 2.51 Irradiation from an anisotropic sky on a tilted collector aperture

conditions) and partly as beam radiation (for clear skies with small CSR). Hay and Davies modified the simple isotropic in an easy way to correct the underestimation of directional radiation by introducing an anisotropy index A_i. A_i is a function of the transmittance of the atmosphere for beam radiation and given as

$$A_i = \frac{I_{bn}}{I_{0n}}, \tag{2.90}$$

where I_{bn} is the hourly terrestrial beam radiation and I_{0n} the extraterrestrial value. The fraction A_i of diffuse irradiation is now considered as circumsolar radiation and treated like beam radiation with an (approximate) incidence angle θ. Therefore, the total irradiance on the tilted surface (G_t) is modified in this model to:

$$G_t = \left(G_{b,n} + G_{d,h} \cdot A_i\right) \cdot \cos\theta + (1 - A_i) \cdot G_{d,h} \cdot \frac{1 + \cos\beta}{2} + \rho \cdot G_h \cdot \frac{1 - \cos\beta}{2}. \tag{2.91}$$

This is a slight improvement, but the horizon brightening is still not covered. An extension of the Hay and Davies model by Reindl et al. (1990) (the so-called HDKR-model) tries to also incorporate this feature.

$$G_t = \left(G_{b,n} + G_{d,h} \cdot A_i\right) \cdot \cos\theta + (1 - A_i) \cdot G_{d,h} \cdot \frac{1 + \cos\beta}{2} \cdot \left(1 + f \cdot \sin^3\left(\frac{\beta}{2}\right)\right) + \rho \cdot G_h \cdot \frac{1 - \cos\beta}{2}, \tag{2.92}$$

where f corrects for the horizon brightening,

$$f = \sqrt{G_{bh}/G_h}. \tag{2.93}$$

Fig. 2.52 The sky representation of Perez's model (Perez et al., 1986)

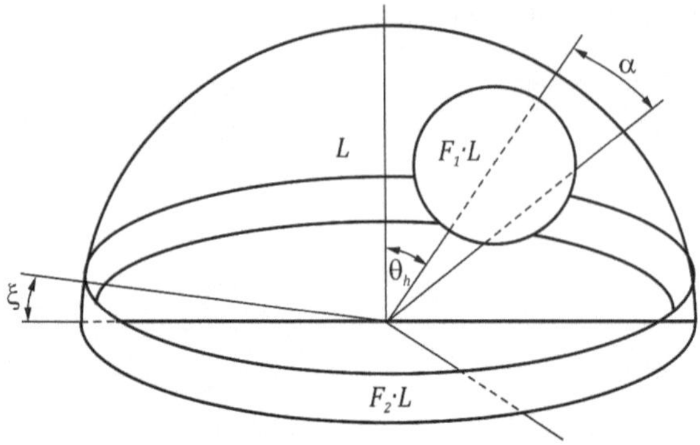

A different approach was undertaken by Perez et al. (1986) when dividing up the sky hemisphere in certain defined regions for circumsolar and horizon brightening using factors for the enhanced brightness (see Fig. 2.52). Here, L is considered as the sky radiance originating from the main portion of the dome, $(F_1 \cdot L)$ comes from the circumsolar zone, with $\alpha = 15°$ the half-angle of the circular region centered on the sun's position, and $(F_2 \cdot L)$ comes from the horizon with the horizon band angular thickness of $\xi = 6.5°$.

The diffuse part of the irradiation calculation for a tilted surface with this model can be written as

$$G_{d,t} = G_{d,h} \cdot \left[F_1 \cdot \frac{\max[\cos\theta, 0]}{\max[\cos\theta_z, \cos 85°]} + (1 - F_1) \cdot \frac{1 + \cos\beta}{2} + F_2 \cdot \sin\beta \right]. \tag{2.94}$$

The factor for the first circumsolar part does consider the incidence angle, but also the reduction of circumsolar area during sunset. Even for zenith angles larger than 90° some intensity is still available. The brightness coefficients are functions of three parameters zenith angle θ_z, a sky clearness ε and a brightness Δ. They are given by:

$$\varepsilon = \frac{G_{bn} + G_d}{G_d}, \tag{2.95}$$

$$\Delta = \frac{AM \cdot G_d}{G_0}. \tag{2.96}$$

The equations for calculating F_1 and F_2 are:

$$F_1 = |f_{11} + f_{12} \cdot \Delta + f_{13} \cdot \theta_z|^+, \tag{2.97}$$

$$F_2 = f_{21} + f_{22} \cdot \Delta + f_{23} \cdot \theta_z, \tag{2.98}$$

Table 2.10 Perez brightness coefficients f_{ij} for the anisotropic sky (Perez et al., 1988)

Range of ε	f_{11}	f_{12}	f_{13}	f_{21}	f_{22}	f_{23}
1.000–1.065	−0.008	0.588	−0.062	−0.060	0.072	−0.022
1.065–1.230	0.130	0.683	−0.151	−0.019	0.066	−0.029
1.230–1.500	0.330	0.487	−0.221	0.055	−0.064	−0.026
1.500–1.950	0.568	0.187	−0.295	0.109	−0.152	0.014
1.950–2.800	0.873	−0.392	−0.362	0.226	−0.462	0.001
2.800–4.500	1.132	−1.237	−0.412	0.288	−0.823	0.056
4.500–6200	1.060	−1.600	−0.359	0.264	−1.127	0.131
6.200 → ∞	0.678	−0.327	−0.250	0.156	−1.377	0.251

where the zenith angle θ_z has to be given in radians. Table 2.10 with the statistically derived coefficients f_{ij}, which depend on the brightness range ε, has been published in Perez et al. (1988). The development of anisotropic sky models took into account the statistical contribution sky brightness conditions on the irradiance of tilted apertures of collectors. If however, for some reason the exact distribution of brightness due to clouds should be incorporated, a more detailed knowledge on the sky radiance is needed. This cannot be provided by simple irradiance measurments, but pictures of the whole clouded sky should be taken by sky cameras. However, in most applications the foregoing analysis gives sufficient accuracy to energy calculations of solar thermal systems.

2.5 Statistical Distribution of Irradiation

The frequency of periods with various radiation levels of clear, overcast, and cloudy days is important for a realistic assessment of the performance of solar systems. We know from our daily experience that the intensity of beam radiation varies substantially over "good" and "bad" days. However the categories cannot be assigned only by measuring the global irradiance level as this varies in both day categories due to solar position and time as well. A better characterizing parameter is the so-called clearness index which relates the measured terrestrial irradiation level to the incident irradiation for the same time period at the same location outside the atmosphere, to the extra-terrestrial irradiation also on a horizontal plane. We differentiate between hourly, daily, and monthly clearness indices k_T, K_T and $\overline{K}_T$:

$$k_T = \frac{I}{I_0}; \tag{2.99}$$

$$K_T = \frac{H}{H_0}; \tag{2.100}$$

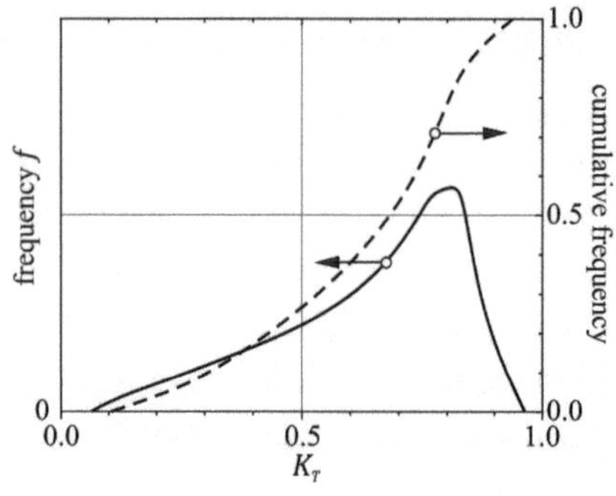

Fig. 2.53 Example of frequency of occurrence of days with various clearness index for a specific monthly average $\overline{K}_T$ (dashed: frequency distribution, solid: cumulative distribution)

$$\overline{K}_T = \frac{\overline{H}}{\overline{H}_0}; \tag{2.101}$$

in which I, H an $\overline{H}$ are the hourly, daily, or monthly sums of the irradiation on a horizontal plane, with the index 0 for extraterrestrial values. This clearness index is therefore a kind of atmospheric transmittance and quantifies the extinction in the atmosphere. For a location with a particular monthly clearness index $\overline{K}_T$ we find that the frequency distribution of days with various values of K_T is depending on the monthly average. If $\overline{K}_T$ is low we must have a large number of days also with cloudy and overcast sky and hence small K_T, if $\overline{K}_T$ is high, then there have to be much more days with clear sky and high K_T. A typical example of such a frequency distribution is shown in Fig. 2.53.

Liu and Jordan (1960) found that the cumulative distribution curves for locations having the same values $\overline{K}_T$ are very close to each other, even though the locations varied widely in latitude and elevation. Based on this they have developed generalized distribution curves of K_T versus fractional time. Bendt et al. (1981) have developed equations to represent the Liu and Jordan distributions, based on additional 20 years of data for 90 locations in the United States. These equations represent the Liu and Jordan curves very well for values of $f(K_T)$ less than 0.9–above 0.9 the correlations overpredict the frequency. The equations have the form:

$$f(K_T) = \frac{\exp(\gamma \cdot K_{T,min}) - \exp(\gamma \cdot K_T)}{\exp(\gamma \cdot K_{T,min}) - \exp(\gamma \cdot K_{T,max})}, \tag{2.102}$$

where γ is can be determined implicitly from the condition that the average of the distribution is the known $\overline{K}_T$. An explicit solution from a curve fit is given by

$$\gamma = -1.498 + \frac{1.184\,\xi - 27.182 \cdot \exp(-1.5\xi)}{K_{T,max} - K_{T,min}}, \tag{2.103}$$

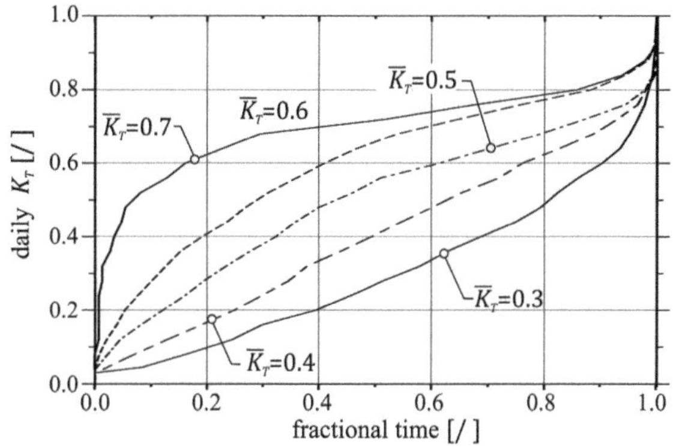

Fig. 2.54 Generalized functions of $f(K_T)$ over the monthly time fraction as a function of the monthly mean clearness (Liu & Jordan, 1960) in comparison to the functional representations (Bendt et al., 1981) (also applicable to hourly averages)

with $\xi = (K_{T,max} - K_{T,min})/(K_{T,max} - \overline{K}_T)$. For the extreme limits of the distribution the following values are recommended:

$$K_{T,max} = 0.6313 + 0.267 \cdot \overline{K}_T - 11.9(\overline{K}_T - 0.75)^8,$$
$$K_{T,min} = 0.05. \tag{2.104}$$

The universality of the Liu and Jordan distributions has been questioned, and in fact, for tropical climates an alternative formulation is suggested. And deviations were found for locations at high altitudes.

Figure 2.54 shows the K_T-values as a function of the time fraction. It has been shown that the same set of curves can also be used to derive $\overline{K}_T$ hourly values of K_T from daily values. Here, for a specific $\overline{K}_T$ a statistical distribution of K_T-values for individual days or subsets of a month can be given from the monthly mean values (e.g. for $\overline{K}_T = 0.7$, 7 days (25%) have a daily value of $K_T \approx 0.4-0.7$; 14 days (50%) have a ratio of $K_T \approx 0.7-0.8$ and the remaining 25% are above $K_T \approx 0.8$).

It has been observed however, that the generalized distribution curves for hourly and daily curves for a location are very similar (Whillier, 1965). Therefore, the distribution curves above may also be used for hourly clearness indexes K_T using the same monthly average $\overline{K}_T$.

The importance of these findings is that for many historical weather data only monthly values have been recorded and are still available. On the basis of the statistical distributions we can, however, construct artificial data for shorter time scales. Of course, for a realistic generation of weather data from long-term averages one needs not only know the statistical distribution of values but also time-correlations of time series. Stochastic modelling takes into account also the perseverance of weather situations. A good introduction into the topic is given by Aguiar et al. (1988) and more refined methods to construct with high confidence synthetic data can be found in Ngoko et al. (2014).

Similarly to the question, which temporal distribution of days with clear, overcast, and cloudy skies exist for a location, from the clearness indexes in turn it can be answered on a statistical basis, what the diffuse fraction of the corresponding hour, day or month is. The variable from which the diffuse fraction x_{dif} of the global irradiation can be derived, is the clearness index.

As the methods which present diffuse and beam fractions are derived from statistical evaluations, the individual results may differ substantially from measured data, especially when short time intervals like hours are considered. Individual cloud formations and changes in physical meteorological conditions (temperature, humidity) will certainly influence the reality whilst the statistics will only average over many similar clearness values.

The usual approach for hourly values is to correlate $x_{dif} = I_d/I$, the hourly fraction of the diffuse part in the horizontal global irradiation versus the hourly clearness index K_T, as illustrated in Fig. 2.55. Although the scatter of data is substantially, a tendency can be recognized. Erbs et al. (1982) have suggested the following correlation to be used based on US and Australian weather stations:

$$x_{dif} = \begin{cases} 1.0 - 0.249 \cdot k_T & \text{for } k_T < 0.35 \\ 1.5571 - 1.84 \cdot k_T & \text{for } 0.35 \le k_T < 0.75 \\ 0.177 & \text{for } k_T \ge 0.75 \end{cases}, \quad (2.105)$$

A similar correlation has been developed by Reindl and Beckman (1990) based on different European and US stations. In this correlation a small influence of the solar altitude angle has been incorporated:

$$x_{dif} = \begin{cases} 1.020 - 0.254 \cdot k_T + 0.0123 \cdot \sin\alpha_s & \text{for } k_T < 0.3 \\ 1.400 - 1.749 \cdot k_T + 0.1770 \cdot \sin\alpha_s & \text{for } 0.3 \le k_T < 0.78 \\ 0.486 \cdot k_T - 0.182 \cdot \sin\alpha_s & \text{for } k_T \ge 0.78 \end{cases}. \quad (2.106)$$

Reindl and Beckman (1990) investigated the addition of three more variables solar altitude, ambient temperature, and relative humidity in the predictor, and reduced then the numbers of variable. The single variable correlation is very close to the curve of Orgill and Hollands (1977), and the solar altitude yields the largest improvement in reduction of residuals (as given in Eq. (2.106), see Fig. 2.56). However, adding temperature and humidity even improved the result.

Similarly to correlation of hourly values, many scientific papers in the past have investigated the correlations of daily and monthly averaged diffuse fraction versus clearness index H_d/H_0 versus K_T and $\overline{H}_d/\overline{H}_0$ versus $\overline{K}_T$ (Table 2.11). The longer the averaging period, the smaller the range of observed and realized values. The correlation functions for monthly averages are typically in the range of $0.3 < \overline{K}_T < 0.7$. And the curves are usually linear or close to linear functions, with the scattering of measured values becoming closer. This is a consequence of statistics when we deal with averaged values. The outliers are singled out.

In some situations it is interesting to know how a solar system would perform on an average "typical" day. We are not interested in fluctuations and variations. Then we may construct a typical day based on a diffuse fraction from a correlation of monthly averages. How can the diffuse and global radiation be distributed of

Table 2.11 Monthly mean values of irradiation $\overline{H}$, clearness index $\overline{K}_T$, temperature $\overline{T}_a$ and degree days DD for two cities in the Federal Republic of Germany

		Jan	Feb	Mar	Apr	May	Jun	Jul	Aug	Sep	Oct	Nov	Dec
Bochum	51.17°												
$\overline{H}$	MJ/m^2	2.29	4.81	9.56	16.14	18.84	20.95	19.07	17.02	12.03	6.81	3.05	1.76
$\overline{K}_T$		0.274	0.356	0.441	0.524	0.497	0.51	0.484	0.51	0.498	0.428	0.321	0.255
$\overline{T}_a$	°C	1.7	2.5	4.9	8.3	13.1	15.7	17.2	16.5	13.9	9.6	5.1	2.7
DD	K_d	536	462	437	32[illegible]	60	0	0	0	14	291	417	505
Hamburg	53.63°												
$\overline{H}$	MJ/m^2	2.29	5.99	11.56	18.56	24.18	25.35	23.59	19.83	14.96	8.04	3.23	1.94
$\overline{K}_T$		0.432	0.497	0.569	0.625	0.645	0.619	0.602	0.607	0.636	0.556	0.402	0.351
$\overline{T}_a$	°C	0.3	1.0	3.5	7.5	13.3	15.4	17.1	16.2	13.6	8.8	4.2	1.6
DD	K_d	580	504	481	345	56	0	0	0	35	316	444	539

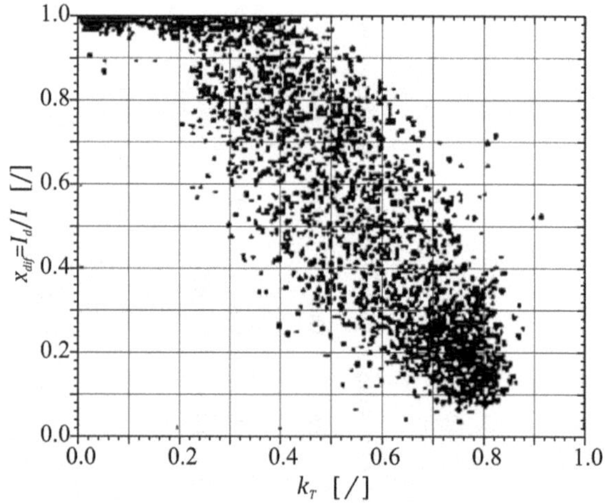

Fig. 2.55 Example of measured diffuse radiation fraction $x_{dif} = I_d/I$ as a function of global radiation in relation to extraterrestrial radiation on a horizontal surface as hourly mean value, adapted from Reindl (1988)

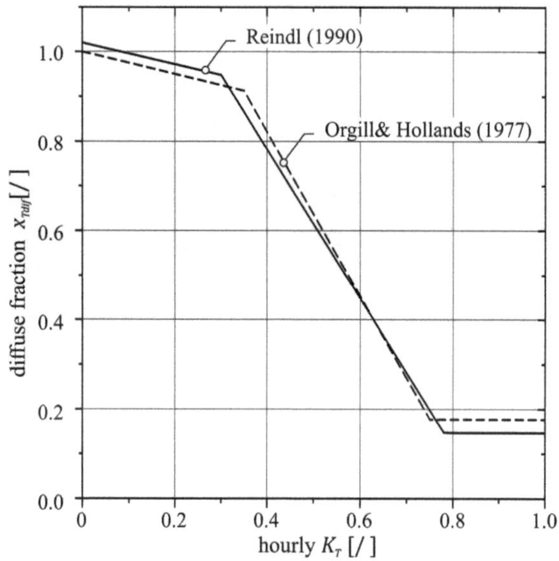

Fig. 2.56 Example of two correlations for diffuse fraction x_{dif} versus clearness index on an hourly basis (Orgill & Hollands, 1977; Reindl, 1988)

the hours of such a day? Already Liu and Jordan have given an answer to such a question. They developed from statistical studies a model how to distribute the hourly global and diffuse radiation over the day. The ratio of hourly to daily total global radiation is called r_t, the ratio of diffuse hourly to daily total diffuse r_d:

$$r_t = \frac{I}{H} \quad \text{and} \quad r_d = \frac{I_d}{H_d}. \tag{2.107}$$

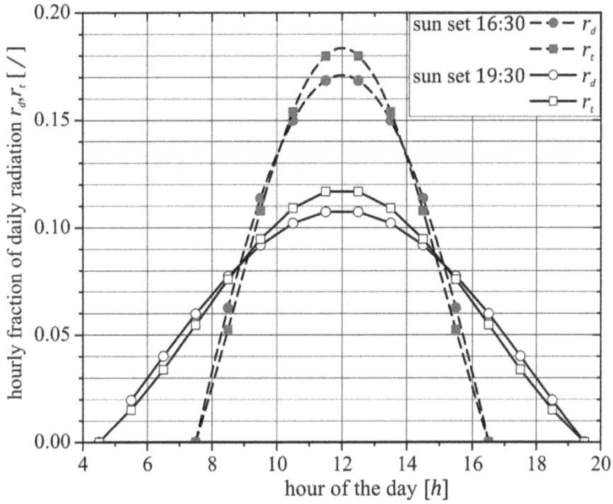

Fig. 2.57 Relationship between hourly and daily global irradiation on a horizontal surface for two days with sun set at 16:30 h and 19:30 h solar time

Collares-Pereira (1979) developed a functional form describing these ratios:

$$
\begin{aligned}
r_d &= \frac{\pi}{24} \cdot \frac{\cos\omega - \cos\omega_s}{\sin\omega_s - \omega_s \cdot \cos\omega_s};\\
r_t &= (a + b\cos\omega) \cdot r_d;\\
a &= 0.409 + 0.5016 \cdot \sin\left(\omega_s - \frac{\pi}{3}\right);\\
b &= 0.6609 - 0.4767 \cdot \sin\left(\omega_s - \frac{\pi}{3}\right);
\end{aligned}
\tag{2.108}
$$

where ω_s is the sunset hour angle. The curves of r_t of the relationship between hourly and daily global radiation is shown for hours in the afternoon as a function of the sunset hour angle in Fig. 2.57.

With the simple means described in the last paragraph we are able to model even starting from simple monthly irradiation values solar hourly irradiation values which we need at least to model the performance of a solar thermal system over the day. Modern computer programs use much shorter time steps of 1–5 min, but the input data there are usually still based on hourly data. If not measured, of course we do not know the statistical distribution over time, the correct time series which is dependent on many weather variables, the prominent ones being clouds, rain fronts and temperature drops. However, we can do better than just producing an average monthly day. This anyway would not give a representative result for a solar system performance but a conservative underestimation. But, at least using the distribution of days in a month we can produce by throwing the dices—in science terms with random Monte Carlo methods—a realistic distribution of clear, overcast, and cloudy days producing the correct monthly average. And similarly methods have been developed to create hourly or sub-hourly data of irradiation. Of course, the limits of the physical reality have to be considered, so global terrestrial irradiation might not exceed extra-terrestrial at any time. The computer program Meteonorm (Remund et al., 2010) has exactly incorporated such stochastic methods to create time series of beam and diffuse irradiation, and the values are even correlated with other variables like humidity and temperature. It is unlikely that a decrease in irradiance will lead to an immediate temperature drop.

But, although artificial weather data are in fact a big help, we need at least a good experimental basis of this, and the best solution anyway is to directly measure the real weather data in a time resolution which is needed. With this question we will deal in the next paragraph.

2.6 Quantitative Measurement of Solar Radiation

Acquiring information on the typical solar irradiation available at potential plant sites is of great importance to the design of solar power plants. The climatological average annual solar irradiance is most important in the planning process, but the characteristic frequency distribution also plays an important role. We talk here about solar resource assessment, in contrast to solar irradiation forecasting, which is the process of predicting solar radiation conditions on different time scales (from minutes to days) at a given time in the future. Ground based measurements are essential both at pre-operational and operational stage. However, satellite data provides broader temporal and special coverage, and therefore long-term estimates are commonly based on satellite-derived solar resource assessment. For a complete assessment of solar thermal plant design and operation, auxiliary meteorological parameters need to be analysed, e.g., ambient air temperature, wind speed and relative humidity. Although these quantities are important regarding plant performance and malfunction, we will not deal in detail with their measurement. Wind for example may influence heat loss, especially of uncovered collectors, wind gusts may be a risk for the safe integration of roof-mounted collectors or might harm mirror structures of concentrating mirrors. Ambient air temperature and humidity affect the necessity of climatization, have an impact on drying substances or determine the choice of heat-rejecting coolers.

The term *irradiance* is used to describe the solar power (instantaneous energy flux) falling on a unit area per unit time, e.g., in $\mathrm{W/m^2}$. The term *irradiation* relates to the amount of solar energy falling on a unit area over a specified time interval such as a day, a month, or a complete year. In climatology, solar irradiation typically is given in $\mathrm{J/m^2}$ summed over the period of a day, whereas for solar engineering, solar irradiation is often reported in units of $\mathrm{kWh/m^2}$.

According to the standard (ISO 9488, 2021), the symbol for the irradiance is G and for the irradiation is H. In practice, often the same symbol (e.g. DNI) is used for irradiance and irradiation. In that case, they have to be differentiated by context or by the attached units.

2.6.1 Terrestrial Measurements

Measuring the Sun's energy incident to the Earth's surface is a difficult field measurement exercise. The measurement technology applied today is based on the conversion of electromagnetic radiant energy to heat or electricity. The sensors for such measurement instrumentation are e.g. photodiodes and thermopiles. The first utilizes the photoelectric effect to convert direct sunlight, and the second one the thermoelectric effect to convert generated temperature differences into an electrical current, which then can be measured.

Instruments for measuring solar radiation are of two basic types:

- A pyranometer is an instrument for measuring total hemispherical solar (beam plus diffuse) radiation with a hemispherical (180°) field of view, usually on a horizontal surface (global irradiance). If shaded from the beam radiation by a shade ring or disc, a pyranometer measures diffuse radiation.
- A pyrheliometer is an instrument using a collimated detector for measuring solar radiation at normal incidence from the sun and from a small portion of the sky (typically 5°) around the sun (i.e., beam radiation) with narrow field of view.

Pyrheliometers operate with a sun-tracking system that positions the sensor adequately towards the sun. Consequently, when considering the measurement of direct sunlight, the requirements for solar measurement stations are advanced since tracking systems bring high technical and mechanical requirements.

2.6.1.1 Thermal Sensors

With thermal solar radiation sensors, the radiant energy is initially converted into thermal energy by means of a black absorbing surface. A temperature difference is created in comparison with a white, reflecting or a shaded surface, as shown in Fig. 2.58 left. The small difference creates an electrical signal by a thermopile working with the thermo-electric effect. This electrical output can be measured with a voltmeter.

Due to the functional principle of thermopile radiometers, they are sensitive in a wide spectral range. Ideally, the black surface of thermopile instruments should absorb all incoming irradiance, like an ideal black body. However, only 97–98%

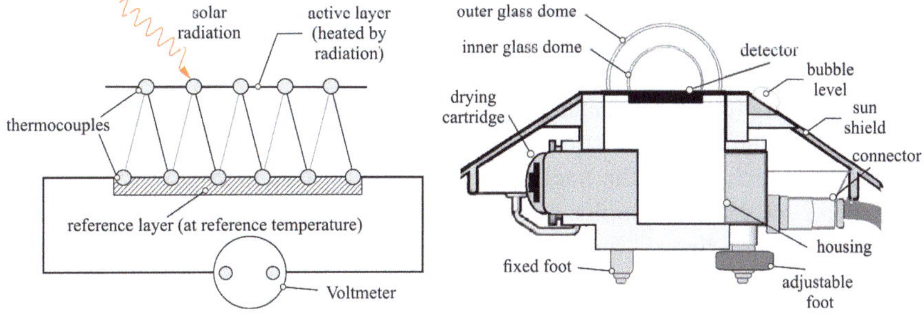

Fig. 2.58 Pyranometer principle (left) schematical functional principle: (right) set-up functional device

of the total solar energy is absorbed by the thermal detector. In addition, thermal losses by convection and radiation have to be minimized so that the temperature of the black surface is a good measure for the absorbed radiation and is not influenced by wind or sky conditions. Glass domes and glass windows are used for that purpose, as illustrated in Fig. 2.58 right. In summary, the sensors do not behave as ideal blackbodies but rather as small solar thermal collectors, so the instrumentation needs to be calibrated and compensation corrections are used to deduce the solar irradiation.

A thermopile element consists of a large number of thermocouple junction pairs, electrically connected in series. A thermocouple consists of two dissimilar metals connected together. The absorption of thermal radiation by the active (or 'hot') thermocouple junctions increases its temperature to T_1 and the reference (or 'cold') junction is kept at a fixed temperature T_2. The differential temperature between the active ('warm') and the inactive ('cold') junction produces an electromotive force directly proportional to the differential temperature created. This is called the thermoelectric effect. The magnitude and direction of the electromotive force is affected by the type of metal and the temperature difference between the hot and cold ends. The relationship between the temperature difference and the output voltage of a thermocouple is nonlinear and is approximated by a polynomial interpolation.

Thermal processes need some time to reach stationarity, therefore the instruments have response times to abrupt changes in irradiation. This means that the sampling time Δt_s of high-quality measurements should be short enough (smaller than response times) in order to track adequately the signal and to be able to correctly average the incoming radiation. The averaged measurements using a good number of sampling data then can be stored after a recording interval $\Delta t_r \gg \Delta t_s$. Recording intervals of 5 min are certainly adequate for solar thermal collector applications. When producing the irradiation data for experimental validation or for high-quality simulations of solar thermal systems, other inaccuracies like zero point accuracy or tilt errors may affect the result. For example zero point errors can be often estimated looking at recorded signals at nighttime where no solar radiation should reach the sensor (but watch out for light sources in the neighbourhood!). Negative values often can be found then which indicate that especially for clear skies with low sky temperatures the thermal radiation exchange with sky produces some uncompensated losses. Of course, these thermal losses might also exist during daytime but are obscured by the large solar irradiance. If this is the case the solar signal might be systematically underestimated in such conditions. Therefore, the quality of the instrumentation is assessed by a number of factors listed in the Table 2.12, which affects the final accuracy of the measurement.

Figure 2.59 shows examples of thermopile pyranometers designed to measure hemispherical irradiance. The hemispherical glass domes allow the capture of GHI without a bias of incidence angles, when installed on a horizontal plane. For solar energy applications often the insolation on a tilted plane is of interest. If this is the case, pyranometers are installed for example in the installation plane of solar thermal collectors with tilt angles from 15° to 60° depending on latitude and

Table 2.12 Uncertainties and pyranometer classes according to ISO 9060 (2018)

Category	Secondary standard	First class	Second class
Response time (95% final value)	<15 s	<30 s	<60 s
Zero point accuracy			
(a) response 200 W/m^2 thermal radiation	±7 W/m^2	±15 W/m^2	±30 W/m^2
(b) response 5 K/h temperature change	±2 W/m^2	±4 W/m^2	±8 W/m^2
Stability/drift per year	±0.8%	±1.5%	±3.0%
Non-linearity (related to 1000 W/M^2)	±0.5%	±1.0%	±3.0%
Tilt error (0°–90°)	±0.5%	±1.0%	±5.0%
Spectral response deviation	±3%	±5%	±10.0%
Influence of temperature range of 50 K	2%	4%	8%
Overall measurement uncertainty (daily sums)	±3%	±5%	±10.0%

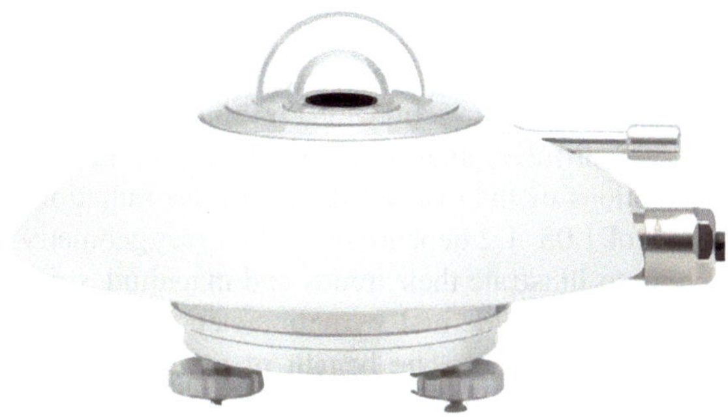

Fig. 2.59 Set-up of a pyranometer

application. These instruments directly capture the global tilted irradiance (GTI) mostly including a part of the irradiation reflected from the ground.

A disadvantage of thermopile pyranometers in comparison to photoelectric pyranometers is a higher price and frequent soiling of the glass dome. Also, measurements might be biased due to weather impact. Common phenomena to be considered here are dew or frost on the glass dome. To overcome this, advanced pyranometer systems have integrated ventilation and heating. Figure 2.59 shows an example of a heated, ventilated pyranometer next to a standard pyranometer. Heavy frosting on the glass dome will falsify the data and thereby generate data loss. To avoid such losses, the instrumentation must be selected depending on meteorological conditions. A disadvantage of air-conditioned systems is the higher price and the higher power consumption. Therefore power supply must be adjusted to the needs of the measurement system.

Although the DNI and diffuse irradiance is defined in ISO-9488 (2021) it is quite vague due to the lack of specification of the angular distribution of the solar radiation. Experimental instruments include or block not only the sun disc, but also a part of the sky. This leads to the requirement of a correction factor if direct radiation only from the solar disc or diffuse radiation including circumsolar radiation is needed. The measured DNI on site e.g. by conventional thermopile pyrheliometers is not necessarily larger than the effective DNI for a concentrator system. This originates from the larger aperture angle of pyrheliometers or equivalent pyranometric systems when compared to that of concentrating collectors leading to an overestimation of measured irradiance since it includes the irradiance from the solar disc and a larger circumsolar region. Also the diffuse radiation blocked by a shadowband has to be corrected.

Measurements of diffuse radiation can be made with pyranometers by shading the instrument from beam radiation. This is usually done by means of a shading ring, as shown in Fig. 2.60. The ring is used to allow continuous recording of diffuse radiation without the necessity of continuous positioning of smaller shading devices; adjustments need to be made for changing declination only and can be made every few days. The ring shades the pyranometer from part of the diffuse radiation, and a correction for this shading must be estimated and applied to the observed diffuse radiation (LeBaron et al., 1990). The corrections are based on assumptions of the distribution of diffuse radiation over the sky and typically are factors of 1.05–1.2 depending on the exact geometry. Typical shade ring correction factors, to illustrate their trends and magnitudes, is shown in Fig. 2.61.

Examples of thermal pyrheliometers that are designed to measure DNI are shown in Fig. 2.62. The benefit of pyrheliometers is the direct measurement of DNI and therefore such instrumentation complies with ISO 9060:2018 Class A standards. In order to capture direct sunlight, pyrheliometers are mounted to a sun tracker as can be seen in Fig. 2.62. For redundant measurements, a station usually

Fig. 2.60 Pyranometer with shadowband

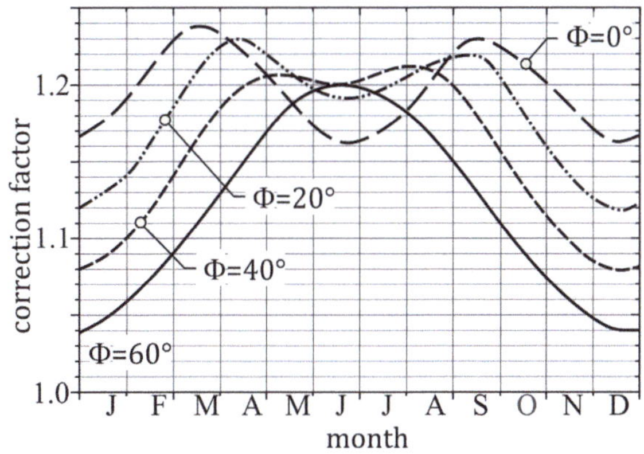

Fig. 2.61 Typical shade ring correction factors to account for shading of the detector from diffuse radiation. Adapted from Coulson (1975)

Fig. 2.62 Pyrheliometers (*Source* NREL (2009), Solar Radiation Research Laboratory)

holds two pyrheliometers and thereby reduces the risk of data loss. In order to avoid dew or frost to affect the measurements, Class A pyrheliometers have integrated window heating, such as the Hukseflux DR30-D1, shown in Fig. 2.62. For the needs of data quality monitoring, it is recommended to consider pyrheliometers with inbuilt tilt sensors, which allow the orientation of the instrument to be monitored.

The definition of the DNI according to Eq. (2.86) is an experimental definition and relates the DNI to the specific measurement device being used. The amount of measured circumsolar scattered irradiance depends on the state of the atmosphere and on the specific penumbra function of the instrument. The corresponding penumbra function can be approximated geometrically according to Fig. 2.63 by introducing three angles: the opening half-angle α, the slope angle α_s, and the limit angle α_l, defined as

$$\alpha = a\tan(R/L);\, \alpha_s = a\tan((R-r)/L);\, \alpha_l = a\tan((R+r)/L), \qquad (2.109)$$

where R, r and L are characteristic dimensions of the instrument. Since the angles are small one approximate for α, so that $\alpha \approx (\alpha_s + \alpha_l)/2$. This definition results in the so-called geometric penumbra function, which describes the fraction of collected radiant flux by an optical aperture depending on the off-axis angles. For off-axis angles $0° \leq \alpha \leq \alpha_s$, the penumbra function $P(\alpha)$ is unity; if $\alpha \geq \alpha_l$ correspondingly $P(\alpha) = 0$. The penumbra function monotonically drops from 1 to 0 in the interval $[\alpha_s, \alpha_l]$. The opening half-angle α is the main characteristic for the description of the field of view of a pyrheliometer, and it corresponds approximately to the centre of the transition range $[\alpha_s, \alpha_l]$. The viewing angle Ω is defined as the solid angle of a cone with apex angle 2α:

$$\Omega = 2\pi(1 - \cos\alpha). \qquad (2.110)$$

The detailed procedure, which is based on a multi-step approach, to arrive finally at a consistent DNI measurement value is described in Blanc et al. (2014).

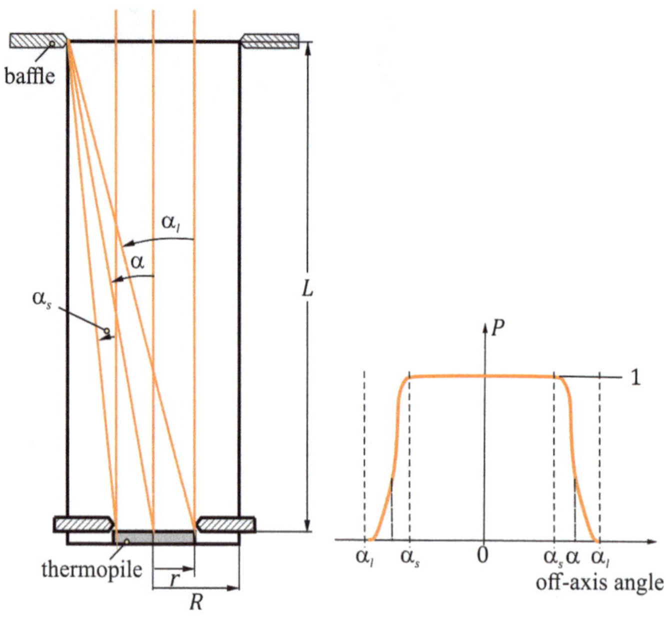

Fig. 2.63 Circular opening of a thermopile pyrheliometer to acquire DNI with definition of the slope angle α_s, opening half-angle α and limit angle α_l and corresponding penumbra function $P(\alpha)$ as afunction of the off-axis angle α (right)

2.6.1.2 Photoelectric Sensors

A photodiode sensor is usually a silicon semiconductor with $p - i - n$ structure or $p - n$ junction. These sensors convert radiant energy directly into an electrical signal. When a photon of sufficient energy strikes the diode, it creates an electron–hole pair. This mechanism is also known as the inner photoelectric effect. If the absorption occurs close to the junction, the pair is separated by the inner field: holes move toward the anode and electrons toward the cathode, and a photocurrent is produced. And in first order the photocurrent is linearly proportional to the irradiance. Photodiodes can be used under either reverse bias (photoconductive mode) or zero bias (photovoltaic mode), as depicted in Fig. 2.64.

Diodes have extremely high resistance when reverse biased. This resistance is reduced when light of an appropriate frequency shines on the junction. Hence, a reverse biased diode can be used as a light detector by monitoring the current running through it. A photodiode signal can be measured as a voltage or a current. Current measurement demonstrates far better linearity, offset, and bandwidth performance. The generated photocurrent is proportional to the incident light power.

Under zero bias, solar radiation incident on the diode causes a current across the device, leading to forward bias, which in turn induces 'dark current' in the opposite direction to the photocurrent. This is called the photovoltaic effect, which is the basis for solar cells. Figure 2.65 shows an exemplary photoelectric pyranometer.

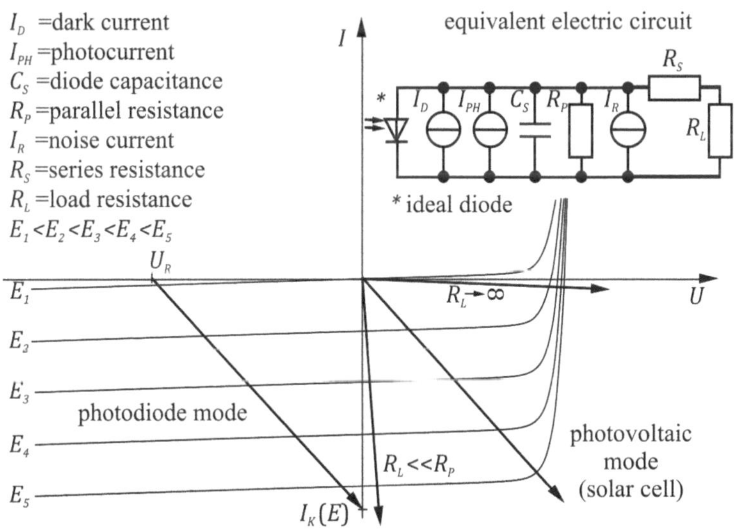

Fig. 2.64 I-V characteristic of a photodiode. The linear load lines represent the response of the external circuit: I = (Applied bias voltage − diode voltage)/total resistance. The points of intersection with the curves represent the actual current and voltage for a given bias, resistance and illumination (modified from Hess (2011))

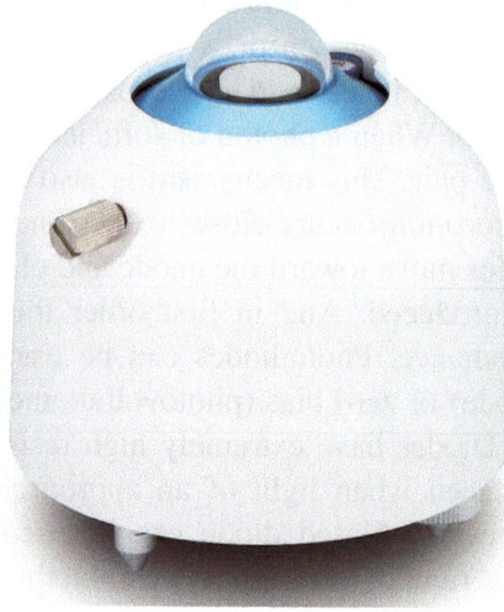

Fig. 2.65 Photoelectric pyranometers (Jrloginov, 2016)

The white diffusing layer on top of the Si diode cell shall improve the angular sensitivity.

They are less precise instruments than the thermopile instruments and have some limitations on their use. They are also less expensive than thermopile instruments and are easy to use. The main disadvantage of photovoltaic detectors is their spectrally selective response. Figure 2.66 shows a typical terrestrial solar spectrum and the spectral response of a silicon solar cell. Spectral response typically is in the range between 0.4 and 1.1 μm, and it is not uniform. If the spectral distribution of incident radiation was fixed, a calibration could be established that would remain constant; however, there are some variations in spectral distribution with clouds and atmospheric water vapor. One may estimate that the error introduced because of spectral response is ±5% maximum under most conditions of natural daylight and is ±3% under typical conditions.

Photovoltaic detectors have additional characteristics of interest. Their response to changing radiation levels is essentially instantaneous and is linear with radiation. The temperature dependence is ±0.15%/°C maximum. In order that the

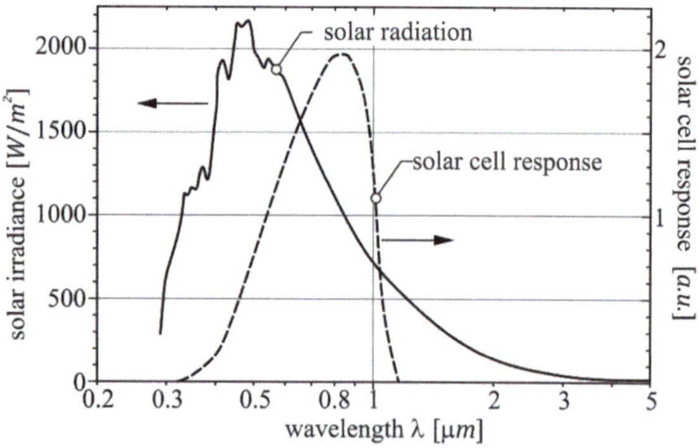

Fig. 2.66 Spectral sensitivity of photoelectric sensors

dependence of response on the angle of incidence is removed, the sensor is often fitted with a white diffusing cover that substantially removes the incidence of the radiation. The response of the detectors is independent of its orientation.

The advantages especially in the robustness and fast response led to the development of commercial Rotating shadowband irradiometers or pyranometers (RSI or RSP respectively) usually employ such a photoelectric radiometer (Fig. 2.65) to measure incident solar radiation. As shown in Fig. 2.67, the pyranometer is mounted on a 'head unit' apparatus, which permits unobstructed measurement of global horizontal irradiance. DNI can be determined from a single photoelectric pyranometer, if it is assembled in an arrangement that periodically blocks the direct beam radiation and causes it to measure global irradiance and diffuse irradiance alternately. The black shadowband rotates around the radiometer and thereby shades it off. While shaded, the radiometer is measuring horizontal diffuse irradiance (DHI). Accordingly, both GHI and DHI can be measured with one instrument. Since the shadowband is motor driven, and RSP's power consumption is demanding for stand-alone off-grid solar measurement stations sufficient power supply must be ensured.

RSP operation is realized by a program coding, which drives a motor control module. The code is usually run by a datalogger, which processes and collects the measured values. Typically, the shadowband rotates over the photodiode once every 30 s, taking approximately 1 s for this motion. During this period, the photoelectric pyranometer signal is sampled approximately 1,000 times. When the sensor is completely shaded from the sun by the shadowband, the lowest pyranometer readings occur. The software determines the DHI by finding the average of minimum values. For these instruments narrow-to-broadband corrections need to be applied to derive the full solar range. Ideally, these consider the spectral effect of lower air masses and also that of various contents of atmospheric trace gases and aerosols.

Fig. 2.67 RSP instrument operated by National Renewable Energy Laboratory-NREL (2009)

DNI measurements derived by RSPs are not reaching the exact same accuracy compared to pyrheliometer measurements, as long as the pyrheliometer tracking is accurate and the measurement device is maintained and properly cleaned. Soiling of measurement devices is often an issue with pyrheliometers, as these are easily affected by dust and dirt accumulation. Accurate pyrheliometer-based DNI measurements require daily and careful cleaning. In real life, the cleaning of pyrheliometers is often not practical since daily station visits can be very expensive and challenging for remote areas—the typically preferred areas for CSP application. RSP-based measurements are less susceptible to soiling since the impact is reduced for photoelectric effect. Accordingly, RSP measurements require less attendance and maintenance. In compliance with internationally recognized Tier2-standards, station maintenance and cleaning of the RSP should be performed twice a week. In case low soiling rates are given, station visits may be reduced down to once per week. Little soiling can be expected, e.g., in remote areas with sufficient ground cover by grass. For a desert environment, however, high soiling rates must be considered. Not only station maintenance but also the purchase price of RSP instruments is reduced compared to pyrheliometer systems. For maximum data reliability, solar monitoring stations with both sorts of instruments are typically configured with redundant sensors, which also allow comparison of both sensors regarding parallel measurement of the same DNI intensity. However, highest standards for solar measurements require pyrheliometer-based measurements of DNI. Such standards are classified as Tier1.

2.6.1.3 Sky Cameras and Measurement of Circumsolar Radiation

The first measurements of the solar aureole were carried out during total solar eclipses at the end of the nineteenth century. Between 1920 and 1955, the Smithsonian Institution measured the circumsolar irradiance coming from an annular region concentrically positioned around the sun. Their instruments had ring-shaped apertures and measured the irradiance coming from the angular region between 3.5° and 14.5° and later between 8.5° and 18.5° (half angles) (Hoyt, 1979). Also pyrheliometers with different collimators have been used.

Measurements of the solar radiance profile including the solar disk and the circumsolar region have been carried out by LBL. The measurements of LBL are of special importance, as approximately 180,000 measurements collected at 11 different sites between 1976–1981 were later digitally published in the LBL reduced data base (Noring et al., 1991). The instrument had a small circular aperture and measured the radiance coming from nearly point-like regions around and inside the solar disk.

Other groups used analogue photographic techniques. In the 90s CCD cameras with a highly linear response were used by Schubnell (1992) and Neumann et al. (2002). The data acquired with this approach agreed well with the LBL data base.

2.6.2 Satellite Data Evaluation

In cases where ground-based measurements are not available, satellite-derived solar radiation values are used. Satellites measure reflected radiation from the Earth's surface in several wavelength bands. Known albedo values by location and complex models and algorithms can be used to determine global, diffuse, and direct beam irradiance components. Raw data are available from various satellite operators. In turn, these are processed by several different organizations providing solar resource satellite data services. In some cases, these are commercial services and in other cases, research, or government. One of the best known and most extensive data sets is provided by NASA. Today, their data is provided through multiple platforms.

The methodology for calculating direct and global irradiance data from satellite data is differing from one institution to the other. From the early development of look-up tables already in the early years scientists have tried to model surface solar irradiance (SSI) data based on satellite observations, but after the year 2000, satellite-based estimation of SSI became increasingly mature thanks to the modern satellite sensors that have onboard calibration for both narrow and broadband channels, as well as further development of remote sensing methods. With quantitative remote sensing information regarding aerosols, water vapor, ozone etc. have become or are becoming available, which increasingly supported the accurate modelling of the radiative process. Therefore, quite sophisticated algorithms have been developed in which detailed radiative transfer processes are considered (Huang et al., 2019).

One major step is usually the determination of irradiation for clear sky conditions (see Fig. 2.68a) in one channel of the transfer calculations. Of the various atmospheric constituents under clear-sky conditions, ozone, water vapor, and aerosols are three critical components because of their influences on SSI and their variability in the atmosphere. Ozone and water vapor absorb solar radiation at certain wavelengths, which means that their loadings can be effectively retrieved by the sharp contrasts between the absorption windows. Aerosols are more challenging to retrieve, particularly over land. Factors like cloud contamination and the difficult differentiation of light reflected by aerosols and surfaces makes the determination of AOD (aerosol optical depth) over bright surfaces prone to large uncertainties.

When the sky is not clear, but clouds are partly covering shading the Earth's surface, solar irradiation is effectively altered (see Fig. 2.68b). Clouds play a critical role in solar radiation. This corresponds to our common perception that clouds obscure the direct sun but may provide a different diffuse light than the clear sky. This is dependent on cloud shape, cloud cover and cloud properties. In view of the decisive influence of clouds on SSI, to some extent, SSI satellite estimation is conducted around how to accurately account for radiative attenuation by clouds (cloud scattering and absorption) in the atmosphere. Towards this end, there are two different strategies:

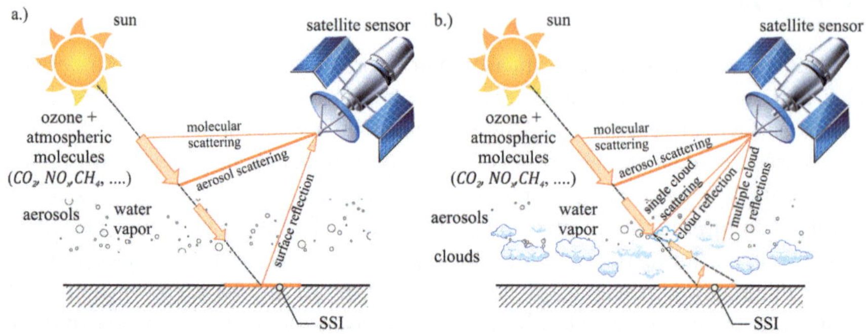

Fig. 2.68 Simplified relations between satellite observations and SSI according to one-dimensional radiative transfer theory for **a** clear sky and **b** cloudy sky modified from Huang et al. (2019)

- directly inferring the attenuation of clouds and subsequently SSI from top-of-atmosphere satellite pictures, or
- estimating the attenuation of clouds from cloud models.

Figure 2.69 and the following short description outlines one of the first modern methodologies to use satellite data for the determination of global and direct solar irradiance. It is taken from the ISIS model description, a method developed by DLR in the ISCCP (International Satellite Cloud Cover Programme). The radiative transfer through the Earth's atmosphere is calculated as 1-dimensional transport through a horizontally layered, plane-parallel atmosphere. The two-stream approximation of the diffuse und direct radiative transfer equation is used to derive the spectrally integrated diffuse und direct irradiances at the surface. Absorption of atmospheric gases is included by a 32 band calculation covering the whole solar spectrum from 250 to 4300 nm.

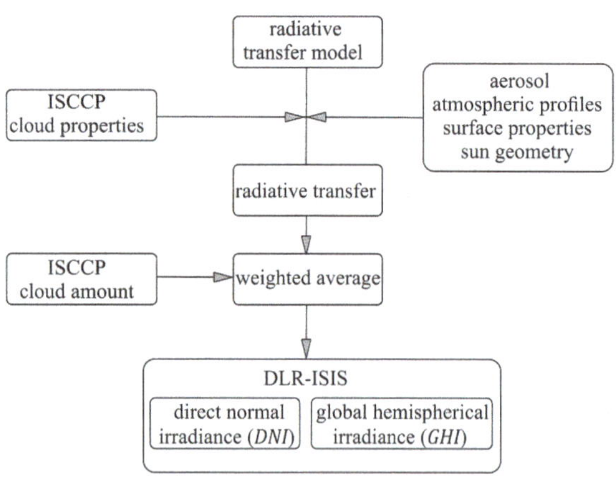

Fig. 2.69 Methodology for calculating direct normal and global irradiance from satellite data as developed in the ISIS project modified from German Aerospace Center (DLR) (2006)

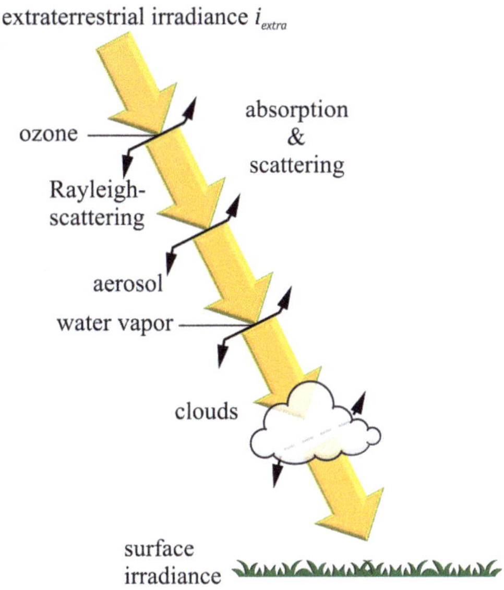

Fig. 2.70 Schematic modelling process in a radiative transfer model for clear sky

Cloud optics are included as two parametrizations for the optical properties of cloud particles. For water clouds Mie-calculations of liquid water droplets are used, for ice clouds a different parametrization is chosen. The clear sky radiative transfer algorithm and parameterizations are included in the radiative transfer library libRadtran. For the derivation of the DLR-ISIS data set, a program was built around this library to ensure the correct combination of atmospheric parameters, initiate radiative transfer calculations (Fig. 2.70), and handle the results; further details may be taken from Lohmann et al. (2006).

For clouds the models are based on the ISCCP D1 cloud data (see ISCCP homepage for more information about cloud data sets). It provides 3-hourly cloud observations on a 280 km × 280 km equal area grid, see Fig. 2.71. ISCCP differentiates between 15 cloud types. The classification includes three intervals of optical thickness in three cloud levels: low, middle, and high clouds. Low and middle cloud types are further divided into water and ice clouds, high clouds are always ice clouds.

For DLR-ISIS, optical thickness, cloud top pressure and cloud phase given in the ISCCP data set are processed to generate clouds for the radiative transfer calculations. One radiative transfer calculation is carried out for each occurring cloud type assuming 100% cloud coverage, plus one calculation for clear sky. For the final result, irradiances are weighted with the cloud amount for each cloud type and for clear sky, respectively, and summed.

The input of the aerosol optical thickness is split into two parts: tropospheric aerosol and stratospheric aerosol. The optical thickness of the tropospheric aerosol is taken from the NASA-GISS data set. Optical properties of the most common aerosol types were measured, and the distribution of aerosol particles are modelled

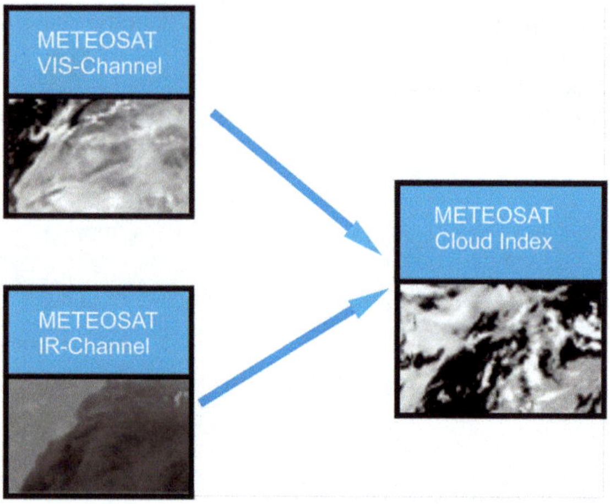

Fig. 2.71 Principle scheme of calculation of cloud cover index from image evaluation of different spectral channels

for individual months with the help of emission data and transport models. Stratospheric aerosol is mostly caused by large volcanic eruptions. The data set here also consist of monthly averages of optical thickness of the stratospheric aerosol. Ozone and water vapor profiles are also provided in the ISCCP input data set, based on the Total Ozone Mapping Spectrometer (TOMS) and the TIROS Operational Vertical Sounder (TOVS). Together with the surface albedo and the mean elevation above sea level for each grid box included in the ISCCP data set, these are used to derive irradiances for the DLR-ISIS data set.

Although the methodologies have been developed further and new versions of the semi-empirical approach with improved accuracy are being used in the engines such as SolarGIS, in the NSRBD of NREL and the Indian SUNY project, the basic principle is similar.

The derivation of solar irradiation values from satellite data is realized by radiative transfer calculations. Irradiation is calculated from infrared and visible channels of satellites with a nominal spatial resolution of 3 km × 3 km at ground level and 5 km × 5 km at ground level for the second and first generation of Meteosat, respectively. The annual sums of satellite-derived data can coincide with ground-measured data within a range of ± 5%. However, in some cases, deviations of around 20% have been observed (Gueymard, 2010). For shorter time periods like single months, days or hours, the deviation of satellite-derived and measured solar irradiance values can increase significantly (Zelenka et al., 1999). Due to insufficient accuracy of satellite data, precise ground-measured data are essential for validation and calibration of long-term satellite-derived data. The minimum measurement period should cover a duration of 1 year, so that at least one complete seasonal cycle can be evaluated (Suri & Cebecauer, 2014). Where to place the contribution of satellite data in a determination of statistical solar irradiance at a site is depicted in Fig. 2.72. Table 2.13 illustrates the input data used for the solar radiation model.

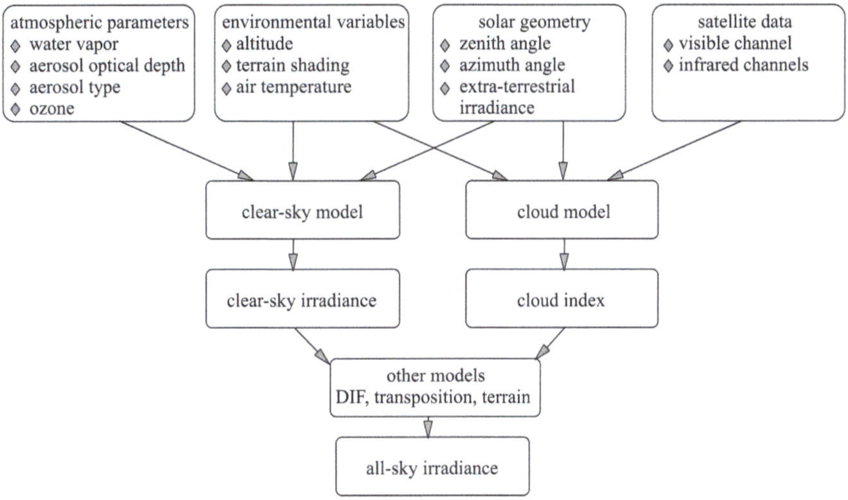

Fig. 2.72 Evaluation path to determine all-sky irradiance, with elements, data and models contributing to the result

Table 2.13 Inputs to a satellite model from Cebecauer and Suri (2016)

Inputs into the SolarGIS model	Source of input data	Time period	Original time step	Approximate grid resolution
Cloud index	Meteosat MFG (PRIME) satellites (EUMETSAT)	1994–2004	30 mins	3.0 km × 3.1 km
	Meteosat MSG satellites (EUMETSAT)	2005–present	15 mins	3.6 km × 3.8 km
Atmospheric optical depth AOD	MACC (ECMWF)	2003–present	6 h (monthly avg. before 2003)	85 and 125 km
Water vapor	CFSR/GS (NOAA)	1994–present	1 and 3 h	35 and 55 km
Elevation and horizon	SRTM-3 (SRTM)	–	–	250 m
SolarGIS primary data outputs (GHI and DNI)	–	1994–present	15 mins	250 m

At present, there is no standard procedure for processing solar radiation data or a set of procedures to be followed for solar resource assessments. As a result, and due to pressing deadlines for projects, project developers do not always follow a strict approach for solar resource assessment. The majority of the decisions for site selection of CSP plants are still based on the usually rough initial site assessments.

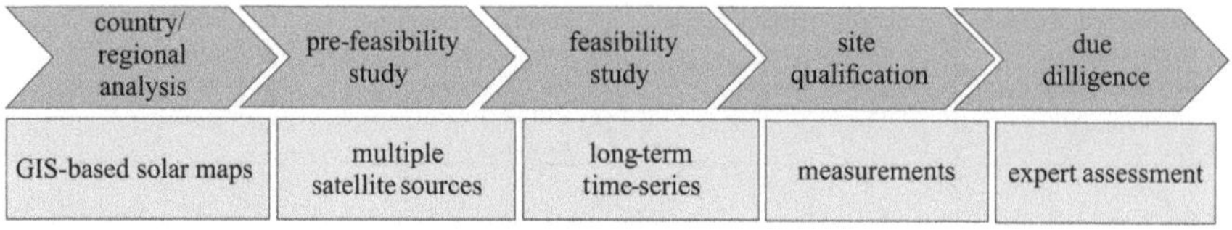

Fig. 2.73 Recommended steps for solar resource analysis during project development from Meyer (2010)

In such cases, solar radiation conditions such as annual cycle of DNI, frequency distribution of DNI, inter-annual variability, and the uncertainty have not been assessed in detail. As a result, the current knowledge of solar resource available at the foreseen location remains limited.

Based on practical experience and analysis of various data products and methods, the best practices to achieve high-quality assessments with reasonable effort are proposed. Following Meyer (2010) or Sengupta et al. (2017) the recommended procedure is illustrated in Fig. 2.73 and involves the following steps:

- When no detailed assessment of a project location has been made, multiple satellite data sources and, if available, ground measurements, should be taken into consideration. At this point, long-term average values of DNI from these sources may suffice.
- Calculate a quality-weighted best average and determine the resulting uncertainty by Gaussian error propagation.
- Once a project site has been confirmed for further development of the CSP project, a suitable measurement station should be installed at the site in order to reach the 1 year of measurement data in parallel with the project approval process.
- Determine the long-term best estimate for the specific project site (in 1 km $\times$ 1 km resolution) and based on as many years of satellite data as available. For DNI, a minimum of 10 consecutive years should be considered, because inter-annual variability is high.
- Multiple site-specific data sets from satellites and ground-based measurements, with overlapping time periods, can provide independent information and increase accuracy when combined. Only reliable data sets shall be considered. The uncertainty of the data is determined individually for each data set.

Site-specific satellite data time series are required to reflect the long-term history of irradiation and must be adapted based on overlapping time periods between different satellite and measurement data sets. Section 2.6.1 introduced the principles of solar ground-based measurements and pointed out the benefit of accurate measurements, providing less risk to power performance estimates and control. However, ground-based measurements come at great expense, preventing from long-term measurement campaigns. To derive long-term best estimates, it is necessary to consider data series exceedingly more than a decade. Thus, the lowest level of risk in solar resource assessment is achieved when long-term satellite data is recalibrated to ground-based measurements. This will reduce uncertainty. Furthermore, when simulating power performance of a CSP plant, data of sub-hourly temporal resolution should be considered (Hummon et al., 2012). Highest temporal resolution is given for ground based data, where common sampling rates are 1 min. Thus, adapting satellite data to ground data holds both advantages of long-term and high-temporal resolution. For spatial resolution of satellite data, at least a $10\,\text{km} \times 10\,\text{km}$ spatial resolution is required. Preferably, it should be $1\,\text{km} \times 1\,\text{km}$ resolution.

The accuracy and reliability of the satellite data provider is important as random and systematic errors can result in data of poor quality. Satellite providers often have a regional focus, so that one provider is not always recommended for all global regions. The current state of satellite-derived methods leaves room for improvement. Also, ground-based measurements can be of poor quality depending on the individual maintenance of the measurement station over the complete measurement period. If a location is seriously considered during the site qualification phase, the satellite-derived time series should be adapted to the ground-measured data by, e.g., using the procedure of Chhatbar and Meyer (2011). However, such adaptions require temporal overlap of ground- and satellite-based data.

2.6.3 Composition of Representative Time Series of Solar Irradiation

A Typical Meteorological Year (TMY) is a data set giving values in high time resolution. It shall well represent the characteristic meteorological conditions of average years at a site.

From the long-term satellite data, which are corrected with the measurement data, the average DNI can be derived, representing the so-called $P50$ value. $P50$ means that the annual solar irradiation has a probability of 50% being above and 50% below this value. It is the arithmetic average of all annual values. Often, $P50$ is referred to as best estimate and represents the statistical level of confidence to which the estimated solar resource exceeds with 50% probability. In other words, the $P50$-value is given as the solar irradiance, which is expected to be exceeded with a probability of 50%. From the corrected satellite data, the climatological average of DNI can also be derived.

Generally in statistics PXX-measures represent a value that is exceeded by XX% of the population of a data set, meaning that $P70$ or $P90$ values for annual solar irradiation represents a data set where 70% or even 90% of the years exhibit better results. $P50$ refers to 50% of the years exceeding the value and is the highest of these quantities. $P70/P90$ are pessimistic values regarding the representative irradiation. Therefore, they are a common quantity in the financial world, and widely applied for PV and wind energy projects. Thus, P-values of solar radiation (DNI) are provided to judge the reliability of the solar resource of a project, a typical example is illustrated in Fig. 2.74. Accordingly, the DNI value related

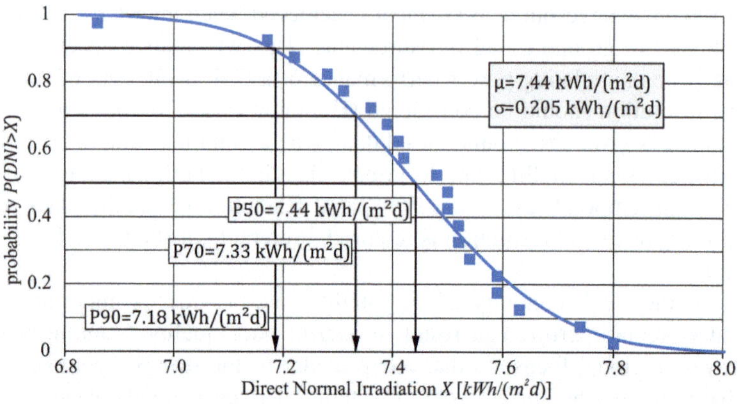

Fig. 2.74 Example of a fitted normal distribution to daily average DNI values of 20 different years. The values for P50, P70 and P90 are indicated

to $P90/P70$ for a given site are lower compared to $P50$, which represents the long-term average. Such values are derived based on the uncertainty of the data basis. A normal distribution is fitted to all available annual values, and from the determined parameter the critical value for $P70$ or $P90$ is determined.

A typical meteorological year (TMY), data set, is a compilation of extracts of data from various real years, assembled such that the annual DNI matches the long-term average for that site. Such data sets are a valuable input for modelling typical system output. For the creation of the most reliable CSP-specific typical meteorological years, site-specific measurements are of great value. According to Hoyer-Klick et al. (2009) data selection for TMY building should be done in a way that for each month only data that deviates by less than 5% from the climatological average is considered. The annual average in the TMY should be within 1% of the climatological average.

Finally, for bankable expert opinions on meteorological conditions for CSP, usually data with a lower probability of underperformance is required in order to satisfy conservative approaches from banks and lenders. This said, additional TMY s based on a more conservative approach are required to assess the risk of lower irradiance values. For this purpose, typically either $P70$ or $P90$ data sets are computed. These represent the data of a typical year with values that are expected to be exceeded in 70% or 90% of all years, respectively.

Alternatively, a risk assessment using performance simulation based on several good and bad years could give sufficient comfort to the banks. Such time series can be derived from satellites but should also be adapted to site-specific characteristics based on ground-measured data. This approach would allow a more detailed assessment of the influence of variable DNI conditions.

Also, by processing of at least 10 years of data, $P70$ or $P90$ values could be derived, which usually are used as basis to calculate the financial base case for a project. Compared to the simpler approach of only using $P70$ or $P90$ years, the advantage of using multiple years is manifold. First, the effects of meteorological variability and uncertainty are considered. Additionally, uncertainties resulting from technical parameters (describing the plant or the uncertainty of the performance simulation models) can be assessed in more detail.

2.6.4 Data Sources for Direct and Global Irradiation

Ground based solar and other meteorological data are available nowadays from many sources, however with different accuracy and temporal resolution. Measurement stations are operated by national meteorological services, airports, academic institutions, and even commercial entities, for instance when the run a large PV power plant. In addition interpolation algorithms are available to approximate meteodata for locations close to existing measurement stations.

Scientific measurement campaigns on solar irradiation in a large global scale started in the 1960s with the creation of the World Radiation Data Centre by the World Meteorological Organisation (WMO) in 1965. Since then a number of quality stations measuring also global solar irradiation have been maintained by national meteorological services around the world for many years. The archived data are still available (World Radiation Data Centre-WRDC (World Radiation Data Centre, 2022)) and the number of stations is indicated in Fig. 2.75.

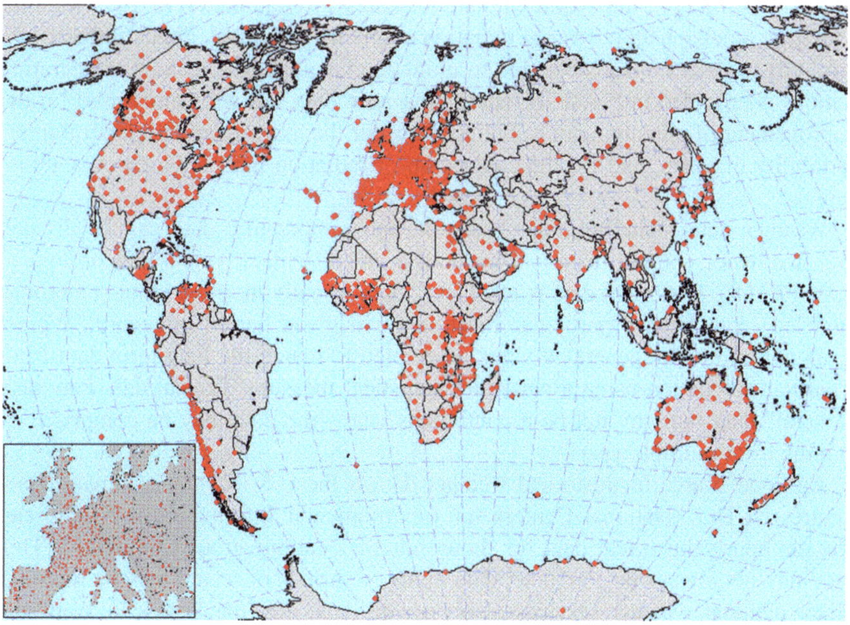

Fig. 2.75 WRDC stations worldwide from World Radiation Data Centre (2022)

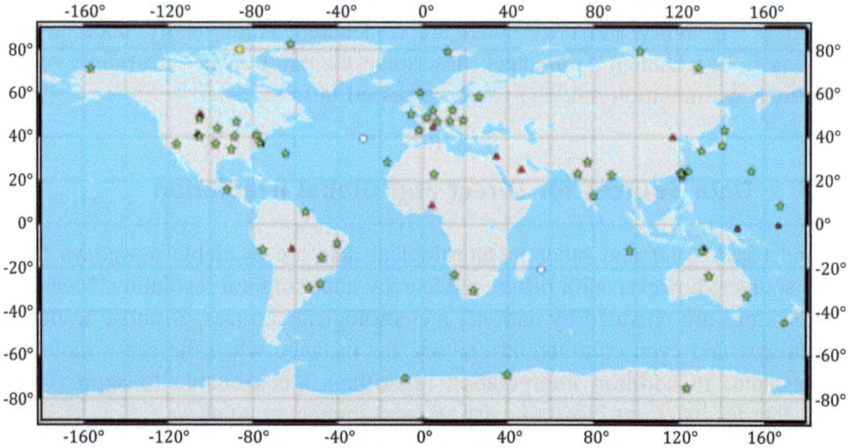

Fig. 2.76 BSRN network of currently 64 stations worldwide from Base Surface Radiation Network (BSRN) (2023)

Compared to the many years of daily, monthly and yearly sums of global irradiation from this network, systematic world-wide scientific measurements of direct normal irradiation DNI has been started only 1992 with the Baseline Surface Radiation Network (BSRN, (Base Surface Radiation Network (BSRN), 2023)), which is a project of the World Climate Research Programme (WRCP) to monitor with high precision all terrestrial radiation components (global, direct, diffuse, UV, Infrared) to detect differences due to climatic change; an indication of operating stations is provided in Fig. 2.76. These long-term data sets are mainly used for the improvement of climate models, but also e.g. for the assessment of temporal inter-variability of years, and for the calibration of radiation models based on satellite data.

Weather data including solar radiation and temperature data can be divided into two major classes: historical meteodata and so–called typical meteorological years (TMY). Historical data is just "real" data: usually measured (but sometimes modeled) data from a particular location for a given period of record. Typical years are artificially generated years assembled to match the long term data from a particular location using a particular statistical measure. Typical data may also be "real" data but may not be a contiguous year—the data may be comprised of months from multiple years.

Ag good source for historical weather data is the U.S. National Climatic Data Centre (NCDC, (National Centers for Environmental Information)). NCDC can provide hourly historical data for thousands of locations around the world. This data may not always be complete; data items or periods of record may be missing. Many national weather services also provide data, e.g. the Deutsche Wetterdienst DWD for Germany, the China Meteorological Bureau, the India ISHRAE Weather Data files. Some of the data provided are typical years, others are historical records. Internationally, the Solar and Wind Energy Resource Assessment (SWERA) project, funded by the United Nations Environment Program, is developing high quality information on solar and wind energy resources in 14 developing countries. Currently typical year hourly data are available for 156 locations.

A tool used widely as meteorological input for simulation of solar applications is the global climatological database Meteonorm (www.meteonorm.com). It's a combination of a climate database, a spatial interpolation tool, and a stochastic weather generator. In this way typical years with hourly or minute time resolution can be calculated for any site. The input of Meteonorm for global radiation is the Global Energy Balance Archive (GEBA, (Global Energy Balance Archive-GEBA)). All other meteorological parameters are taken from databases of WMO and NCDC (periods 1961–90 and 1996–2005). The stochastic generation of global radiation is based on a Markov chain model for daily values and an autoregressive model for hourly and minute values (Aguiar et al., 1988 or Aguiar & Collares-Pereira, 1992). The generation of temperature is based on global radiation and measured distribution of daily temperature values of approx. 5000 sites. Meteonorm generates also additional parameters like precipitation, wind speed or radiation parameters like diffuse and direct normal irradiance.

Satellite-derived solar and meteorological data are provided for example by NASA (NASA—National Aeronautics and Space Administration). Also commercial providers present maps on a monthly or annual basis e.g. SolarGIS. PV-GIS (SOLARGIS) is a free tool to find solar data worldwide as well. Programs have been sponsored by the World Bank to provide good data for developing countries. For instance, NASA's Land Processes Distributed Active Archive Centre (LPDAAC) offers DNI data to the public without charge for any location across the globe. However, freely available data only holds monthly averages and thereby does not hold information on fluctuations throughout the day. The data is derived from 22 consecutive years of satellite data with a basic resolution of approximately 30 km, which has been processed to a $1° \times 1°$ geographical grid.

All satellite data providers have different temporal and spatial coverage, different temporal, and spatial resolution, use different algorithms and different inputs. As a result, solar radiation values derived for a given place from different satellite providers can differ significantly. There are many advantages associated with satellite-derived values:

- Satellites have high spatial resolution covering most of the sites of interest.
- Compared to ground-based measurement campaigns, the temporal coverage of meteorological satellites is extensive. Some data date back to the late 1970s and hold almost continuous coverage since the 1980s. Since satellites cover such long terms of data, the historic data basis can be applied to estimate future irradiation, and thus can serve as basis for planning and sizing of solar systems. Furthermore, inter-annual variations in energy yield can be estimated. Such estimates are vital for risk analysis of CSP plants.
- Satellite-derived solar radiation values (real-time data) can be utilized for monitoring and managing dispersed solar power in the grid.

References

Aguiar, R., & Collares-Pereira, M. (1992). Statistical properties of hourly global radiation. *Solar Energy, 48*(3), 157–167. https://doi.org/10.1016/0038-092X(92)90134-V

Aguiar, R., Collares-Pereira, M., & Conde, J. (1988). Simple procedure for generating sequences of daily radiation values using a library of Markov transition matrices. *Solar Energy, 40*(3), 269–279. https://doi.org/10.1016/0038-092X(88)90049-7

Base Surface Radiation Network (BSRN). (2023, February 16). *Global baseline network for surface radiation for the Global Climate Observing System* (J. R. European Commission, Ed., & D. C. Lanconelli, Compiler). Alfred-Wegener-Institut. https://bsrn.awi.de/contact-persons/

Bendt, P., Collares-Pereira, M., & Rabl, A. (1981). The frequency distribution of daily insolation values. *Solar Energy, 27*(1), 1–5. https://doi.org/10.1016/0038-092X(81)90013-X

Blanc, P., Espinar, B., Geuder, N., Gueymard, C., Meyer, R., Pitz-Paal, R., & Wilbert, S. (2014). Direct normal irradiance related definitions and applications: The circumsolar issue. *Solar Energy, 110*, 561–577.

Buie, D., Monger, A., & Dey, C. (2003). Sunshape distributions for terrestrial solar simulations. *Solar Energy, 74*(2), 113–122. https://doi.org/10.1016/S0038-092X(03)00125-7

Cebecauer, T., & Suri, M. (2016). Site-adaptation of satellite-based DNI and GHI time series: Overview and SolarGIS approach. *AIP Conference Proceedings, 1734*, 150002-1–150002-11. https://doi.org/10.1063/1.4949234

Chhatbar, K., & Meyer, R. (2011). The influence of meteorological parameters on the energy yield of solar thermal plants. In *Proceedings of the SolarPACES 2011 Conference* (p. 8). https://doi.org/10013/epic.37915.d001

Collares-Pereira, M. A. (1979). The average distribution of solar radiation-correlations between diffuse and hemispherical and between daily and hourly insolation values. *Solar Energy, 22*(2), 155–164. https://doi.org/10.1016/0038-092X(79)90100-2

Coulson, K. (1975). *Solar and terrestrial radiation-methods and measurements*. Academic Press.

Duffie, J. A., & Beckmann, W. A. (2020). *Solar engineering of thermal process, photovoltaics and wind*. Wiley.

Erbs, D., Klein, S., & Duffie, J. (1982). Estimation of the diffuse radiation fraction for hourly, daily and monthly-average global radiation. *Solar Energy, 28*(4), 293–302. https://doi.org/10.1016/0038-092X(82)90302-4

Fröhlich, C. (1977). The solar output and its variation. In O. R. White (Ed.), *Contemporary measures of the solar constant* (p. 93). Colorado Associated University Press.

Fröhlich, C., Foukal, P. V., Hickey, J. R., Hudson, H. S., & Willson, R. C. (1992). Solar irradiance variability from modern measurements. In *Sun in time* (3rd ed.). Univ. of Arizona Press.

German Aerospace Center (DLR). (2006). *Derivation of irradiances from ISCCP cloud data*. Linder Höhe, 51147 Cologne, Germany. Retrieved February 27, 2023, from http://www.pa.op.dlr.de/ISIS/method.html

Gerthsen, C., Kneser, H. O., & Vogel, H. (1995). *Physik. Ein Lehrbuch zum Gebrauch neben Vorlesungen* (17. Auflage ed.). Springer.

Global Energy Balance Archive-GEBA. (n.d.). *Central database for the worldwide instrumentally measured energy fluxes at the surface* (E. Z. Institute for Climate and Atmospheric Sciences, Ed.). Retrieved March 20, 2022, from https://geba.ethz.ch/

Gueymard, C. (2010). Variability in direct irradiance around the Sahara: Are the modeled datasets of bankable quality? In *Proceedings of the 16th SolarPACES Conference* (p. 9).

Hess, G. (2011, June 22). *I-V curve of a photodiode and the equivalent circuit*. Retrieved February 27, 2023, from https://commons.wikimedia.org/w/index.php?curid=15579778

Hippler, R. (2005). *Atmosphärenphysik*. Universität Greifswald.

Hoyer-Klick, C., Hustig, F., Schwandt, M., & Meyer, R. (2009, September 15–18). Characteristic meteorological years from ground and satellite data. In *Proceeding of the 15th SolarPACES Symposium.*

Hoyt, D. (1979). Pyrheliometric and circumsolar sky radiation measurements by the Smithsonian Astrophysical Observatory from 1923 to 1954. *Tellus, 31*(3), 217–239.

Huang, G., Li, Z., Li, X., Liang, S., Yang, K., Wang, D., & Zhang, Y. (2019). Estimating surface solar irradiance from satellites: Past, present, and future perspectives. *Remote Sensing of Environment, 233*, 111371. https://doi.org/10.1016/j.rse.2019.111371

Hummon, M., Ibanez, E., Brinkman, G., & Lew, D. (2012). Sub-hour solar data for power system modeling from static spatial variability analysis. In *Conference Paper, 2nd International Workshop on Integration of Solar Power in Power Systems*, Lisbon, Portugal, 12–13th November.

Iqbal, M. (1983). *An introduction to solar radiation.* Academic Press Inc.

ISO 9060. (2018). *Solar energy—Specification and classification of instruments for measuring hemispherical solar and direct solar radiation.*

ISO 9488. (2021). *Solar energy—Vocabulary.* ISO/TC 180 Solar energy.

Johnson, F. (1954, December). The solar constant. *Journal of the Atmospheric Sciences*, 431–439. https://doi.org/10.1175/1520-0469(1954)011<0431:TSC>2.0.CO;2

Jrloginov. (2016, July 16). *Class A secondary standard pyranometer*. Retrieved February 27, 2023, from https://commons.wikimedia.org/w/index.php?curid=86344332

Kasten, F., & Young, A. T. (1989). Revised optical air mass tables and approximation formula. *Applied Optics, 28*(22), 4735–4738.

LeBaron, B., Michalsky, J., & Perez, R. (1990). A simple procedure for correcting shadowband data for all sky conditions. *Solar Energy, 44*(5), 249–256. https://doi.org/10.1016/0038-092X(90)90053-F

Liu, B. Y., & Jordan, R. C. (1960). The interrelationship and characteristic distribution of direct, diffuse and total solar radiation. *Solar Energy, 4*(3), 1–19. https://doi.org/10.1016/0038-092X(60)90062-1

Lohmann, S., Schillings, C., Mayer, B., & Meyer, R. (2006). Long-term variability of solar direct and global radiation derived from ISCCP data and comparison with reanalysis data. *Solar Energy, 80,* 1390–1401. https://doi.org/10.1016/j.solener.2006.03.004

McCracken, G., & Scott, P. (2005), *Fusion: the Energy of the Universe.* Elsevier Academic Press, Burlington MA.

Meyer, R. (2010). Recommendations for bankable meteorological site assessments for solar thermal power plants. In *Proceedings of SolarPACES 2010* (p. 8). http://www.solarpaces2010.org/

Monteith, J., & Unsworth, M. (2013). *Principles of environmental physics-plants, animals, and the atmosphere* (4th ed.). Elsevier Science. ISBN 9780123869937.

NASA—National Aeronautics and Space Administration. (n.d.). *The POWER Project-Prediction of worldwide Energy Resources* (P. Stackhouse, Compiler). 300 E. Street SW, Suite 5R30, Washington, DC 20546, United States of America. Retrieved March 20, 2022, from https://power.larc.nasa.gov/

NASA. (2022, March 14). *EOS Project Science Office—Earth Observatory* (N. G. Center, Editor). https://earthobservatory.nasa.gov/features/SORCE/sorce_03.php

National Centers for Environmental Information. (n.d.). *Archives of atmospheric, coastal, geophysical, and oceanic research* (N. O. (NOAA), Ed.). 1401 Constitution Avenue NW, Room 5128, Washington, DC 20230, United States of America. Retrieved March 20, 2022, from https://www.ncei.noaa.gov/

National Renewable Energy Laboratory-NREL. (2009). *NREL pyrheliometer comparison: NPC2009.* Solar Radiation Research Laboratory. https://midcdmz.nrel.gov/apps/npcphotos.pl?year=2009

Neumann, A., Witzke, A., Jones, S., & Schmitt, G. (2002). Representative terrestrial solar brightness profiles. *Journal of Solar Energy Engineering, 124*(2), 198–204. https://doi.org/10.1115/1.1464880

Ngoko, B., Sugihara, H., & Funaki. (2014). Synthetic generation of high temporal resolution solar radiation data using Markov models. *Solar Energy, 103*, 160–170. https://doi.org/10.1016/j.solener.2014.02.026

Noring, J., Grether, D., & Hunt, A. J. (1991). *Circumsolar radiation data: The Lawrence Berkeley Laboratory reduced data base*. National Renewable Energy Lab.

Orgill, J., & Hollands, K. (1977). Correlation equation for hourly diffuse radiation on a horizontal surface. *Solar Energy, 19*(4), 357–359. https://doi.org/10.1016/0038-092X(77)90006-8

Paetzold, H. K. (1952, May). Ein Beitrag zur atmosphärischen Extinktion. *Astronomische Nachrichten, 281*, 17–22. https://doi.org/10.1002/asna.19522810106

Pecker, J.-C. (1994). The sun: A synoptic view. *Vistas in Astronomy, 38*(2), 111–164. https://doi.org/10.1016/0083-6656(94)90028-0

Perez, R., Stewart, R., Arbogast, C., Seals, R., & Scott, J. (1986). An anisotropic hourly diffuse radiation model for sloping surfaces: Description, performance validation, site dependency evaluation. *Solar Energy, 36*(6), 481–497. https://doi.org/10.1016/0038-092X(86)90013-7

Perez, R., Stewart, R., Seals, R., & Guertin, T. (1988). *The development and verification of the Perez diffuse radiation model*. Sandia National Lab.

Rapp-Arrarás, Í., & Domingo-Santos, J. M. (2011). Functional forms for approximating the relative optical air mass. *Journal of Geophysical Research, 116*, D24308-13. 0148-0227/11/2011JD016706

Reindl, D. (1988). *Estimating diffuse radiation on horizontal surfaces and total radiation on tilted surfaces*. Solar Energy Laboratory. University Wisconsin Madison: Faculty Mechanical Engineering. http://digital.library.wisc.edu/1793/47660

Reindl, D., & Beckman, D. J. (1990). Diffuse fraction correlations. *Solar Energy, 45*(1), 1–7. https://doi.org/10.1016/0038-092X(90)90060-P

Reindl, D., Beckman, W. A., & Duffie, J. (1990). Evaluation of hourly tilted surface radiation models. *Solar Energy, 45*(1), 9–17. https://doi.org/10.1016/0038-092X(90)90061-G

Remund, J., Müller, S., Schilter, C., & Rihm, B. (2010). The use of Meteonorm weather generator for climate change studies. In *Proceedings of the 10th EMS Annual Meeting; 8th European Conference on Applied Climatology* (Vol. 7, pp. EMS2010-417).

Schubnell, M. (1992). Sunshape and its influence on the flux distribution in imaging solar concentrators. *Journal of Solar Energy Engineering, 114*(4), 260–266. https://doi.org/10.1115/1.2930015

Sengupta, M., Habte, A. G., & Wilbert, S. R. (2017, December). *Best practices handbook for the collection and use of solar resource data for solar energy applications: Second edition*. National Renewable Energy Laboratory (NREL).

SOLARGIS. (n.d.). *Online data and apps*. Bottova 2A, 811 09 Bratislava, Slovakia. Retrieved March 20, 2022, from https://solargis.com/

Stöcker, H. (1994). *Taschenbuch der Physik*. Harry Deutsch.

Suri, M., & Cebecauer, T. (2014). *Satellite-based solar resource data: Model validation statistics versus user's uncertainty*. Environmental Science, ASES SOLAR 2014 Conference, San Francisco, 7–9 July 2014. https://solargis2-web-assets.s3.eu-west-1.amazonaws.com/public/publication/2014/1f0b376723/Suri-Cebecauer-ASES-Solar2014-Satellite-Based-Solar-Resource-Data-Model-Validation-Statistics-Versus-User-Uncertainty.pdf

Thekaekara, M. (1976). Solar radiation measurement: Techniques and instrumentation. *Solar Energy, 18*(4), 309–325. https://doi.org/10.1016/0038-092X(76)90058-X

Van de Hulst, H. C. (1981). *Light scattering by small particles*. Dover Publications Inc.

Whillier, A. (1965). Solar radiation graphs. *Solar Energy, 9*(3), 164–165.

Willson, R. C. (1978). Accurate solar 'constant' determinations by cavity pyrheliometers. *Journal of Geophysical Research, 83*(C8), 4003–4008. https://doi.org/10.1029/JC083iC08p04003

World Radiation Data Centre. (2022). *Meteorological Services world solar radiation data* (V. M. Observatory, Ed., & H. o. Dr. Anatoly V. Tsvetkov, Compiler). Retrieved March 20, 2022, from http://wrdc.mgo.rssi.ru/wrdc_en_new.htm

Zelenka, A., Perez, R., Seals, R., & Renné, D. (1999). Effective accuracy of satellite-derived hourly irradiances. *Theoretical and Applied Climatology, 62*, 199–207. https://doi.org/10.1007/s007040050084

Optical Conversion

3

Abstract

Solar collector systems always need to maximize the solar gain from the short-wave solar radiation of the sun and hence minimize optical losses. Many components and different materials are used in a solar collector to convert solar radiation into thermal heat. It is crucial to understand the principles and limitations of the interactions of transparent, reflecting, and absorbing components with the electromagnetic radiation. The concepts of spectral selectivity and optical concentration are introduced and several methods for achieving that are described.

Keywords

Radiation transport • Transparent media • Reflector • Absorber • Spectral selectivity • Directional optical property • Anti-reflection • Optical concentration • Imaging concentrators • Non-imaging optics • Fresnel optics

3.1 Electromagnetic Radiation

Solar radiation is electromagnetic radiation emitted from the sun. Due to the internal processes in the hot sun the electromagnetic waves emitted have energies mainly corresponding to a wavelength range between approximately $0.2\mu m$ and $2.6\mu m$ as discussed in Chap. 2. However, when dealing with solar technology, we also find electromagnetic radiation with other wavelengths. Notably, we also consider the thermal radiation transport with waves of wavelengths depending on the temperature of the emitting body. The typical range where the associated black-body or Planck spectra of thermal radiation have their energy is between 5 and $100\mu m$ for ambient temperatures, and between 2 and $40\mu m$ for $600^\circ K$; the spectrum is depicted in Fig. 3.1. Although ultraviolet UV, visible and near infrared solar radiation and thermal radiation all is transported by electromagnetic waves, due to the different energy densities the interaction with matter is different. We will deal with that in a separate chapter. At first, however we want to introduce

R. Stieglitz and W. Platzer, *Solar Thermal Energy Systems*,
https://doi.org/10.1007/978-3-031-43173-9_3

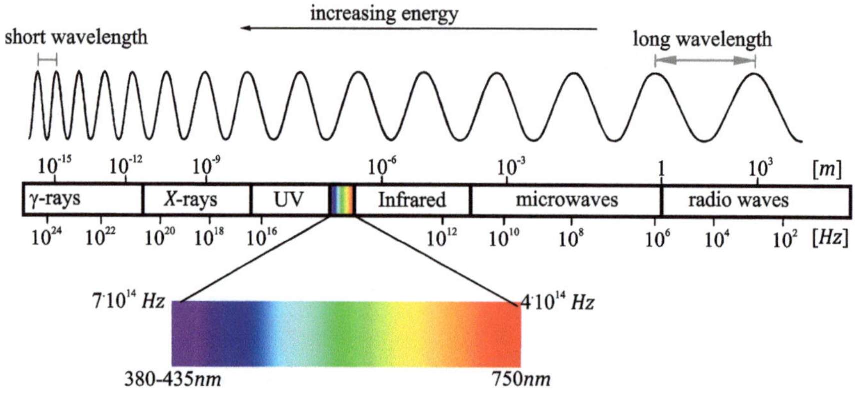

Fig. 3.1 Overview over energies and wavelength spectrum for different electromagnetic waves

the basic physical laws which govern the propagation of electromagnetic waves, the Maxwell equations.

An electromagnetic wave, such as a light wave, is described by the electric field strength vector ($\overrightarrow{E}$), the magnetic induction vector ($\overrightarrow{B}$) and the wave propagation vector ($\overrightarrow{k}$), which are perpendicular to each other, thus:

$$\vec{E} \perp \vec{B}, \ \ \vec{E} \perp \vec{k} \ \text{ and } \ \vec{k} \perp \vec{B}. \tag{3.1}$$

In classical electromagnetic theory the Maxwell equations describe how electric and magnetic fields are related to each other and to electric charges and electric current under given boundary conditions. Together with the Lorentz force, they explain all phenomena of classical electrodynamics. They, therefore, also form the theoretical basis of optics and electrical engineering.

If we only want to discuss the propagation of electromagnetic waves, of light and thermal radiation, in vacuum and in a continuous medium, the equations for the electric and magnetic fields simplify, because we need not to consider source terms. Also, we need only solve the equations for one field as the other one is coupled directly to that. Using the so-called wave equation, we can deal with a wide range of optical phenomena which are important for solar energy technologies; may it be the transmission of solar radiation in transparent glass panes or the scattering of light due to aerosol particles. So essentially, we have to find a solution for the electric field vector ($\overrightarrow{E}$) within a given domain and for pre-scribed boundary conditions for the following equation

$$\frac{d^2}{dt^2}\vec{E} = \frac{1}{\varepsilon_0 \mu_0 \varepsilon_r \mu_r} \cdot \Delta \vec{E} = \frac{1}{\varepsilon \cdot \mu} \cdot \Delta \vec{E}, \tag{3.2}$$

in which ε is the absolute permittivity of a medium characterizing the polarizability of the medium, μ the absolute magnetic permeability of a medium related to the magnetisation of a medium and ε_0 the vacuum permittivity or electric constant given by $\varepsilon_0 = 8.8541878128 \cdot 10^{12}\ As/(Vm)$ and μ_0 the magnetic permeability of vacuum given by the constant $\mu_0 = 4 \cdot \pi \cdot 10^{-7}\ H/m$.

For vacuum the relative permittivity and permeability is unity (1). The product of permeability and permittivity is related to the Index of refraction n and the absolute speed of light in the medium c resp.in vacuum c_0.

$$c = \frac{1}{\sqrt{\varepsilon.\mu}}, \tag{3.3}$$

$$c_0 = \frac{1}{\sqrt{\varepsilon_0 \mu_0}} = 2.997893 \cdot 10^8 m/s, \tag{3.4}$$

$$n = \frac{c}{c_0}. \tag{3.5}$$

One solution of the wave Eq. (3.2) are plane waves with a linear propagation direction. The wave can be represented with respect to the electric field vector $\overrightarrow{E}$ in the following form (similar for the magnetic field $\overrightarrow{H}$):

$$\overrightarrow{E}(t, z) = \overrightarrow{E}_0 \cdot \cos(\omega \cdot t - k \cdot z), \tag{3.6}$$

where ω is the angular frequency, t is time and $k = 2\pi/\lambda = \omega \cdot \sqrt{\varepsilon \cdot \mu}$ is the absolute value of the wave vector $\overrightarrow{k}$. A projection of the wave in the $x - y$-plane provides a line, as illustrated in Fig. 3.2.

The polarization of light is defined by the direction of the electric field vector. If the electric field vector of a wave always points in the same direction perpendicular to the z-axis (see Fig. 3.2), the wave is called linearly polarized. If a wave propagates along the z-axis $\overrightarrow{k} = k \cdot \overrightarrow{e}_z$ and the contributions $\overrightarrow{E}_{0x}$ and $\overrightarrow{E}_{0y}$

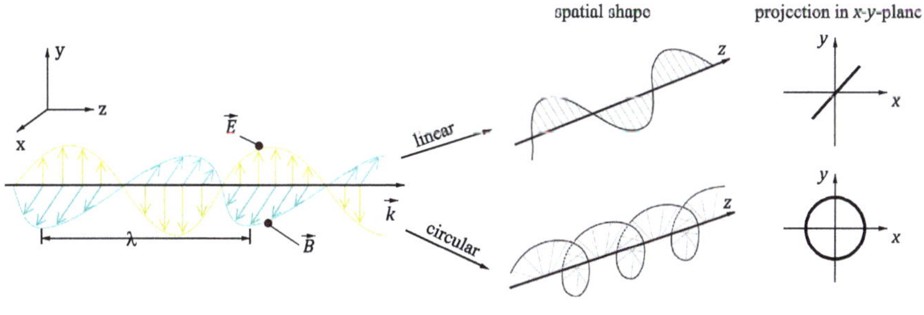

Fig. 3.2 Light as an electromagnetic wave. Polarization as a form of the spatial propagation of the electric field vector and the projection into the $x - y$-plane of a linearly polarized wave (top) and a circularly polarized wave (bottom)

of the wave vector are equal but phase shifted by $\pi/2$, it is called a circularly polarized wave. The electric field vector describes a spiral and it applies:

$$\vec{E}_x(t,z) = \vec{E}_{0x} \cdot \cos(\omega t - kz) \text{ and } \vec{E}_y(t,z) = \vec{E}_{0y} \cdot \cos\left(\omega t - kz - \frac{\pi}{2}\right). \quad (3.7)$$

Both forms of polarization are sketched in the Fig. 3.2.

There is a multitude of polarization types depending on the orientation in time of the resulting amplitude vector (E_x, E_y) perpendicular to the propagation direction.

In the following paragraph we want to deal with the physical reasons for the interaction of electromagnetic waves with matter, which in classical continuum theory results in the functions of permeability and permittivity. As we are dealing in solar mostly with non-magnetic materials, we assume in the further development $\mu_r = 1$.

More general for a continuum the electromagnetic field can be written as a complex function. This is important for a medium where due to energy dispersion the waves can be also attenuated and extinguished (absorption of energy and conversion into thermal energy). We write the solutions of the Maxwell equations in complex form:

$$\vec{E}(t,z) = \vec{E}_0 \cdot \exp\left(i \cdot \omega \cdot \left(t - \frac{n}{c_0} \cdot z\right)\right) \cdot \exp\left(-\omega \cdot \frac{k}{c_0} \cdot z\right). \quad (3.8)$$

Here, the second part describes the attenuation and damping of an electromagnetic wave due to interaction with a dispersive medium. The intensity decreases exponentially and is described by the imaginary part of the complex index of refraction. The attenuation of the electromagnetic wave depends on its complex refractive index, into which the electrical and magnetic permittivities of the media ε_r and μ_r between the interfaces enter in a complex form. The complex refractive index $\tilde{n}$ is:

$$\tilde{n} = n + i \cdot k. \quad (3.9)$$

Please note that the imaginary part of the complex index of refraction is denoted $k-$ as the absolute value of the wave vector characterizing wavelength and direction of a plane wave. It depends on the context which quantity should be referred to. The imaginary part of the index of refraction causes absorption. It leads to an attenuation of electromagnetic radiation in the direction of propagation. The attenuation coefficient α is related to k by

$$\alpha = \frac{4\pi}{\lambda} k \text{ and is given as } \alpha = \frac{1}{I}\frac{d}{dz} I, \quad (3.10)$$

The spatial reduction of the radiant intensity in dependence of the layer thickness z

$$I(z) = I(z = 0)e^{(\alpha \cdot z)} \approx |E|^2, \tag{3.11}$$

is proportional to $|E|^2$ and falls off exponentially with increasing$z$. For transparent dielectric media the penetration depth must be large compared to its thickness$z$, whereas for metals it is exceedingly small. The energy flux of the radiation is given by the Poynting vector, and the time-average is obtained by squaring the amplitudes of the electric field.

$$I = \left|\vec{S}\right| = \left|\vec{E} \times \vec{H}\right| = \frac{n}{\mu \cdot c_0}\left|\vec{E}\right|^2 = \frac{n}{\mu \cdot c_0}\left(E_x^2 + E_y^2 + E_z^2\right). \tag{3.12}$$

For a complex formulation the complex conjugate product is needed to obtain the (real) intensity:

$$I = \frac{\tilde{n}}{\mu \cdot c_0}\vec{E} \cdot \overrightarrow{E^*} = \frac{\tilde{n}}{\mu \cdot c_0}\left(E_x \cdot E_x^* + E_y \cdot E_y^* + E_z \cdot E_z^*\right). \tag{3.13}$$

3.2 Radiation Transport in Transparent Media

3.2.1 Basic Definitions and Nomenclature

When we talk about the processes of transmission, reflection, absorption and scattering in transparent media, we are interested in the impact of components like polymeric films and glass panes on the solar radiation in different wavelength ranges, generally speaking on radiative power or energy. Radiometric terms describe the distribution of the radiation's power in space. Here, several quantities are being used, describing on the one side the radiation, and on the other side the material optical properties. In experimental measurements these properties are determined. Therefore, we use here the nomenclature of radiometry (Table 3.1).

A basic quantity is the radiant power or radiant flux Φ, defined as the amount of radiant energy dQ per unit time dt.

$$\Phi = \frac{dQ}{dt} \quad [\mathrm{W}]. \tag{3.14}$$

The radiant intensity is defined by:

$$I = \frac{d\Phi}{d\Omega} \quad [W/sr]. \tag{3.15}$$

And expresses the radiant power per unit solid angle and does not include any information about the emitting source or receiving element. To describe the radiant

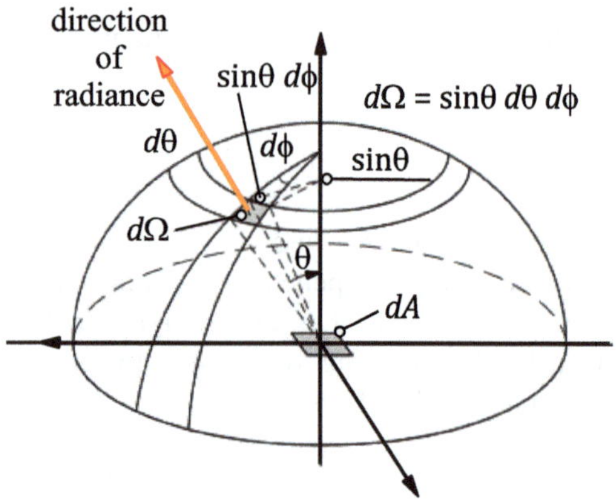

Fig. 3.3 Visualization of radiance L, the radiant flux per unit solid angle $d\Omega$ per projected area dA

flux emitted by a surface A into a specific solid angle Ω, the term radiance is defined, as depicted in Fig. 3.3. The radiance L in $W/(sr \cdot m^2)$ is the radiant power per unit solid angle per unit *projected* source area and thus results in:

$$L = \frac{d^2\Phi}{\cos\theta dA d\Omega}. \tag{3.16}$$

To distinguish between radiance leaving and incident on a surface, reflected radiance is in the following L_r or L_2 and incident radiance L_i or L_1.

The term irradiance E in W/m^2 describes the radiant power incident on a surface. It can be expressed as the integrated radiance over a defined solid angle in space as:

$$E = \frac{d\Phi}{dA} = \int_0^{2\pi} L_i d\Omega. \tag{3.17}$$

The index 'e' denotes the radiometric range. For photometry in the visual wavelength range the index 'v' is used. The optical properties of components now are defined in the Table 3.2. As an example, the mathematical definition of the reflectance is given in the following equation (the definition for transmittance and absorptance is analogous).

Reflectance R describes the ratio of radiant flux Φ_r reflected from a surface dA to the incident radiant flux Φ_i. Often it is additionally relevant to analyse radiant flux Φ_i incident from a specific solid angle Ω_i onto a surface dA and the radiant flux Φ_r reflected into a specific solid angle Ω_r. The reflectance is given by:

$$\rho = \frac{d\,\Phi_r}{d\Phi_i} = \frac{\int_{\Omega r} L_r d\Omega_r}{\int_{\Omega_i} L_i d\Omega_i}. \tag{3.18}$$

The reflection and transmission coefficients describe the ratio of magnitude of the amplitudes of the considered electromagnetic field to the incident field. When

Table 3.1 Definition of radiometric quantities (ISO 80000–7, 2019)

Quantity	Symbol	Unit	Description
Flux	Φ_e	W	Radiative flux from a source (EM radiation)
Irradiance	E_e	W/m^2	Radiative specific flux falling onto a surface (EM radiation)
Radiosity/spec. radiance	M_e	W/m^2	Radiative specific flux emitted by a surface (EM radiation)
Intensity	I_e	W/sr	Radiative flux per unit solid angle (from or into a cone) (EM radiation)
Radiance	L_e	$W/(sr \bullet m^2)$	Radiative specific intensity emitted by a surface into a solid angle (EM radiation)

Table 3.2 Definition of optical properties in radiometry according to (ISO 80000–7, 2019); all dimensionless

Property	Symbol	Description
Reflection coefficient	r	Magnitude of reflected amplitude divided by amplitude of incident radiation
Transmission coefficient	t	Magnitude of transmitted amplitude divided by amplitude of incident radiation
Transmissivity	τ	Transmitted power through a surface or a medium divided by incident power
Reflectivity	ρ	Reflected power from a surface divided by incident power
Absorptivity	α	Absorbed power at a surface divided by incident power
Emissivity	(ε)	Emitted power from a surface divided by maximum blackbody emitted power
Transmittance	τ	Transmitted power through a component divided by incident power
Reflectance	ρ	Reflected power from a component divided by incident power
Absorptance	α	Absorbed power by a component divided by incident power
Emittance	ε	Emitted power from a component divided by maximum blackbody emitted power

we speak about "-ivity" in the context of optical properties, we speak about intrinsic material or surface properties. The terms ending on "-ance", however, describe the properties of complete components, for instance a pane of glass where the reflectivity on the front and the back surface adds to the reflectance, and where the thickness of the materials has an impact on that as well. Also, inhomogeneous materials are described by transmittance and reflectance terms. Sometimes the difference is minute and mainly shows the emphasis we want to give, for example when we talk about emittance or emissivity of a solar absorber. Strictly speaking the emissivity describes the emissive power of a material, and the emittance the emissive power of the absorber material on, say the copper substrate.

We will see in many literature texts, that not always a clear distinction has been kept. This makes it sometimes complex to compare results from different authors (coming maybe from different academic background) when they write about the same things. Symbols moreover have always been interpreted in the context they are used. A symbol a for example might denote a multitude of physical quantities, from solar absorptivity to thermal diffusivity to heat transfer coefficient. In the optical realm, moreover, the conditions for in- and outgoing radiation must be understood as well. Is the incoming radiative flux directional or diffuse, is the outgoing flux considered integral or is the angular distribution of radiation considered? Is the flux measured spectrally or for a certain wavelength range? Very often indexing may give some indication what is meant, but although quantities may be mathematically better defined using multiple indexing, the comprehensibility of text usually decreases. In doubt the context is important. Therefore, we hope that the reader will not be completely confused, when sometimes we do not stick slavishly to a certain connotation.

3.2.2 Origin of Absorption and Reflection in Transparent Media

Solid matter can absorb electromagnetic radiation. The degree of absorption depends on its electronic band structure (see for a thorough treatment e.g. (Ashcroft, Mermin, Wei, and Mermin, 2016). Electromagnetic radiation (or photons, if you want to use the particle representation of light!) provide the energy to excite quantum states of the solids with higher energy. With solar and thermal radiation, the excitation of electrons and phonons is possible, whereas for the nuclei the energy is not sufficient. In solids, the outer electrons are no longer firmly bound to the atoms of the solid but can move freely in electron bands. Two bands are important for absorption:

- The valence band and,
- The conduction band.

The classification of substances into band models is described herein with the help of quantum mechanics. At temperature zero, the valence band is the highest fully occupied energy band. Bands which are energetically above, unoccupied, or partially occupied are called conduction bands. In semiconductors and insulators (dielectrics), there is a forbidden area between the two bands called the band gap. The electrical and thermal conductivity of the solid depends on the one hand on whether the conduction band at temperature zero is occupied, partially occupied or unoccupied, and on the other hand on the temperature of the crystal. A fully occupied band, just like an unoccupied band, does not contribute to charge transport. The reason for this is that electrons in fully occupied bands cannot absorb energy, for example, through an electric field, because for them there are no unoccupied

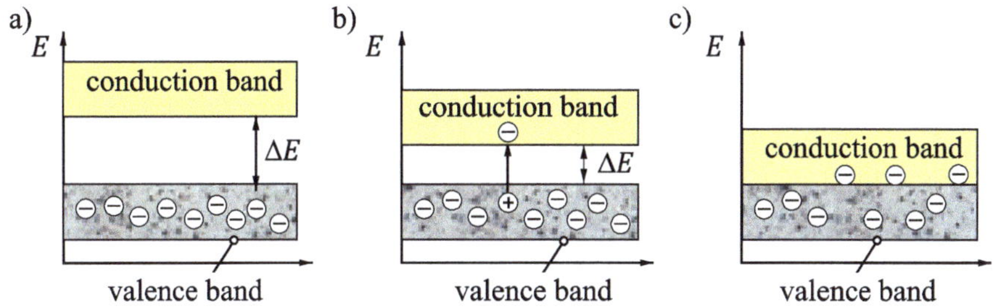

Fig. 3.4 Energy bands in an electrical non-conductor (**a**), a semiconductor (**b**) and a metal (**c**)

states with suitably higher energy. Only a partially occupied band enables a net current induced by the electric field.

Semiconductors and insulators have a band gap ΔE, semiconductors in the range of $1eV$ and for insulators well above $3eV$. Since the valence bands are fully occupied at $0K$, no electrical charge transport is possible. If sufficient energy is supplied to a semiconductor by increasing the temperature or by irradiating with light, electrons can overcome the band gap and be lifted into the conduction band. In this way, the unoccupied band can be partially occupied. These electrons and the holes remaining in the valence band both contribute to the conductivity of the electric current.

According to this idea, insulators like glasses have an unoccupied conduction band and a very large band gap ($\Delta E > 3eV$, see Fig. 3.4a). At room temperature and even at much higher temperatures, only a few electrons can be transferred from the valence band to the conduction band by thermal excitation. As a result, they have a high electrical resistance. However, the conduction band can be partially filled by electrons from the valence band overcoming the energy gap between the valence and conduction band by absorbing light. This requires light with an energy that exceeds a certain lower limit, which is given by the size of the energy gap.

Like for insulators, the conditions for a semiconductor are similar, but here the band gap is relatively small, typically ($1eV < \Delta E < 2eV$ corresponding to $0.6 - 1.2$ μm wavelength, see Fig. 3.4b), this can be overcome more easily by thermal or optical energy supply. An electron can be lifted into the conduction band and is freely movable here. At the same time, it leaves a gap in the valence band which can be filled by neighbouring electrons. Semiconductors, therefore, have pronounced absorption bands with a sharp edge at a maximum wavelength. The Fig. 3.5 illustrates this dependence of the absorption coefficient of various semiconductors on the wavelength in the visible and near infrared range. The figure exhibits a sharp cut-off edge of the absorption coefficient of all semiconductors when a certain wavelength is exceeded.

However, the following properties might make actual absorption impossible: If the semiconductor has an ideal interface with air, due to the large complex part of the index of refraction, its reflectivity is particularly high at wavelengths for which it has a large absorption coefficient (extinction coefficient). To avoid reflection, the

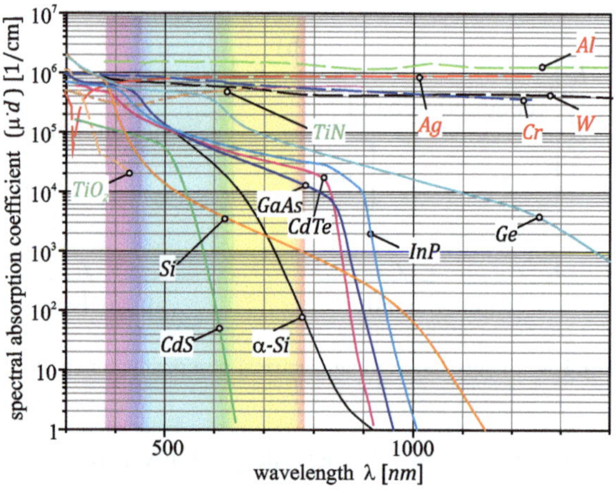

Fig. 3.5 Spectral absorption coefficients of various semiconductors in the visible light range and in the near infrared as a function of wavelength λ (McNaught and Wilkinson, 1997)

refractive index must not change abruptly at the interface. One possibility is to use a transparent coating with intermediate index of refraction, or to emboss a surface texture on the surface with structures in the nanometre range. Then, the structure cannot be resolved by light with the wavelength of visible light and the effective index of refraction is changing continuous from air to solid material. Thus, the surface of a photocell requires special treatment. If the surface looks red when viewed, then this treatment is missing or insufficient. This is because the cell reflects exactly the light that should actually be absorbed.

In a metal one usually does not speak of a conduction or valence band. For monovalent metals, the highest occupied energy band is half filled, as shown in Fig. 3.4c. In polyvalent metals, the energy bands partially overlap. Electrons can therefore change to a higher energy state when small electric field strengths are applied and can thus move more or less freely and contribute to the flow of electric current. Metals are, therefore, good electrical conductors. An increase in temperature leads to a decrease in conductivity because the increased scattering of the electrons causes a lower average velocity (Wiedemann–Franz law). Electrons can also be raised to higher energy states in the conduction band by absorbing light. Metals are therefore in principle ideal light absorbers, as indicated in Fig. 3.5 exhibiting no cut-off wavelength at all. However, similar to semiconductors, the optical constants again prohibit that for an ideal smooth surface, because light is almost completely reflected ($\rho \to 1$) exactly when the absorption coefficient is large. Metallic conductors with ideal polished surfaces therefore do barely absorb light. This contradiction is resolved by the fact that the permittivity leading to high absorption inside the body also lead to high reflectivity of the interface air-metal which prohibits an intrusion of light into the volume.

3.2.3 Polarization, Refraction, and Reflection

When light, that is electromagnetic radiation, crosses the border between one medium and another, then the boundary conditions for the EM-radiation according to the Maxwell equations determine the interaction with the interface. If the incident wave encounters the interface, it is split into two new waves, as depicted in Fig. 3.6. One is refracted while proceeding into the second medium and the other is reflected, see (Born and Wolf, 1999).

In the continuum theory the media are characterized by the permissivity and permeability. Valid solutions for the dominating wave equation must fulfil boundary conditions at the interface. And derived from this mathematical theory the laws of refraction, of reflection and transmission at the boundary result. The change in optical constants ε and μ of the medium (or$\tilde{n}$) determines the relation between frequency (the colour of light) and its propagation (the velocity of light). Whereas the frequency does not change, the velocity of light changes and hence the energy density. This is the reason for the change of direction when a wave crosses the boundary between two media. The resulting law in optics then is the law of refraction.

Although the following topics can be reformulated also in complex numbers, we will stick to the formulation in real numbers. The complex number formulation is useful for metals and the laws of Snellius and Fresnel have formally the same shape as the simpler real formulation for dielectrics, however the interpretation of a complex incidence angle is less straight-forward.

Snellius' law of refraction is used to describe the transition of an electromagnetic wave at an interface where both media have a different wave velocity. The changed phase velocity causes a deflection of the wave, i.e. a change in direction.

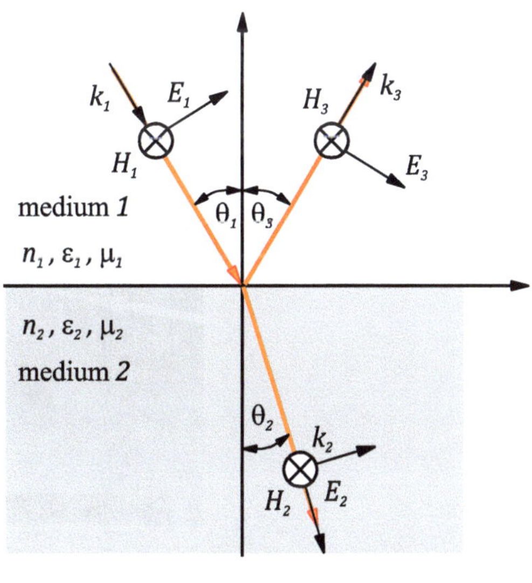

Fig. 3.6 Reflection and refraction of a plane wave with p polarization

In the field of optics, when passing from one transparent medium to another, this means that the light beam is refracted.

If you look at a light beam that is incident from medium 1 into medium 2 at an angle of θ_1, the non-reflected portion continues to travel at an angle of θ_2 in medium 2 relative to the perpendicular. Light travels in any isotropic medium at the velocity of propagation dependent on the refractive index n_i, which indicates the ratio of the velocity of light in a vacuum c_0 to that of the respective medium (c_i).

$$n_1 = \frac{c_0}{c_1} \text{ and } n_2 = \frac{c_0}{c_2}. \tag{3.19}$$

Here, the refractive index is a material constant dependent on the wavelength, which causes a splitting of the different wavelengths of the incident beam during refraction. In a dispersion prism, this wavelength dependence allows an incident light beam to be deflected from its original direction of propagation by refraction at two non- parallel surfaces. Since the strength of the deflection depends on the wavelength, it is possible to spectrally decompose polychromatic light, as shown in the Fig. 3.7.

The Table 3.3 shows typical refractive indices of some liquids and glasses as averaged over the solar spectrum. An example of the wavelength dependence, also called dispersion, is shown in Fig. 3.8. This functional form is dependent on the proximity of a resonance frequency in the medium which determines the shape of the complex permittivity function.

If two parallel incident light beams meet an ideal interface between two media of different optical density, an additional distance $\lambda_1 = (c_1 \cdot t)$ in medium 1 (Fig. 3.9) is geometrically obtained for the second beam. The first light beam must travel an additional distance $\lambda_2 = (c_2 \cdot t)$ in medium 2. At the interface, one finally obtains, via geometrical relations and using the Eq. (3.19) Snellius' law of refraction

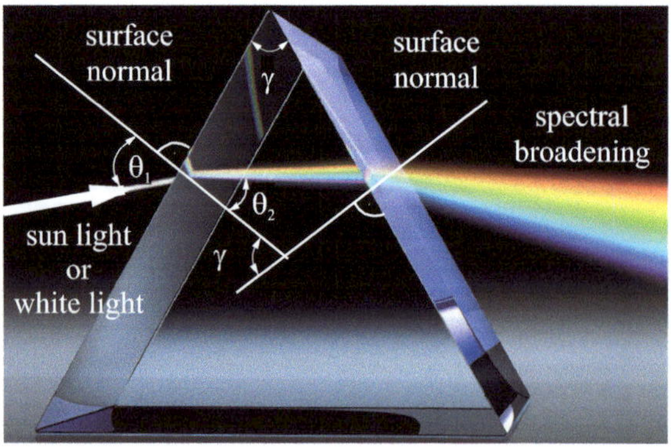

Fig. 3.7 Spectral decomposition of polychromatic light in a dispersion prism

Table 3.3 Refractive indices (n) of some selected transparent liquids and solids

Material	Refractive index n [/]	Material	Refractive index n [/]
Dry air (1 bar)	1.000278	Acrylic glass (PMMA)	1.49
Ice (−8 °C)	1.309	Optical lenses	1.51 – 1.58
Water (0 °C)	1.3346	Quartz glass	1.544 – 1.553
Ethanol	1.3614	Sapphire/Ruby	1.757 – 1.779
Ethyl glycol	1.4318	Lead glass	1.805
Carbon tetrachloride	1.46	Cubic zircon	2.17 – 2.21
Glycerine	1.47	Zinc sulphide	2.37
Benzene	1.49	Diamond	2.419
Epoxy resin	1.55 – 1.63	Silicon carbide	2.65 – 2.69

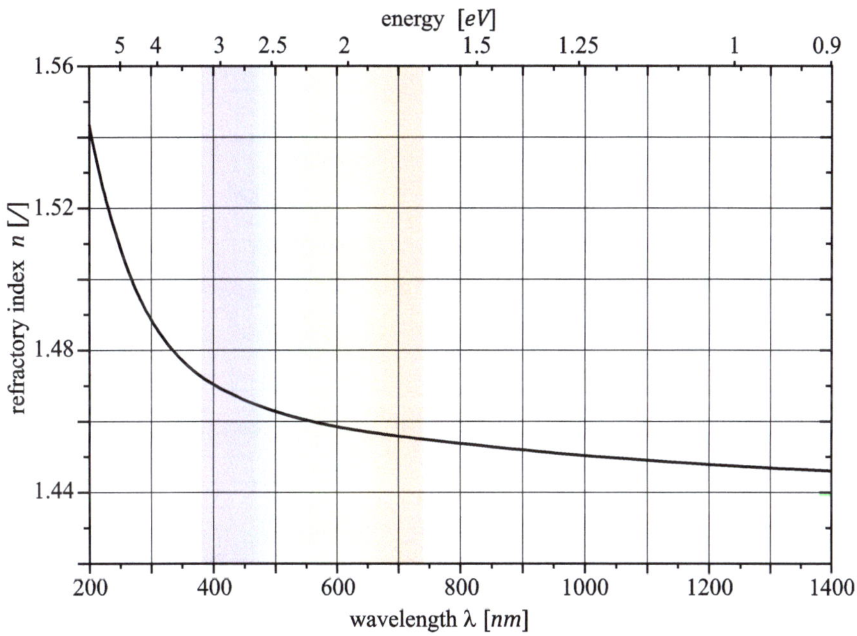

Fig. 3.8 Example of wavelength dependent dispersion in SiO_2

$$\frac{\sin\theta_1}{\sin\theta_2} = \frac{n_2}{n_1} = \frac{c_1 \cdot t}{c_2 \cdot t} = \frac{\lambda_1}{\lambda_2}. \tag{3.20}$$

For absorbing media in which the refractive index n is defined as a complex number, the law of refraction can also be applied in a similar modified form, however

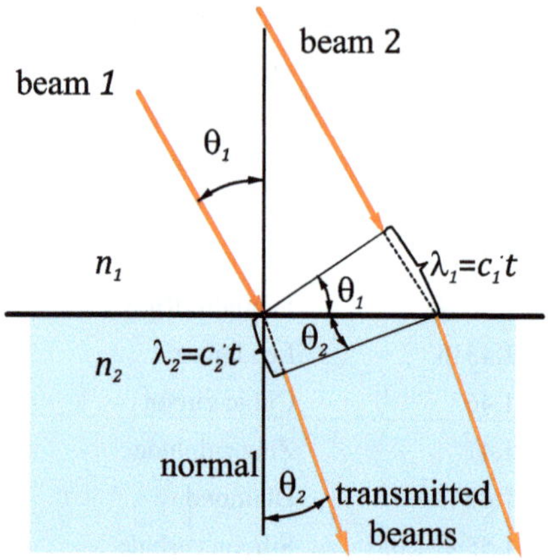

Fig. 3.9 Sketch for the derivation of Snellius' law of refraction

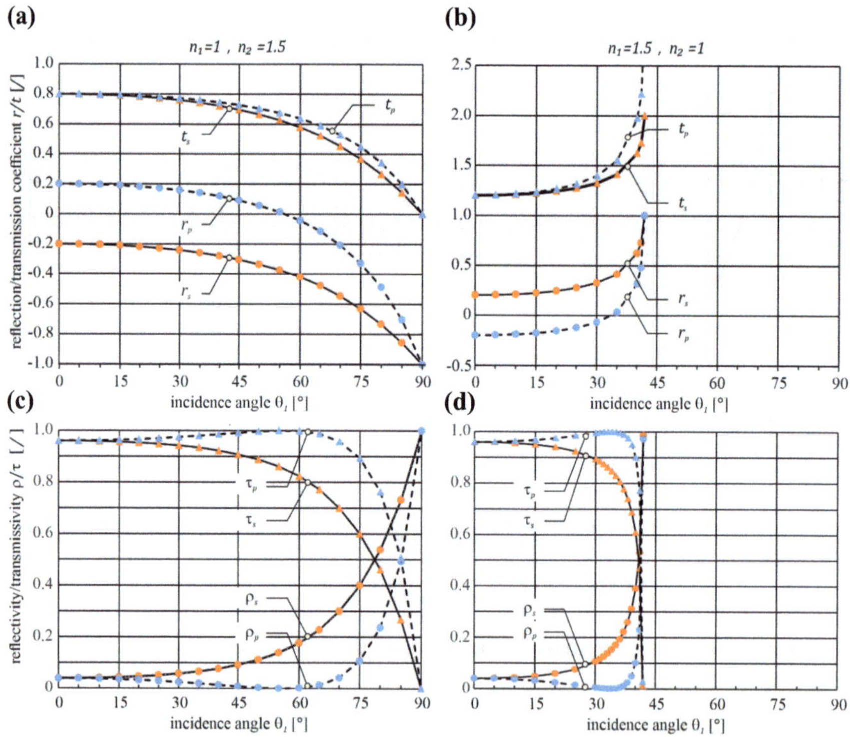

Fig. 3.10 Calculated amplitude ratios r_i, t_i (top) and reflectivity/transmittanceρ_i, τ_i (bottom) for the interface air ($n = 1$) and glass ($n = 1.5$). Light incident on interface with angle θ_1 from the air side (**a, c**) and the glass side (**b, d**)

the interpretation of the complex quantities is not straight-forward. It describes the relation between propagation direction, propagation velocity and attenuation. It should be remembered that the even the resulting "angle of refraction" is a complex quantity.

When the wave passes from a medium with smaller optical density into a medium with larger density ($n_2 > n_1$) then the refraction is towards the surface normal $\theta_2 < \theta_1$. Contrary, when the wave passes from the dense medium to the thinner medium, the refraction is away from the surface normal. The optical laws are valid in forward and backward direction equally. An example is the situation when a diver below water is looking to the water surface into the environment above. One might see the full landscape within a smaller cone of view. But there is obviously a limiting angle where the refracted beam already is travelling parallel to the surface, the water surface in air in our case. A larger incidence angle on the interface then leads to a transmitted wave with propagation direction into the source medium, i.e. the beam is reflected. The solution of the wave equation for these boundary conditions leads to the case of total reflection of an electromagnetic wave, where the wave is completely reflected back into the output medium. The angle where total reflection starts is referred to as the critical angle θ_c, for which

$$\theta_c = \arcsin\left(\frac{n_2}{n_1}\right). \tag{3.21}$$

If the critical angle ($\theta > \theta_c$) is exceeded, a total reflection is obtained in which $\theta_1 = \theta_2$ Snellius' law merely states in which direction a wave (or beam of light) is deflected, but not how much of the wave is transmitted or reflected at the transition between the two media. The total reflection can be easily observed by holding a mirror under a water surface. If the mirror is turned towards the water surface, total reflection sets in from a certain angle, the critical angle ($\theta = \theta_c$).

When we deal with the question which part of the intensity of light is transmitted or reflected, we need information also for the amplitudes of the electromagnetic waves, not only the directions. This is determined with the help of Fresnel's formulas.

Since an electromagnetic wave field vector can be represented as a linear combination of component perpendicular and a component parallel to the incident plane. Treating the boundary conditions at the interface, the individual components of the amplitudes can thus be determined. For this purpose, amplitude transmission coefficients t_s, t_p and amplitude reflection coefficients r_s, r are defined in perpendicular and parallel directions, which indicate the ratio of the amplitude of the outgoing or reflected electric field to the amplitude of the incoming component of the incident electric field $(E_{0e})_s$ or $(E_{0e})_p$. These coefficients thus describe the amplitude damping in the respective direction, i.e. in perpendicular or parallel direction.

A well-known case of the Fresnel equations is for dielectric materials in which the complex refractive index k in Eq. (3.9) is negligible and close to zero. However, the Fresnel equations can be solved in a universal form for all materials, i.e. for glasses, metals, etc., in which then the amplitude ratios t_s, t_p or r_s, r_p are given as a function of the complex refractive index $\tilde{n}$, the relative angles (θ_1, θ_2) and the magnetic permeability (μ_{r1}/μ_{r2}) between the media, compare for example (Nolting, 2009) or (Demtröder, 2004).

A dielectric material may absorb to some extent light, in fact otherwise one would not have coloured dielectrics. However, the imaginary part of the index of refraction $k \ll 1$ is so small that a simplified extinction coefficient can be derived (Beer's law). For this case of dielectrics, for example the pairing air-glass, the Fresnel formulas can be simplified considerably. The relations between electromagnetic radiation on either side of the interface can be reduced as the Fresnel laws for real numbers use only the real part of the refractive index n_i of the respective medium. First of all, the component which is polarized linearly perpendicular (index:s) to the plane of incidence shall be considered. It is also often referred to in the literature as the transversal electrical (TE) component. The following transmission amplitude coefficients t_s and amplitude reflection coefficients r_s result for the vertical polarization:

$$t_s = \left(\frac{E_{0t}}{E_{0e}}\right)_s = \frac{2n_1 \cdot \cos\theta_1}{n_1\cos\theta_1 + n_2\cos\theta_2} = \frac{2\sin\theta_2\cos\theta_1}{\sin(\theta_1+\theta_2)},$$
$$r_s = \left(\frac{E_{0r}}{E_{0e}}\right)_s = \frac{n_1 \cdot \cos\theta_1 - n_2\cos\theta_2}{n_1\cos\theta_1 + n_2\cos\theta_2} = -\frac{\sin(\theta_1-\theta_2)}{\sin(\theta_1+\theta_2)}. \quad (3.22)$$

Correspondingly, for parallel polarization (TM) the correlations are:

$$t_p = \left(\frac{E_{0t}}{E_{0e}}\right)_p = \frac{2n_1 \cdot \cos\theta_1}{n_2\cos\theta_1 + n_1\cos\theta_2} = \frac{2\sin\theta_2\cos\theta_1}{\sin(\theta_1+\theta_2)\cos(\theta_1-\theta_2)},$$
$$r_p = \left(\frac{E_{0r}}{E_{0e}}\right)_p = \frac{n_2 \cdot \cos\theta_1 - n_1\cos\theta_2}{n_2\cos\theta_1 + n_1\cos\theta_2} = -\frac{\tan(\theta_1-\theta_2)}{\tan(\theta_1+\theta_2)}. \quad (3.23)$$

The equations after the second equals sign are based on Snellius' law of refraction and the addition theorem. The calculated amplitude coefficients r_i, t_i for the substance pairing light from air in glass (a) and light from glass in air (b) are shown in Fig. 3.10a, b as a function of the angle of incidence.

Looking at the vertical incidence of light on the interface between media 1 and 2, the angle of incidence is $\theta_1 = \theta_2$, and the amplitude reflection coefficient r_s or r_p is then:

$$r_s = \left(\frac{E_{0r}}{E_{0e}}\right)_s = \frac{n_1 - n_2}{n_1 + n_2} = -r_p = \left(\frac{E_{0r}}{E_{0e}}\right)_p. \quad (3.24)$$

Using the fact that the intensity I is proportional to the square of the amplitude of the electric field E, one can calculate the intensity of the portion reflected at the interface. In the vertical transition of visible light from air ($n_1 = 1$) to quartz glass ($n_1 = 1.46$), 3.5% of the incident intensity is reflected regardless of the polarization.

Figure 3.10a and b show that there are amplitude coefficients for both transmission and reflection which have a negative sign. In this case a phase shift of the wave by 180° (180° $= \pi$) occurs physically. The amplitude reflection coefficient r_p has a zero crossing, therefore, the light polarized parallel to the plane of incidence is completely refracted and only the part polarized perpendicular to the plane of incidence is reflected. An analysis of Eq. (3.22) shows that the following must apply to this case:

$$r_p = -\frac{\tan(\theta_1 - \theta_2)}{\tan(\theta_1 + \theta_2)} = 0 \rightarrow \theta_1 + \theta_2 = 90°. \tag{3.25}$$

This means that the beam refracted at the angle (θ_2) is perpendicular to the reflected beam.

If polarized light were also reflected parallel to the plane of incidence, this would result in a longitudinally oscillating light wave. Due to the properties of the electromagnetic field, such a field cannot exist, since, for example, no radiation can occur in the direction of the dipole axis. Physically, the incoming light beam causes an excitation of the electrons of the material, which leads to an oscillation called displacement polarization. This results in a collection of atomic dipoles that oscillate in the direction of polarization and thus emit secondary waves. Inserting the relation (3.25) into Snellius' law of refraction (Eq. (3.20)) gives the following angular relation for the angle referred to as Brewster angle θ_B:

$$\theta_B = \arctan\left(\frac{n_2}{n_1}\right). \tag{3.26}$$

This means that light polarized parallel to the plane of incidence and incident at Brewster angle θ_B is not reflected (see also Fig. 3.11).

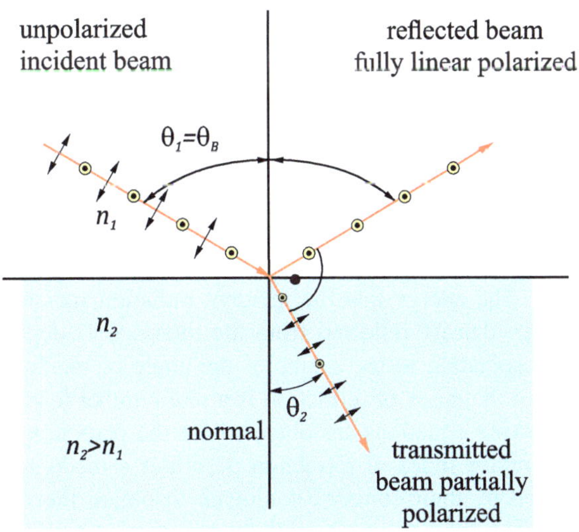

Fig. 3.11 Schematic sketch to explain the Brewster angle θ_B

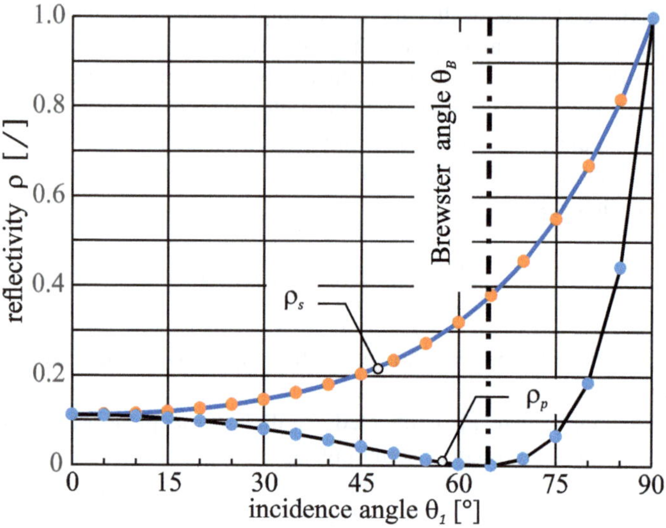

Fig. 3.12 Position of the Brewster angle at the transition from an optically thinner ($n_1 = 1$) to a denser medium ($n_2 = 2$) and calculated reflection amplitude as a function of the angle of incidence θ_1

This parallel polarized light penetrates the surface completely. Light with perpendicular polarization, on the other hand, is also partially reflected, as can be seen in the Fig. 3.12. Therefore, the reflected light is completely linearly polarized. One may observe polarisation of natural unpolarized light when looking at obliquely reflected light approximately incident under the Brewster angle e.g. from a low sun reflected by a water surface of a smooth lake, or by window shields from a car, when one uses sunglasses with polarizers. Rotating the sunglasses by 90° changes the transmitted intensity substantially, because these glasses are selectively transmitting only one linear polarization.

If light enters from an optically denser medium into an optically thinner one, i.e. $n_1 > n_2$, the amplitude ratios become complex from a certain angle onwards. From this critical limit angle θ_c, (cf. Eq. (3.21), the total reflection occurs, cf. Figure 3.13 for the pairing glass ($n_1 = 1.5\ 5$)$\rightarrow$ air ($n_2 = 1$) the Brewster angle is $\theta_B = 33.7°$ and the angle for total reflection is $\theta_c = 41.8°$.

So far, only amplitude coefficients have been considered. Much more important in application is the change in relative intensity as it passes through the interface. The energy that the incident radiation radiates onto an interface I_e and the energy density reflected from the interface I_r depends on the angle of incidence corresponding to the cosine of the angle of incidence. For the case of dielectrics where the index of refraction is a real number leading to real field amplitudes, the following equations are obtained for the perpendicular or parallel components (if a complex index of refraction describes a medium, the amplitudes must be multiplied by their complex-conjugate to obtain the real reflection and transmission

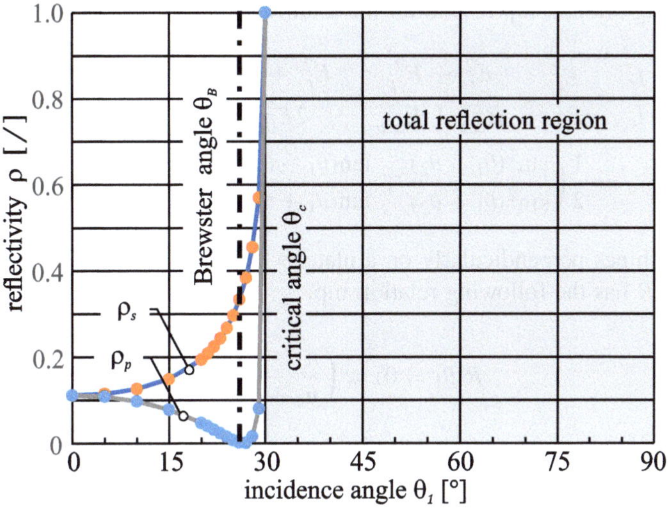

Fig. 3.13 Reflection amplitudes in both directions at the transition of light from an optically denser medium ($n_1 = 2$) to a thinner one ($n_2 = 1$), as well as the position of the Brewster angle (θ_B) and the critical angle (θ_c)

coefficients):

$$R_{s/p} = \left(\frac{E_{0r}}{E_{oe}}\right)^2_{s/p} = r^2_{s/p} \text{ and } T_{s/p} = \frac{n_2 \cdot \cos\theta_2}{n_1 \cdot \cos\theta_1} \cdot \left(\frac{E_{0t}}{E_{oe}}\right)^2_{s/p} = \frac{\tan\theta_1}{\tan\theta_2} \cdot t^2_{s/p}, \tag{3.27}$$

wherein i represents the s- or p-polarized component, respectively. Without absorption, the following energy current balance applies for the individual polarization directions:

$$T_s + R_s = 1 \text{ and } T_p + R_p = 1. \tag{3.28}$$

We denote here the reflectivity and transmissivity with capital R and T to avoid confusion when we talk about reflectance and transmittance of components like glass sheets or polymer films having two interfaces with surrounding air. However, strictly speaking also in this case the symbols ρ and τ could be used if the context is clear.

According to Fresnel's equations, unpolarized light on a smooth surface is partially reflected when passing from a material with refractive index n_1 to a material with n_2. If one assumes this unpolarized light, the individual components must be determined separately according to the proportions parallel (p) and perpendicular (s) to the surface. If one additionally postulates that the size of the perpendicular and parallel light portion of the incident beam was identical, i.e. $(E_{0e})_s = (E_{0e})_p$, the following relationship results for the total reflection coefficient R:

$$\begin{aligned} R = \frac{I_r}{I_e} = \frac{E_r^2}{E_{0e}^2} = \frac{E_{rs}^2 + E_{rp}^2}{E_{0es}^2 + E_{0ep}^2} = \frac{E_{rs}^2 + E_{rp}^2}{2E_{0es}^2} = \frac{1}{2}\left(R_s^2 + R_p^2\right) \\ = \frac{1}{2}\left(\frac{\sin^2(\theta_1 - \theta_2)}{\sin^2(\theta_1 + \theta_2)} + \frac{\tan(\theta_1 - \theta_2)}{\tan(\theta_1 + \theta_2)}\right) \end{aligned} . \quad (3.29)$$

If light shines perpendicularly on a plate, which is $\theta_1 = 0$, then the reflection coefficient R has the following relationship:

$$R(\theta_1 = 0) = \left(\frac{n_1 - n_2}{n_1 + n_2}\right)^2 . \quad (3.30)$$

If, for example, visible unpolarized light falls perpendicularly on the interface between air ($n_1 = 1$) and quartz glass ($n_1 = 1.46$), then according to the Eq. (3.30) the portion of 3.5% of the incident intensity is reflected independently of the polarization.

3.2.4 Scattering and Absorption in Media

As a rule, when light passes through a body, partial absorption occurs in the transparent material itself, reducing the intensity of the light beam. This attenuation (or extinction) of the radiation intensity is a function of the path length of the beam traveling through the material. The intensity I in the transparent body when the beam passes through it can be described as follows:

$$\frac{dI}{dx} = -\mu \cdot I(x), \quad (3.31)$$

in which the extinction coefficient μ depends on the material. The extinction may be caused by a mixture of absorption processes, where radiation energy is converted into heat—or in the molecular picture into kinetic energy of the molecules, and scattering, where the direction of the radiation is changed. Scattering may occur by interaction of light with phonons (vibrational waves of the solid) or with small particles embedded in the medium. These particles may be much smaller than the wavelength of light and hence may not be visible. Depending on their chemical composition the particles can be also absorbing and scattering. In such a case the attenuation depends essentially on the concentration c_k of the absorbing substance or particles in the transparent medium and the molar wavelength-dependent extinction coefficient ε_λ of all elements contained in the transparent body. Thus, the extinction coefficient is a function of both quantities, and the relation is $\mu = f(c_k, \varepsilon_\lambda)$.

This extinction coefficient (which may be spectral or broadband, depending on the accuracy needed in the calculations) in general is a sum of scattering coefficient σ and absorption coefficient κ:

$$\mu = \kappa + \sigma. \tag{3.32}$$

In contrast to colour dispersion in refraction, the dependence of then extinction on wavelength cannot be described by a simple function. The dependence on wavelength is complex and determined by the transparent medium and it complex index of refraction, and by molecular properties of the dissolved substances or embedded particles. There are theories which try to derive the complex index of refraction and the extinction properties for mixtures of substances (e.g. for metallic or oxidic particles embedded in a dielectric matrix), the so-called effective medium theories (Bruggeman, Maxwell-Garnet, etc., see e.g. (Wang and Pan, 2008)). It is not only important which particles sizes and chemistry we have for a specific case, but also which grain structure and topology is found for the mixture. Depending on that, different theories have to be applied to give good estimates.

Generally, coloured substances show one or several defined absorption maxima on the scale of wavelengths. Other dyes show a continuously increasing or decreasing absorbance with the wavelength in the visible range. These properties are based on the molecular structure (especially the electron configurations in the molecules). Therefore, the spectral reflectance or absorptance in the visible and infrared spectral range can be used for chemical analysis, where many substances can be directly identified and quantitatively measured due to their specific spectral extinction patterns, may it be iron oxide, metallic particles, organic chlorophyll, or haemoglobin.

Even for a dielectric medium with a very small imaginary part k of the index of refraction for relatively thick transparent media the spectral absorption is non-negligible. For example, if we assume $k = 10^{-6}$ and want to calculate the reduction of the intensity for a wavelength $\lambda = 1\ \mu m$ and a medium thickness of $d = 4\ mm$ the spectral attenuation according to electromagnetic wave theory is:

$$I(\lambda, d) = I_0 \cdot \exp\left(-\frac{4\pi}{\lambda} \cdot k \cdot d\right). \tag{3.33}$$

Evaluating the exponential yields 95.1 or 4.9% absorption with the thickness of 4 mm. For real dielectric materials it is easier to provide a solar absorption coefficient $\kappa(\lambda)$ which includes already the full dependence on wavelength. Using the broadband value is critical. Especially for polymers with a very pronounced spectral absorption bands the broad band approximation fails when treating different film thicknesses and spectral calculation is preferable.

When we consider the absorption in a dielectric layer, where the intensity travels with a propagation direction $\theta\prime$ across a layer with thickness d, we must take into account also the longer path length. Using the extinction law of Lambert (3.33) we define an internal attenuation factor or "internal transmissivity" x_i:

$$x_i = \exp\left(\frac{-\mu \cdot d}{\cos\theta\prime}\right). \tag{3.34}$$

The transmission behaviour of conventional glasses, which consist of a multitude of individual chemical elements, is much more inconsistent. The spectral dependence of transmission is essentially determined by the composition of the glass. For solar covers, a high transmission in the visible spectrum up to the near infrared (approx. $2-3\ \mu m$) is required in particular, since the intensity of the solar radiation must be utilized in this wavelength range. In the solar range, it is mainly the iron oxide content of the glass that influences the energy transmission most. The ultraviolet part of the spectrum is nearly blocked completely but carries only a very small fraction of the solar energy. Figure 3.14 shows the spectral transmittance of glasses with different iron oxide contents. As the iron content decreases, the transparency of glass increases, especially in the near infrared part of the solar radiation spectrum. To thermal radiation, which occupies the wavelength range of $3\ \mu m$ and above, glass is practically opaque. It is therefore not possible to create a thermal image with a conventional infrared camera through a window. Figure 3.14 shows that for optimum glass coverage, glass with as little iron oxide as possible should be used. However, with iron oxide contents of less than 0.1%, the production costs of the glasses increase considerably. Table 3.4 provides for some glass types the absorptivity and transmissivity in the visible spectrum ($\lambda < 3\ \mu m$) of a $4\ mm$ thick glass plate.

An alternative to a glass cover may be a polymeric film. Figure 3.15 shows the spectral transmission behaviour of a polyvinyl fluoride Tedlar film. Like most plastic films, this film shows considerable transparency in the long-wave infrared range. This transparency leads to large radiant heat losses, especially at high absorber temperatures. As a transparent cover for a collector, it is therefore

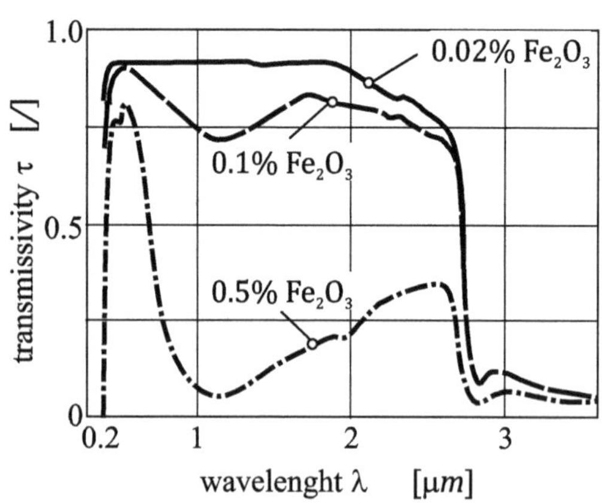

Fig. 3.14 Spectral transmittance of glasses of different iron oxide content from (Duffie and Beckmann, 2020)

Table 3.4 Absorption and transmission coefficients in the visible spectrum ($\lambda < 3\mu m$) for some transparent materials

Material	$\kappa[1/m]$	x_i $(d = 4mm)$	$\tau_{vis}[/]$
White solar glass	4	0.984	0.94
Float glass	16	0.938	0.861
Greenish glass	32	0.88	0.808
Green glass	80	0.726	0.666

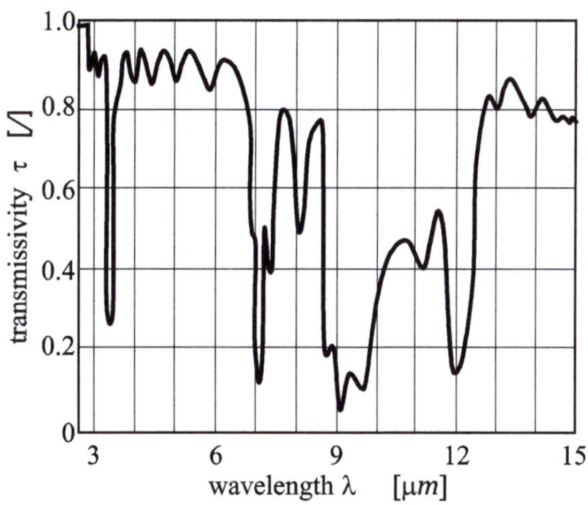

Fig. 3.15 Spectral transmission coefficient of a polyvinyl fluoride Tedlar film modified from (Duffie and Beckmann, 2020)

only slightly suitable. Films often show this only slightly wavelength-selective behaviour.

Colours and spectral absorptance patterns play a decisive role in the natural environment. Reflections from natural environment have an impact on the efficiency of some solar energy applications which also accept these radiative reflections. For example, the albedo, the reflection form of the ground has a smaller effect on the output of a tilted solar thermal collector placed in this environment. The absorption of solar in water, in swimming pools or lakes, has an impact on the temperature distribution in the water.

Let us take the first example. Especially at low sun elevations ($\alpha_s < 40°$) the diffuse scattering and thus the environment plays a larger role in the generation of solar energy. The efficiency of a solar thermal system can be significantly improved by including the immediate vicinity of the collector in an efficiency calculation for the design and alignment of the collector. Adding white or light grey gravel on the ground increases the irradiation on a collector row. In cold climates such effects may be noticed in winter when snow is covering the ground. For some solar systems a low-cost technical installation increasing ground albedo might even improve economics. Figure 3.16 depicts the spectral reflectance of several ground types.

In the case of water, the following can be said. The reflection coefficient of water decreases monotonically with increasing wavelength. Clear and deep waters

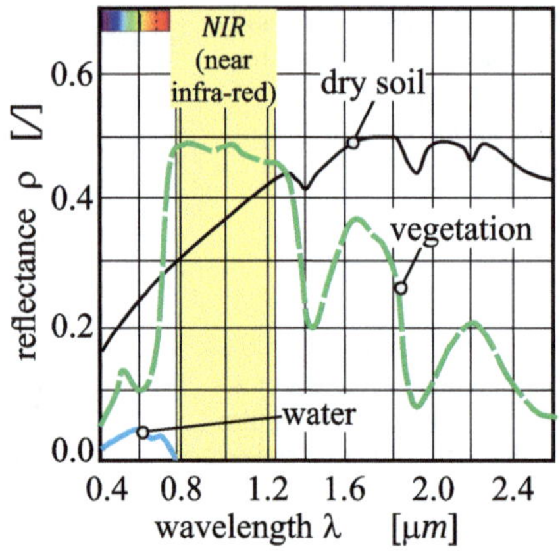

Fig. 3.16 Spectral reflectance of different soil formations according to (Thiemann, 2000)

reflect only visible light; in the infrared range their reflectance is practically zero. With increasing turbidity and decreasing depth, however, the degree of reflection rises, so that turbid and shallow waters in the near infrared range show quite measurable reflection values, as seen in Fig. 3.17.

Spectral reflection is determined in detail by the presence and concentration of dissolved and suspended organic and inorganic constituents. Suspended solids act as a scatterer, dissolved substances absorb selectively. The chlorophyll content is an important indicator of the status of water bodies. The Fig. 3.18 shows the spectral reflectance at different chlorophyll contents of the water. In waters low in chlorophyll, the reflection is large at small wavelengths and decreases monotonically with increasing wavelength. A high chlorophyll content, on the other hand, shows a homogeneous almost constant reflection up to a wavelength of $\lambda \approx 550\ nm$ (green) and then drops sharply.

Figure 3.19 shows a distorted water transmission photograph up to a depth of about 5 m with optical selective absorption. In water, the longer wavelengths are absorbed by the water body already in the uppermost 20 cm. The result is a relative

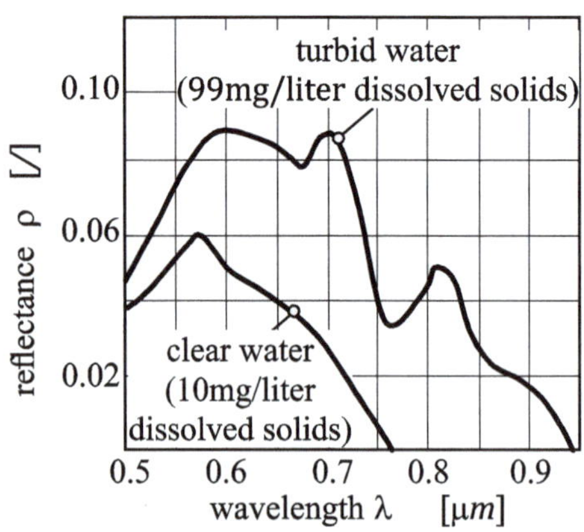

Fig. 3.17 Spectral reflection of clear and turbid water from (Jarmer et al., 2005)

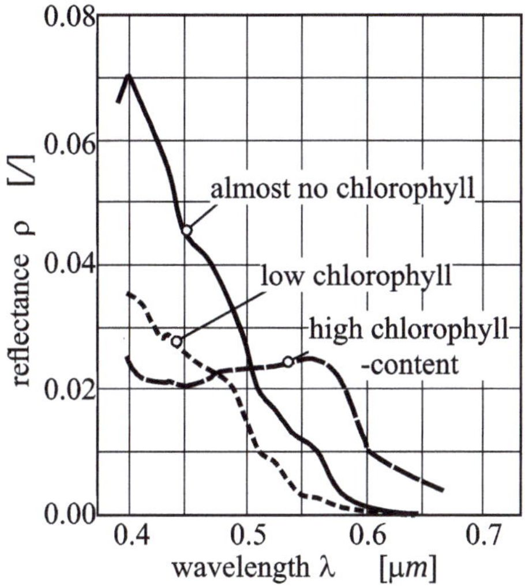

Fig. 3.18 Spectral reflectance at different chlorophyll contents of water from (Jarmer et al., 2005)

energy maximum at 0.5 m (limit blue/green). From a depth of about two meters, no radiation with a wavelength greater than 0.8 μm penetrates the water. Depths of about 20 meters are only reached by the blue and green spectral range.

However, other ground configurations also exhibit a special spectral behaviour that allows remote sensing of often difficult to access areas of the earth's surface with the help of satellites. Figure 3.16 shows the different spectral behaviour for dry soil, vegetation, and water. It can be clearly seen that in vegetation a high proportion of the incident radiation is already reflected in the near infrared, whereas in dry soil the maximum is shifted towards large wavelengths. Water differs significantly from this behaviour and shows only a low reflectivity.

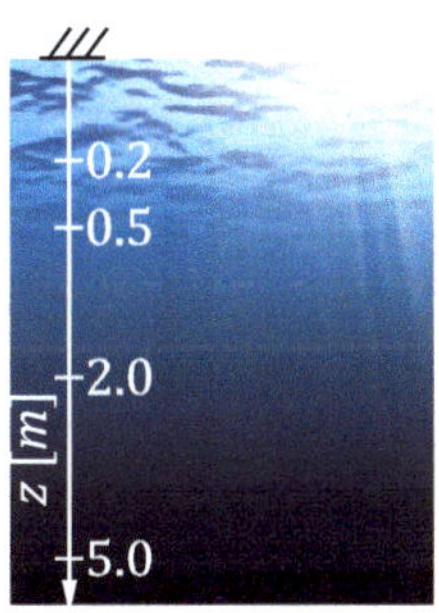

Fig. 3.19 Transmission photograph of unpolarized light entering water

3.2.5 Transmission and Reflection in (Semi-)transparent Components

In this subchapter we want to discuss the transmittance, reflectance and absorptance of common collector elements with respect to solar radiation. The most accurate treatment will consider the spectral properties of the materials, variable incidence angles and polarization effects. In practical situations however, a quick assessment using broadband properties and average conditions for polarization and incidence angle also are helpful. The formulae given can be applied to do the complex calculations, however they can be applied also in a simplified way.

The first basic physical law with respect to the optical properties, transmission, reflexion, and absorption, is energy conservation. It states that the sum of transmittance, reflectance and absorptance is unity. This is true for each wavelength and each incidence angle:

$$\tau(\lambda, \theta, \phi) + \rho(\lambda, \theta, \phi) + \alpha(\lambda, \theta, \phi) = 1. \tag{3.35}$$

This is of course also valid when all the quantities are integrated over wavelength or over a solid angle region or over both. In other words, any radiation flux is either transmitted, reflected, or absorbed.

The second basic optical law is given by Kirchhoff's radiation law which describes the relation between absorption and emission of a body. It is equivalent to the second law of thermodynamics (whereas Eq. (3.35) is equivalent to the first law of thermodynamics, and therefore, sometimes is called Kirchhoff's first radiation law):

$$\varepsilon(\lambda, \theta, \phi, T) = \alpha(\lambda, \theta, \phi, T). \tag{3.36}$$

It is important to state here that the temperature of the body in question also may play a role, not because a body at higher temperature radiates generally more emissive power (that is covered by the Planckian function of blackbody radiation), but because optical constants might change with temperature (e.g. when a phase change of the material occurs). This 2nd law will play a role in the subchapter on selective absorbers further below.

Let us now consider the case of a transparent body, for example a glass pane. How can the optical properties as defined in the Sect. 3.2.1 and as described in Eq. (3.35) be calculated form the optical constants?

For this we consider a single pane optical material, a layer with a certain thickness d. When a beam of light passes from a medium into the transparent layer, e.g. a sheet of glass, a large number of reflections occur. The incident beam is partially reflected at the upper entrance interface while another part penetrates into the layer towards the second interface. There at the exit from the transparent medium the light beam is again partially reflected at the interface. The part not reflected at the interface leaves the layer. This process repeats infinitively but the energy contained in the layer becomes smaller and smaller, as illustrated in Fig. 3.20. The energy balance should account also for absorption in the (semi-)transparent medium.

Fig. 3.20 Schematic diagram of the transmission of a light beam at the interfaces of an ideal transparent medium in which the internal attenuation factor is x_i

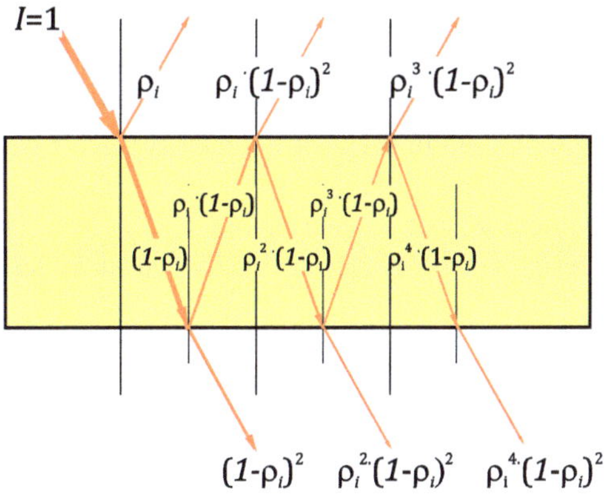

The reflectivity and transmissivity at the border between the two media are given by the Eqs. (3.23) and (3.22) for the parallel and perpendicular polarization states. In order not to confuse these values with the resulting transmittance and reflectance values for the complete layer, we denote reflectivity in the following paragraphs with the symbol R and transmissivity with the symbol T (instead of ρ and τ).

The transmittance of one single sheet $\tau_{s/p}$ emerging on the underside of the panel in one of the two directions is then calculated as the sum of the individual passing components. In each passage the internal extinction in the material reduces the intensity by a factor of x_i (see Eq. (3.38)). In many conditions the infinite sum can be approximated with good accuracy using the first zero and first order components. Only for large reflection values for instance for oblique incidence the multiple reflections gain in importance. Since, according to the equations, the reflectivity values $R_{s/p}$ are less than unity, the transmitted energy fraction decreases with each reflection. The following relationship results for the transmittance τ_i:

$$\tau_{s/p} = \left(1 - R_{s/p}\right)^2 \cdot x_i \sum_{j=0}^{\infty} x_i^{2j} R_{s/p}^{2j} = \frac{\left(1 - R_{s/p}\right)^2 \cdot x_i}{1 - R_{s/p}^2 x_i^2}, \tag{3.37}$$

where s/p represents the variable for the perpendicular (s) and parallel (p) component.

The reflectance ρ is similarly calculated by summing up the infinite series of reflected parts:

$$\rho_{s/p} = R_{s/p} + \left(1 - R_{s/p}\right)^2 \cdot R_{s/p} \cdot x_i^2 \sum_{j=0}^{\infty} x_i^{2j} R_{s/p}^{2j} = R_{s/p} \cdot \left[1 + \frac{\left(1 - R_{s/p}\right)^2 \cdot x_i^2}{1 - R_{s/p}^2 x_i^2}\right], \tag{3.38}$$

For incident unpolarized light, the arithmetic average of both results for the two polarization states has to be considered.

A fundamental result of this consideration is that the passage of unpolarized light even through an ideal glass pane without any absorption leads to an angle-dependent transmittance. We want to illustrate the magnitude of this effect in an example, the passage of light from the air ($n_1 = 1$) through a conventional glass pane with the refractive index $n_2 = 1.526$ should be viewed at the angle $\theta_1 = 0°$ (vertical incidence) and the angle$\theta_1 = 60°$. The reflectance at the first air in glass interface gives the result for vertical incidence according to the Eq. (3.39):

$$R(\theta_1 = 0°) = \frac{1}{2}\left(r_s^2 + r_p^2\right) = \left(\frac{n_1 - n_2}{n_1 + n_2}\right)^2 = 0.04336. \tag{3.39}$$

For the incidence angle 60°, Eq. (3.20) yields the exit angle of $\theta_1 = 34.57°$; thus the corresponding amplitude reflection amplitudes can be determined according to Eqs. (3.23) and (3.22). With an irradiation under the angle $\theta_1 = 60°$ one obtains a reflectivity for unpolarized radiation of:

$$\begin{aligned} R(\theta_1 = 60°) &= \frac{1}{2}\left(r_s^2 + r_p^2\right) = \frac{1}{2}\left[\left(-\frac{\sin\,(\theta_1 - \theta_2)}{\sin\,(\theta_1 + \theta_2)}\right)^2 + \left(\frac{\tan\,\sin\,(\theta_1 - \theta_2)}{\tan\,(\theta_1 + \theta_2)}\right)^2\right]. \\ &= 0.075523 \end{aligned} \tag{3.40}$$

This means that at an angle of $\theta_1 = 60°$ a share of 7.5% of the incident radiation intensity is already reflected on the first surface of a glass pane, while only about 92% passes the pane and is available for an absorber to convert into thermal energy.

At the second surface of the pane, the calculations repeated this time with a reversed order of media yield exactly the same result. We will arrive with the Fresnel formulae at the same values, not matter whether we consider the interface where radiation enters the glass from air, or whether we consider the one where it enters air from glass. Simply summing up the two reflectivities gives a first approximation of the reflectance of the glass pane. When we want to be exact, we must also consider the multiple reflections, which becomes more important for oblique incidence. Figure 3.21 illustrates this dependence of reflectivity as a function of incident angle θ_1 for different types of glass from Table 3.4.

Calculation of optical properties for a several transparent layers

We want to consider now a second pane of glass added. How is the transmittance and reflectance of a glazing consisting of several glass sheets calculated? The following procedure is giving a very general approach which is iterative and can be applied to as many layers one may think of.

Consider a stack of N layers with transmittance τ_N and reflectance ρ_N. As the stack may not be symmetrical the reflectance on the back side which we denote by ρ_N^* might be different. This could be due to coatings on the individual layers, or due to different layers in the series. For example, a white bright layer on the back side and a black dark layer on the front side would be asymmetric in reflectance.

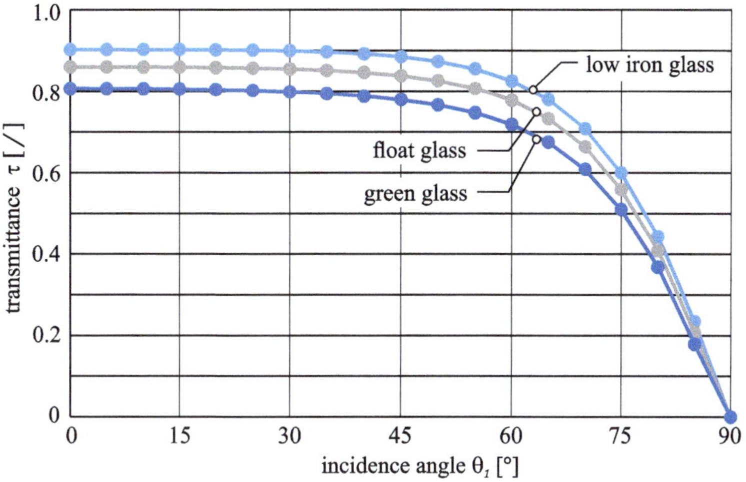

Fig. 3.21 Transmittance of different glass panes, 4 mm thickness and different iron-oxide content, i.e. low-iron glass ($\kappa = 4m^{-1}$), normal float glass ($\kappa = 16m^{-1}$) and green glass ($\kappa = 32m^{-1}$) from Table 3.4

Now we would like to put in front of this stack another set of M layers, having the optical properties transmittance τ_M, reflectance ρ_M and back reflectance ρ_M^*. The combined stack $M + N$ then has the following properties:

$$\tau_{M+N} = \frac{\tau_M . \tau_N}{1 - \rho_N \cdot \rho_M^*}; \tag{3.41}$$

$$\rho_{M+N} = \rho_M + \frac{\rho_N \cdot \tau_M^2}{1 - \rho_N \cdot \rho_M^*}; \tag{3.42}$$

$$\rho_{M+N}^* = \rho_N^* + \frac{\rho_M^* \cdot \tau_N^2}{1 - \rho_N \cdot \rho_M^*}; \tag{3.43}$$

$$\alpha_{M+N}^{(M)} = \alpha_M \cdot \left(1 + \frac{\tau_M \cdot \rho_N}{1 - \rho_N \cdot \rho_M^*}\right); \tag{3.44}$$

$$\alpha_{M+N}^{(N)} = \frac{\tau_M \cdot \alpha_N}{1 - \rho_N \cdot \rho_M^*}. \tag{3.45}$$

Simplification for no absorption

For the case where there is absolutely no absorption in the pane, we reach a simplified formula:

$$\tau = 1 - \rho = 1 - \frac{1}{2}\left(r_s^2 + r_p^2\right) = \frac{1}{2}\left[\left(1 - r_s^2\right) + \left(1 - r_p^2\right)\right] = \frac{1}{2}\left(t_s^2 + t_p^2\right), \tag{3.46}$$

where r_s and r_p describe the amplitude reflection coefficients and t_s and t_p are the amplitude transmission coefficients, respectively. Assuming an originally unpolarized irradiation on the transparent pane, the total transmission coefficient, which includes both directions, is:

$$\tau = \frac{1}{2}\left[\frac{1 - R_p}{1 + R_p} + \frac{1 - R_s}{1 + R_s}\right]. \tag{3.47}$$

An analogous procedure can also be carried out for the case of N several panes arranged parallel to each other. If there is no absorption within the layer, the following equation can be derived:

$$\tau = \frac{1}{2}\left[\frac{1 - R_p}{1 + (2N - 1)R_p} + \frac{1 - R_s}{1 + (2N - 1)R_s}\right]. \tag{3.48}$$

If, for example, one looks again at the vertical incidence of light ($\theta_1 = 0°$) from the air on a double pane, one obtains for the coefficient of transmission:

$$\tau(\theta = 0) = \frac{1}{2}\left[\frac{1 - R_p}{1 + (2N - 1)R_p} + \frac{1 - R_s}{1 + (2N - 1)R_s}\right] = \frac{1 - R_s}{1 + 3R_s} = 0.85. \tag{3.49}$$

This already means a loss of intensity of about 15% for an ideal double glazing without absorption. An oblique incidence angle different from the vertical incidence leads to a further loss of light intensity. At an angle of incidence of $\theta_1 = 60°$, two panes without any absorption only result in a total transmittance of $\tau(\theta_1 = 60°) = 0.76$. The result of the observation of ideal glass panes shows that the intensity available after a glass pane is sensitive with respect to the refractive index and angle of incidence and even without absorption a significant loss of intensity can already occur. Since absorption is a passive damping process, glasses in solar thermal systems must have a low absorption; at the same time, care must be taken to ensure that radiation is as perpendicular as possible. The Fig. 3.22 shows the total transmission coefficient τ as a function of the angle of incidence θ_1 for a different number of ideal absorption-free panes arranged in parallel one behind the other. A value of $n_2 = 1.526$ was used as the refractive index for the glasses.

The calculation is of course simplified and does not consider the effects of polarization and multiple reflection as addressed in Sect. 3.2.3. Both effects lead to the fact that for incident angles of $\theta_1 > 45°$ both the transmission through the pane becomes slightly larger and the reflectivity of the pane slightly decreases. This effect of the real values for τ and ρ is illustrated in Fig. 3.23, showing that the simplification is conservative but relatively accurate.

In order to quantify the attenuation, we will first look at the Fig. 3.24. There, a beam of intensity I_0 hits a glass pane at an entrance angle θ_1. Assuming a homogeneous isotropic transparent body with a constant extinction coefficient, the

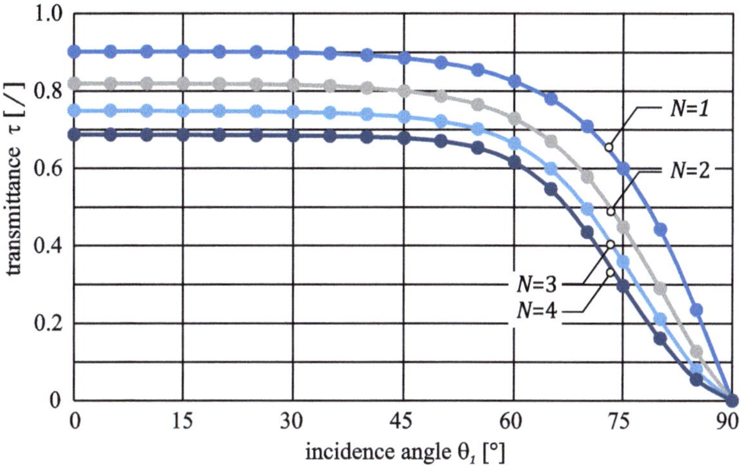

Fig. 3.22 Calculated ideal transmission coefficient τ (without absorption $\mu = 0$) as a function of the angle of incidence θ_1 for a different number N of parallel arranged panes for the pairing of air ($n_1 = 1$) and quartz glass ($n_2 = 1.526$)

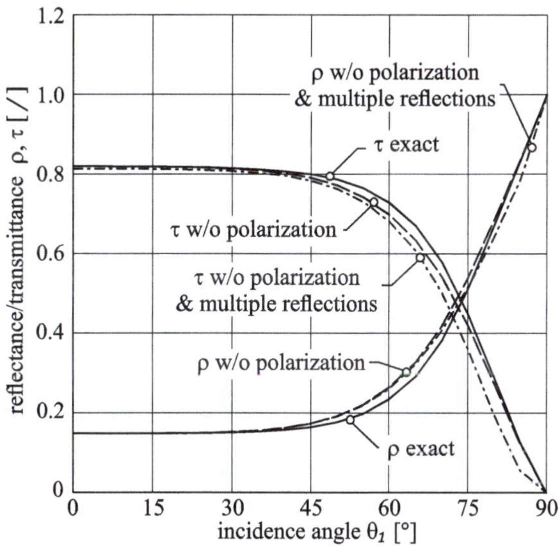

Fig. 3.23 Transmittance and reflectance of double glazing (stack of two identical float glasses), using different approximations (exact: polarization and multiple reflections both considered, w/o polarization: multiple reflections considered, calculation of double stack with average properties of single float glass, w/o pol and multiple reflections: calculations as before but neglecting multiple reflections between 2 panes)

Eq. (3.31) can be integrated over the disc thickness d and one obtains the following result for the intensity for the single passage of the beam through the plate:

$$I(x = d) = I_0 \cdot (1 - R_i)^2 \cdot \exp\left(-\mu \cdot \frac{d}{\cos\theta_2}\right). \qquad (3.50)$$

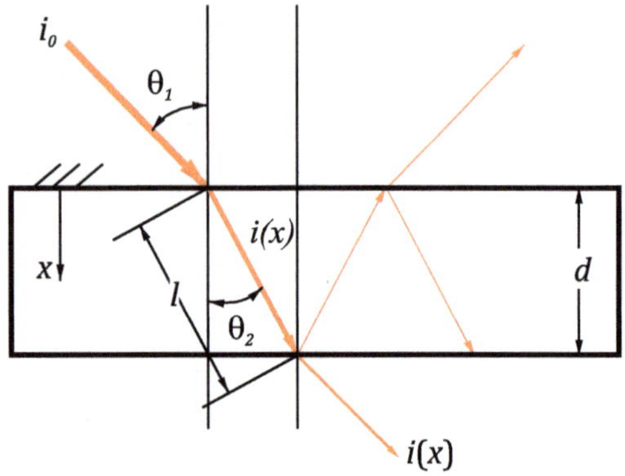

Fig. 3.24 Schematic diagram of the transmission with absorption

The factor $(1 - R_i)^2$ describes the entrance and exit from the material and the exponential function describes the light absorption along the path. The index i stands for the parallel and perpendicular reflection in the material. Since many reflections occur in the material, the total intensity i emerging from the layer is i_τ:

$$I_\tau(x = d) = I_0(1 - R_i)^2 \cdot \exp\left(-\mu \cdot \frac{d}{\cos\theta_2}\right) \cdot \sum_{j=1}^{\infty}\left[R_i^2 \exp\left(-2\mu \cdot \frac{d}{\cos\theta_2}\right)\right]^j. \tag{3.51}$$

If the transparent body reflects only slightly, i.e. i is small ($R_i \ll 1$, attention i depends on the angle θ_1) and absorbs only slightly ($\mu \to 0$), the exponential expression $\left[\exp(-2\mu d/\cos\theta_2)\right]$ approaches unity (1). If, in addition, the perpendicular and parallel reflection coefficients are approximately the same, i.e. $R_s \approx R_p$, then one obtains i_τ for the intensity of the outgoing (transmitted) beam:

$$I_\tau(x = d) = I_0 \cdot \frac{(1 - \rho)}{(1 + \rho)} \cdot \exp\left(-\mu \cdot \frac{d}{\cos\theta_2}\right). \tag{3.52}$$

The solution of the exponential attenuation of the intensity thus obeys the Lambert–Beer law, which was already discussed in Chap. 2 in connection with absorption in the atmosphere. The transmission coefficient in the case of absorption is reduced compared to the ideal case by the exponential expressionexp$(-2\mu d/\cos\theta_2)$. An analogous procedure for N layers arranged in parallel one behind the other can be carried out in the same way as for the case without absorption losses, so that the following results are obtained for τ_a andτ_{aN}:

$$\tau_a = \underbrace{\frac{(1 - \rho)}{(1 + \rho)}}_{\text{reflection}} \cdot \underbrace{\exp\left(-\mu \cdot \frac{d}{\cos\theta_2}\right)}_{\text{absorption}} \text{ and } \tau_{aN} = \left(\frac{1 - \rho}{1 + \rho}\right)^N \cdot \exp\left(-N\,\mu \cdot \frac{d}{\cos\theta_2}\right). \tag{3.53}$$

The transmittance defined in Eq. (3.53) has a reflection part τ_ρ and a pure absorption part τ_a and is only valid for small absorption coefficients. It is often common in literature to combine both terms multiplicatively, so that absorption is separated from reflection, i.e. $\tau = \tau_\rho \cdot \tau_a$.

The absorption in the glass leads to a significant reduction of the transmission intensity. If, for example, the material pairing air and quartz glass ($n_2 = 1.526$) is selected again and $(\mu d) = 0.05 m^{-1}$ is applied for the absorption, the intensity of the light after passing through the pane is reduced from 93 to 88%, considering the absorption and with vertical irradiation ($\theta_1 = 90°$). The Fig. 3.25 shows this reduction of the transmission coefficient as a function of the angle of incidence with several panes placed one behind the other for the ideal case and taking absorption into account. The figure clearly shows that even with two panes of glass connected in series, taking absorption into account $(\mu d) = 0.064\ m^{-1}$, the intensity behind the pane assumes a value which corresponds to the order of magnitude of four ideal glass panes arranged one behind the other. This illustrates that even a small proportion of reflections and absorption can lead to a considerable loss of intensity. This loss of intensity is no longer available for use by an absorber.

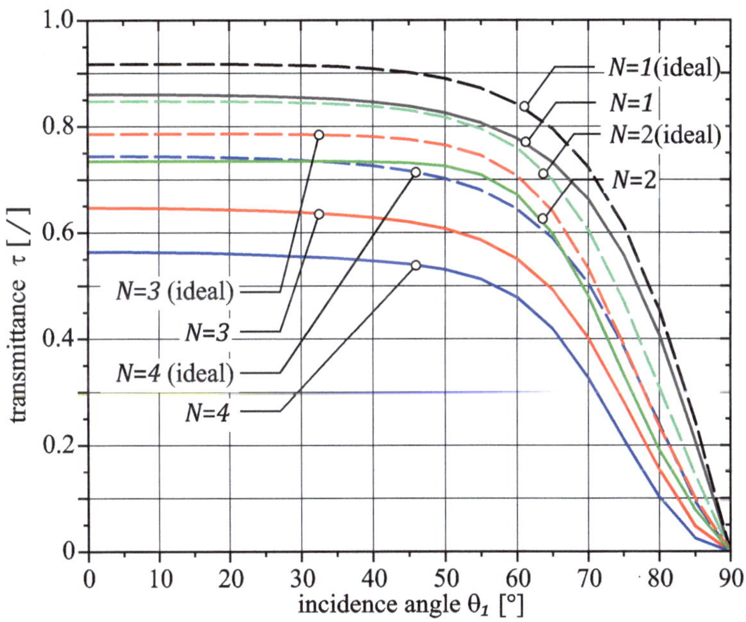

Fig. 3.25 Calculated transmission coefficient τ_a as a function of the angle of incidence θ_1 for a different number N of parallel panes of air-quartz glass ($n_1 = 1, n_2 = 1.526$) arranged one behind the other, assuming an absorption $(\mu d) = 0.05 m^{-1}$. The respective curves of an ideal glass pane arrangement $(\mu d) = 0$ show the dashed lines

3.2.6 Anti-Reflection Coatings

When electromagnetic waves pass through the interface between two media with different refractive indices, a part of the radiation is reflected. For the transition $n_1 < n_2$ (for example $n_1 = 1$ as ambient medium air and n_2 as substrate material), the larger the refractive index difference and the larger the angle of incidence θ_1 the more radiation of the incident light is reflected at the interface. For the example of float glass $n_2/n_1 = 1.5$ a reflection at perpendicular incidence ($\theta_1 = 0°$) results already in 4% and at $\theta_1 = 60°$ in 9%, i.e. a reflection loss more than twice as large.

It is extremely desirable to reduce the occurring reflections for solar energy systems in general, as these reflections reduce the efficiency of the optical conversion. But reflections are also disturbing in other aspects. Reflections from solar energy systems might disturb visually in sensitive areas e.g. public places, critical traffic areas or airports. With the help of anti-reflection layers, disturbing reflections can be reduced, for example in technical optics e.g. in spectacle lenses as well as for solar components like windows and collector covers, thus increasing the functionality. Depending on the application different technical solutions for anti-reflection coating must be applied. It is important whether a certain wavelength region needs to be anti-reflected or a broadband measure for the whole solar spectrum is looked for. Also, weatherability, sensitivity against scratching etc. also impacts the choice of technology. The various technical methods will be described and discussed in the following paragraphs.

3.2.6.1 Interference Coatings

One of the most common methods is interference coatings. In these layers, the degree of reflection on the coated surface is reduced by means of destructive interference of the reflected rays.

The use of thin layers is nowadays an established process for the anti-reflection coating of surfaces. To reduce the reflected light at the interface of two optical media with different refractive indices (n_1, n_s), an optical layer is applied at this interface whose refractive index (n_s) lies between that of the two optical media, i.e. $n_1 < n_s < n_2$. The Fig. 3.26 shows schematically the course of a light wave when passing through a thin transparent layer of thickness d. At each interface (air → layer and layer → substrate), the penetration of the light beam causes part of the incident light to be transmitted and another part to be reflected. The intensity remaining in each partial beam is calculated using Fresnel's formulae. In most cases, the multiple reflections occurring within the thin layer as well as the absorption occurring within them are neglected in such processes. The two reflected partial waves R_1 and R_2 come from the same light source and therefore oscillate synchronously for a certain time, i.e. they are coherent. Since the wave reflected at the interface layer-substrate covers a longer optical path than the wave reflected at the front interface air-layer, R_1 and R_2 have a phase difference between them. Depending on the optical thickness ($n_s \bullet d$) of the layer and the direction

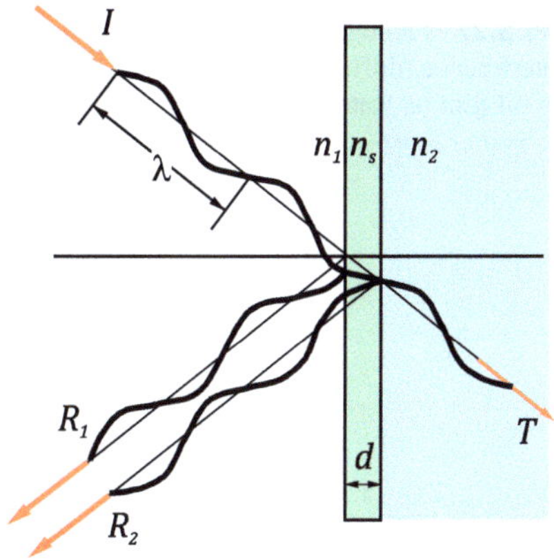

Fig. 3.26 Sketch to describe the destructive interference using $\lambda/4$ layers for $n_1 < n_s < n_2$

of the refractive index transition to the substrate ($n_s < n_2$,$n_s > n_2$), these partial reflections can now intensify or cancel each other out.

The dependence of the phase position of the partial beams on the light wavelength λ as well as the optical thickness of the layer *d* is described by the following equation assuming a perpendicular incidence of light,

$$\delta = \frac{2 \cdot \pi}{\lambda} \cdot n_s \cdot d, \tag{3.54}$$

where δ is the path difference, *d* is the thickness of the layer and n_s is its refractive index.

If the optical thickness of a low-refractive layer is a quarter of the light wavelength $\lambda/4$) or an odd multiple of $\lambda/4$, the phase difference between the two reflected partial waves after travelling back and forth through the layer or an odd multiple of π. The total reflection is the sum of the two reflected waves. With an optical layer thickness d of $\lambda/4$, the partial waves are shifted in such a way that the signs of the amplitudes are opposite. As a result, these values partially cancel each other in the sum (at $n_1 < n_s < n_2$). In the case of weakening there is a so-called destructive interference. The $\lambda/4$-thickness of a single layer is the basis for the design of most interference layer systems, whose detailed calculation and generation in (Willey, 2002) is described.

The opposite effect is achieved by using a highly refractive layer, as can be seen in oil films floating on water. The refractive indices for air, oil and water are $n_{air} = 1$, $n_{oil} = 1.5$ and $n_{water} = 1.33$. Since oil is denser than air, the light is refracted and broken down into its spectral colours when it enters the gasoline. At the water surface under the oil it is then reflected, penetrates the oil again and is broken again. At the boundary with the air it is reflected again, sent back into the oil. If the gear difference is $\lambda/2$, the waves cancel each other out with this refractive index combination ($n_{oil} > n_{water}$). If the path difference is a quarter of the light wavelength $\lambda/4$) or an odd multiple of $\lambda/4$, the amplitudes are amplified, and a colour film can be seen (see Fig. 3.27). Soap bubbles also show colourful

Fig. 3.27 Photograph of the interference (diffraction) of an oil film on water $n_s = n_{oil} > n_2 = n_{water}$

effects due to interference. If the soap film thickness is less than $5\mu m$, the light reflected on the inside interferes with that on the outside.

For a complete antireflection coating of the substrate surface ($\rho \to 0$) not only the phase condition described above must be fulfilled, but also the amplitude condition, i.e. the amplitudes of both partial waves must be equal (amplitude condition: $R_1 = R_2$). This requirement can only be met if the following condition is met for the refractive index of the layer according to the Fresnel formulae:

$$n_s = \sqrt{n_1 \cdot n_2}, \tag{3.55}$$

when light passes from glass ($n_2 = 1.5$) to air ($n_1 = 1$), this means that the layer must have a refractive index of $n_s \approx 1.22$. However, this refractive index, which is necessary according to theoretical analyses, cannot be realized industrially, since dielectric solid materials with this refractive index are not known in the visible spectral range (VIS = Visual spectral range). The substance magnesium fluoride (MgF_2) with a refractive index of $n \approx 1.38$ comes relatively close to this value. For an anti-reflection coating with a single layer, therefore, a MgF_2 layer is usually used, which brought Smakula to industrial application as early as 1935 and in 1957 (Smakula, 1957) fully published; details are available in (Anders, 1967) to find. With a $\lambda/4$-MgF_2 layer, the glass surface can only be anti-reflective from $\rho = 4\%$ to1.41%, as the Fig. 3.28 shows: the reflection coefficient ρ is a function of the wavelength λ and, therefore, the effect is limited to a limited wavelength interval. In addition to the significantly reduced reflection compared to the glass pane, a wavelength dependence of the reflection coefficient can also be observed.

If the light beam does not hit the interface perpendicularly but at an angle, the optical path length in the coating changes according to Snellius' law of refraction, so that a different optimum coating thickness d results, which is calculated

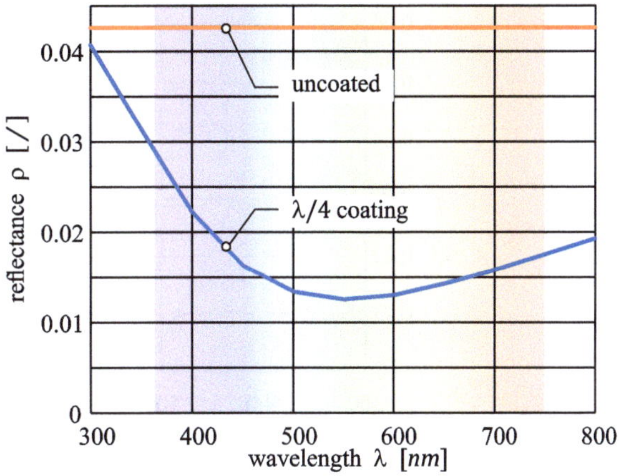

Fig. 3.28 Reflectance ρ as a function of wavelength once with $\lambda/4$ coating and once without

according to equation

$$d = \frac{\lambda}{4 \cdot n_s} \cdot \sqrt{1 + \left(\frac{n_1}{n_s}\right)^2} \cdot \sin\theta_1, \tag{3.56}$$

The design wavelength is the wavelength for which the $\lambda/4$-layer was calculated. A zero-percent reflection is generally conceivable for any wavelength. For an antireflection coating in the visible range, the design wavelength is usually specified as 550 nm, as this is approximately the middle of the visible spectrum. This results in a layer thickness d of 137.5 nm for $\lambda/4$ at $\lambda = 550$ nm.

Since in practice a single layer does not allow a reflection factor of 0%, the use of layer stacks is used. When using two $\lambda/4$-layers, a more effective reduction of reflection at one wavelength can be achieved. After applying a medium refractive $\lambda/4$-layer ($n_s > n_2$) as the first layer on the substrate, it is possible to achieve a reflection of 0% at the design wavelength by applying another layer with a low refractive index. These systems are also called V-reflection or $V - coat$ due to the shape of the spectral reflection curve. The Fig. 3.29 shows this using a glass coated with a medium refractive λ/4-layer Al_2O_3 ($n_s = 1.7$) and then with a low refractive λ/4-layer MgF_2 ($n_s = 1.38$). At the design wavelength, almost no reflection occurs now.

To obtain an anti-reflection effect over a wider wavelength range, anti-reflection coatings with three layers have been developed. By introducing a highly refractive $\lambda/2$-layer ($n_s = 2.3$) into the two-layer mirror coating, the reflection at two wavelengths can theoretically be reduced to zero, which is why this design is also called $W - Coat$, see (Pulker, 1999). Figure 3.30a illustrates this type of layer stacking. For very broadband anti-reflective coating systems, the use of at least 4 to 6 layers (and sometimes even more) are necessary, which results in much more complex technical designs. Thus, the reflection can be eliminated in almost the entire visible wavelength range as Fig. 3.30b represents. However, all types of

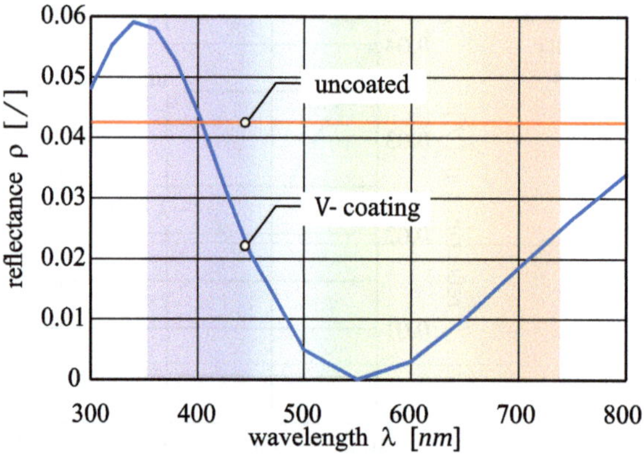

Fig. 3.29 Reflectance of glasses without and with a $V - Coat$ ($2 \bullet \lambda/4$-layers with Al_2O_3 ($n_s = 1.7$) and then with MgF_2 ($n_s = 1.38$)

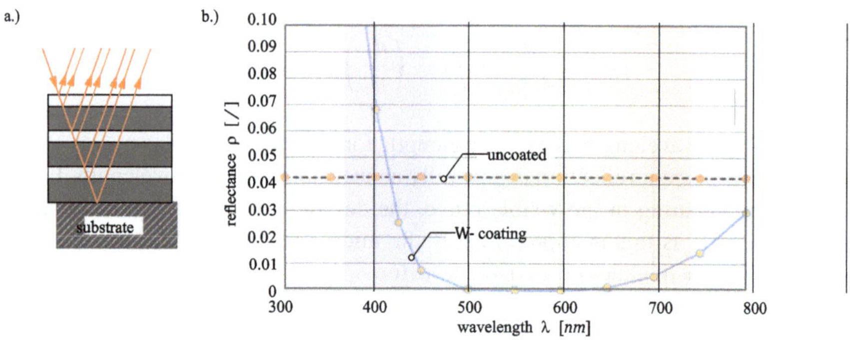

Fig. 3.30 (**a**) layer stacking of a $W - Coat$; (**b**) spectral reflectance without coating and with a $W - Coat$ according to (Pulker, 1999)

destructive interference also limit the transmission of light intensity, so that there is a certain optimum of the number of layers.

3.2.6.2 Porous Coatings

An alternative to anti-reflection coating by interference coatings is the use of porous coatings, such as the introduction of a surface structuring as used by (Gombert and Rommel, 1997) can be described. If a solid is mixed with air in a size dimension below the wavelength of light, very low refractive indices can be achieved. In fact, such media have a refractive index between that of air and that of the material. In the case of porous layers, this can be determined by the volume fractions of different media (e.g. by the size or density of pores) according to (Bläsi, 2000) can be specifically influenced.

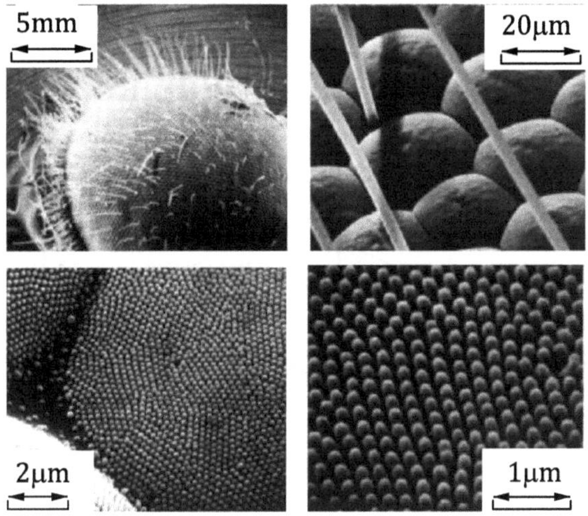

Fig. 3.31 Scanning electron microscope images of the surface of a moth eye (MacLeod, 2001)

A reflex reduction of the surface can also be achieved by introducing a suitable microstructure, which is called "moth eye structure" in reference to the surface structure of insect eyes. The model for this anti-reflection principle was the eye of nocturnal moths on their cornea (see Fig. 3.31). The individual compound eyes have structures that are very similar to a periodic hexagonal surface relief lattice with periods of $230nm$. The introduction of this structure leads to a refractive index gradient at the interface of an optically transparent medium, which increases continuously towards the substrate (Bläsi et al., 2002).

3.2.6.3 Micro-Structured Surfaces

Surfaces with "moth-eye structures" are also known as periodic surface profiles, which can be described as optical gratings. Optical gratings are locally periodic structures that act on the amplitude or phase of light radiation. The lattice properties are determined by the material properties, the structural shape and structural period. The latter can be used for a classification. It is divided into three areas:

- Structure period less than the wavelength of the electromagnetic radiation (a),
- Structural period approximately equal to the wavelength of the electromagnetic radiation (b),
- Structural period larger than the wavelength of the electromagnetic radiation (c).

Different diffraction efficiencies result, which correspond to the share of redistributed energy for the respective diffraction orders. The Fig. 3.32 shows schematically the beam path with different ratios of structure period to wavelength.

If the structure period is much larger than the wavelength (case c.), significantly more diffraction orders are capable of propagation than in the case of approximately the same order of magnitude (case b.). It can propagate for $\lambda \gg \Lambda$ in

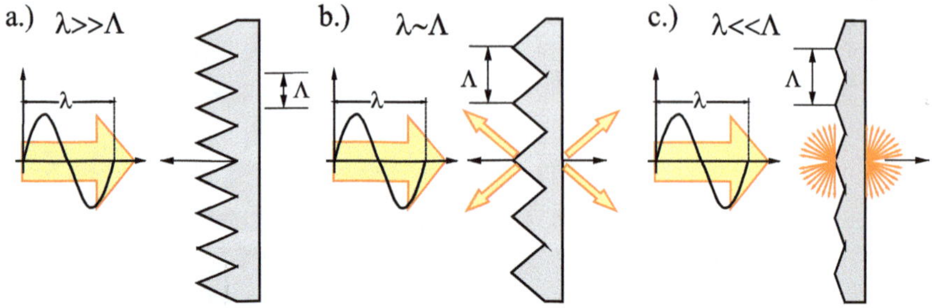

Fig. 3.32 Ratio of grating period Λ to wavelength λ of the irradiated electromagnetic wave determines the optical behaviour of the grating. (**a**) $\lambda \gg \Lambda$, (**b**) $\lambda \approx \Lambda$ und (**c**) $\lambda \ll \Lambda$

Fig. 3.33 Scanning electron microscope image of a section through the moth eye structure on the cornea of nocturnal insects after (Miller et al., 1966) and the resulting refractive index curve from Gäbler (Gäbler, 2005)

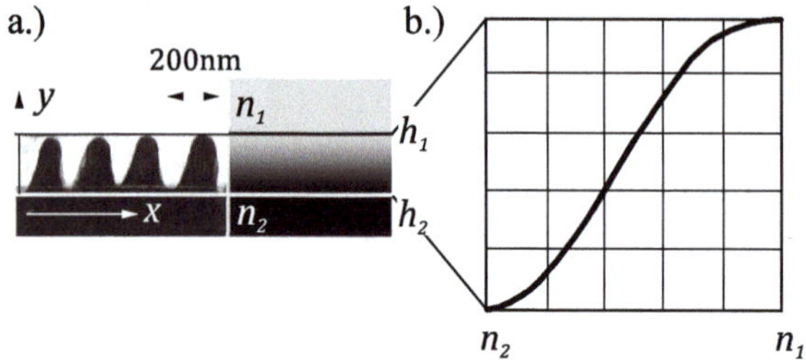

each case only the zeroth diffraction-order in reflection and transmission. The light passes through the structured interface as through a plane interface, which means that the region can be considered optically homogeneous and described with an effective refractive index (Gombert, et al., 1999). Figure 3.33a show a section through a moth's eye made with a scanning electron microscope. Above the height y of the eye there is a continuous increase in material density. The resulting refractive index is shown in Fig. 3.33b.

Thus, for $\lambda \gg \Lambda$, i.e. at long wavelengths, one obtains hardly any reflections and a high transmission. At the other extreme (case c.), small wavelengths and large structures result in very many reflection orders and less transmission.

For the description of the optical properties of material mixtures or the moth-eye structure, there are now many methods, models and simulations that seem suitable for the different aspects. For example, RCWA (Rigorous Coupled Wave Analysis) is well suited for the simulation of periodic gratings. For the description of the structures, the model of the effective media approach (EMA) is often favoured, which physically essentially uses material mixture under consideration of two constituents. Three mixing models are technically applied: the Maxwell–Garnett, the Bruggeman and the Lorentz-Lorenz mixing model.

3.3 Spectral Selectivity

3.3.1 Introduction

The concept of *spectral selectivity* is a key concept in solar thermal energy conversion. We deal with solar radiation basically in a wavelength range between 0.2 and 2.6 μm, which is responsible for our system energy gains, and with thermal radiation a different wavelength region between about 2 μm and 50 μm (depending on the temperatures considered), which is responsible for system energy losses. Optical properties of materials and components used therefore are searched for which take different values in these spectral regions. And this is called spectral selectivity:

3.3.1.1 Selective Transmittance

One spectral selectivity we discuss about is the *selective transmittance* of a transparent element. Here, we distinguish two important classes of selective transmission:

Visual selectivity: high light transmittance in the visual part of the spectrum ($0.38 - 0.78$ μm, weighted with the eyes sensitivity curve$V(\lambda)$, as defined in (ISO 2353, 2005) combined with minimum solar transmittance ($0.3 - 2.5$ μm, weighted with a standard solar spectrum (ISO 9845–1, 1992), (ISO 9050, 2003):

- $\tau_{vis} \gg \tau_e$;
- $\rho_{vis} \ll \rho_e$.

This kind of selective glazing does not transmit the near infrared invisible part of the solar spectrum and hence may reduce the cooling load of buildings. As in windows the heat transport shall be minimized also, the visual selectivity often is produced by a coating of the base glass with low thermal emittance. In principle however, visual selectivity could also be enhanced by layers transparent to the visual light and absorbing in the near infrared range (NIR) between 0.8 μm and 2.5 μm . Water or polymers could be used theoretically for such purpose.

Thermal selectivity: high solar transmittance is combined with low thermal transmittance and minimized heat transport. Ordinary float glass and many transparent polymers are already an example for this selective transmission properties, as only a very small part of the thermal radiation is transmitted directly through these media (of course depending on the thickness of the polymer film or glass). The heat transport however may be further improved by using suitable coatings with low thermal emissivity on the transparent element, which reduce further the thermal radiative transport.

- $\tau_e \gg \tau_{th}$;
- $\rho_e \ll \rho_{th}$.

The performance of a candidate high-performance selective glazing can be characterized by its visual and solar transmittance and its thermal emittance. Using Kirchhoff's law, spectral transmittance and emittance are dependent very much on reflectance $\rho(\lambda)$ for transparent materials,

$$\tau(\lambda) = 1 - \rho(\lambda) - \alpha(\lambda), \tag{3.57}$$

$$\varepsilon(\lambda) = \alpha(\lambda). \tag{3.58}$$

Using standard spectrophotometers, solar reflectance and transmittance are usually measured in the $0.3 - 2.5\mu m$ wavelength range at near-normal angle of incidence. Similarly, in the infrared range above $2\mu m$ emittance is determined by measuring near-normal hemispherical reflectance and transmittance with an integrating sphere, and by using Eqs. (3.57) and (3.58). This may lead to unrealistically low predictions of radiative heat transport because the effective emittance relevant for that is systematically underestimated. The reason is that the property determined from normal-hemispherical reflectance is the normal emittance, which is usually smaller than hemispherical emittance relevant for the overall radiative heat loss of a surface. An alternative is the calorimetric measurement of radiative heat loss of a surface in vacuum at the relevant temperature. Whereas spectral reflectance can be measured at room temperature, total emittance must be measured at elevated temperatures. An average emittance over a wavelength range is frequently reported from spectral data weighted with blackbody curves

$$\varepsilon(T) = \frac{\int_{\lambda_{\min}=0}^{\lambda_{\max}=\infty} \varepsilon(\lambda, T) \cdot e_B(\lambda, T) d\lambda}{\sigma \cdot T^4}, \tag{3.59}$$

where $\sigma = 5.67 \cdot 10^{-8} W/(m^2 K^4)$ is the Stefan-Boltzmann constant and $e_B(\lambda, T)$ is the spectral irradiance of a blackbody curve from

$$e_B(\lambda, T) = \frac{c_1}{\lambda^5 \cdot \left[\exp\left(\frac{c_2}{\lambda \cdot T}\right) - 1\right]}, \tag{3.60}$$

where $c_1 = 3.7405 \cdot 10^8 W/(\mu m^4 m^2)$ and $c_2 = 1.43879 \cdot 10^4 \mu m K$ are Planck's first and second radiation constants, respectively (Howell, Mengüc, Daun, and Siegel, 2020).The actual performance of an absorber at high temperatures may not correspond to the calculated emittance.

Using the spectral data as described in the previous section we may determine similarly the solar and visual transmittance and reflectance by integration using a standard solar spectrum as weighting factor:

$$\tau_e = \frac{\int_{\lambda_{\min}=0}^{\lambda_{\max}=2.6} \tau(\lambda) \cdot e_s(\lambda)\, d\lambda}{\int_{\lambda_{\min}=0}^{\lambda_{\max}=2.6} e_s(\lambda)\, d\lambda}, \tag{3.61}$$

$$\tau_{vis} = \frac{\int_{\lambda_{min}=0}^{\lambda_{max}=2.6} \tau(\lambda) \cdot \mathrm{V}(\lambda) \cdot e_s(\lambda)\, d\lambda}{\int_{\lambda_{min}=0}^{\lambda_{max}=2.6} \mathrm{V}(\lambda) \cdot e_s(\lambda)\, d\lambda}, \tag{3.62}$$

where $e_s(\lambda)$ is a suitable solar standard spectrum and $V(\lambda)$ is the eye sensitivity. For window application a global irradiation spectrum is taken depending on the standard. A recommended standard global spectrum here is ISO 9845 solar global spectrum $AM1.5$ (ISO 9845–1, 1992).

3.3.1.2 Selective Absorptance

Another spectral selectivity is the relation of solar absorption and thermal emission for *spectrally selective absorbers*. Here, the purpose is a high absorption in the solar range combined by a low emittance in the thermal range to reduce thermal radiation losses.

The surface of an ideal selective absorber is black (i.e. completely absorbing) in the solar wavelength range between $0.3\mu m$ and $2.6\mu m$ approximately. To reduce the thermal heat losses via emitted radiation from the absorber tubes, the surfaces on the other hand should ideally have a very low emittance for the wavelengths that they emit.

Absorber surfaces are governed by the basic principles of optical behaviour of materials. Key to this are two main principles:

- The absorptance of a surface at a particular wavelength (i.e. the fraction of incident radiation that it absorbs) is equal to its emittance at the same wavelength (the amount of radiation it emits expressed as a fraction of the amount that an ideal black body would emit).
- All the radiation incident on a surface must either be absorbed, reflected, or transmitted, so absorptance, transmittance and reflectance must sum to 1.

The key to the idea of a selective surface is that the wavelengths that a hot solar receiver will emit are different to the solar spectrum wavelengths. Whilst absorptance must equal emittance at each specific wavelength, they can and do vary with wavelength.

As a consequence, the ideal reflectance function of an effective so-called selective solar absorber is a step function between a minimum value of reflectance (hence high absorptance) in the solar wavelength range and a high reflectance (hence low emittance) in the infrared (IR) thermal radiation range (see Fig. 3.34). Ideally the lower reflectance value limit would approach 0% while the upper reflectance limit would be 100%, however in reality these values cannot be achieved. Perfect gold surfaces have reflectance values in the IR of nearly 98%, whilst blackened surfaces with 98% absorptance in the solar wavelength range are also possible. For lower temperatures there is no energetically significant overlap between the thermal emission spectrum of a body (with the blackbody spectrum according to Planck being the upper limit) and the solar spectrum. A critical wavelength λ_{crit} for the step function can be defined which separates the two respective

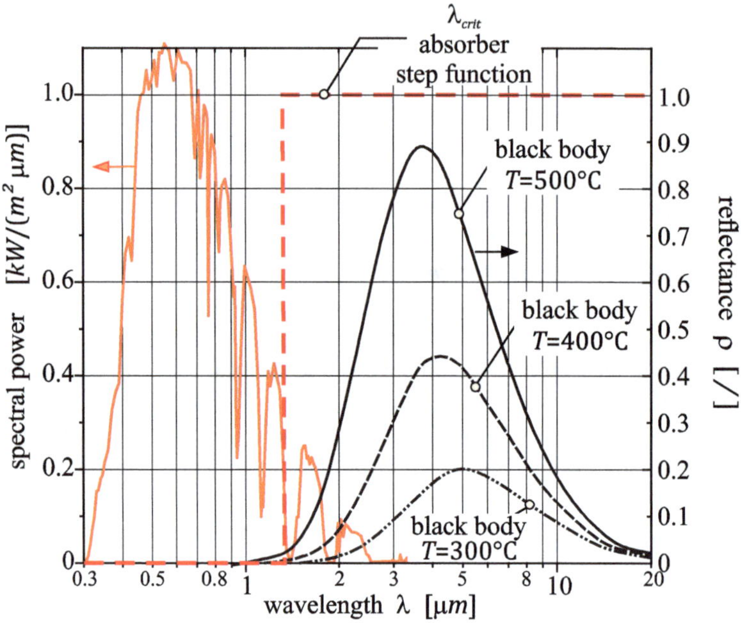

Fig. 3.34 Reflectance of an ideal selective solar absorber together with standard solar spectrum ($AM1.5$ direct, not concentrated) and blackbody spectra (300 $°C$, 400 $°C$, 500 $°C$)

wavelength regions. For higher temperatures however, the thermal emission spectrum is shifted towards shorter wavelengths which implies an increasing overlap of the spectra. Therefore, the critical wavelength λ_{crit} must be optimized with respect to solar absorptance and thermal emittance for different operation temperatures.

When the Critical wavelength λ_{crit} of the ideal reflectance step function is varied (see Fig. 3.34), the integrated values of solar absorptance α_s and of thermal emittance ε_{IR} for a specific temperature change due to a different overlap with solar and thermal spectra. For a low value of the critical wavelength a part of the solar spectrum overlaps with the high reflectance, and therefore solar reflectance increases—and due to the basic principles solar absorptance decreases. When the critical wavelength exceeds the maximum wavelength of significant solar radiation (at about $2.6\mu m$) a further increase leaves the integrated value constant, as illustrated in Fig. 3.35.

In contrast, the infrared emittance rises with increasing critical wavelength, as a part of the low reflectance/high absorptance curves enters the infrared region where the thermal spectrum is significant. Consequently, shifting the critical wavelength to higher values in order to gain a higher absorption will always result in a higher emittance for a given temperature. Similar to a transparent window coating, the performance of a candidate solar absorber can be characterized by its solar absorptance and thermal emittance, both depending on the spectral reflectance. For opaque (zero transmittance) materials, Eq. (3.57) can be simplified setting $\tau = 0$.

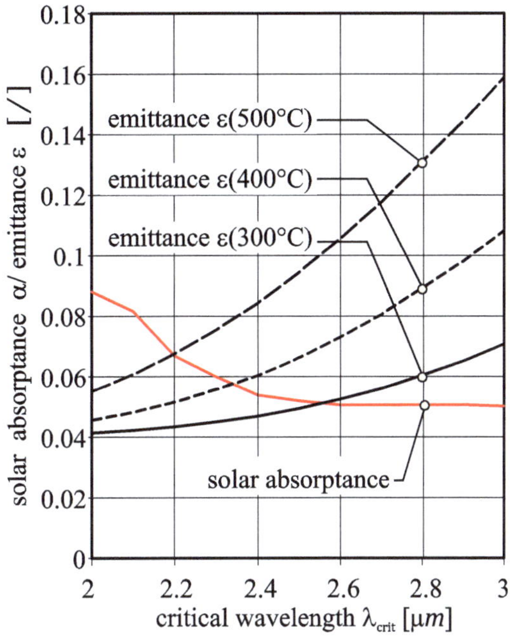

Fig. 3.35 Solar absorptance and thermal emittance for absorber temperatures 300 °C, 400 °C *and* 500 °C versus critical wavelength λ_{crit} of an ideal selective absorber (step function) with short wavelength reflectance 0.04 and long wavelength reflectance 0.96

Development of spectrally selective materials depends on reliable characterization of their optical properties. The determination of the thermal emittance is as described in the last chapter on coated glass, only setting transmittance to zero and using Eqs. (3.59) and (3.60).

The actual performance of an absorber at high temperatures may not correspond to the calculated emittance. This is because small errors in measured ρ can lead to large errors in small values of ε. Calculating the emittance from spectral data taken at room temperature assumes that the spectral characteristics do not change with increasing temperature. This is only valid if the material is invariant for example does not undergo a phase change (as do some titanium containing materials), break down or undergo oxidation (as do paints and some oxide coatings) at higher temperatures. It is important before using high-temperature emittance calculated from room temperature data, that the calculated data is verified with high-temperature emittance measurements for each selective coating. Therefore, it is preferable to also measure the total emittance at the operating temperatures and conditions (Brunotte, Lazarov, and Sizmann, 1992). In addition, measurement errors might lead to differences to the direct calorimetric measurement.

Emittance is a surface property and depends on the surface condition of the material, including the surface roughness, surface films, and oxide layers. Coatings typically replicate to some degree the surface roughness of the substrate. Therefore, to facilitate development, it is important to measure the emittance of each coating-substrate combination as well as the uncoated substrate when developing a solar selective coating. Using the spectrally measured reflectance as described before we may determine similarly the solar absorptance α_e by integration using a standard solar spectrum as weighting factor.

$$\alpha_e = \frac{\int_{\lambda_{\min}=0}^{\lambda_{\max}=2.6} [1 - \rho(\lambda, T)] \cdot e_s(\lambda) d\lambda}{\int_{\lambda_{\min}=0}^{\lambda_{\max}=2.6} e_s(\lambda) d\lambda}, \tag{3.63}$$

where $e_s(\lambda)$ is a suitable solar standard spectrum. For concentrating collectors only direct irradiation is used, hence a direct solar spectrum should be taken. A recommended standard spectrum here is ASTM G173-03 $AM1.5$ Direct Normal (ASTM G173-03, 2020).

3.3.2 Selective Glazings

In order to produce spectrally highly selective glazings with better performance than the ordinary float glass transparent coatings are applied on the glass. As for all glazings the reduction of U-value is a target, a low emissivity coating is the key element. For commercial coatings two different coating types are used: the sputtering of ultra-thin noble metal films or the spray pyrolysis of doped semiconductor oxides. In both materials the number of free electrons determine the spectral limit between transparent and reflective properties of the coating. The properties of these coating types are described in the next paragraphs.

3.3.2.1 Noble Metal Coatings (Soft Coatings)

Very thin noble-metal coatings produced by sputtering are, at least in principle, the simplest solution for reaching a significant short wavelength transmittance combined with long-wavelength reflectance. The best optical properties are obtainable with copper, silver, and gold (Valkonen et al., 1984). Alternative materials may be TiN (Claesson et al., 1990) and aluminium. Thin silver coatings stand out as the superior material on account of their low absorption of luminous and solar radiation (Valkonen and Karlsson, 1987) and are mainly used for commercial coatings.

It is necessary to improve the transmittance by additional antireflection layers. Dielectrics with high refractive indices—such as Bi_2O_3, In_2O_3, SnO_2, TiO_2, ZnO and ZnS - give the largest enhancement of the transmittance. By selecting the thicknesses of the three-layer configuration properly, one can optimize for a warm climate (minimum solar gain with high visible transmission) or a cold climate (maximum solar gain combined with low emissivity). Figure 3.36 shows the spectral optical properties of two silver-based coated glasses, one optimized for high solar gain, the other one for reasonable solar rejection with similar visible transmission.

3.3.2.2 Doped Oxide Semiconductor Coatings

Doped oxide semiconductor coatings by spray pyrolysis offer an alternative to the earlier discussed noble-metal based coatings. The materials which are known

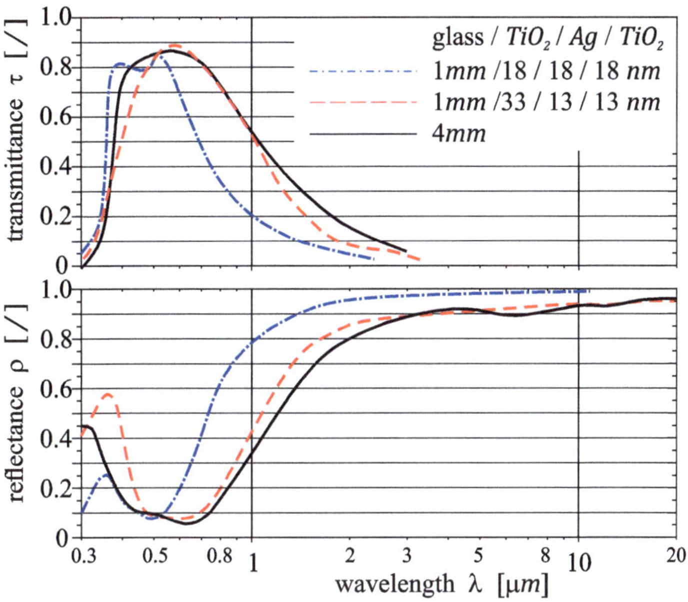

Fig. 3.36 Spectral transmittance and reflectance of glass with silver-based coatings, solar control coating: high visible transmittance with maximal solar reflectance, Solar gain coating: increased solar gain in the solar infrared (above $0.7\mu m$), (Granqvist, 1991a, 1991b)

to be useful are oxides based on zinc, cadmium, indium, tin, thallium, lead, and alloys of these. The required doping is often achieved by the addition of a foreign element; particularly good properties have been obtained with $SnO_2 : F$, $SnO_2 : Sb$, $In_2O_3 : Sn$ and $ZnO : Al$ (Granqvist, 1990), see Fig. 3.37. The industrial products are based on tin oxide, where the coating is directly done in the float process. A specific and important advantage of the doped oxide semiconductors is their excellent chemical and mechanical durability, which allows their use on glass surfaces exposed to the air.

The required coating thickness t is usually around $t \approx 1\mu m$, much thicker than the multilayer soft coatings! This is done to avoid iridescence with unpleasant coloured fringes. The human eye then senses a uniform coloration. However, the use of a much larger thickness than the one demanded for a low τ is clearly inefficient in terms of materials utilization, coating time, and cost. Further, thick oxide semiconductor coatings can display some light scattering, sometimes referred to as haze.

This phenomenon is associated with the occurrence of large crystallites as well as surface roughness, see cf. (Schade and Smith, 1985).

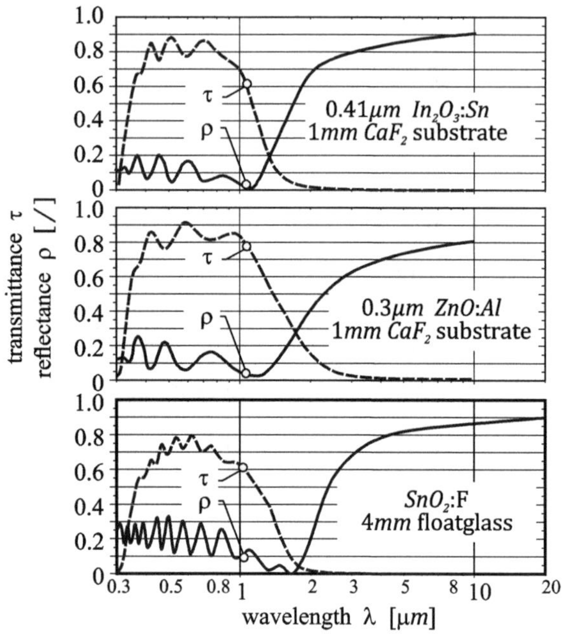

Fig. 3.37 Transmittance and reflectance for thick pyrolithic semiconductor oxide coatings, (Granqvist, 1990)

3.3.3 Design of Selective Absorbers

Selective absorber surface coatings can be categorized into several distinct types based on different physical principles:

a. Intrinsic,
b. Semiconductor–metal tandems,
c. Multilayer absorbers,
d. Multi-dielectric composite coatings,
e. Textured surfaces, and
f. Selectively solar-transmitting coating on a blackbody-like absorber.

Whereas the first two principles need only one homogeneous material, the other ones work with layers of different optical properties.

(a) Intrinsic absorber coatings use a material having intrinsic properties that result in the desired spectral selectivity.
(b) Semiconductor–metal tandems absorb short wavelength radiation where photon energy is above the semiconductor bandgap and have low thermal emittance because of the metal layer.
(c) Multilayer absorbers use multiple reflections between layers to absorb light and can be tailored to be efficient selective absorbers.

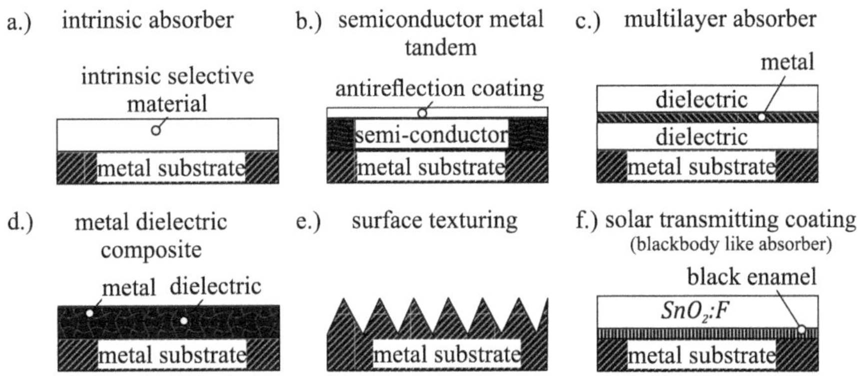

Fig. 3.38 Schematic designs of six types of coatings and surface treatments for selective absorption of energy, adapted from (Kennedy, 2002)

(d) Metal-dielectric composites—cermets—consist of fine metal particles in a dielectric or ceramic host material.
(e) Textured surfaces can produce high solar absorptance by multiple reflections among needle-like, dendritic, or porous microstructure.
(f) Additionally, selectively solar-transmitting coatings on a blackbody-like absorber are also used but are typically used in low-temperature applications.

These constructions are shown schematically in Fig. 3.38a-f, respectively, and are discussed in more detail below.

3.3.3.1 Intrinsic Absorbers

Intrinsic selective absorbers consist of a single material with intrinsic selective property of the material. No naturally occurring material exhibits intrinsically ideal solar-selective properties, but some have mild selective properties. Intrinsic properties are found in transition metals and semiconductors, but both need to be greatly modified to serve as an intrinsic absorber.

Examples include the transition metal tungsten W (Agnihotri and Gupta, 1981) and a series of semiconducting materials, for example MoO_3-doped Mo (Seraphin, 1979), Si doped with B, HfC (Seraphin and Meinel, 1976), ZrB_2 (Randich and Allred, 1981), SnO_2 and others. Hafnium carbide (HfC) could be useful as an absorbing selective surface at elevated temperatures because of its high melting point. However, the absorptance of HfC has to be increased e.g. by an antireflective (AR) layer. Using Zirconium bromide ZrB_2 one can reach the best selectivity with $\alpha = 88\%$ and $\varepsilon(100°C) = 8\%$, which may be even inproved by a $Si_3N_4 - AR$-Layer to $\alpha = 93\%$ and $\varepsilon(100°C) = 10\%$ (Randich and Allred, 1981).

Intrinsic absorber materials are usually structurally very stable but optically less effective than other types of selective absorbers. Therefore, historically, research in intrinsic absorbers has not been very productive because there are no ideal intrinsic materials; but the intrinsic materials are finding increasing use as a component in high-temperature absorber multilayers and composite coatings.

3.3.3.2 Surface Texturing

Surface texturing is a common technique to obtain higher absorption due to optical trapping of solar energy. When surfaces are textured with a lateral dimension close to the wavelength to be absorbed, they appear as a series of cavities. Thus, structures smaller than $1\mu m$ absorb solar energy whereas for longer wavelengths the texture cannot be resolved and appears highly reflective and mirror-like to thermal energy. The emittance can be adjusted (higher or lower) by modifying the microstructure of the surfaces with ion-beam treatments. The selective properties depend on the ratios of mean height deviations and the autocorrelation distance to the wavelength (Cuomo et al., 1975).

Needle-like, dendrite, or porous microstructures on the same scale as the wavelength of the incident radiation exhibit both wavelength and some directional selectivity. The surface of the microstructure must be protected from damage caused by surface contact or abrasion. Selection of a material having a high intrinsic absorption coefficient can further optimize the absorptance. A number of methods exist to prepare textured microstructures, (Seraphin and Meinel, 1976) for example:

- Liquid–solid phase separations.
- Lithography in combination with ion-etching.
- Unidirectional solidification of eutectic alloys.
- Lithography with X-rays.
- Ion-exchange reactions between metals—isothermal transport occurs between two metals where the difference in the work function (ΔE_w) is $> 0.2eV$, e.g. for $Cu - Ni$ alloy.
- Vapor–liquid-solid (VLS) mechanism grown in a controlled way, whiskers on substrates from the liquid alloy zone at the interface (e.g., Si, Ge or group $III - V$ whiskers).
- Vapor deposition—the condensation of a metal or compound from the gas phase onto a substrate by Chemical Vapour Deposition (CVD) or Physical Vapour Deposition (PVD) (e.g., $Ni - Al2O3$, Ni).
- Oxidation of metals at high temperature—the growth of whiskers on metals by the oxidation process in air or O_2 at high temperature ($400° - 850\ °C$), e.g. $Fe_2O_3 - Fe$, steel, $CuO - Cu$, phosphor bronze, $ZnO - Zn$, brass, W; Ni or Mo.

These techniques can be useful in texturing selective surfaces. Examples for absorber surfaces are dendritic tungsten or structured Copper, Nickel or stainless steel surfaces (Zhao, 2007). Typical values are $\alpha = 92\%$ and $\varepsilon(100°C) = 22\%$; using Nickel which is not stable at high temperatures the emittance can be as low as $\varepsilon(100\ °C) = 10\%$, see (Kennedy, 2002).

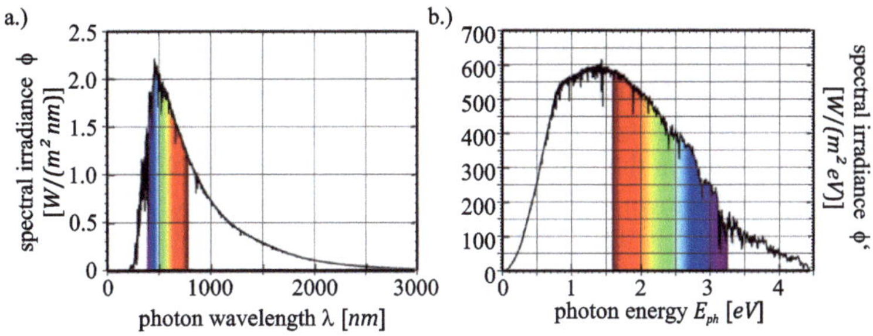

Fig. 3.39 Spectral irradiance as function of wavelength λ in nm (left) and photon energy in eV (right), visible light indicated in colour

3.3.3.3 Semiconductor–metal Tandems

Semiconductors with bandgaps from about $\sim 0.5eV$ to $1.26eV$, corresponding to wavelengths of $2.5\mu m$ to $1.0\mu m$ (cf. Figure 3.39), absorb short-wavelength radiation and are transparent to longer wavelengths. Hence, an underlying metal provides low emittance to give the desired spectral selectivity to semiconductor–metal tandems. Semiconductors of interest include Si $(1.1eV)$, Ge $(0.7eV)$, and PbS $(0.4eV)$ (Agnihotri and Gupta, 1981). An antireflection treatment (coatings or porous surfaces) is needed because the useful semiconductors have high refractive indices resulting in large detrimental reflectance. Si-based designs produced by Chemical-Vapor Deposition (CVD) are well known that are suitable for mid-to high-temperature applications. Best values to be reached are quoted as $\alpha = 91\%$ and $\varepsilon(100°C) = 10\%$ as mentioned in (Kennedy, 2002).

3.3.3.4 Multilayer Absorbers

Multilayer absorbers or multilayer interference stacks consist of many alternating thin layers of transparent dielectric layers and semi-transparent metallic or semiconducting layers, for which the schematic design is illustrated in Fig. 3.40. The lowest opaque thick coating must be metallic to provide a low emittance. The alternating thin-layer stack is optimized for high solar absorption while being transparent to longer wavelengths. Thus, the stack can be designed to become an efficient selective absorber.

The basic physics of multilayer absorbers is well understood, and computer modelling can predict the optical properties given by an optimum multilayer design of candidate materials (Born and Wolf, 1999), provided adequate optical constants

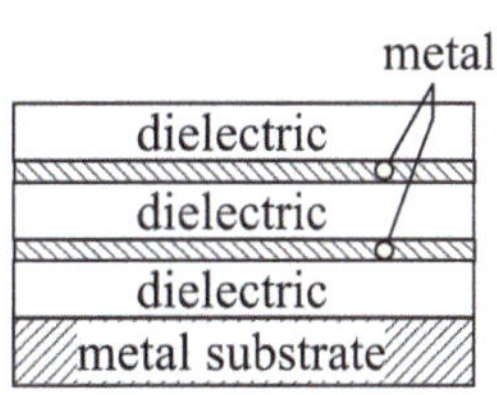

Fig. 3.40 Schematic designs of multilayer absorber film structure adapted from (Kennedy, 2002)

for the thin films are available. Multilayer interference stacks have high solar absorption, low thermal emittance, and can be stable at elevated temperatures ($\geq 400°C$) depending on the materials used. Several multilayer absorbers using different metals (e.g. Mo, Ag, Cu, Ni) and dielectric layers (e.g., Al_2O_3, SiO_2, CeO_2, ZnS) have been cited in the literature for high-temperature applications (Kennedy, 2002). A very well-known example is the AMA-coating ($Al_2O_3 - Mo - Al_2O_3$) with an absorption $\alpha = 92 - 95\%$ and $\varepsilon(100°C) = 6 - 10\%$ (Thornton et al., 1980).

3.3.3.5 Metal-Dielectric Composite Coatings (Cermets)

These absorbers consist of an Anti-Reflection (AR) layer on top of a composite layer, which is highly absorbing in the solar region (i.e., black) and transparent in the IR, deposited onto a highly IR-reflective metal substrate, for which two types are depicted in Fig. 3.41. The highly absorbing metal-dielectric composite, or Cermet, consists of fine metallic nanoparticles in a dielectric or ceramic matrix. These films are transparent in the thermal IR region, while the absorption in the solar region is either due to inter-band transitions in the metal or due to small particle resonances. The cermet layer may have uniform metal content, but graded profiles are preferable to enhance absorption especially towards the substrate. The metal-dielectric concept offers a high degree of flexibility, and the solar selectivity can be optimized by proper choice of constituents, coating thickness, particle concentration, size, shape, and orientation. The solar absorptance can be boosted with a suitable choice of substrates and AR layers, which can also provide protection (for example, from thermal oxidative degradation). A variety of techniques, such as electroplating, anodization, inorganic pigmentation of anodized aluminium, Chemical Vapour Deposition (CVD), and co-deposition of metal and insulator materials by Physical Vapor Deposition (PVD), can produce the composite coatings.

In cermets, solar absorptance is mainly determined by the response of the absorbing particles. There is a shift of the absorption and scattering cut-offs to larger wavelengths when the particle radius, r, increases. This effect is accompanied by a reduction in the scattering and absorption efficiencies roughly proportional to $1/r$ (Arancibia-Bulnes, Estrada, and Ruiz-Suarez, 2000).

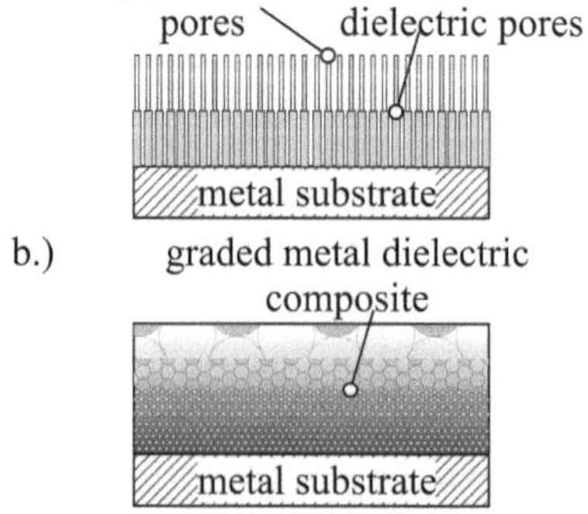

Fig. 3.41 Schematic designs of two different metal-dielectric solar selective coatings adapted from (Kennedy, 2002)

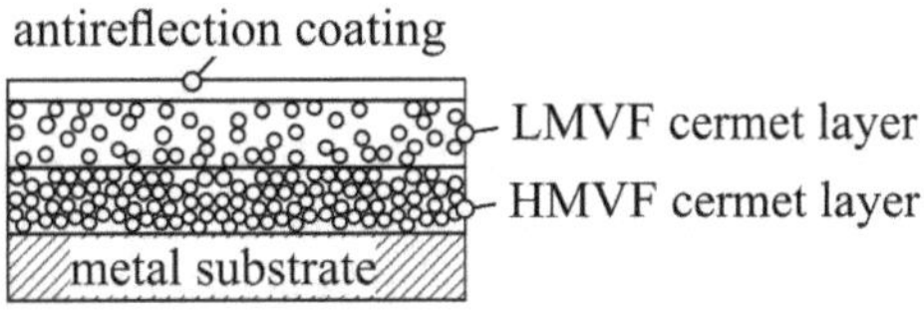

Fig. 3.42 Schematic design of double-cermet film structure adapted from (Kennedy, 2002)

In a graded cermet (Fig. 3.41b), the reflectance from the cermet is reduced by gradually increasing the metal volume fraction, hence the refractive index, as a function of depth from the surface to the base of the film. A double-cermet film structure has been developed similarly increasing the absorption of a homogeneous cermet layer like the graded cermets (Zhang and Mills, 1992). Solar radiation is effectively absorbed internally and by phase interference in double-cermet solar coatings.

Further, it is easier to deposit the double-cermet selective coating than graded-cermet layer selective surfaces. The typical double-cermet layer film structure from surface to substrate consists of the following arrangement:

- An AR layer that enhances solar absorption.
- An absorbing layer composed of two homogenous cermet layers,
- A low-metal-volume fraction (LMVF) cermet layer on
- A high-metal-volume fraction (HMVF) cermet layer; and
- A metallic infrared reflector layer to reduce substrate emittance (Zhang and Mills, 1992),

which is shown in Fig. 3.42. High selectivity can be achieved with these absorber types. Typical values are $\alpha = 95\%$ and $\varepsilon(100°C) = 5\%$, see (Kennedy, 2002).

3.3.3.6 Selectively Solar-Transmitting Coating on a Blackbody-Like Absorber

A selectively solar-transmitting coating on a blackbody-like absorber is the last concept. Here the low emitting surface is the top layer, which transmits radiation in a wavelength selective manner. This solar-transmitting coating can be a highly doped semiconductor (e.g., $SnO_2 : F$, $SnO_2 : Sb$, $In_2SO_3 : Sn$, and $ZnO : Al$) over an absorber with a proven long-term durability. Some low-temperature flat-plate collectors have used black enamel as the absorber material (Granqvist, 1991a, 1991b). Highly doped semiconductors may be useful with high-temperature black absorber materials. The selectivity, however, is limited for this type of coatings.

3.3.4 Production Methods for Selective Layers

The typical layer thicknesses are in the micrometre and nanometre range, up to monomolecular layers. This also makes the use of expensive materials economical if the desired effect can be achieved despite small quantities. By using thin-film technology, material efficiency can be combined with energy resource conservation in a suitable way.

A thin layer applied to a substrate must not only have predetermined optical properties (refractive index, absorption), but must also meet additional technical requirements so that the product can withstand everyday stress. This includes in particular the outdoor use of solar thermal systems:

- Mechanical strength (adhesion, hardness, abrasion and scratch resistance, centring, ultrasound),
- Thermal resistance (cold, heat),
- Chemical resistance (solvents, detergents, sweat, UV radiation, humidity).

The thermal expansion coefficient of the coating material must not normally differ from that of the substrate. In most cases, the coatings already have much higher internal stresses during production than are caused by temperature differences during operation, for example. Flaking of the layer therefore occurs (especially with thicker layers) when the internal stresses are superimposed on the load or the adhesion of the layer to the substrate is too low.

The combination of all the above-mentioned requirements means that only a few substances are even considered as coating materials. In the field of optical thin films, for example, not every refractive index n is available due to the small selection of coating materials. The Table 3.5 lists the refractive index of some common optical coating materials.

In principle, there are several processes for solar thermal purposes to produce thin films. The most common methods are the following:

- Chemical Vapor Deposition (CVD) is a gas phase reaction (usually at or near the substrate surface). The reaction gases are simultaneously fed into the reaction chamber with the substrate to be coated. The mostly preheated gases are thermally activated by the heated substrate and react with each other. The desired material is separated and chemically bound (chemisorption). Besides countless CVD variants, which differ in working pressure and other process

Table 3.5 Average refractive indices n of some selected optical coating materials

Material	Molecular formula	$n[/]$
Magnesium fluoride	MgF_2	1.38
Silicon dioxide	SiO_2	1.46
Aluminium oxide	Al_2O_3	1.7
Zirconium dioxide	ZrO_2	2.05
Praseodymium	$PrTiO_3$	2.1
Titanium oxide	TiO_2	2.3
Zinc sulphide	ZnS	2.3

Table 3.6 Overview of process parameters (temperature, pressure) of common CVD processes and their application areas

CVD-process	Temperature $T[°C]$	Pressure $p[mbar]$	Activation type	Intended use
APCVD	$400 - 1300$	1000	Thermal	Poly silicon deposition for conductor tracks, gate oxide, epitaxy of silicon wafers
LPCVD	$500 - 1000$	$0.1 - 10$	Thermal	Conductor paths, gate oxide SiO_2, silicon nitride Si_3N_4, polysilicon
HFCVD	$150 - 750$	$0.01 - 200$	Thermal	Carbon-based deposition (diamond, amorphous carbon, carbon nanotubes), silicon-based layers (amorphous / crystalline Si, Si_3N_4, ...
PECVD	$200 - 500$	1	Thermal plasma	SiO_2

parameters and are listed in the Table 3.6, there are also some coating processes which represent strongly modified CVD processes.

- Plasma polymerization: In this process, gaseous monomers excited by a plasma form a highly crosslinked layer on a substrate.
- Atomic layer deposition (ALD): Atomic layer deposition is a strongly modified CVD process in which the reaction or sorption on the surface stops automatically after the surface is completely covered. This self-limiting reaction is carried out in several cycles (with intermediate rinsing steps), so very good aspect ratios and exact layer thicknesses can be achieved.

The chemical principles of the CVD process are described in Powell et al. (Powell, Oxley, Blocher, and Klerer, 1966) in detail.

In contrast to CVD processes, physical vapour deposition (PVD) is based on purely physical processes. PVD is also a vacuum-based coating process in which, in contrast to the CVD process, the layer is formed directly by condensation of a material vapour on the starting material. The procedure is characterised by three successive processes. First, a gas (vapor) generation of the layer forming particles takes place; in the second stage the vapor is transported to the substrate and finally the vapor is condensed on the substrate which then forms the layer.

- Thermal evaporation

In thermal evaporation as shown in Fig. 3.43, the evaporation material is heated until it evaporates at a suitable evaporation rate. There are three "subprocesses" depending on the type of evaporator used (inductive, resistive or electron beam evaporators are possible). To ensure the deposition of high-quality and homogeneous layers, it is necessary to keep the space between the evaporator and the substrate as free of material (i.e. vacuum) as possible. The interactions (usually impacts) of the particles with residual gas atoms can bind or scatter them in such

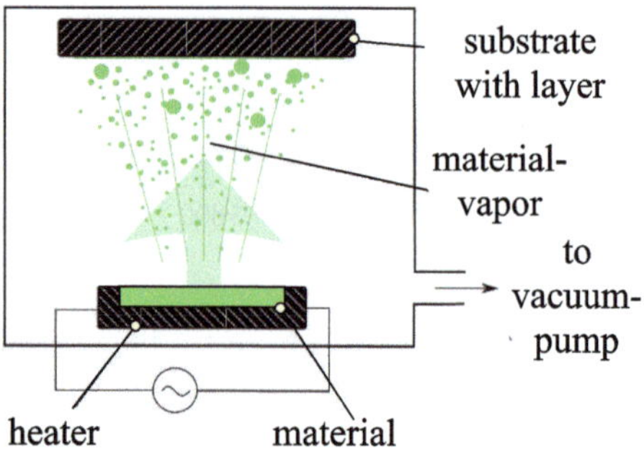

Fig. 3.43 Schematic representation of surface coating by thermal evaporation using a resistance heater

a way that the reproducibility of the coating cannot be guaranteed. Oscillating crystals are often used to measure and control the evaporation rate and layer thickness (alternatively, optical monitoring is also used).

- Sputter deposition:

During sputtering (also called cathode sputtering) particles are removed from the surface by ion bombardment. This process can be used to clean the surface, e.g. of oxides or water that have entered the material through production, processing, or storage. In thin-film technology, this physical process is also used to atomize material from the target, i.e. to transfer it into the gas phase. The resulting gaseous material is then fed onto the substrate to be coated and condenses there, as shown in the Fig. 3.44. Compared to evaporation, sputter deposition has the advantage that alloys can also be transferred to the wafer in "the same ratio". However, it must be considered that different materials have different sputtering coefficients, i.e. they can be atomized to different degrees. The layer thickness is often controlled by means of a time switch-off.

- *Ion plating.* Ion plating is a vacuum-based and plasma-assisted PVD process for metals and metal compounds. Vaporized metal (e.g. by arc discharge) is fed into a plasma. There a part of the metal vapour cloud ionises and is accelerated towards the substrate. The metal ions form a layer on the substrate surface, which is sputtered back together with the substrate material at the beginning by the constant bombardment by metal ions.
- *Ionized cluster beam deposition (ICBD) technology.* The cluster beam technique is also a modified evaporation process. The crucible used for evaporation is kept closed. The heating of the evaporation material creates an overpressure in the sealed crucible. If this steam is discharged through a nozzle, an adiabatic expansion causes sudden cooling. Neutral atomic clusters are formed, which partially dissolve when they hit the substrate surface and are deposited distributed over the surface.
- Molecular beam epitaxy (MBE).
- Electrochemical process: Galvanization.

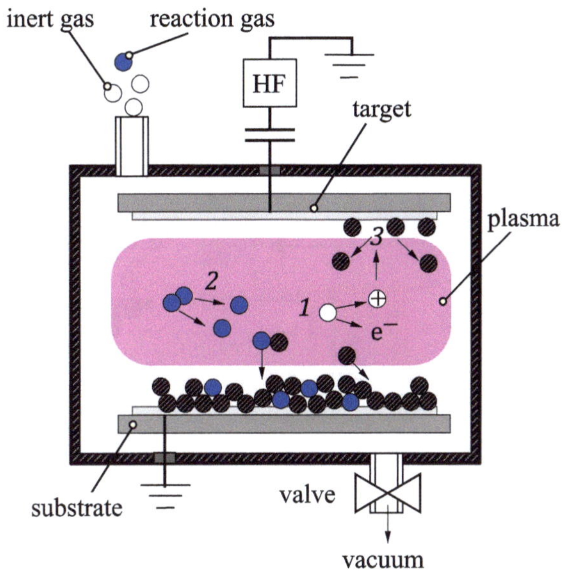

Fig. 3.44 Diagram of a sputter deposition system (reactive HF sputtering) for surface coating

- Electroplating is the electrochemical deposition of metallic deposits (coatings) on objects. In electroplating, an electric current is sent through an electrolytic bath. At the positive pole (anode) is the metal to be applied, e.g. copper or nickel, at the negative pole (cathode) is the object to be coated. The electric current detaches metal ions from the consumption electrode and deposits them on the target object by reduction. In this way, the object to be finished is evenly coated on all sides with the target metal. The longer the object is in the bath and the higher the electric current, the thicker the metal layer (e.g. copper layer) becomes.
- Paint coatings and dipping method.
- Rotation coating (spin coating).
- Self-growing oxide films.

Microlayers are often produced by combined processes.

One of the best known is the sol–gel process, which is used to produce microporous layers of homogeneous thickness. For the production of coatings, the brines used as coating solutions must first be applied to the substrate surface. In principle, all coating processes that can also be used for paint systems are suitable for this purpose. The coating technology used has a decisive influence on the desired properties such as layer thickness, homogeneity, or low defect density. In sol–gel processes, many process parameters must be considered, such as viscosity of the coating solution, flammability of the solvent, volatility of the solvent, geometry of the substrate (planar, curved, cylindrical, complex shapes), size of the surface, flexibility, and fracture behaviour of the substrate (flat glass or metal strip) and many

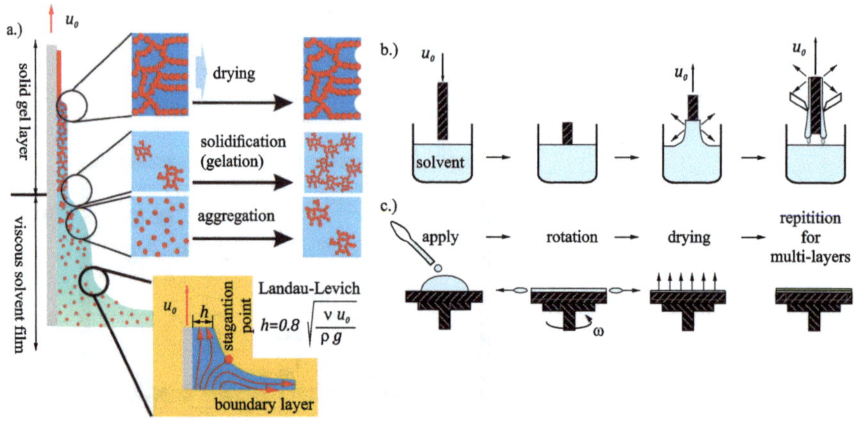

Fig. 3.45 (**a**) Schematic representation of the continuous solidification of a thin surface coating along a vertical flat plate using the sol–gel process; (**b**) dip-coating and (**c**) spin-coating

more. During the entire coating process and the drying of the sol, the hydrolysis and condensation reactions of the precursor molecules (or precursor) used will continue until the aggregation of the sol particles results in a solid gel film. The process is sketched in Fig. 3.45a.

The pre-treatment and cleaning of the substrate is of central importance. Contamination can lead to inadequate wetting with the coating solution. Particulate impurities such as dust also often lead to local defects. In order to improve the adhesion of the coating, metal surfaces are often provided with a thin oxide layer by brief heating (tempering of the surface). The simplest way to evenly wet a surface with the substrate is to immerse and pull it out at a constant speed. This process is known as *dip coating* and is shown in Fig. 3.45b. In *spin coating*, which is shown in Fig. 3.45c, the coating solution is distributed by fast rotation, which is why planar or slightly convex surfaces are particularly suitable. Compared to dip coating, the maximum substrate size is mechanically more limited.

Hybrid polymer coating materials may contain little or no solvent, while the proportion of volatile components in inorganic systems is relatively high. In general, however, the microscopic processes during drying play an important role. During film production, a gel network forms before the solvent has completely evaporated. During further drying from the gel film, liquid menisci are formed which allow capillary forces to act on the pore walls. The magnitude of these forces can be described by a modified form of the Kelvin equation. The forces increase strongly with decreasing meniscus radius and reach high values for gel films with typical pore diameters well below$100 nm$.

(Sakka, 2005) describes that the solid gel layers can only condense perpendicular to the substrate plane during drying. In addition, an unhindered shrinkage of the film in the substrate plane would lead to a "pulling back" of the film from the edges of the substrate and thus to damage. This tendency is counteracted by the adhesion of the layer to the surface. If the tensile stresses occurring in the plane exceed the internal stability of the gel film, drying cracks will occur.

For the production of micro-structured surfaces, the use of mostly multi-stage, more complex processes are required. Structure generation can be done, for example, as in semiconductor manufacturing, using masks, lithography or even electron beam processing.

In holographic lithography, two expanded laser beams are superimposed at a certain angle. The interference image of this superposition is a sinusoidal intensity modulation with which a photoresist (light-sensitive material) is exposed. During development, this photoresist is ablated, creating a surface grid that can be moulded into metals with virtually no structural loss. This impression, in turn, serves as a pattern for the structuring of polymers with which large areas of anti-reflective, transparent surfaces can be produced at low cost.

In this context, a process to produce micro-structured surfaces, the so-called *LIGA* process, which was specially developed in Karlsruhe, will be discussed. The German acronym *LIGA* (stands for the process steps: lithography, electroplating and impression taking) describes a process based on a combination of deep lithography, electroplating and micro impression taking. The *LIGA* method was developed in the early 1980s at the then Nuclear Research Centre Karlsruhe by a team led by Becker and Ehrfeld (Becker, et al., 1982) was developed as part of the development of the separation nozzle process for uranium enrichment in order to be able to produce extremely small separation nozzles. The process is carried out in five steps:

- The starting material is a flat substrate, for example a silicon wafer or a polished disk made of beryllium, copper, titanium, or other materials. The substrate, if not already electrically conductive, is provided with a metallic "seed layer", usually by sputter deposition or evaporation. A thick photo or X-ray sensitive positive resist (often PMMA) is applied to the starting layer Fig. 3.46a).
- The resist is exposed (Fig. 3.46b).
- After developing, a negative mould of the metal structure remains (Fig. 3.46c).
- In an electroplating process, a metal is deposited on the substrate in those areas where the resist has been removed during development (i.e. where the seed layer has been exposed) (Fig. 3.46d).
- After removal of the resist, the substrate, the seed layer, and the electrodeposited metal remain.

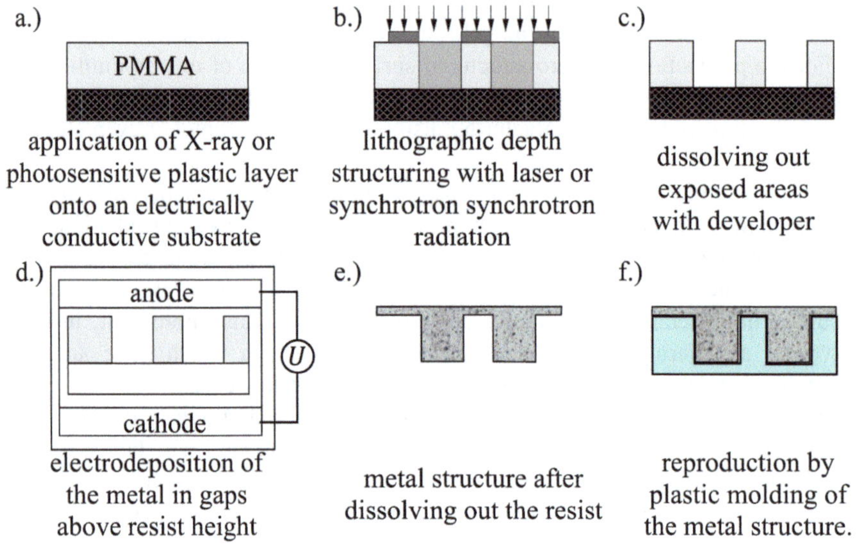

Fig. 3.46 Individual process steps of the LIGA process for the production of microstructured surfaces (Becker, et al., 1982)

Now there are different possibilities for the further procedure:

a. By etching the seed layer (which now functions as a sacrificial layer) and possibly the substrate, (small) metallic components can be produced directly.
b. By further electroplating ("overgrowth") and subsequent removal of substrate and germ layer, a mould insert can be produced from the lithographically produced microstructure, which is then built into a moulding tool, with which in turn the ultimately desired plastic component is moulded, for example by injection moulding or hot stamping (Fig. 3.46e and f).
c. Alternatively, the mould insert can be cut directly from the substrate with the lithographically produced microstructure, for example by spark erosion, and incorporated into a moulding tool.

If the steps of exposure, development and electroplating are repeated several times, more complex structures develop, but these must taper towards the substrate (the wafer), otherwise the component would not be able to be released from the mould. This limits the complexity of the structure to be produced. If the mould produced by *LIGA* is used for injection moulding, a special feature results in contrast to the production of macroscopic parts. There is no need to provide holes for the air present in the mould to escape, as the unevenness of the contact surface between the mould and its counterpart is sufficient for the air to escape when manufacturing components several hundred micrometres in size. Only the holes for feeding the material to be sprayed must be created.

3.3.5 Low and Medium Temperature Absorber Layers in Air/ Vacuum

Selective absorbers for flat-plate and vacuum tube collectors were developed in the 1980's to improve the performance of these collectors. The coatings were mainly based on the Cermet principle. They had to be stable in air for stagnation temperatures up to $250 - 300\ ^\circ C$. Commercially the first products were produced with galvanic deposition. Until about 1995 essentially two fundamentally different types of absorber layer were prepared:

- The first absorber layer called "black chrome" proved to be very stable with good selectivity. The layer is applied by a galvanic bath process, which means inherently wastewater containing heavy metals and especially chromium is produced. The layer thickness of a black chrome layer is about $5 - 10\ \mu m$ and must be removed in the recycling of used absorbers.
- The second layer called "black nickel" is a type of "nickel-pigmented alumina": It has a more limited lifetime due to sensitivity against moisture in the collector. Moreover, the optical selectivity is modest. The production of these layers is, however, less energy-intensive and polluting. A very successful product in the first years of selective collectors was the coated strip-absorber which could be applied from a roll.

In 1995, the TiNOX GmbH was the first company with a highly selective Cermet absorber coating for solar thermal collectors on the market, produced by physical vapour deposition (PVD) in vacuum. The TiNOX absorber layer is composed of several functional layers: a metal substrate is used as a carrier material having good IR reflection. A thin titanium carbide compound serves as an adhesion-promoting layer and diffusion barrier. The absorber layer is made of a titanium compound with oxygen and nitrogen. This patented The Titanium-oxynitride gave the "Ti-NOX" company its name. On top of this layer there is an additional anti-reflection and protective layer of silicon oxide. This minimizes the surface reflection. Due to interference effect the optimized anti-reflection layer reflects small amounts of blue light, so that the absorber appears "blueish". The properties of these commercial products are given amongst others in Table 3.7.

Decisive for selection of a suitable selective absorber layer is the range of the absorber temperature achieved in the collector, and the maximum expected exposure, which is dependent on the location (climate, roof pitch, ...), the type of collector (flat, tubular storage collector) and the intended use (process heat or

Table 3.7 Commercial selective absorber coatings and their specific characteristics

Product	TiNOX	Sunselect	Diamanta	Sunstrip	Gibo
Functional layer	TiN_xO_y	Cermet	–	$Ni-NiO$ black nickel	$Cr-CrO$ black chrome
Substrate	Cu	Cu	$Cu(Al$, stainless steelt	Al	Cu
α_s [/]	0.95	0.95	$0.9-0.96$	$0.92-0.96$	0.96
ε_{IR} [/]	0.05	0.05	$0.02-0.05$	0.1	0.1
$\alpha_s/\varepsilon_{IR}$ [/]	19	19	$12.85-19.2$	$9.2-9.6$	9.6
Production route	PVD	Sputtering	Combined sputtering CVD	Sputtering	Sputtering or electro-plating

power use). Although there is no clear distinction between low and high temperature applications, for the absorbers, an imaginary border at about $250°C$ is considered for example in (Kennedy, 2002) for the low to intermediate temperature range and above $250°C$ for the high temperature range.

For the absorber layers, the temperature range is essential, because chemical reactions like oxidation are strongly temperature-dependent, and because due to different thermophysical properties of substrate material and coating mechanical stresses might develop.

Several physical principles and a wide range of different production methods can be used to produce solar selective absorbers. The selectivity and corresponding optical properties are very important for assessing the relevance for solar thermal applications. However, other factors like stability and cost of production certainly come into consideration. Multilayer cermet absorber coatings seem to be the most adaptable and selective class of high-temperature absorber coatings used up to now. Although the inherent properties of the coatings (materials chosen, number of barrier layers and contact layers, thicknesses, and number of different layers as well as gradients) are important for the optical properties and the stability of a material, other factors add to that. The roughness and purity of the substrate surface, the substrate material and phase transitions have large impacts on the practical applicability of an absorber coating type. Production processes must be stable and controlled to guarantee homogeneous properties and long-term performance of selective absorbers at elevated temperatures.

As a consequence, good laboratory results cannot automatically be transformed into good products. Nevertheless, laboratory research and numerical optimization and modelling of absorber coatings can stimulate development. A large number of historical approaches for selective absorbers in the medium temperature $(150-250°C)$ have been described in a comprehensive review (Kennedy, 2002). The various approaches are listed in the overview Table 3.8. Especially the sol-gel production technology might be interesting commercially, especially for the coating of tubular absorbers for solar process heat collectors.

Table 3.8 Different absorber layer systems for the mid temperature range when operating in a vacuum and air (Kennedy, 2002), $TSSS$=thickness sensitive spectrally selective layer, SS=stainless steel

Material	Substrate	Fabrication process	Absorptance α_s	Emittance $\varepsilon(100°C)$	Stability limit [°C] vacuum air	Comm. product
$TSSS$	Al	Painted	0.92	0.015	<135	Solaris Select
PbS	Al	Evaporation pyrolysis	0.93 – 0.99	0.1 – 0.21	~200	
$NiCrO_x$	SS Al polyamide	Sputtering	0.8 0.92 – 0.93	0.14 0.06	<200 <200	
coloured SS	SS	Chem. convection	0.62 – 0.93	0.1	<200	SEL INCO
Black Nickel NiS/ZnS $NiSn$	Ni / steel Cu / steel	El.-deposition sol–gel	0.88–0.96 0.92–0.98	0.03–0.1 0.08–0.25	<200 300	Maxorb black crystal
Graphite films $aC:H/Cr$ $A-C:H$	Cu Al	MF-pulse PVD/PEVD	0.92 0.876	0.025 0.061	250 250	
Ge in Si Si PbS	Cu, SS	Painted	0.91 0.83 0.96	0.7 0.7 0.7	300	
Ag diel	Al	Sol–gel			300	
Black copper $Cu/Cu_2O/Cu$	Cu	El.-deposition	0.97 - 0.98	0.02	370250	
Black chrome Cr/Cr_2O_3 Mo/Cr_2O_3	$Ni-Cu$ Cu / steel	El.-deposition	0.97 0.95	0.09 < 0.3	400 3 50 425	MTI, Chrome Coat, Energie Solaire
TiN_xO_y	Cu	El. beam vapor deposition / sputtering	0.92	0.06	400	TiNOX, Almeco
$CuFeMnO_4/Si$	glass,Si	Sol–gel	0.6	0.29–0.39		
Cr, Fe, Mo, SS, Ta, Ti, Wsilicates	Cu	DC reactive sputtering	0.76 – 0.82 0.81 – 0.86	0.02 – 0.3	400 250	

(continued)

Table 3.8 (continued)

Material	Substrate	Fabrication process	Absorptance α_s	Emittance $\varepsilon(100°C)$	Stability limit [°C] vacuum air	Comm. product
$Cr, Fe, Mo,$ $SS, Ta,$ $Ti, WCarbide$	Cu	DC reactive sputtering	0.76 – 0.81 0.84 – 0.90	0.02 0.035 – 0.06	400 250	
$Ni - NiO$	Al	Reactive sputtering	0.96	0.10	300	Sunstrip
Ni-pigmented Al_2O_3	Al	Anodization	0.85 – 0.97	0.08–0.21	400	Showa

3.3.6 Absorber Layers for High Temperature Applications

In concentrating systems higher temperatures are achieved. Here, it is very difficult to produce absorber coatings stable in air. Therefore, the development of vacuum receivers came first, which also had the advantage of reducing even further the heat losses when compared to a system filled with air. However, the tightness of the vacuum tube over many years of lifetime is another challenge. Especially for organic heat transfer media this is an issue.

Historically the first commercially successful providers of vacuum receivers and corresponding absorber coatings for parabolic trough collectors after the pioneers LUZ were Solel Solar Systems, Israel, and Schott Solar CSP GmbH, Germany. The absorber tube is covered by a borosilicate glass tube to keep the vacuum and to reduce the heat losses. Polished stainless steel tubes (mean roughness index $R_a < 0.2\mu m$) are heated in vacuum for 10 minutes at a temperature of $600°C$. Then the coating system consisting of Al_2O_3 diffusion barrier, the $Mo-$ IR-mirror, the Mo/Al_2O_3-cermet and the SiO_2 AR layer are deposited on the tube (Lanxner and Elgat, 1990). Solel and Schott for a long time dominated the market with their coatings stable up to about $500°C$.

In 2005 in Italy a DC (Direct Current) sputtered absorber system of a $500nm$ thick tungsten (W) or ZrN IR-mirror, a cermet of TiN_x, ZrN_x or HfN_x in AlN. The AR-layer is made from AlN or Al_2O_3. The absorber coating is stable in vacuum up to $580°C$ and reaches $\alpha > 95\%$ and $\varepsilon(580°C) < 12\%$. It is specially developed and used at the moment for receivers with molten salt as a heat transfer fluid (HTF) exceeding the temperature of the HTF thermo-oil at $390°C$.

The market of the suppliers of vacuum receiver has undergone major changes in recent years. While the Spanish company Rioglass now has taken over the technologies of Solel and Schott, in China other producers developed. Table 3.9 provides an overview of the properties of some commercial products.

An overview of various laboratory development of high-temperature selective coatings stable in vacuum or even in air is given in Table 3.10. Many systems use the expensive metals platinum (Pt), Molybdenum (Mo) or Tungsten (W) as

Table 3.9 Comparison of absorptance and emittance of commercial high-temperature vacuum tube receiver coatings

Product	Rioglass Solar S.A.U (Rioglass Solar SA, 2022)	Archimede Solar Energy S.r.l (Tagle-Salazar and Nigam, 2020)	Royal Tech CSP Limited	Himin Solar Co. Limited (Hi-min SOLAR, 2022)
α $AM1.5$ direct (ASTM)	0.9452	0.95	0.93	0.96
$\varepsilon(T = 400°C)$	0.0953	0.073	0.098	0.11
τ ASTM	0.96	0.965	0.961	0.96
Max. operation temperature [$°C$]	350	580	400	400

infrared mirror and Pt in Al_2O_3 as Cermet. Material and the slow high frequency sputtering process make these designs costly. A commercially successful development could be based on silver (Ag) in a DC sputtering process using polished stainless steel as a substrate. Theoretically this would lead to lower emissivity at high temperatures and therefore better solar thermal efficiencies. But a problem of silver is the agglomeration in vacuum, but that can be overcome by a special multi-layer design using barrier layers.

For use at high temperatures in air no selective absorber coatings are commercially available. There are concepts for absorbers for temperatures above 350°C at laboratory scale. One of the most stable material candidates for absorber systems is platinum (Pt). Many authors describe it as a stable IR-mirror, and embedded in Al_2O_3 it forms a stable cermet (Thornton et al., 1980; Vien et al., 1985; and Sella, et al., 1989).

The most extensive investigations describe the influence of different process parameters on the stability of the system $Pt-$ IR-mirror, Pt/Al_2O_3 -cermet and Al_2O_3 AR-layer on glass and on stainless steel with an Al_2O_3 diffusion barrier. Sputtering with RF (Radio Frequency) produces better results than DC. Also sputtering with higher substrate temperatures seems to be beneficial for stability and low emittance. However, platinum is an extremely expensive material, and secondly RF sputtering is slow, which makes a commercial production process expensive. Also, the substrate heating requires longer production times and additional investment costs. Highly alloyed substrates are necessary which are too expensive for receiver tubes.

Therefore, less expensive materials and processes are still required. Substrates used should be only stainless steel tubes and not extremely expensive materials, at least for linear reflectors. The suitability of many other materials Cr, Mo, Ni, and Ta have therefore been investigated as IR-Mirrors. One problem is the relatively fast oxidation in air for these materials, although embedding in the cermet layer helps. Silver (Ag) is an excellent IR-mirror and on the other hand has, as a noble metal, a relatively low tendency for oxidation. The main degradation mechanism is the formation of Ag_2S through reaction with H_2S in air, see (Graedel,

Table 3.10 Common absorber layer systems for high temperature applications in air and vacuum adapted from (Kennedy, 2002), SS=stainless steel

Material	Substrate	Fabrication process	Absorpt α_s	Emittance $\varepsilon(100°C)$	Stability limit $[°C]$ vacuum	Stability lim $[°C]$ air
$Ni - Al_2O_3$	$Mo-Ni-SS$	RF sputtering	0.94	0.07	500	350 – 400
$SiO_2 - AR$	$Ni - Al$	RF sputtering	0.94	0.04	350 – 500	600
$Co - Al_2O_3$	$steel$	RF sputtering	0.96	0.16(350)	500	600
$Mo - Al_2O_3$	Cu	CVD	0.97 – 0.98	0.1 – 0.07(400)		
$W - Al_2O_3$		RF sputtering	0.85	0.04		
$Pt - Al_2O_3$			0.90 – 0.98	0.08		
$Al_2O_3 - Pt - Al_2O_3$						
$Mo - Al_2O_3$	Cu	DC sputtering	0.96	0.06(350)	500	
$SS - AlN$			0.95	0.10(350)	350 – 500	
$Mo - AlN$			0.92 – 0.94	0.08 – 0.1(350)	500	
$W - AlN$					500	
Quasicrystals	Cu, Si		0.90	0025	500	400
Multi-layer cermet			0.86 – 0.92	0.031 – 0.05	550	
$Si_3N_4/Si - Ge/Ag$	SS, Al	CVD	0.89	0.0389 (300)0.0545 (500)		650(He)
$Ni - -SiO_2$ $Cr - SiO.$	Al, Cu	DC reactive sputtering	0 .90 – 0.96.	0.03 – 0.14		400 – 800(Ar)
$Al - AlN_x - AlN$	SS	DC reactive sputtering	0.97	0.10	500	
CuO	Cu, SS	Electroplating	0.91	0.18	700	< 400
$Ag/CuO/Rh_2O_3/CeO_2$	SS	Organ.-metal.	0.90	0.10	775	500
$CeO_2/Ag/Pt/CuO,$		spray	0.86 – 0.88	0.1	700	550
$Rh/Rh_2O_3/Ag/Pt$			0.88 – 0.92	0.06 – 0.12		500
$CeO_2/CuO/CoO, Mn_2O_3/Pt$						

(continued)

Table 3.10 (continued)

Material	Substrate	Fabrication process	Absorpt α_s	Emittance $\varepsilon(100°C)$	Stability limit [°C] vacuum	Stability lim [°C] air
Black cobalt			$0.96-0.92$	$0.71-0.017$		$400-650$
Co_3O_4/Co			0.95	0.10		
$Ni-Co_3O_4/Co$						
Black molybdenum $Mo-MoO_2$		CVD	0.94	0.3(500)	500	350
Black tungsten $W-WO_x$		CVD	0.83	0.15	800	
$Au-TiO_2$	SS	Sol–Gel	0.85	0.01(400)	> 500	400
$Au-MgO$	Mo, SS	RF sputtering	$0.90-0.93$	$0.04-0.1$		
ZrC_xN_y	Al		0.85	0.074(325)	600	125
$Al_2O_3/ZrC_xN_y/Ag$	S S		0.91	0.05(325)	700	175
$ZrO_x/ZrC_x/Zr$			0.90	0.05(20)	700	
TiN	Cu, Al	DC reactive sputtering	0.80	$14-0.4$		500
$Ti_{1-x}Al_xN$						$750-900$
$M_bO_c+M'Fe_2O_4$	$Ni-Mo$ alloy	Brushing, plasma	> 0.90	> 0.45		700 1060

(continued)

Table 3.10 (continued)

Material	Substrate	Fabrication process	Absorpt α_s	Emittance $\varepsilon(100°C)$	Stability limit $[°C]$ vacuum	Stability lim $[°C]$ air
VB_2, NbB_2, TaB_2 TiB_2, ZrB_2LaB_6, $WSi_2, TiSi_2$, $Si_3N_4, AR - ZrB2$	Glass ZrB_2	DC reactive sputtering CVD	0.99 $0.88 - 0.93$	$0.95 - 0.97$ $0.08 - 0.12$	$2300 - 3040$ $(Melt)$	500
C-textured Cu	Cu		0.9	0.04(20)	400	< 300
textured Ni	Ni		0.92 ± 0.2	0.09 ± 0.02	> 440	> 500
textured SS	SS		0.93 ± 0.02	0.22 ± 0.02	550	> 600
textured Cr	Cr		$0.8 - 0.9$	$0.1 - 0.3$		800
W whiskers,Mo, Rh, Pt, W, HfC, Au			0.98	>0.26		
NiO, CoO_3						

1992). Also, the thin layers tend to agglomerate for temperatures above $200°C$, cf. (Hollingsworth Smith and Gurev, 1977). Therefore, protection and adhesion layers are needed to stabilize Ag in the system. Using that approach absorber coatings on stainless steel tubes with $\alpha > 95\%$ and $\varepsilon(450°C) < 12\%$ could be produced but do not have long-term stability for testing temperatures of $500°C$ (Hildebrandt, 2009). However, based on the emittance of polished stainless steel a CrO_x cermet in combination with a $SiO_x - AR$ layer was stable $3000h$ at $500°C$. The optical properties reached for such a system were $\alpha = 94\%$ and $\varepsilon(450°C) = 18\%$ (Hildebrandt, 2009). For solar towers even, higher temperatures have to be reached, but the requirement on selectivity is lower due to the higher concentration. Here absorption is the key parameter.

High solar absorptance (non-selective) coatings can be easily applied by spray coating. The coating equipment is cheap and the possibility of being able to recoat the receiver on the tower after a performance loss due to degradation is a big advantage. During the coating process attention must be paid to the homogeneity of application as the rather low heat conductivity of the thick coatings (around $100\mu m$) might lead to hot spots and result in a degradation of the receiver. Currently developments are ongoing.

3.3.7 Degradation Mechanisms of Selective Absorber Layers

Absorber coatings of the various sorts that have been discussed are precisely formulated to achieve the optical properties that are desired. They must operate at elevated temperatures, subject to temperature fluctuations, and need to achieve long lifetimes to be practically useful.

Surface temperatures experienced during operation which exceed the production temperatures of the absorber systems may lead to structural reordering within the materials, to diffusion and redox reactions. The optical properties will usually change as a consequence. High temperatures above $400°C$ pose a special challenge because kinetics for most processes is much faster at higher temperatures. A short overview of degradation processes is given here based on reference (Hildebrandt, 2009).

3.3.7.1 Diffusion Processes

There are different kinds of diffusion processes that can contribute to performance degradation. Diffusion of atoms along interstitial lattice sites or lattice vacancies plays a minor role if the material in question exhibits lattice disorder like grain boundaries or dislocations. The diffusion along these defects is several orders of magnitude faster than the first process and is also called "short circuit diffusion" (Heumann and Mehrer, 1992). Sputtered layers may also have a high porosity. The diffusion of molecules, for example of oxygen from air, may take place.

Diffusion processes may take place between substrate and absorber systems or within the individual layers. In cermet layers even, the diffusion within the cermet itself changes the gradient of constituents and therefore the optical properties of the system.

To prevent or at least substantially reduce the diffusion of atoms e.g. from the substrate (Fe, C) into the IR-mirror layers, barrier layers can be prepared. Oxides, for example Al_2O_3 have been used for these layers. High density of the barrier layer and continuous layer formation is important. These layers must cope also with thermal expansion and stresses of the substrates because cracks will destroy the barrier functionality. A different approach is taking a thick IR- mirror metal layer as barrier.

3.3.7.2 Oxidation

The oxygen in air is obviously a threat for degradation due to oxidation in high temperature absorbers operated in air. Metallic layers of the IR-mirror and the metallic particles in the cermet layer are subject to oxidation. Barrier layers between surrounding air and the absorber system must prevent oxygen diffusion without destroying the desired optical properties of the absorber system.

3.3.7.3 Redox Reactions

Redox reactions are important in cermet layers even in vacuum. During the production of a reactively sputtered cermet Me_1/Me_2-oxide,[1] e.g. a Mo/Al_2O_3-cermet, the metallic constituent Me_1 (Mo) usually is also partially oxidized whereas the second metal Me_2 (Al) is not completely oxidized. During a tempering process redox reactions drive the system towards chemical equilibrium, i.e. the oxidized Mo will be reduced and give the oxygen to the metallic aluminium. The size and shape of metallic particles in a cermet might also change. All these processes provoke strong spectral changes in the reflection spectrum and hence a change in solar absorptance and thermal emittance.

3.3.7.4 Thermo-Mechanical Stresses

The temperature differences occurring within high temperature receivers and / or, unequal thermal expansion coefficients of adjacent layers may lead to thermally induced stresses on the individual layers. This in turn may induce flaking or chipping off of layers. Cracks develop and barrier layers are destroyed locally. The problems may be ameliorated by introducing special adhesion layers between problematic materials.

3.3.7.5 Other Environmental Stresses

Apart from the mechanisms described above, a number of other environmental stresses may deteriorate the optical properties of an absorber system reversibly or irreversibly. For instance, for non-hermetically sealed receivers open to air, humidity and even dust, several reactions may occur. For example, dust on a hot receiver

[1] Me_1 And Me_2 references general combinations of metal 1 and metal 2.

may fasten on the surface irreversibly. Condensation of humidity during cold periods (night-time) in combination with gases like sulphur or carbon dioxide may produce acids attacking surface layers of the absorber. Sulphur itself can be a problem e.g. for silver IR-mirrors, if the oxide is reduced. For plants built in the proximity of an ocean, salt may react aggressive with both steel and absorber system.

Although accelerated testing (Brunold, Frei, Carlsson, Möller, and Köhl, 2000a, 2000b) may try to incorporate all conceivable effects, there will be always a risk that the exposure in the real environment will have additional effects. The conditions at a specific location may be different from what is expected in a standard atmosphere, and interactions between different mechanisms may not be the same as in a test. Nevertheless, it is important to develop methods for estimating long-term stability and predicting service lifetime as far as possible to reduce the risks for degradation and failure.

In addition to the initial efficiency, long-term stability is an important requirement for absorber coatings. Complex designs of absorber coating systems are frequently susceptible to changes upon exposure to elevated temperature. However, these changes may even help to improve the performance of a freshly produced absorber system. Tempering processes may improve the optical properties but should stop after a short period of time. Slow continuous changes on the other hand are likely to lead to long-term performance reduction and must be quantified and eliminated.

In order to quantify "degradation" of an absorber system a so-called Performance Criterion (PC) is required. For flat-plate collector selective absorber testing (i.e., non-concentrating, $1-2X$ sunlight intensity) such a criterion has been developed (Brunold, Frei, Carlsson, Möller, and Köhl, 2000a, 2000b). The PC describes the influence in the change of solar absorption ($\Delta\alpha_s$) and emittance ($\Delta\varepsilon$) on the solar fraction:

$$PC = -\Delta\alpha_s + 0.25 \cdot \Delta\varepsilon \leq 0.05. \tag{3.64}$$

It provides a single quantifiable parameter for maximum acceptable degradation. Service lifetime testing for this criterion is performed by exposing the absorber coatings of flat-plate collectors for $200h$ at $250°C$. If the material survives, it is then exposed for $75h$ at $300°C$, followed by $600h$ at $40°C/95\%$ relative humidity (RH), then $85h$ at $60°C/95\%RH$, for more details see (Brunold, Frei, Carlsson, Möller, and Köhl, 2000a, 2000b). After exposure testing, the emittance is typically measured at 100 °C. A similar criterion has been developed for testing the service lifetime of high-temperature absorbers for CSP applications. It is obvious that for concentrating collectors due to the concentration a change in the absorptance is even more important than for non-concentrating collectors. On the other hand, emittance at higher temperatures is more important due to the Stefan-Boltzmann law. Therefore, we would suggest as a first approach to use a PC for concentrating collectors:

$$PC = -\Delta\alpha_S + \frac{0.25 \cdot f}{C} \cdot \Delta\varepsilon \leq 0.05 \text{ with } f = \frac{\sigma \cdot \left(T_{\text{op,CSP}}^4 - T_{\text{amb}}^4\right)}{\sigma \cdot \left(T_{\text{op,FPC}}^4 - T_{\text{amb}}^4\right)} \text{and}$$
$$T_{op,CSP} = 273.2 + 340\ K;\ T_{op,FPC} = 273.2 + 60\ K;\ T_{amb} = 273.2 + 25\ K. \tag{3.65}$$

where C is the concentration factor, f is a factor describing the higher operating temperatures compared to flat-plate hot water collectors, T_{amb} is ambient temperature; T_{op} is the operating temperature of the solar system and the subscripts FPC denote a flat-plate collector and CSP a concentrating collector for solar thermal power plant.

Thermal stability is sometimes determined by the thermal properties of the individual materials or the processing temperature parameters. Durability or thermal stability is typically tested by heating the selective coating, typically in a vacuum oven or in air at elevated temperatures, for a relatively short duration (thousands of hours) compared to the desired lifetime ($25 - 30 years$). This procedure of accelerated indoor testing may pose problems if cascaded processes and interactions during exposure occur (Kennedy, 2002). Degradation of high-temperature absorbers usually changes the reflectance spectrum significantly and causes increasing emittance; therefore, optical properties are sensitive indicators to monitor degradation with exposure.

3.4 Optical Concentration

3.4.1 Motivation for Optical Concentration

The solar collector energy balance is strongly dependent on the absorber or fluid temperatures reached. The thermal loss as compared to an ordinary black absorber plate can be reduced by many means, but in fact they limit the maximum temperatures which can be achieved. Concentration of solar radiation is an elegant possibility to extend the temperature range of solar collectors. When the low power density of solar radiation with maximum around $1000 W/m^2$ can be focused from a large collecting area, the aperture, to a much smaller absorber area, then the losses for a specified absorber temperature can be reduced dramatically. And although the thermal losses increase non-linearly with absorber temperature, the break-even temperature where losses and gains for the small area are equal (for collectors usually called stagnation temperature), increases also very much. Absorber temperatures of more than $1000°C$ can be achieved in practical systems, and very often the limitation is rather the stability of the collector material rather than the energy balance. This is the reason why concentration of solar radiation is such a powerful tool.

The ratio between collecting aperture A_a and reradiating receiver area A_r is called the geometric concentration ratio C_g, independently of the optical efficiency of the optical concentrator system trying to achieve that:

$$C_g = \frac{A_a}{A_r}. \tag{3.66}$$

Optical efficiency is influenced by the optical properties of real materials like mirrors or lenses which do not have 100% reflectance or transmittance, respectively, even for perfect concentrator geometry. On top of that the shapes of real optical concentrators deviate to some extent from the ideal geometry, on the nano level due to surface irregularities up to the macro level due to mechanical stresses, imperfect support structures, thermal expansion, wind load sensitivity and many other factors. In some applications optical efficiency also might be compromised by design, for example if a homogeneous irradiation distribution on the receiver area is a target. Also, economic considerations play a role, and therefore often concentrator designs lead to a non-perfect concentration with some radiation not directed towards the receiver area. The optical efficiency (η_{opt}) is a measure for the irradiation collected by the aperture compared with the radiation reaching the receiver and is thus defined by.

$$\eta_{opt} = \frac{\iint^{A_a} I_0(x, y)dxdy}{\iint^{A_r} I(x, y)dxdy} = \frac{A_a \bullet I_0}{\iint^{A_r} I(x, y)dxdy}. \tag{3.67}$$

The intercept factor, which also frequently used especially for imaging concentrators (the ones which have a focal point or line as the image of the sun), is very similar to that definition, as it is a measure for the irradiation redirected towards the receiver and reaching the receiver. In the case of a flat receiver area one could write the intercept factor γ as:

$$\gamma = \frac{\iint_{-\infty}^{+\infty} I(x, y)dxdy}{\iint^{A_r} I(x, y)dxdy}. \tag{3.68}$$

The optical efficiency relates the received radiation to the input of the optical concentrator system, the intercept to the (redirected) output of the concentrator. When we look at a heliostat for example, the average reflectance of the heliostat surface should be multiplied by γ to get optical efficiency η_{opt}.

An optical concentration ratio is defined for any part of the receiver area as the intensity at the receiver $I(x, y)$ divided by the incoming solar intensity I_0 hitting the aperture. The optical concentration is often reported as a factor X or as X suns. It may be specified as peak optical concentration for the focal area or averaged over the whole receiver:

$$C_{opt} = \frac{I_r}{I_0}. \tag{3.69}$$

The average optical concentration ratio is obviously related to the intercept factor by:

$$\overline{C_{opt}} = \eta_{opt} \cdot C_g. \tag{3.70}$$

Concentrating systems only make use of direct beam radiation. They do not collect diffuse radiation, except some concentrators with rather small concentration ratio in the range of 1.3 to 5. The reason is that the acceptance angle due to fundamental physics is becoming smaller when the concentration ratio increases. This leads to a smaller and smaller view angle of the concentrator system to the sky, resulting in negligible diffuse contribution. Therefore, generally a part of the global irradiation is not used in concentrating systems. The higher temperatures possible can be offset by small energy gains in climates with little direct beam radiation. This is often a reason for selecting a concentrating or a non-concentrating system design for solar thermal applications. For very high point concentrating systems one has even to consider that a part of the circum-solar radiation (CSR) may be cut out due to a small acceptance angle. So, these limits to concentration must be understood and considered in the system design and selection.

3.4.2 Limits of Optical Concentration of Solar Radiation

The concentration ratio C_g determines the absorber temperature and thus also the long-term performance of the solar thermal system. It is easy to see, through simple considerations, that there are some limits to the concentration ratio. It's these:

- Divergence of solar radiation.
- Technical design of the imaging optics (focal point or focal surface design).
- Precision of the image (surface quality, cleanliness, temperature, …).

In addition, there are also material-technical questions concerning the mirrors or lenses as well as the type of tracking and its quality.

The sun produces direct divergent irradiation on the Earth's surface where the solar radiation comes within a cone divergence angle θ_s as illustrated in Fig. 3.47, the angle of view of the sun disk. This has a consequence for the maximum optical concentration possible. Assume a two-dimensional, perfectly concentrating system, which concentrates the incident radiation with a circular aperture of the surface A_a onto a receiver (or absorber) of the surface A_r. At the same time, both surfaces should be ideally perpendicular to the direction of irradiation. According

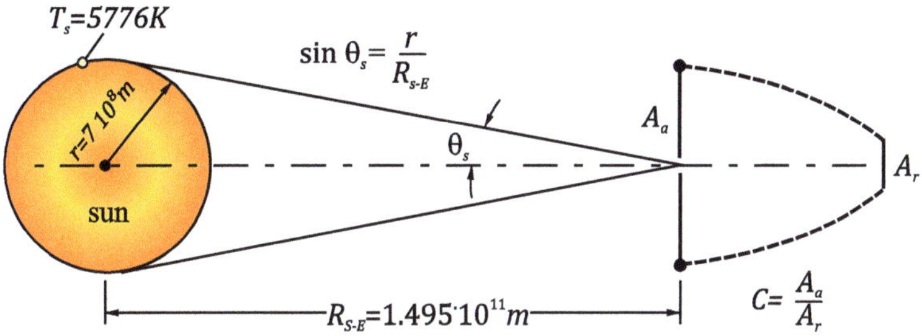

Fig. 3.47 Derivation of the angular distribution of the solar radiation

to the sketch, the solar radiation power $\dot{Q}_{S-a}$. indicated in the Eq. (3.71) falls on the aperture and further on the receiver.

$$\dot{Q}_{S-a} = A_a \frac{r^2}{R_{s-E}^2} \sigma T_s^4, \tag{3.71}$$

where A_a is the aperture area, R_{s-E} the average distance of the earth from the sun with $R_{s-E} = 1.495 \cdot 10^{11}\, km$, r is the radius of the sun with $r = 7 \cdot 10^8\, m$ and σ the Stefan-Boltzmann constant ($\sigma = 5.67 \cdot 10^{-8} W/(m^2 K^4)$). T_s describes the surface temperature of the sun. The angle of view of the sun θ_s directly results as:

$$\sin\theta_S = \frac{r}{R_{s-E}} \text{ with } \theta_s = 0.27° = 16'. \tag{3.72}$$

Assuming that the receiver is a "black radiator", the following relationship results for the radiation power of the absorber (receiver) in the direction of the sun $\dot{Q}_{r-s}$:

$$\dot{Q}_{r-s} = A_r \cdot \sigma \cdot T_r^4 \cdot \varepsilon_{r-s}. \tag{3.73}$$

In Eq. (3.73) T_r describes the temperature of the absorber (receiver). ε_{r-s} is the emission coefficient between absorber (receiver) and sun reflects the radiation exchange between the receiver and the sun. The maximum emission can be the radiation of a black body, i.e. the maximum emission coefficient is $\varepsilon_{r-s} = 1$. According to the second law of thermodynamics, in a borderline case the receiver temperature (T_r) can be at most the radiation temperature of the sun, i.e. in a limiting case $T_r = T_s$. It follows directly from this:

$$\dot{Q}_{s-E} = \dot{Q}_{r-s} \rightarrow A_r \cdot T_r^4 = A_a \cdot \left(\frac{r}{R_{S-E}}\right)^2 T_S^4. \tag{3.74}$$

With the result of the viewing angle considerations resulting in Eq. (3.75) one obtains a maximum possible concentration ratio for C for the two-dimensional case of:

$$C_{\max,2-\dim} = \frac{A_a}{A_r} = \frac{R^2_{s-E}}{r^2} = \frac{1}{\sin^2\theta_s}. \tag{3.75}$$

Assuming the maximum angle on the aperture of $\theta_s = 0.27° = 16'$ the maximum concentration ratio of a two-dimensional concentrator is $C_{\max,2-\dim}$:

$$C_{\max,2-\dim} \approx 4.5613 \cdot 10^4. \tag{3.76}$$

For a linear concentrating mirror system that is one-dimensional, it can be shown with the help of similar considerations that the maximum achievable one-dimensional concentration ratio $C_{\max,1-\dim}$ is limited to the following value:

$$C_{\max,1-\dim} = \frac{1}{\sin\theta_s} \approx 213. \tag{3.77}$$

In practice, however, such values cannot be achieved due to surface defects and adjustment errors of the mirrors. Higher concentration ratios can be achieved with spherically symmetrical mirrors than with cylinder-symmetrical ones. Paraboloids with a C_g -value of $C = 5 \cdot 10^3 - 8 \cdot 10^3$ have been realized with great effort; however, the values are usually around $C_g \approx 10^3$. In line-focusing systems, technical values around 30 are achieved.

Another important parameter for concentrating solar systems is the so-called acceptance angle θ_A. In a real solar receiver, the position of the sun and thus the solar elevation angle α_s as well as the direction of irradiation and thus the azimuth angle γ_s changes due to the sun's daily motion. The acceptance angle θ_a is now defined as the angle at which the solar radiation is just about geometrically able to hit the absorber. (This assumption implies that a complete reflection occurs on the contour surface ($\rho = 1$) and the sun is a point source, which is also idealized). In fact, the solar disk is seen on Earth at an angle of $(2 \cdot \theta_s) \approx 0.54°$ (see Fig. 3.47).

If the acceptance angle θ_A now decreases below the viewing angle θ_s of the solar disk S, the radiation power emitted by the sun can no longer be fully utilized. This effect is known as radiation loss. If the angle of acceptance corresponds to the angle of view of the sun ($\theta_a = \theta_s$), maximum use is made of the irradiation power. From this, a maximum possible geometric concentration ratio C_g can be derived, as described by (Kreider, 1979). Through geometric considerations, it is easy to consider that with smaller absorber dimensions, the acceptance angle θ_a of a collector also decreases monotonically. However, a decreasing acceptance angle means an increasing concentration ratio and thus increasing absorber temperatures when the collector is ideally adjusted towards the sun.

The maximum theoretical concentration as discussed above and given in Eqs. (3.76) and (3.77) can be derived also from the optical theorem of conservation of etendue in a perfect optical system. In simple terms, the Étendue describes how a bundle of rays expands in geometrical optics. To calculate this, the cross-section and the solid angle are required, as well as the direction vector of the radiation. For an infinitesimal area immersed in a medium with refractive index n the Etendue is given by

$$dG = n^2 \cos\phi \, dA \, d\Omega, \tag{3.78}$$

in which dA is the surface, $d\Omega$ is the solid angle extended by the ray bundle with a propagation at angle ϕ with respect to the surface normal. This may be interpreted also a volume in phase space $d\phi d\theta$ in Hamiltonian optics. If a light cone has a divergence angle of θ_{in}, the integration over the solid angle and a finite area A yield

$$G_{in} = \pi n^2 \sin^2 \theta_{in} \, \mathrm{A}. \tag{3.79}$$

The quantity $(n\sin\theta_{in})$ in optics is called the numerical aperture NA. The Etendue is conserved not only in free space but also when refraction or reflection occurs. Consider now a concentrator system immersed in a medium with refractive index n_2 and focusing all radiation on a receiver area A_r and incoming radiation on an aperture A_a coming within a cone of θ_1 in a medium with refractive index n_1. Then the conservation of Etendue requires that $G_1 = G_2$ which implies that

$$\pi \, n_1^2 \sin^2 \theta_{in} \, \mathrm{A}_a = \pi \, n_2^2 \sin^2 \theta_{out} \, \mathrm{A}_r. \tag{3.80}$$

The situation is sketched in Fig. 3.48. The concentration C is given then by

$$C = \frac{\mathrm{A}_a}{\mathrm{A}_r} = \frac{{n_2}^2 \sin^2\theta_{out}}{{n_1}^2 \sin^2\theta_{in}}, \tag{3.81}$$

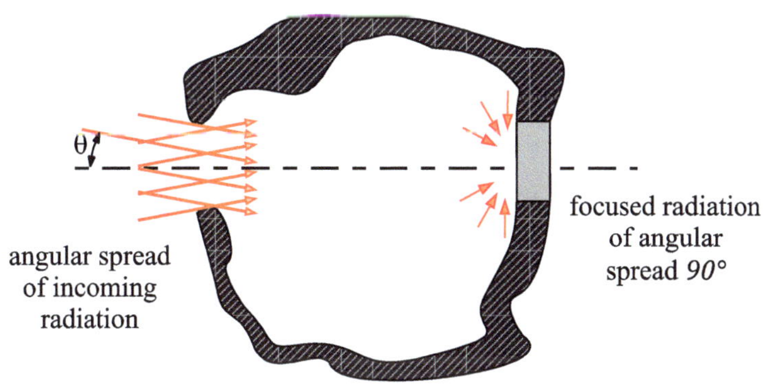

Fig. 3.48 Maximum concentration from Etendue conservation (considering media with refractive index n_i)

and the maximum concentration for incoming solar radiation within an angle θ_s in air ($n_1 = 1$) is reached with $\theta_{out} = 90°$ leading to

$$C_{\max,2-\dim} = \frac{A_a}{A_r} = \frac{n^2}{\sin^2\theta_S}. \tag{3.82}$$

The difference of Eqs. (3.82) and (3.75) is the factor n_2 which gives a larger concentration when the optical concentrator is immersed in a medium with contact to the receiver. In optical microscopy this is used when the lens is immersed in a liquid in which the object is immersed thus providing a larger numerical aperture. In concentrator optics a secondary concentrator made of glass or acrylic glass connected to the receiver surface would utilize this effect.

As can be seen from burning mirror experiments, the temperature on the imaging surface increases with increasing concentration ratio. To derive the maximum absorber temperature, therefore, the energy incident on the absorber is first determined, which results from the product of the aperture area A_a and the radiant power on the aperture ($\dot{Q}_{s-E}$). If the relation (3.75) is inserted into Eq. (3.74), the radiant power incident on the aperture is $\dot{Q}_{s-E}$:

$$Q_{s-E} = A_a \cdot \sigma \cdot T_S^4 \cdot \frac{1}{C_{\max,2-\dim}}. \tag{3.83}$$

Since the absorber temperatures also increase with increasing concentration ratio, radiation losses make up an ever larger share of the total losses and dominate the loss share at high temperatures. If only radiation losses of the absorber are considered as losses, the greatest losses are obtained when the absorber is a blackbody radiator, i.e. the emission coefficient $\varepsilon_{a-S} = 1$. Thus, the power loss $\dot{Q}_{a-s}$ is given by the Eq. (3.72). The highest absorber (receiver) temperatures $T_{r,max}$ are obtained when the gains of irradiation correspond to the losses to the outside, i.e. $\dot{Q}_{s-E} = \dot{Q}_{a-s}$. Equating the gains with the losses physically means that the collector efficiency is zero. The highest possible absorber temperature $T_{r,max}$ thus is:

$$T_{r,max} = T_s \cdot \sqrt[4]{\frac{C}{C_{\max,2-\dim}}}, \tag{3.84}$$

where $T_{r,max}$ in this sense refers to the stagnation temperature of a concentrating system.

3.4.3 Reflector Materials

Solar thermal concentrators are in most cases constructed with reflecting materials and not with glass or polymeric lenses. One of the reasons is certainly that over the wide solar wavelength range of solar thermal collectors the dispersion of dielectric materials is substantial, leading to different focal lengths when using dielectric materials. This is not the case for reflector materials as the reflection law is independent on wavelength.

The basis of reflectors are metallic layers. The ones with the highest solar reflectance are using Aluminium (Al) or Silver (Ag). Contrary to dielectric materials the reflectivity of metals is even high for very thin layers. This is cause by the substantial imaginary part of the complex index of refraction as introduced in Eq. (3.9). For calculation of the reflectivities of layers the Fresnel laws (3.22) and (3.23) as well as Snell's law (3.20) have to be used with complex mathematics. For example, a complex angle of refracted electromagnetic waves is given by

$$\sin\theta_1 = (n - ik)\sin\theta_2. \quad (3.85)$$

For the common solar reflector materials aluminium and silver Fig. 3.49 shows the reflectance R_p and R_s for thick layers exhibiting no interference effects as a function of angle of incidence θ_1.

The optical properties of media described by complex indices of refraction vary with wavelength λ due to specific excitations of the material involved. Within the solar spectral range these excitations are typically electronic excitations leading to resonant absorption in the UV, metallic excitations (e.g. plasmon excitations) in the Vis–NIR range and molecular excitations (e.g. vibration absorption bands) in the NIR and IR. With the spectral variation of the complex index of refraction, transmittance and reflectance are also a function of wavelength. In Fig. 3.50 the reflectance $R(\lambda)$ at normal incidence ($\theta = 0°$) is plotted versus the wavelength for a thick layer of pure silver and aluminium (Palik, 1985). The reflectance of silver increases steeply from UV to an almost constant high reflectance over the

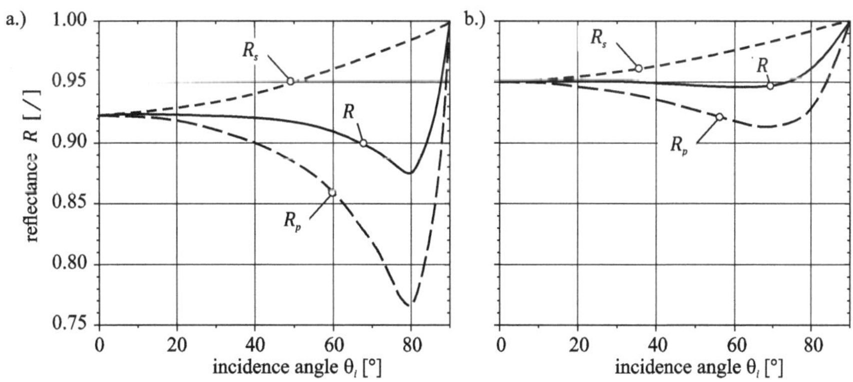

Fig. 3.49 Simulated reflectance R_p, R_s and mixed reflectance R at metallic surfaces as function of the incidence angle θ_1 at a wavelength $\lambda = 510nm$ for a thick aluminum layer **a** and a thick silver layer **b** using optical constants from (Palik, 1985)

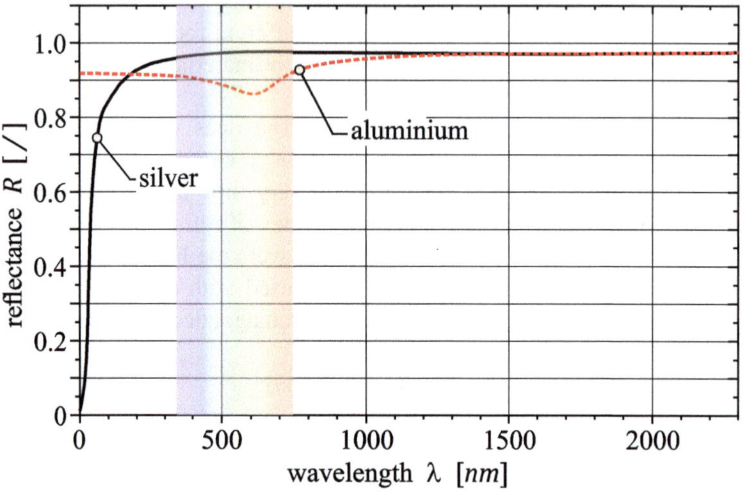

Fig. 3.50 Computed reflectance ρ of silver (Ag, straight) and aluminium (Al, dashed) versus the wavelength λ for 'unpolarized' radiation and near-normal incidence

solar wavelength range, whereas aluminium displays a material-specific minimum in the near infrared between $800nm$ and $900nm$ due to an absorption band.

Although silver is more expensive than the abundant aluminium most reflectors in concentrator devices are based on silver. We must distinguish between first surface and second surface reflectors, as illustrated in Fig. 3.51.

The classical bathroom mirror is a second surface reflector, because the front glazing here protects the thin silver layer from the humid environment. Silver

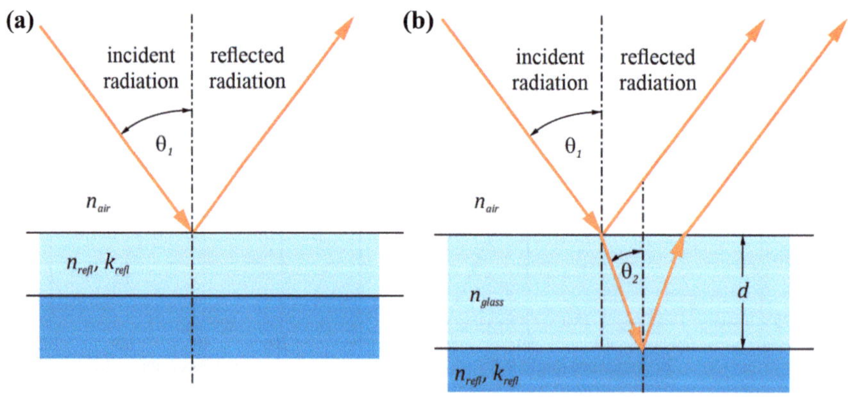

Fig. 3.51 First surface mirror with reflecting metallic material on a glass substrate **a** and second surface mirror with reflecting metallic material behind a glass substrate **b**

Table 3.11 Solar reflectance of Ag and Al refector materials (normal incidence)

Material	Solar reflectance
Silver, fresh, electroplated	0.96
Aluminium, fresh, high-purity	0.91
Aluminium, foil 99.5% purity	0.86
Silver, back surface, anti-reflected glass	0.94
Silver, back surface, white glass	0.91
Aluminium, front surface, sputtered	0.89
Aluminium, front surface, added SiO coating	0.87

is very sensitive to humidity and other components in the atmosphere (e.g. sulphides). Also, the back and edges of a silver mirror must be protected, usually some layers of protecting lacquer. First surface mirrors are not protected by a thick dielectric layer, but very often ultra-thin protecting layer of sputtered oxides are applied. Aluminium is better suitable for first surface mirrors as it is self-passivating with the formation of an aluminium-oxide layer. Also, high-reflectance reflectors can be optically improved by adding stacks of dielectric layers. The principle of constructive interference is applied to the dielectric coating of aluminium-based solar reflectors, (see Table 3.11).

The key requirement for a mirror surface is a reflectance close to the value of one in the solar wavelength spectrum, where reflectance indicates the fraction of incident radiation intensity that is reflected by the surface. All solar thermal mirrors aim at specular reflection, where the direction of the incident solar radiation according to Snellius' law of refraction and the direction of the outgoing light form the same angle with the normal of the mirror surface. However, since the reflectance of a metallic surface is different for different wavelengths, it must be specified for a particular wavelength or range of wavelengths.

The fraction of sunlight that is concentrated on the receiver only by the mirror material can be described by the "solar weighted reflectance" $\rho_r(SW, \theta_i, \theta_c)$ without considering other possible imperfections such as shape, receiver orientation, and tracking errors addressed within the context of the optical defects. An acceptance half-angle of $\theta_c = 7.5 - 12.5\ mrad$ (usually $12.5\ mrad$) as indicated in Fig. 3.52 is considered appropriate for typical parabolic trough designs. The solar weighted reflectance $\rho_r(SW, \theta_i, \theta_c)$ cannot be determined directly because spectrophotometers can measure specular reflectance at a given θ_i only in a narrow wavelength range. Therefore, $\rho_r(SW, \theta_i, \theta_c)$ must be approximated.

The hemispherical reflectance $\rho_h(\lambda, \theta_i, h)$ is measured over a wavelength range representative of the terrestrial solar spectrum ($\lambda = 250 - 2500\ nm$) using a spectrophotometer covering the solar spectrum wavelength domain and an integrating sphere (typically for an incidence angle $\theta_i = 8°$) relative to standards. The spectrum is then weighted by the standard terrestrial solar spectrum to calculate a sun-weighted hemispherical reflectance, $\rho_h(250 - 2500\ nm)$, as a meaningful

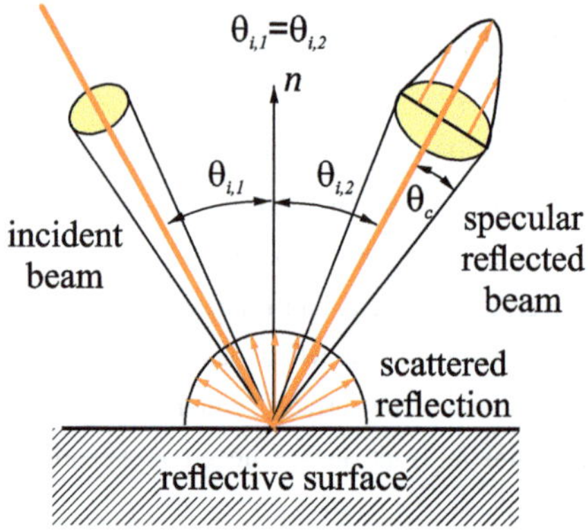

Fig. 3.52 Sketch to determine solar weighted specular reflectance using specific acceptance half angles

single measure of optical performance. The procedure and measurement are specified in standards (e.g. (ASTM G173-03, 2020), at an air mass factor $AM = 1.5$. The specular reflectance $\rho_s(\lambda, \theta_i, \theta_c)$ is measured with a specular reflectometer at $\theta_c = 3.5, 7.5, 12.5mrad$ with an incidence angle θ_i=15° for $\lambda = 660nm$.

For highly specular glass or front-surface reflectors, the specular reflectance can be described by the ratio between the specular reflectance $\rho_s(\lambda, \theta_i, \theta_c)$ and the hemispherical reflectance $\rho_h(\lambda, \theta_i, h)$ at the same wavelength, which is assumed to be constant and equal to the ratio of the solar-weighted values. The solar-weighted specular reflectance is then approximated by:

$$\rho_r(SW, \theta_i, \theta_c) = \frac{\rho_r(\lambda = 660nm, \theta_i, \theta_c)}{\rho_h(\lambda = 660nm, \theta_i, \pi)} \cdot \rho_h(SW, \theta_i, \pi). \tag{3.86}$$

The solar-weighted reflectance thus indicates the proportion of solar energy that is reflected by a mirror and is therefore the decisive quality criterion for efficient mirrors. This data provided by the manufacturer refer always to clean mirror.

3.4.4 Linear-Imaging Concentrators

The simplest case of a concentrator is a parabola of focal length f in two-dimensional space. A translational extension of the two-dimensional shape then forms a linear concentrator, a parabolic trough. The light rays incident parallel to the y-axis are focused by the parabola contour to the focal point F of the parabola, as shown in the Fig. 3.53. Provided that the lines FP and PQ are of equal length, the tangent is always correct to focus all vertical beams into the focal point F. According to geometric algebra, the following relationship is obtained for the contour shape of the parabola:

$$y + f = \sqrt{x^2 + (y - f)^2} \rightarrow y = \frac{1}{4f}x^2. \tag{3.87}$$

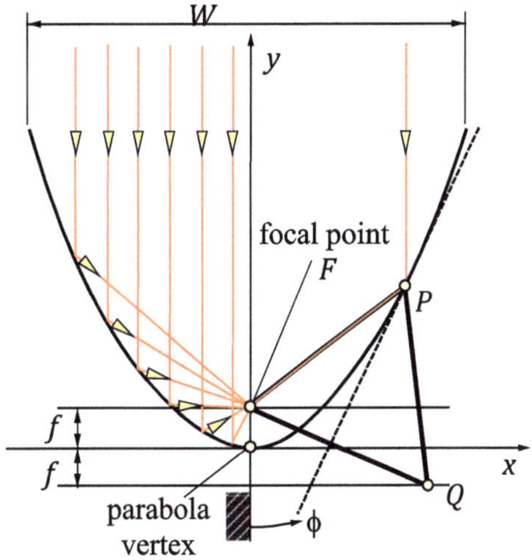

Fig. 3.53 Concentration of incident radiation in a parabolic concentrator with focal length f

The result of this simple observation is that as the focal length f of the parabola increases; a flatter parabola becomes necessary to focus the light in a point-like manner. In practice, this means a smaller contour length and thus less material requirement for the same mirror width (as will be seen later, constant concentration ratio). On the other hand, the distance to the focal point increases, so that mirror errors have a greater effect.

Although modern manufacturing machines allow the production of parabolic contours, it is technically easier to produce precise circular segments. Under certain conditions the parabola can be replaced by a segment of a circle. The following relationship results from the circular equation:

$$x^2 + (y - R)^2 = R^2 \rightarrow y = R \pm R\sqrt{1 - \left(\frac{x}{R}\right)^2}, \tag{3.88}$$

where R denotes the radius of the circle and x, y represent the coordinates from the Fig. 3.53.

If the surface segment DE is small in comparison to the radius R ($DE/R \ll 1$), it is also true for all coordinate ratios x/R that these are much smaller than one ($x^2/R^2 \ll 1$). In this case, a Taylor series expansion of the root expression in Eq. (3.88) yields:

$$\sqrt{1 - \left(\frac{x}{R}\right)^2} = 1 - \frac{1}{2} \cdot \left(\frac{x}{R}\right)^2 - \frac{1}{8}\left(\frac{x}{R}\right)^4 - \frac{1}{16}\left(\frac{x}{R}\right)^6 - \ldots \approx 1 - \frac{1}{2} \cdot \left(\frac{x}{R}\right)^2. \tag{3.89}$$

Inserting the result in Eq. (3.88) we obtain:

$$y = \frac{1}{2} \cdot \left(\frac{x}{R}\right)^2. \tag{3.90}$$

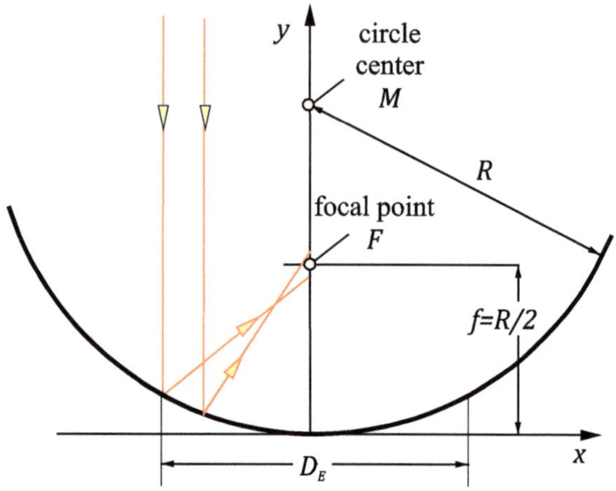

Fig. 3.54 View of the focal length of a circle segment as a replacement for a parabolic contour

Thus, the focal length f of a circle segment is $f = R/2$, as illustrated in Fig. 3.54.

Theψ_r rim angle ψ_r of a parabola is defined as the angle between the axis and a line from the focus to the physical edge of the concentrator. Together, the focal length f and rim angle ψ_r of a parabolic concentrator completely define its cross-sectional geometry. The rim angle of a parabola is given by

$$\tan\psi_r = \frac{W/2}{f - h} = \frac{W/2}{4f^2 - (W/2)^2}, \tag{3.91}$$

where W is the width and $h = y_r$ is the full depth of the parabola,

$$h = \frac{W^2}{16f}. \tag{3.92}$$

Assuming W as the width of the aperture, one can find a relation between the rim angle and the diameter to focal length ratio (d/f) and it results after some transformations:

$$\frac{f}{W} = \frac{1}{4\tan(\psi/2)}. \tag{3.93}$$

It follows directly from the relationship that a small rim angle corresponds to a large focal length (for a given aperture diameter) and vice versa. The rim angle is an important design feature of collectors, especially because it affects the concentration ratio.

The parabolic effect of focusing on a single focal line only occurs with perfectly parallel incoming radiation. As noted above, sunlight has a range of incident angle due to the physical extent of the solar source. This reduces the optical concentration achievable at the focus of parabolas, as discussed in the following sections.

3.4.4.1 Limits with Flat Receivers

Each point on a parabolic mirror will reflect a cone of rays that matches the angular distribution of the solar source (half-angle size θ_s). Consider the size of the spot formed by the cone of rays reflected from the points on the mirror when incident on a flat target placed in the focal plane (see Fig. 3.53). The rays from the rim will form the widest such spot. The distance x_r of the rim reflection point from the vertical axis is

$$x_r = r \cdot \sin\psi_r, \tag{3.94}$$

and the width of the focal spot on the focal plane due to reflection from this point will be

$$D = \frac{2r \cdot \sin\theta_S}{\cos\psi_r}. \tag{3.95}$$

The reflected cone from a single spot on the mirror will form an elliptical spot on the target with a major axis length of d. This applies equally for dishes and troughs. Clearly, the size of the focal-plane spot depends on the received incidence angle θ, so the accumulated effect of all the spots across the mirror will not be a true image of the sun formed at the focal plane. This is the reason that solar concentrator optics are referred to as non-imaging optics. If the rim angle ψ_r is small, then the distance mirror to the focus r is approximately equal to f over the whole mirror, and then, to a good approximation, it will form a true image of the sun in the focal plane with the image diameter being

$$\mathrm{D} = \frac{2f \cdot \sin\theta_s}{\cos\psi_r} \approx 2f \cdot \sin\theta_s. \tag{3.96}$$

On the other hand, for a dish or trough with a flat receiver and a rim angle of 90°, the mirror elements at the very edge will make an infinitely long image, spreading their radiation over the entire focal plane. If the receiver is large enough to accept reflected spots from the entire mirror surface, then the diameter of the receiver will be defined by the reflected spot size from points at the edge of the mirror with $x = W$ and $\phi = \psi_r$. The geometric concentration ratio for a parabolic trough with a flat receiver will then be,

$$C_g = \frac{A_a}{A_r} = \frac{W}{D}, \tag{3.97}$$

leading to

$$C_g = \frac{\sin\psi_r \cdot \cos\psi_r}{\sin\theta_S} = \frac{\sin 2\psi_r}{2\sin\theta_S}. \tag{3.98}$$

The optimal rim angle $\psi_{r,opt}$ can be found by setting the derivative of Eq. (3.98) to zero, i.e.

$$\frac{dC_g}{d\psi_r} = 0, \tag{3.99}$$

which yields the maximum concentration at $\psi_{r,opt} = 45°$ for the flat receiver and ideal shape of the parabola. When using the divergence of the sun with $\theta_s = 4.65 mrad$ we get as maximum concentration $C_{g,max,flat}$,

$$C_{g,max,flat} = \frac{1}{2\sin\theta_s} \approx 108. \tag{3.100}$$

When the beam is reflected from an imperfect mirror with imperfect shape, an additional dispersion half-angle δ must be added to θ_s in the formulae. Then the equivalent beam spread θ_s^* with

$$\theta_s^* = \theta_s + \delta \tag{3.101}$$

should be used, resulting, of course, in a smaller maximum concentration.

3.4.4.2 Limits for Cylindrical Receivers

More useful for solar energy applications is the consideration of a receiver with circular cross-section as shown in Fig. 3.55. In this case, the diameter of the target should be at least

$$\mathrm{D} = 2r \cdot \sin\theta_S. \tag{3.102}$$

For a trough with a cylindrical receiver, then

$$C_g = \frac{A_a}{A_r} = \frac{2 \cdot r \cdot \sin\psi_r}{2\text{ßr} \cdot \sin\theta_S} = \frac{\sin\psi_r}{\pi \sin\theta_S}. \tag{3.103}$$

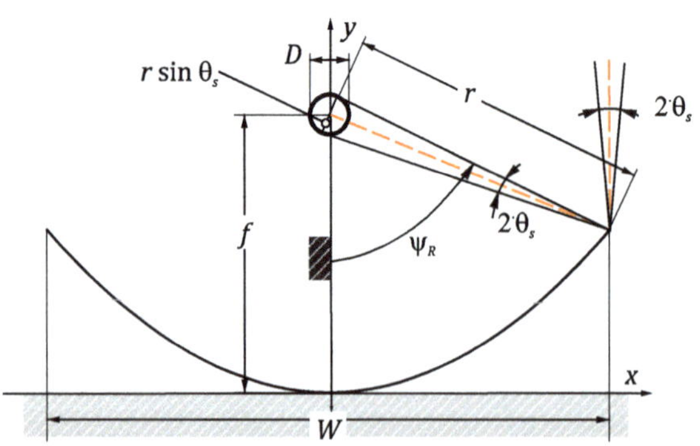

Fig. 3.55 Parabolic concentrator with circular target (cross section of parabolic trough)

Again, the shading fraction of the receiver tube on the mirror is neglected. Solving for maximum geometric concentration ratio as before, the optimal trough rim angle now is calculated as 90°, and for this angle the maximum concentration is

$$C_{g,max,tube} = \frac{1}{\pi \sin\theta_s} \approx 68. \tag{3.104}$$

Note that many companies quote however the concentration of their product as described by Eq. (3.97) for a flat receiver – arguing that the projection of the tube is the relevant receiver area. However, the area of the receiver emitting thermal radiation is undoubted ($\pi \bullet D \bullet L$) and not the projection ($D \bullet L$) when L is the receiver length oriented normal to the cross-sectional plane. Also, it should be emphasized that this analysis only considers the geometric concentration ratio, and the above derivation indicates that the local contributions to focal spot size vary as a function of reflection radius x, so the geometric concentration ratio limits are less than the optical concentration ratio limits at the centre of the focal spot.

3.4.5 Point-Imaging Concentrators

When we rotate the parabola in Fig. 3.53 around its focal axis, we get a paraboloid, a concentrator with a point focus. The width of the parabola W then is the diameter of a circular aperture. Instead of focusing on a focal line as for a trough there is a focal point for this three-dimensional concentrator. This rotational concentrator is called parabolic dish.

It follows directly from the relationship (3.93) that a small rim angle corresponds to a large focal length (for a given aperture diameter) and vice versa. The rim angle is an important design feature of parabolic dishes, especially because it affects the concentration ratio. A perfect paraboloid mirror shows the highest concentration ratio on a flat receiver (as shown in 3.4.4.1) at a rim angle of $\psi = 45°$, and we now consider how the concentration ratio behaves when deviating from this ideal rim angle. If the rim angle becomes very small, the reflector is far from the focal point for a given diameter, so that the divergence of radiation, which is unavoidable due to the spread of the solar disk, produces a larger solar image. This leads to a lower concentration ratio at small edge angles for a given collector diameter. Therefore, reflector geometries with small rim angles are unfavourable for large concentration ratios.

Assuming a receiver with a flat circular aperture in the focal plane and a large rim angle ($\psi > 45°$), as shown in Fig. 3.56 for beam 2, large rim angles lead to an expansion of the focal spot exceeding the aperture plane.

The surface area of a mirror with a large edge angle is larger than the surface area of a mirror with the same diameter and smaller edge angle. This affects the material requirements and the weight of the collector. The absolute mirror surface

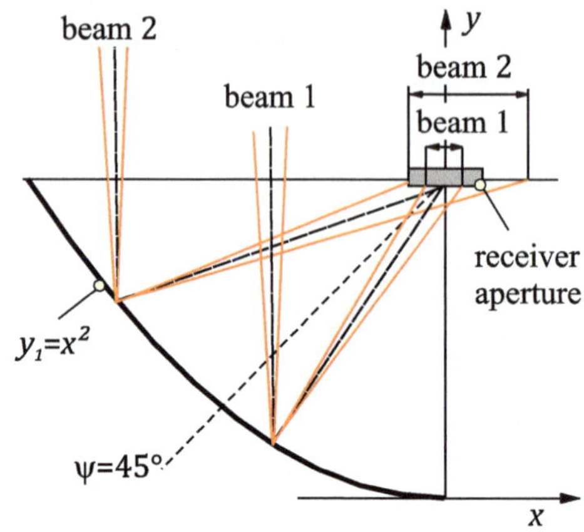

Fig. 3.56 Radiation path of an angle-variant irradiation on a parabolic mirror at different rim angles ψ

(A_{mir}) is

$$A_{mir} = 4\pi f^2 \frac{\sin^2\psi}{(1 + \cos\psi)^2}. \tag{3.105}$$

Therefore, not only parabolic mirrors with very small edge angles, but also mirrors with very large edge angles are not favourable for a high concentration ratio. In addition to these considerations about a sufficiently large concentration ratio of parabolic mirrors, there are other aspects related to the rim angle. One advantage of mirrors with a very small rim angle is that as the rim angle decreases, the parabolic shape can be better approximated by a spherical shape, which is more favourable for manufacturing reasons, since many mirrors are segmented and a larger paraboloid with sufficiently large rim angles can be composed of many spherical shaped, small facets. Moreover, there is a relationship between the rim angle of the collector and the receiver type. Small rim angles mean low angular tolerance of incident radiation at the receiver, while large rim angles have the opposite effect, meaning that cavity receivers are used at rim angles ($\psi < 50°$), while external receivers allow larger rim angles. Since dish systems target high temperatures, thermal radiation losses play an excellent role, so cavity receivers are in the development focus.

The diameter of the focal point in the receiver (d_{rec}) can be calculated by using Eq. (3.105), which affects the size of the receiver. Here, the acceptance angle (θ_A) to the parabolic dish appears, which describes the angle of worst condition that collect all reflected solar radiation from the concentrator at small time for tracking,

$$d_{rec} = \frac{f\theta_A}{\cos\psi(1 + \cos\psi)}, \tag{3.106}$$

with θ_A in (rad). If one considers the radiation distribution of the sun, large concentration ratios such as those of a dish receiver result in an extreme flux maximum present at the actual focal point, which exceeds the permissible temperatures of a possible absorber material. Therefore, cavity geometries are usually chosen for the receiver targeting in the design to achieve the most uniform wall radiation flux distribution possible. Here, spherical receiver shapes offer a good starting point for shape optimization with which a desirable shape (inverted bulb) with nearly uniform distribution can be achieved; if necessary, the central receiver can also be supplemented by a secondary reflector.

The geometrical concentration factor for flat circular receivers can be derived similarly to the paragraphs 3.4.4.1 and 3.4.4.2. For flat receivers the concentration is

$$C_{g,flat} = \frac{\sin^2\psi_r \cdot \cos^2(\psi_r + \theta_S)}{4 \cdot \sin^2\theta_S} - 1. \tag{3.107}$$

For a spherical bulb receiver (which is different from the inverted bulb mentioned above) the concentration is given by

$$C_{g,sphere} = \frac{\sin^2\psi_r}{4 \cdot \sin^2\theta_S} - 1. \tag{3.108}$$

The optimum rim angles here again are 45° for flat disks and 90° for the spherical receiver. As described above, the formulae are valid for perfect mirrors, otherwise the equivalent beam spread θ_s^* (3.101) has to be used.

3.4.6 Non-Imaging Concentrators

Concentration ratios corresponding to the ideal concentrator reach the levels achieved by (Winston, 1974) proposed Compound-Parabolic Concentrators, which are often abbreviated as CPC in the technical literature. Rabl investigated (Rabl A., 1976) different absorber geometries with adapted CPC contours and compares them with parabolic trough concentrators (PTC).

In combination with a flat absorber, the contour of a CPC is composed of two parabolic branches whose focal points are located at the opposite ends of the absorber strip. The parabola axes run parallel to the shadow lines that start from the contour ends and meet the opposite absorber ends. The angle between these two shadow lines gives twice the acceptance angle. The Fig. 3.57 shows schematically the geometrical structure of a Compound-Parabolic Concentrator (CPC), whose contour is composed of two parabolic branches.

In the case of a round absorber, the shadow lines from the upper contour ends touch the absorber as a tangent, as shown in the Fig. 3.58. Below the absorber between the points where the shadow lines meet, the contour has the shape of two involutes. Above the points of impact, it is composed of a sequence of intersections of parallels to the shadow lines and tangents to the absorbers. Further combinations of absorber geometries and CPC contours are described in Rabl (Rabl A., 1976) as well as in (Kreith and Kreider, 1978).

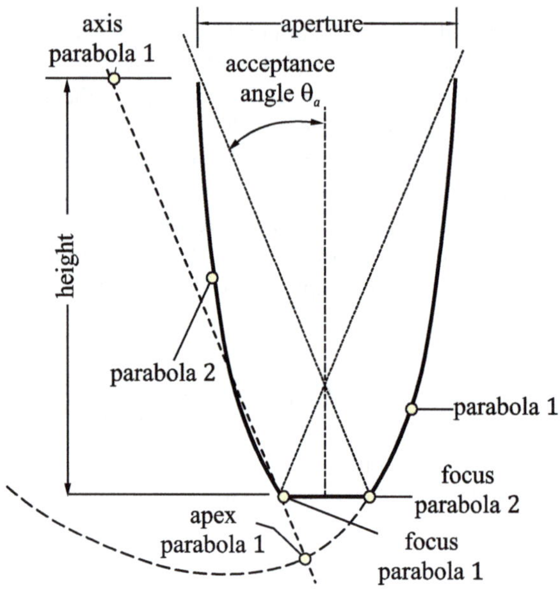

Fig. 3.57 Schematic diagram of the contour of a Compound Parabolic Concentrator (CPC) with a flat absorber strip; its mirror contour is composed of two parabolic branches

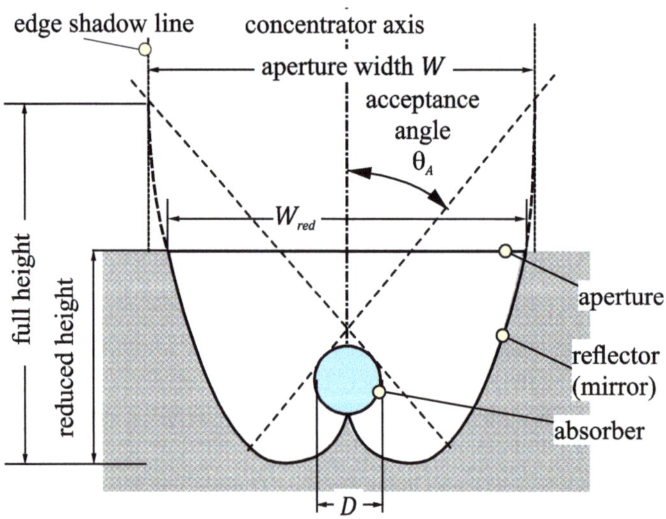

Fig. 3.58 Schematic sketch of the contour geometry of a compound parabolic concentrator (CPC) with round absorber tube at full or reduced height

Figure 3.59 shows the ratio of the reflector contour length to the aperture width W as a function of the concentration ratio C at various specified acceptance angles θ_a. The graph illustrates two disadvantages of CPC concentrators (shown here for flat absorbers) As the concentration ratio C_g increases, the contour length grows faster than the aperture width W. This would lead to very large reflector heights and therefore collectors of the CPC type are generally only built with a shortened or reduced height.

Figure 3.60 illustrates a comparison of a full height CPC and reduced height $CPC's$. In addition, at high concentration ratios C the CPC level, the acceptance angle θ_a decreases very rapidly due to the large height. This in turn requires a large number of tracking steps to ensure a high daily operating time.

The design of full-height CPC reflectors leads to complex multiple reflections on the mirror surface at high concentration ratios. Figure 3.61 shows the mean number of reflections n_ρ as a function of the concentration ratio C_g at different

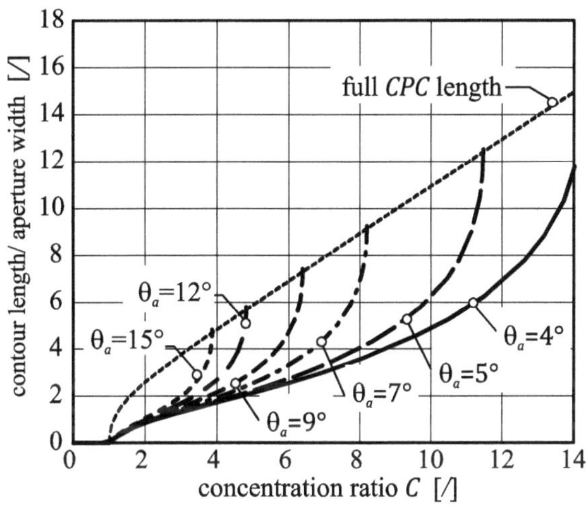

Fig. 3.59 Ratio of the reflector contour length to the aperture width W as a function of the concentration ratio C at a given acceptance angle θ_a in (°) for CPC collector types with full and reduced height

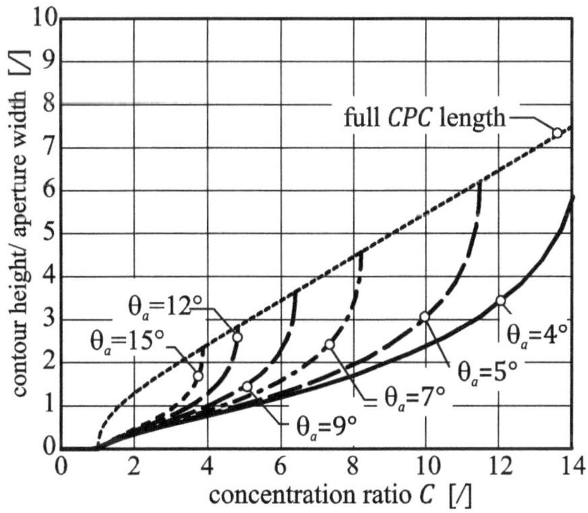

Fig. 3.60 Ratio of the concentrator height to the aperture width W as a function of the concentration ratio C at a given acceptance angle θ_a in (∘) for CPC collector types with full and reduced height

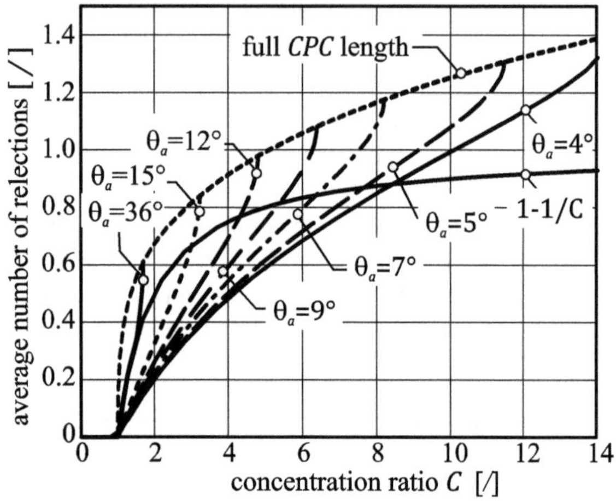

Fig. 3.61 Mean number of reflections n_ρ (right) as a function of the concentration ratio C at a given acceptance angle θ_a for CPC collector types with full and reduced height

given acceptance angles θ_a. This figure shows that the average number of reflections until the radiation from the reflector edge areas reaches the absorber increases with increasing concentration ratio C_g. Since the reflection coefficient is $\rho < 1$, the optical efficiency decreases with the number of reflections proportionally with ρ^n.

Most applications of CPC concentrators are therefore limited to small concentration ratios, typically $C_g \leq 2$. There are several collectors on the market with fixed CPC concentrators as back-up for vacuum tubes or they are built into such tubes. On the other hand, (Niemann and Schreitmüller, 1993) showed that double concentrating $CPC's$ in combination with good vacuum tubes offer efficiencies of $\eta \approx 50\%$ at operating temperatures of $200°C$ and $\Delta T/i = 0.1K/\left(Wm^{-2}\right)$.

In the concentration range of $C_g \leq 2$, CPC collectors compete economically with other non-imaging concentrators that are easier to manufacture. These include, for example, V-shaped channels with flat mirrors on both sides of the absorber, whose reflector cross section forms a V-shape. The Fig. 3.62 schematically shows the geometrical structure of a V-shaped channel. Rabl (Rabl A., 1976) shows, using equations of linear algebra, that up to an angular deviation of δ^* between the direction of irradiation and the normal, all irradiation hitting the aperture surface comes to the absorber. At angular deviations greater thanθ_C, no radiation at all reaches the absorber. The limit angle θ_C is thus the sum of the angle at the tip of the V-groove with $(2 \cdot \theta_V)$ and δ^*:

$$\theta_C = 2\theta_V + \delta^*. \tag{3.109}$$

Between θ_C and δ^*, the proportion of the incident radiation decreases approximately linearly with the angle. With this approximation the possible concentration

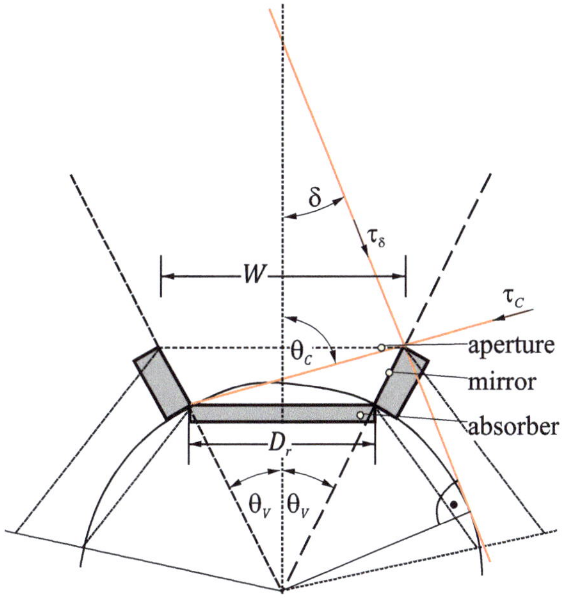

Fig. 3.62 V-shaped trough with a flat absorber and two flat lateral mirrors; up to an angular deviation δ^* between the direction of irradiation and the normal, all the irradiation hitting the aperture surface reaches the absorber; at angular deviations greater than θ_c, no more radiation at all hits the absorber

ratio $C_{g,V}$ can be given:

$$C_{g,V} = \frac{1}{\sin(\delta^* + \theta_V)}. \tag{3.110}$$

It is difficult to draw a comparison between the CPC and the V-channel, as several sizes have to be considered. (Rabl A., 1976) contrasts examples where the average number of reflections is equal. If this number is 0.75 and the acceptance angle $\theta_a = 19.5°$ a truncated CPC reaches a concentration of$C_{g,CPC} = 2.9$. In contrast, the V-trough with half the trough angle $\theta_V = 11°$ reaches a concentration of$C_{g,V} = 2$. The ratio of mirror length to receiver width is 2.8 for the CPC and 2.5 for the V-trough. Only with small C_g and trough angles of $\theta_V \leq 20°$ do both concentrator types reach comparable values. For direct evaporation tests of water, flat plate collectors with a V-trough concentrator (location Karlsruhe,$\Phi = 49°$) were used, on which the collector mirrors were mounted on the long sides so that they could be swivelled. The long sides and thus also the axes of rotation of the mirrors were in east–west direction. A tracking system was dispensed with to avoid constructively complex flexible steam lines, and a rod system connected the mirror ends. This meant that one lock was sufficient for simultaneous adjustment of both mirrors during tracking.

3.4.7 Fresnel Concentrators

3.4.7.1 Primary Mirror Field

The primary mirror or reflector array consists of multiple rows of slightly curved mirrors, rotating with half the angular speed of the sun in the transversal projection path over the rows. The basic geometry (see Fig. 3.63) has variables; as primary mirror width $w_{pr,i}$, gap Δx_i, focal length f_i and height h_i which all may be functions of distance x_i of mirror row i from the receiver unit.

There are several optical loss mechanisms related to geometry. Even for perfect reflectors, cosine losses occur due to off-normal incidence on the reflectors most of the time. Blocking considers the partial interception of reflected rays by mirror rows closer to the receiver tower. In addition, mirror rows may partially shade rows behind them as seen from the sun (shading). Additionally, any real mirror has a small but finite absorption and—if dust or a rough surface must be considered – scattering extinction both reduce the direct radiation reflected towards the receiver. Blocking and shading is dominant for low solar altitudes.

According to (Abbas et al., 2018) there are four principal classes of LFC primary mirror designs:

•

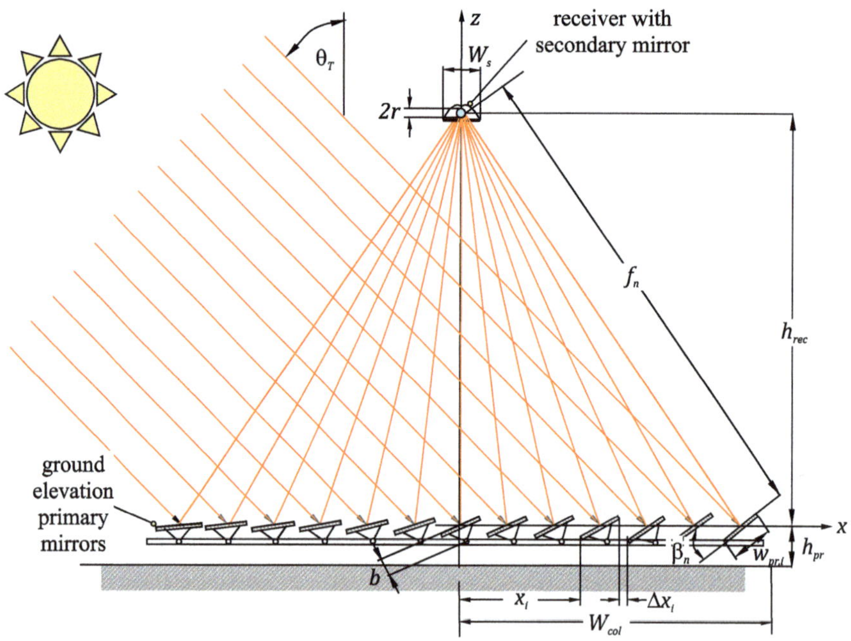

Fig. 3.63 Geometrical transversal cross-section of a primary reflector array of a Fresnel collector (w_i=mirror width, Δx_i= gap between neighbouring mirrors, b=distance between rotational axis and mirror axis, f_i=focal length)

Constant mirror width and constant gap (CW-CG),
- Constant mirror width and variable gap (CW-VG),
- Variable mirror width and constant gap (VW-CG),
- Variable mirror width and variable gap (VW-VG).

The advantage of constant mirror width is of course that the production line does not need to deal with different mirror sizes. So, it is a question whether variable mirror width options will be chosen by a commercial development. Providing a variable gap to minimize blocking and shading losses however may be solved by mounting options—but of course at the expense of additional assembly effort and more careful design. This design however will result in a somewhat larger field area for the same reflector area.

A further design option is that the focal lengths of individual mirror rows can be constant or variable. The focal lengths should be adapted to the distance of the mirror rows to the receiver aperture. In practice, however, it is too costly to produce different focal lengths for every row, so practical designs typically utilize one to four different focal lengths. In addition to losses due to optical design there are other factors which degrade the quality of the solar image on the receiver. Rough, dusty, or scratched mirror surfaces as well as shape or contour inaccuracies widen the reflected beam. Also tracking inaccuracies must be avoided.

Optical design tries to optimize the size and efficiency of the reflector array (to increase concentration) with respect to economics. The further away a reflector row is from a receiver tower, the lower is its time-averaged optical efficiency. In financial terms every additional investment of adding a row has less and less addition to overall output. This is one reason for the CLFR approach where the mirror rows can be reoriented, according to sun position, to reflect to a receiver in better position.

3.4.7.2 Secondary Optics

For identical receiver area large primary reflector fields result in larger concentration, however, to enhance the efficient collection of the reflected radiation secondary concentrator optics may be used. This allows larger fields, as the receiver aperture will be larger than the absorbing area. It is important that radiation impinging on the receiver aperture should be directed towards the absorber tube or tubes. Non-imaging optics provides us with the means to design such secondary concentrators. Several principles have been proposed, from truncated compound parabolic concentrators (CPC) to elliptic or hyperbolic concentrators. Typically, there is slightly more than one reflection on average in a secondary concentrator, which means that reflectivity should be very high, because energy is lost with any reflection. In addition, any absorption by the reflector heats up the secondary concentrator due to the high intensity. This must be considered carefully in the choice of secondary material and shape accuracy. Deformation and degradation may be caused by high temperatures.

***CPC* Receiver**

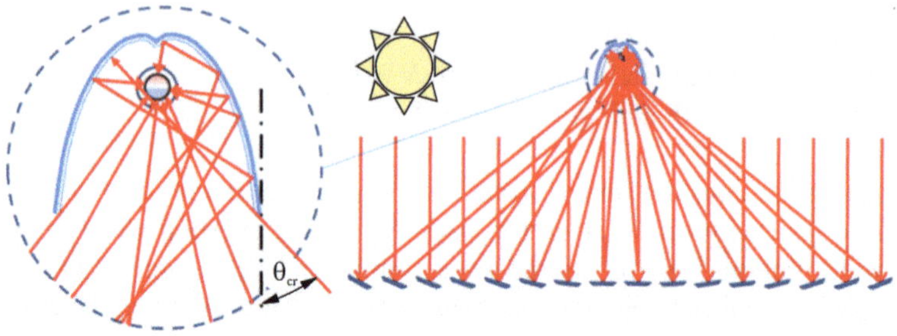

Fig. 3.64 CPC Receiver within a linear Fresnel collector (LFR)

The geometry used in most commercial approaches is a truncated CPC, as illustrated in Fig. 3.64. A CPC reflector ideally redirects all rays from a certain field of view (within the critical acceptance angle θ_{cr} which should cover the primary reflector field) towards the receiver surface. Truncation reduces the length of the CPC without much sacrifice in concentration (Rabl and Winston, 1976). The advantage of this design is a relatively shallow geometry using little reflector material. An ideal CPC with a cylindrical absorber would have a cusp nearly touching the absorber tube. To avoid that, the cusp often is flattened out, allowing some rays within the theoretical acceptance range to escape the concentrator. Experimental methods have been developed to assess the quality of the secondary concentrator early in development, see (Platzer, Heimsath, Hildebrandt, and Georg, 2008).

XX-SMS Fresnel concentrator
A similar concept from non-imaging optics using the Simultaneous Multiple Surface method yields a taller secondary concentrator but resulting in higher concentration (Canavarro et al., 2014) or (Guerreiro et al., 2015), which is depicted schematically in Fig. 3.65. It is called the LFR XX-SMS concept. This advanced optical design obviously has consequences on other issues like construction of the tower structure (due to higher wind loads) or concentrator material costs. It must be shown that the additional optical concentration leading eventually to better thermal performance at low irradiation, justifies the effort. According to (Guerreiro et al., 2015). a concentration of 73.7 can be reached using 16 mirrors 1.06 m wide within a field 20.1 m wide. The tower has to be 9.8 m high. The optical efficiency is close to 70%.

TERC concentrator
A further secondary concentrator type has been designed by Chaves and Collares-Pereira also based on etendue conservation using non-imaging optics maximum (Chaves and Collares-Pereira, 2010). However, in this case the optics is designed for a modified primary mirror field where the rotational axes of the mirrors are not on one altitude level. The primary field mirrors of one segment between receiver towers, point to different receivers alternatingly, similarly to the CFLR concept as

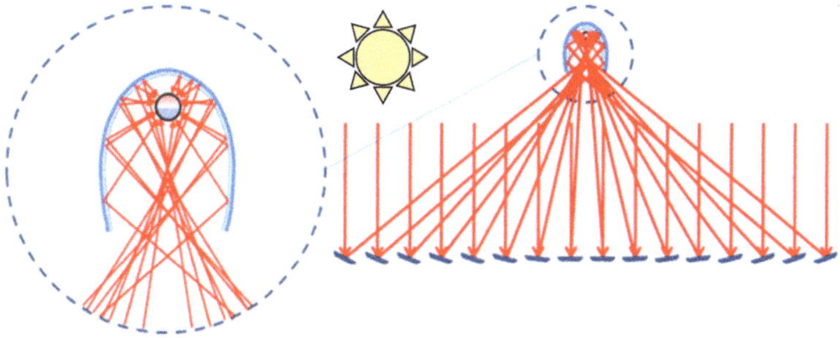

Fig. 3.65 LFR XX SMS Concept according to (Guerreiro et al., 2015)

depicted in Fig. 3.66. In this way the primary mirror field can be extremely compact minimizing the gaps. The mirrors further away from the receiver are elevated to avoid blocking. The receiver in this case is a V-shaped absorber with two wings of mirrors on the sides. These concentrators are called Tailored Edge Ray Concentrators (TERC). From a theoretical point this is an ideal optical concentrator with highest possible optical efficiency for infinitesimally small mirrors. Practically viewpoints lead to finite size mirrors and mirrors in one plane which brings the efficiency down.

Adaptive Design Concentrator (ADC)
NREL's adaptive design concentrator is not defined by a set of equations but by a repetitive procedure to determine the next segment of the reflector from a previously defined one (Zhu, 2017). The resulting secondary shape looks very similar to a truncated CPC and is also determined by the design variables as indicate in Fig. 3.67a; absorber tube diameter, aperture width and height position of the tube above the secondary aperture plane. A recent comparative optical study concludes that the optical performance is very similar to that of a CPC, but however may have some benefits at higher flux intensities (Abbas et al., 2018).

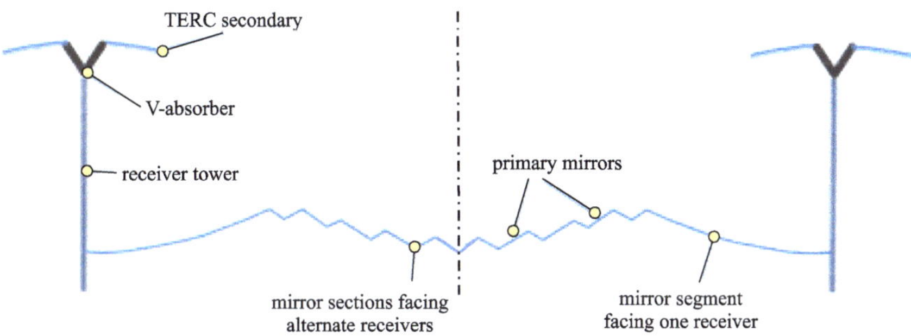

Fig. 3.66 Concentrator for several receivers with Fresnel primary and truncated secondary mirrors (TERC) after (Chaves and Collares-Pereira, 2010)

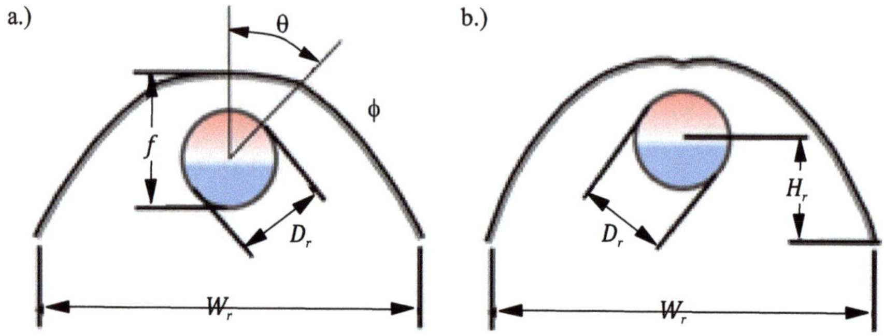

Fig. 3.67 Design parameters in an ADC (**a**) and a SPSC (**b**)

Segmented Parabolic Secondary Concentrator (SPSC)
This shape is composed of a parabolic trough broken at a given point to change its tilt. The result is a kind of reflecting cavity with 3 segments of curved walls which is depicted in Fig. 3.67b. The design parameters are the receiver diameter, the rim angle of the bottom segment where the parabola is cut, and then the tilt angle for the wing segments. In the comparison of (Abbas et al., 2018) the SPSC has a clearly lower performance which could be accepted if cost would be lower.

3.4.7.3 Comparison of Design Options

A combination of primary concentrator with a receiver and possibly a secondary concentrator results in an angle-dependent optical efficiency of the collector. The situation is depicted in Fig. 3.68. Let us define the angles transversal, the longitudinal and the solar longitudinal incidence angles θ_T, θ_L and θ_{LS} as

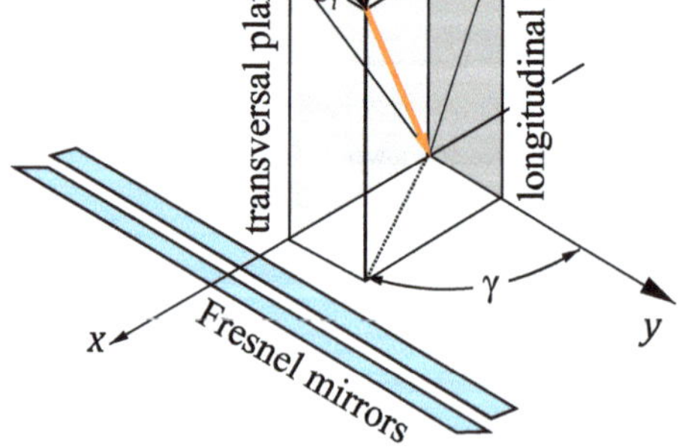

Fig. 3.68 Incidence angles for a Linear Fresnel Collector (LFC) from (Mertins, 2006)

$$\theta_T = \tan^{-1}(|\sin\gamma| \cdot \tan\theta_z); \tag{3.111}$$

$$\theta_L = \tan^{-1}(\cos\gamma \cdot \tan\theta_z); \tag{3.112}$$

$$\theta_{LS} = \sin^{-1}(\cos\gamma \cdot \sin\theta_z). \tag{3.113}$$

The incidence angle modifier function (IAM) describes the optical efficiency for off-normal incidence angles normalized by optical efficiency at normal incidence. It is usually written as the product of a transversal and a longitudinal function assuming transversal and longitudinal angle dependence to be independent, which is a reasonable assumption for most incidence angles θ and reads mostly to:

$$IAM = K(\theta_{LS}, \theta_T) = K_T(\theta_{LS} = 0, \theta_T) \cdot K_L(\theta_{LS}, \theta_T = 0), \tag{3.114}$$

where the coefficients are provided by the manufacturer of a collector assembly. A typical IAM distribution as a function of the incidence angle θ_i is shown in Fig. 3.69 for a Linear Fresnel collector (LFC) in comparison with a parabolic trough (PTC), which exhibit significant differences further discussed in Chaps. 7 and 10.

In a comparison of design options theoretical maximization of optical efficiency should not be the target. This approach certainly leads to designs with high concentration and preserving etendue, however, it must be kept in mind that at the end manufacturing, construction and operating cost will also determine the final economic performance. Complex solutions with variable component geometries, with changing altitude of mirror rows, with complicated construction manuals will produce extra costs and sources of errors. The decision which approaches to use and which to abandon due to cost certainly depends very much on experience and production technology of an individual commercial entity. So, a purely academic design recommendation will probably not meet the final expectations. The optimized design is always a dialogue between science and commercial considerations.

3.4.8 Practical Concentrators

So far, we assumed error-free concentrators. However, practical designs of mirrors will have tolerances in the contours (roughness radius waviness), tracking or surface contamination. Kreider (Kreider, 1979) propose to assume a Gaussian distribution for the radiation distribution of the sun. In fact, the sun is about 2.5 times brighter in the center of the solar disk than at the edge (Giovanelli and Beck, 1987). The defects of the concentrator fan out the beam reflected in the direction

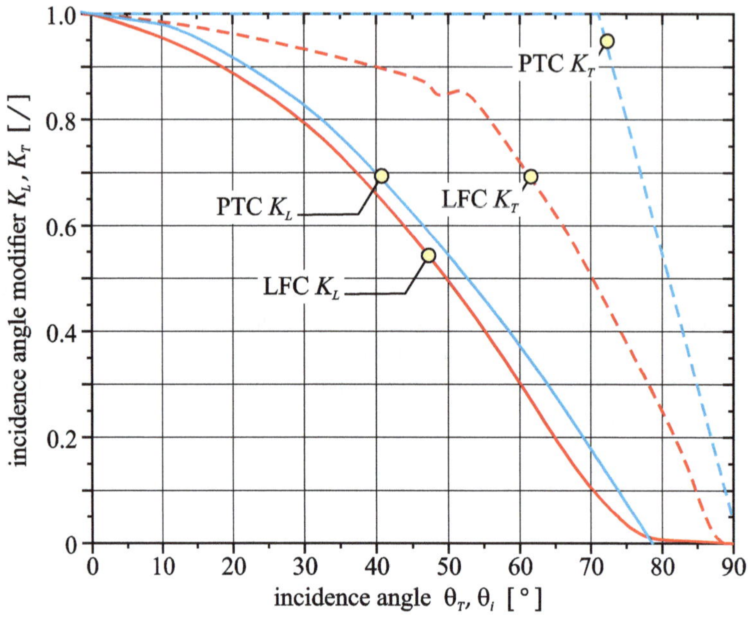

Fig. 3.69 Simulated incidence angle modifiers (IAM, relative to DNI) for the Linear Fresnel Collector (LFC) for transversal (K_T) and longitudinal (K_L) directions, in comparison to the Parabolic Trough Collector (PTC) (here transversal K_T describes the row-to-row shading) from (Morin et al., 2012)

of the absorber. With standard distributions of the individual errors (from geometric ripples, imaging, adjustment, and tracking), their sum forms the total optical error σ_{opt} of the reflector:

$$\sigma_{\text{opt}}^2 = 4\sigma_{\text{shape}}^2 + \sigma_{\text{rough}}^2 + \sigma_{adjust}^2 + \sigma_{\text{track}}^2. \tag{3.115}$$

Together with the assumed radiation distribution of the sun, this gives the standard deviation of the reflected ray, which is represented by a total standard deviation σ_{tot} in the way,

$$\sigma_{\text{tot}}^2 = \sigma_{\text{opt}}^2 + \sigma_{\text{sun}}^2. \tag{3.116}$$

This will be elaborated further in the context of the collectors in Chaps. 7 and 10. For the linear Fresnel reflector e.g. the tracking angle error has to be added as $4\sigma_{\text{track}}^2$.

3.5 Measurement Methods

3.5.1 Spectral Measurements—Reflectance and Transmittance

Optical materials should be characterized spectrally if detailed investigations are needed for optimization. Broadband measurements over the whole solar spectrum using solar simulators are possible, however, the quality of the simulator spectrum as well as other criteria like homogeneity of irradiation and temporal stability lead very often to lower accuracy. The classical characterization instrument for scientific spectral measurements are double-beam spectrometers, where a light beam is split up in to two parts, one being used as the 100% reference beam and the other with the sample in the light path for determination of the transmitted or reflected radiation. The beam path of a double-beam spectrometer is shown in Fig. 3.70.

Single beam spectrometer are useful tools in practice when spectral information has to be collected in mobile applications or outdoors. The accuracy is lower because of the inherent problem that the reference of 100% is less secure. On the one hand side, an unstable light source might lead to artefacts in the spectral curve as it takes a certain time to collect the information. Although nowadays with CCD-detector arrays and very fast data acquisition a scan of a total spectrum might need only milliseconds, nevertheless the exact determination of the reference can be a problem. If the materials are transparent and non-scattering, the direct beam transmittance or reflectance can be collected very precisely using a double-beam design using gratings and lock-in-techniques with shutters as shown in Fig. 3.70. When diffuse parts of the light also shall be collected (for determining the hemispherical reflectance or transmittance, so-called integrating spheres are used).

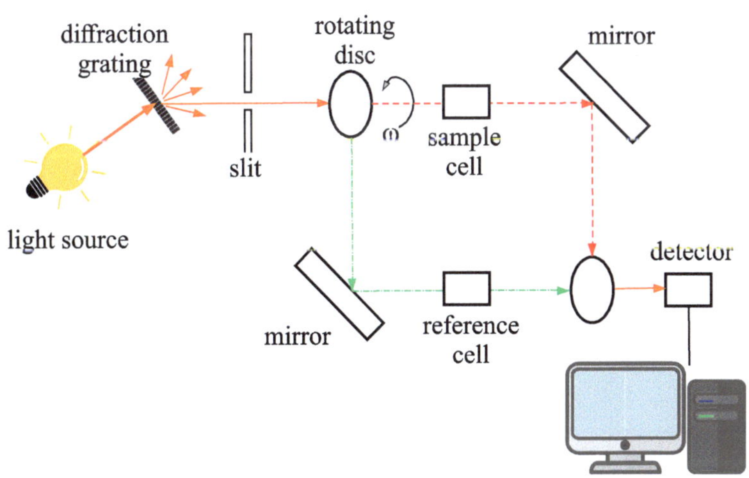

Fig. 3.70 Schematical design of double-beam spectrometer

3.5.2 Scattering and Surface Characterization

Scattering of light due to surface roughness or soiling particles reduces the amount of reflected intensity in the direction predicted by the reflection law. Therefore, a straightforward measurement technique is the determination of direct reflectance with a defined aperture acceptance angle as described in the guideline developed by an international working group of IEA SolarPACES (Fernandez-Garcia, et al., 2018). An equipment measuring the integrated total scattering signal TIS can be a so-called Coblentz sphere (with specularly reflecting walls redirected all scattering to a sensor) or an Ulbricht sphere (with diffusely reflecting walls and a sensor detecting the brightness level) (Kadkhoda, et al., 2000). However, the directionality of the scattering should be known for concentrating devices especially. The bidirectional reflectance distribution function (BRDF) is a comprehensive characterization, where narrow forward or wide-angle scattering can be quantified depending on incidence and exit angles. The geometrical definition of the $BRDF$ is given in Fig. 3.71.

In the photometric measurement set-up, a sensor with a narrow acceptance angle measuring only a cone with width $d\Omega_2$ is moved over the complete hemisphere above the reflector to measure the distribution for a specific incidence angle. The mathematical description is given in Eq. (3.117).

$$BRDF = \frac{\text{differential radiance}}{\text{differential irradiance}} = \frac{dL_2(\theta_1, \varphi_1; \theta_2, \varphi_2; E_1)}{dE_1(\theta_1, \varphi_1)}, \tag{3.117}$$

where the incident solid angle (θ_1,φ_1) and the scattered solid angle (θ_2,φ_1) are given in spherical coordinates. For practical application the $BRDF$ is often approximated by a simplified term [41]

$$BRDF|_{\theta_1,\varphi_1} = \frac{dL_2(\theta_1, \varphi_1; \theta_2, \varphi_2)}{dE_1(\theta_1, \varphi_1)} = \frac{d\Phi_2(\theta_2, \phi_2)}{d\Omega_2 dA \cos\theta_2} \frac{dA}{d\Phi_1} \approx \frac{\Phi_2}{\Omega_2 \cos\theta_2} \frac{1}{\Phi_1}, \tag{3.118}$$

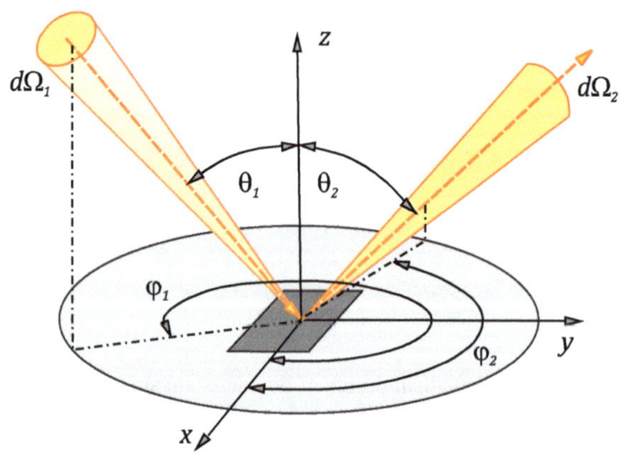

Fig. 3.71 Geometrical definition of bidirectional reflectance distribution function

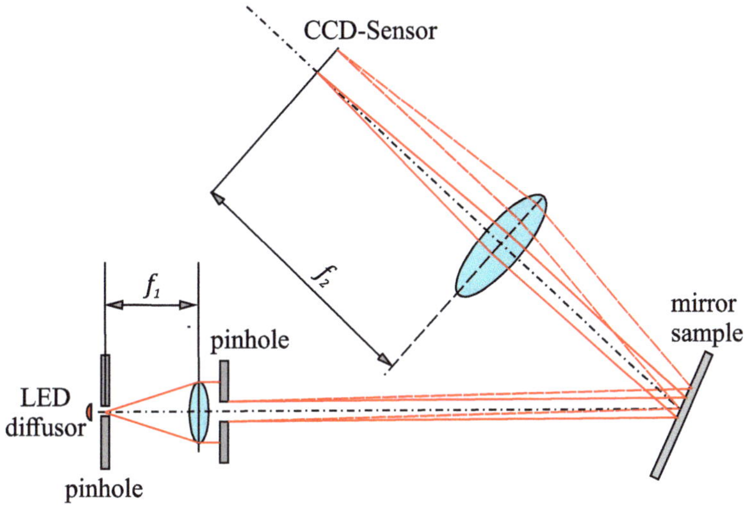

Fig. 3.72 Schematic representation of set-up for specular reflectance measurement. Sample measurement; adapted from (Heimsath et al., 2015)

valid for a nearly collimated beam of incident light reflected from a small differential surface elementdA. The $BRDF$ is relating the radiant power Φ_2 scattered into a solid angle Ω_2 to the incident power Φ_1For analysis of surface scatter it is sometimes useful to drop the cosine correction, the resulting term is then called angle resolved scatter function (ARS), which is the cosine corrected$BRDF$.

$$ARS|_{\theta_1,\varphi_1} = \frac{\Phi_2}{\Omega_2}\frac{1}{\Phi_1}. \tag{3.119}$$

Experimental design in the past used collimated light or laser light to illuminate a reflecting small sample material, and a sensor with a high sensitivity over several decades of intensity e.g. a PIN diode were used to detect the scattered light by moving it over the hemisphere over the complete hemisphere. Such an instrument is described for example in (Platzer and Bergkvist, 1993). A more modern version using a highly linear CCD camera has been developed at Fraunhofer able to detect a forward scattering range in one shot (Fig. 3.72).

As seen in Fig. 3.73b the (logarithmic) intensity plot of a scattering surface may show a pattern without rotational symmetry. Integrated over the azimuth angle the fast-declining intensity over few $mrad$ can be better seen in Fig. 3.73a.

3.5.3 Deflectometry

Other methods are needed when the geometrical shape of the reflector is not ideal. Then deviations from the ideal concentrator will result in an optical error, i.e.

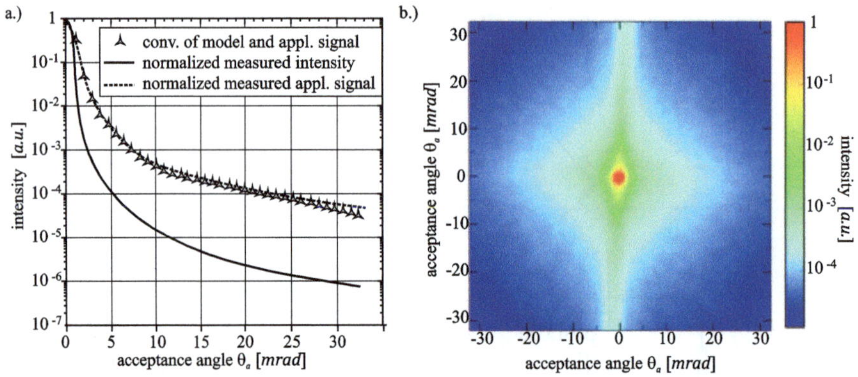

Fig. 3.73 Measured angle resolved scattering distribution of an aluminium based reflector. (left) Measurement data (blue) and convolution (red-crosses), as well as apparatus signal (green) obtained by integration (right) 2-D plot of the measured intensity. Data were obtained at $\lambda = 455nm$ and $\theta_i = 15°$ (Heimsath, 2020)

the beam reflected towards the receiver will have a wider divergence spread as compared to the ideal situation. The situation is depicted in Fig. 3.74.

The large deviations might have an origin in torsion, thermo-mechanical stress, and gravity forces, which due to limited stiffness of the reflector construction lead to macroscopic distortions. on thermal movements. Waviness or indentations on a smaller scale may be due to mechanical damages in the production process. The microscopic surface roughness of the material discussed above adds to these errors. If the curved cross section of the concentrator lies in the $x-z$-plane, the local slope deviation is defined as the difference between the measured local surface gradient

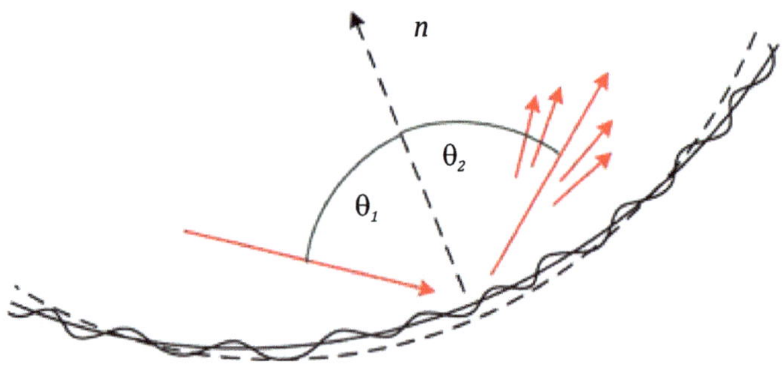

Fig. 3.74 Large and medium scale geometrical error of a reflector surface. The dashed line shows the ideal surface, the solid line a distorted average surface and the thin solid line shows the actual surface, including small surface undulations or scratches. The angular scattering of the reflected solar rays at the surface normal n, the angle of reflectance θ_2 and the indicdent solar ray at the incidence angle θ_1 are sketched

dz/dx_m and the local surface gradient of the design shape dz/dx_d:

$$sd_x = \frac{\partial z}{\partial x}_m - \frac{\partial z}{\partial x}_{id} with \overline{sd_x} = \sqrt{\frac{1}{A}\int\int_A sd_x dxdy}. \quad (3.120)$$

Experimentally large areas have to be analysed, with only sometimes parts of a millimetre deviation from the original shape. This needs an experimental instrumentation working exact even for large samples like parabolic trough mirrors. Contour accuracy of solar reflectors is measured either indirectly, by analysis of the focal line or focal spot, or directly. For direct measurement techniques either the reflection of laser profilometry (Maccari and Montecchi, 2007) or camera-based image processing techniques such as photogrammetry (Pottler, Lüpfert, Johnston, and Shortis, 2005), video processing or use of other reflected images are used. (Xiao et al., 2012).

A very powerful technique working with the reflected image of a pattern is the fringe reflection technique (FRT) or deflectometry (Heimsath, Platzer, Bothe, and Li, 2008) (Kammel, 2005). The advantage is its fine lateral resolution and its ability to provide the desired information for the entire reflector area. The principle is based on a phase-shifting method in which multiple sinusoidal patterns are reflected on the reflector surface to determine the surface gradients of the specular reflecting surface by analysing the reflected image mathematically, see Fig. 3.75.

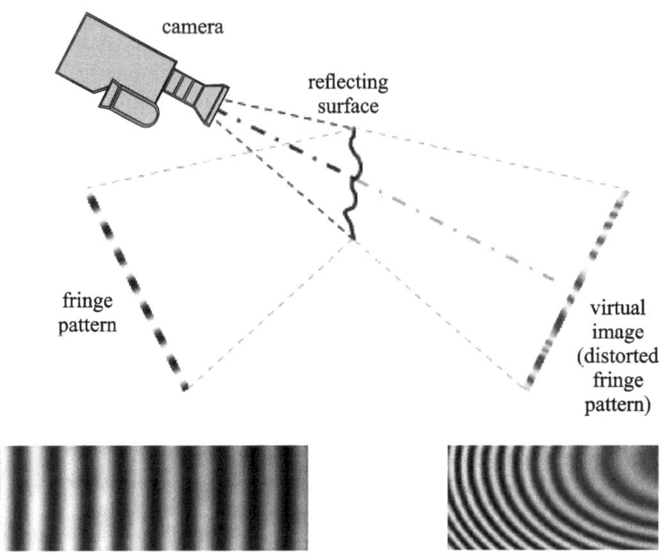

Fig. 3.75 Deflectometry measurement principle. A regular fringe pattern is displayed on a monitor (TFT) or projected onto a target. A camera records the reflected pattern. Irregularities on the specular surface result in a distorted fringe pattern which is evaluated by a phase-shifting technique, adapted from (Burke, Li, Heimsath, Kopylow, and Bergmann, 2013)

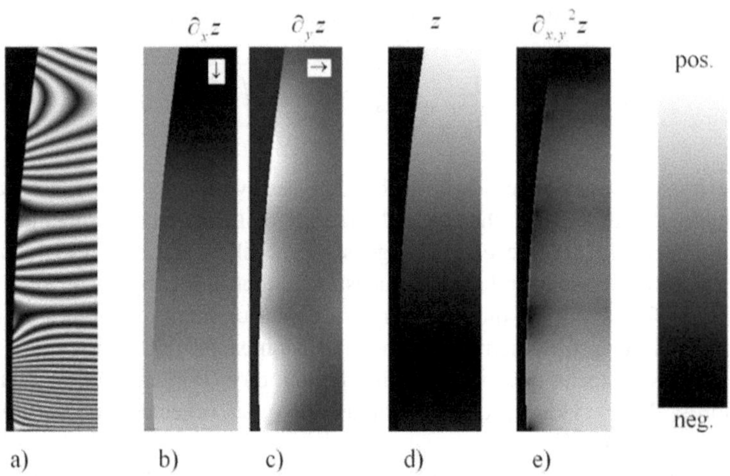

Fig. 3.76 Principal results of fringe reflection or deflectometry techniques. **a** reflected pattern image **b** gradient in x- and **c** y- direction, **d** surface height map and **e** local curvature (Heimsath, Platzer, Bothe, and Li, 2008)

Results of FRT measurements are shown as examples in the following figures. Figure 3.76a display the distorted fringe pattern on part of a reflector. Slope distortions in the range of $mrad$ are clearly visible and demonstrate the sensitivity of the method. The fringe distortions are evaluated by phase measurement and deliver surface gradients in two directions, Fig. 3.76b and c. A height map of the reflector may be calculated by integration, Fig. 3.76d. The curvature of the reflector in Fig. 3.76e shows the slope change per lateral distance and is obtained by differentiation.

In Fig. 3.76 one may see how the image of the reflected and distorted sinusoidal pattern (a) is evaluated to generate local surface gradients. The gradients are calculated in the $x-$ (b) and $y-$ (c) directions. The surface height map (d) is generated by integration and the local curvature (e) by further differentiation. If mirror concentrators are described by such position resolved data, which may be imported into raytracing models, then the effect of the localized slope deviations on the intensity distribution on an absorber geometry can be simulated. This interference-based technology (constructive for $\lambda/2$ and destructive for $\lambda/4$) provides resolution accuracy on the order of wavelength (λ), but it includes only limited directional information. These disadvantages can be compensated by multi-wavelength interferometry, light polarization, white light interferometry (low coherence length) or acousto-optical modulation.

In addition to interferometry, there are a number of other methods for measuring specular surfaces, such as the time-of-flight method (ToF) and triangulation, which are based on the use of coherent light. If unstructured light is used, intensity methods are the obvious choice. Based on the basic principles shown in Fig. 3.77, different signal coding methods can also be used via the operating mode of the

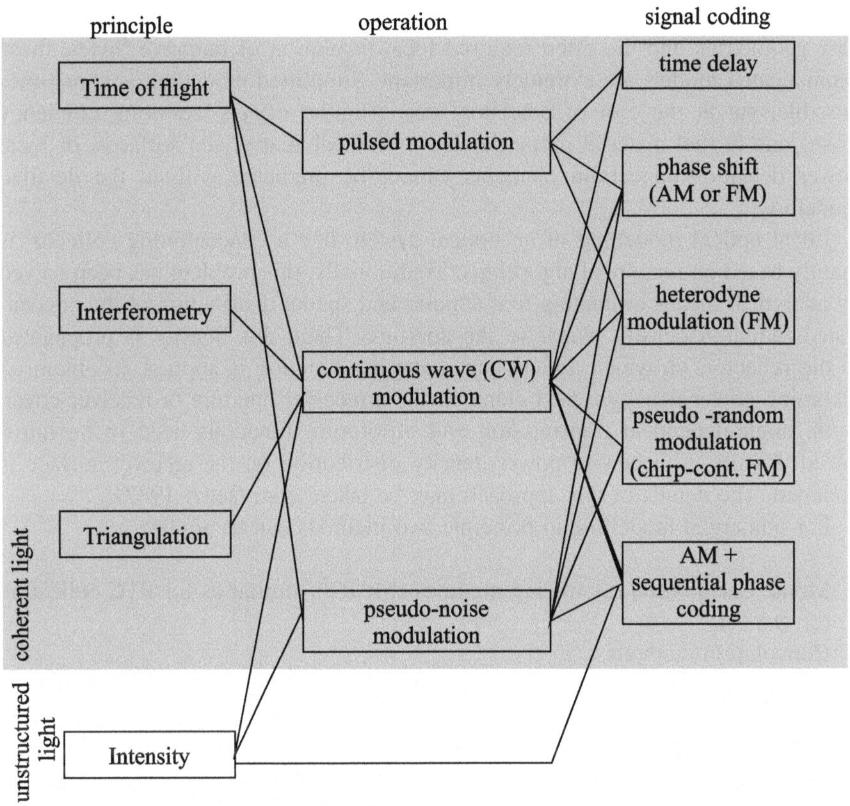

Fig. 3.77 Classification of measurement methods for measuring specular surfaces, with different modes of operation and signal coding options

source (continuous, pulsed, or pseudo-noise methods), based on either time offset or amplitude modulation (AM) or phase modulation (FM) or mixed methods. Successfully also with high time resolution were used so far, the double layer projection method (DLP- (Hillenbrand et al., 2012) and/or the time-of-flight method (ToF), see (Brenneis, 2022).

3.6 Modelling of Optical Systems

For the optical modelling of simple geometries as found in windows or non-concentrating collectors, as treated in Chap. 6 usually a simplified one-dimensional treatment of optical radiation fluxes with balancing equations is sufficient. Only in the case of local shading effects depending on the exact three-dimensional position of transparent and non-transparent components more sophisticated modelling is needed. For concentrating collectors, however, as treated in Chap. 10, the complex geometries and the often required local resolution of radiative fluxes, these sophisticated models are extremely important. Simplified modelling is sometimes possible, but at the cost of precision loss. Angular effects, resulting efficiency losses due to real material properties and non-ideal shapes and surfaces, or local power densities on certain elements cannot be predicted without the detailed modelling.

Local optical modelling of an optical system like a concentrating collector is mainly based on numerical algorithms. Traditionally, this problem has been solved by analytical means evaluating first angular and spatial distribution of the concentrated radiation density planar to the aperture. Then, this density is propagated to the reflector, on which a Gaussian scattering function is applied, to obtain an incoming power density a next element like a receiver aperture or receiver cover. Both, angle-dependent transmission and absorption functions need to be introduced. Finally, a piecewise power density distribution on the receiver surface is obtained. The details of this approach may be taken from (Jeter, 1987).

For numerical modelling in principle two methods can be applied.

- Monte Carlo based ray tracing methods (MCRT), similar as for PTC collectors (cf. 9.3.2.1).
- Convolution methods.

Both methods MCRT and convolution methods require a source description, i.e. a modelling of the sun disk accounting for the non-uniformity of the flux distribution on the solar disk and the widespread solar half-angle of $4.66 mrad$ (0.267°). Almost all simulations assume that the rays pass through air with 100% transmittance, and the reflectance of the reflectors (primary and secondary mirrors is constant; the same is true for transmittance and refractive index of the receiver glass cover. However, in MCRT modelling also the angular-dependent optical properties may be included exactly. A difference between both modelling approaches is that MCRT methods are probabilistic, and only by increasing the number of emitted photons, the resulting power density curve approaches a smooth distribution. The convolution solution, however, is based on the evaluation of a convolution with a Gaussian distribution function, and thus relatively fast. Therefore, this virtual advantage of MCRT methods is not only paid by higher computational effort but also counterbalanced by a fine spatial discretization of all mirror elements to attain a high-quality solution.

In the following two sections we would like to illustrate in detail the two approaches, each time with an example of a specific concentrating collector.

3.6.1 Monte Carlo Raytracing Techniques (MCRT)—Example PTC Collector

Monte-Carlo based Ray Tracing methods ($MCRT$) are based on the probabilistic treatment oft he interaction of a light ray with optical elements like mirrors, absorber, glasses etc.—each interaction of the ray or beam is governed by probability distribution functions. The calculation of a huge number of rays then should reproduce the optical behaviour. The process could also be imagined in a way that a large number of photons are followed on their passage through an optical system. If they hit for example a mirror, a random number and the probability of reflection decides whether the individual photon is reflected or absorbed.

$MCRT$ use randomized rays as indicated in Fig. 3.78 in the example of a PTC, which are traced along their beam path based on a model relying on geometrical optics. The MC simulation first requires a model for the angular distribution of the solar source. The calculation of the ray energy transport from the source to the absorber by means of a $MCRT$ simulation is then structured sequentially as follows:

- First, the position, direction, and energy of each ray on the aperture plane of the PTC are initialized by probability distribution functions.
- Then, based on geometrical optics, the path of each ray, considering reflection at the mirror, transmission and refraction in the glass tube, and absorption at the receiver at each interface, is determined by a series of uniformly distributed random numbers. The ray then is split up or one ray selected with an energy weight, both according to the values of the optical coefficients
- In the third step, the position of the intersection of each ray with the receiver surface with the corresponding energy content is calculated and stored.
- If the relative change in, for example, the energy content for one local target element falls below a pre-determined limit, statistical convergence is assumed

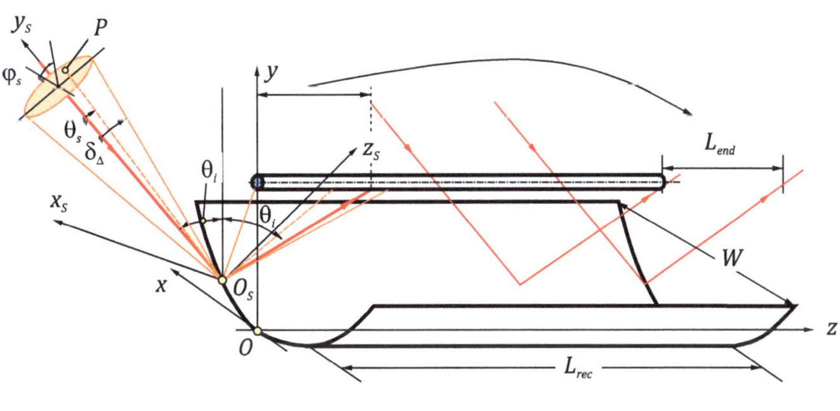

Fig. 3.78 Ray path in a PTC for an irradiation with a longitudinal incident angle of θ_i

for this element and the energy flux density on the absorber surface can be calculated. The entire simulation is finished if all tallies of all target elements have attained statistical convergence, which some time require considerable amounts of rays.

MCRT procedure formulation

In order to trace an arbitrary ray emanating from the sun in as source the world coordinate system $O_s x_s y_s z_s$ must be transformed in the local coordinate system of the PTC, given by $Oxyz$. The coordinate of hitting point (x_0, y_0, z_0) is given by

$$(x_0, y_0, z_0) = \left(W \cdot \kappa_1, x_0^2/4f, L_{rec} \cdot \kappa_2\right), \tag{3.121}$$

in which κ_1, κ_2 are random numbers. Assuming as starting point of a ray P (see Fig. 3.78) on the solar disk its world coordinates are given by,

$$(x_s, y_s, z_s) = (l\sin\theta_s\cos\varphi_s, l\cos\theta_s, l\sin\theta_s\sin\varphi_s), \tag{3.122}$$

where l is the distance $\overline{PO_S}$. θ_s and φ_s denote the radial angle of a sun point and the circumferential one. From this the unit direction vector of the incident ray in the world coordinate system is $\overrightarrow{S} = (\sin\theta_s\cos\varphi_s, -\cos\theta_s, \sin\theta_s\sin\varphi_s)$. Further, the radial ($\theta_s$) and the circumferential angle (φ_s) of the ray´s starting point in the sun are described

$$(\theta_s, \varphi_s) = \left(\arctan\sqrt{\kappa_3}\tan\delta, 2\pi\kappa_4\right), \tag{3.123}$$

withκ_3, κ_4 being again random numbers. The transformation of the world-coordinate system $O_s x_s y_s z_s$ to the PTC coordinate system by translation and rotation is expressed by a transformation matrix$M$, for details see (Zou et al., 2017). The inner normal unit vector of the parabolic reflector ($\overrightarrow{n_r}$) and the glass envelope ($\overrightarrow{n_g}$) are given by

$$\overrightarrow{n_r} = \left(-\frac{x}{\sqrt{x^2+4f^2}}, \frac{2f}{\sqrt{x^2+4f^2}}, 0\right), \overrightarrow{n_g} = \left(\frac{-2x}{d_g}, -\frac{2(y-f)}{d_g}, 0\right), \tag{3.124}$$

in which d_g is the outer diameter of the glas tube. For the unit vectors of the angle of incidence ($\overrightarrow{\lambda}$), the normal unit vector ($\overrightarrow{n}$) and the unit reflected vector ($\overrightarrow{r}$) on the surface (mirror or glass cover), using Fresnel's law, we obtain

$$\vec{r} = \vec{\lambda} - 2 \cdot \left(\vec{\lambda} \cdot \vec{n}\right) \cdot \vec{n}. \tag{3.125}$$

The optical behaviour independently whether reflection, transmission or absorption on a surface is determined as a function of a uniformly generated random number. For example, a uniformly generated random number (κ_3) is compared to the reflection coefficient (ρ) of the mirror. When a ray hits the surface of the mirror the ray is reflected if $\kappa_3 < \rho$, otherwise it is discarded. Other optical quantities along the beam path such as transmission and reflection in the glass cover and absorption at the receiver tube are determined in a similar manner.

In order to initialize the position of the ray on the Sun, a probability distribution function is required, which can be derived using a sunshape model. This describes the intensity distribution in transversal direction perpendicular to the receiver tube. It is given by

$$F(x) = \frac{\int_0^x \phi(\theta) \sin\theta \, d\theta}{\int_0^{\delta_\Delta} \phi(\theta) \sin\theta \, d\theta}, \tag{3.126}$$

in which $\phi(\theta)$ is the solar intensity profile assuming radial symmetry and δ_Δ the outer extent of the solar aureole, finally yielding a probability density function

$$f(x) = \frac{\phi(x) \sin(x)}{\int_0^{\delta_\Delta} \phi(\theta) \sin\theta \, d\theta}. \tag{3.127}$$

Since the direct sampling method is barely applicable mostly acceptance-rejection sampling methods (or hit-and-miss method) are used for random variable sampling, as in other Monte-Carlo techniques as well.

The sun is imaged using countless individual line light sources. These sources are reflected at the mirror surface and those that reach the absorber surface provide the intensity for optical conversion. The image of any line light source in the optical cone with one unit height can be derived using a sunshape model and the intensity $\psi(\theta)$ is obtained by,

$$\psi(\theta) = 2 \int_0^{\sqrt{\delta_\Delta^2 - \theta^2}} \phi\left(\sqrt{\theta^2 + \tau_s^2}\right) d\tau_s, \tag{3.128}$$

where θ is the radial angular displacement of line light source. From this, the total energy in any angle span ($\Delta\theta$) in the light cone can be calculated and further the fraction of energy limited by a given radial angular displacement (θ) of a given line light source can be determined. It can now happen that only a part of the rays located around the light cone are intercepted after reflection on the absorber tube, as shown in Fig. 3.79. This radial intercept angle (θ_γ) can be determined analytically for each position of a point A, x_A,

$$\theta_\gamma = \arcsin\left(\frac{d_{rec}}{\left(2\sqrt{x_A^2 + \left(x_A^2/4f - f^2\right)}\right)}\right), \tag{3.129}$$

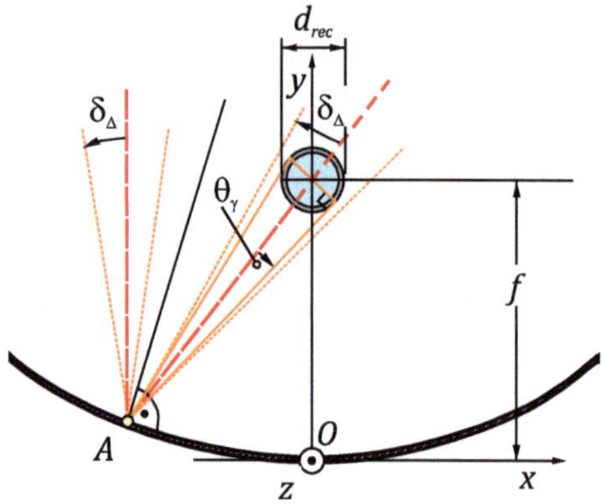

Fig. 3.79 Partially intercepted rays in a PTC collector

in which d_{rec} describes the outer receiver diameter. Again analytically, this allows to define a minimum receiver diameter (d_{min}) given by (3.130) which ensures that all reflected rays reach the receiver. However, in the probabilistic MCRT result, there will be an intensity profile decreasing strongly within the cone of θ_γ, but also outside this cone intensity will be found. Therefore, quantitatively it can be evaluated for which receiver radius how much intensity is lost and spilled.

$$d_{min} = 2\left(W^2/16\ f + f\right) \cdot \sin\delta_\Delta\ . \tag{3.130}$$

However, it is also clear from Fig. 3.78 that part of the incident rays at the angle of incidence (θ) hits the receiver tube after reflection offset on the z-axis. If the entire reflector length L is scanned, a part of the rays no longer reaches the absorber; end losses occur. If the collector row continues infinitely in direction z, the reflected rays can be intercepted by the neighboring collector, but at the end of a collector row, end losses are unavoidable. With the help of trigonometry, the final loss length (L_{end}) as shown in Fig. 3.78 can be calculated:

$$L_{end} = \left(x_A^2/4f + f\right) \cdot \tan\theta_i. \tag{3.131}$$

Due to the longitudinal incidence angle (θ_i) the effective solar radiation intensity (DNI_{eff}) falling on the aperture is reduced on the basis of Lambert's cosine law according to:

$$DNI_{eff} = DNI\ \cos\theta_i. \tag{3.132}$$

This reduction of effective intensity is called cosine-loss and unavoidable. Moreover, it constitutes the largest fraction of losses.

Using the locally evaluated radiance on the receiver tube, one can determine a local concentration factor (C_L), which is defined as the ratio of the local energy flux density ($I(\varphi_{rec})$) to the effective solar irradiance, incident on the aperture.

$$C_L = \frac{I(\varphi_{rec})}{DNI_{eff}}. \tag{3.133}$$

By sampling enough rays to match the statistic convergence criterion set in advance of the computation, which may require in complex designs up to 10^8 rays, which is not a computational challenge since the Monte-Carlo technique allows in a simple manner for massive parallelisation. Despite the fact that the MCRT method describes in a quasi-exact manner the transport along the ray path, it stills contains simplifications. A drawback is that the ray transport at the interface in many codes is considered to be governed by constants as for reflectivity of the mirror, transmissivity of the receiver glass hull and absorptivity of the receiver. Of course, this distorts the results to some extent as all quantities are dependent on the incidence angle (θ_i). It is not a fundamental problem, but a problem of calculation time, when probabilistic distribution functions for the angular optical properties are built into a MCRT code. Especially the transmittance of cover glass (τ_{cover}) depends on the beam direction of its intersection with the absorber surface, so that a simple counting of the accumulated intensity as done in most MCRT-codes yields an overestimation of the yield.

3.6.2 Modelling LFR Collectors -Convolution Methods

Shading and blocking modelling can be done as for a PTC with simple calculation methods, however, requiring in contrast to PTC also consideration of transversal angle dependence (θ_T) in LFC. Details may be taken from (Sharma et al., 2015), (Canavarro et al., 2014). On the other hand, the distribution of incident power on the receiver tube is a localized effect, which can be modelled by MCRT or convolution methods.

The radiance distribution of the sun as well as the optical errors are regarded as statistically independent distributions. The distribution of the optical errors σ_{tot} and the sunshape s_s can be combined by convolution to a resulting distribution L_2 by means of a convolution integral:

$$L_2(\alpha) = \int_{\alpha^*=-\frac{\pi}{2}}^{\frac{\pi}{2}} \sigma_{opt}(\alpha - \alpha^*) \cdot s_s(\alpha^*)\, d\alpha^*. \tag{3.134}$$

This approach is used in the analysis of heliostat fields for tower power plants (Walzl, 1977). Also, the mapping of the heliostat into the image plane of the absorber at an idealized parallel irradiation is taken as a probability density distribution. The convolution of the distributions of sun, heliostat and optical errors results again in a probability density distribution in the image plane of the absorber. The determined function corresponds to the radiant flux density distribution in this image plane. To keep the computational effort of the convolution low, the probability density distributions are approximated by Gauss-Hermite polynomials (Winter et al., 1991). Due to the orthogonality of the polynomials, the convolution can be determined from the moments of the distribution using arithmetic operators. Occlusion or shading losses are not explicitly included in the convolution and are only considered proportionally. The flux density distribution is location dependent.

For Fresnel collectors, the angular distribution of the secondary reflector must be additionally determined from the spatial relationship of heliostat and absorber.

For line-focusing systems, the use of the Gauss-Hermite polynomials can be considerably simplified, since angles of incidence deviating from the transverse plane ($\theta_i > 0°$) change only the length (l) of the reflected rays. But the angular relationships with respect to the transverse plane are preserved. The actual length ($l\prime$) then results from:

$$l' = \frac{l}{\cos\theta_i}. \tag{3.135}$$

First, the primary mirror is discretized into small mirror elements. Further, the reflected radiation fraction at a point P of the receiver aperture is determined (compare Fig. 3.80a). This depends exclusively on the angles α_1 and α_2 at which the mirror element is seen by point P deviating from the ideal reflection. If α_1 and α_2 have the same sign and at the same time the angular range of the radiance distribution L_2 is larger, the point P cannot receive radiation from the mirror element under consideration. With different angular sign but also larger amount than the angular range of L_2, P receives the total irradiance of the source. The radiance distribution $L_2(\alpha)$ includes the source i.e. the sun with its distribution (sunshape) as well as the optical error σ_{opt} of the primary mirrors and is calculated with the help of Eq. (3.135). The convolution must be determined only once because the radiance distribution does not change at the location of the mirror element. Due to the geometrical arrangement of the primary mirrors, a partial shading of the mirror element is possible, which can be described by a flat reduction factor. However, in order to reflect the distribution correctly, the shading angle β (cf. Fig. 3.80b) must be included in the convolution. This results in $L_2(\alpha)$ for the radiance distribution,

$$L_2(\alpha) = \int_{-\frac{\pi}{2}}^{\frac{\pi}{2}} \begin{cases} \alpha* < \beta\ \sigma_{opt}(\alpha - \alpha*)\, L(\alpha*)\, d\alpha* \\ \alpha* > \beta\ 0 \end{cases}, \tag{3.136}$$

which Fig. 3.81 illustrates as a function of the angle α for an unshaded and a shaded primary mirror element.

In principle, this radiance distribution does not change with the location of the mirror element, which is why it is sufficient to determine the convolution only

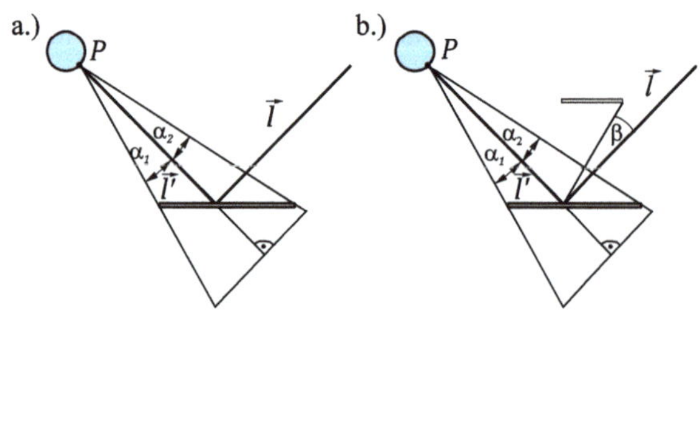

Fig. 3.80 Sketch for determining the reflected radiation fraction at a primary mirror element; (a) without shading and (b) with shading

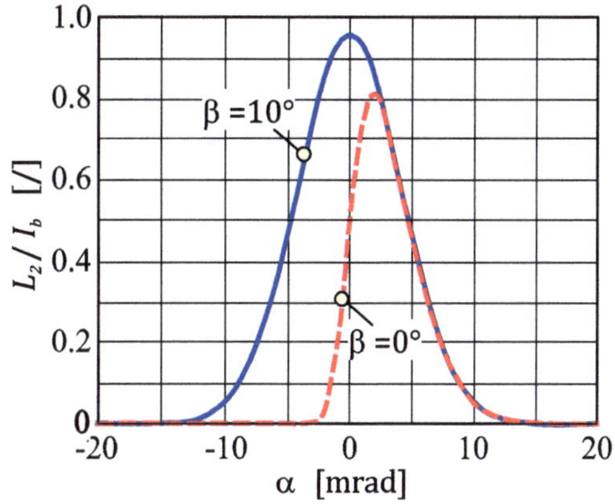

Fig. 3.81 Calculated normalised convolution $L_2(\alpha)$ for unshaded ($\beta = 10°$) and partially shaded primary mirror element ($\beta = 0°$)

once. The radiance transmitted through the mirror element to the point P can be calculated by integrating the convolution $L_2(\alpha)$ from α_1 to α_2 in the form of

$$E(\alpha_1, \alpha_2, \beta) = \int_{\alpha_1}^{\alpha_2} L_2(\alpha, \beta)\, d\alpha, \tag{3.137}$$

where the reflected radiance is the area below the normalised convolution L_2 in the boundaries of α_1 and 2, which is depicted in Fig. 3.82a. Equation (3.137) actually only applies to infinitely long primary mirrors, but this restriction is only relevant when considering the row end losses. The consideration also applies to a sun position outside the transverse plane ($\theta_i > 0$). The limiting angles $\alpha_{1,2}$ are given by the points of the primary mirror element boundaries x_i:

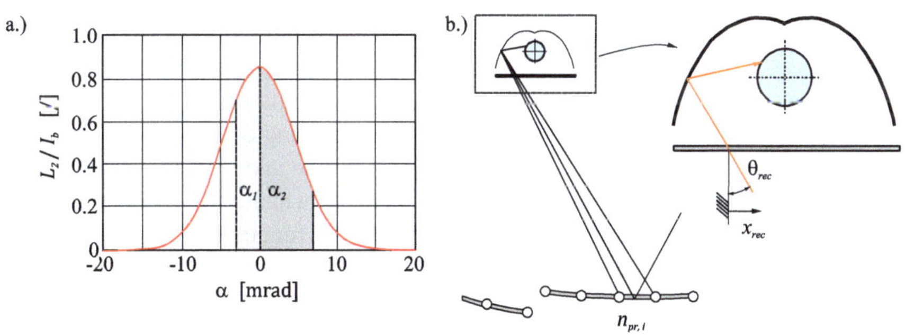

Fig. 3.82 **a** Calculated reflected radiation fraction as area below the normalised convolution in the limits from α_1 to α_2; **b** Coordinate definition for the acceptance range of the secondary reflector of a Fresnel collector

$$\alpha_{1,2} = \cos^{-1}\left(\frac{\left(\vec{P} - \vec{x}_{i,1,2}\right) \cdot \vec{l}'}{\left|\left(\vec{P} - \vec{x}_{i,1,2}\right)\right| \cos\theta_i}\right). \tag{3.138}$$

The total illuminance $E(x_{rec})$ at location P of the receiver results from the summation of the proportions of all mirror elements $n_{pr,i}$ with:

$$E(x_{rc}) = I_b \cdot \sum_{i=1}^{n_{pr}} E(\alpha_1, \alpha_2, \beta). \tag{3.139}$$

For Fresnel systems, the procedure must be extended, since up to now only the depicted portion of the direct radiation of a mirror element at a location on the receiver is determined. For a Fresnel collector, however, the radiation reaching the absorber is important, which is not identical to the above result. The coordinates of a light beam incident on the aperture are determined by the angle θ_{rec} and the coordinate x_{rec} (cf. Fig. 3.82b). Whether the beam hits the absorber tube can be calculated using the forward ray tracing method. The result can be summarised in a matrix representing the acceptance range $\Gamma_{rec}(x_{rec}, \theta_{rec})$ of the secondary reflector. The total radiation E_{abs} reaching the absorber results from the summation over all receiver locations (x_{rec}) and each primary mirror element ($x_{pr,i}$) by:

$$E_{abs} = \rho_{pr} \cdot \sum_{n_{xrec}} \sum_{n_{pr}} \cos(\theta_{rec})\ E(\alpha_1, \alpha_2, \beta)\ \Gamma_{rec}(x_{rec}, \theta_{rec})\ \tau_{rec}(\theta_{rec})\ \Delta x_{rec}. \tag{3.140}$$

A typical result of such a calculation is shown in Fig. 3.83. Plotted is the relative illuminance as the sum of the mirror elements of a primary mirror over the location x_{rec} and the influence of the utilisation of the receiver, which is also a function of θ_{rec} and x_{rec}. In the centre, the mirror directly hits the absorber tube. At the edges of this area, there is a strong attenuation due to the angle-dependent absorption, as the absorber tube is hit at increasingly unfavourable angles. This is followed by the area that reaches the target, i.e. the receiver, via one and then two reflections.

Another central result is depicted in Fig. 3.84a, in which the acceptance range $\Gamma_{rec}(x_{rec}, \theta_{rec})$ of a secondary reflector is shown as a function of the length x_{rec}. The solar radiation that reaches the input aperture of the receiver within the black bordered areas directly or indirectly hits the absorber tube. Figure 3.84b illustrates the phase diagram of the radiation incident from the primary mirror field. The area shadowed by the receiver at $\theta_{rec} = 0°$ is clearly visible. Due to their greater distance, the outer mirrors smear their image over a wider area. In order to lose as little radiation as possible, both images should overlap as well as possible.

Although the convolution method provides both temporally and spatially resolved data, its implementation is more complex and time consuming.

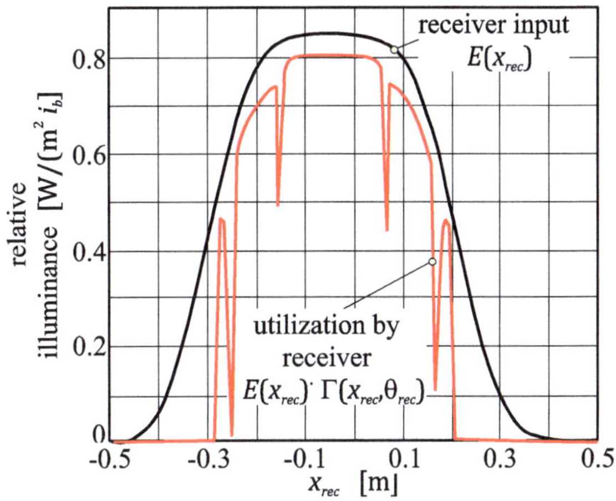

Fig. 3.83 Relative illuminance distribution at the input aperture of the receiver $E(x_{rec})$ of a linear Fresnel collector and its utilisation by the receiver $E(x_{rec})\Gamma_{rec}(x_{rec}, \theta_{rec})$

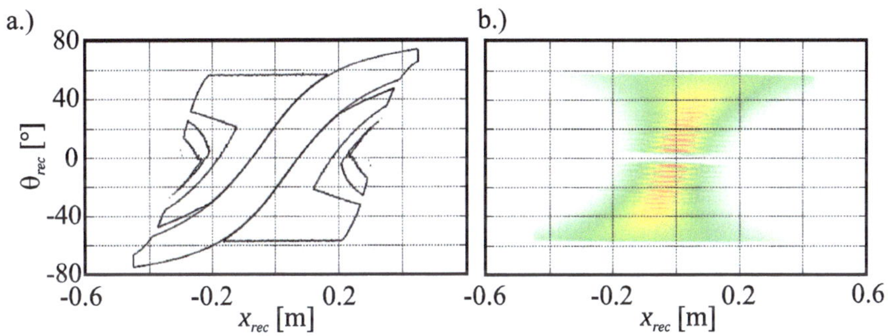

Fig. 3.84 Phase diagrams of the acceptance range of the secondary reflector Γ_{rec} **a** and the radiation **b** as a function of the transverse coordinate x_{rec} and the angle θ_{rec} (Mertins, 2006)

3.6.3 Conclusion

For the modelling of geometrical optics, the MCRT and convolution methods are versatile and flexible. They can provide detailed results with spatial resolution. Although commercial MCRT tools often use simplified assumptions to reduce simulation time, such as optical properties which are constant with incident angle, or not spectrally dependent, in principle all these properties can be included in the code by providing an addition probability distribution function. However, the more properties are included in the probabilistic code, the longer takes a calculation of a ray path. And more over, much more photons with different (spectral) properties have to be emitted and followed in order to get a decent statistic with a representative result.

Other optical simulation software than presented here, e.g. on spectral properties of transparent glasses, coatings, and thin-films, therefore are more sophisticated in the optical calculations, but deterministic and they do not take into account spatially resolved interactions but calculate only a one-dimensional radiation balance in one propagation direction.

References

Abbas, R., Sebastián, A., Montes, M., & Valdés, M. (2018). Optical features of linear Fresnel collectors with different secondary reflector technologies. *Applied Energy, 232*, 386–397. https://doi.org/10.1016/j.apenergy.2018.09.224

Agnihotri, O. P., & Gupta, B. K. (1981). *Solar selective surfaces*. Wiley-Interscience.

Anders, H. (1967). *Thin Films in Optics*. Focal Press Ltd.

Arancibia-Bulnes, E., & C., & Ruiz-Suarez, J. (2000). Solar absorptance and thermal emittance of cermets with large particles. (I. P. Ltd, Ed.). *Journal of Physics. D. Applied Physics, 33*, 2489–2496. https://doi.org/10.1088/0022-3727/33/19/320

Ashcroft, N., Mermin, N., Wei, D., & Mermin, N. (2016). *Solid State Physics* (2nd ed.). Cengage Learning.

ASTM G173–03. (2020). *Standard Tables for Reference Solar Spectral Irradiances: Direct Normal and Hemispherical on 37° Tilted Surface.* West Conshohocken, PA, 19428–2959, USA: ASTM Headquarters . doi:DOI: https://doi.org/10.1520/G0173-03R20

Becker, E. W., Ehrfeld, W., Münchmeyer , D., Betz, H., Heuberger, A., Pongratz, S., . . . von Siemens, R. (1982). Production of separation-nozzle systems for uranium enrichment by a combination of X-ray lithography and galvanoplastics. *Naturwissenschaften*, Vol 69(11), pp. 520-523.

Bläsi, B. (2000). *Holographisch hergestellte Antireflexoberflächen für solare und visuelle Anwendungen.* Dissertation, Albert-Ludwigs-Universität Freiburg i. Br., Freiburg im Breisgau.

Bläsi, B., Aufderheide, K., & Abott, S. (2002). Entspiegeln mit Mottenaugenstrukturen. *Kunststoffe, Jahrgang, 92*, 50–53.

Born, M., & Wolf, E. (1999). *Principles of Optics-Electromagnetic Theory of Propagation.* Cambridge University Press. https://doi.org/10.1017/CBO9781139644181.

Brenneis, B. (2022). Long Range Optical Distance Sensors for Liquid Metal Free Surface Detection. *Journal of Nuclear Engineering, 3*(4), 364–372. https://doi.org/10.3390/jne3040023.

Brunold, S., Frei, U., Carlsson, B., Möller, K., & Köhl, M. (2000). Accelerated life testing of solar absorber coatings: Testing procedure and results. *Solar Energy, 68(4)* , pp. 313-323. doi:https://doi.org/10.1016/S0038-092X(00)00034-7.

Brunold, S., Frei, U., Carlsson, B., Möller, K., & Köhl, M. (2000). Round robin on accelerated life testing of solar absorber surface durability. *Solar Energy Materials and Solar Cells*, 61(3), pp. 239-253. doi:https://doi.org/10.1016/S0927-0248(99)00113-0.

Brunotte, A., Lazarov, M., & Sizmann, R. (1992). Calorimetric measurements of the total hemispherical emittance of selective surfaces at high temperatures. *Proc. SPIE 1727, Optical Materials Technology for Energy Efficiency and Solar Energy Conversion XI, 1727*. doi:https://doi.org/10.1117/12.130497.

Burke, J., Li, W., Heimsath, A., Kopylow, C. v., & Bergmann, R. B. (2013). Qualifying parabolic mirrors with deflectometry. *Journal of the European Optical Society - Rapid publications; Vol 8.*

Canavarro, D., Chaves, J., & Collares-Pereira, M. (2014). Simultaneous multiple surface method for linear fresnel concentrators with tubular receiver. *Solar Energy, 110*, 105–116. https://doi.org/10.1016/j.solener.2014.09.002.

Chaves, J., & Collares-Pereira, M. (2010). Etendue-matched two-stage concentrators with multiple receivers. *Solar Energy, 84*(2), 196–207. https://doi.org/10.1016/j.solener.2009.10.022.

Claesson, Y., Georgson, M., Roos, A., & Ribbing, C.-G. (1990). Optical characterisation of titanium-nitride-based solar control coatings. *Solar Energy Materials, 20*(5–6), 455–465. https://doi.org/10.1016/0165-1633(90)90035-Y.

Cuomo, J., Ziegler, J., & Woodall, J. M. (1975). A new concept for solar energy thermal conversion. *Applied Physics Letters, 26*(10), 557–550.

Demtröder, W. (2004). *Experimentalphysik 2 (2* (Auflage). Springer Verlag.

Duffie, J. A., & Beckmann, W. A. (2020). *Solar Engineering of Thermal Process, Photovoltaics and Wind*. John Wiley & Sons Inc.

Fernandez-Garcia, A., Sutter, F., Montecchi, M., Sallaberry, F., Heimsath, A., Heras, C., & Le Baron, E. a.-G. (2018). *International Guideline Version 3.0: Parameters and method to evaluate the solar reflectance properties of reflector materials for concentrating solar power technology.* www.solarpaces.org: IEA SolarPACES.

Gäbler, D. (2005). *Entspiegelung von Silizium-Photodioden nach dem Vorbild der Nanooptik von Mottenaugen*. Diplomarbeit TU Ilmenau, Illmenau.

Giovanelli, R., & Beck, R. (1987). *Geheimnisvolle Sonne*. VCH Verlagsgesellschaft.

Gombert, A., & Rommel, M. (1997). Breitbandige Antireflexbeschichtungen. (F. Sonnenenergie, Ed.) *Solare Gebäudetechniken*, 81–86.

Gombert, A., Glaubitt, W., Rose, K., Dreibholz, J., Bläsi, B., Heinzel, A., . . . Wittwer, V. (1999). Subwavelength-structured antireflective surfaces on glass. *Thin Solid Films*, Vol 351(1), pp. 73-78.

Graedel, T. E. (1992). Corrosion Mechanisms for Silver Exposed to the Atmosphere. *J. Electrochem. Society, 139*(7), 1962–1970.

Granqvist, C. (1990). Window Coatings for the Future. *Thin Solid Films*, 193-194, pp. 730-741. doi:dx.doi.org/https://doi.org/10.1016/0040-6090(90)90225-3.

Granqvist, C. (1991). Energy-efficient windows: Present and fortcoming technology. In C. (. Granqvist, *Materials Science for Solar Energy Conversion Systems* (pp. 106–167). Oxford: Pergamon Press plc.

Granqvist, C. (1991). Solar energy materials-Overview and some examples. *Appl. Phys. A*, 52, pp. 83-93.

Guerreiro, L., Canavarro, D., & Collares Pereira, M. (2015). Efficiency improvement and potential LCOE reduction with an LFR-XX SMS plant with storage. *Energy Procedia*, 69, pp. 868-878. doi:doi: https://doi.org/10.1016/j.egypro.2015.03.112.

Heimsath, A. (2020). *Optical efficiency of concentration solar collectors - Investigation of loss mechanisms of solar reflectors*. Dissertation, Universität Bremen, Bremen.

Heimsath, A., Platzer, W., Bothe, T., & Li, W. (2008). Characterization of optical components for linear Fresnel collectors by the fringe reflection method. *Proc. 14th SolarPACES International Symposium, Las Vegas, 3–7 Mar. 2008.*

Heimsath, A., Schmid, T., & Nitz, P. (2015). Angular Resolved Specular Reflectance Measured with VLABS. *Energy Proc., 69*, 1895–1903.

Hildebrandt, C. (2009). *Hochtemperaturstabile Absorberschichten für linear konzentrierende solarthermische Kraftwerke*. Dissertation, Universität Stuttgart, Institut für Thermodynamik und Wärmetechnik, Stuttgart.

Hillenbrand, M., Stieglitz, R., & Neitzel, G. (2012). Detection of liquid–metal, free-surface flow using the DLP measurement technique. *Experiments in Fluids*, 52, pp. 179–192. doi:doi.org/https://doi.org/10.1007/s00348-011-1214-z.

Hi-min SOLAR. (2022). Parabolic Trough Receiver data sheet. Sun-Moon Mansion, Solar Valley Road, Dezhou, Shandong, PR China. Retrieved March 22, 2022, from http://himinsun.com/1-3-parabolic-trough-receiver.html.

Hollingsworth Smith, P., & Gurev, H. S. (1977). Silicon Dioxid as a High Temperature Stabilizer for Silver Films. *Thin Solid Films, 45*, 159–168.

Howell, J., Mengüc, M. P., Daun, K., & Siegel, R. (2020). *Thermal Radiation Heat Transfer* (7 ed.). CRC Press ,Taylor & Francis Group, LLC. Retrieved from https://doi.org/10.1201/9780429327308.

ISO 2353. (2005). *Photometry—The CIE system of physical photometry.*

ISO 80000–7. (2019). *Quantities and units—Part 7: Light and radiation.*

ISO 9050. (2003). *Glass in building—Determination of light transmittance, solar direct transmittance, total solar energy transmittance, ultraviolet transmittance and related glazing factors.*

ISO 9845–1. (1992). *Solar energy—Reference solar spectral irradiance at the ground at different receiving conditions — Part 1: Direct normal and hemispherical solar irradiance for air mass 1,5.* ICS:27.160 Solar energy engineering.

Jarmer, T., Serwaty, J., & Knote, C. (2005). *Grundlagen und Praxis radiometrischer; Abschnitt: Fernerkundliche Erfassung photosynthetisch aktiver Pigmente und weiterer Blattinhaltsstoffe.* Universität Trier.

Jeter, S. M. (1987). Analytical determination of the optical performance of practical parabolic trough collectors from design data. *Solar Energy, 39*(1), 11–21.

Kadkhoda, P., Müller, A., Ristau, D., Duparré, A., Gliech, S., Lauth, H., Bennett, J. M. (2000). International round-robin experiment to test the International Organization for Standardization total-scattering draft standard. *Applied optics*, 39(19), pp. 3321–3332.

Kammel, S. (2005). *Deflektometrische Untersuchung spiegelnd reflektierender Freiformflächen.* Dissertation, Universität Karlsruhe.

Kennedy, C. E. (2002). *Review of mid- to high-temperature solar selective absorber materials.* Technical Report , National Renewable Energy Laboratory, Golden, Colorado.

Kreider, J. F. (1979). *Medium and high temperature solar processes.* Academic Press Inc.

Kreith, F., & Kreider, J. F. (1978). *Principles of solar engineering.* Hemisphere Publishing Corporation.

Lanxner, M., & Elgat, Z. (1990). March 12–13). Solar selective absorber coating for high service temperatures, produced by plasma sputtering. (e. G. C., Ed.) *SPIE- Proc. Optical Materials Technology for Energy Efficiency and Solar Energy Conversion IX, 1272*, 240–249.

Maccari, A., & Montecchi, M. (2007). An optical profilometer for the characterisation of parabolic trough solar concentrators. *Solar Energy, 81*(2), 185–194.

MacLeod, A. (2001). *Thin Film Optical Filters.* Bristol: Adam Hilger Ltd.

McNaught, A. D., & Wilkinson, A. (1997). *IUPAC. Compendium of Chemical Terminology (2nd* (edition). Blackwell Scientific Publications.

Mertins, M. (2006). *Technische und wirtschaftliche Analyse von horizontalen Fresnel-Kollektoren.* Uinversität Karlsruhe: Fakultät Maschinenbau.

Miller, W. H., Möller, A. R., & Bernhard, C. G. (1966). The corneal nipple array. In C. G. Bernhard (Ed.), *Functional organization of the compound eye* (pp. 21–33). Pergamon Press.

Morin, G., Dersch, J., Platzer, W., Eck, W., & Häberle, A. (2012). Comparison of Linear Fresnel and Parabolic Trough Collector plants. *Solar Energy*, 86, pp. 1-12. doi:doi:https://doi.org/10.1016/j.solener.2011.06.020

Niemann, M., & Schreitmüller, K. R. (1993). Investigations on high efficiency collectors for industrial process heat generation up to 200°C. *Proc. of ISES World Congress Budapest, Hungary.*

Nolting, W. (2009). *Grundkurs Theoretische Physik 7: Viel-Teilchen-Theorie.* Springer Verlag.

Palik, E. (1985). *Handbook of optical constants of solids* (8th ed.). Academic Press.

Platzer, W. J., & Bergkvist, M. (1993). Bulk and Surface Light-Scattering from Transparent Silica Aerogel. *Solar Energy Materials and Solar Cells, 31*(2), 243–251.

Platzer, W., Heimsath, A., Hildebrandt, C., & Georg, A. (2008). Quality Control of concentrating collector components for the optimization of performance. *Proc. 14th SolarPACES International Symposium, Mar 4–7, 2008.* Las Vegas, USA.

Pottler, K., Lüpfert, E., Johnston, G. H., & Shortis, M. R. (2005). Photogrammetry. A Powerful Tool for Geometric Analysis of Solar Concentrators and Their Components. *J. Sol. Energy Eng. 127 (1)*, pp. 94–101.

Powell, C. F., Oxley, J. H., Blocher, J. M., & Klerer, J. (1966). Vapor Deposition. *Journal of The Electrochemical Society,* 113.

Pulker, H. K. (1999). *Coatings on Glass* (2nd ed.). Elsevier Science B.V.

Rabl, A. (1976). Comparison of solar concentrators. *Solar Energy, 18*, pp. 93-111.

Rabl, A. (1976b). Optical and thermal properties of compound parabolic concentrators. *Solar Energy, 18*(6), 497–511.

Rabl, A., & Winston, R. (1976). Ideal concentrators for finite sources and restricted exit angles. *Applied Optics, 15*(11), 2880–2883. https://doi.org/10.1364/AO.15.002880

Randich, E., & Allred, D. (1981). Chemically vapor-deposited ZrB2 as a selective solar absorber. *Thin Solid Films, 83*(4), 393–398. https://doi.org/10.1016/0040-6090(81)90646-5

Rioglass Solar SA. (2022). Receiver Tubes for Linear CSP. Poligono Industrial Villallana Pola de Lena, 33695, Spain. Retrieved March 22, 2022, from https://www.rioglass.com/our-products/hce-tubes.html

Sakka, S. (2005). Applications of Sol-Gel Technology. In S. Sakka (Ed.), *Handbook of Sol-Gel Science and Technologie (Vol* (*III*, pp. 599–760). Kluwer Academic Publishers.

Schade, H., & Smith, Z. (1985). Mie scattering and rough surfaces. *Applied Optics, 24*(19), 3221–3226. https://doi.org/10.1364/AO.24.003221

Sella, C., Bichri, A., Martin, J. C., Lafait, J., Driss-Khodja, K., & Berthier, S. (1989). Adjustable optical properties of coatings based on cermet thin films near the percolation threshold. *Physica a: Statistical Mechanics and Its Applications, Vo., 157*(1), 555–560.

Seraphin, B. (1979). Spectrally selective surfaces and their impact on photothermal solar energy conversion. In B. Seraphin, *Solar Energy Conversion* (pp. 5–55). Springer,. doi:https://doi.org/10.1007/3-540-09224-2_2.

Seraphin, B., & Meinel, A. (1976). Optical Properties of Solids — New Developments. *In B. Seraphin. Amsterdam: North-Holland.* https://doi.org/10.1016/0030-3992(76)90029-3.

Sharma, V., Nayak, J., & Kedare, S. B. (2015). Comparison of line focusing solar concentrator fields considering shading and blocking. *Solar Energy, 122*, 924–939. https://doi.org/10.1016/j.solener.2015.10.011.

Smakula, A. (1957). *Acta Crystallographica, Vol. 10*, 735–863.

Tagle-Salazar, P., & Nigam, K. R.-S. (2020). Parabolic trough solar collectors: A general overview of technology, industrial applications,energy market, modeling, and standards. (D. Gruyter, Ed.) *Green Processing and Synthesis, 9*, pp. 595–649. doi:https://doi.org/10.1515/gps-2020-0059.

Thiemann, S. (2000). *Das Prinzip Fernerkundung.* Scientific Technical Report 00/04, Geo-Forschungs Zentrum Potsdam, Potsdam.

Thornton, J. A., Penfold, A., & Lamb, J. L. (1980). Sputter-deposited Al2O3/Mo/ Al2O3 selective absorber coatings. *Thin Solid Films, 72*, 101–109.

Valkonen, E., & Karlsson, B. (1987). Optimization of metal-based multilayers for transparent heat mirrors. *Energy Research, 11*(3), pp. 397-403. doi:doi.org/https://doi.org/10.1002/er.4440110309.

Valkonen, E., Karlsson, B., & Ribbing, C.-G. (1984). Solar optical properties of thin films of Cu, Ag, Au, Cr, Fe Co. *Ni and Al. Solar Eneregy, 32*(2), 211–222. https://doi.org/10.1016/S0038-092X(84)80038-9

Vien, T. K., Sella, C., Lafait, J., & Berthier, S. (1985). Pt-Al2O3 selective cermet coatings on superalloy substrates for photothermal conversion up to 600°C. *Thin Solid Films, 126*, 17–22.

Walzl, M. D. (1977). A Solar Flux Density Calculation for a Solar Tower Concentrator Using A Two-Dimensional Hermite Function Expansion. *Solar Energy, 19*, 239–253.

Wang, M., & Pan, N. (2008). Predictions of effective physical properties of complex multiphase materials. *Materials Science and Engineering* R, 63, pp. 1-30. doi:doi:https://doi.org/10.1016/j.mser.2008.07.001.

Willey, R. (2002). *Practical Design and Production of Thin Films* (2nd ed.). Marcel Dekker Inc.

Winston, R. (1974). Principles of solar concentrators of a novel design. *Solar Energy, 16*(2), 89–95.

Winter, C. J., Sizmann, R. L., & Vant-Hull, L. L. (1991). *Solar power plants: Fundamentals, technology, systems, economics.* Springer Verlag.

Xiao, J., Wei, X., Lu, Z., Yu, W., & Wu, H. (2012). A review of available methods for surface shape measurement of solar concentrator in solar thermal power applications. *Renewable and Sustainable Energy Reviews, 16*(5), 2539–2544.

Zhang, Q. C., & Mills, D. (1992). New cermet film structures with much improved selectivity for solar thermal applications. *Appl. Phys. Lett.*, 60, pp. 545-547. doi:doi.org/https://doi.org/10.1063/1.106602.

Zhao (2007). *Spectrally selective solar absorbing coatings prepared by dc magnetron sputtering.* Uppsala Schweden. Acta Universitatis Upsaliensis.

Zhu, G. (2017). New adaptive method to optimize the secondary reflector of linear Fresnel collectors. *Solar Energy, 144*(1), 117–126. https://doi.org/10.1016/j.solener.2017.01.005.

Zou, B., Yao, Y., & Jiang, Y. (2017). A detailed study on the optical performance of parabolic trough solar collectors with Monte Carlo Ray Tracing method based on theoretical analysis. *Solar Energy, 147*, 189–201. https://doi.org/10.1016/j.solener.2017.01.055.

Heat Transfer

4

Abstract

Thermal technologies and thus also solar thermal always involve heat transfer within a body or between bodies. In some cases the heat transfer has to be enhanced in order to achieve a highly efficient optical conversion of solar radiation into heat, in other cases a minimal heat loss over the system boundaries to the environment is the target. In particular, the optical conversion of short wave solar irradiance (electromagnetic waves) to heat occurs in solids. Within the solid body the heat is released mostly volumetrically in thin layers or layer stacks causing not only temperature rise but also differential thermal elongation. This heat has to be transported via passive means towards a heat sink for utilization. At the same time, however, a body that is warmer than its surroundings loses this heat in the form of thermal radiation and convection, and as the temperature rises, this proportion increases rapidly. This conflicting feature emphasizes the excellent importance of understanding passive heat transfer mechanisms in solar thermal applications. In this context we consider as passive heat transfer processes such as heat conduction and heat radiation, which are not associated with a medium transport. Convection, however, which enhances heat transport through molecular heat conduction in fluid media by a movement of fluid volumes, will be treated in the Chap. 5. The driving mechanisms of passive heat transport are scalar quantities such as pressure and temperature, which characterize the thermodynamic state in a continuum, as well as waves, which propagate in a medium. The propagation of the waves in a solid or fluid medium, in turn, depends on its molecular thermo-physical properties and in cases of solids even its lattice structure. For this reason, any temperature difference leads unavoidably to a heat transfer changing the thermal properties of a solar thermal system, but also represents a key variable in component design. An important technical parameter here is, among others, the maximum permissible wall temperature. After an introductory description of the distinction between heat transfer and transport processes, the steady and transient heat conduction in gases, fluids and solids is covered. Successively, radiative heat

R. Stieglitz and W. Platzer, *Solar Thermal Energy Systems*,
https://doi.org/10.1007/978-3-031-43173-9_4

transport in different configurations as well as methods to evaluate to transferred quantities are elaborated. Finally, as a result of a heat transport by conduction and radiation across interfaces mixed situations may occur, in which density differences in the adjacent fluid/gas are induced yielding to natural buoyant convective fluid motion increasing the heat transfer. Some selected cases of combined heat transfer problems are covered in the last section.

Keywords

Steady and transient heat conduction • Conduction in solid and gases • Heat radiation • Radiosity • Irradiance • Radiation transport

4.1 Definition of a Passive Heat Transport

In thermodynamics, energy that crosses the boundary of a system is called heat if the energy transport is caused solely by a temperature difference between the system and its environment. According to the second main theorem of thermodynamics, heat always flows in the direction of lower temperature within a body and across the body interface. Adopting this approach the term passive heat transfer describes a heat energy transport that occurs exclusively via microscopic collisions within a continuum or interactions of waves (phonons, photons, electrons), i.e. molecular or electromagnetic interactions. Thus, it is not a process necessarily associated with a matter transport such as a fluid flow, however, it describes the heat transport within a continuum and across a continuum interface. This passive type of heat transport is either influenced by the thermo-physical material properties or via the surface properties of the interface. Thus, the solution of a passive heat transfer problem in terms of thermodynamics answers the following questions:

- heat transfer though molecular conduction in a continuum and,
- heat exchange across interface by radiation between bodies.

General thermodynamics considers macroscopically a continuum in equilibrium or quasi-equilibrium states called phases, while heat transfer considers the exchange of heat. This exchange of heat changes the internal energy of the volume under consideration. Since thermodynamics considers only the initial and final states, it is not able to make a statement about how fast or intense the heat transfer occurs. In contrast to thermodynamics, which operates with homogeneous systems having the same properties, the so-called phases, heat transfer is a continuum theory in which spatially extended fields with variable properties occur, which may also depend on time. In most cases the driving temperature differences play the dominant role in heat transfer. Therefore, temperatures differences ΔT are used, where the temperature T is described in relation to a fixed reference temperature T_0:

$$\Delta T = T - T_0. \tag{4.1}$$

Using as reference temperature $T_0 = 273.15$ K, ΔT corresponds to the temperature in degree Celsius. Temperature differences are independent of the reference temperature, so that its specification can usually be omitted. The continuum theory of steady-state or transient heat transfer at an interface deals with all possible forms of heat energy exchange between and within media. In multi-body problems, for example, there is a superposition of heat conduction and radiation. Typical cases are packed beds, insulating materials or the gap between the glass cover and the absorber of a solar collector.

However, the temperature differences along the interface of a body generated by passive heat transport can also induce density gradients that trigger macroscopic fluid motion (buoyant or natural convection), so that thermal conduction, thermal radiation, and convective heat transport interact. Since the radiation and moreover convection are non-linear physical phenomena, only a few cases allow for analytical access. Some such coupled issues are addressed in a simplified manner at the end of the chapter, here especially emphasis is put on the derivation of dimensionless numbers to characterize an engineering phenomenon in order to decide in which manner it can be solved. The detailed treatment of convective heat transport by natural convection is provided in Chap. 5.

4.2 Heat Conduction in a Continuum

Heat conduction is a net transfer of kinetic energy of unordered movements (Brownian motion) between neighbouring molecules in otherwise non-moving media due to a temperature gradient. In metals, the free electrons also transfer energy. In opaque media, energy is transported by thermal conduction alone; in gases and liquids, energy transport by the flowing motion of a fluid (convection) and by thermal radiation is often superimposed on the thermal conduction process. Depending on the medium in which conductive heat transport occurs, the carriers of energy transport are atoms, molecules, electrons, or phonons. The latter are energy quanta of elastic waves, which are responsible for the transport of thermal energy in non-metals and—besides electrons—also in metals.

Let us consider the simplest case of a flat, homogeneous wall (or a gap homogeneously filled with a continuum at rest) with temperature T_1 on one side and temperature T_2 on the other side as depicted in Fig. 4.1. The energy transfer in the heat-conducting material due to the temperature difference can be described by the vector field of the heat flux density $\dot{q}$, where

$$\dot{q} = \overrightarrow{\dot{q}}\,(\vec{x}, t), \tag{4.2}$$

is a vector and indicates the strength and direction of an energy flow, which can also depend on the time t, at a location characterized by the vector $\vec{x}$. The corresponding temperature field T is in contrast to the heat flux a scalar but also a function of location $\vec{x}$ and time t and is given by

$$T = T(\vec{x}, t). \tag{4.3}$$

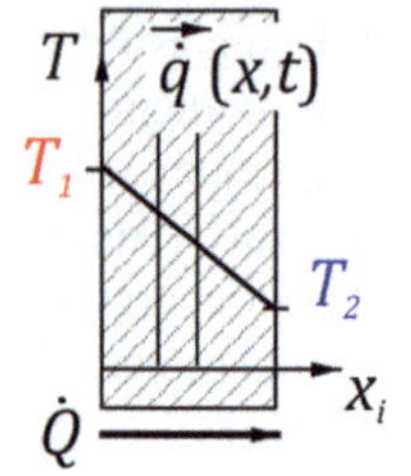

Fig. 4.1 Sketch of heat transport by heat conduction in a continuum

For a steady process ($\partial/\partial t = 0$) the time dependence of t disappears. All points of the body, which have the same temperature T at a certain time, can be thought of as connected by a surface a so-called isotherm. The strongest temperature change occurs normal to the isotherms and is defined by the temperature gradient ∇T:

$$\nabla T = \frac{\partial T}{\partial x} + \frac{\partial T}{\partial y} + \frac{\partial T}{\partial z}. \tag{4.4}$$

The heat flux density $\vec{q}$ is defined such that the heat flux $d\ \dot{Q}$ through an arbitrarily oriented element dA is given by

$$d\dot{Q} = \vec{q}\,(\vec{x}, t) \cdot \vec{n} dA = \left|\vec{q}\right| \cdot \cos\beta \cdot dA, \tag{4.5}$$

where $\vec{n}$ denotes the plane normal vector forming the angle β with $\vec{q}$. The heat flux is thus a time-related energy, i.e. a power, and has the dimension J/s = W and the heat flux density accordingly an area-related reading given in W/m_2. If we now assume the temperature gradient as the cause of the occurring heat flux it is obvious to formulate a proportionality between cause and effect and the first Fourier law results:

$$\vec{q} = -k \cdot \nabla T = -k \cdot \frac{dT}{d\vec{x}}. \tag{4.6}$$

The minus sign appearing in the Eq. (4.6) considers the second law of thermodynamics, wherein heat always flows in the direction of the lower temperature. The proportionality constant k is a material property and is called specific thermal conductivity and has the SI-Units W/(mK). It depends on the temperature and pressure. Only if the material is isotropic, the specific thermal conductivity is a scalar quantity, otherwise it is a tensor of the second order.

For the heat flow $d\dot{Q}$ through an arbitrarily oriented surface element dA, substituting the Eq. (4.6) into (4.5) yields the following result:

$$d\dot{Q} = -k\ \nabla T\ (\vec{n}\ dA) = -k \cdot dA \cdot \left(\frac{\partial T}{\partial n}\right). \tag{4.7}$$

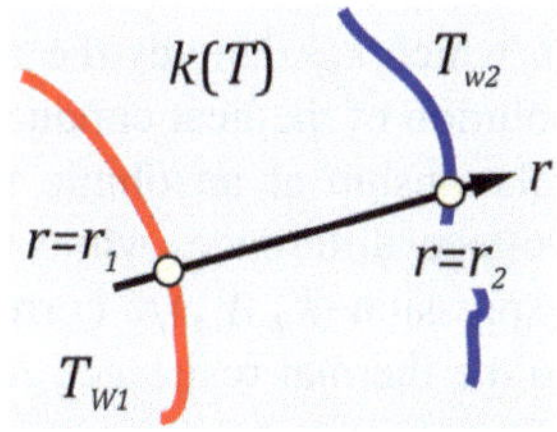

Fig. 4.2 Sketch of steady conductive heat transport in an arbitrary continuum

The basis of solving complicated heat conduction problems is the differential equation for the temperature field in a medium at rest, called the heat conduction equation. The solution of a heat conduction problem means to determine the temperature field $T(\vec{x}, t)$ in its spatial and temporal dependence. Once it is known, the associated heat flux density field can be calculated using Fourier's law and the heat fluxes transported by heat conduction can be determined at arbitrary locations of the body.

In the stationary case of a one-dimensional heat conduction, the temperature field T depends only on one spatial coordinate $\vec{r}$. The same applies to a hollow sphere, which is therefore included in the following considerations for the sake of completeness. If we consider first a stationary one-dimensional heat conduction, then the temperature is only a function of the spatial coordinate r, where at the place r the temperature $T = T(r)$ appears and for each $r = const.$ an isothermal surface is obtained. Assume that for $r = r_1$ the continuum has the temperature $T = T_{W1}$, respectively for $r = r_2$ a temperature $T = T_{W2}$, see Fig. 4.2.

If $T_{W1} > T_{W2}$, then a quantity of heat flows by heat conduction from r_1 over the surface A_1 to r_2 into the surface A_2. According to Fourier's law in Eq. (4.6), the heat flow in [W/m^2] through a surface $A(r)$ along r is given in the following way:

$$\dot{Q} = \dot{q}(r)A(r) = -k(T)\frac{dT}{dr}A(r), \tag{4.8}$$

or

$$-k(T)dT = \dot{Q}\frac{dr}{A(r)}. \tag{4.9}$$

Integration of Eq. (4.9) over the wall thickness from r_1 to r_2 ($\delta = r_2 - r_1$) leads to

$$-\int_{T=T_{W1}}^{T=T_{W2}} k(T)dT = \dot{Q}\int_{r=r_1}^{r=r_2} \frac{dr}{A(r)}, \tag{4.10}$$

and with the mean value theorem of the integral calculus we obtain:

$$-k_m(T_{W2} - T_{W1}) = \dot{Q}\frac{\delta}{A_m} \text{ or } \dot{Q} = \frac{k_m}{\delta}A_m(T_{W1} - T_{W2}), \tag{4.11}$$

in which k_m denotes the average specific thermal conductivity. In this form, the solution of the heat conduction equation corresponds to that of the current–voltage relationship at an ohmic resistor. The temperature difference is equivalent to a voltage difference, while the heat flow is equivalent to an electric current. The expression $(k_m A_m)/\delta$ corresponds to the thermal conductance G and its reciprocal to the thermal resistance R_H

$$R_H = \frac{\delta}{k_m A_m} = \frac{1}{G}\left[\frac{K}{W}\right]. \quad (4.12)$$

By analogy with Ohm's law in electricity, the following is then obtained for the heat flow $\dot{Q}$:

$$\dot{Q} = \frac{(T_{W2} - T_{W2})}{R_H}. \quad (4.13)$$

This analogy allows the use of further methods from electrical engineering. The average thermal conductivity k_m is given by:

$$k_m = \frac{1}{(T_{W2} - T_{W1})} \int_{T=T_{W1}}^{T=T_{W2}} k(T)dT, \quad (4.14)$$

which can be simplified often considering small temperature differences. If the temperature dependence of thermal conductivity in the considered temperature interval is negligible a constant thermal conductivity $k_m = k$ can be used, while in case of a linear temperature dependence of the thermal conductivity, we obtain:

$$k_m = \frac{1}{2}[k(T_{W1}) + k(T_{W2})]. \quad (4.15)$$

For a flat wall, area A normal to the heat flux does not change with the spatial coordinate r, so $A \neq A(r)$ holds. In the case of a hollow cylinder or a hollow sphere this is different, and one obtains for the area A:

$$A(r) = \begin{cases} A_1 = A_2 & \text{for a planar wall} \\ 2\pi L r & \text{for a hollow cylinder of length } L \\ 4\pi r^2 & \text{for a hollow sphere} \end{cases}. \quad (4.16)$$

Accordingly, the following result for the mean area A_m is obtained:

$$A_m = \begin{cases} A_1 = A_2 = \frac{1}{2}(A_1 + A_2) & \text{for a planar wall} \\ \frac{(A_2 - A_1)}{\ln\left(\frac{A_2}{A_1}\right)} & \text{for a hollow cylinder} \\ \sqrt{A_1 A_2} & \text{for a hollow sphere} \end{cases}. \quad (4.17)$$

If one assumes a constant temperature-independent specific thermal conductivity $k_m = const.$, then the following results:

$$-dT = \frac{\dot{Q}}{k}\frac{dr}{A(r)} \tag{4.18}$$

and integration over r yields the sought temperature distribution $T(r)$:

$$\frac{T(r) - T_{W2}}{T_{W1} - T_{W2}} = \begin{cases} \frac{r_2 - r}{r_2 - r_1} & \text{for a planar wall} \\ \frac{\ln\left(\frac{r_2}{r}\right)}{\ln\left(\frac{r_2}{r_1}\right)} & \text{for a hollow cylinder} \\ \frac{\frac{1}{r} - \frac{1}{r_2}}{\frac{1}{r_1} - \frac{1}{r_2}} & \text{for a hollow sphere} \end{cases} . \tag{4.19}$$

Thus, for $k_m = const.$ the central result for the temperature distribution given by Eq. (4.19) is:

- in a flat wall, the dimensionless wall temperature scales linearly with the thickness r,
- in a hollow cylinder, the temperature changes logarithmically with the cylinder radius r,
- in and inside the wall of a hollow sphere, temperature is hyperbolically dependent on the coordinate r.

To derive the general equation of conductive heat transfer, we first apply the first law of thermodynamics to a closed system. To do this, we consider a continuous region of any size, which we imagine to be cut out of the heat-conducting body, as shown in Fig. 4.3.

The first law of thermodynamics provides the power balance in the differential Volume $dV = dx \cdot dy \cdot dz$ and reads to:

$$\frac{dU}{dt} = \sum_i \dot{Q}_i + \dot{W} + q_V dV. \tag{4.20}$$

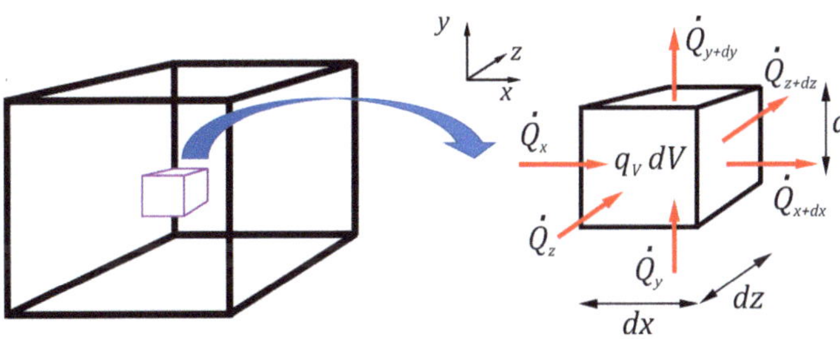

Fig. 4.3 Conductive heat transport in a differential control volume

It attributes the temporal change of the internal energy U within the volume element dV to three sources; the heat flows $\dot{Q}_i$ and the technical power flows $\dot{W}_i$ in the volume element dV over the system boundary and the released heat power $q_V dV$ present in the volume element dV with the volume specific source term q_V. In a quiescent incompressible medium the mechanical power flow is zero ($\dot{W}_i = 0$). The specific volumetric power release q_V acting as source term can be of different types (either electric, chemical reactions or nuclear radiation), but in solar thermal energy it is especially given by optical conversion of solar radiation into heat within the optically selective absorber. Thus q_V depends on the local coordinates and the time:

$$q_V = q_V(\vec{x}, t) \quad \left[W/m^3\right]. \tag{4.21}$$

If we consider heat conduction in a solid, the density changes due to temperature and pressure changes are marginal and we can apply the material model of an incompressible body with constant specific density ($\rho = const.$), resulting in the following:

$$m = \rho \cdot V_{\mathrm{a}} \text{ and } U = u \cdot m = u \cdot \rho \cdot V, \tag{4.22}$$

in which m denotes the mass und u the specific internal energy. Under this assumption, the change in internal energy can transform in the following way:

$$\frac{dU}{dt} = \frac{d}{dt}\int_V \rho u dV = \rho \int_V \frac{du}{dt} dV, \tag{4.23}$$

where the integral extends over the volume element dV. The hereby introduced mass-specific internal energy u of the incompressible body depends on its temperature. The change of internal energy with temperature is related to the specific heat capacity c_V in the following way:

$$du = c_V(T) \quad dT, \tag{4.24}$$

so that insertion in (4.24) yields

$$\frac{dU}{dt} = \rho \int_V c_V(T) \frac{\partial T}{\partial t} dV. \tag{4.25}$$

In order to calculate the heat flux $\dot{Q} = \sum_i \dot{Q}_i$ that crosses the system boundary first consider an element dA of the surface whose normal $\vec{n}$ is directed outwards. The heat flux $d\dot{Q}$ flowing into the area through dA is calculated to:

$$d\dot{Q} = -\overrightarrow{q}\,\vec{n} dA. \tag{4.26}$$

By integrating all heat fluxes and using the integral theorem of Gauss, we obtain:

$$\dot{Q} = -\int_A \vec{q} \cdot \vec{n} dA = -\int_V \nabla \cdot \vec{q}\, dV. \tag{4.27}$$

Substituting in Eqs. (4.25) and (4.27) the power balance (Eq. 4.20) then yields:

$$\frac{dU}{dt} = \sum_i \dot{Q}_i + \underbrace{\dot{W}}_{=0} + q_V dV,$$

$$\rho \int_V c_V(T) \frac{\partial T}{\partial t} dV = -\int_V \nabla \cdot \vec{q}\, dV + \int_V q_V \quad dV. \tag{4.28}$$

This volume integral vanishes for arbitrarily chosen balance domains only if the integrand itself equals zero and it results:

$$\rho c_V(T) \frac{\partial T}{\partial t} = -\nabla \cdot \vec{q} + q_V. \tag{4.29}$$

In the last step we use Fourier's law (Eq. 4.6) and combine the heat flux density $\vec{q}$ with the temperature gradient and obtain the differential equation for the sought temperature field:

$$\rho c_V(T) \frac{\partial T}{\partial t} = \nabla \cdot (k(T) \nabla T) + q_V. \tag{4.30}$$

Equation (4.30) often called second Fourier equation is a first order differential equation in time and second order in space and a linear equation if c_V and k are invariant (constant) in the temperature interval considered. The assumption of constant thermophysical material properties also yields that the square of the Nabla operator becomes the Laplace operator Δ. The Laplace operator has a different form depending on the formulation of the geometric problem. The corresponding forms are listed in Table 4.1.

If no source terms are considered and constant material values are assumed, the Eq. (4.30) is simplified to:

$$\frac{\partial T}{\partial t} = \frac{k}{\rho \cdot c_V} \cdot \Delta T = a \cdot \Delta T. \tag{4.31}$$

Table 4.1 Mathematical forms of the Laplace operator ΔT for different coordinate systems

Cartesian	Cylindric	Spherical
$\frac{d^2T}{dx^2} + \frac{d^2T}{dy^2} + \frac{d^2T}{dz^2}$	$\frac{1}{r}\frac{\partial}{\partial r}\left(r\frac{\partial T}{\partial r}\right) + \frac{1}{r^2}\frac{\partial^2 T}{d\varphi^2} + \frac{\partial^2 T}{\partial z^2}$	$\frac{1}{r^2}\frac{\partial}{\partial r}\left(r^2\frac{\partial T}{\partial r}\right) + \frac{1}{r^2 sin^2\Theta}\frac{\partial^2 T}{d\varphi^2} + \frac{1}{r^2 sin\Theta}\frac{\partial}{\partial\Theta}\left(\sin\Theta\frac{\partial T}{\partial\Theta}\right)$

Table 4.2 Typical examples of the magnitude of thermal diffusivity a of some materials

Material	a[m^2/s 10^{-6}]
Metals	5–10
Glass	0.6–0.8
Stone, ceramics	0.5–1.5
Water	0.143
Organic liquids	0.09–0.1
Air 20 °C, 0.1 MPa	21.4

The expression $a = k/(\rho c_V)$ denotes a substance size combination, which is referred to in the literature as the temperature coefficient, thermal diffusivity, or thermal diffusion coefficient. The thermal diffusivity has the unit $\left[m^2/s\right]$ analogous to the dynamic viscosity. Typical values of thermal diffusivity of some substances are listed in Table 4.2. Metals and gases have the high thermal diffusivities. This means that the temperature equalization in metals and gases is about the same. Liquids and non-metallic substances have lower thermal diffusivities, so temperature equalization is slower there.

The following generally valid statements can be derived directly from the second Fourier equation without presence of volumetric heat sources.

- For $(\partial^2 T/\partial x_i^2 > 0)$, the immediate result is $(\partial T/\partial t) > 0$. In other words, in areas of positive curvature of the temperature curve, there is an increase in temperature.
- If $(\partial^2 T/\partial x_i^2 < 0)$, we get $(\partial T/\partial t) < 0$. A negative curvature means a decrease of the temperature with increasing time.

All in all, it can be stated that the transient influence works towards a straightening of the temperature distribution, because for $t \to \infty$ the linear relationship must be achieved.

4.2.1 Initial and Boundary Conditions, Dimensionless Numbers

Initial conditions define the temperature distribution in the volume at a reference time, often referred to as $t = 0$, i.e. $T(x_i, t = 0) = T_0(x_i)$. The boundary conditions either define the temperature at the boundaries of the considered volume, or they make statements about its time derivative. There are essentially three types of thermal boundary conditions, whose physical meaning are depicted in Fig. 4.4:

- The Dirichlet (or first-type) boundary condition specifies the values that a solution needs to take along the boundary $\partial\Omega$ of the domain Ω and reads to:

$$T(x_i, t)|_{\partial\Omega} = T_{\partial\Omega}(x_i, t)|_{\partial\Omega}, \tag{4.32}$$

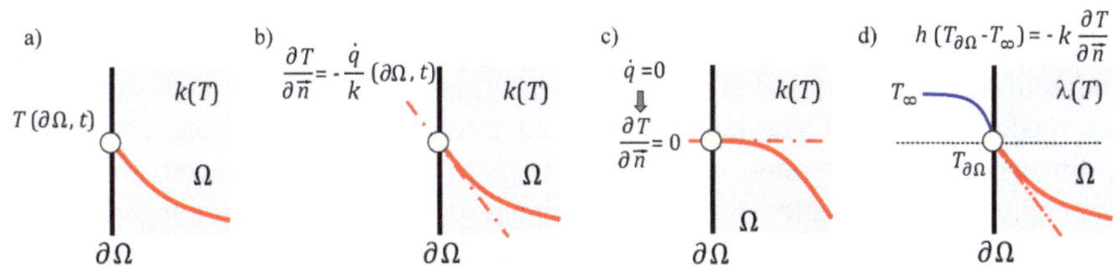

Fig. 4.4 Different thermal boundary conditions to the heat transport equation. **a** Dirichlet; **b** general Neumann condition; **c** homogeneous Neumann condition corresponding to adiabatic wall; **d** Robin or mixed boundary condition

where $T_{\partial\Omega}$ is a known value defined on the boundary $\partial\Omega$.

- The Neumann (or second-type) boundary condition specifies the values in which the derivative of a solution is applied within the boundary of the domain and is given by:

$$\left.\frac{\partial T(x_i, t)}{\partial n}\right|_{\partial\Omega} = T_{\partial\Omega}(x_i, t)|_{\partial\Omega}. \tag{4.33}$$

Regarding the thermal problem considered a Neumann condition means a prescribed heat flux from/to surface boundary $\partial\Omega$, which is in solar thermal applications given by solar radiation but could also be the electrical power provided e.g. by trace heating elements. One specific case is the homogeneous Neumann boundary condition denoting an adiabatic boundary which reads to $(\partial T/\partial n)|_{\partial\Omega} = 0$.

- The Robin boundary condition or third type boundary condition specifies a combination of the temperatures *and* the values of its derivative on the boundary of the domain. The linear Robin condition,

$$-k\left.\frac{\partial T(x_i, t)}{\partial n}\right|_{\partial\Omega} = h(T_{\partial\Omega} - T_\infty). \tag{4.34}$$

applies to problems where heat conduction in the domain is connected with a convective heat transfer at the boundary $\partial\Omega$, expressed by a heat transfer coefficient h, while the non-linear Robin condition

$$-k\left.\frac{\partial T(x_i, t)}{\partial n}\right|_{\partial\Omega} = \varepsilon\left(T_{\partial\Omega}^4 - T_S^4\right) + h(T_{\partial\Omega} - T_\infty). \tag{4.35}$$

addresses the heat transfer with combined surface radiation and convection in which the surface radiates with heat to surrounding surface of temperature T_S with a rate ε proportional to its emissivity current temperature. In Eqs. (4.34) and (4.35) T_∞ refers to the fluid temperature far away the boundary $\partial\Omega$.

To solve the heat conduction differential equation, it is useful to introduce dimensionless quantities, with the help of which the number of variables is reduced and uniform solutions for similar problems (geometric and thermal similarity) are made possible. If problems without an internal heat source are considered a dimensionless temperature $\Theta = (T - T_C)/(T_0 - T_C)$ is introduced, where T_C is a temperature characteristic for the problem and T_0 is the body temperature at the beginning. If internal heat sources are considered, the normalization is done in such a way that the prefactor of the second spatial derivative is omitted, so that the dimensionless temperature assumes the following scaling $\Theta \sim k(T - T_0)/\left(q_V L^2\right)$. If, for example, a flat absorber strip with the homogeneous initial temperature T_0 cools down steadily in a colder environment until the plate temperature corresponds to the ambient temperature T_{amb}, the ambient temperature T_{amb} is to be selected as the characteristic temperature T_C. The length scale on the x-coordinate is obtained by normalization to a characteristic length L, for example the plate thickness, which leads to $\xi = x/L$.

Furthermore, it is useful to introduce a dimensionless time, the so-called Fourier number (Fo):

$$Fo = \frac{k}{\rho c_V} \cdot \frac{t}{L^2} = \frac{at}{L^2}, \tag{4.36}$$

which can be interpreted as the ratio of the transport rate to the storage rate. For transient heat conduction, it describes the ratio of the rate at which sensible heat is transported by conduction to the rate at which it is absorbed.

Problem specific boundary conditions of the differential equation may also be made dimensionless. In many practical cases the heat, which is transported in a body by heat conduction, is transferred at its surface by convection into a surrounding fluid of temperature T_{amb}. At the body surface-fluid interface (index W) the continuity of the heat flux must be fulfilled. If one introduces the previous scaling for the de-dimensioning, Eq. (4.34) read in dimensionless form:

$$-\frac{1}{\Theta_W} \frac{\partial \Theta}{\partial \xi}\bigg|_W = \frac{hL}{k} = Bi, \tag{4.37}$$

in which $\Theta = (T - T_{amb})/(T_0 - T_{amb})$ and $\Theta_W = (T_W - T_{amb})/(T_0 - T_{amb})$. The boundary condition for the heat conduction problem in the body is now a function of a dimensionless quantity $Bi = (hL)/k$ called Biot number. In the case of heat transfer through the surface of a body, the Biot number indicates the ratio of the body's thermal resistance to the heat transfer resistance of the surrounding medium. Thus, the Biot number is formally formed equal to the Nusselt number, where, however, the specific thermal conductivity k_F of the fluid is used instead of the solid at rest k, and the length scale L has a different meaning. A large Biot number means that temperature differences within the solid body are larger than in the boundary layer of the fluid, so that an improvement of the external heat transfer, e.g. by forced instead of free convection, does not significantly accelerate the heat transfer process. From the solution of this heat conduction problem in dimensionless form, the dimensionless temperature of the body is calculated. This has the form:

$$\Theta = f(\xi, Fo, Bi). \tag{4.38}$$

For some practical problems with one-dimensional, unsteady heat conduction, solutions in analytical form are known. Some of them are presented in the subsequent subchapters as examples.

Normally, the Nusselt number Nu is used to describe the heat transfer of a body to flowing fluids. The Nusselt number is also a dimensionless parameter from similarity theory that measures the improvement in heat transfer from a surface. It relates the intensity of a convective heat transfer at a solid surface to one imagined at a fluid at rest when pure heat conduction is acting. However, the Nusselt number can also be understood as a dimensionless gradient of the temperature at a surface. If one uses the fluid and the dimensions of the fluid space as a reference system, Nu results in:

$$Nu = \frac{hL}{k_F}. \tag{4.39}$$

However, the Nusselt number, in addition to the geometry due to its different reference frame, depends essentially on flow parameters of the fluid such as the Reynolds number and the Prandtl number, as shown in Chap. 5. If one uses the fluid system with a normalization of the temperature $\Theta = (T - T_\infty)/(T_W - T_\infty)$. and the length normalization of the fluid system $\xi = x/L_F$ the reference system for de-dimensioning, then the dimensionless temperature gradient corresponds to that of the Nusselt number:

$$-\left.\frac{\partial \Theta}{\partial n}\right|_{\mathrm{W}} = Nu. \tag{4.40}$$

4.2.2 Stationary Heat Conduction

Stationary or steady state heat conduction treats problems where the temperature field does not change with time. A constant temperature difference between surfaces or a constant heat flux into a boundary cause such stationary conditions.

4.2.2.1 Stationary Heat Conduction in a Solar Absorber

In any solar thermal collector, heat gain occurs through optical conversion in a continuum volume being at rest. The resulting increase in temperature causes heat not only to be transported to the desired heat sink, but also to be lost by conduction through structures or radiation exchange with all collector boundaries.

Let there be an absorber strip as shown in Fig. 4.5, which absorbs the solar radiation i in [W/m^2] over the entire upper surface and conducts the heat to a pipe

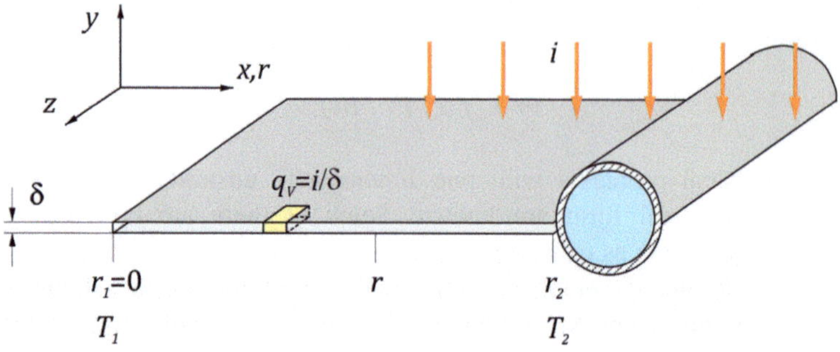

Fig. 4.5 Schematics of a "volumetric" heat release in an absorber strip and the conductive heat transport to the heat transfer medium

in which the heat transfer fluid is located. Dividing the irradiation power by the thickness of the absorber strip δ gives a volumetric power $q_V(x, y, z)$ in [W/m^3]. In most cases, the absorber plates are so thin that the temperature difference normal to the plate can be neglected ($\partial/\partial y \approx 0$). Likewise, the heat conduction parallel to the heat transfer tubes can be neglected ($\partial/\partial z \approx 0$). This results in a one-dimensional case with a volumetric source $q_V = q_V(r)$ as a first approximation. The heat conduction with an internal heat source also occurs in the glass covers, but in terms of magnitude it is only significant for concentrating systems in which large temperature differences can occur.

For the absorber plate, the irradiance would have to be converted to the cross-section of the absorber strip to equal a power density. Assuming a stationary case ($\partial/\partial t \approx 0$), Eq. (4.29) simplifies to:

$$\frac{d^2T}{dr^2} + \frac{q_V(r)}{k} = 0. \tag{4.41}$$

The general solution with a volumetric power density independent of r is thus given:

$$T = -\frac{q_V}{2\lambda} r^2 + C_1 r + C_2, \tag{4.42}$$

where C_1 and C_2 are integration constants. Their value depends on the respective boundary condition, i.e. whether, for example, the absorber strips are adjacent to a heat transfer tube on one or both sides. For a strip attached to a heat transfer tube on one side only and the maximum temperature $T_1(r_1 = 0)$ at the free end, as sketched in Fig. 4.5, the temperature curve is obtained:

$$T(r) = T_1 - \frac{q_V}{2k} \cdot r^2. \tag{4.43}$$

The temperature then decreases with increasing r_2. The average temperature of the absorber strip can be calculated if r_2 is the interface to the heat transfer tube and $r_1 = 0$:

$$\overline{T} = \frac{\int_{r=r_1}^{r=r_2} T(r)dr}{\int_{r=r_1}^{r=r_2} dr} \text{ and taking } r_1 = 0 \rightarrow \overline{T} = T_1 - \frac{1}{3}\frac{1}{2}\frac{q_V}{k} r_2^2. \tag{4.44}$$

Thus, the average temperature differs from the maximum temperature at the free end by one third (1/3) of the temperature difference between both boundaries of the strip $\Delta T = T_2 - T_1$:

$$\overline{T} = T_1 - \frac{1}{3}\Delta T. \tag{4.45}$$

In the case of a cylinder, the steady-state heat conduction equation assuming again constant thermal conductivity (k = const.) and a power density q_V independent of the coordinate r takes the following form:

$$\frac{1}{r}\frac{d}{dr}\left(r\frac{dT}{dr}\right) + \frac{q_V}{k} = 0. \tag{4.46}$$

The first integration provides the following relationship

$$r\frac{dT}{dr} = -\frac{q_V}{2k}r^2 + C_1, \tag{4.47}$$

and the solution reads to:

$$T(r) = -\frac{q_V}{4k} r^2 + C_1 \cdot \ln(r) + C_2. \tag{4.48}$$

Considering a solid cylinder with $r_1 = 0$ as an example, the boundary condition at this point is:

$$\left.\frac{dT}{dr}\right|_{r=r_1=0} = 0. \tag{4.49}$$

If $T_{1,2}$ are the temperatures in the centre of the cylinder and at the surface, respectively, the boundary condition (Eq. 4.49) leads to $C_1 = 0$. Assuming T_1 as the temperature at $r_1 = 0$, the second integration constant is obtained and as a solution for the temperature distribution $T(r)$ the result:

$$C_2 = T_1 + \frac{q_V}{4k}r_1^2 \text{ and } T(r) = \frac{q_V r_1^2}{4k}\left(1 - \frac{r^2}{r_1^2}\right) + T_1. \tag{4.50}$$

The temperature also changes here with r, however, decreasing quadratically.

For a hollow cylinder with an inner radius $r_1 \neq 0$ and an outer radius r_2, as well as the assumption that the inner side is adiabatic ($\partial T/\partial n = 0$), the following boundary condition results for the inner side:

$$\left.\frac{dT}{dr}\right|_{r=r_1} = 0. \tag{4.51}$$

The solution then leads to the following temperature distribution:

$$T(r) = T_1 - \frac{q_V}{4k}\left(r^2 - r_1^2\right) + \frac{q_V}{2k} r_1^2 \ln \frac{r}{r_1}. \tag{4.52}$$

Equations (4.45), (4.50) and (4.52) illustrate that the temperature depends essentially on q_V and thus on the power density generated volumetrically in the absorber by optical conversion, i.e., i_{glob}/δ. In addition to conductive heat conduction in the absorber strip, the temperature of the absorber is also determined by its thermal conductivity a, its ability to transfer heat in the form of radiation, which is given by emissivity ε and discussed in the following chapters of thermal radiation, and convective heat transfer to the heat sink, so that the temperature is a function of several influencing parameters in the following form:

$$\Delta T = f\left(i_{glob}, \varepsilon, a, \delta \ldots\right). \tag{4.53}$$

4.2.2.2 Stationary Heat Transfer Through a Wall

In practical heat transfer problems, heat conduction in a solid and convective heat transfer in a fluid boundary layer at the surface of the solid are often coupled. For large temperature differences, coupling with heat transfer by radiation is added. If the temperature is considered as a potential, as already mentioned, a total thermal resistance can be calculated from the individual thermal resistances in analogy to electrical engineering.

Let us consider the heat transfer through a plane wall of thickness δ surrounded on one side by fluid 1 of temperature T_1 and on the opposite side by fluid 2 of temperature T_1 with $T_1 > T_2$, as shown in Fig. 4.6. Due to the higher temperature, heat is transferred from fluid 1 to fluid 2. The heat flow is tfirst convective from fluid 1 to the wall surface, then the heat is transferred conductive in the wall to the opposite wall surface and there it is transferred again to fluid 2 by means of convective heat transfer. The total process consisting of heat conduction and convective heat transfer is called heat transfer. The resulting qualitative temperature profile from Fluid 1 to Fluid 2 is shown in Fig. 4.6.

The heat flow which is transferred due to the temperature difference between fluid 1 and fluid 2 can be calculated as

$$\dot{Q} = \frac{T_1 - T_2}{R_H} \tag{4.54}$$

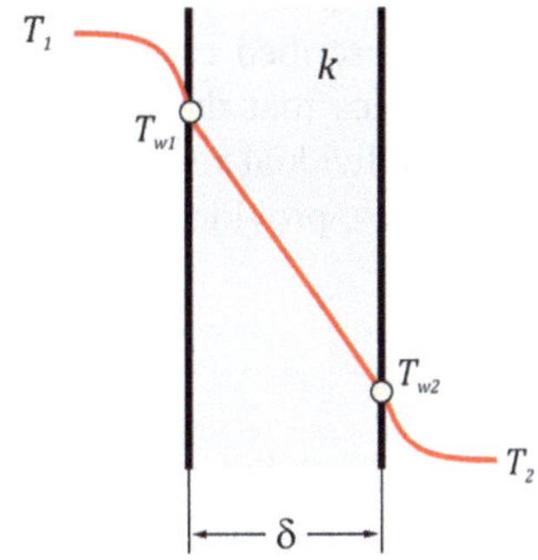

Fig. 4.6 Qualitative temperature profile during heat transfer through a planar wall. flat wall

in analogy to Ohm's law. If one assumes a one-dimensional heat transport normal to the wall surface, one obtains a total thermal resistance R which is the sum of the individual thermal resistances in the following way:

$$R_H = \frac{1}{h_1 A} + \frac{\delta}{kA} + \frac{1}{h_2 A} \quad \left[\frac{K}{W}\right], \tag{4.55}$$

where the convective thermal resistances are of the form $R_{conv,i} = 1/(h_i A)$ and the thermal resistances are given by $R_{cond,i} = \delta/(\lambda A)$. The heat flux transferred from fluid 1 to fluid 2 can then be given by a temperature difference $(T_1 - T_2)$,

$$\dot{Q} = k_{tot} A (T_1 - T_2), \tag{4.56}$$

in which k_{tot} is the total heat transfer coefficient in $\left(W/\left(m^2 K\right)\right)$, which combines in a single coefficient the specific heat conduction and heat transfer quantities and is related to the total thermal resistance R_H as follows:

$$k_{tot} = \frac{1}{R_H A}. \tag{4.57}$$

In a cylindrical tube of length L and thickness δ, through which fluid 1 flows on the inside and fluid 2 flows over it on the outside, one can also assume one-dimensional, now radial heat transport, resulting in the following total thermal resistance R_H:

$$R_H = \frac{1}{k_{tot} A} = \frac{1}{h_1 A_1} + \frac{\delta}{k A_m} + \frac{1}{h_2 A_2}, \tag{4.58}$$

where $A_1 = (2\pi r_1 L)$ denotes the inner shell area, $A_2 = (2\pi r_2 L)$ is the outer shell area, $\delta = (r_2 - r_1)$ is the pipe wall thickness, and $A_m = (A_2 - A_1)/\ln(A_2/A_1)$ is the logarithmically averaged cross-sectional area already introduced in Eq. (4.17). If we solve Eq. (4.58) for the total heat transfer coefficient k_{tot}, we see that the value for k_{tot} for a pipe must be specified in terms of the cross-sectional area A to which the value relates, which is not necessary for a flat wall because the areas are the same.

The described case of heat transfer from fluid 1 through a plane wall to fluid 2 illustrates that the total thermal resistance is composed of a series connection of the individual resistances and can be represented as the sum of the individual resistances, provided that one-dimensional heat transfer is present.

$$R_H = \sum_i R_i. \tag{4.59}$$

If the planar wall between the fluids consists of several layers j of thickness δ_j with different thermal conductivities k_j, the total thermal resistance when the individual resistances are connected in series is given by:

$$R_H = \frac{1}{k_{tot}A} = \frac{1}{h_1 A_1} + \sum_j \frac{\delta_j}{k_j A} + \frac{1}{h_2 A_2}. \tag{4.60}$$

The summation term changes for cylindrical bodies with the number n of layers of different thermal conductivity k_i or spheres of radius r in the following way:

$$\frac{1}{R_H} = \frac{1}{\sum_{i=1}^{n} \frac{1}{k_i} \ln\left(\frac{r_{i+1}}{r_i}\right)} \text{ (cylinder)}; \quad \frac{1}{R_H} = \frac{1}{\sum_{i=1}^{n} \frac{1}{k_i}\left(\frac{1}{r_i} - \frac{1}{r_{i+1}}\right)} \text{ (sphere)}. \tag{4.61}$$

Conversely, the analogy of dealing with electrical resistors can also be applied to the parallel connection of thermal resistances, provided that one-dimensional heat transport is given by the parallel thermal resistances.

4.2.2.3 Variable Thermal Conductivity and Kirchhoff Transformation

Often the thermal conductivity of the bodies analysed is temperature dependent, and a constant heat conductivity is normally present temperature only if small temperature intervals are considered. Moreover, the thermal conductivity substantially varies between the different types of continua appearing in in heat transfer problems. The Fig. 4.7 illustrates the temperature dependence of some gases, solid crystalline substances and metals appearing in solar thermal engineering and shows clearly that there are orders of magnitude differences depending on the material type, but also on the material structure. The physics is for this behaviour will be discussed in Sect. 4.2.4.

In case the difference between a temperature level T_1 and T_2 is in the range of a few percent, an average value can be determined for the thermal conductivities k_m in this range:

$$k_m = \frac{1}{(T_1 - T_2)} \int_{T=T_1}^{T=T_2} k(T)dT, \tag{4.62}$$

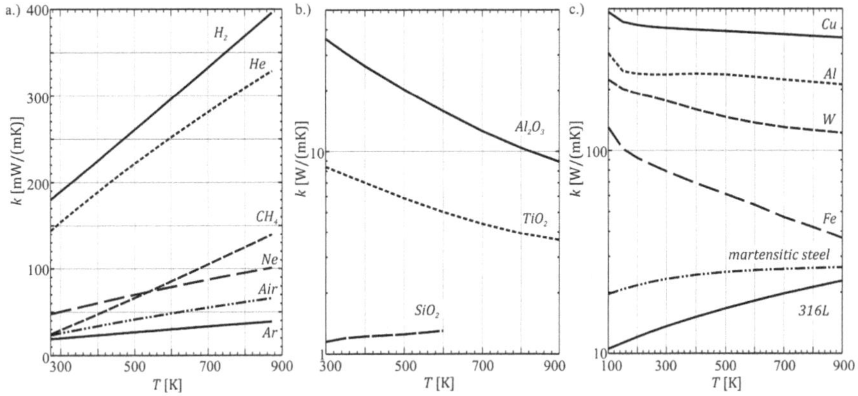

Fig. 4.7 Temperature dependent molecular heat conductivity k of **a** some gases, **b** ceramic materials based on ion bonding and **c** pure metals together construction metals

which can be reasonably used in calculations. Nonetheless, in this case, even in a flat wall, the temperature distribution is no longer linear.

The Kirchhoff transformation can be used to reduce the partial differential equation (Eq. 4.30), which becomes non-linear due to the temperature dependence of the thermal conductivity, to an equivalent linear equation that can be solved for steady-state problems. To do this, one introduces a new temperature variable T^* in the form:

$$k^* = \int_{T=T_1}^{T=T_2} k(T)dT \quad \text{and} \quad T^* = \frac{1}{k_m} \int_{T=T_1}^{T=T_2} k(T), \tag{4.63}$$

which adjusts the "normal" temperature to the variable thermal conductivity. This yield $k^* dT^* = k(T)dT$. From the definition of T^* we get for $T^*(T = T_1) = 0$ and for $T^*(T = T_2) = 1$. The transformed temperature T^* is monotonically increasing and $T(T^*)$ is the inverse function of $T^*(T)$. Rewriting the nonlinear differential equation yields for the stationary case:

$$k^* \frac{\partial^2 T^*}{\partial x_i^2} + q_V = 0, \tag{4.64}$$

and thus, a linear equation, provided that q_V does not depend on the temperature.

4.2.2.4 Two-Dimensional Heat Conduction

The heat flow from an isothermal surface A_1 to another isothermal surface A_2 in arbitrary geometrical arrangements is calculated in a general way by:

$$\dot{Q} = -\lambda \iint_{A_1} \left(\frac{\partial T}{\partial n}\right)_1 dA_1 = k \iint_{A_2} \left(\frac{\partial T}{\partial n}\right)_2 dA_2. \tag{4.65}$$

By introducing a form coefficient S of the type,

$$S = \frac{\iint_{A_1} \left(\frac{\partial T}{\partial n}\right)_1 dA_1}{(T_2 - T_1)} = \frac{\iint_{A_1} \left(\frac{\partial T}{\partial n}\right)_1 dA_1}{(T_2 - T_1)}, \tag{4.66}$$

and normalizing it to a characteristic length L, e.g. S/L, one obtains a dimensionless form factor S^* and ultimately a heat flux per unit length:

$$\frac{\dot{Q}}{L} = kS^*(T_2 - T_1). \tag{4.67}$$

In engineering data bases these dimensionless form factors (S^*) are tabulated for different geometrical arrangements.

4.2.3 Transient Heat Conduction

Time-varying temperature fields $T(\vec{x}, t)$ play a decisive role in many technical applications. Their calculation allows, for example, to answer questions about the duration of a heating or cooling process, about the heat dissipated or absorbed, or about the location inside the body where a maximum permissible temperature is reached at a certain time. For geometrically simple body shapes, a large number of analytical exact solutions are known, see e.g. (Carslaw & Jaeger, 1986) among others. Here, only some practically relevant and simple solutions shall be discussed, which are relevant for solar thermal systems. Transient heat transfer methods play an excellent role not only in evaluating the performance of collectors, but also in determining the thermal conductivity of materials, heating cooling of buildings, temporal resolution of thermocouples, etc., so an understanding of these effects is essential.

4.2.3.1 One-Dimensional Heat Flow with Abrupt Change of the Ambient State

Let us first consider a flat plate which is located in an environment of temperature T_∞. Normalizing the problem using the following scaling

$$\xi = \frac{x}{L}; \quad \Theta = \frac{T - T_\infty}{T_B - T_\infty} \quad \text{and} \quad \tau = \frac{at}{L^2}, \tag{4.68}$$

where L denotes the characteristic length, a the heat condcutance and T_B the temperature at the beginning, yields:

$$\frac{\partial \Theta}{\partial \tau} = \frac{\partial^2 \Theta}{\partial \xi^2}. \tag{4.69}$$

The direction of the heat flow vector in this case always coincides with a coordinate direction ξ and the planes $\xi = const.$ are then isothermal surfaces. The initial condition is given by $\Theta(\xi, \tau = 0) = 1$, while as boundary condition we assume the general case of a Robin boundary condition, in which the ratio of the internal thermal resistance of the plate to the external heat transfer resistance from the plate to the environment is expressed using a Biot number (Bi). This yield:

$$\left(\frac{1}{Bi}\frac{\partial\Theta}{\partial\xi} + \Theta\right)\bigg|_{|\xi|=1} = \begin{cases} 1 & \text{for } \tau = 0 \\ 0 & \text{for } \tau > 0 \end{cases}. \tag{4.70}$$

The formulation has the advantage that both a heating ($\Theta_\infty > \Theta_B$) and a cooling ($\Theta_\infty < \Theta_B$) of a body is described and tends towards zero for $\tau \to \infty$.

A consequence of the normalization of the boundary condition is that all tangents to the temperature curves at the surface of the body ($\xi = 1$) intersect at one point. The coordinates of this reference point are ($\Theta = 0, \xi = 1 + 1/Bi$). If the heat transfer from the plate to the environment is very small (i.e. the thermal resistance is high), the reference point is far away; conversely, an excellent heat transfer means that it is close to the wall and the Robin boundary condition mutates to a boundary condition of the first kind, i.e. $\Theta(\xi = 1, \tau > 0) = 0$.

With the help of a geometrical consideration, the maximum temperature difference $\Delta\Theta_{\max}$ occurring in the body can be estimated as follows:

$$\Delta\Theta_{\max} < \lambda\frac{Bi}{1 + Bi}. \tag{4.71}$$

From the figure it can be seen that at small times (e.g. the time $\tau = \tau_1$ in Fig. 4.8), for which $Fo \leq 0.01$ holds, the temperature in the centre of the plate has not yet noticed the temperature change at the interface. This behaviour is called boundary layer character. Conversely, this means that the thickness with its heat capacity has no influence on the temperature field and the plate body can be regarded as semi-infinitely extended. By introducing dimensionless wall distance $\xi' = (1 - \xi)/(2\sqrt{\tau})$ and a modified Biot number formed with the dimensionless wall distance $Bi' = Bi\sqrt{\tau}$ the following special solution of the differential Eq. (4.69) applies:

$$\Theta(\xi', Bi') = \text{erf}(\xi') + \exp(\xi'^2)\cdot\exp(\xi' + Bi')^2\cdot\left[1 - \text{erf}(\xi' - Bi')\right], \tag{4.72}$$

where erf is the Gaussian error integral defined as

$$\text{erf}(x) = \frac{2}{\sqrt{\pi}}\cdot\int_{x=0}^{z}\exp(-x^2)dx. \tag{4.73}$$

The latter is usually solved numerically and Fig. 4.9 shows the evaluation of the Gaussian error integral. For short times, the temperature in the centre of the body corresponds to the initial temperature and the temperature of the surface $\Theta_S(\xi' = 0)$ is obtained from Eq. (4.72) in dimensionless notation by

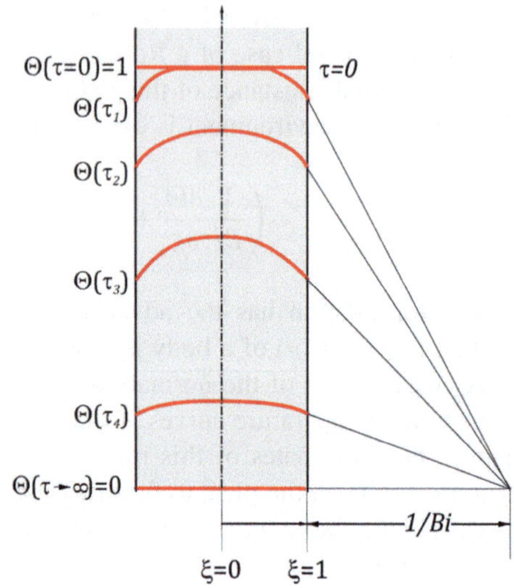

Fig. 4.8 Schematic dimensionless temperature profile $\Theta(\xi, \tau, Bi)$ for the transient heat transfer in a plate using Robin boundary conditions

$\Theta_S = \exp(Bi'^2) \cdot [1 - \text{erf}(Bi')]$. For $Bi' > 2$, Eq. 4.72 yields a dimensionless surface temperature of $\Theta_S = 1/(\sqrt{\pi} Bi')$ with an error less than 1%. For $Bi' \to \infty$ the surface temperature gets zero ($\Theta_S = 0$), i.e. there is a boundary condition of the first kind, and from Eq. (4.72) only the first summand remains:

$$\Theta_S = erf(x') = \text{erf}\left(\frac{x}{2 \cdot \sqrt{a \cdot t}}\right). \tag{4.74}$$

The instantaneous heat flux density through the surface is then obtained by:

$$\dot{q}_S(\xi = 1, t) = \frac{k}{\sqrt{\pi \cdot a \cdot t}}(T_B - T_S) = \frac{\sqrt{\lambda \cdot \rho \cdot c_V}}{\sqrt{\pi \cdot t}}(T_B - T_S). \tag{4.75}$$

Thus, the heat flux decreases with inverse of the square root of time ($1/\sqrt{t}$) and is proportional to the quantity $b = \sqrt{k \cdot \rho \cdot c_V}$ called thermal effusivity or heat penetration coefficient. The integral mean value of the heat flux density at the

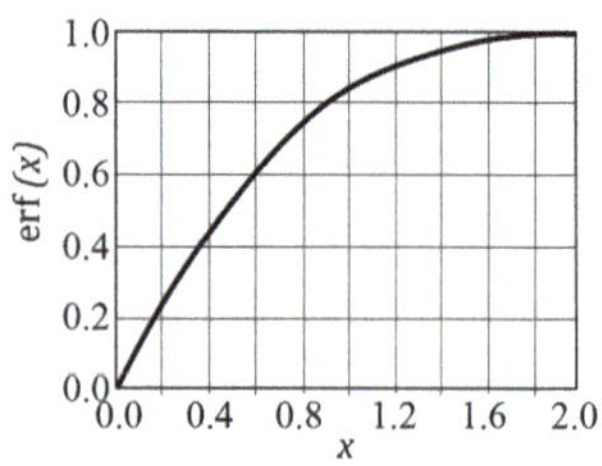

Fig. 4.9 Schematic evaluation of the Gaussian error integral

surface from $t = 0$ to t is calculated as:

$$\dot{q}_S(\xi = 1, t) = \int_{t=0}^{t} \dot{q}_S(t)\, dt = \frac{2 \cdot k \cdot t}{\sqrt{\pi \cdot a \cdot t}} (T_B - T_S). \tag{4.76}$$

The mean temperature $\overline{T}$ can be used to calculate the heat added or removed in a given time. For this purpose, the integral caloric mean temperature $\overline{T}$ is required, which is defined by:

$$\overline{T} = \frac{\int_V (\rho \cdot c_V \cdot T) dV}{\int_V (\rho \cdot c_V) dV} [K] \quad \text{or} \quad \overline{\Theta} = \frac{\left(\overline{T} - T_\infty\right)}{\left(T_B - T_\infty\right)} [/]. \tag{4.77}$$

With the initial and the mean temperature, the change of the heat content is determined. This results in:

$$Q = \rho \cdot V \cdot c_V \cdot \left(T_B - \overline{T}\right) = m \cdot c_V \cdot \left(T_B - \overline{T}\right)\left(1 - \overline{\Theta}\right) \text{in}[W], \tag{4.78}$$

where $\overline{T}$ denotes a dimensional mean temperature. Finally, for the mean temperature one obtains:

$$\left(T_B - \overline{T}\right) = \frac{2 \cdot A \cdot k \cdot t}{V \cdot \rho \cdot c_V \cdot \sqrt{\pi \cdot a \cdot t}} (T_B - T_S) = \frac{2 \cdot A}{V} \cdot \frac{\sqrt{a \cdot t}}{\pi} \cdot (T_B - T_S). \tag{4.79}$$

The ratio of the surface area A to the volume V is $1/L$ for a plate (, where it should be noted that L is half the thickness of the plate), $4/L$ for a circular cylinder and $6/L$ for a sphere. Thus, the surface temperature calculated from Eq. (4.74) can be used to determine the mean temperature of the body. Equations 4.74 and 4.79 are valid only if the Biot number Bi is very large ($Bi \gg 1$). Physically, this assumption means that the heat transfer coefficient h tends to infinity ($\rightarrow \infty$), i.e. there is perfect heat transfer. For sufficiently long times τ, the temperature profiles inside the body of finite thickness are similar to each other and can all be described by a location function $f(\xi)$, which is reduced in scale with increasing time.

Introducing a multiplicative approach to the differential Eq. (4.69) for the temperature in the following way $\Theta = g(\tau) \cdot f(\xi)$, we obtain in each case for the time scale function $g(\tau) = \exp\left(-m^2\tau\right)$. Therein, m is a constant to be determined from the boundary conditions. The form of the location function $f(\xi)$ depends on whether it is a plate, a cylinder, or a sphere. Unfortunately, these functions alone are not able to represent the temperature field for short times, especially the initial temperature distribution. Therefore, a representation in the form of an infinite series as:

$$\Theta = \sum_{i=1}^{n} C_i(m_i) \cdot f(m_i\xi), \tag{4.80}$$

is necessary, whose coefficients C_i have to be determined in such a way that a finite series describes the initial temperature distribution $\Theta(\xi, \tau = 0) = 1$ with a minimum of the error sum of squares and converges to this initial temperature distribution for $n \to \infty$.

For the one-dimensional plate of thickness L, the general solution of the differential equation for the temperature can be represented as a Fourier series of in the form:

$$(x, t) = \sum_{n=1}^{\infty} [C_1 \cos(B \cdot x) + C_2 \sin(B \cdot x)] \cdot \exp(-C_3 \cdot a \cdot t), \tag{4.81}$$

in which the constants B and C_i can be determined with the help of the boundary conditions. If the heat transfer coefficient is infinite on the outside (i.e., the external wall temperature is equal to the temperature of the surrounding fluid corresponding to $1/Bi \to 0$), the solution in the case of the infinite plate is:

$$\begin{aligned}\Theta = \frac{T(x,t) - T_\infty}{T_A - T_\infty} &= \frac{4}{\pi} \sum_{n=1}^{\infty} \frac{1}{n} \sin\left(\frac{n \cdot \pi \cdot x}{2 \cdot L}\right) \exp\left(-\frac{n^2 \pi^2 a t}{4L^2}\right) \\ &= \frac{4}{\pi} \sum_{n=1}^{\infty} \frac{1}{n} \sin\left(\frac{n \cdot \pi \cdot x}{2 \cdot L}\right) \exp\left(-\frac{n^2 \pi^2 Fo}{4}\right),\end{aligned} \tag{4.82}$$

where n is a series of odd values with $n = 1, 3, 5, \ldots, (2n - 1)$. For $\tau > 0.5$ the local and the mean temperature Θ and $\overline{\Theta}$ are:

$$\Theta = \frac{4}{\pi} \cos\left(\frac{\pi x}{2L}\right) \exp\left(-\frac{n^2 \pi^2 a t}{4L^2}\right) \text{ and } \overline{\Theta} = \frac{8}{\pi} \exp\left(-\frac{n^2 \pi^2 a t}{4L^2}\right), \tag{4.83}$$

and the heat flux density at the surface results to:

$$\dot{q}_S = \frac{2k}{L} (T_B - T_\infty) \exp\left(-\frac{\pi^2 a t}{4L^2}\right). \tag{4.84}$$

4.2.3.2 Approximations for Transient Heat Conduction Problems in Practice

From the analytical solutions (of Eq. 4.81), the temperature field $\Theta = \Theta(\tau, \xi, Bi, a_V^*)$ can be calculated as a function of time, location, external heat transfer coefficient and body shape. Herein, a_V^* is a dimensionless ratio in the from $a_V^* = (AL)/V$, where A is the surface area, V the volume and L half the thickness of a plate. If one defines for the transient heat conduction in the solid an internal heat transfer coefficient h_i with the heat flux through the surface $\dot{Q}$ and the difference between the caloric mean temperature $\overline{T}$ and the surface temperature T_S being $h_i = \dot{Q}/\left[A\left(\overline{T} - T_S\right)\right]$ it can be expressed in asymptotic equations

Table 4.3 Constants for different body shapes of the long term asymptotic in one-dimensional transient heat transfer

	Plate	Cylinder	Sphere
a_V^*	1	2	3
m_1	$\pi/2$	2,4048	π

for long and short times using additionally the non-dimensional Nusselt number $Nu = (h_i L)/\lambda$, for more details see e.g. (von Boeckh & Wetzel, 2011). Accordingly, the following results for the long-term asymptote with the values from the Table 4.3:

$$Nu(t \to \infty) = Nu_\infty = \frac{2 + a_V^* + Bi}{1 + \left(a_V^* Bi\right)/m_1^2}, \tag{4.85}$$

whereas the mean value of the short-term asymptote gives the relationship:

$$Nu(t \to 0) = Nu_0 = \frac{1}{\sqrt{\tau}} \frac{\sqrt{\pi} + 10Bi\sqrt{\tau}}{1 + 5Bi\sqrt{\pi\tau}}. \tag{4.86}$$

By superimposing the asymptotes, a mean Nusselt number Nu:

$$Nu = \sqrt{Nu_\infty^2 - 0.04 + (Nu_0 + 0.2)^2}, \tag{4.87}$$

is obtained, which can be used to approximately calculate a caloric mean temperature $\overline{\Theta}$:

$$\overline{\Theta}(\tau) = exp\left(-\frac{a_V^* \quad \tau}{1/Bi + 1/Nu(\tau)}\right). \tag{4.88}$$

The argument of the preceding exponential function represents a dimensionless time. If one analyses the contributing dimensions from the numerator and denominator $(kAt)/(\rho c_V V)$, one recognizes that they correspond to the NTU (Number of Transfer Units) values of a heat exchanger.

To determine the surface temperature, which results from the thermal resistance ratio k/h_o, with $1/k = 1/h_o + 1/h_{i,t}$, one needs the instantaneous heat transfer coefficient instead of the time average values (Eq. 4.86), which is:

$$Nu_{0t} = \frac{1}{\sqrt{\tau}} \frac{2.3\sqrt{\pi} + 2Bi\sqrt{\tau}}{2\left(2.3 + Bi\sqrt{\pi\tau}\right)}. \tag{4.89}$$

The superposition of the asymptotes gives the instantaneous value of the Nusselt number Nu_t,

$$Nu_t = \sqrt{Nu_\infty^2 - 0.04 + (Nu_{0t} - 0.2)^2}, \tag{4.90}$$

from which the surface temperature $\Theta_0(\tau, \xi = 1)$ can be calculated:

$$\Theta_0(\tau, \xi = 1) = \frac{\overline{\Theta}}{1 + Bi/Nu_t}. \tag{4.91}$$

The temperature in the centre of the body $\Theta_m = \Theta(\tau, \xi = 0)$ exhibits a time offset with respect to the time change of the caloric mean temperature, which can be taken into account by means of the introduction of a new time scale $\tau_m = (\tau - \Delta\tau)$ with the variable dead time $\Delta\tau$:

$$\Delta\tau = \left(1/\Delta\tau_\infty^m + 1/\tau^m\right)^{-\frac{1}{m}} \text{ with } m = \tfrac{30+Bi}{10.8+Bi/4} + \tfrac{0.16}{\sqrt{\tau}} \text{ and} \\ \tfrac{1}{\Delta\tau_\infty} = 4 + \tfrac{2a_V^*(6+a_V^*+2Bi)}{6+a_V^*+Bi(2.71+0.03a_V^*)}. \tag{4.92}$$

The temperature at the centre of the body now reads using $\tau_m = (\tau - \Delta\tau)$:

$$\Theta_m = \Theta(\tau, \xi = 0) = \exp\left(-\frac{a_V^* \tau_m}{1/Bi + 1/Nu(\tau_m)}\right). \tag{4.93}$$

The approximate Eqs. (4.85)–(4.93) show a maximum difference of less than 3% compared to the exact solutions, also for the caloric mean temperature and for the surface temperature.

4.2.3.3 Coupled Interface Boundaries

If two bodies 1 and 2 of different temperature are brought into contact with each other at time $t = 0$, as sketched in Fig. 4.10, the contact temperature T_C is established after an arbitrarily short time at the surface of both bodies. The two bodies can have different material properties, but since they have the same contact area, the temperature at the interface of the two bodies is the same, so that the boundary condition at the contact point is as follows:

$$T_1(x = 0) = T_2(x = 0). \tag{4.94}$$

In the case of real interfaces of solid bodies, this condition does not necessarily have to be fulfilled; for example, roughness, oxide layers can be produced by imperfect contact, so that an additional contact resistance occurs. This type of contact leads to a boundary condition that describes the continuity of the heat flux at the contact point.

$$k_1 \left.\frac{\partial T}{\partial n}\right|_{1,x=0} = k_2 \left.\frac{\partial T}{\partial n}\right|_{2,x=0}. \tag{4.95}$$

Fig. 4.10 Sketch for the temporal temperature evolution of the contact of two bodies of different temperature

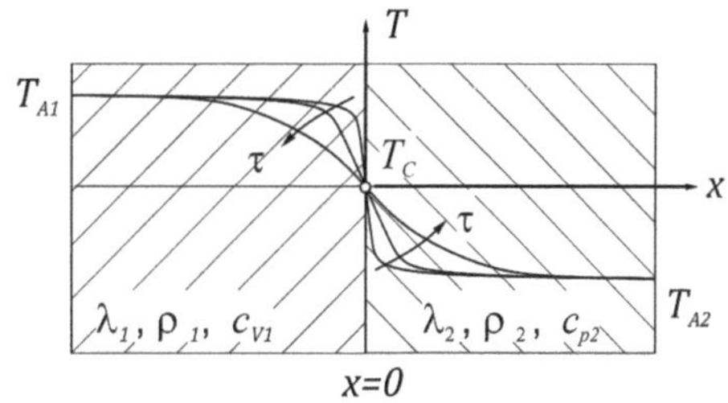

Assuming an ideal contact, we have a boundary condition of the first kind at the contact point $x = 0$, the solution of which is Eq. (4.74):

$$\begin{aligned} \frac{T-T_C}{T_{A1}-T_C} &= -\mathrm{erf}\left(\frac{x}{2\cdot\sqrt{a_1\cdot t}}\right) \text{ for } x < 0 \text{ and} \\ \frac{T-T_C}{T_{A2}-T_C} &= \mathrm{erf}\left(\frac{x}{2\cdot\sqrt{a_2\cdot t}}\right) \text{ for } x > 0. \end{aligned} \tag{4.96}$$

This condition for the continuity of the temperature at the interface can be inserted into Eq. (4.94)

$$\sqrt{k_1 \cdot \rho_1 \cdot c_{p1}}(T_{A1} - T_C) = \sqrt{k_2 \cdot \rho_2 \cdot c_{p2}}(T_C - T_{A2}), \tag{4.97}$$

and can be resolved according to the contact temperature T_C so that the following is obtained:

$$T_C = \frac{\left(T_{A1} + \sqrt{\frac{k_2\cdot\rho_2\cdot c_{p2}}{k_1\cdot\rho_1\cdot c_{p1}}} \cdot T_{A2}\right)}{\left(1 + \sqrt{\frac{k_2\cdot\rho_2\cdot c_{p2}}{k_1\cdot\rho_1\cdot c_{p1}}}\right)} = \frac{\left(T_{A1} + \frac{b_2}{b_1} \cdot T_{A2}\right)}{1 + \frac{b_2}{b_1}}. \tag{4.98}$$

Thus, the contact temperature T_C depends on the ratio of the thermal effusivities $b_i = \sqrt{(k\rho c_V)_i}$ and it is always closer to the initial temperature of the body that has the higher thermal effusivity. Equation (4.98) can be used to explain why different bodies feel differently warm and the contact temperature is determined by the ratio of the thermal effusivity of the two bodies.

A copper plate has a much larger thermal effusivity than the human hand. Therefore, one feels approximately the temperature of the copper plate. With a Styrofoam plate it is opposite, here one feels rather the temperature of the hand. As long as the temperature change does not reach deeper layers of the body, the contact temperature remains constant. If larger layers are covered, a gradual temperature equalization occurs, which is determined by the heat content of the two bodies. This can be demonstrated, for example, on an aluminium foil. If one touches a very thin, hot aluminium foil, a contact temperature arises at the first moment which corresponds approximately to the temperature of the aluminium foil. Since the foil is very thin and thus has little mass to store the heat, it cools down very quickly. It is not able to heat the skin in such a way that the temperature increase reaches the nerves with which the temperature is registered. As a

Table 4.4 Thermal effusivities b of selected matter in $\left[J/\left(m^2 K\sqrt{s}\right)\right]$ (some take from https://thermtest.com/materials-database)

Material	$b\left[J/\left(m^2 K\sqrt{s}\right)\right]$	Material	$b\left[J/\left(m^2 K\sqrt{s}\right)\right]$
Graphite	16318	Cu (liquid)	27700
Si (solid)	7660	Fe (liquid)	16100
Fe (solid)	12600	Al (liquid)	17000
Stainless-steel (316L)	7230	$NaNO_3$	1921
Aluminium (solid)	23688	Na (liquid)	1381
Copper (solid)	36983	Si (liquid)	11600
Concrete stone	1260	Water (H_2O)	1588
MgO	5990	Oil (motor 10W30)	519
Al_2 $O3$	5890	Diesel	491
SiO_2	3610	Ethanol	589
CaO	7530	Methane (CH_4)	7
Granite	2490	He	12
SiC(SiC76Si24)	11266	Hydrogen (H_2)	272
Glass (avg. property)	1419	Paper	305
Hard wood	380	Human skin	1120

result, the foil feels only moderately warm. If you touch an aluminium plate that is equally hot, you will burn your fingers. Some typical thermal effusivities of solid matter are listed in Table 4.4.

The importance of the thermal effusivity especially appears in in the storage of sensible heat, where it characterizes the ability to exchange thermal energy with its surroundings. Since the effusivity is related to the surface heat exchange the most favourable medium at high heat loaded surfaces is that which can absorb (dissipate) more heat without raising temperature. Thermal effusivity is also of importance to measure heat fluxes by means of calorimetric methods, where the transient heat conduction plays a prominent role.

4.2.3.4 Limiting Cases with $Bi = 0$ and $Bi \to \infty$

If the Biot number approaches zero ($Bi = 0$), this means that the external heat transfer coefficient h is very small or the thermal conductivity λ of the material is very large. In this case, the temperature in the body is independent of the location and therefore only dependent on time. For very large Biot numbers ($Bi \to \infty$), i.e. for very large external heat transfer coefficients h or very small thermal conductivity of the body under consideration, the surface temperature T_S approaches the external temperature T_∞, which is depicted in Fig. 4.11a. The surface temperature is equal to the external temperature. Figure 4.11c shows the case for $Bi = 0$. Here, the temperature inside the body is independent of the location. Figure 4.11b illustrates the temperature behaviour for finite Biot numbers, where the temperature depends on both location and time.

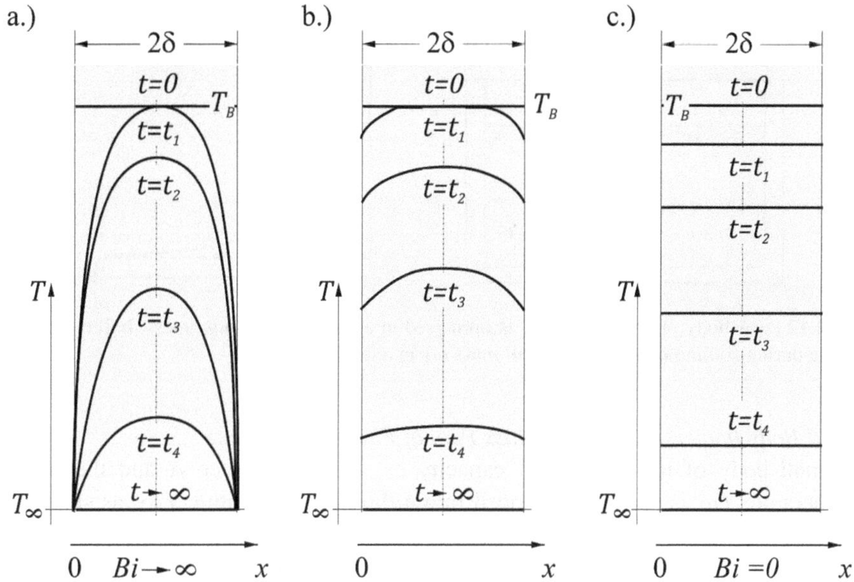

Fig. 4.11 Qualitative temperature profile of transient heat conduction at $Bi \to \infty$ (**a**), real Biot number (**b**) and $Bi = 0$ (**c**)

4.2.3.5 Solar Thermal Relevant Cases of Temperature Changes for $Bi \ll 1$

In many technical processes, it is not the temperature distribution in the body that is of interest during cooling or heating, but the mean temperature change. If the Biot numbers are small (the external heat transfer coefficient is smaller than that within the body), there is only a small temperature difference between the average temperature and the temperature in the centre of the body between the surfaces. If the Biot numbers are less than 0.5, the calculations can be done with fairly good accuracy such that the temperature in the whole body is taken as the mean temperature. For Biot numbers smaller than unity, a calculation performed in this way is still acceptable for estimations.

- A small body heats/cools transiently in a large environment with homogeneous temperature. This occurs when an absorber foil heats up quickly due to solar radiation, but the air in the collector box is still cold (danger of thermal shock!).
- Unsteady heating or cooling for bodies of approximately the same mass. The volumetric part of an absorber in a collector is relatively small, but its mass is of similar size compared to the air surrounding it (density air !).
- Heat transfer through a flowing heat transfer medium. This case occurs, for example, during collector operation, loading and unloading of thermal storage.

Subsequently, these cases and their results are discussed.

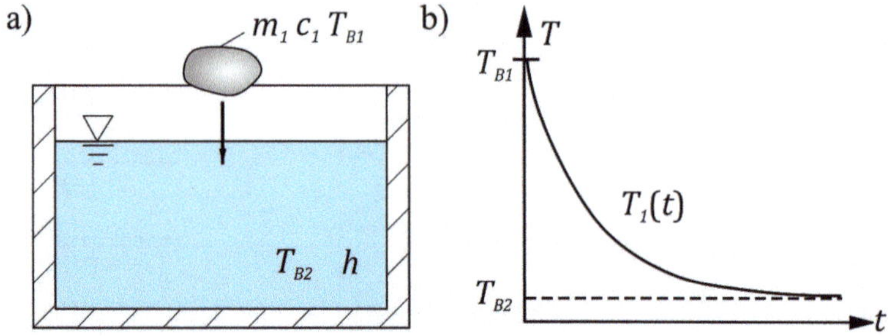

Fig. 4.12 **a** A body of small mass m_1 is immersed in a fluid pool of large mass. **b** Temperature profile during cooling of a body of small mass m_1 in a fluid pool

Small Body Immersed in Large Mass Pool of Fluid

A small body of mass m_1, heat capacity c_1 with the surface A and the initial temperature T_{B1} is suddenly dropped into a fluid of temperature T_{B2}, as sketched in Fig. 4.12a. Assume now that $T_{B1} > T_{B2}$ and the mass of the fluid is so large that no temperature change occurs in the fluid due to the immersion of the body. At the surface of the body, the heat transfer coefficient h is considered to have a given value and the walls of the vessel towards the ambient are supposed to be isolated. Then the heat flow transferred from the body to the fluid is given by the change in the heat content of the body over time as:

$$\dot{Q} = -m_1 \cdot c_1 \cdot \frac{dT_1}{dt}. \tag{4.99}$$

The transferred heat flow is determined by the heat transfer coefficient h and the temperature difference between the fluid and the body $(T_1 - T_{B2})$. If we note that when the body cools, its temperature change is negative and the temperature difference is positive, the result for the heat flow at the interface from the body to the fluid is:

$$\dot{Q} = h \cdot A \cdot (T_1 - T_{B2}). \tag{4.100}$$

Since the heat flow out of the body must equal the heat flow into the fluid, Eqs. (4.99) and (4.100) can be equated and an ordinary differential equation of the form is obtained:

$$\frac{dT_1}{dt} = -\frac{h \cdot A \cdot}{m_1 \cdot c_1}(T_1 - T_{B2}). \tag{4.101}$$

Replacing the temperature T_1 by the temperature difference $(T_1 - T_{B2})$, the following expression is obtained after separating the variables:

$$\frac{d(T_1 - T_{B2})}{(T_1 - T_{B2})} = -\frac{h \cdot A \cdot}{m_1 \cdot c_1} \cdot dt. \tag{4.102}$$

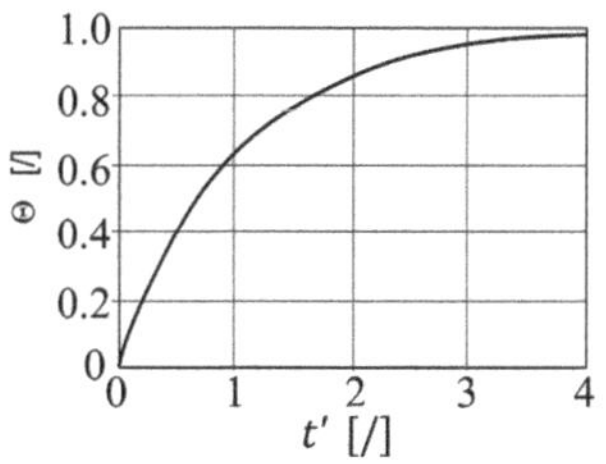

Fig. 4.13 Dimensionless temperature profile Θ as a function of dimensionless time t' when a small body cools in a large pool

If we assume a constant heat transfer coefficient h and a constant heat capacity (c_1), we can integrate this equation and obtain:

$$(T_1 - T_{B2}) = (T_{B1} - T_{B2}) \cdot exp\left(-\frac{h \cdot A}{m_1 \cdot c_1} \cdot t\right). \tag{4.103}$$

Figure 4.12b shows schematically the temporal evolution the body cooling. The temperature of the small body approaches the fluid temperature negatively exponentially asymptotically with increasing time. For the dimensionless representation, the dimensionless temperature and dimensionless time t' can be introduced

$$\Theta = \frac{(T_{A1} - T_1)}{(T_{A1} - T_{A2})} = 1 - \exp(-t') \text{ with } t_0 = \frac{h \cdot A}{m_1 \cdot c_1} \text{ and } t' = \frac{t}{t_0}. \tag{4.104}$$

Here, the time t_0 denotes the time in which the immersed body is heated by $1°K$ at $1°K$ temperature difference with respect to the surrounding fluid. The dimensionless representation allows to depict the solution for the temperature in a universal form, as it is shown in Fig. 4.13.

Body Immersed in Fluid of Similar Mass

In this limiting case, we will consider the immersion of a body in a basin thermally isolated from the outside and filled with a fluid with a relatively small fluid mass (and small dimension L) as illustrated in Fig. 4.14a. At immersion, the body has temperature T_{B1} and the fluid has initial temperature T_{B2}. The mass and the specific heat capacity of the body have index 1, respectively that of the fluid index 2. Since the fluid mass is no longer large, the temperature of the fluid is changed by the immersion of the body. The heat flux released by the body corresponds to the change over time of its heat content and one obtains:

$$-\dot{Q} = m_1 \cdot c_1 \cdot \frac{dT_1}{dt}. \tag{4.105}$$

For reasons of energy conservation, the heat flow absorbed by the fluid must be of the opposite magnitude to the change in the heat content of the fluid over time, so that:

$$-\dot{Q} = m_2 \cdot c_2 \cdot \frac{dT_2}{dt}. \tag{4.106}$$

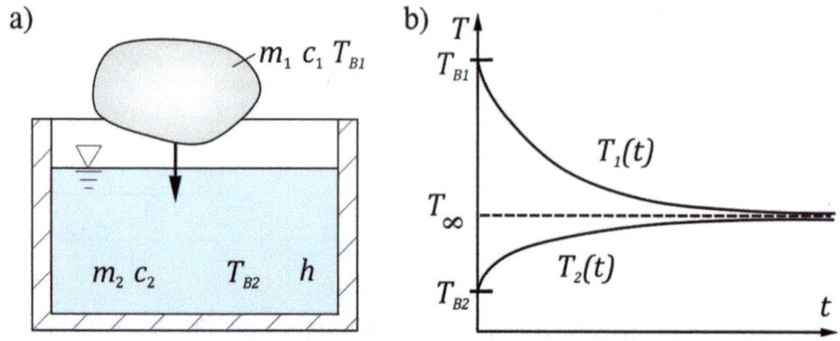

Fig. 4.14 **a** A body of small mass is immersed in a fluid of comparable mass. **b** Temporal temperature profile during cooling of a body of small mass m_1 in a fluid pool of comparable mass

The heat flux is determined by the heat transfer coefficient h and the driving temperature difference between fluid and body $(T_1(t) - T_2(t))$. The heat flow is thus given by:

$$\dot{Q} = h \cdot A \cdot (T_1 - T_2). \tag{4.107}$$

For the change of the temperature difference between fluid and body, we obtain from Eqs. (4.106) and (4.107) the relation:

$$d(T_1 - T_2) = -\dot{Q} \cdot \left[\frac{1}{m_1 \cdot c_1} + \frac{1}{m_2 \cdot c_2}\right] dt. \tag{4.108}$$

The heat flow from Eq. (4.107) can thus be substituted into Eq. (4.108), yielding the following differential equation

$$\frac{d(T_1 - T_2)}{(T_1 - T_2)} = -h \cdot A \cdot \left[\frac{1}{m_1 \cdot c_1} + \frac{1}{m_2 \cdot c_2}\right] dt. \tag{4.109}$$

Provided that the masses m_i, the heat transfer coefficient h and the specific heat capacities c_i are constant, Eq. (4.109) can be integrated. The integration is performed from time $t = 0$ to t and from initial temperatures T_{B1} and T_{B2} to temperatures $T_1(t)$ and $T_2(t)$ and results in the time dependent temperature difference $(T_1(t) - T_2(t))$:

$$(T_1(t) - T_2(t)) = (T_{B1} - T_{B2}) \cdot \exp\left[-h \cdot A \cdot \left(\frac{1}{m_1 \cdot c_{p1}} + \frac{1}{m_2 \cdot c_{p2}}\right) \cdot t\right]. \tag{4.110}$$

Equation (4.110) can be used to determine the temperature difference between the fluid and body, but not the temperature of the body or fluid. The temperature of the body and fluid can be determined from Eqs. (4.105) and (4.106) for any time.

Since in the equations the masses m_i and specific heat capacities c_i are considered as constant, the heat removed and added in time t can be calculated

$$Q(t) = m_1 \cdot c_1 \cdot (T_{B1} - T_1(t)); \quad Q(t) = m_2 \cdot c_p \cdot (T_2(t) - T_{B2}). \qquad (4.111)$$

Equation (4.110) is also valid for infinite times at which temperature equilibrium is reached between the body and the fluid. The temperature after equilibration for $t \to \infty$ is calculated with the aid of equilibrium to be

$$T(t \to \infty) = T_\infty = \frac{m_1 \cdot c_1 \cdot T_{B1} + m_2 \cdot c_2 \cdot T_{B2}}{m_1 \cdot c_1 + m_2 \cdot c_2}. \qquad (4.112)$$

For the temporal changes of the temperature one obtains from the two previous equations:

$$\frac{T_{B1} - T_1}{T_2 - T_{B2}} = \frac{T_{B1} - T_\infty}{T_\infty - T_{B2}} = \frac{m_2 \cdot c_2}{m_1 \cdot c_1}. \qquad (4.113)$$

Equation (4.113) can be solved for T_1 or T_2 and inserted into Eq. (4.110). The result is the temperature of the body and fluid, the time course of which is shown schematically in Fig. 4.14b. Finally, the time negative exponentially changing temperatures of the bodies 1 and 2 reads to:

$$\begin{aligned} T_1 - T_\infty &= (T_{B1} - T_\infty) \cdot \exp\left[-h \cdot A \cdot \left(\frac{1}{m_1 \cdot c_1} + \frac{1}{m_2 \cdot c_2}\right) \cdot t\right]; \\ T_2 - T_\infty &= (T_{B2} - T_\infty) \cdot \exp\left[-h \cdot A \cdot \left(\frac{1}{m_1 \cdot c_1} + \frac{1}{m_2 \cdot c_2}\right) \cdot t\right]. \end{aligned} \qquad (4.114)$$

Transient Heat Transport Through a Flowing Heat Transfer Medium
Figure 4.15 shows an externally insulated container filled with a fluid 2. This container is heated or cooled by a heat exchanger flowing with the mass flow $\dot{m}_1$. The fluid 2 in the container is separated from the heat transfer fluid 1 by a wall. The temperature of the fluid in the vessel is T_{B2} at the beginning of heating or cooling. The heat transfer fluid has the temperature T_{B1} at the inlet and $T_1(t)$ at the outlet. It is cooled or heated by the fluid 2 in the vessel. The heat transfer takes place with a constant heat transfer coefficient h. After an infinitely long time, the fluid in the vessel reaches the inlet temperature of the heat transfer medium T_{B1}, after which the fluid temperature of the heat transfer medium no longer changes. Assuming that the mass flow rate of the heat transfer fluid $\dot{m}_1$, the heat transfer coefficient h, the specific heat capacities c_i and local fluid temperatures are constant, the time temperature changes of the fluid 2 (T_2) and those of the heat transfer fluid 1 at the outlet T_2 can be calculated. The result is without going into the derivation:

$$T_{B1} - T_2 = (T_{B1} - T_{B2}) \cdot \exp\left[\frac{\dot{m}_1 \cdot c_1}{m_2 \cdot c_2} \cdot \left(\exp\left(\frac{-h \cdot A}{\dot{m}_1 \cdot c_1}\right) - 1\right) \cdot t\right]. \qquad (4.115)$$

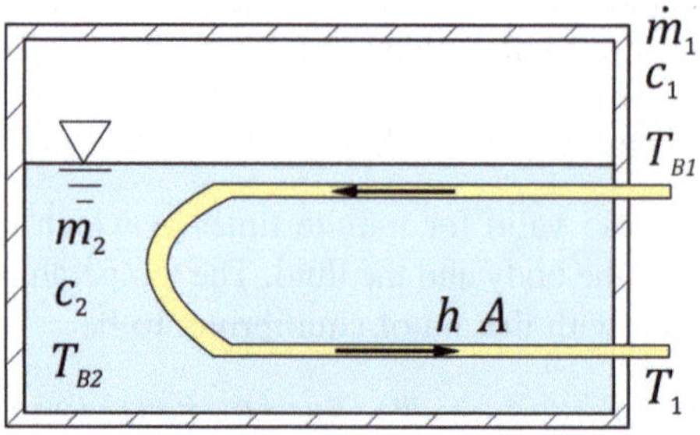

Fig. 4.15 Heating/cooling of a fluid 2 in an insulated container by means of a heat exchanger 1 through which the fluid 1 and the mass flow $\dot{m}_1$

4.2.3.6 Periodic Temperature Changes

Another boundary condition that occurs in many applications is the periodic change of the environmental condition, such as the daily change of solar radiation on a collector or a building structure.

When a periodic external temperature variation hits a plate, a heat wave propagates through the plate. Due to thermal conduction in the solid, the amplitude of the imposed external temperature variation is reduced. The damping of the wave, i.e. the attenuation of the amplitude, is determined by the thermal effusivity coefficient b. However, not only the amplitude of the temperature changes as the wave passes through the wall, but also the time response. A large specific thermal storage coefficient $s = (\rho \cdot c)$ in (J/(m^{3}K)) causes a slower progression of the temperature wave. The imposed temperature variation is therefore modulated not only in amplitude but also in phase, as shown in Fig. 4.16a, where a sinusoidal heat wave is assumed on the outside of the wall. As a rule of thumb, the higher the bulk density, the longer the phase shift and the higher the heat storage coefficient.

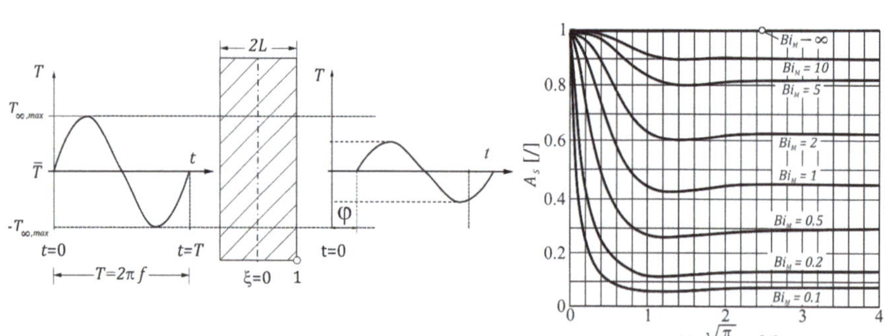

Fig. 4.16 **a** Schematic response behaviour of the establishing outer wall temperature $(T_S(\xi = 1))$(right) of a wall with the thermal conductivity λ for a periodically changing outer temperature T_∞ (left) of period T. **b** Dimensionless amplitude of a plate surface temperature as a function of M and Bi

The simplest form of this type of an imposed external boundary condition is the specification of the ambient temperature as a harmonic oscillation around a mean value $\overline{T}$ with the amplitude $A = \left(T_{\infty,\,\max} - \overline{T}\right)$:

$$T_\infty - \overline{T} = \left(T_{\infty,max} - \overline{T}\right) \cdot \sin\left(2\pi \frac{t}{T}\right), \tag{4.116}$$

in which T specifies the period of the imposed temperature disturbance. Assuming that this boundary condition has existed for such a long time that any disturbances from an initial temperature distribution have decayed, the following temperature distribution results inside a body

$$T - \overline{T} = \left(T_{\infty,\,\max} - \overline{T}\right) \cdot A \cdot \sin\left(2\pi \frac{t}{T} + \varphi\right), \tag{4.117}$$

where A denotes the amplitude and φ the the phase shift of the temperature wave, which are functions of the normalized distance $\xi = x/L$ from the centre, the Biot number Bi and a parameter M:

$$M = L \cdot \sqrt{\pi/(at)}, \tag{4.118}$$

which can be interpreted as the reciprocal value of Fourier number. In case of slow changes of ambient temperature (with a large period duration $T \gg \left(\pi L^2\right)/a$) corresponding to a small M, the heat wave is hardly damped and the amplitude A is close to unity; at the same time there is hardly any phase shift, i.e. $\varphi \approx 0$. The temperatures inside the body therefore follow the fluctuation of the of the ambient temperature almost completely. For large values of M, which imply a rapid change of the ambient temperature, the amplitude A decreases ccontinuously from the surface ($\xi = 1$) to the centre ($\xi = 0$) of the body. Thereby, a negative phase shift occurs, increasing in the same direction. For sufficiently large values of M, only a narrow edge zone of the body follows the external temperature variation corresponding to a boundary layer character. The analytical expressions for the amplitude and phase shift can be universally represented for a plate in the following way, (details for the cylinder and sphere can be obtained from (von Boeckh & Wetzel, 2011):

$$\begin{gathered}
A(\xi, M, Bi) = \sqrt{\frac{F_R^2(M\xi)+F_I^2(M\xi)}{G_R^2(M,Bi)+G_I^2(M,Bi)}} \text{ and } \varphi(\xi, M, Bi) = \arctan\left(\frac{F_I G_R - F_R G_I}{F_R G_R - F_I G_I}\right), \\
\text{in which } F_R(M\xi) = \cosh(M\xi) \cdot \cos(M\xi), \\
F_I(M\xi) = \sinh(M\xi) \cdot \sin(M\xi), \\
\left.\frac{dF_R}{d(M\xi)}\right|_{\xi=1} = \mathrm{F}'_R = \sin M \cdot \cos M - \cosh M \cdot \mathrm{sinM}, \\
\left.\frac{dF_I}{d(M\xi)}\right|_{\xi=1} = \mathrm{F}'_I = \cosh M \cdot \sin M - \sinh M \cdot \cos M \text{ with} \\
G_R(M, Bi) = \left(F_R(M\xi) + \frac{M}{Bi} \cdot \frac{dF_R}{d(M\xi)}\right)_{\xi=1} \text{ and} \\
G_I(M, Bi) = \left(F_I(M\xi) + \frac{M}{Bi} \cdot \frac{dF_{RI}}{d(M\xi)}\right)_{\xi=1}.
\end{gathered} \tag{4.119}$$

The Fig. 4.16b illustrates the amplitude $A_S(\xi = 1)$ at the surface of the plate as a function of M and Bi_M, which is defined as $Bi_M = Bi/M = (h/\lambda)\sqrt{(aT)/\pi}$.

Due to the external heat transfer resistance 1/h, the harmonic oscillation of the ambient temperature (according to Eq. 4.116) is transmitted to the surface T_S with a smaller amplitude A_S and a time offset φ_S, which is called negative phase shift, see schematic sketch in Fig. 4.16b so that:

$$T_S - \overline{T} = \left(T_{\infty,max} - \overline{T}\right) \cdot A_S \cdot sin\left(2\pi \frac{t}{T} + \varphi_S\right). \quad (4.120)$$

The time shift φ_S of the surface temperature T_S with respect to the ambient temperature can reach for thin bodies ($M \to 0$) and simultaneously a small external heat transfer coefficient ($Bi/M \to 0$) a quarter of the period, i.e. $\varphi_S = -\pi/2$. Then, however, the amplitude of the surface temperature oscillation A_S is also almost zero as well, in other words the wall does not notice anything from the surroundings.

For large values of $M\xi$ the plate can be considered as a semi-infinite body and Eq. (4.119) simplify and a boundary layer character is obtained. By introducing a boundary layer coordinate η and a modified Biot number Bi_M in the following form,

$$\eta = M(1 - \xi) = \sqrt{\frac{\pi}{(aT)}} y \text{ and } Bi_M = \frac{h}{a}\sqrt{\frac{(aT)}{\pi}}, \quad (4.121)$$

in which y denotes the coordinate from the plate surface towards the plate center the following result is obtained for the amplitude $A(\eta, Bi_M)$ and the phase shift $\varphi(\eta, Bi_M)$:

$$A(\eta, Bi_M) = \frac{\exp(-\eta)}{\sqrt{\left(1 + 2/Bi_M + 2/Bi_M^2\right)}} \text{ and } \varphi(\eta, Bi_M) = -\left(\eta + \arctan\left(\frac{1}{1 + Bi_M}\right)\right). \quad (4.122)$$

With the imposed sinusoidal boundary condition (4.116), the direction of the heat flow is reversed after half a period, i.e. heat is stored in the plate during the first half-wave and released to the environment during the next half-wave. Thus, the instantaneous heat flux is constantly changing. If we integrate over a half period ($t = T/2$), we obtain the amount of heat that the plate stores in the structure during the first half-wave in the form of internal energy and releases again during the next half-period. Using the abbreviations F'_R and F'_I from Eq. (4.119), one obtains the amount of heat stored after one half period $Q_{T/2}$:

$$Q_{T/2} = \frac{2}{\pi} A^* \cdot k \cdot \frac{T}{2} \sqrt{\frac{\pi}{(aT)} \cdot \left(\frac{2F'_R + 2F'_I}{G_R^2 + G_I^2}\right)} \cdot \left(T_{\infty,\,\max} - \overline{T}\right), \quad (4.123)$$

in which A^* represents the surface area of the plate. If we define a heat transfer coefficient $k_{T/2}$ in the way

$$Q_{T/2} = A^* \cdot \frac{T}{2} \cdot k_{T/2} \cdot \left(T_{\infty,\max} - \overline{T}\right), \tag{4.124}$$

in which the time-averaged temperature difference $\Delta\overline{T}$ is obtained from Eq. (4.116) by,

$$\Delta\overline{T} = \frac{1}{T/2} \int_{t=0}^{T/2} \left(T_\infty - \overline{T}\right) dt = \frac{2}{\pi} \cdot \left(T_{\infty,\max} - \overline{T}\right), \tag{4.125}$$

and the comparison of variables of Eqs. (4.123) and (4.124) lead to:

$$k_{T/2} = k \cdot \sqrt{\frac{\pi}{(aT)} \cdot \left(\frac{2\mathrm{F}'_R + 2\mathrm{F}'_I}{G_R^2 + G_I^2}\right)}. \tag{4.126}$$

The right-hand side of Eq. (4.126) is a function of M and Bi_M and for $Bi_M \to 0$, meaning a negligible internal heat transfer resistance compared to the external one, one gets

$$\lim_{Bi_M \to 0} \left(\frac{k_{T/2}}{k} \cdot \sqrt{\frac{(aT)}{\pi}}\right) = Bi_M \text{ or } \lim_{Bi_M \to 0} \left(k_{T/2}\right) = h. \tag{4.127}$$

In the other limiting case $Bi_M \to \infty$, an internal heat transfer coefficient h_i is obtained for $k_{T/2}$, which is

$$h_i = k \cdot \sqrt{\frac{\pi}{(aT)} \cdot \left(\frac{2\mathrm{F}'_R + 2\mathrm{F}'_I}{G_R^2 + G_I^2}\right)}. \tag{4.128}$$

Knowing the internal heat transfer coefficient, one can determine an average internal Nusselt number $Nu_{i,m} = (h_i/\lambda) \cdot \sqrt{(aT)/\pi}$. For large values of M, $Nu_{i,m}$ tend towards the constant value of $Nu_{i,m} = \sqrt{2}$, while in the opposite case for small values of M has the limit value $Nu_{i,m} = 2M$. If the internal heat transfer coefficient h_i is analyzed at a fixed period T for a plate, it passes through a maximum with increasing plate thickness, which is located at $M = 1.2$.

Practically occurring periodic boundary conditions usually deviate more or less strongly from the simple harmonic oscillation; however, they can be represented by Fourier analysis as sums of harmonic oscillations.

4.2.3.7 Application of Transient Heat Conduction to Building Walls

Based on the findings of the previous section, various conclusions can be drawn regarding the application in building technology and collector development. For example, the smaller the heat penetration coefficient b (see e.g. Table 4.4) of the room boundary surfaces, the faster the heating of a room with a certain heat output of the heating device. Rapid heating over time may be desirable for reasons of comfort in winter; this requires a propagation of a temperature field in a material, i.e. a high thermal diffusivity a. In the heating period, for example, heating energy can be saved by lowering the night temperature; in this case, a high thermal diffusivity a is disadvantageous. In the building sector, the storable energy is given by a storage coefficient s, which is the product of the density ρ and the heat capacity c of the material, and thus $s = (\rho c)$ in [J/(m^{3}K)]. However, large storage masses (high heat storage coefficients s) are desired for thermal inertia in summer, this means only a small heating of the building at high solar radiation. These storage masses can be used efficiently if there is an exterior thermal insulation with a small thermal conductivity. This creates a small contact temperature T_C and the building heating is significantly delayed.

This result, obtained on the basis of heuristic considerations, can be analysed on the basis of the temporal heating and cooling behaviour of typical building walls as a function of the selected building materials. Let us therefore analyse the heating and cooling behaviour of two walls for a given outside air temperature $T_{A,o} = 5\ °C$ and a fixed inner room air temperature of $T_{A,i} = 20\,°C$. The wall 1 consists of $\delta = 0.24\,m$ thick aerated concrete ($\rho = 500\,kg/m^3$) a thermal conductivity $k = 0.16$ W/(mK), resulting. in a thermal resistance of $R_H = 1.5$ K/W (or U-value of $\Lambda = k/\delta = 0.66$ W/(m^{2}K)) according to Eq. (4.55). Wall 2 is composed of two layers at first a $\delta = 0.06\,m$ thick polystyrene layer ($\rho = 30\,kg/m^3$, $k = 0.04$ W/(mK)) followed by a 10 cm wide heavy concrete layer ($\rho = 2500\,kg/m^3$, $k = 2.1$ W/(mK)) corresponding to $\Lambda = 0.646$ W/(m^{2}K) being almost equal to that of wall 1. Figure 4.17 shows the heating (left) and cooling (right) behaviour of the two different walls as a function of time for the same given external air temperature $T_{A,o}$.

Both walls reveal a completely different time behaviour of the surface temperature on the inner wall side $T_{S,i}$. Wall 1 offers advantages during the heating period since the inside wall temperature $T_{S,i}$ increases faster than in wall 2. At the same time, however, this wall exhibits a higher heat loss to the outside because the wall-normal temperature gradient $\partial T/\partial n$ is larger (compare Eq. 4.7). These high heat losses are also evident in the cooling of the wall. The massive wall 1 loses heat much faster than wall 2, and an uncomfortable living feeling quickly results after a winter night. Conversely, in summer the interior air in wall 1 heats up rapidly, so that an uncomfortably warm climate can establish. In this context, the question immediately arises as to the most energy-efficient way of insulating a wall. Which type of insulation is more favourable in a house depends on the way the room is used.

Figure 4.18 shows two identically thick aerated concrete walls on which the insulation is applied either externally or internally. With the exterior insulation (left

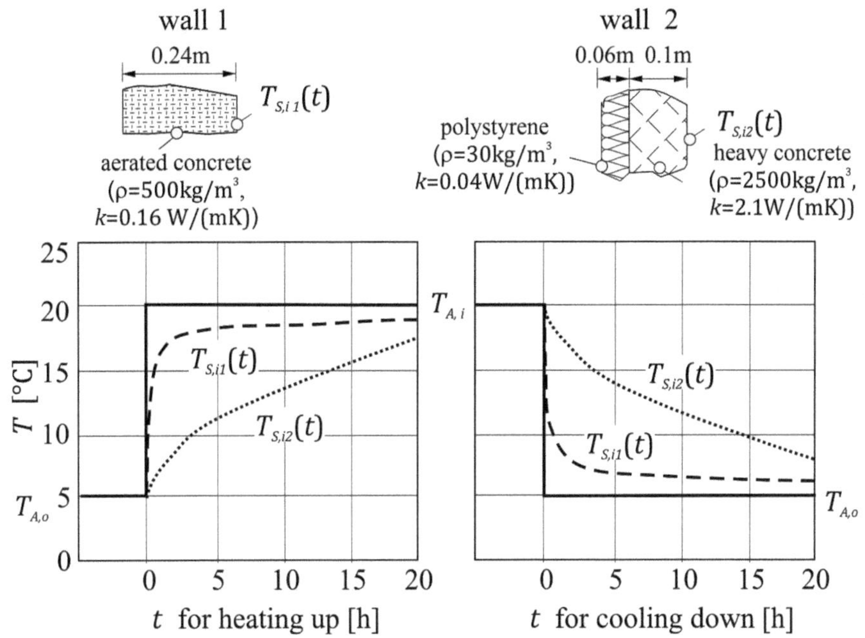

Fig. 4.17 Heating and cooling curves of the internal wall temperatures $T_{S,i}$ of two different walls as a function of time at for a fixed outside air temperature of $T_{A,o} = 5$ °C and a set internal room air temperature $T_{A,i} = 20$ °C

figure, a), the temperature gradient within the insulation layer is very high. This means that both at high temperatures in summer and low temperatures in winter, the building mass temperature hardly changes. The storage effect of the concrete walls is thus significantly increased. This type of insulation is disadvantageous in winter, if a "cold house" has to be heated. The heat loading of the solid storage wall by solar radiation is low and lasts for a relatively long time. In the case of interior insulation, rapid heating/cooling of the rooms by the direct solar radiation occurs, this is advantageous in the case of temporary use of the room or intermittent use. However, the storage effect of the relatively thick concrete wall is lost. In addition, at the interface concrete insulation in the case of internal insulation can fall below the dew point in winter, which can lead to undesirable meld growth. This type of insulation is not very useful for residential buildings.

If, for example, strong fluctuations in the outside temperature occur on the outer shell of a building on several consecutive summer days (24-h daily cycle), then a heat wave propagates through the outer components towards the interior every day. Heat conduction in the building material reduces the amplitude of the outside temperature fluctuation. The attenuator, i.e. the attenuation of the amplitude, is determined by the heat penetration coefficient b as discussed in Sect. 4.2.3.6. However, not only the amplitude of the temperature changes as it passes through the wall but also the time response. A large specific thermal storage coefficient s

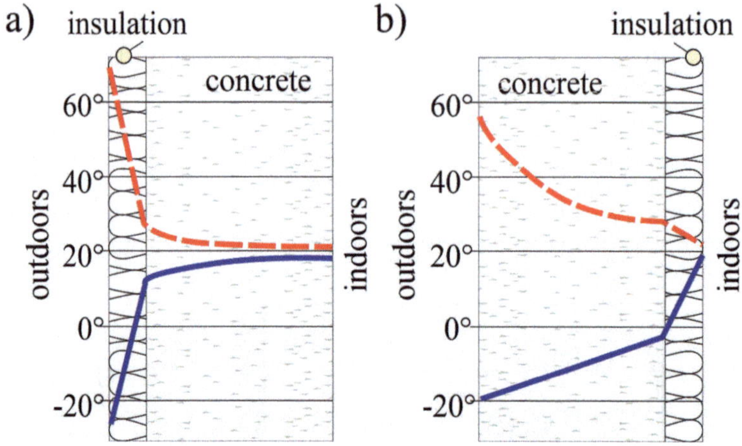

Fig. 4.18 Schematic temperature curve for exterior insulation (**a**) and interior insulation (**b**) of two identically thick building material layers of concrete and insulation material

causes a slower advance of the temperature wave. Thus, the temperature variation is modulated not only in amplitude but also in phase. Insulation materials based on renewable raw materials usually have a higher density and heat storage capacity and thus offer better thermal protection in summer. The ratio of the surface wall temperatures outside to the value inside is referred to in building technology as the temperature-amplitude ratio or TAR. The phase shift φ describes the time delay of the occurrence of the maxima on the outside and inside due to the storage of heat in the structural material. Figure 4.19 shows the effect of TAR on the indoor air temperature in a building with the same k-value but different heat storage capacities indexed as lightweight construction and heavy construction on ten consecutive summer days, determined by numerical simulation. It can be clearly seen that the TAR is significantly smaller for the heavy construction method.

A cooling time t_c is often specified in building services engineering. It indicates in hours how long it takes for a building component to cool down to the same temperature level in all layers under constant heat extraction. The longer the cooling time t_c of the outer walls, the more comfortable the indoor climate. It is defined as the ratio of the specific thermal heat storage coefficient s and the equivalent heat transfer coefficient h_{eq}, which scopes the thermal resistance of the insulation layer (k/δ), and the external heat transfer coefficient h. It has as SI unit the value in seconds, i.e. $t_c = s/h_{eq}$ $[s]$.

$$h_{eq} = \left(\frac{1}{h} + \frac{\delta}{k}\right)^{-1}. \tag{4.129}$$

Values for t_c of more than 120 h are considered very good. In general, the greater the density of the material, the lower the specific heat (c). In the case of solid building materials, it varies so little that the bulk density provides sufficient

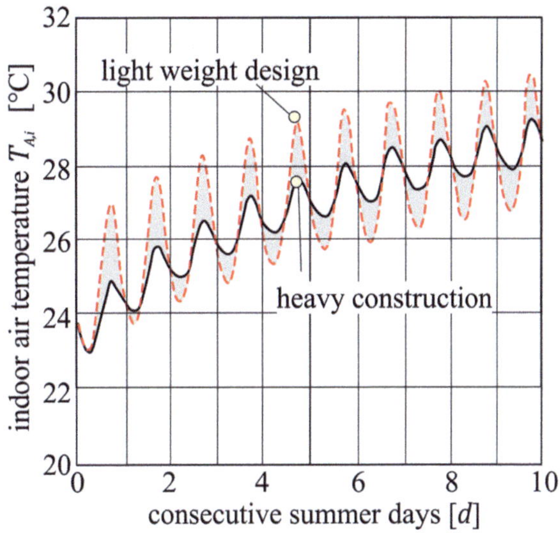

Fig. 4.19 Simulated temperature amplitude ratio (TAR) for 10 consecutive days for a light construction and a heavy construction

orientation for the heat storage capacity (thickness × bulk density = mass per unit area). The simple rule is that the more heat a material can store, the more sluggishly it reacts during heating and cooling, which is often referred to as so-called amplitude damping. Table 4.5 shows typical values of a solid wall of constant thickness 17.5 cm with different insulation thickness and the resulting h_{eq}-values. This constellation results in different cooling times, temperature-amplitude ratios (TAR) and phase shifts φ.

4.2.4 Molecular Heat Conductivity

In the context of the Kirchhoff-Transformation in the section of steady heat transfer we recognized large variations of the thermal conductivity in a continuum. Both, the order of magnitude differences in the temperature-dependent thermal conductivity of different materials, illustrated in Fig. 4.20, and their nature of temperature dependence vary considerably. Thermal conductivity is obviously a transport property of a body and a high thermal conductivity of a structural and/or functional material allows large heat fluxes; moreover, different materials exhibit a substantial temperature dependent thermal conductivity. Functional materials as the spectral selective solar thermal absorbers target at high temperatures to attain high collector efficiencies, while as thermal insulation aim at low heat losses to the ambient requiring low thermal conductivities. According to Fourier's law determining the heat flux magnitude, this flux through matter is associated is associated with the heat conductivity. Relevant for this heat transfer is obviously an exchange of energy between neighbouring particles. Apparently, in gases and liquids where the particles can freely move and are not bound to a lattice structure this energy transfer is established by collisions. In solids that is different because the atoms

Table 4.5 Solid wall of constant thickness 17.5cm made of different materials with different insulation thickness, the resulting h_{eq}-values. Resulting cooling times t_c, TAR and phase shifts φ in [h]

Material	Insulation thickness δ [cm]	h_{eq} [W/(m^{2}K)]	Cooling characteristic time t_c [h]	TAR [–]	Phase shift φ [h]
Clay wall	0	2.59	11.08	0.492	5.5
	2	1.13	48.58	0.082	8.2
	4	0.72	86.2	0.043	8.5
	8	0.42	161.7	0.022	8.8
	12	0.30	237.8	0.015	9.3
	16	0.23	314	0.011	9.5
Brick wall	0	3.26	4.2	0.867	2.5
	2	1.14	38.39	0.106	7.5
	4	0.73	68.33	0.057	7.8
	8	0.42	128.5	0.029	8.2
	12	0.30	189.2	0.019	8.7
	16	0.23	250.4	0.014	9.2
Aerated	0	1.23	5.7	0.672	4.1
	2	0.61	25.0	0.161	8.5
	4	0.47	37.3	0.100	9.0
	8	0.32	62.4	0.056	9.6
	12	0.24	87.6	0.039	10.2
Lightweight	0	1.59	3.14	0.861	2.5
	2	0.88	6.14	0.597	4.1
	4	0.61	9.31	0.422	5.0
	8	0.36	15.9	0.253	6.0
	12	0.27	22.9	0.176	6.7

are integrated in a lattice structure; nonetheless the atoms are elastically linked to each other by bonding forces. Thereby, they are able to transfer vibrational energy from one particle to the next in the lattice structure. For this reason, the transport of vibrational energy in non-metallic solids can also be thought of as a motion of quasi-particles. The quasi-particles are called phonons, which have a dual character as a wave and as a particle similar as photons. Since the phonon energy transfer occurs quantised, the lattice arrangement and therefore the atom distance plays an essential role. This can be fundamentally derived by solid state physics.

By this description the heat conductivity in gases, liquids as well as solids can be formulated in a general way as an energy transfer from colliding particles by an energy diffusion. An exception are metals where an additional effect

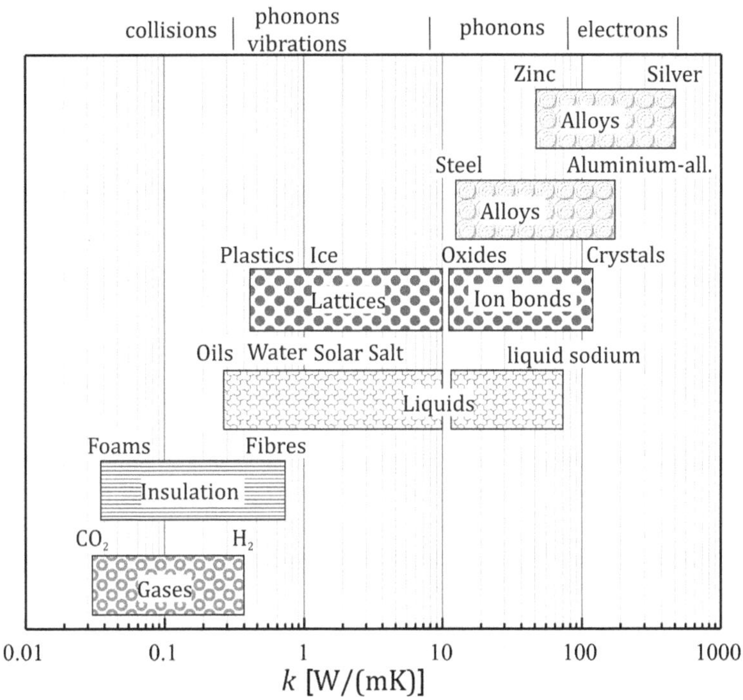

Fig. 4.20 Thermal conductivities k of various matter and leading processes for the heat conduction

is superimposed. Aside from the phonons also freely moving particles, the electrons, contribute to the energy transfer. These free conduction electrons of the electron gas outweigh the phonon contribution especially in pure metals is and is responsible for their good thermal conductivity.

4.2.4.1 Thermal Conductivity of Solids

Generally, in solids the energy propagation is described by the motion of phonons and these phonons transfer energy via collisions. This collisional energy transfer in non-electrically conducting solids makes heat conduction processes similar to that in liquids and gases. In the meantime between two phonon collisions of a phonon, the phonon travels on average a distance ℓ, which is called the mean free path as illustrated in Fig. 4.21. This mean free path refers to the motion of phonons in a three-dimensional frame of reference. But, relevant for the heat transfer in the one-dimensional case is only the motion of phonons in the direction of the heat flux and here in x-direction. The mean free path length ℓ is related to the mean free path components in x-, y- and z-direction. If we assume an isotropic substance, then the phonons can move in all spatial directions similarly and hence between two collisions the phonons travel the same distance on time average. If the phonons are assumed to be the thermal energy carriers, they cool down by the amount of

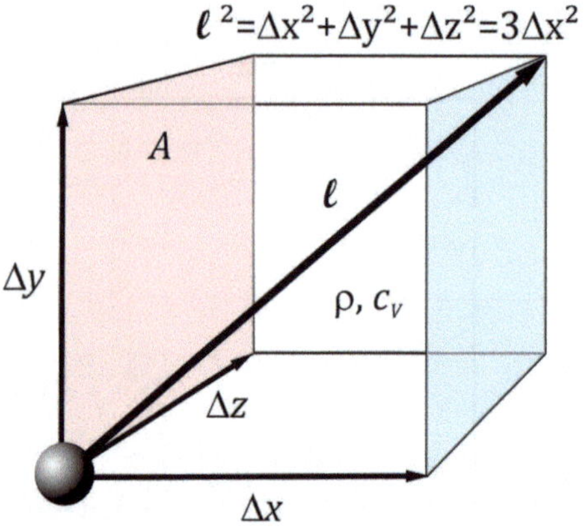

Fig. 4.21 Definition of mean free path ℓ in threedimensional space

ΔT due to the collisions when passing the distance Δx. The phonon itself losses energy through this but has transferred it to the next phonon. The magnitude of the energy transfer Q through a surface A at a temperature change ΔT results from the specific heat capacity c_V and the mass of the domain m ($m = \rho \cdot V$), which is given by:

$$Q = m \cdot c_V \cdot \Delta T = \rho \cdot V \cdot c_V \cdot \Delta T. \tag{4.130}$$

The average distance is given by the time between two collisions τ during which the heat Q is transferred. The thermal power $\dot{Q}$ through the area A is the given by $\dot{Q} = Q/\tau$. By inserting Fourier's law $\dot{Q}/A = \lambda \Delta T/\Delta x$ in Eq. (4.130) and replacing Δx by the expression for the mean free path 0 ℓ, we obtain

$$k = \frac{1}{3} c_V \cdot \rho \cdot v \cdot \ell, \tag{4.131}$$

where $v(= \ell/\tau)$ is the mean velocity of phonons which is approximately equal to the speed of sound. The existence of a certain finite value of ℓ is the result of phonon scattering on phonons, on defects of crystal lattice and, in particular on grain boundaries of the entire specimen. The temperature dependence of $k(T)$ is determined by that of ℓ and c_V on temperature. The Eq. (4.131) states that for no-metallic solids, the greater the mass specific heat capacity and the greater the density, the greater the thermal conductivity. Thus, for a high thermal conductivity, it is advantageous if the matter is composed of very many, but light particles. The specific heat capacity is not a constant value but depends on the temperature and the same holds for the mean free path, while the phonons, which are essentially sound waves, can be considered to be independent of temperature to a first approximation. In contrast to gases, atoms in solids cannot move freely but are bound to specific locations, also rotation is not possible. The movements of the atoms are thus limited to oscillation around an equilibrium position, whereby the oscillations can take place in all three spatial directions. This results in a degree of freedom for the atoms (f) in solids, which must be counted twice because of the potential and kinetic energy inherent in an oscillation. With effectively six degrees of freedom $(f = 6)$, the molar heat capacity of solids is theoretically three times the value of the universal gas constant $(\mathcal{R})$, which is called the Dulong-Petit-law, which reads to:

$$c_V = \frac{f}{2}\mathcal{R} = 3 \cdot \mathcal{R} = 3 \cdot N_A \cdot k_B, \tag{4.132}$$

where $\mathcal{R}$ can be expressed by the product of the Avogadro number N_A and the Boltzmann constant k_B, with $k_B = 1.380649 \cdot 10^{-23}$J/K. At high temperatures, it is mainly the mean free path length impacting the thermal conductivity and relatively strong lattice vibration occur. The many and violent lattice oscillations then influence each other, i.e. phonons are scattered by other phonons. Then the phonons travel only short distances before they are scattered again by other phonons. As the temperature increases, there are more phonons and the mean free path declines. Since the phonon density or population is proportional to the temperature, the mean free path length decreases inversely proportional to the temperature and thus in the temperature domain relevant for solar thermal engineering the thermal conductivity scales with $k \sim 1/T$.

The deduction explains the dependence $k(T)$ in crystal dielectric in general but does not account for deviations from a constant lattice due to atomic impurities of other elements. At a high temperature the resistance offered by impurities to heat transport is independent of temperature. This allows to estimate the degree of crystal purity from variation of its thermal conductivity with temperature. The same fact enables us to determine the distribution between a glassy and crystal substance in the natural mineral if the glass is considered as an impurity in a crystal.

In amorphous solids, too, a dependence of the thermal conductivity with increasing temperature is observed. However, the mean free path length is relatively independent of the temperature, since these do not have a regular lattice structure, but consist almost entirely of lattice perturbations. Therefore, the mean free path length of these materials is of the order of the atomic distance. The fact that the thermal conductivity of amorphous materials also changes with increasing temperature must therefore originate from the change in the heat capacity. In case of amorphous materials, the heat capacity is often greater than in crystalline materials, but the mean free path length is substantially smaller. This usually results in lower overall thermal conductivity values compared to crystalline matter.

4.2.4.2 Thermal Conductivity of Metals

A high thermal conductivity inherent appears in metals as shown in Fig. 4.20. Additionally, k for nearly all pure metals declines with increasing temperature. Another effect also observed is, that the coefficient of thermal conductivity for steel is substantially lower than for pure iron (see also Fig. 4.7), an effect which is observed in nearly all alloys. Also, for similar metal alloy compositions, different thermal conductivities are obtained. Obviously, this depends on the lattice configuration, which can be modified by mechanical and thermal treatment; e.g. quenching and cold treatment of metals yield a decrease of the thermal conductivity k by introducing additional lattice defects, while reversely pre-heating up to a high temperature anneals of defects and diffusion of alloy constituent to grain boundaries and thus an increasing thermal conductivity. Therefore, the knowledge of the treatment of construction alloys is of crucial importance especially for high thermally loaded materials.

In contrast to the non-metallic solids the thermal conductivity in metals depends not only on phonons but to essentially on the motion and interaction of current carriers, i.e., conduction electrons. Thus, the thermal conductivity k of metals is a superposition of lattice thermal conductivity by phonons k_{ph}, and conduction electrons k_e, where $k_e \gg k_{ph}$. If we assume a flow of free electrons and postulate that the length of their free path does not depend on velocity, then the Wiedemann–Franz law is obtained, postulating that the ratio of specific thermal conductivity to that of the specific electrical conductivity (σ) equals the Lorenz constant (L) timest he temperature, i.e.:

$$\frac{k}{\sigma} = L \cdot T \text{ with } L = \frac{k}{\sigma\, T} = \frac{\pi^2}{3} \cdot \left(\frac{k_B}{e}\right)^2 = 2.44 \cdot 10^{-8} \frac{WV}{AK^2}, \tag{4.133}$$

in which k_B is the Boltzmann constant and e is the elementary charge with $e = 1.602175 \cdot 10^{-19} As$. The Wiedemann–Franz law is used to calculate k_e from the measured electrical conductivity. The experimental findings show that the thermal conductivity of metals at high temperatures is a slightly decreasing function of temperature, however, there are numerous exceptions; e.g. k of iron strongly decreases with temperature while niobium exhibits a positive gradient.

The heat transport in semiconductors is more complex than in dielectrics, although they are of general interest as selective absorbers for solar thermal application since the band gap corresponds to the energy of the solar spectrum. The reason for the complex derivation of the thermal conductivity is the same order of magnitude of the phonon and electron part of the thermal conductivity. Moreover, impurities, bipolar diffusion, and other little studied factors considerably affect the heat transfer.

4.2.4.3 Thermal Conductivity of Liquids

The thermal conductivity in liquids is mostly based on the Bridgman model, relying that the sound speed v in a liquid exceeds by an order of magnitude the speed of the molecules determined from the kinetic theory. Another difference is that the coefficient for thermal conductivity k is based on the heat capacity at a constant pressure c_p, so that the following relation is obtained:

$$k = \rho \cdot c_p \cdot v \cdot \mathcal{L}, \tag{4.134}$$

where $\mathcal{L}$ denotes the intermolecular spacing with $\mathcal{L} = \Delta - d$ being Δ the center-to-center distance of molecules of diameter d. The coefficients of thermal conductivity and viscosity of liquids obey more, and no simple proportionality characteristic of gases exists between λ and their viscosity μ. One reason for this is the presence of intermolecular force fields in liquids as e.g. van der Waals forces, which prevent to some extent the relative motion of neighbouring molecules. For increasing temperature, the kinetic energy of the thermal motion of the molecules rises and thus yields a decreasing the attraction forces. As a consequence, the thermal conductivity k decreases. For many liquids the temperature dependence can be approximated by a linear function $k(T)$ with the proportionality factor from -0.0005 to -0.002 for different types of liquids. The sound speed for liquids is mainly experimentally determined; however, semiempirical relations as the one suggested by Missenard (see e.g. in Grifoni & Passerini, 2006) are used; which read to:

$$k = B \cdot c_p \cdot \sqrt{\rho \cdot T_m}, \tag{4.135}$$

where B is assumed as a constant for all the liquids with the same number of atoms in the molecule N, being $B = N^{-1/4}$ and T_S the boiling point of the liquid. All liquids exhibit an increasing thermal conductivity with increasing pressure, which is based on the fact that compression increases the mutual attraction of the molecules, however, this effect is marginal for pressure below 5 MPa.

Contrary to gas mixtures in which components are mixed at the atomic and molecular level, the structure of systems with liquid components can differ essentially and reliable assessments should be based on manufacturer data. In solar thermal applications this is of particular importance, as water–glycol mixtures are used for frost protection in domestic applications or solar salt mixtures for power plant as well as sensible heat storage purposes. Based on generalized experimental data Filippov (1968) suggested an empirical relation for estimating thermal conductivity of solutions 1 and 2:

$$k = k_1(T) \cdot C_1 + k_2(T) \cdot C_2 - 0.72/k_1(T) - k_2(T)/(C_1 \cdot C_1), \tag{4.136}$$

where C_1 and C_2 are the mass concentrations of components with thermal conductivities k_1 and k_2 respectively. For the calculation of the thermal conductivity of electrolytes (salt solutions) often the relation suggested by Missenard is recommended:

$$k = 9.3 \cdot 10^{-2} \cdot N \cdot m^{-5.6} \cdot \sqrt{\rho \cdot T_m}, \tag{4.137}$$

in which m represents the molar mass of the molcules, N the number of atoms in a molecule, ρ the electrolyte density and the T_m melting temperature.

4.2.4.4 Thermal Conductivity of Insulation Materials and Gases

Table 4.6 shows some typical thermal conductivities of solids, such as metals, liquids, composites, and gases. From Table 4.6 and Fig. 4.20 it is clear that there is a significant difference between the values of solids and especially gases, the causes of which have been discussed in the previous sections. It is interesting to compare the thermal conductivity values of the gases with those of insulating materials and loose fills, which is given in Table 4.7. Both, exhibit similar values being substantially smaller than those of liquids and metals.

Insulating materials are inhomogeneous agglomerations of gas-filled structures (fibres, beads, porous bodies). In these inhomogeneous materials there is a mixture of radiation and conductive-convective heat transport occurring on a microscopic scale, which constitutes in an equivalent "heat conductivity k" on a macroscopic scale. As long as there are many interactions of radiative or conductive exchange in a layer of the materials (say within a few cm), the Fourier law of conduction holds. However, if the layers become very thin, the exact transport equations of in the structures (e.g. between the fibres in mineral wool) have to be solved. In the case of pebble bed fills, there are only a few surfaces and mostly only point contacts between particles occur. These small thermal bridges cause a large thermal conductance resistance and there are long paths of the heat transport by conduction processes compared to the layer thickness of the medium, as illustrated by Fig. 4.22. In order to calculate for this type of pebble beds as they e.g. appear

Table 4.6 Typical specific thermal conductivity k in [W/(mK)] of some materials at standard pressure (1013 m bar) and temperature (25 °C)

Metal	k [W/(mK)]	Liquids	k [W/(mK)]	Solids	k [W/(mK)]	Gases	k [W/(mK)]
Aluminium	221	Gasoline	0.14	Concrete	2.1	Ar	0.016
Bismuth	8.4	Glycerine	0.286	Glass	0.76	He	0.143
Lead	35	Fuel oil	0.14	Limestone	2.2	CO_2	0.015
Iron	80.2	Machine oil	0.124	Marble	2.8	CH_4	0.034
Gold	314	Ice at 0 °C	2.2	Sand	0.58	O_2	0.023
Potassium	135	H_2O (10 °C)	0.58	Sandstone	2.3	N_2	0.02
Copper	393	H_2O (60 °C)	0.644	Mild steel	40	H_2	0.18
Brass	120	H_2O (100 °C)	0.679	Stainless steel	15	Vacuum	0.0
Sodium	133			Brick	0.5–1.4	Xe	0.0051
Silver	429			Diamond	2300	Air	0.0245
Tungsten	167			Silicon	148		
Zinc	110			Sulphur	0.269		
Tin	67				2.1		

Table 4.7 Specific thermal conductivity k in $[W/(mK)]$ of some insulation materials and loose fill beds

Insulating materials	k $[W/(mK)]$	Insulating materials	k $[W/(mK)]$
Expanded polystyrene (EPS)	0.035–0.05	Aerated concrete	0.08–0.25
Extruded polystyrene	0.032–0.04	Poroton	0.08–0.45
Glass wool	0.035–0.05	Foam glass crushed stone	0.08
Cork	0.035–0.046	Perlite	0.04–0.07
Foam glass	0.04	Straw bales	0.038–0.07
Wool	0.035	Sand loose	0.04
Vacuum insulation panel	0.004–0.006	Wood pressed	0.1–0.2

Fig. 4.22 Schematic sketch of heat conduction in a pebble

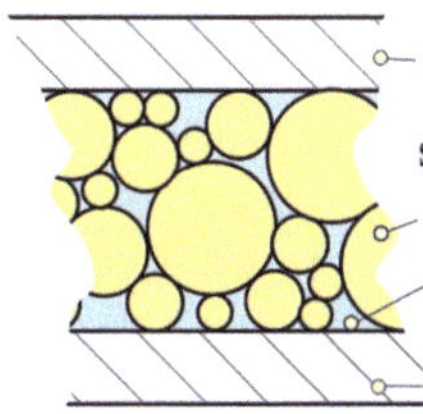

in filler beds of thermal energy storages the effective thermal conductivity, the mixed form between gas and solid conduction as well as thermal radiation must be considered, which we will discuss at the end of this chapter.

In gases, heat transport does not occur within a fixed lattice; rather, in gases, the gas atoms or their molecules encounter each other as they move randomly, transferring energy from particles with higher energy to those with lower energy. The gas molecules can move both translationally and rotationally. Thus, the energy transfer in gases is determined by the changes in translational as well as rotational energy as the gas atoms collide. Higher temperatures are equivalent to higher molecular energies.

For the gas to be conceived as a continuum, many collisions of the gas atoms with each other must occur. This is dealt within the idealized treatment of kinetic gas theory. The requirement for a high collision frequency directly implies that the mean free path length ℓ of the gas atoms or molecules between each other must be small compared to the dimension of the control space (vessel or gap in which the heat exchange is to take place). The mean free path length ℓ of the gas atoms between two collisions is given by the collision cross-section σ_{12} (in cm^2) and the particle or atom density n of the gas. A collision between two particles with particle radii r_1 and r_2 occurs when the particles approach each other to the distance of $(r_1 + r_2)$. For "point" gas atoms, a collision would occur when one of two particles fly around the second particle through a disk of cross-section $\pi(r_1 + r_1)^2 = \sigma_{12}$, as outlined in Fig. 4.23.

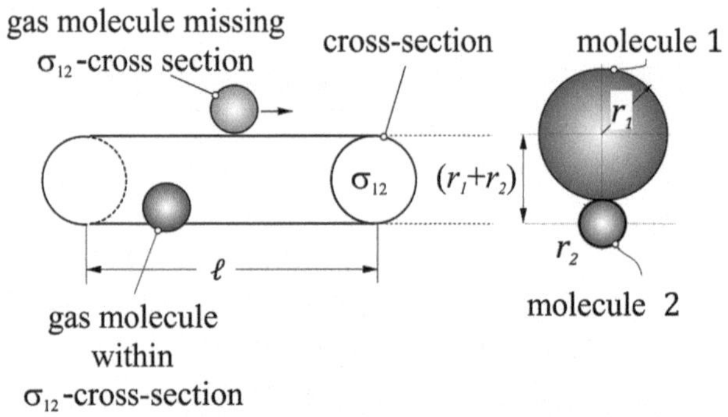

Fig. 4.23 Concept of the effective cross section σ_{12} for the collision between two gas molecules

If one imagines a particle stream with particle density n_1 flying through a volume with particle density n_2, collisions on a path element dx reduce the original particle stream n_1 by dn_1:

$$dn_1 = -n_1 \cdot \sigma_{12} \cdot n_2\, dx. \tag{4.138}$$

Integration along path x yields a negative exponential function for the decrease in particle density n_1 along path x relative to the original particle density $n_{1,0}$ at location $x = 0$. This yield:

$$n_1(x) = n_1(x = 0) \cdot \exp(-\sigma n_2 x). \tag{4.139}$$

The mean distance ℓ between two collisions is then given by

$$\ell = \frac{\int_{x=0}^{x\to\infty} x dn_1}{\int_{x=0}^{\infty} dn_1} = \frac{\int_{x=0}^{x\to\infty} x dn_1}{n_1(x = 0)}. \tag{4.140}$$

Also important is the number of collisions, which is described with the help of the collision frequency z_{12} between the two types of particles. The collision frequency is the number of collisions per time, which the incoming particles n_1 have with the particle 2 located in the volume V. The particles 1 have an average velocity $\overline{v}_1$, so that the following relationship results for the number of collisions between the particles 1 and 2:

$$z_{12} = \frac{n_2 \cdot V}{\Delta t} = \frac{\sigma_{12} \cdot \ell \cdot n_2}{\Delta t} = \sigma_{12} \cdot \overline{v}_1 \cdot n_2. \tag{4.141}$$

If the particles of type 2 are also moving, the mean velocity is replaced by the mean relative velocity of both particles $\overline{v}_{12}$ and a modified impact number z_{12} is obtained. Finally, the impact density Z_{12} results from the multiplication of the impact number z_{12} by the particle density n_1 of the particles 1 entering the control volume V:

$$z_{12} = \sigma_{12} \cdot \overline{v}_{12} \cdot n_2 \text{ and } Z_{12} = \sigma_{12} \cdot \overline{v}_{12} \cdot n_2 \cdot n_1 \left[\frac{1}{(m^3 s)} \right]. \tag{4.142}$$

As a rule, the gas molecules do not all have the same velocity, rather there is a statistical distribution of velocities in both types of particles, which is given by the Boltzmann distribution. For the sake of simplicity, we consider further only one type of particle which has a statistical velocity distribution, and as a result we obtain the following relation for the mean free path length ℓ:

$$\ell = \frac{\overline{v}_1 \cdot \Delta t}{z_{11} \cdot \Delta t} = \frac{1}{\sqrt{2}} \cdot \frac{1}{\sigma \cdot n_1}. \tag{4.143}$$

In a plane gas-filled gap as sketched in Fig. 4.24, which is bounded by two walls of temperature T_1 and T_2, where $T_1 > T_2$, heat transport occurs from the hot to the cold side. For this purpose, at an arbitrary point x_1 in the gap, one can think of an intermediate plane with the thickness of the free path length ℓ below and above the plane x_1. If we first consider the number of particles that pass over the interfaces at $(x_1 + \ell)$ and $(x_1 - \ell)$, we obtain a net particle flux ΔN:

$$\Delta N = N(x_1 - \ell) + N(x_1 + \ell) = [n(x_1 - \ell) - n(x_1 + \ell)] \cdot A \cdot \frac{\ell}{6}, \tag{4.144}$$

where A denotes the cross-sectional area. The factor $1/6$ results from the fact that all particles in the control volume can move in the three translational directions (x, y and z) as well as the three rotational directions. The boundary transition from N to dN yields the following result using Eq. (4.139):

$$dN = -2 \cdot A \cdot \frac{\ell}{6} \cdot \ell \cdot \frac{\partial n}{\partial x} = \frac{\ell^2}{3} \cdot A \cdot \frac{\partial n}{\partial x}. \tag{4.145}$$

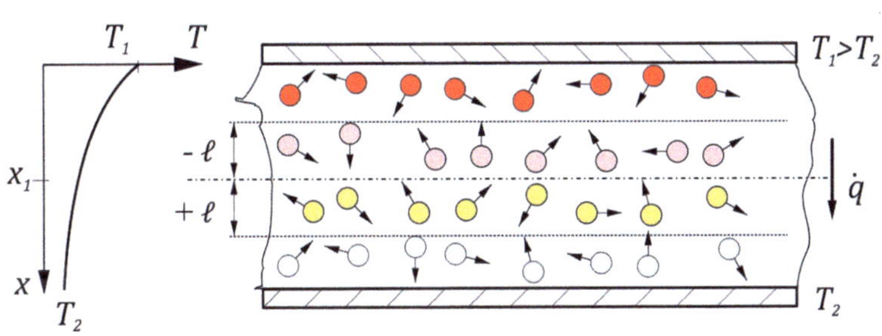

Fig. 4.24 Sketch of the derivation of the heat conduction in a gap

Thus, the number of particles exchanged over the imaginary control volume is proportional to the square of the mean free path length ℓ and the gradient of the particle density $\partial n/\partial x$. Further, one can define a particle flow in the x-direction J_x and using $\ell = \overline{v} \cdot t$ assuming a constant mean velocity yield:

$$J_x = \frac{dN}{A \cdot dt} = - \underbrace{\frac{\ell}{3} \cdot \overline{v}}_{\text{diffusion constant}} \frac{\partial n(x,t)}{\partial x}. \tag{4.146}$$

Each molecule has a certain amount of heat Q, which is given by the number of degrees of freedom f of its motion, and is accordingly calculated as follows

$$Q = \frac{1}{2} \cdot f \cdot k_B \cdot T. \tag{4.147}$$

Therein, f is the number of degrees of freedom—3 of translation and 3 of rotation, depending on the type of molecule, k_B is the Boltzmann constant, and T is the absolute temperature in Kelvin. The heat flux transported from the hot to the cold side in the gap is then given by the product of the particle flux and the heat stored in each particle to give:

$$\dot{q} = -\frac{\ell}{3} \cdot \frac{\partial}{\partial x}\left(\overline{v} \cdot n \cdot \frac{f}{2} \cdot k_B \cdot T\right). \tag{4.148}$$

If we now express the mean kinetic energy of the gas particles E_{kin}, given by $E_{kin} = m\overline{v}^2/2$, by the temperature T of the gas, given by $E_T = 3/2(k_B T)$, we obtain the ideal gas relation.

$$p = n \cdot k_B \cdot T \text{ with } n = 1/V, \tag{4.149}$$

in which n denotes the particle density $\left[1/\text{m}^3\right]$. Thus, if the temperature T changes but the pressure p remains constant, according to the general gas law, the particle density n must change in the opposite sense so that the ratio $(T/V) \sim (n \cdot T)$ remains constant. Assuming the validity of the gas law, only the mean velocity $\overline{v}$ can change in the term for the derivative in Eq. (4.148). This gives the following relationship for the specific heat flux

$$\dot{q} = -\frac{\ell}{6} \cdot n \cdot f \cdot k_B \cdot T \frac{\partial \overline{v}}{\partial x}. \tag{4.150}$$

However, the mean velocity is also related to the temperature. As already shown in connection with the probability of collisions among each other (Eq. 4.142), a frequency distribution of velocity exists in a gas, which is described with the help of the Maxwell–Boltzmann distribution. The distribution F of the gas molecules with velocity v is given in the one-dimensional case by the relation:

$$F(v) = \sqrt{\frac{m}{2 \cdot \pi \cdot k_B \cdot T}} \cdot \exp\left[-\frac{m \cdot v^2}{2 \cdot k_B \cdot T}\right], \tag{4.151}$$

where m denotes the particle mass. If the mean velocity is defined as the sum of the individual velocities of the particles and divided by their total number N, the mean velocity $\overline{v}$ of the gas-molecule composite is given by the following equation

$$\overline{v} = \frac{1}{N}\sum_{i=1}^{i\to\infty} v_i = \int_{v=0}^{v\to\infty} v \cdot F(v)\, dv = \sqrt{\frac{8}{\pi} \cdot \frac{k_B \cdot T}{m}}. \tag{4.152}$$

This allows us to convert the derivative of the mean velocity in Eq. (4.150) into a differential of the temperature.

$$\frac{\partial \overline{v}}{\partial x} = \frac{\partial}{\partial x}\sqrt{\frac{8}{\pi} \cdot \frac{k_B T}{m}} = \frac{\partial}{\partial T}\sqrt{\frac{8}{\pi} \cdot \frac{k_B T}{m}} \cdot \frac{\partial T}{\partial x} = \frac{1}{2} \cdot \frac{1}{\sqrt{T}}\sqrt{\frac{8}{\pi} \cdot \frac{k_B}{m}} \cdot \frac{\partial T}{\partial x} = \frac{1}{2} \cdot \frac{\overline{v}}{T} \cdot \frac{\partial T}{\partial x}. \tag{4.153}$$

This expression for the change in mean velocity can now be substituted into Eq. (4.150) and the result obtained for the heat flux density $\dot{q}$ transferred.

$$\dot{q} = -\frac{n}{12} \cdot f \cdot k_B \cdot \ell \cdot \overline{v} \cdot \frac{\partial T}{\partial x}. \tag{4.154}$$

The equation for the heat flow corresponds to that of the 1st Fourier law (Eq. 4.6) at the beginning of this chapter. By a comparison of variables, the following relation is obtained for the thermal conductivity $k(T)$ of an ideal gas:

$$k_{Gas} = -\frac{\overline{v}}{12} \cdot \ell \cdot n \cdot f \cdot k_B \left[\frac{W}{(mK)}\right]. \tag{4.155}$$

Thus, the thermal conductivity of the gas can be analysed as a function of the individual contributions. If we consider the average velocity of the molecules $\overline{v}$, we get a higher conductivity of the gases scaling with $\sqrt{T}$ (from Eq. 4.152). The higher the temperature, the more collisions occur, and thus more heat can be transferred. Another result of the molecular velocity analysis is that the thermal conductivity of the gases decreases correspondingly with $1/\sqrt{m}$ as the molecular mass increases. Table 4.8 gives the thermal conductivity of some gases at standard pressure and temperature. The decrease of the specific thermal conductivity with increasing molecular mass can be seen particularly well for the noble gases; for example, for xenon it is only about one third of that of helium.

In Eq. (4.143), it was found that the mean free path length ℓ is inversely proportional to the particle density n, thus $k \sim 1/n$. Using the ideal gas equation, the particle density n is directly proportional to the pressure p, thus $p \sim n$.) If we insert this relationship into the conductivity relationship, we obtain the striking result that the thermal conductivity of a gas does not depend on the particle density. Thus, the thermal conductivity of a gas is independent of the pressure p; likewise, it can be shown that the viscosity of gases is independent of the pressure

Table 4.8 Measured specific thermal conductivity k in $[W/(mK)]$ of some gases at standard pressure (1013 m bar) and temperature (25 °C)

Gas	k [W/(mK)]	Gas	k [W/(mK)]
Hydrogen (H_2)	0.168	Helium (He)	0.144
Methane (CH_4)	0.030	Neon (Ne)	0.046
Air	0.0262	Argon (Ar)	0.016
Carbon dioxide (CO_2)	0.0146	Krypton (Kr)	0.0088
SO_2	0.0086	Xenon (Xe)	0.0051

as well. This statement is true as long as the mean free path length ℓ is small compared to the width of the considered gap or the vessel dimension. Therefore, for a calculation of the gas conductivity, it is mandatory to first determine the mean free path length ℓ_{gas} of the gas molecules at a certain residual gas pressure p_{gas}. As a rule of thumb, the following relationship can be used here:

$$\ell_{gas} = \frac{C_1(p_{gas})}{1 + C_2/T}[m], \tag{4.156}$$

where T is the absolute temperature in Kelvin and C_1 and C_2 represent two gas-specific constants. Table 4.9 provides some data for selected gases.

For air at normal pressure and room temperature, $\ell_{gas} \approx 0.6\mu m$. Next, the vessel dimensions δ must be checked. This can be the distance of the walls over which the driving temperature gradient occurs. For insulation materials, this can be the dimensions of the pores or the spacing of foils. For most insulation materials, $\ell_{gas} \gg \delta$. This means that the mean free path ℓ_{gas} in these materials is not so much determined by molecule–molecule collisions but is rather limited by the mean foil spacing δ. According to (Kaganer, 1969), the following relationship then applies

Table 4.9 Constants for determining the mean free path of various gases ℓ_{gas}. C_1 applies for a pressure of $p_0 = 133\ Pa$; for $p_{gas} \neq p_0$: $C_1(p_{gas}) = (p_0/p_{gas}) \cdot C_1(p_0)$, (Stephan et al., 2019)

Gas	$C_1(p_0)[\mu m]$	$C_2[K]$
Hydrogen (H_2)	105.6	76
Nitrogen (N_2)	61	112
Oxygen (O_2)	68.7	132
Helium (He)	160	79
Neon (Ne)	111.9	56
Argon (Ar)	70.3	169
Krypton (Kr)	59.6	142
Xenon (Kr)	48.7	252
Water (H_2O)	95	600
Carbon monoxide (CO)	60.2	100
Carbon dioxide (CO_2)	57	273

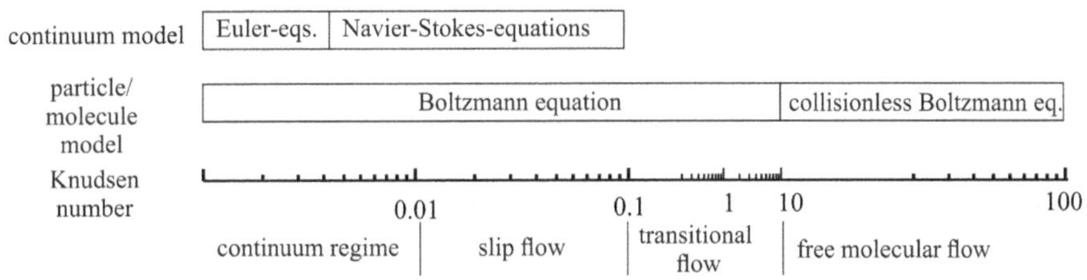

Fig. 4.25 Classification of flow regimes and associated continuum models as a function of the Knudsen number (Bird, 1994)

to the residual gas thermal conductivity k_{gas}:

$$\ell_{gas}(p_{gas}) = \frac{k_0}{1 + 2\beta Kn}\left[\frac{W}{(mK)}\right]. \tag{4.157}$$

In Eq. (4.157), k_0 denotes the thermal conductivity of the gas at standard pressure, β a weighting factor and Kn the Knudsen number. The Knudsen number Kn indicates the ratio of the mean free path length in the gas ℓ_{gas} to the typical characteristic dimension δ of the problem e.g. the pipe diameter or wall distance and is defined as follows:

$$Kn = \frac{\ell_{gas}}{\delta} \cdot . \tag{4.158}$$

The magnitude of the Knudsen number serves as a criterion to characterize the flow regime and the continuum model to be applied for computations. As Fig. 4.25 illustrates continuum models can be applied up to roughly $Kn \approx 0.1$.

The weighting factor β contains a correction quantity k_c and the adiabatic index γ, which is the ratio of the isochoric heat capacity c_V and the isobaric heat capacity c_p. The adiabatic index γ of an ideal gas is,

$$\gamma = \frac{c_p}{c_V} = \frac{f + 2}{f} \quad [/], \tag{4.159}$$

where f denotes the degree of freedom of the gas. The accommodation coefficient α is also included in the weighting factor β, so that the exact formulation is as follows:

$$\beta = \frac{2 \cdot k_c}{\kappa + 1} \cdot \frac{2 - \alpha}{\alpha}. \tag{4.160}$$

Some correction values k to the specific heat capacity c_V for the calculation of β in Eq. (4.160) are listed in Table 4.10. The accommodation coefficient α is usually only imprecisely known. For the light gases (H_2, He) $\alpha \approx 0.3$, for the heavy gases for example nitrogen (N_2) α is about unity (Wakao & Vortmeyer, 1971). For air $\beta \approx 1.63$.

Table 4.10 Correction values k_c (to specific heat c_V) to calculate β from Eq. (4.160)

Gas	$T[°C]$ −182.6	−78.4	0	100	200	300	400
Hydrogen (H_2)	2.35	–	2.03	–	2.12	–	2.19
Nitrogen (N_2)	–	–	1.95	–	–	–	–
Oxygen (O_2)	–	–	1.915	–	–	–	–
Air	–	1.94	1.95	1.945	1.943	–	–
Carbon monoxide (CO)	–	–	1.865	–	–	–	–
Carbon dioxide (CO_2)	–	1.805	1.67	1.60	1.56	1.55	–
Water (H_2O)	–	–	–	1.41	1.395	1.41	1.425
Helium (He)	–	–	2.43		–	–	–
Neon (Ne)	–	–	2.50		–	–	–
Krypton (Kr)	–	–	2.49		–	–	–
Xenon (Xe)	–	–	2.49		–	–	–

According to Eq. (4.160), the gas thermal conductivity k_{gas} becomes small if k_0 is chosen to be as small as possible, as in the case of noble gases, for example, or if the Knudsen number Kn is aimed for to be as large as possible. Noble gases with their small conductivity are used in the construction industry for very high-quality multiple-pane glazing. In some cases, the cavity between the panes is filled with krypton or even xenon.

Technically, the more common method is to reduce the pressure to increase the Knudsen number. At low pressures, heat conduction then depends only on the number of particles n and then decreases proportionally with the pressure. There is a significant increase in the free path length at pressures lower than 100 Pa. However, in turn this means a rough vacuum is therefore insufficient to reach the dimensions of a macroscopic system with the free path length. Since the evacuation of a volume causes the mean free path length ℓ_{gas} to increase, the lower the pressure, the larger the Knudsen numbers (- or the smaller the λ_{gas}− values). The mean free path of some gases ℓ_{gas} as a function of pressure p_0 in $[Pa]$ is given in Table 1.11. This method has been standard for thermos cans or, for example, Dewar vessels in cryogenics for over 100 years. In the building industry, it has been pursued as a vision for some time, but here problems often arise in the practical implementation in the area of sealing surfaces, the mechanical stability of large glass surfaces and, last but not least, the technical implementation on site at the construction site. Therefore, vacuum insulation elements for building applications have only recently become available. It also follows from Eq. (4.158) that large Knudsen numbers can be obtained by specifying very small film (or wall) spacings δ or, as described in Table 4.11, by using fine-grained fills with very fine pore/particle diameters δ.

By evacuation and by means of small δ-values, it is finally possible on a technical scale to completely eliminate the thermal conductivity component k_{gas}. Plates

Table 4.11 Mean free path length of some gases ℓ_{gas} in $[m]$ as a function of pressure p_0 in $[Pa]$ at standard temperature $T = 25\,°C$

Gas	$p_0[Pa]$ 10^{-6}	10^{-3}	1	10^3	10^5
Air	$6.8 \cdot 10^3$	6.8	$6.8 \cdot 10^{-3}$	$6.8 \cdot 10^{-6}$	$6.8 \cdot 10^{-8}$
Argon (Ar)	$7.2 \cdot 10^3$	7.2	$7.2 \cdot 10^{-3}$	$7.2 \cdot 10^{-6}$	$7.2 \cdot 10^{-8}$
Carbon dioxide (CO_2)	$4.5 \cdot 10^3$	4.5	$4.5 \cdot 10^{-3}$	$4.5 \cdot 10^{-6}$	$4.5 \cdot 10^{-8}$
Water (H_2O)	$4.2 \cdot 10^3$	4.2	$4.2 \cdot 10^{-3}$	$4.2 \cdot 10^{-6}$	$4.2 \cdot 10^{-8}$
Helium (He)	$19.6 \cdot 10^3$	19.6	$19.6 \cdot 10^{-3}$	$19.6 \cdot 10^{-6}$	$19.6 \cdot 10^{-8}$
Nitrogen (N_2)	$6.7 \cdot 10^3$	6.7	$6.7 \cdot 10^{-3}$	$6.7 \cdot 10^{-6}$	$6.7 \cdot 10^{-8}$
Neon (Ne)	$14 \cdot 10^3$	14	$14 \cdot 10^{-3}$	$14 \cdot 10^{-6}$	$14 \cdot 10^{-8}$
Oxygen (O_2)	$7.2 \cdot 10^3$	7.2	$7.2 \cdot 10^{-3}$	$7.2 \cdot 10^{-6}$	$7.2 \cdot 10^{-8}$

filled with mineral or polymer foam cores are commonly used for vacuum insulation. Figure 4.26 shows the thermal conductivity of some insulation materials as a function of the gas pressure in the fill. It is clear from the figure that as the dimensions of the voids decrease (smaller δ-values), the specific thermal conductivity is lower at the same pressure. Furthermore, a fine-grained bed reaches a saturation value already at higher gas pressures. This saturation value is given by the dimension of the voids. The drop of the thermal conductivity with decreasing gas pressure is often referred to in literature as Knudsen effect.

The better the insulation capacity is at low gas pressures, the more the local contact areas within the insulation material play a role. For example, they are smaller in spheres than in fibre-like composites. Figure 4.27a illustrates this by comparing

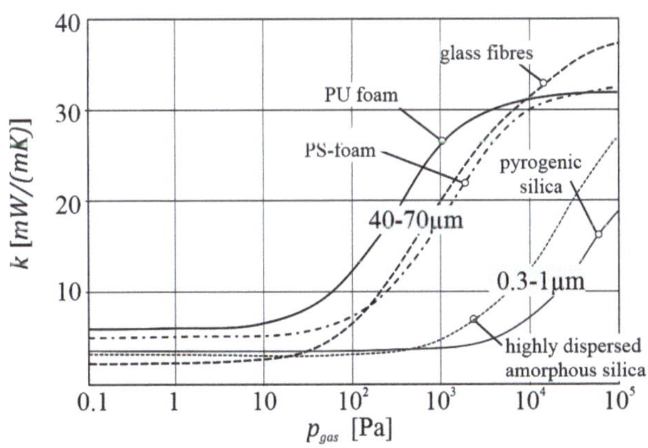

Fig. 4.26 Specific thermal conductivity k of several high insulation materials in [mW/(mK)] as a function of gas pressure p_{gas} in the bulk in $[Pa]$ at 20 °C (dry nitrogen) for different mean effective pore diametera. PU = polyurethane, PS = polystyrene, silica materials (specific surface, 300 m^2/g), (IfB GmbH Wolfgang Sorge, 2009)

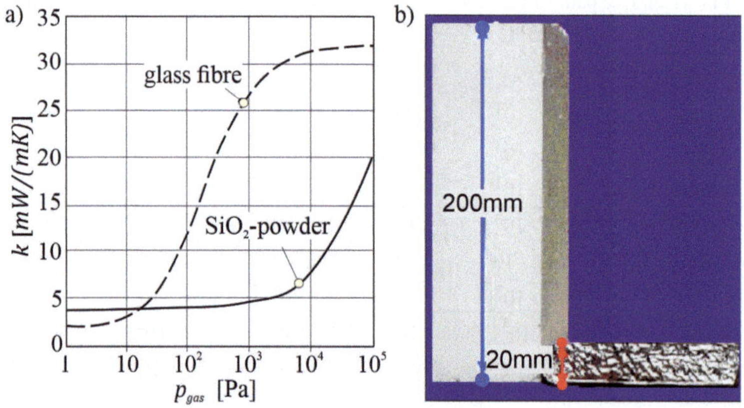

Fig. 4.27 Thermal conductivity of glass fibre and silicon oxide powder fill k in [mW/(mK)] as a function of gas pressure in nitrogen at 25 °C. **b** Thickness comparison of a solid polystyrene (PS) board and a vacuum insulated PS fill with the same heat transfer coefficient

a glass fibre with a silicon oxide spherical fill. The bulk material reaches the value of thermal conductivity of the pure material already at relatively large pressures, whereas in the case of the glass fibre composite, depending on the contact areas and gap dimensions, an optimum insulation effect is only achieved at much lower gas pressures. The performance of such insulation is illustrated by Fig. 4.27b, which shows the comparison of a solid polystyrene (PS) sheet with an amorphous vacuum-insulated pyrogenic silica fill. The solid structure builds 10 times thicker with the same heat transfer coefficient.

Note, the vacuum filled beds are rather exceptional limiting cases. For insulation material the effective heat conductivity (k) is usually a mixture of heat conduction of (residual) gas and insulation materials; moreover, for high emissivities and especially at higher temperatures also radiative effects contribute to the effective heat conductivity k, which is then constituted by a super-position of heat transfer coefficients arising from heat conduction, radiation, and convection.

4.3 Radiation Heat Transfer

From a microscopic physics point of view, there are only two mechanisms of heat transfer; either momentum transfer due to local collisions or diffusion (the transfer of heat via molecular interactions) and radiation, where the transfer of heat is realized via photons, which can be conceived as quasi-particles similar as phonons or as electromagnetic waves. Convection, being the bulk transport of a fluid, is not precisely a heat transfer mechanism. The physics of radiation transport is distinctly different than those of diffusion transport. The diffusion of particles in gases or fluids and of quasi-particles (phonons) in solids is a local phenomenon, meaning that the rate of heat transfer, at a point in space, precisely depends only on the local nature about the point, i.e., the temperature gradient and its thermo-physical properties at this point. Of course, the temperature field also depends on the boundary and initial conditions imposed on the system. Radiation, on the other hand, is not confined to a local point; the flux of radiation at a point, in general, directly, and instantaneously dependents on the radiation flux from all other visible points in a system. Unlike diffusion, radiation thus acts over a distance. Accordingly, the mathematical description of radiation transport employs an integral equation for the radiation field balancing all in- and output, as opposed to the differential equation for heat diffusion via thermal conduction.

4.3.1 Fundamental Radiation Properties

Radiation can be conceived as the transport of energy by discrete photons, and the basic relationship between the energy of a photon e, the frequency ν or wavelength $\lambda = c_0/\nu$ of the wave is provided by Planck's relation,

$$e = h \cdot \nu = \frac{h \cdot c_0}{\lambda}\cdot, \tag{4.161}$$

where h is Planck‘s constant $\left(h = 6.6261 \cdot 10^{-34}\mathrm{J/s}\right)$ and c_0 the speed of ligth in vacuum $c_0 = 2.9979 \cdot 10^{8}\mathrm{m/s}$. Thus, radiation is quantized. Thermal radiation occurs in a range of the electromagnetic spectrum of energy emission, and the full electromagnetic spectrum includes an enormous range of energy-bearing waves, of which the thermal radiation covers only a small part parts as Fig. 4.28 illustrates. From the graph it becomes visible that thermal radiation refers to the spectrum of radiation from the ultra-violet range via the visible spectrum ($\lambda = 0.4 - 0.7\,\mu\mathrm{m}$) to the infrared (IR) or far-IR wavelengths. Comparing these wavelengths with the atomic ones it becomes clear that by solar radiation the rotational and vibrational energy levels of molecules can be excited and thus energy is transferred by the solar spectrum to the molecules in the form of heat. This has been subject of Chap. 3 (optical conversion). There it could be demonstrated that by selective absorption of thermal radiation a raise of the temperature is achieved. In turn this means, thermal radiation emission is a direct result of vibrational and rotational motions of molecules, atoms, and electrons of a substance and temperature is a measure of these activities. Thus, the rate of thermal radiation emission increases with increasing temperature. Additionally, thermal radiation is a volumetric phenomenon. However, for opaque solids such as metals, radiation is considered to be a surface phenomenon, since the radiation emitted by the interior region never reaches the surface.

A body that emits the maximum possible radiation power is called a "blackbody". The radiation emitted by a blackbody has a wavelength distribution. In addition to the extra-terrestrial formulation, se Chap. 2, in which the gas density does not play a role, the refractive index n occurs in a body and the specific

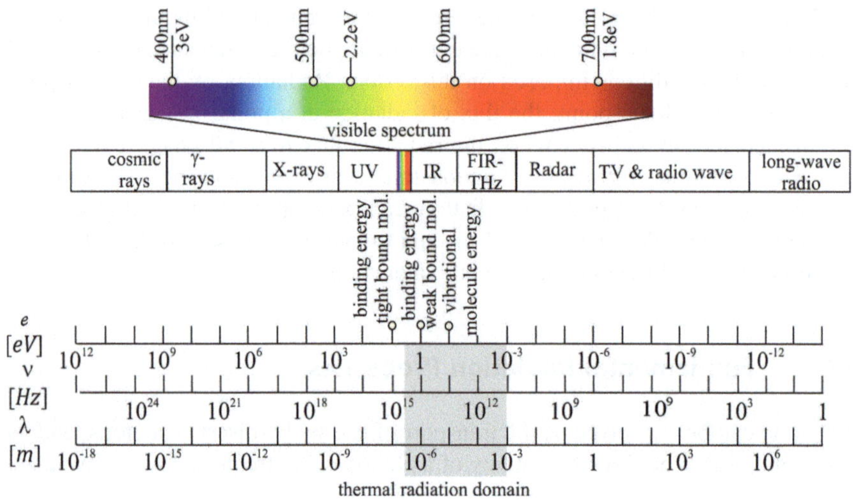

Fig. 4.28 Thermal radiation spectrum as function molecule energy e $[eV]$, wavelength λ $[m]$ and frequency ν $[Hz]$

spectral radiosity $e(\lambda, T)$ in $[\mathrm{W}/(\mathrm{m}^2\mu\mathrm{m})]$ is given in complete form by:

$$e_{b\lambda}(\lambda, T) = \frac{2\pi\ h\ c^2}{n^2 \cdot \lambda^5}\ \frac{1}{\exp\left(\frac{hc}{\lambda k_B T}\right) - 1}. \tag{4.162}$$

in which k_B is the Boltzmann constant ($k_B = 1.36 \cdot 10^{-23} J/K$).

4.3.2 Blackbody and Its Concept

In the following, a “blackbody” is considered. Physically, such a “blackbody” is an idealized body that completely absorbs electromagnetic radiation incident on it at any wavelength; it is not transparent to any radiation through and does not reflect or scatter anything. At the same time the emissivity is maximal at all wavelengths. Thus, any real body cannot emit more thermal radiation than a blackbody at any wavelength. It is the ideal thermal radiation source. Moreover, its spectrum depends *only* on its temperature. In particular, material properties have no influence. Thus, it is the reference source for numerous theoretical and practical purposes. The total radiation emitted by the surface a blackbody e_b per unit area into half-space follows from the integration of Planck’s relation over all wavelengths from $\lambda = 0$ to $\lambda \to \infty$ and yields the Stefan-Boltzmann law:

$$e_b = \int_{\lambda=0}^{\lambda\to\infty} e_{b\lambda}(\lambda, T) d\lambda = \sigma T^4 \left[\frac{W}{m^2}\right], \tag{4.163}$$

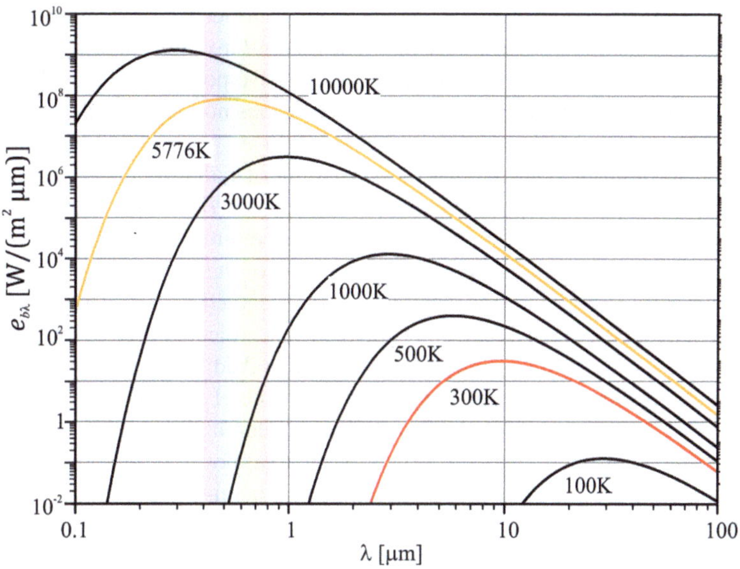

Fig. 4.29 Blackbody spectral power distribution $e_{b\lambda}$ as a function of the wavelength at different tempratures

in which $\sigma = 5.67 \cdot 10^{-8} \mathrm{W}/\left(\mathrm{m}^2\mathrm{K}^4\right)$ is the Stefan-Boltzmann constant. A plot of $e_{b\lambda}$ as a function of the wavelength λ is illustrated in Fig. 4.29 on a double logarithmic scale and reveals to essential aspects. An increase of the temperature T yields a a rise of the spectral emissive power at all wavelengths and moreover the wavelength corresponding to the maximum power shifts towards the shorter wavelengths. The value of the wavelength at which the maximum occurs $\lambda_{\max}$ is predicted by the Wien displacement law,

$$\lambda_{\max} = \frac{2898}{T} \quad [\mu \mathrm{mK}]. \tag{4.164}$$

An ideal black body cannot be realized technically. There are no materials that completely absorb the electromagnetic waves independent of frequency. Even the production of a body that comes close to the ideal black body is difficult. A sooty surface has an absorptivity of about $\alpha \approx 0.96$ and is sufficient for simple purposes, amorphous graphite absorbs about 99% of the incoming radiation. For practical applications only, the absorption and emission properties of the radiation source are of interest, but not its shape. Therefore, instead of a surface, the aperture of a cavity radiator or simply a long blind hole is used. This allows a better representation of the ideal properties of a blackbody radiator, even if the inner surfaces have a low emissivity. Most real substances, on the other hand, reflect a portion of the incident radiation ($\rho > 0$) and absorb only a portion described by the absorptance or absorption coefficient $\alpha < 1$. Likewise, the emissivity ε differs from 1.

Although the blackbody emissive power from a surface has as quantity the units of [W/m^2] being the same as a heat flux, but it is *not* a vector quantity which can be resolved into directional components. The emissive power describes the energy leaving a surface per unit area of the surface and does not describe the directional characteristics of the radiation as it leaves the surface.

In order to describe the directional propagation of radiation through space, we have to introduce a quantity referred to as the radiance I. The radiance is defined by means of the Fig. 4.30 in which radiation is emitted from a small surface element ΔA and travels in all directions. A fraction of the radiation hit a small area element on a hemisphere enveloping ΔA, being denoted as dA. The net amount of energy $d\dot{Q}$ falling on dA is proportional to the projected area of the emitting surface which is $(\Delta A \cos\theta)$, and to the target dA. Additionally, it scales inverse square of the distance between source and target. i.e. $1/r^2$. The latter originates from the fact that the area of the hemisphere scales as r^2 and the total power emitted to the hemisphere, which is equal tot the total power emitted by the source is constant. Hence the power per unit area on the hemisphere goes as $1/r^2$. The target area dA is given by the distance r and the polar coordinates θ, ϕ:

$$dA = r^2 \sin\theta d\theta d\phi. \tag{4.165}$$

Now, $d\dot{Q}$ is proportional to dA/r^2, and

$$\frac{dA}{r^2} = \sin\theta d\theta d\phi \equiv d\Omega \tag{4.166}$$

defines a differential solid angle $d\Omega$ with steradian (sr) as unit. Because the surface area A of a sphere is $\left(4\pi r^2\right)$, the definition implies that a sphere subtends 4π steradians ($\approx 12.56637 sr$) at its centre. By means of the solid angle definition, the power emitted towards the target gets proportional to the projected area of

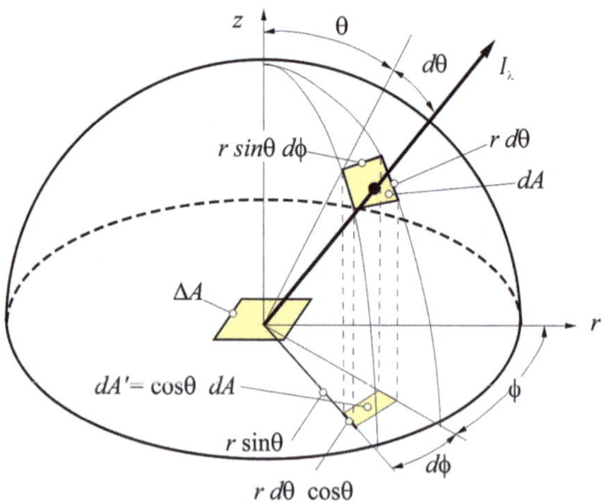

Fig. 4.30 Sketch do define the radiation intensity I

the source and the solid angle subtended by the target with respect to the source. Thereby, the power is not only be proportional to the magnitude but also the directionality of the radiation leaving the source and propagating through space. Since the power $d\dot{Q}$ is in $[W]$, and $d\dot{Q}$ is proportional to $(\Delta A \cdot \cos\theta \cdot d\Omega)$, it follows that radiation propagating through space must have units of $[W/(m^2 sr)]$. This quantity is called radiance L and is defined

$$L = \frac{d\dot{Q}}{\Delta A \cdot \cos\theta \cdot d\Omega}. \tag{4.167}$$

Although the radiance L displays the directional distribution of the propagating radiation, it is not a vector rather the intensity is a scalar that is dependent on the directional coordinates θ and ϕ together with the spatial coordinates and time.

Inherently, we assumed that the radiance L (Eq. 4.167) is a wavelength–integrated quantity. Thus, the spectral radiance L_λ is defined similarly but on a per–unit–wavelength basis in the way,

$$L_\lambda = \frac{d\dot{Q}}{\Delta A \cdot \cos\theta \cdot d\Omega \cdot d\lambda}, \tag{4.168}$$

and the integral of L_λ over all wavelength corresponds to L.

If the ΔA emits as a perfect blackbody the radiance distribution leaving the surface is independent of direction, which means that the blackbody radiance L_b landing on the hemisphere in Fig. 4.30 isn´t a function of θ. (Note that the power per unit area landing on the hemisphere $(d\dot{Q}/dA)$ is a function of θ.) If the total power leaving the black source is $\dot{Q} = e_b \Delta A$ and Eq. (4.167), then the total power landing on the target the hemisphere is,

$$\begin{aligned} \Phi = \dot{Q} = e_b \Delta A = \int d\dot{Q} = L_b \Delta A \int\limits_{\Omega=4\pi} \cos\theta d\Omega \\ = L_b \Delta A \int\limits_0^{4\pi} d\phi \int\limits_0^{\pi/2} \cos\theta \sin\theta d\theta = \pi L_b \Delta A. \end{aligned} \tag{4.169}$$

Here, we used $L = L_b = const.$ Since all blackbody radiators emit diffusely, the power density of a blackbody is,

$$e_b = \pi L_b, \tag{4.170}$$

and in the case of monochromatic radiation:

$$e_{b\lambda} = \pi L_{b\lambda}. \tag{4.171}$$

Assuming an ideally diffuse surface ΔA with an isotropic distribution of radiance L in all directions in half-space, i.e. that of a blackbody, and substituting

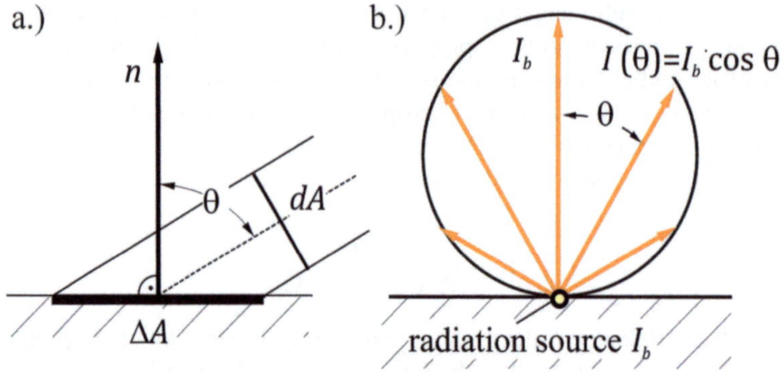

Fig. 4.31 **a** Independence of the emitted radiant intensity (radiant intensity I) from the direction of observation. **b** Radiance I over a diffusely radiating surface according to Lambert's cosine law

this into Eq. (4.167) using (4.166), we obtain the intensity $I(\theta)$ in $[W/sr]$ of the black surface ΔA in direction θ around the solid angle $d\Omega$, which is illustrated in Fig. 4.31a.

$$I(\theta) = \frac{d\dot{Q}}{d\Omega} = \Delta A \cdot L_b \cdot \cos\theta = I_b \cdot \cos\theta. \tag{4.172}$$

This law is called Lambert's cosine law. The radiation intensities I plotted as correspondingly long arrows above an area element of such a radiator come to rest with their tips on a sphere, as Fig. 4.31 shows. A good approximation for a Lambertian radiator is paper (for visible light!). Small air pockets between the fibres form scattering centres for the visible light. If they are missing, for example by impregnating the paper with water, paper loses its reflective properties and becomes transparent.

The intensity being a scalar describes the directional distribution of radiation energy transfer, while the radiative heat flux $\vec{q}_R$, is a vector representing the net direction and magnitude of radiant energy transfer in space. Consider a small surface element in space, dA, oriented so that its normal vector points in the positive z-direction, and radiation, arriving from all directions with an intensity distribution $I(\theta, \phi)$, passes through the element, then the net rate at which energy dQ is transported across the surface, is

$$d\dot{Q} = dA \int\limits_{\phi=0}^{\phi=2\pi} \int\limits_{\theta=0}^{\pi} I(\theta, \phi) \cos\theta \sin\theta d\theta d\phi. \tag{4.173}$$

It is important that the term $\cos\theta$ appears because $(dA\cos\theta)$ is the projected area of the element with respect to the θ-direction. The quantity $d\dot{Q}$ can be either positive, negative, or zero. The sign indicates the direction of a net transfer across dA being positive in positive z-direction. It is essential to note that a zero value does not necessarily imply that $I(\theta, \phi) = 0$, it rather means that all positive and negative contributions annihilate. This occurs if the intensity distribution is isotropic, i.e., independent of direction.

If one calculates for a black body according to Stefan-Boltzmann law (Eq. 4.163) the power of the emitted radiation and sets for the temperature $392.8° K$, one gets for the solar constant a value of $i_{SC} = 1357 W/m^2$, which nearly corresponds to the value of the solar constant (see Chap. 2). A black-body on an extra-terrestrial position with one side facing perpendicular to the sun and the other surfaces ideally insulated could thus not reach a temperature above about $120°C$. However, there are two ways to exceed that temperature limitation. The first is concentrating mirror systems that increase the irradiance on the target body; the second is to use selective surfaces that absorb well in the solar radiation wavelength range but emit little in the thermal radiation wavelength range at solar collector operating temperatures.

4.3.3 Radiation Balance

Most real substances, reflect part of the incident radiation expressed by coefficient named reflectance ρ being $\rho > 0$ and absorb only a portion described by the absorptance or absorption coefficient $\alpha < 1$. Similarly, the emissivity ε differs from unity. A grey radiator is a body whose surface does not completely absorb the incident radiation and therefore does not radiate the full radiation back into space at a given temperature. The numerical value of how "grey" the surface is expressed by the absorptivity α (- wich can equal the emissivity ε, see later Kirchhoff's law). An ideal gray radiator, which does not exist in reality, describes a body whose emissivity ε does not change with the wavelength l (see also in the following section).

Moreover, if a body is transparent, a part of the radiation will pass through it, as depicted in Fig. 4.32, which is characterized by the transmittance τ. The part of the radiation that is incident on the body will be the same as the part of the radiation that is not incident on the body. The fraction of the incident radiation that is reflected from the surface is given by the reflectance. Since only an equal or lesser intensity can be reflected back, the reflectance ρ is always $\rho < 1$. In transparent bodies, some electromagnetic (or thermal) radiation can pass through the body. This transparency by the transmittance τ characterizing the part of radiation passing the body. An energy balance of the surface element in Fig. 4.32 states that the energy incident on the surface element $(I \cdot dA)$ must correspond to the portions that leave the surface element by reflection and transmission and lead to heating by absorption. Assuming thermodynamic equilibrium, the first law of thermodynamics leads to the following relationship:

$$\alpha + \tau + \rho = 1. \tag{4.174}$$

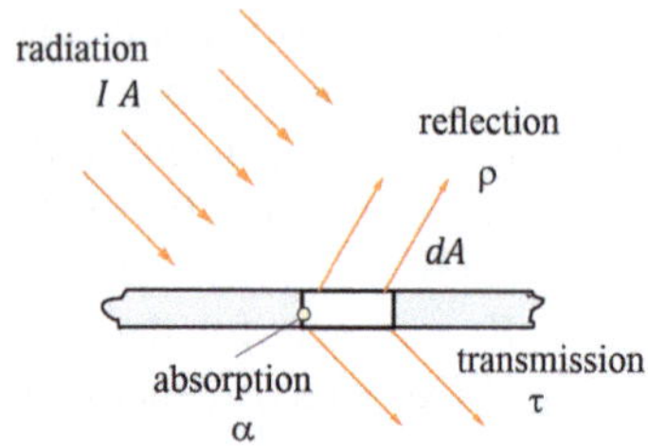

Fig. 4.32 Sketch of the radiation balance on a real body

The absorption coefficient α is similar in its behavior to the emission coefficient ε. For example, a surface with an emission or absorption coefficient of $\alpha = 0.5$ absorbs half of the incident radiation, but at a given temperature also emits only half of the thermal radiation compared to a blackbody. Heat radiation can be reduced by using bare metal surfaces. i.e. metal layers on rescue blankets and insulating bags, mirroring in thermos flasks and super insulation. To increase the thermal radiation of a metallic body that reflects light well, it can be covered with a "dark" matter coating in the relevant wavelength range of thermal radiation (infrared range). Technically, this is used, for example, in painting radiators with paint (which emits radiation well in the infrared range), in black anodizing aluminium heat sinks in computers and electrical equipment, or in enamelling stove pipes and metal furnaces.

The relation (4.174) suggests that the absorptance α, the transmittance τ and the reflectance ρ are constants, which is unfortunately not the case. Since thermal radiation, like solar radiation, has a spectral behaviour, it is also dependent on location, wavelength, and temperature.

Opaque bodies do not let any radiation through, so that for their transmittance $\tau = 0$ is valid. Depending on the body, the radiation can be reflected at the interface to different degrees. This reflection at the surface can be specular or diffuse, as outlined in Fig. 4.33. If the angle between the surface normal and the incident ray is equal to the angle of the outgoing ray and surface normal (incidence angle = emerging angle, i.e. $\theta_1 = \theta_2$), it is a specular reflection. In the case of ideal diffuse reflection, the reflected radiation is isotropically distributed in intensity in all directions in the half-space above the reflecting surface (compare Fig. 4.33b). Real surfaces reflect are neither ideally specular nor ideally diffuse, and a space-dependent reflection results, as illustrated by Fig. 4.33c. Well-polished surfaces come very close to the ideal case with a specular reflection behaviour, while very rough surfaces tend to a homogeneous diffusive reflection.

Thus, an analysis of the thermal radiation exchange requires a characterization of the surface properties of the surfaces involved in the radiation problem.

4.3.4 Surface Properties and Radiation

As recognized in the context of the radiation balance of a body surface can emit, absorb, and reflect radiation and we have assigned to the surface coefficients

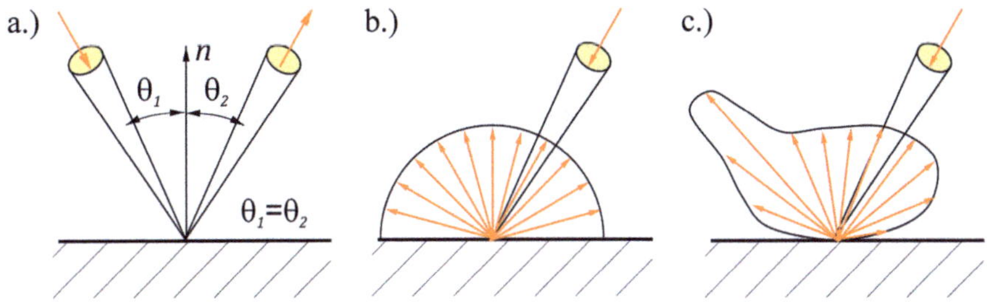

Fig. 4.33 Reflection at an ideally specular (**a**), at an ideally diffuse (**b**) and at a real, mixed reflecting surface (**c**)

describing absorptivity and reflectivity, i.e., the absorptivity represents the fraction of incident radiation on a surface that is absorbed by the surface. Nonetheless, more precise definitions need to be developed to account for the directional and spectral dependencies of the radiation falling on or emitted by a surface.

4.3.4.1 Emissivity

One of the advantages of a blackbody is that the intensity emitted is independent of direction and given by $I_b = e_b/\pi$. Likewise, the spectral (or wavelength–dependent) blackbody intensity is independent of direction and given by $I_{b\lambda} = e_{b\lambda}/\pi$. However, a real body surface emits a spectral intensity into some direction θ and ϕ and within a small wavelength interval about λ, whichs always be less than the spectral blackbody intensity for a surface at the same temperature. We can therefore define the spectral directional emissivity $\varepsilon'_\lambda = I_{e\lambda}(\theta, \phi)/I_{b\lambda}$, in which $I_{e\lambda}$ is the emitted intensity from the surface. This spectral directional emissivity represents the most irreducible information on the emission properties of a surface and a function of direction, wavelength, temperature as well as the physical and chemical properties of the surface, but much too complex for technical application. By integration of ε'_λ over the unit sphere the spectral hemispherical emissivity ε_λ is obtained which represents the directional average. It is defined by:

$$\varepsilon_\lambda = \frac{\int_{\phi=0}^{\phi=2\pi} \int_{\theta=0}^{\pi/2} I_{e\lambda}(\theta, \phi) \cos\theta \sin\theta d\theta d\phi}{\int_{\phi=0}^{\phi=2\pi} \int_{\theta=0}^{\pi/2} I_{b\lambda} \cos\theta \sin\theta d\theta d\phi}$$
$$= \frac{\int_{\phi=0}^{\phi=2\pi} \int_{\theta=0}^{\pi/2} \varepsilon_\lambda\prime(\theta, \phi) \cos\theta \sin\theta d\theta d\phi}{\pi} = \frac{e_\lambda}{e_{b\lambda}}. \tag{4.175}$$

That is,ε_λ is the ratio of the emitted spectral power from the surface, e_λ to the blackbody power $e_{b\lambda}$. Likewise, the total directional emissivity ε' is obtained by an integration over all wavelength to obtain a wavelength average by

$$\varepsilon' = \frac{\int_{\lambda=0}^{\infty} I_{e\lambda} d\lambda}{\int_{\lambda=0}^{\infty} I_{b\lambda} d\lambda} = \frac{\int_{\lambda=0}^{\infty} \varepsilon'_\lambda I_{b\lambda} d\lambda}{I_b}. \tag{4.176}$$

This total directional emissivity is of technical importance and observed in many practises. Figure 4.34 illustrates the directional emissivity as a function of the incidence angle for various electrically non-conducting materials. The deviation of the emissivity from that of a Lambertian radiator is particularly evident for the non-conductors in the region of large angles θ. Because of the comparatively low radiation power in this angular range, the technical impact remains marginal. The assumption of ideally diffuse radiating surfaces is therefore reasonably well fulfilled for electrical non-conductors.

Metallic bare surfaces are far from a Lambert radiator behaviour as Fig. 4.35 illustrates, where a clear dependence of the total directional emissivity can be seen.

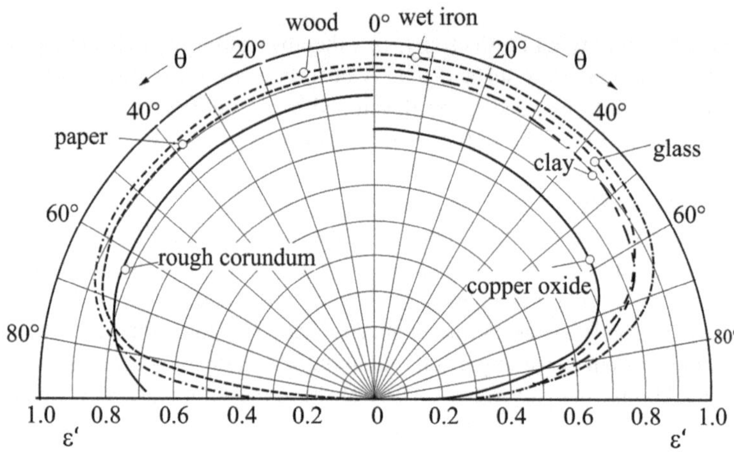

Fig. 4.34 Total directional emissivity ϵ' as a function of the incidence angle θ for various electrically non-conducting material modified from (Stephan et al., 2019)

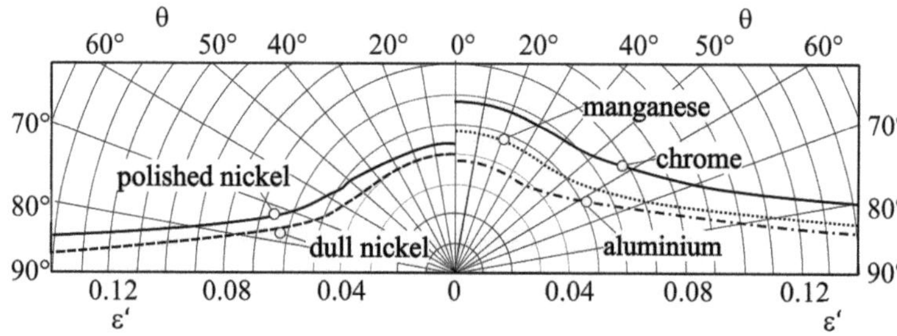

Fig. 4.35 Total directional emissivity ϵ' as a function of the incidence angle θ for some metals modified from (Stephan et al., 2019)

Finally, the total hemispherical emissivity ε is obtained from either a directional average of ε' or a wavelength average of ε_λ;both yield the same result, which is:

$$\varepsilon = \frac{\int_{\lambda=0}^{\infty}\int_{\phi=0}^{\phi=2\pi}\int_{\theta=0}^{\pi/2} I_{e\lambda}(\theta,\phi)\cos\theta\sin\theta d\theta d\phi d\lambda}{\int_{\lambda=0}^{\infty}\int_{\phi=0}^{\phi=2\pi}\int_{\theta=0}^{\pi/2} I_{b\lambda}\cos\theta\sin\theta d\theta d\phi d\lambda}$$
$$= \frac{\int_{\phi=0}^{\phi=2\pi}\int_{\theta=0}^{\pi/2} \varepsilon'(\theta,\phi)\cos\theta\sin\theta d\theta d\phi}{\pi} = \frac{\int_{\lambda=0}^{\infty}\varepsilon_\lambda I_{b\lambda}d\lambda}{I_{b\lambda}} = \frac{e}{e_b}. \tag{4.177}$$

4.3.4.2 Absorptivity

Let us assume an opaque surface ($\tau = 0$) exposed to an incident source of spectral intensity given by $I_\lambda(\theta,\phi)$ shining on its surface.When the radiation strikes the surface a fraction of it will be absorbed by the surface, the remainder will be reflected. Defining the absorbed intensity as $I_{\alpha\lambda}$ the spectral directional absorptivity α'_λ is defined by:

$$\alpha'_\lambda = \frac{I_{\alpha\lambda}}{I_\lambda}, \tag{4.178}$$

and it describes the fraction of incident spectral intensity absorbed by the surface. The spectral directional absorptivity describes the fundamental absorption properties of the surface. Further, similar with as the emissivity averages can be introduced with the direction, wavelength, or both by the corresponding integrations.

The spectral hemispherical absorptivity α_λ writes to:

$$\alpha_\lambda = \frac{\int_{\phi=0}^{\phi=2\pi}\int_{\theta=0}^{\pi/2} I_{\alpha\lambda}(\theta,\phi)\cos\theta\sin\theta d\theta d\phi}{\int_{\phi=0}^{\phi=2\pi}\int_{\theta=0}^{\pi/2} I_\lambda\cos\theta\sin\theta d\theta d\phi}$$
$$= \frac{\int_{\phi-0}^{\phi=2\pi}\int_{\theta-0}^{\pi/2} \alpha'_\lambda(\theta,\phi)\mathrm{I}_\lambda(\theta,\phi)\cos\theta\sin\theta d\theta d\phi}{q_\lambda} = \frac{q_{\alpha\lambda}}{q_\lambda}, \tag{4.179}$$

in which q_λ ist he spectral flux in direction of the surface and $q_{\alpha\lambda}$ the spectral absorbed flux. The total directional absorptivity α' is:

$$\alpha' = \frac{\int_{\lambda=0}^{\infty} I_{\alpha\lambda}d\lambda}{\int_{\lambda=0}^{\infty} I_\lambda d\lambda} = \frac{\int_{\lambda=0}^{\infty}\alpha'_\lambda I_\lambda d\lambda}{I}, \tag{4.180}$$

and, finally, the total hemispherical absorptivity α is:

$$\alpha = \frac{\int_{\lambda=0}^{\infty}\int_{\phi=0}^{\phi=2\pi}\int_{\theta=0}^{\pi/2} I_{\alpha\lambda}(\theta,\phi)\cos\theta\sin\theta d\theta d\phi d\lambda}{\int_{\lambda=0}^{\infty}\int_{\phi=0}^{\phi=2\pi}\int_{\theta=0}^{\pi/2} I_\lambda\cos\theta\sin\theta d\theta d\phi d\lambda}$$

$$= \frac{\int_{\phi=0}^{\phi=2\pi} \int_{\theta=0}^{\pi/2} \alpha'(\theta,\phi) I(\theta,\phi) \cos\theta \sin\theta d\theta d\phi}{q} = \frac{\int_{\lambda=0}^{\infty} \alpha_\lambda q_\lambda d\lambda}{q} = \frac{q_\alpha}{q}. \tag{4.181}$$

At a first glance the definitions look quite similar to those of the emissivity, however, there are especially with respect to the solar thermal application fundamental differences. In the emissivity relations for ε_λ and ε Eqs. (4.175) and (4.177) definition the blackbody intensity I_b appears which is not a function of the direction, which allowed for a removal from integral over direction. But, with respect to the absorptivity the incident intensity irrespective if it is spectral (I_λ) or total (I) is, in general, a function of direction, and it cannot be removed from the corresponding definition (Eqs. 4.179, 4.181) for the hemispherical absorptivity.

Thus, it is important to emphasize that the thermal emission by radiation is a function of the surface only, while regarding the absorption the directional intensity distribution falling on a surface as well as the wavelength dependent absorptivity α_λ play a role. As illustrated already in Chap. 3, a surface illuminated by the sun from the normal direction ($\theta = 0°$) has a different hemispherical absorptivity than the same surface illuminated by the sun at an oblique angle ($\theta > 60°$), or by the sun on a cloudy day, which is dominated by diffuse illumination. Likewise, the total absorptivity (either directional or spectral) is dependent on the spectral distribution of the incident radiation. A spectral selective of a solar thermal receiver exposed to visible light exhibits a total absorptivity and correspondingly a high temperature, whereas than the same surface illuminated by IR radiation source will barely heat up, since the absorptivity in this wavelength domain is marginal.

4.3.4.3 Reflectivity

The complexity to describe the reflection is one step higher than for the absorption, because not only the incident radiation direction plays a role but also the direction of the outgoing (reflected) radiation. We therefore introduce the superscripts „−" for the incident and „+" to assign the reflected intensity. By means of this notation one can define a spectral bi-directional reflectivity ρ_λ'':

$$\rho_\lambda''(\Omega, \Omega') = \frac{I_\lambda^+(\Omega')}{\pi I_\lambda^-(\Omega)}. \tag{4.182}$$

The spectral directional–hemispherical reflectivity ρ_λ' is obtained by integration of the reflected intensity over the entire hemisphere

$$\rho_\lambda' = \frac{\int_{\phi'=0}^{2\pi} \int_{\theta=0}^{\pi/2} I_\lambda^+(\theta', \phi') \cos\theta' \sin\theta' d\theta' d\phi'}{\pi I_\lambda^-(\theta,\phi)}$$
$$= \frac{1}{\pi} \int_{\phi'=0}^{2\pi} \int_{\theta=0}^{\pi/2} \rho_\lambda''(\theta,\phi,\theta',\phi') \cos\theta' \sin\theta' d\theta' d\phi'. \tag{4.183}$$

All of the radiation incident on the surface is either absorbed by the surface or reflected. Assuming an opaque body ($\tau = 0$), we get:

$$\alpha'_\lambda + \rho'_\lambda = 1. \tag{4.184}$$

4.3.5 Kirchhoff's Law and Grey Bodies

To derive Kirchhoff's radiation law, first assume that a body is in a cavity of temperature T. The body is supposed to be in thermal equilibrium with the cavity. If the cavity walls behave inwardly like a perfect blackbody radiator, they emit radiation e_b in the direction of the blackbody according to the Stefan-Boltzmann law (Eq. 4.163). In the state of equilibrium, a blackbody for $t \to \infty$ attains the same temperature as the cavity walls. The incident radiation at the location of the blackbody with area A is thus $(\dot{q} \cdot A)$. In equilibrium, the blackbody emits radiation also according to Eq. (4.163), which is given by the product $(e_b \cdot A)$. Because of the conservation of energy, the power absorbed by the body must be equal to its emitted power, so that holds:

$$e_b \cdot A = \dot{q} \cdot A. \tag{4.185}$$

If we now consider a real body in the same cavity, it will absorb a lower radiation power by the absorptivity α ($\alpha < 1$). Furthermore, the real body emits a radiation power lower by the factor ε ($\varepsilon < 1$) than a blackbody. In temperature equilibrium, the absorbed radiation must be equal to the emitted one, so that:

$$\underbrace{\varepsilon \cdot e_b \cdot A}_{\substack{\text{emitted} \\ \text{power}}} = \underbrace{\alpha \cdot \dot{q}}_{\substack{\text{absorbed} \\ \text{power}}} \cdot A. \tag{4.186}$$

Up to this point, no restrictions have been made with regard to the spectral, spatial and temperature behaviour of the body. As discussed in the context of the reflectivity ρ, both emissivity and absorptivity depend on the wavelength (λ), the solid angles (θ, ϕ) and the temperature (T). In order to maintain the thermal equilibrium, however, the real body must re-emit the absorbed amount of energy at the same wavelengths in the same directions in order to replace the energy taken from the black cavity ("Hohlraum"). The division of Eq. (4.179) with Eq. (4.179) thus results in,

$$\frac{e_\lambda}{e_{b\lambda}} = \alpha'_\lambda(\theta, \phi, \lambda, T) = \varepsilon'_\lambda(\theta, \phi, \lambda, T), \tag{4.187}$$

where θ, ϕ are the solid angles in sphere. The key statement of Kirchhoff's law is thus:

In thermal equilibrium, for the same frequencies and directions, the directional spectral absorptivity and the directional spectral emissivity are equal.

Technically, this means that good absorbers are also good emitters. Kirchhoff's radiation law initially applies only in thermal equilibrium; for the most part, however, it is also applicable to a good approximation for bodies that are not in thermal equilibrium with their surroundings, as long as their directional spectral absorption and emission coefficients do not change under these conditions.

The identity of absorption and emission coefficients applies *only* to the directional spectral absorptivity and the directional spectral emissivity. These quantities, which describe the explicit directional and wavelength dependence of the absorption and emission processes, are not known or available for many materials. Usually, for a specific material, the hemispherical spectral emission coefficient $\varepsilon_\lambda(\lambda, T)$ integrated over all directions of the half-space or the total hemispherical emissivity ε, which is still a function of the temperature. The equality of the integrated absorptivity with emissivity holds only in a few special cases, since the integrated absorptivities also depend on the directional and wavelength distribution of the incident radiation. This means that total absorptivities and emissivities are not pure material properties.

Kirchhoff's radiation law is valid in the following cases:

- For ideally diffusely radiating surfaces the emissivity is direction-independent and we get (Baehr & Stephan, 2008):

$$\alpha_\lambda(\lambda, T) = \varepsilon_\lambda(\lambda, T), \tag{4.188}$$

- for ideal grey bodies $\varepsilon \neq \varepsilon(\lambda)$ and thus

$$\alpha(\theta, \phi, T) = \varepsilon(\theta, \phi, T), \tag{4.189}$$

- for diffuse and grey radiating bodies

$$\alpha(T) = \varepsilon(T). \tag{4.190}$$

Real bodies are often diffuse radiators to a good approximation, while the requirement for grey radiating surfaces can be poorly met. Non-metals (i.e. electrical insulators, dielectrics) behave approximately like diffuse emitters with few exceptions. Moreover, in many cases their directional spectral emissivity is reasonably constant for wavelengths above about $1 - 3\mu m$. Thus, for radiative transfer in the long-wavelength region (especially thermal radiation at temperatures that are not too high), dielectrics can often be treated approximately as diffuse grey radiators, and Eq. (4.190) applies. In the case of metals, on the other hand, the directional dependence of the emissivity usually does not allow an approximation

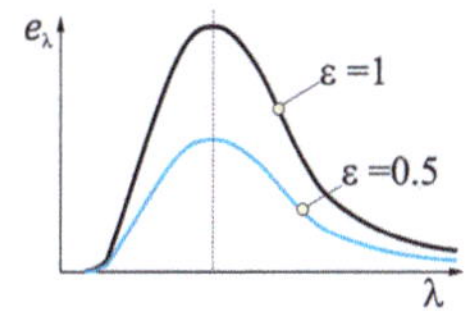

Fig. 4.36 Schematic representation of the radiation intensity ε_λ as a function of the wavelength λ for a black and a diffuse and gray radiator ($\varepsilon = 0.5$)

by a diffuse radiator. In addition, their spectral emissivity is not constant even at long wavelengths, so they are not grey radiators either; thus, Eq. (4.190) is not satisfied. Oxide layers or contamination, however, can strongly change the radiation properties of metals and make them approach those of dielectrics. Important for the validity of Kirchhoff's law is the assumption of thermodynamic equilibrium as well as the consideration of the radiation properties of the body. If it is fulfilled, the radiation emission of a diffuse grey radiator is reduced by the factor ε compared to a black radiator; it results:

$$e_{\varepsilon<1} = \varepsilon \cdot \sigma T^4. \tag{4.191}$$

For the case of a diffuse and grey emitting radiator, the spectrum is shifted downward by a factor e over all wavelengths, as schematically sketched in Fig. 4.36. However, real surfaces not only emit less than a blackbody, but their emissivity varies with wavelength. It is therefore often necessary to start from the monochromatic properties and normalize to the blackbody as shown in Eq. (4.175) and then to integrate over all wavelengths.

The grey assumption can fail miserably, however, when the incident radiation has a significantly different spectra than the emitted radiation. In the context here, as an example sunlight is falling on a solar collector. The radiation spectrum of sunlight is similar to that of a blackbody with a temperature of $T \approx 5800K$, which is concentrated to a considerable fraction in the visible wavelengths. Solar collectors, on the other hand, typically operate at a temperature of around $T \approx 350K$, and correspondingly their thermal emission at this temperature is concentrated in the mid IR wavelengths. For such conditions, a collector with a high spectral absorptivity in the visible yet a small spectral emissivity in the IR is desirable.

4.3.6 Radiosity and Irradiance

By now we considered only the radiation leaving or accepted by a surface, which is of rather academic nature. Technically relevant, however, is the knowledge of the net rate of heat transfer to each surface in the domain considered. Thus, in the next step we assume N surfaces which can exchange radiation heat, and for simplicity we consider their surfaces to be diffuse so that the emissivity and absorptivity are not a function of the direction, but to maintain generality, the properties are initially taken to be wavelength–dependent, so that $\alpha \neq \varepsilon$.

Fig. 4.37 Incoming and emerging energy for a diffuse surface A_1

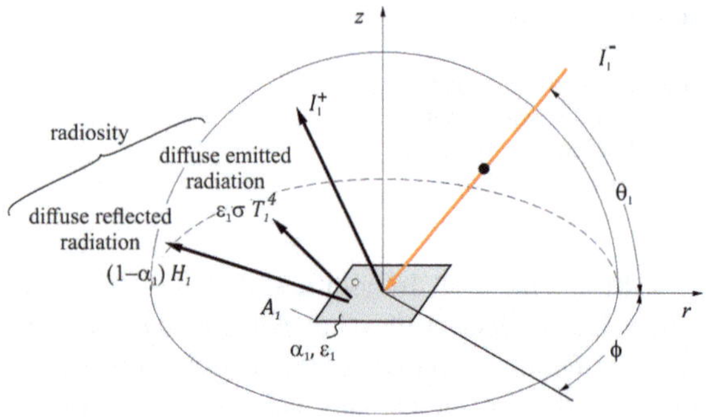

In order to define the basic quantities of radiation heat exchange we consider a surface 1, with an area A_1 and a uniform constant total hemispherical emissivity ε_1 and absorptivity α_1, see Fig. 4.37.

The irradiance H_1 in $[W/m^2]$ is the flux of radiant energy reaching the surface averaged over the surface area and the total radiant energy falling on the surface is thus

$$H_1 \cdot A_1 = \int\limits_{A_1}\int\limits_{\Omega} I_1^- \cos\theta_1 d\Omega\, dA_1, \tag{4.192}$$

where I_1^- represents the intensity falling on the surface A_1 depending on the direction (θ, ϕ) of the incident radiation. The radiosity J_1 decribes the surface averaged radiant energy flux emerging the surface:

$$J_1 \cdot A_1 = \int\limits_{A_1}\int\limits_{\Omega} I_1^+ \cos\theta_1 d\Omega dA_1, \tag{4.193}$$

where I_1^+ denotes the intensity leaving the surface A_1. The radiosity is connected to the irradiance H_1 by:

$$J_1 = \varepsilon_1 \cdot e_{b1} + \rho_1 \cdot H_1 = \varepsilon_1 \cdot e_{b1} + (1 - \alpha_1) \cdot H_1. \tag{4.194}$$

Thus, the radiosity consists of two parts, the emission from its own surface, which is given by $e_1 = \varepsilon\sigma T_1^4$, complemented by the reflected $(\rho_1 = (1 - \alpha_1))$ fraction of the incident radiation. The irradiance H_1 depends on the incoming radiation field, which is incident on surface A_1 from all surrounding surfaces at a "viewing angle". Hence, radiosity J_1 represents the total radiation energy streaming from a surface, per unit area per unit time, and is the summation of the reflected and the emitted radiation.

The net average heat flux of the surface $(\dot{q}_1)$ is the difference of the flux arriving and the one emerging from the surface,

$$\dot{q}_1 = J_1 - H_1 = \varepsilon_1 \cdot e_{b1} - \alpha_1 \cdot H_1, \tag{4.195}$$

and thus, relates the emissive flux from the surface to the absorbed incident flux. The net balance ($\dot{q}_1 \cdot A_1$) corresponds to the net rate of the heat transfer from/to the surface similar as in heat conduction. If ($\dot{q}_1 \cdot A_1) = 0$ the surface is adiabatic otherwise it describes power required tobe supplied to the surface to keep it at the constant Temperature T_1. Eliminating H_1 from the previous equations one obtains

$$\dot{q}_1 = \frac{1}{(1 - \alpha_1)}(\varepsilon_1 \cdot e_{b1} - \alpha_1 \cdot J_1). \tag{4.196}$$

In case of blackbody radiation ($\alpha = \varepsilon = 1$), the radiosity J_1 corresponds to e_{b1} and thus the solution of (4.196) is trivial, however, the essential message here is that the net transfer is associated with a balance equation accounting for emission and absorption (or reflectance).

For surface which are opaque, diffuse, and grey ($\varepsilon_1 = \alpha_1$) and considering them to be isothermal Eq. (4.194) simplifies

$$J_1 = \varepsilon_1 \cdot e_{b1} + (1 - \varepsilon_1) \cdot H_1. \tag{4.197}$$

Using an energy balance, the net rate of radiation heat transfer from a surface 1 of surface area A_1 can be expressed as

$$\dot{Q}_1 = \dot{q}_1 \cdot A_1 = A_1 \cdot (J_1 - H_1) = A_1\left(J_1 - \frac{J_1 - \varepsilon_1 \cdot e_{b1}}{1 - \varepsilon_1}\right) = \frac{A_1\varepsilon_1}{1 - \varepsilon_1}(e_{b1} - J_1). \tag{4.198}$$

In electrical analogy to Ohm's law, a thermal resistance can be defined as:

$$\dot{Q}_1 = \frac{(e_{b1} - J_1)}{R_1} \text{ with } R_1 = \frac{1 - \varepsilon_1}{A_1\varepsilon_1}, \tag{4.199}$$

where R_1 can be conceived as "*surface resistance*" to radiation. Note, that the surface resistance to radiation for a blackbody is zero. For insulated or adiabatic surfaces, the net heat transfer through them is zero. In these cases, the surface is called re-radiating surface and thus there is no net heat transfer to a re-radiating surface.

4.3.7 Radiation Transport Between Bodies

Radiation transport differs from conduction and convection as heat transport mechanisms in that the observed net heat transport cannot be divided into two heat transports only mathematically. The net heat transport from surface 1 to surface 2 is given by the heat transport from 1 to 2 minus the heat transport from 2 to 1. Thus, the thermal radiation transports the heat not only in one direction, as in a diffusive energy transport in a continuum of, for example, heat conduction (from warm to cold), but at the same time the thermal radiation emitted by the colder body also hits the warmer body and can be absorbed by it. In technical heat transfer problems, usually only the effective heat exchange by thermal radiation between two or more bodies with known surface temperatures is of interest. Several methods are available to determine this heat exchange, such as the net radiation method, the zone method, Monte Carlo methods or the method of discrete ordinates. For the sake of clarity, the following derivations are limited to the net radiation method, more details of the complementary calculation methods or a more refined description of the radiative heat exchange can be found in (Modest, 2014).

4.3.7.1 Radiation Exchange Between Gray, Diffusely Radiating Infinite Surfaces

If we consider two plane, grey, equally sized parallel plates 1 and 2, both will emit thermal radiation according to their temperature according to Eq. (4.190) and according to their emissivity ε_i. The radiation emitted by plate 1 will be partly absorbed and partly reflected by plate 2. The radiation emitted by plate 1 will be partly absorbed and partly reflected by plate 2. In the simplest case, the intensity of the reflected beam is lower by the reflection coefficient $\rho_2 = (1 - \varepsilon_2)$. The reflected part of the beam again hits plate 1, where it is in turn partially absorbed and reflected. With each further reflection, another part of the radiation is absorbed. As a result, the radiation energy decreases by $(1 - \varepsilon_i)$ each time, as schematically sketched in Fig. 4.38.

The sum of all absorbed radiation fractions results in a geometric series, which yields for radiated heat flux $\dot{q}_{12}$ from plate 1 to plate 2:

$$\dot{q}_1 = \sum_{k=1}^{k\to\infty} \varepsilon_1 \cdot \varepsilon_2 \cdot (1 - \varepsilon_1)^k \cdot (1 - \varepsilon_2)^k \cdot \sigma \cdot T_1^4. \tag{4.200}$$

If the areas A_i of the two plates are equal and the length dimension (L) is considerably larger than the gap width, the summation of Eq. (4.200) yields the following result:

$$\dot{q}_1 = \sigma \cdot T_1^4 \cdot \frac{1}{\frac{1}{\varepsilon_1} + \frac{1}{\varepsilon_2} - 1}. \tag{4.201}$$

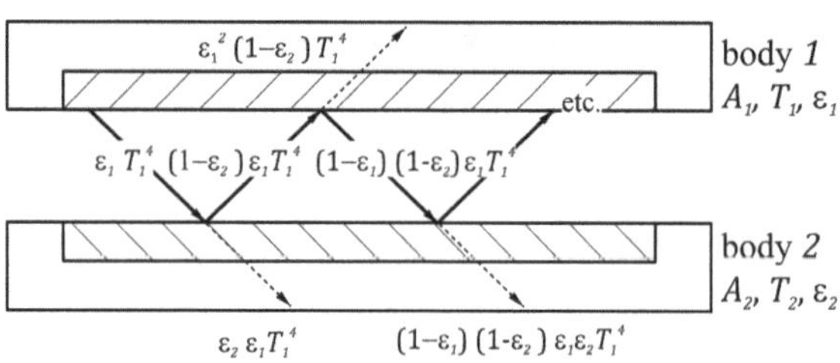

Fig. 4.38 Radiation exchange between two planes, parallel, gray radiating plates

Conversely, there is a radiation flux from the body 2 to the body 1 with the result:

$$\dot{q}_2 = \sigma \cdot T_2^4 \cdot \frac{1}{\frac{1}{\varepsilon_1} + \frac{1}{\varepsilon_2} - 1}. \tag{4.202}$$

Thus, the net radiation balance is the difference of the radiation intensities absorbed by both surfaces:

$$\dot{q}_{12} = \sigma \cdot \frac{\left(T_1^4 - T_2^4\right)}{\frac{1}{\varepsilon_1} + \frac{1}{\varepsilon_2} - 1}. \tag{4.203}$$

In the case of radiation heat exchange between two concentric tubes, the exchange areas are different in size, so that a geometry factor is added if the axial extension of the tubes is much larger than the gap between the tubes. Thus, all emitted radiation is transferred from tube 1 to tube 2 and vice versa. For this purpose, assume that the area of the inner tube is A_1 and A_2 is that of the cladding tube. In this simple case, the heat flow from the inner tube A_1 to the outer tube can be given by the equation below, where the heat flow is related to the area of the inner tube with A_1.

$$\dot{q}_{12} = \frac{\sigma\left(T_1^4 - T_2^4\right)}{\frac{1}{\varepsilon_1} + \frac{A_1}{A_2}\left(\frac{1}{\varepsilon_2} - 1\right)}. \tag{4.204}$$

Obviously, the relative orientation of the radiating surfaces plays a role, which is depicted here by an area ratio.

4.3.7.2 Radiation Exchange Between Diffuse Radiating Bodies

To determine the magnitude of the relative orientation of two radiating bodies, we consider two radiating surfaces A_i and A_j, each with homogeneously distributed surface temperatures T_i and T_j, as shown in Fig. 4.39. Let H_i be the total irradiance in $\left[W/m^2\right]$ incident on A_i and J_i the total radiosity in $\left[W/m^2\right]$ outgoing from A_i (uniformly). Since J_i includes the reflected portion of H_i in addition to the radiation emitted by A_i $\left(\varepsilon_i \sigma T_i^4\right)$, we obtain immediately according to Eq. (4.194).

$$J_i = \varepsilon_i \cdot e_{bi} + \rho_i \cdot H_i = \varepsilon_i \cdot e_{bi} + (1 - \alpha_i) \cdot H_i, \tag{4.205}$$

with $e_{bi} = \sigma T_i^4$. Now only a fraction of the emitted radiosity J_i reaches the surface A_j. The dimensionless number $F_{ij} \in [0, 1]$, which has yet to be determined, is called the view factor (alternatively often used are expression as configuration factors, form factors, angle factors or shape factors). This view factor depends exclusively on the shape and relative position of the surfaces A_i and A_j with

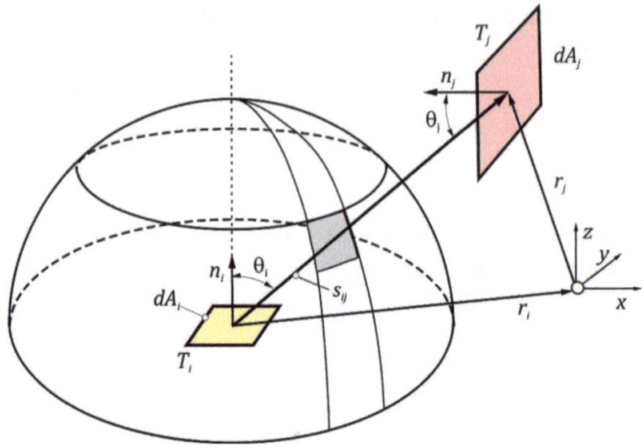

Fig. 4.39 Two surfaces with radiation exchange

respect to each other. The distance vector $\vec{s}_{ij} = \vec{r}_j - \vec{r}_j$ between the two surface elements dA_i and dA_j depends on the angles θ_i and θ_i in the following form:

$$\cos\theta_i = \frac{\vec{n}_i \cdot \vec{s}_{ij}}{\left|\vec{s}_{ij}\right|} \text{ and } \cos\theta_j = \frac{\vec{n}_j \cdot \vec{s}_{ij}}{\left|\vec{s}_{ij}\right|}, \tag{4.206}$$

in which $\vec{n}_i$ and $\vec{n}_j$ denote the area normal vectors of dA_i and dA_j.

The total radiant energy emitted from the surface element dA_i is $(J_i dA_i)$ with the radiosity according to (4.205), of which the fraction $(J_i dA_i \cos\theta_i)$ is emitted in the direction of $\vec{s}_{ij}$. The radiant energy incident on the surface element dA_j is calculated to be $\left(J_i dA_i \cos\theta_i \cos\theta_j\right)$. Thus, the radiant energy incident on the entire surface A_j and emitted from the entire surface A_i is calculated to be:

$$\mathrm{J}_{ji} = \frac{J_i}{\Gamma} \int\limits_{A_i} \int\limits_{A_j} \cos\theta_i \cos\theta_j dA_j dA_i, \tag{4.207}$$

in which Γ represents a normalization factor yet to be determined. Since for a hemisphere of the surface A_j with the radius $\left|\vec{s}_{ij}\right|$ the incident radiation energy J_{ji} must correspond to the radiation energy J_i emitted by the surface A_i for energy conservation reasons the following must apply:

$$\frac{1}{\Gamma} \int\limits_{A_i} \int\limits_{A_j} \cos\theta_i \cos\theta_j dA_j dA_i = 1. \tag{4.208}$$

Evaluating Eq. (4.208) for the hemisphere with radius $\left|\vec{s}_{ij}\right|$, it follows that $\theta_j = 0$ and we obtain:

$$\int\limits_{A_i} \cos\theta_i dA_i = \int\limits_0^{2\pi} \int\limits_0^{\pi/2} \mathrm{s}_{ij}^2 \cos\theta_i \sin\theta_i d\theta_j d\phi = \Gamma. \tag{4.209}$$

From Eq. (4.207) together with (4.209) we immediately obtain the definition equation for the view factor as the ratio of J_{ji} (radiant energy incident on the surface A_j and emitted from the surface A_i) to $(J_i A_i)$ (total radiant energy emitted from the surface A_i):

$$F_{ij} = \frac{J_{ji}}{J_i A_i} = \frac{1}{A_i} \int\limits_{A_i} \int\limits_{A_j} \frac{\cos\theta_i \cos\theta_j}{\pi \left|\vec{s}_{ij}\right|^2} dA_j dA_i. \tag{4.210}$$

Provided that no other heat processes occur, in thermodynamic equilibrium the radiative heat fluxes from i to j and from j to i must correspond to each other for energy conservation reasons and one obtains the reciprocity rule:

$$A_i \cdot F_{ij} = A_j \cdot F_{ji}. \tag{4.211}$$

Since a surface can emit at most the power it has also received, the sum of the visibility factors from surface i to all other surfaces j must equal unity. This results in the summation rule, which in its general form reads to:

$$\sum_j F_{ij} = 1. \tag{4.212}$$

Since the view F_{ij} factor is of fundamental importance for a thermal radiation exchange calculation some general aspects shall be summarized here:

- The view F_{ij} factor ranges between zero and one.
- $F_{ij} = 0$ indicates that two surfaces do not see each other directly. $F_{ij} = 1$ indicates that the surface j completely envelopes surface i.
- The radiation that strikes a surface does not need to be absorbed by that surface.
- F_{ii} is the fraction of radiation leaving surface i that strikes itself directly. $F_{ii} = 0$ for plane or convex surfaces, and $F_{ii} \neq 0$ for concave surfaces, meaning that they see each other for example the surface of a cylinder inner side.
- Radiation analysis of an enclosure consisting of N surfaces requires the calculations of N^2 view factors, which form a matrix. However, all of these calculations are not necessary. Once a sufficient number of view factors are available, the rest of them can be found using the reciprocity and summation rule.
- The summation rule can be applied to each surface of an enclosure by varying i from 1 to N (number of surfaces). Thus, the summation rule gives N equations. Also, reciprocity rule gives $(0.5N(N-1))$ additional equations. Therefore, the total number of view factors that need to be evaluated directly for an N-surface enclosure becomes $(0.5N(N-1))$.
- The superposition rule states that the view factor from a surface i to a surface j is equal to the sum of the view factors from surface i to the parts of surface j and reads to $F_{i(j+k)} = F_{ij} + F_{ik}$.

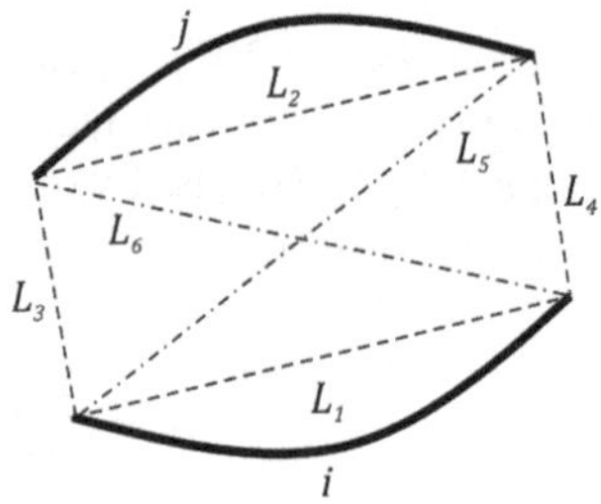

Fig. 4.40 Cross-string method to determine a view factor in 2D-geometries

- A good way to check the correctness of a thermal radiation calculation is to assume blackbody radiation everywhere in the geometry under consideration. The product $\left(A_i \cdot F_{ij}\right)$, which can be understood as a radiation exchange factor matrix, must always be symmetrical, otherwise there is a calculation error.

Geometries such as channels and ducts that are very long in one direction can be considered two-dimensional (since radiation through end surfaces can be neglected). The view factor between their surfaces can be determined by the cross-string method as shown by the example in Fig. 4.40 in the following way:

$$F_{ij} = \frac{\sum(\text{crossed strings}) - \sum(\text{uncrossed strings})}{2 \cdot (\text{string on surface} i)} = \frac{(L_5 + L_6) - (L_3 + L_4)}{2L_1}. \tag{4.213}$$

What is finally missing is to determine the exchange?

For the simple problem stated with two bodies j and i we had $(H_i \cdot A_i) = \left(J_j \cdot F_{ji} \cdot A_j\right)$, expressing the irradiance emitted by i corresponds to the radiosity from surface j considering the view factor of both. This can be generalized to a N–surface system:

$$H_i \cdot A_i = \sum_{j=1}^{N} J_j \cdot F_{ji} \cdot A_j. \tag{4.214}$$

Using the reciprocity rule (Eq. 4.211) allows to eliminate A_i and inserting this into Eq. (4.194) we obtain system of equations for the radiosities J_i:

$$J_i = \varepsilon_i \cdot e_{bi} + (1 - \alpha_i) \cdot \sum_{j=1}^{N} J_j \cdot F_{ij}, \tag{4.215}$$

with $i = 1, 2, \ldots N$. Knowing the temperature of each surface i ($e_{bi} = \sigma T_i^4$) and determining the view factors the system of equations can be solved for the radiosity at each surface and using Eqs. (4.195) or (4.196) we get the heat flux $\dot{q}_i$. Unfortunately, in many technical applications the heat transfer to a surface is known, and the temperature of the surface are an unknown and of relevance. A typical example is the insulated surface leading to $q_i = 0$. If the temperatures are not known also e_{bi} becomes an unknown. Then the path follows first from $\dot{q}_i$:

$$\dot{q}_i = J_i + \sum_{j=1}^{N} J_j \cdot F_{ij}, \tag{4.216}$$

to obtain the set of radiosities. Once the radiosities are obtained, the temperature of the surface can be obtained from Eq. (4.195).This equation has a network interpretation, since multiplying J_j using the summation rule $\left(\sum F_{ij} = 1\right)$ yields:

$$\dot{Q}_i = \dot{q}_i \cdot A_i = \sum_{j=1}^{N} \left(J_i - J_j\right) \cdot F_{ij} \cdot A_i. \tag{4.217}$$

The general procedure is as follows: say our system has $M < N$ surfaces on which the temperature is prescribed, and $N - M$ surfaces on which $\dot{q}_i$ is prescribed. We apply Eq. (4.215) to the M surfaces with specified temperature, and Eq. (4.216) to the $N - M$ surfaces with specified $\dot{q}_i$. Thereby, N linear equations for the radiosities $J_1, J_2, \ldots .J_N$ are obtained. And once we have solved for these quantities, we can calculate either the heat flux or the temperature of the surface.

Similar as for heat conduction it is possible to treat the radiation problem as an electrical circuit. If we consider two diffuse, grey, and opaque surfaces of arbitrary shape maintained at uniform temperatures, then the net rate of radiation heat transfer from surface i to surface j can be expressed as:

$$\dot{Q}_{ij} = A_i \cdot J_i \cdot F_{ij} - A_j \cdot J_j \cdot F_{ji} = A_i \cdot F_{ij} \cdot \left(J_i - J_j\right), \tag{4.218}$$

where in the second part the reciprocity rule has been applied. In analogy with Ohm's law, a resistance R_{ij} can be defined as:

$$\dot{Q}_{ij} = \frac{\left(J_i - J_j\right)}{R_{ij}} \text{ with } R_{ij} = \frac{1}{\left(A_i \cdot F_{ij}\right)}, \tag{4.219}$$

in which R_{ij} is called the *space resistance* to radiation. In a cavity consisting of N-surfaces, the conservation of energy requires that the net heat transfer from surface i to be equal to the sum of the net heat transfers from i to each of the N surfaces of the cavity leading to:

$$\dot{Q}_i = \sum_{j=1}^{N} \dot{Q}_{ij} = \sum_{j=1}^{N} \frac{\left(J_i - J_j\right)}{R_{ij}}. \tag{4.220}$$

Using the derived a relationship for the net radiation from a surface (Eq. 4.199) and inserting this in previous relation, we get:

$$\frac{(e_{bi} - J_i)}{R_i} = \sum_{j=1}^{N} \frac{(J_i - J_j)}{R_{ij}}. \tag{4.221}$$

Thus, the heat transfer from surface i can therefore be interpreted as a sum of currents flowing from i to allother surfaces, with $\left(J_i - J_j\right)$ being the potential (or voltage) difference and R_{ij} the resistancebetween i and j.

To solve simple radiation problems necessitates that either the temperature or the net rate of heat transfer must be given for each of the surfaces to obtain a unique solution for the unknown surface temperature and heat transfer rates. Using the electrical network method, the following procedure is subsequently conducted:

- form an enclosure; consider fictitious surface(s) for openings, room, etc.;
- draw a surface resistance associated with each surface of the enclosure;
- connect the surface resistances with space resistances;
- solve the radiation problem (radiosities) by treating it as an electrical network problem.

Note, that this method is not practical for enclosures with many surfaces.

For very large problems in which many surfaces are involved a different approach is used. Considering N grey emitting surfaces ($\varepsilon_i = \alpha_i, \rho_i = (1 - \varepsilon_i)$, then all absorptions and reflections among the N surfaces are participating in the radiation problem, as sketched in Fig. 4.41 by the surfaces i, j and k. For this purpose, it is useful to introduce Gebhart factor which describe the ratio of radiation absorbed by any other surface versus the total emitted radiation from the given surface; thus, it is defined by:

$$B_{ij} = \frac{\dot{Q}_{ij}}{\varepsilon_i A_i \sigma T_4^4}. \tag{4.222}$$

As such, the Gebhart factor becomes the radiation exchange factor between a number of surfaces: It has to be emphasized that the Gebhart factor approach assumes that the surfaces are *grey* and are illuminated *diffusely* and *uniformly.* The total heat flux of surface i is $\left(\varepsilon_i \sigma T_4^4\right)$, but only $\left(\varepsilon_i \sigma T_4^4 F_{ij}\right)$ reaches surface j with F_{ij} the dimensionless view factor. Thus, the surface j absorbs the fraction $\left(\alpha_i F_{ij} = \varepsilon_i F_{ij}\right)$. This fraction is called primary exchange. From the emissions from surface i to surface k, $(\alpha_k F_{ik})$ is absorbed by k, from which surface k reflects again $(\rho_k F_{ik} = (1 - \varepsilon_k) F_{ik})$, which is absorbed by surface j $(\alpha_k F_{ik} B_{kj})$. This fraction coming from i must be traced over all surfaces and thus the Gebhart factors can be mathematically represented in the following way.

$$B_{ij} = \varepsilon_j \cdot F_{ij} + \sum_{k=1} \rho_k \cdot F_{ik} \cdot B_{kj} = \varepsilon_j \cdot F_{ij} + \sum_{k=1} (1 - \varepsilon_k) \cdot F_{ik} \cdot B_{kj}. \tag{4.223}$$

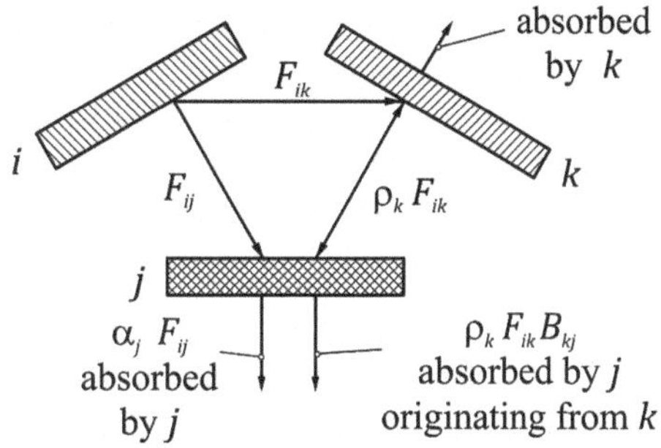

Fig. 4.41 Radiation path of thermal rays in a three-body problem

Since $B_{ij} \leq 1$, Eq. (4.223) can be solved iteratively, that is, starting with $B_{ij}^0 = 0$ and obtaining $B_{ij}^1 = \left(\alpha_j F_{ij}\right)$ as a first approximation. The calculation is repeated until the relative change of the Gebhart factors falls below a given threshold. Another possibility is to set $B_{ij} = \left(\delta_{ik} B_{kj}\right)$, where δ_{ik} is the Kronecker tensor. This allows to transform the Eq. (4.223) into a matrix equation which reads to

$$\sum_{k=1} \left(\delta_{ij} - \rho_k F_{ik}\right) \cdot B_{kj} = \alpha_j \cdot F_{ij}. \tag{4.224}$$

Thereby one obtains a modified space resistance $R'_{ij} = \left(\varepsilon_i A_i B_{ij}\right)$. It is obvious that for all $\varepsilon_i = \alpha_i = 1$ and thus omitting reflection ($\rho_i = 0$) the Gebhart factor become identical to the view factor, i.e. $B_{ij} = F_{ij}$ and thus the results for the blackbody are obtained, which is a recommended consistency check in any radiation calculation.

4.3.8 Radiation in the Presence of the Sun

If a surface A_1 is enclosed by a very large surface A_2, the view factor $F_{12} \rightarrow 1$. The surface A_2, however, hardly sees the surface A_1 and therefore receives almost no thermal radiation from it; as a consequence, $F_{12} \rightarrow 0$ is obtained. In this special case, the radiation exchange is independent of the surface properties of the large surface A_2, and the result is:

$$\dot{Q}_{12} = \varepsilon_1 \cdot A_1 \cdot \sigma \cdot \left(T_2^4 - T_1^4\right). \tag{4.225}$$

In other words, the large surface A_2 behaves like a blackbody. This special case corresponds to the situation of a collector which sees the sky. The radiation exchange of a collector with the area $A_C = A_1$ the emissivity $\varepsilon_C = \varepsilon_1$ and the temperature T_C with the sky of the temperature T_{sky} is thus given by the following equation:

$$\dot{Q}_{12} = \varepsilon_C \cdot A_C \cdot \sigma \cdot \left(T_C^4 - T_{sky}^4\right). \tag{4.226}$$

However, the atmosphere is not a black body, because at wavelengths between 8 and 14 μm it is almost transparent (cf. Chap. 2). So how the effective sky temperature is obtained?

If we assume the atmosphere to be a grey body and then the atmospheric long-wave irradiance in the range of 3 to 100 μm (I_{sky}) can be approximated as a fraction of the blackbody emissive power, termed as effective sky emissivity ε_{sky}, at the surface level ambient temperature (T_a).

$$\varepsilon_{sky} = \frac{I_{sky}}{\sigma \cdot T_a^4}, \tag{4.227}$$

where σ ist he Stefan-Boltzmann constant. Assuming the sky as blackbody we can define a sky temperature $I_{sky} = \sigma \cdot T_{sky}^4$. If we insert this in Eq. (4.227) we obtain

$$T_{sky} = \varepsilon_{sky}^{1/4} \cdot T_a, \tag{4.228}$$

and thereby, we recognize that the effective sky temperature is lower than the surface level air temperature, since ε_{sky} cannot exceed unity. As a consequence, the apparent radiation temperature is several degrees below the sensible temperature reading of a thermocouple in air, which is not of relevance for non-concentrating systems or low temperature applications, because the thermal radiation losses are low compared to other loss mechanisms, however, for high concentrating applications ranging from solar cookers to solar towers, we require a closer look.

The sky emissivity relates the atmospheric radiation, which emanates from the gases in the air, some of which absorb and emit in the far infrared range. Although oxygen and nitrogen constitute by far the majority (about 99% by volume) both gases are transparent to such radiation and do not absorb or emit in the far infrared. Also, carbon dioxide, ozone, and other asymmetrical molecules do not have much influence on the variation of atmospheric radiation received at the ground, see (Lienhard, 2001). Thus, the variation in the water vapor in the atmosphere near the ground is a major contributor to the atmospheric radiation and about 90% of the atmospheric radiation received at the ground level originates at an altitude less than 1 km (Bliss, 1961). Although the effective sky temperature is always lower than the ambient air temperature due to a decrease in elevation, but the presence of clouds in the sky usually tend to increase the effective sky temperature, causing it to approach the clear-sky temperature. In general, the effective sky emissivity of the atmosphere can be expressed as a function of ambient temperature (T_a), relative humidity ($\phi[/]$), the partial pressure of water vapor (p_v) and dew-point temperature (T_{dp}). The partial pressure of water vapor (p_v) in $[Pa]$ and dew point temperature T_{dp} $[K]$ als well as the moisture content d $(kg/(kgdryair))$ can be expressed as a function of ambient temperature (T_a) the Magnus expression (Alduchow & Eskridge, 1996):

$$\begin{aligned} p_v &= 610.94\left(\tfrac{\phi}{100}\right)\exp\left(\tfrac{17.625(T_a-273.15)}{T_a-30.11}\right), \\ T_{dp} &= \tfrac{243.04\ln(p_v/610.94)}{17.625-\ln(p_v/610.94)} + 273.15, d = \tfrac{0.622p_v}{p_a-p_v}, \end{aligned} \tag{4.229}$$

in which p_a is the air pressure in $[Pa]$. For energy balance applications in solar engineering, the effective emissivity during the daytime is more favourable while for night-time passive cooling applications, the effective emissivity during the night-time is more favourable. Therefore, often parametric clear-sky Brunt models for both daytime and night-time are used to calculate the clear sky emissivity $\varepsilon_{sky,c}$ (Alados, 2012):

$$\varepsilon_{sky,c} = 0.598 + 0.057\sqrt{p_v}\,\text{day};\ \varepsilon_{sky,c} = 0.633 + 0.057\sqrt{p_v}\,\text{night}. \tag{4.230}$$

Apart from these primary parameters, the sky emissivity additionally also depends on the cloudiness of the sky, moisture content, and so forth, which have to be taken from statistical meteorological data on site. The presence of clouds modifies the longwave irradiance I because the radiation emitted by water vapor and other gases in the lower atmosphere is supplemented by the emission from clouds. Therefore, under cloudy conditions, the effective sky emissivity is higher compared to clear-sky value ($\varepsilon_{sky,c}$). Parametric models validated by numerous experiments are then used to estimate the all-sky condition for the longwave sky irradiance I in $[W/m^2]$ for an all-day model, see (Li & Yiang, 2017):

$$I = I_{sky,c}(1 - 0.78CF) + 0.380\sigma T_a^4 CF^{0.95}\phi^{0.17}, \tag{4.231}$$

in which CF (%) is the cloud cover fraction in the sky dome. For a simple assessment of the performance of solar thermal components the relationship of Berdahl and Fromberg (1982) is sufficient, which reads to:

$$T_{sky} = T_a\left[0.711 + 0.0056 \cdot T_{dp} + 0.00004T_{dp}^2 + 0.013\cos(15 \cdot t)\right]^{1/4}, \tag{4.232}$$

where T_{sky} is the radiation temperature of the sky in $[K]$ and t is the number of hours after midnight. The relation is derived from measured data at temperatures from $-20°$ to $+ 30$ °C. The deviations of the radiation temperatures of the sky from the air temperatures can be 5 °C in hot and humid climate and up to 30 °C in cold and dry conditions.

The following example is intended to demonstrate the mode of operation of a radiation shield which is used, among other things, in high-temperature receivers for solar towers as night cover. If several such shields, which are technically designed as, for example, thin aluminium or steel foils, are arranged one behind the other, this is referred to as multi-layer insulation, or MLI for short. Figure 4.42 shows a schematic diagram of this. A spacer is usually placed between two foils so that the foils do not come into contact with each other. It shall be considered in the following how the number of foils increases the insulation performance. The walls and the foils are assumed to have an emissivity ε and the spacers transfer the heat flux by means of heat conduction corresponding to its heat conduction resistivity according to $C = (k \cdot A)$. If the areas are equal, the radiation resistance is obtained for $R = (\varepsilon \cdot A)/(2 - \varepsilon) \sim (\varepsilon \cdot A)/2$ (see e.g. Eq. 4.203). In the steady

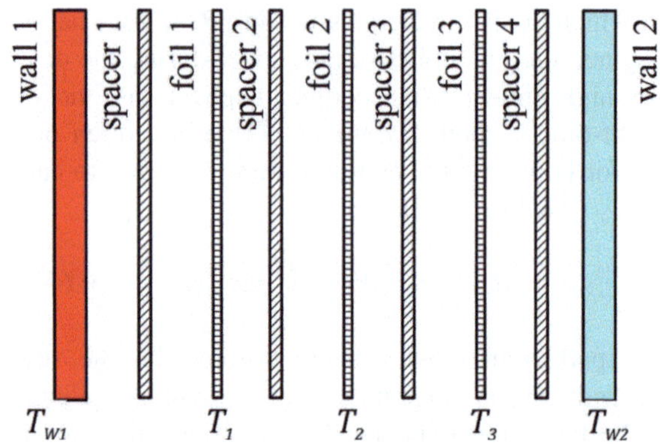

Fig. 4.42 Sketch for determining the insulation effect in a multi-layer insulation with three foils and four spacers

state, the same heat flux flows in each interstitial space, which can be calculated as follows:

$$\begin{aligned}\dot{q}_{W1\to T1} &= C\cdot(T_{W1}-T_1)+\sigma\cdot R\cdot\left(T_{W1}^4-T_1^4\right)=\dot{q}_{1\to 2},\\ \dot{q}_{1\to 2} &= C\cdot(T_1-T_2)+\sigma\cdot R\cdot\left(T_1^4-T_2^4\right)=\dot{q}_{2\to 3},\\ \dot{q}_{2\to 3} &= C\cdot(T_2-T_3)+\sigma\cdot R\cdot\left(T_2^4-T_3^4\right)=\dot{q}_{3\to W2},\\ \dot{q}_{3\to W2} &= C\cdot(T_3-T_{W2})+\sigma\cdot R\cdot\left(T_3^4-T_{W2}^4\right).\end{aligned} \tag{4.233}$$

Summing up all heat fluxes the temperatures of the foils are eliminated and we obtain:

$$\dot{q}_{W1\to W2}=\frac{C}{4}\cdot(T_{W1}-T_{W2})+\frac{\sigma\cdot R}{4}\cdot\left(T_{W1}^4-T_{W2}^4\right). \tag{4.234}$$

The main result of the MLI structure is that the insulation effect, which is composed of heat conduction and radiation losses, reduces by a factor of $(N+1)$ with each foil layer.

In the case of radiation exchange outside a cavity under the influence of the sun, Kirchhoff's law with $\varepsilon_i=\alpha_i$ no longer applies. The emissivity differs from absorptivity and thus $\varepsilon_i\neq\alpha_i$. The reason lies in the definition of the absorptivity α_i, which is related to the spectrum of the sun. The radiant heat flux from the sun or irradiance is,

$$I_s=\int_{\lambda=0}^{\lambda\to\infty} I_{s\lambda}(\lambda,T)d\lambda, \tag{4.235}$$

while the absorptivity of a surface is given by the following relation:

$$\dot{q}_\alpha(T)=\int_{\lambda=0}^{\lambda\to\infty}\alpha(\lambda,T)\cdot I_{s\lambda}(\lambda,T)\,d\lambda. \tag{4.236}$$

The ratio of the previous two equations yields the temperature dependent absorption coefficient:

$$\alpha(T) = \frac{\int_{\lambda=0}^{\lambda\to\infty} \alpha(\lambda, T) \cdot I_{s\lambda}(\lambda, T)\, d\lambda}{\int_{\lambda=0}^{\lambda\to\infty} I_{s\lambda}(\lambda, T)\, d\lambda}. \tag{4.237}$$

Equation (4.237) is the basis of "passive" solar thermal energy. Only this validity allows to manipulate the surface temperatures by a specific choice of the surface properties. This will be shown by means of an example.

A plate with the area A floats in space and is ideally insulated on the side facing away from the sun ($\partial\dot{q}/\partial n = 0$). On the other side the sun shines with solar constant $i_{sc} = 1370 W/m^2$. The temperature of the space is the temperature of the universe $T_s = 0°K$. Searched and unknown is now the plate temperature T. The plate has the radiation exchange factor $R = (\varepsilon A)$ and absorbs thermal power from the sun according to $\dot{Q} = (\alpha \cdot A \cdot i_{sc})$. The power absorbed by the plate must equal the power received from the sun and thus:

$$\dot{Q} = \alpha \cdot A \cdot i_{SC} = \sigma \cdot R \cdot \left(T^4 - T_{Space}^4\right). \tag{4.238}$$

Solving the equation for temperature finally yields:

$$T = \left(\frac{\alpha}{\varepsilon}\frac{i_{Sc}}{\sigma}\right)^{\frac{1}{4}}. \tag{4.239}$$

This dependence on (α/ε) is used in so-called SSMs (second surface mirrors), for example in radiators, to emit heat despite incident solar radiation. For this purpose, a metal film (e.g. silver or aluminum) is vapor-deposited on the back of a carrier material (e.g. polyester film). In front of this is a Teflon or plastic film that is transparent to solar radiation, i.e. permeable. The metal layer reflects most of the solar radiation and absorbs only a little ($\alpha = 0.08 - 0.14$). The outer layer emits with a high emission coefficient ε ($\varepsilon = 0.6 - 0.8$) and prevents excessive heating. Table 4.12 shows some temperatures of different materials according to Eq. (4.239).

Table 4.12 Temperatures of a sunlit plate in space

Gas	$\alpha[-]$	$\varepsilon[-]$	$\alpha/\varepsilon[-]$	$T[K]$
Gold	0.24	0.03	8	661
Aluminium	0.13	0.05	2.6	499.08
Black paint	0.97	0.9	1.08	400.46
White paint	0.18	0.85	0.212	366.62
SSM	0.08	0.8	0.1	221.02

4.4 Practical and Coupled Radiative-Convective Heat Transfer Problems

In solar energy and building physics, the heat exchange between surfaces separated by an infrared transparent medium (usually air) is a very common problem. Examples are the heat transfer from the absorber to the collector casing, or the heat exchange of walls with different temperatures in a room or the heat exchange between heating surfaces (radiator/convector) with its surrounding, or the heat exchange between collector surface with the environment and the sky. As air has a minimal infrared absorption coefficient. When considering the radiation exchange between neighbouring buildings or walls in a room with only a few meters distance, the absorption can be neglected. This is not the case for the complete atmosphere layer of several kilometres. Here infrared radiation is effectively hold back – on of the basic pre-requisites of our life on earth.

In an engineering treatment the radiation exchange between surfaces can be described as described in Sect. 4.3.7. But there is another mode of exchanging heat: the convective heat transfer from a surface to the air. Also here a convective heat transfer coefficient to air can be defined:

$$h_{conv} = \frac{\dot{q}_{conv}}{T_s - T_{air}} \tag{4.240}$$

in which T_s denotes the surface temperature, and T_{air} the air temperature outside the thin boundary layer. If the surface temperatures in a problem are given, radiative and convective heat transfer can be separately calculated and then the two solutions add to a net outgoing heat flux for each surface. This quasi-linearization can be applied if the temperature differences are not too large, usually in the range of $\Delta T < 100K$, unless non-linear effects come into play. However, if for some surfaces not temperatures but heat fluxes ($\dot{q}$) are given, a coupled problem appears requiring non-linear equations be solved.

Additional challenges appear, if the gas itself is absorbing and emitting radiation, or within a cavity a solid structure is in radiative interaction with the boundaries creating temperature differences in the gas leading to convection. In such a context the superposition of radiation and convection collapses and the combined heat transfer problem to be treated in a fully coupled way.

In solar collector problems, a coupling between convection and the radiation transport is possible if for example polymer films as internal glazings are coming too close to a selectively emitting absorber, or if transparent insulation materials (see Chap. 6) like honeycomb structures or aerogel, which are partially transparent to infrared radiation are used to improve the heat losses in the air gap. Simplified methods to solve the coupled equations have been developed alongside these developments, and some details are discussed for example in (Pflüger & Wittwer, 1984; Platzer, 1988, 1992).

One important result is that by using an appropriate air gap in this case between the selective absorber and the absorbing-emitting structure, the heat transport modes may be effectively decoupled, with the results that a much lower combined heat transfer can be observed, which is close to the one calculated with the additive simplification. Other research where coupled heat transfer of both modes, radiation and convection occurs, is the energy loss of stellar atmospheres. Although combined heat transfer is needed in many practical problems in engineering, the treatment still seems to be a challenge in many cases (Dombrovsky, 2022). The separation of the individual effects is often not easy, because apart from the given collector or storage dimensions and the material specifications, the environmental conditions such as solar irradiation (direction and intensity) also external parameters such as wind overflow of the collector or flow through the storage or the wall panels determine which of the heat transfer mechanisms plays an excellent role. Thus, the key to an engineering assessment and the subsequent decision on how a problem can be solved is the evaluation of dimensionless number characterizing the problem, which require some engineering assessment. In the next step a simple analytical example of pure radiative heat transfer within a cylinder is addressed to illustrate how a radiation heat transfer calculation is executed Based on this we subsequently address two practical combined problems appearing in solar thermal systems such as

- the heat transfer between two glass panes in buildings.
- the effective thermal conductivity of packed beds, which are used as thermal energy storages in gas-cooled systems, among others, both in low-temperature systems such as packed beds in air collectors and in high-temperature systems.

4.4.1 Non-Dimensional Numbers Characterizing Combined Problems

Despite the complexity of conjugate radiation-convection and heat-conduction problems, it is possible to assign certain dominant phenomena by forming characteristic dimensionless parameters. This allows the selection of suitable solution approaches and methods, either described in the literature such as (Howell et al., 2020a, 2020b) or being depicted by commercially available numerical solution packages like COMSOL Multiphysics.

Thus, when a considerable heat conduction or convection occurs simultaneously with thermal radiation in the presence of an absorbing emitting medium, a more complex mathematical treatment is needed. In principle, if none of the modes is negligible, a non-linear integrodifferential equation will result for the energy equation for the general problem having combined modes of heat transfer. Sometimes the formulation can be simplified, for example when the absorbing-emitting medium can be considered as optically thick, so that emitted infrared photons may travel very short before being absorbed again. Under such conditions the so-called diffusion approximation in radiation transport is applicable, leading to differential terms only instead of the integral equations. The other extreme case, an optically very thin medium also allows the simplification of additive solutions. However, as combined mode problems in general are mathematically complex, it is usually not possible to obtain an analytical solution even for geometrically simple cases.

Nevertheless, there is the possibility to characterize the coupled problem with the help of dimensionless numbers before starting the solution procedure. The Stark number (N), also called the conduction and radiation parameter, weights the energy transport by heat conduction versus radiation and reads:

$$N = \frac{(\alpha + \sigma_s) \cdot k_s}{4 \cdot \sigma \cdot \Delta T^3}, \tag{4.241}$$

in which α is the total, linear absorption/emission coefficient, σ_s the total isotropic scattering coefficient, k_s the thermal conductivity of the solid, σ the Stefan Boltzmann-constant and $\Delta T^3 = T_a^3 - T_b^3$ is determined from two characteristic temperatures (T_a, T_b) of the problem considered. If N falls, thermal radiation effects increase. Roughly, three characteristic regimes for N can be distinguished, i.e. $N = 1, 0.1, 0.01$. The dimensionless numbers for the other heat transfer modes can also be derived as the ratio between the heat fluxes due to convection ($\dot{q}_{conv}$), radiation ($\dot{q}_{rad}$) and conduction ($\dot{q}_s$). These heat fluxes read to

$$\dot{q}_{conv} = u_0 \rho c_p \Delta T; \; \dot{q}_{rad} = (\alpha + \sigma_s)\sigma \Delta T^4; \; \dot{q}_s = k_s \Delta T / d, \tag{4.242}$$

in which u_0 is the mean characteristic velocity, ρ the specifc density, c_p the heat capacity and d a characteristic length scale. The Boltzmann number (Bo) determines the relative importance of the energy transported by forced convection to thermal radiation and is given by

$$Bo = \frac{\dot{q}_{conv}}{\dot{q}_{rad}} = \frac{u_0 \rho c_p \Delta T}{(\alpha + \sigma_s)\sigma \Delta T^4}. \tag{4.243}$$

Typically, three regimes exist at the Boltzmann number, i.e. $Bo = 0.1, 1, 10$, which implies an increasing influence of convection on the versus radiative heat transfer. The convection-conduction number (Cn) characterizes the ratio between heat flux generated by forced convection to that of molecular thermal conduction

$$Cn = \frac{\dot{q}_{conv}}{\dot{q}_s} = \frac{u_0 \rho c_p d}{k_s}. \tag{4.244}$$

where d specifies the characteristic length of the system along the direction of heat conduction, typical domains are separated by $Cn = 1, 10, 100$. In case of buoyancy induced problem, where the temperature differences in a fluid induce natural convection, the characteristic velocity u_0 has to be replaced by the buoyant mean velocity u_b given by $u_b = \sqrt{g \beta \Delta T d}$, in which g is the gravity acceleration and β the volumetric expansion coefficient.

When considering radiative heat transfer problems, two other non-dimensional quantities must be considered in advance. The linear scattering albedo (ω), defined as,

$$\omega = \frac{\sigma_s}{(\alpha + \sigma_s)}. \tag{4.245}$$

is the ratio between the scattering coefficient and the extinction coefficient and describes the relative importance of scattering compared to absorption/emission. Two scattering regimes can be distinguished. For $\omega \approx 0$, a strongly absorbing/emitting material or a non-scattering medium exists; while $\omega \approx 1$ represents a strongly scattering medium.

The optical thickness (or opacity, τ), which we described in Chap. 3 as product of refractory index (n) and homogeneous medium thickness (d), can be conceived as the ability of a medium to attenuate radiation and is defined as

$$\tau = (n \cdot d) = (\alpha + \sigma_s) \cdot d, \tag{4.246}$$

where d is the characteristic length of the medium layer. Essentially, there are four different physical states here:

- non-participating medium $\tau \approx 0$;
- optically thin medium $\tau \ll 1$, where the radiative heat transfer is determined by the interfacial emission and the radiation of the medium is limited;
- self-absorbing medium $\tau \approx 1$, in which the contributions of the interfaces and the internal radiation balance each other;
- optically thick medium $\tau \gg 1$, where radiation is essentially a local phenomenon and radiative transfer behaves like a diffusion process (similar as molecular transport).

For transient heat transfer phenomena, the characteristic diffusion time of the solid (τ_d), has to be considered, which is defined as follows;

$$\tau_d = \frac{d^2}{a} = \frac{(\rho c_p)_s}{k_s} \cdot d^2, \tag{4.247}$$

where a specifies the thermal conductivity. The diffusion time τ_d can be interpreted as a measure of the time required to establish thermal equilibrium in the solid. When heat transfer occurs through the liquid–solid interface, the thermal activity ratio (TAR) given by,

$$TAR = \frac{\sqrt{(\rho k c_p)_f}}{\sqrt{(\rho k c_p)_s}}, \tag{4.248}$$

describes the ratio between the heat transfer coefficients of the liquid to the solid media and can be used to quantify the interfacial heat flux. High values of the TAR values indicate a low heat flux, while small TAR values reflect a large heat flux.

4.4.2 Radiative Transfer Equation and Simple Example

Since the radiative transfer describes a net exchange between the different materials involved in a heat transfer problem, the determination of the heat transfer is fundamentally different from that in solids, which are linked via the thermo-physical material properties, or via interfaces, which are characterized by a media transport such as flows over a wall. To solve energy conservation in a control space, the energy conservation equation must be satisfied in a control volume at the given thermal boundary conditions. For a translucent medium, the energy conservation equation in general form is given by

$$\rho c_p \frac{dT}{dt} = \beta T \frac{dp}{dt} + \nabla \cdot (k \nabla T - \dot{q}_{rad}) + \dot{q} + \Phi_{diss}. \tag{4.249}$$

In it, ρ and c_p describe the specific density and heat capacity, T the temperature, β the volumetric expansion coefficient, p the pressure, k the thermal conductivity, $\dot{q}$ a vo-lumetric heat source/sink, and Φ_{diss} the heat release by volumetric viscous dissipation. d/dt denotes the substantial derivative.

To obtain the radiation contribution $\dot{q}_{rad}$ to the energy equation, radiative transfer relations must be provided. Since some of the radiation terms such as absorption, emission, and reflection depend not only on wavelength but also on temperature, the radiative transfer equation (RTE) must be solved simultaneously with the energy conservation equation to determine the temperature distribution and radiation properties.

To obtain the temperature distribution in a control volume by solving Eq. (4.249), we need $(\nabla \cdot \dot{q}_{rad})$ as function of the temperature distribution. Unfortunately, $(\nabla \cdot \dot{q}_{rad})$ depends on the scattering of the radiation, which must be determined within the medium. There are now two options to obtain the required relations for $(\nabla \cdot \dot{q}_{rad})$.

The first is to derive $\dot{q}_{rad}$ and then differentiate to obtain $(\nabla \cdot \dot{q}_{rad})$. The second option is to directly determine $(\nabla \cdot \dot{q}_{rad})$ by the local radiation interaction within a control volume in the medium, since $(\nabla \cdot \dot{q}_{rad})$ is the difference in absorbed radiation emitted in the same volume element. This in turn depends on the absorption coefficient and the intensity incident from all directions. The radiation energy required for the energy equation (Eq. 4.249) is the total radiation over all wavelengths, so for both absorption and emission relationships are required to obtain the difference of the total radiation Ω_i incident from all solid angles, absorbed locally in the volume, and the locally emitted radiation. These two quantities (absorbed and emitted radiation) are described by the two integral terms over all wavelengths on the right-hand side of the following equation.

$$-\nabla \cdot \dot{q}_{rad} = \underbrace{\int_{\lambda=0}^{\infty} \alpha_\lambda \left(\int_{\Omega_i=0}^{4\pi} I_\lambda(\Omega_i) d\Omega_i \right) d\lambda}_{absorption} - \underbrace{4\pi \int_{\lambda=0}^{\infty} \alpha_\lambda I_{\lambda b} d\lambda}_{emission} . \tag{4.250}$$

To determine the radiative power density ($\dot{q}_{rad}$), we now still need the radiation intensity (I_λ) along a propation direction $\vec{s}$. The radiation intensity is given as balance equation containing all contributions (propagation, emission, absorption, and scattering) in the volume element, which is called the radiative transfer equation (RTE) reading in its general form to

$$\frac{dI_\lambda}{d\vec{s}} = \alpha_\lambda(\vec{s}) I_{\lambda b}(T, \vec{s}) - \beta_\lambda I_\lambda(\vec{s}) + \frac{\sigma_{\vec{s},\lambda}}{4\pi} \int_{\Omega_i=0}^{4\pi} I_\lambda(\vec{s}, \Omega_i) \Phi_\lambda(\Omega, \Omega_i) d\Omega_i , \tag{4.251}$$

in which α_λ, β_λ and $\sigma_{\vec{s},\lambda}$ are the absorption, extinction, and scattering coefficients and Φ_λ is the scattering phase function. The RTE expresses the general conservation of radiant energy along a line of sight in an absorbing, emitting, and scattering medium. The solution of this differential equation is by no means trivial, especially in engineering geometries, but often a number of simplifications can be made, such as the existence of an opaque medium ($\tau = 0$), assumptions of a gray beam ($\varepsilon = const$), and many others that reduce the problem to an interface problem.

There are a number of approximate solvers for the RTE equation, which are essentially based on the assumption of local thermodynamic equilibrium. Mostly, however, the solution is done numerically. Some of these solution methods are applicable to special environments and an overview is given, for example, in (Howell et al., 2020a, 2020b). The numerical methods most commonly used in radiative heat transfer and their main features can be summarized as follows:

Monte Carlo

- based on conceptually simple physics, which represents quasi-exactly the behaviour of a single ray;
- completely general; can solve time-dependent and 3D problems with arbitrary geometry;
- easy to program.
- Challenge: reduction of statistical error, which can be reduced by tracking more initial rays at the expense of computation time (requiring longer computer times)

Discrete Ordinates Systems

- Complex programmability;
- poor convergence on complex scattering phenomena with large gradients.

Invariant Embedding

- Difficult programmability;
- limited to 1D problems;
- includes all orders of multiple scattering and the computed radiances have no statistical errors;
- extremely fast.

Due to their simplicity, quasi-exact representation of the behaviour of a single ray, and most importantly, the extreme increase in computer capacity, Monte Carlo methods are now mostly used to solve $RTEs$.

Radiative heat transfer phenomena are an often underestimated or neglected aspect in the calculation of temperature distributions in technical assemblies. Using a simple example of a cylinder, the radiative heat exchange between finite surfaces will be sketched here to vividly illustrate the calculation procedure presented in Sect. 4.3.7. With the coordinate designations given in Fig. 4.43, one obtains

$$x = r/h;\ y = r/h \rightarrow z = 1 + \left(1 + y^2\right)/x^2,$$

for which we obtain as view factor F_{12} between the area A_1 and A_2:

$$F_{12} = 1/2\left(z - \sqrt{z^2 - (2y/x)^2}\right)$$

We now assume the hollow cylinder with the dimensions $r = 0.1m$ and the height $h = 0.25m$ to be insulating to the outside and there should be no heat conduction within the cylinder, and we want to determine the radiation exchange factors and temperatures.

First, the areas A_1 and the radiation exchange factors F_{ij} must be calculated.

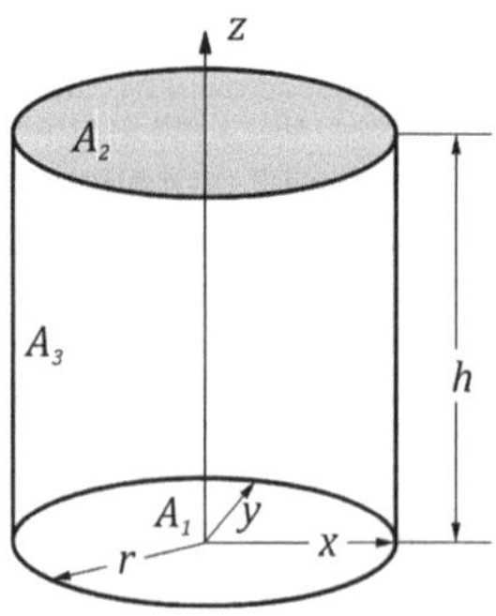

Fig. 4.43 Schematic sketches for the determination of the visibility factors in a closed cylinder

- Areas: The areas of the cylinder under consideration are

$$A_1 = A_2 = \pi \cdot r^2 = 0.031416\,\mathrm{m}^2 and A_3 = 2 \cdot \pi \cdot r \cdot h = 0.15708\,\mathrm{m}^2.$$

- View factors: For $n = 3$ surfaces, there exist are $n^2 = 9$ view factors. Since the ground and top surfaces are considered to plane parallel, they cannot see themselves. This results in $F_{11} = F_{22} = 0$. The view factor from 1 to 2 yields for $x = 0.4$, $y = 0.4$ and $z = 8.25$ to $F_{12} = 0.123$. Since symmetry applies, $F_{12} = F_{21} = 0.123$. The visibility factor F_{31} results from the summation rule (Eq. 4.212), according to which $F_{11} + F_{12} + F_{13} = 1$ must be to $F_{13} = 0.877$.
- Also now the symmetry is applicable, and it is $F_{23} = F_{13} = 0.877$.
- The symmetry also requires $F_{31} = F_{32}$. With the help of the reciprocity law, F_{31} can now be determined. It is $(A_1 \cdot F_{13}) = (A_3 \cdot F_{31})$. Thus, $F_{31} = (A_1/A_3) \cdot F_{13} = (r/2h) \cdot F_{13} = 0.1754 = F_{32}$.
- The inner shell of the cylinder sees itself, so $F_{31} \neq 0$ (!!!). Now the summation law is applied, and it yields $F_{31} + F_{32} + F_{33} = 1$ and thus $F_{33} = 0.6942$. Thus, the view factor matrix is complete, and it results in:

$$F_{ij} = \begin{bmatrix} 0 & 0.123 & 0.877 \\ 0.123 & 0 & 0.877 \\ 0.1754 & 0.1754 & 0.6942 \end{bmatrix}$$

- Accordingly, the following matrix is obtained for the radiation exchange factors.

$$R_{ij} = A_i \cdot F_{ij} = \begin{bmatrix} 0 & 0.03864 & 0.02755 \\ 0.03864 & 0 & 0.02755 \\ 0.02755 & 0.02755 & 0.101945 \end{bmatrix}$$

The radiation exchange factor matrix must always be symmetrical, otherwise there is a calculation error. This is an important checking mechanism for complex technical calculations in difficult geometries. The determination of the radiation exchange factors, and their symmetry behaviour also apply to grey bodies.

In the following it is assumed that all three surfaces are isolated from each other. Let surface 1 be at $T_1 = 550K$ and surface 2 at $T_2 = 300K$. The steady state temperature ($t \to \infty$) of the shell surface is obtained from the heat balance leading to:

$$\sigma \cdot \left[\mathrm{R}_{31} \cdot \left(\mathrm{T}_1^4 - \mathrm{T}_3^4\right) + \mathrm{R}_{32} \cdot \left(\mathrm{T}_2^4 - \mathrm{T}_3^4\right)\right] = 0.$$

From this, after solving for the shell surface temperature T_3, we get

$$T_3^4 = \frac{R_{31} \cdot T_1^4 + R_{32} \cdot T_2^4}{R_{31} + R_{32}} = \frac{T_1^4 + T_2^4}{2} \to T_3 = 472.4K$$

It is noteworthy that without the influence of convection, the temperature T_3 is approximately at the level of the hot surface A_1.

Now the same cylinder should not be "black" radiating but only one black surface A_1 with $\varepsilon_1 = 1$ and two gray radiating surfaces A_2 with $\varepsilon_2 = 0.1$, A_3 with $\varepsilon_3 = 0.75$; the temperatures $T_1 = 550K$ and $T_2 = 300K$ are assumed identical as before. The first two calculation steps for the areas (A_i) and the view factor matrix (F_{ij}) remain the same, however we have now to evaluate the Gebhart factors (B_{ij}) according to Eq. (4.224) leading to

$$B_{ij} = \begin{bmatrix} 0.0727 & 0.176 & 0.90968 \\ 0.1762 & 0.00479 & 0.8190 \\ 0.2426 & 0.0218454 & 0.7356 \end{bmatrix}.$$

This in turn results in the following radiation exchange matrix

$$R_{ij} = \begin{bmatrix} 0.00228 & 0.00554 & 0.028578 \\ 0.00554 & 0.000015 & 0.002573 \\ 0.028578 & 0.002573 & 0.08666 \end{bmatrix}.$$

If we now calculate the temperature of the shell surface A_3, we obtain from the heat balance a value of $T_3 = 539.34$ K and thus approx. 69 °C more (!!!) than in the case of the black body assumption. Even if the example is of an academic nature, it shows that it is possible to manipulate the temperatures on interfaces by controlling the radiation properties of the surfaces, which is of excellent importance for the integrity of components, especially in high-temperature applications.

4.4.3 Coupled Heat Transfer Between Two Glass Panes

Using Fig. 4.44a, let us first consider the heat transfer between two parallel vertical panes of glass, as typically occurs in windows in buildings. When calculating heat losses through a double-glazed window, the heat flow in the individual sections of the window is composed of components of radiative, convective, and conductive heat transfer on the outer and inner surfaces of the double-glazed window and in the gap between the panes. In this case, the value of each of these components at a known temperature difference ΔT in each separate section of heat transfer is calculated using the coefficients of convective h_{conv} and radiative heat transfer h_{rad}, as well as the thermal conductivity of gas (k_{gas}) and the glass panes (k_I, k_{II}), respectively. This interaction of different heat transfer mechanisms can be clearly represented in the form of an electrical equivalent circuit as shown in Fig. 4.44b.

In the first step, we will only consider the radiative heat loss through a single-chamber double-glazed window installed in the opening of the building envelope. In order to consider the problem in a one-dimensional way, we assume that the areas of all four glass surfaces are equal, and the emission coefficients take the values ε_1, ε_2, ε_3 and ε_4. Let the temperature of the inner surface of the room be

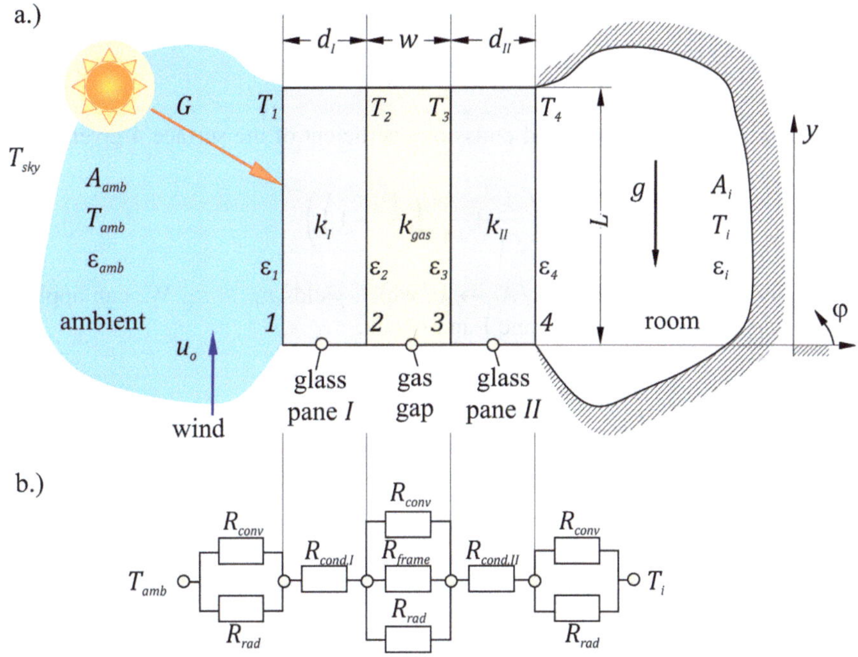

Fig. 4.44 Sketch to calculate the combined convective-radiative and conductive heat transfer across a double-glazed window(a) and corresponding equivalent electric circuit expressed by heat resistanccs (R_i) (b)

constant and be T_i, while that of the surroundings is also constant and has the value T_{amb}, where $T_{amb} < T_i$. Thus, heat is radiative transferred from the room to the surroundings. The objective is to determine the glass temperatures T_I, T_{II} and the radiant heat flux density $\dot{q}_{rad}$ through the window. For the analysis of radiant heat transfer, we assume that there is no temperature gradient across glass thickncss, i.c. $T_I = T_1 = T_2$ or $T_{II} = T_3 - T_4$. This is sufficiently justified since the thermal conductivity of the glass is considerably greater than that of the gas in the gap between the panes.

The temperatures T_{amb} and T_i are the radiation temperatures of the surfaces that differ from the ambient temperature Tamb and the room air temperature T_i. In indoor surfaces the air temperature is usually equal to the radiant temperature, while the equation of the radiant temperature and the air temperature in the outdoor window surface reflects a dense cloudiness of the sky, otherwise the radiant temperature of the sky must be determined, (see cf. Sect. 4.3.8). Here, we further assume the identity of temperatures.

The surface of the space A_i and the surface of the inner glass A_4 form a closed system, and since the surface of the glass A_4 is flat does not radiate on itself. Thus, we can determine the heat flux density $\dot{q}_{rad}$ transferred by radiation from the inner surface of the room to the inner surface of the glass by thermal radiation.

$$\dot{q}_{rad,4\rightarrow i} = \varepsilon_{4i} \cdot \sigma \cdot \left(T_i^4 - T_{II}^4\right), \tag{4.252}$$

in which ε_{4i} denotes the reduced emission coefficient of the surface 4 given by

$$\varepsilon_{4i} = \left(\frac{1}{\varepsilon_4} + \frac{A_4}{A_i}\left(\frac{1}{\varepsilon_i} - 1\right)\right)^{-1}. \tag{4.253}$$

Usually $A_i \gg A_4$ so that $A_4/A_i \rightarrow 0$, which yields $\varepsilon_{4i} \approx \varepsilon_4$. We can apply a similar approach for the glass pane I and obtain:

$$\dot{q}_{rad,amb\rightarrow 1} = \varepsilon_1 \cdot \sigma \cdot \left(T_I^4 - T_{amb}^4\right). \tag{4.254}$$

At last, we calculate the radiative heat flux density of the two panes with temperatures T_I and T_{II}, whose emission coefficients of the surfaces are ε_2 and ε_3, which is given as follows:

$$\dot{q}_{rad,2\rightarrow 3} = \varepsilon_{23} \cdot \sigma \cdot \left(T_{II}^4 - T_I^4\right), \tag{4.255}$$

where the reduced emission coefficient ε_{23} is given by the following relation.

$$\varepsilon_{23} = \left(\frac{1}{\varepsilon_2} + \frac{1}{\varepsilon_3} - 1\right)^{-1}, \tag{4.256}$$

In the steady state of radiative heat transfer through a double-glazed window, the values of all heat fluxes are equal. We apply here the solution method used in the calculation of heat transfer by radiation through a system of screens, since the glasses in a double-glazed window correspond to the role of screens in heat radiation. Thus, this methodology for double-glazed windows is also applicable to any number of parallel arranged glass panes. This results in the radiative heat flux density of:

$$\dot{q}_{rad} = \frac{1}{\left(\frac{1}{\varepsilon_1} + \frac{1}{\varepsilon_2} + \frac{1}{\varepsilon_3} + \frac{1}{\varepsilon_3} - 1\right)}\left(T_i^4 - T_{amb}^4\right). \tag{4.257}$$

After determining $\dot{q}_{rad}$, T_I and T_{II}, the values of the radiative heat transfer coefficients h_{rad} can be calculated for each of the heat transfer points through the window, for which we assume $\dot{q}_{rad,ij} = h_{rad,ij} \cdot \left(T_i - T_j\right)$. This yields $h_{rad,amb1}$ for the outer glass environment; $h_{rad,23}$ for the glass interspace; $h_{rad,4i}$ for the inner pane surface facing the room. The value of the total thermal resistance ($R_{rad,tot}$) for radiant heat transfer is composed of the sum individual thermal resistance, which is given by

$$R_{rad,tot} = \frac{1}{h_{rad,amb1}} + \frac{1}{h_{rad,23}} + \frac{1}{h_{rad,4i}} = R_{rad,amb1} + R_{rad,23} + R_{rad,4i}$$

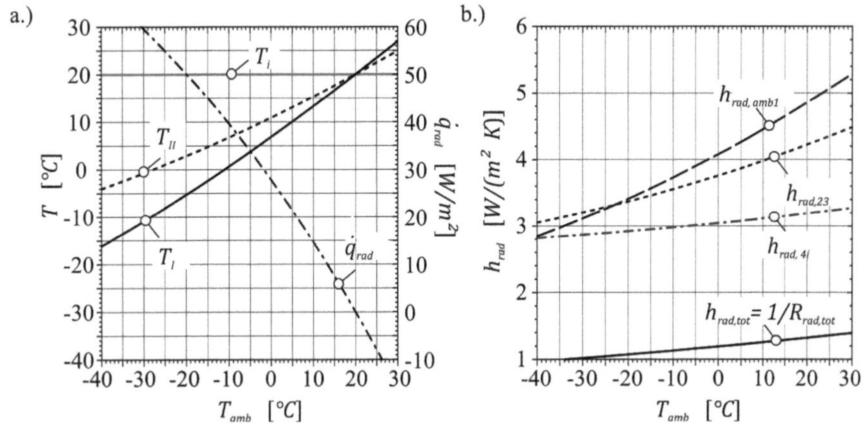

Fig. 4.45 **a** Calculated temperatures T_I and of T_{II} the window panes and radiative heat flux ($\dot{q}_{rad}$) for a given inner room temperature T_i as a function of the ambient temperature T_{amb} for a glas emissivity of $\varepsilon = 0.85$; **b** corresponding total ($h_{rad,tot}$) and individual radiative heat transfer coefficients ($h_{rad,ij}$)

$$R_{rad,tot} = \frac{1}{h_{rad,tot}}. \tag{4.258}$$

Figure 4.45a shows the temperatures of the two panes (T_I, T_{II}) as a function of the ambient temperature (T_{amb}) and the heat flux ($\dot{q}_{rad}$) resulting from radiation exchange alone, assuming an emissivity of $\varepsilon = 0.85$ on each surface of the panes. The corresponding heat transfer coefficient ($h_{rad,ij}$) due to radiation and the total heat transfer $\left(h_{rad,tot}\right)$ coefficient is shown in Fig. 4.45b.

The figure shows that the coefficient of radiative heat transfer decreases both with a decrease in the maximum temperature of two surfaces and with an increase in the temperature difference of these surfaces. The coefficient of radiative heat transfer ($h_{rad,23}$) between the pane surfaces also decreases monotonically with decreasing arithmetic mean temperature of the panes. Thus, with the increase of the temperature difference at a double-glazed window when the ambient temperature is lowered, the total radiation thermal resistance ($R_{rad,tot}$) increases, which in turn means that the radiative heat transfer coefficient decreases as the ambient temperature decreases. Thereby, a reduction of the radiative heat loss can only be realized by means of two methods.

- increase the number of glasses arranged in parallel behind each other.
- the use of glasses with one/more heat-shielding coatings (low emissivity-ε).

While the first method is already known to us from the problem of radiation screens, it makes sense to deal with the second method in particular. For this purpose, let us consider that one or both inner sides of the panes are provided with a low-emissivity coating ($\varepsilon = 0.05$) instead of the prior value of $\varepsilon = 0.05$.

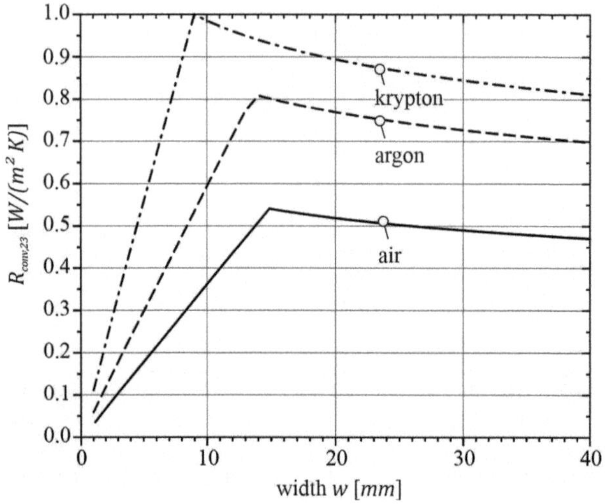

Fig. 4.46 Calculated convective heat resistance ($R_{conv,23}$) as a function of the gas gap width w for temperature difference between the inner pane surfaces of $(T_2 - T_3) = 20°C$ for three gas gap fillings

Since the reduced emission coefficient ε_{23} (see Eq. 4.256) immediately impacts the radiative heat flux exchanged between the panes (Eq. 4.257), the ratio of the radiative heat fluxes for the three cases reads to:

$$\dot{q}_{rad}(\varepsilon_2, \varepsilon_3 = 0.85)/\dot{q}_{rad}(\varepsilon_2 = 0.05, \varepsilon_3 = 0.85)/\dot{q}_{rad}(\varepsilon_2, \varepsilon_3 = 0.05) \approx 11/1.84/1. \quad (4.259)$$

In other words, the specific radiation-related heat loss can be substantially reduced by a low-emissivity coating. The case of emission-reduced coating of the interface 3 is particularly important for solar thermal collectors, since we have seen in the Chap. 3 on optical conversion that emission in the infrared range is a particular development goal for an optically selective absorber.

In the next step, we will consider when heat transfer by conduction and convection is superposed to the radiative heat transfer. The total heat flow ($\dot{q}$) through the glass panes is obtained as a parallel connection of the individual thermal resistances composed of conduction, radiation, and convection, as indicated in Fig. 4.44, and reads to:

$$\dot{q} = U \cdot (T_i - T_{amb}), \quad (4.260)$$

in which U denotes the thermal transmittance in $(W/(m^2 K))$.

This total heat transfer coefficient consists of parts comprising the outer pane surface to into the environment (k_{amb1}), the heat conduction in panes I and II, the convective heat transfer between panes I and II as well as the heat radiation between them termed as the total thermal conductance of the glazing (k_t), and finally the combined radiative-convective heat transport from the pane into the room (k_{4i}), which can be expressed by:

$$\frac{1}{U} = \frac{1}{h_{amb1}} + \frac{1}{h_t} + \frac{1}{h_{4i}}. \tag{4.261}$$

The total thermal conductance of the glazing (h_t) is obtained by assuming continuity of the heat flux at each interface and is calculated for M parallel panes with N number of gas spaces as follows

$$\frac{1}{h_t} = \sum_1^N \frac{1}{h_s} + \sum_1^M \frac{d_j}{k_j}, \tag{4.262}$$

where N is the number of gas spaces, d_j the thickness of the of each pane j and k_j its thermal conductivity. Neglecting the contribution of the frame to the heat losses and considering only the heat flow normal to the pane, the thermal resistance from radiation and that from conduction and convection represents a parallel circuit according to the analogue of the electrical equivalent circuit. Thus, the total heat transfer coefficient of each gap ($h_{s,l}$) can be calculated as the sum of the mixed conductive and convective heat transfer coefficient ($h_{g,l}$) and the radiation heat transfer coefficient ($h_{rad,l}$) as follows.

$$h_{s,l} = h_{g,l} + h_{rad,l}, \tag{4.263}$$

In the next step, the external heat transfer from the pane to the environment (h_{amb1}) is determined, which results from a radiation component ($h_{rad,amb1}$) and the convective component ($h_{conv,amb1}$). These are calculated as

$$h_{rad,amb1} = \sigma \cdot \varepsilon_1 \cdot \left(T_{g,1} + T_{amb}\right)\left(T_{g,1}^2 + T_{amb}^2\right), \tag{4.264}$$

where σ is the Stefan-Boltzmann's constant and $T_{g,1}$ is the outer surface temperature of the glass pane to the ambient, which is not constant anymore as in the case of radiative heat transfer. The convective part is calculated by

$$h_{g,amb1} = 0.86 \cdot \frac{u_0^{3/5}}{L^{2/5}} \quad \left(W/\left(m^2K\right)\right), \tag{4.265}$$

where u_0 describes the mean wind speed along the glazing, and L the lateral dimension of the window. $h_{g,amb1}$ shall not be below $5W/\left(m^2K\right)$ even is wind conditions may yield lower values as recommended by (Kalogirou, 2013). For standard conditions and vertical arrangement the European Norm for glass in buildings (European Standard EN673, 2011) has a standard value of $25W/\left(m^2K\right)$ for the purposes of comparison of glazing U-values, which does not consider the improvement of the U-value due to the presence of externally exposed coated surfaces with an emissivity lower than $\varepsilon_1 = 0.837$ (corresponding to corrected emissivity of soda lime glass).

Similar to the external heat transfer the inter heat transfer coefficient (h_{4i}) from the inner pane to the room can be evaluated, in which the radiative part is:

$$h_{rad,4i} = \sigma \cdot \varepsilon_4 \cdot \left(T_{g,4} + T_i\right)\left(T_{g,4}^2 + T_i^2\right), \tag{4.266}$$

while the European norm here simplifies to $h_{rad,4i} = 4.898 \cdot \varepsilon_4 W/\left(m^2 K\right)$. Also for the convective part ($h_{g,4i}$) the European Norm provides a constant value of $h_{g,4i} = 3.6W/\left(m^2 K\right)$, so that $h_{4i} = h_{rad,4i} + h_{g,4i} = 7.7W/\left(m^2 K\right)$.

The thermal conductivity of the gas in the gap between the panes ($h_{gas,23}$) is given by the following equation:

$$h_{gas,23} = Nu \cdot \frac{k_{gas}}{w}, \tag{4.267}$$

in which k_{gas} is the thermal conductivity of the gas in the gap and w the width of the gap interspace. The Nusselt number is a dimensionless parameter and has the physical meaning of the ratio between the magnitude of the heat flux transferred by convection and the magnitude of the heat flux transferred by thermal conductivity through the stationary gas filling the interlayer. To determine the Nusselt number, we actually need the basic equations of convective transport of a fluid, which are the subject of Chap. 5. Here, we will limit ourselves to quantifying the natural convective heat transfer expressed by the Nusselt number using correlations. The Nusselt number correlations themselves depend on the Rayleigh number (Ra) and the inclination angle of the pane gap described by the angle φ with respect to the gravity field vector g, so that $Nu = f(Ra, \varphi)$.

The Rayleigh number itself is the product of the Grashof number (Gr) and the Prandtl number (Pr), i.e. $Ra = (Gr \cdot Pr)$. The Grashof number (Gr) describes the ratio of the buoyancy force acting on a fluid element and the counteracting viscous fluid forces caused by the temperature difference between the panes ($T_2 - T_3$) or a heat flux ($\dot{q}$) applied

$$Gr = \frac{g \cdot \beta \cdot (T_2 - T_3) \cdot L^3}{\nu^2} = \frac{g \cdot \beta \cdot \dot{q} \cdot L^4}{\nu^2 k_{gas}}, \tag{4.268}$$

with L the length of the panes, ν the kinematic gas viscosity, β its volumetric expansion coefficient (for gases $\beta = 1/T = 1/(273.15 + (T_2 + T_3)/2)$), k_{gas} is thermal conductivity, and g the acceleration due to gravity.

The Prandtl number describes the ratio of the transport coefficient of the velocity field versus the temperature field and is defined as follows:

$$Pr = \frac{\rho \nu c_p}{k_{gas}} = \frac{\nu}{a}. \tag{4.269}$$

in which ρ denotes the specific density of the gas and c_p the specific heat capacity. Following the recommendations for the Nusselt-number correlation of the European- Norm, we get for glazings

$$Nu = A \cdot (Gr \cdot Pr)^n, \tag{4.270}$$

in which for vertical glazings $A = 0.35$ and $n = 0.38$, while for horizontal glazings $A = 0.16$ and $n = 0.28$; for $\varphi = 45°$ values of $A = 0.1$ and $n = 0.31$ are stated. Note, that these values contain quite rough approximations with errors in the range of $\pm 10\%$ with respect to the Nusselt number.

Let us first consider a vertical double glazing. For small temperature differences or narrow gaps between the panes, the temperature gradient is constant, and the temperature drops linearly to the colder side. Thus, the heat flow through the gap corresponds to the heat flow transferred by the thermal conductivity of the quiescent gas filling the gap. This mode, in which the Nusselt number takes the value of unity, i.e. $Nu = 1$, is called conductive. As the Rayleigh number increases, the intensity of the gas circulation increases and gradually deviates from the linear distribution. Thus, the Nusselt grows slowly. Initially, a stable laminar state is established, which is characterized by a single convection cell. From about a Rayleigh number of $Ra = 6 \cdot 10^3$, a second cell forms and the Nusselt number continues to increase. Each further increase in the Ra number leads to a gradually more complex flow structure and both the convection velocity and thus the Nusselt number increase continuously. A higher heat transfer from the hot pane towards the cold one expressed yields also to decrease of the heat resistance ($R_{conv,23}$), which is given by

$$R_{conv,23} = \frac{1}{Nu} \cdot \frac{w}{k_{gas}} \tag{4.271}$$

The resulting thermal heat resistance depends on the Nusselt number relationship, the temperature difference between the glass surfaces ($T_2 - T_3$), the thickness of the gas layer w, and the physical properties (type) of the gas, and the impact of the inclination φ with respect to the vertical plane which we subsequently analyse.

- Effect of the thickness of the gas layer (w) on its thermal resistance

The Fig. 4.44 shows the convective heat resistance ($R_{conv,23}$) using the Nusselt number relationship as a function of the distance w of both panes for a vertical double glazing filled with air, at a mean temperature of $10°C$ and a temperature difference between the inner pane surfaces of $(T_2 - T_3) = 20°C$. The curve consists of two sections. A linear section in which the heat resistance grows linearly with the gap width. This is followed by the section in which the heat resistance decreases by the onset of convection. The transition occurs at the point where the Nusselt-number relationship takes on the value of unity and represents the optimal pane distance (or optimal gas layer thickness) with the smallest convective heat losses. This optimal width is associated with a Rayleigh number domain in the range of $3 \cdot 10^3 \leq Ra \leq 10^4$. From this, one can already derive a measure for reducing convective heat transport. If individual elements are introduced into the glass gap, which reduce the active height for the formation of a natural convective flow, the Rayleigh number, which scales with the third (or fourth) power of the length L, is reduced, and Nu keeps small. Such measures are used, for example, in transparent thermal insulation. These so-called convection suppression devices (CSD) have been investigated very early in the solar collectors in order to suppress the convection in the air gab of a flat-plate collector. Parallel films (slats) or tube / honeycomb structures have been investigated as early as in the 1960s. The aspect ratios ($AR = L/w$) of these devices were in a range of $AR = 3 - 8$ and still showed some convection inside the cells structures. Later on with a further development of the so-called transparent insulation structures larger aspect ratios $AR > 10$ could be produced, which practically suppressed convection even for temperature differences between absorber and glazing of $80K$ and more. Cross section and total thickness of the structures proved also to be important (see Platzer, 1984).

- Impact of the thermophysical gas properties

Thermal resistance can be increased by using fill gases with lower thermal conductivity. Table 4.13 shows the physical properties of some such gases in a temperature range between -10 and $20°C$. The dependency of the heat resistance in relation to changed gas properties assuming a constant temperature difference of $20°C$ between the two glass interior surfaces is also shown in Fig. 4.46. If the heat conductivity of the gas decreases, i.e. from air via argon to krypton, the gradient of the heat resistance increases in the conductive (linear) heat transfer mode. However, the value of the optimal thickness of the gas layer decreases, since the maximum heat resistance is achieved for gases with lower heat conductivity at smaller gap widths. By reducing the optimal thickness of the gas layer, both the amount of expensive inert gas can be reduced, and also multi-chamber double-glazed windows of acceptable thickness can be produced.

- Dependence of the temperature difference of the glass surfaces ($T_2 - T_3$)

The optimum thickness of the gas gap (w_{opt}) between the panes is directly linked to the temperature difference between the disks via the Nusselt number relationship (Eq. 4.270), so that the following relation can be established,

$$w_{opt}^3 \cdot (T_2 - T_3) = Ra_{opt}\left(\frac{\nu^2}{g\beta Pr}\right) = C, \tag{4.272}$$

where the constant C depends on the type of gas. Thereby, an increase in the temperature difference between the glasses leads to a decrease in the optimum distance between them, which is caused by an enhanced convective heat transfer.

Table 4.13 Thermophysical values of some gases from (European Standard EN673, 2011)

Gas $T[°C]$	density ρ $[kg/m^3]$	dynamic viscosity μ $[kg/(ms) \cdot 10^5]$	thermal conductivity k_{gas} $[mW/(mK)]$	spec. heat conductivity c_p $[J/(kgK)]$
Air				
−10	1.326	1.661	23.36	1008
0	1.277	1.711	24.16	
10	1.232	1.761	24.96	
20	1.189	1.811	25.76	
Argon				
−10	1.829	2.038	15.84	519
0	1.762	2.101	16.34	
10	1.699	2.164	16.84	
20	1.640	2.228	17.34	
Krypton				
−10	3.832	2.260	8.42	245
0	3.690	2.330	8.70	
10	3.560	2.400	9.00	
20	3.430	2.470	9.26	

In this case, the corresponding value of $R_{conv,23,opt}$ also decreases.As a result, at $(T_2 - T_3) \geq 25°C$, the distance between the glasses will practically no longer affect the value of thermal resistance. Such undesirable effect of temperature difference on the value of thermal resistance can be reduced if the gas layer with glass is divided into two parts of equal thickness. Then half of the total temperature difference is realized on each of the halves. This is another way to increase the thermal resistance ($R_{conv,23,opt}$) of convective heat transfer through the glass.

- Effect of pane inclination φ on the convective heat transfer resistance

The main effect of inclination of the double glazing φ to the gravity vector g is the reduction of the effective height above which a convection flow is formed. In the case of a horizontally arranged double glazing, i.e. $\varphi = 0°$, the effective height for determining the Rayleigh number corresponds only to the gap width w, while in the purely vertical arrangement considered so far ($\varphi = 90°$) it corresponds to the complete pane height L. Thus, naturally, higher Rayleigh numbers result from an increasing φ according to the sine function of φ, and the convective heat transport between the panes reaches a maximum with a vertical arrangement of the double glazing. This is a complex problem of fluid mechanics, which is usually solved computationally with the help of CFD (see Chap. 5). Nevertheless, there are a number of Nusselt number correlations in the literature, which are supported by numerous experiments. Table 4.14 gives some of the most commonly used relations, more details can be found in Summ et al. (2021), Stephan et al. (2019) and others.

The solution of the whole combined radiation, heat conduction and convection problem consisting of Eqs. (4.260)–(4.263) to determine the combined heat transfer through a single-chambered glass package is iterative. First, we will determine the thermal resistance of the convective heat transfer through the intermediate gas layer $R_{conv,23}$ for which we need the temperatures of the inner surfaces of the glasses T_2 and T_3. First of all, all constant parameters have to be taken into account, such as w, T_i, T_{amb}, $k_{conv,amb1}$, $k_{konv,4i}$, d_j, k_j, ε_i, ε_{ij}, as well as the properties of the filling gas. In a first approximation, we choose arbitrary values for the temperature of the inner surfaces of the glasses T_2 and T_3 from the interval between T_i and T_{amb}. For example, for glasses we can use the condition of equality of temperature drops in all sections: $(T_2 - T_3) = (T_2 - T_{amb}) = (T_i - T_3)$, from which we obtain the average temperature of the gas layer $(T_2 - T_3)/2$. For this temperature we obtain Ra, Nu and the value $R_{conv,23}$. Next, we calculate the individual components and the total thermal resistance of the convective heat transfer and obtain the value of the heat flux from which we obtain the temperatures of all glass surfaces as a first approximation. The values of the temperature of the inner surfaces of the glasses T_2 and T_3 obtained in the first approximation are used in the second approximation for repeated calculations and the loop is repeated until in two successive approximations the values do not change within a previously determined error interval, usually the third decimal digit is used for this.

The double glazing is a similar problem as a solar thermal collector. In the solar collector also a coupling between convection and the radiation transport is possible if for example polymer films as internal glazings are coming too close to a selectively emitting absorber, or if transparent insulation materials (see Chap. 6) like honeycomb structures or aerogel, which are partially transparent to infrared radiation are used to improve the heat losses in the air gap. Simplified methods to solve the coupled equations have been developed alongside these developments, and some details are discussed for example in (Pflüger & Wittwer, 1984; Platzer, 1988, 1992).

One important result is that by using an appropriate air gap in this case between the selective absorber and the absorbing-emitting structure, the heat transport modes may be effectively decoupled, with the results that a much lower combined heat transfer can be observed, which is close to the one calculated with the simplified additive superposition. Other research where coupled heat transfer of both modes, radiation and convection occurs, is the energy loss of stellar atmospheres. Although combined heat transfer is needed in many practical problems in engineering, the treatment still seems to be a challenge in many cases (Dombrovsky, 2022). Simplified superposition is not always correct.

Table 4.14 Nusselt number correlations for inclined cavities modified from Summ et al. (2021), Stephan et al. (2019) von Boeckh & Wetzel (2011), $Ra_\varphi = Racos\varphi$; value of the terms indicated by "+"only considered if > 0, else being 0; $\varphi > 90°$ heated from above

Source	Nu correlation	$Ra/Ra_\varphi/Gr$ [/]	φ [°]	valid AR [/]
Ozoe et al. (1975)	$Nu = 0.109 Ra_\varphi^{1/3}$ for $T = const$	$10^3 < Ra < 10^5$	< 10	8–16
Hollands et al. (1976)	$Nu = 1 + 1.446\left(1 - \frac{1708}{Ra_\varphi}\right) \cdot \left(1 - 1708 \frac{sin(1.8\varphi)^{8/5}}{Ra_\varphi}\right)^+ + \left(\left(\frac{Ra_\varphi}{5830}\right)^{1/3} - 1\right)^+$ for $T = const.$	$Ra < 10^6$	$0 \leq \varphi \leq 60$	50
Randall et al. (1979)	$Nu = 0.118 Ra_\varphi - 4520 - 29$ for $T = const.$	$2 \cdot 10^3 < Ra < 2.2 \cdot 10^5$	$45 \leq \varphi \leq 90$	9–36
Schinkel (1980)	$Nu = C Ra^b$ $\varphi = 0°C = 0.08$; $\varphi = 20°C = 0.075$; $\varphi = 40°C = 0.074$; $\varphi = 60°C = 0.072$; $\varphi = 90°C = 0.062$; $b = 1/3$ for $T = const.$ $b = 1/4$ for $\dot{q} = const.$	$10^5 < Ra < 4 \cdot 10^6$	$0 \leq \varphi \leq 90$	6–27
Inaba (1984)	$Nu = 1 + 1.21\left(1 - \frac{2500}{Ra_\varphi}\right)$ $2.5 \cdot 10^3 < Ra < 6 \cdot 10^3$ $Nu = 0.199 Ra_\varphi^{0.258}$ $6 \cdot 10^3 < Ra < 4 \cdot 10^5$ $Nu = 0.785 Ra_\varphi^{0.33}$ $4 \cdot 10^5 < Ra < 1.2 \cdot 10^6$ $Nu = 0.271 AR^{-0.21} (Rasin\varphi)^{1/4}$ $5 \cdot 10^3 < Rasin\varphi < 1.2 \cdot 10^6$ and $60 \leq \varphi \leq 120$	$2.5 \cdot 10^3 < Ra < 1.2 \cdot 10^6$	$60 \leq \varphi \leq 120$	5–83

(continued)

Table 4.14 (continued)

Source	Nu correlation	$Ra/Ra_\varphi/Gr$ [/]	φ [°]	valid AR [/]
ElSherbiny (1996)	$Nu_{180} = \left(1 + (0.212 Ra^{0.136})^{11}\right)^{1/11}$ $Nu_{120} = \left(1 + (0.0566 Ra^{0.332})^{4.76}\right)^{1/4.76}$ $Nu = Nu_{180} + \frac{180-\varphi}{60}(Nu_{120} - Nu_{180})$	$10^2 < Ra < 10^6$	$120 \leq \varphi \leq 180$	20 – 80
Matuska (2009)	$Nu = (0.1464 - 2.602 \cdot 10^{-4}\varphi - 2.046 \cdot 10^{-6}\varphi^2) Ra^{0.29}$	$10^4 < Ra < 2 \cdot 10^6$	$0 \leq \varphi \leq 90$	n.a
Eismann (2015)	$Nu = 1 + Nu_1 + Nu_2$ $Nu_1 = 1.44\left(1 - \frac{1708}{Ra+1708R_c}\right)\left(1 - (sin 1.8\varphi)^{1.6} \frac{1708}{Ra+1708R_c}\right)$ $Nu_2 = \left(\left(\frac{Ra+5830R_c}{5830}\right)^{0.39} - 1\right)(1 + CR_c)$		$10^3 < Ra < 10^6$	n.a

4.4.4 Effective Thermal Conductivity of Packed Beds

Packed beds consisting of spheres, fibres, or irregular other geometric bodies find engineering applications in many places, including solar thermal issues. Excellent examples are the insulating materials already mentioned in Sect. 4.2.4.4. Other potential applications include thermal heat storage systems, where a solid filling material is used as a heat storage material, which is loaded or unloaded with the help of a heat transfer fluid, for example in a thermocline. As we have seen from the example of insulating materials, the effective thermal conductivity (ETC) does not correspond to that of either the solid or the gas. All these examples are combined effects of heat conduction, convection, and heat radiation.

The most common model for determining the effective thermal conductivity of a packed bed is based on combining all relevant heat transfer mechanisms into a single representative value, the effective thermal conductivity of a packed bed (k_{bed}), which is composed of three contributing components.

The first component is the effective thermal conductivity (k_{eff}) of the bed with a fluid at rest, which in turn includes four different heat transfer mechanisms:

- heat conduction through the solid particle;
- heat conduction through the contact area between adjacent particles, considering the surface roughness;
- heat conduction through the stagnant liquid/gas phase; and
- thermal radiation between solid surfaces.

The second component is the increased effective thermal conductivity of the gas ($k_{g,eff}$), which is due to convective fluid motion flowing through the interstices of the packing parallel to the wall while the solid phase is at rest.

The third component occurs when both the fluid and solid phases are in motion ($k_{s,eff}$), such as during particle transfer in particle receivers. This increases heat transfer in the solid phase due to particle motion of the particles.

As a result, for the effective thermal conductivity of the packed bed (k_{bed} often also referred to as ETC), one obtains a (simplified) superposition of the three components given by:

$$k_{bed} = k_{eff} + k_{g,eff} + k_{s,eff}. \tag{4.273}$$

Due to the increased computer capacities, it is possible to simulate the heat transfer properties of packed beds with the help of the discrete element method (DEM), in which the heat transport over the contact surfaces and the exchange of radiation between the individual particles i and j is determined, whereby the walls and the flow through them are considered as boundary conditions. Since the packed bed is usually a mixture of solid particles of different size and thus porosity and the flowing fluid, however, an exact heat transfer model for the bed is also numerically a very complex and computationally expensive problem.

When considering heat transfer in a packed bed at the micro level, one possible assumption is that each individual particle i has a characteristic temperature T_i. As a consequence, the heat transfer equation for particle i can be written as

$$M_i \cdot c_{p,i} \frac{dT_i}{dt} = \sum_{i}^{n_m} \dot{Q}_{ij} + \dot{Q}_i, \tag{4.274}$$

where M_i and $c_{p,i}$ represent the particle mass and specific heat capacity, respectively, and Q_i denotes the volumetric heat output originating from the internal heater, while $\dot{Q}_{ij}$ is the heat flux transferred through each neighboring particle j. $\dot{Q}_{ij}$ is then summed over all n_m particles adjacent to particle i. Consequently, the heat transfer of the entire granular system results from a system of linear equations for all particles as well as additional thermal equations describing the initial and boundary conditions.

Before that, however, the we´ve to determine some unknown geometrical quantities, like the arrangement (packing density-porosity), the shape of the single particles, their size distribution and contact areas, but also a model for the thermal contact to determine the unknown $\dot{Q}_{ij}$, which scopes conduction, convective and radiative heat transport.

Geometric Parameters of a Packed Bed

One of the first bed quantities is the porosity, i.e. the gas volume fraction within a total volume. This void fraction (μ) determines not only the packing density but also, in convective problems, significantly the pressure drop through the bed and is defined as follows:

$$\mu = V_g / V_T = V_g / \left(V_g + V_s\right), \tag{4.275}$$

where V_g is the volume of void-space (such as fluids/gases) and V_T is the total or bulk volume of material, including the solid and void components $\left(V_g + V_s\right)$.

From the knowledge of the void fraction, the effective heat capacity of the packed bed is automatically calculated as the weighted average between the heat capacity of the liquid and the solid filler, reading to:

$$\left(\rho \cdot c_p\right)_{eff} = \mu\left(\rho \cdot c_p\right)_g + (1 - \mu) \cdot \left(\rho \cdot c_p\right)_s, \tag{4.276}$$

in which ρ is the specific density and c_p the specific heat capacity.

Far from being constant, the porosity of a packed bed varies from the wall to the centre of a bed. In addition, it matters whether the bed is densely packed or loose random packed. One relevant quantity required for the bed type is given by the coordination number which describes the number of particles in contact with the particle under consideration. This is relevant to model the heat transport phenomena through packed beds. Typical coordination numbers for dense sphere packings are between 5 and 6 in the bed bulk with a porosity (void fraction) of

$\mu = 0.39$. Detailed correlations to coordination numbers and the corresponding porosities can be found e.g. in (van Antwerpen et al., 2010).

Heat Transfer Modelling

In general the heating power between the particles i and j, $\dot{Q}_{ij}$ can be expressed as

$$\dot{Q}_{ji} = -\dot{Q}_{ij} = -H_{ij}(T_i - T_j), \tag{4.277}$$

In which H_{ij} is the effective heat conductance $[W/K]$ between particles i and j. The total heat transfer is then described by:

- heat conduction through the solid and fluid phase;
- heat conduction through the contact area of the of solid between particle i and j;
- radiant heat transfer between solid surfaces in case of gases
- radiant heat transfer between adjacent voids in gases; and the
- convective heat transfer, including the solid–fluid convection and the fluid mixing convection.

This allows us to represent the effective thermal conductance (H_{ij}) exactly in the following way,

$$H_{ij} = H_{ij}^{t} + H_{ij}^{c} + H_{ij}^{rad}, \tag{4.278}$$

where H_{ij} is composed of the solid contact heat conduction fraction H_{ij}^{t}, the solid–fluid solid conduction fraction H_{ij}^{c} and the radiation fraction H_{ij}^{rad}. For the exact description, the consideration of thermal contacts between neighboring particles with separating gaps is relevant, since the solid–fluid-solid heat conduction and the radiation part occur, although the heat conduction through contacting solids does not contribute.

Solid–solid Contact of Pebbles

A solid-state heat conduction exists between two contacting particles with a contact area, as shown in Fig. 4.47a. Since the temperature distribution in the region near the contact is significantly different from that in the region far from the contact, the solid contact conduction must be considered. There is an analytical approximate solution for the effective thermal conductivity H_{ij}^{t} between two smooth elastic spheres with a contact area of radius r_t, which can be written as follows

$$H_{ij}^{t} = 2r_t k_s = \frac{4\sqrt{R_e \delta_n}}{1/k_{si} + 1/k_{sj}} = \frac{4}{1/k_{si} + 1/k_{sj}} \left(\frac{3F_n R_e}{4E_e}\right)^{1/3}, \tag{4.279}$$

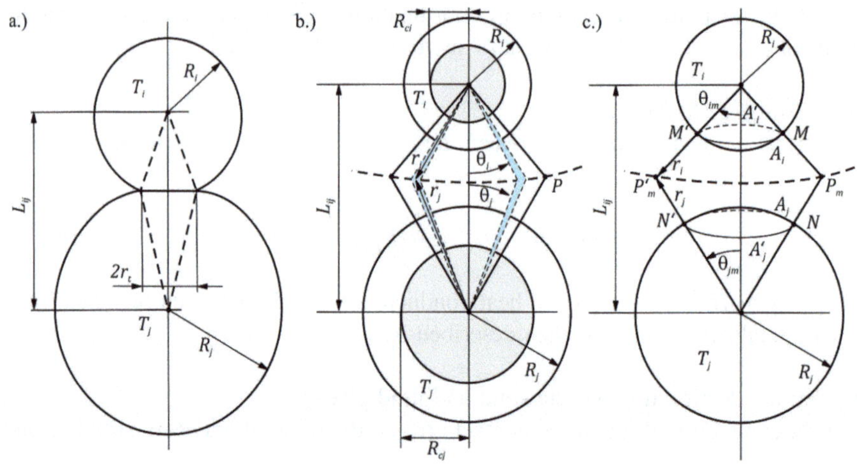

Fig. 4.47 Sketch of the heat transfer between two neighboring particles. **a** solid-state heat conduction via particle contact; **b** heat transfer via solid–fluid-solid conduction; **c** heat transfer between particles via radiation

in which E_e, R_e and δ_n are given by

$$E_e = \left(\frac{1-\nu_i^2}{E_i} + \frac{1-\nu_j^2}{E_j}\right)^{-1}; R_e = \left(\frac{1}{R_i} + \frac{1}{R_j}\right); \delta_n = R_i + R_j - L_{ij}, \quad (4.280)$$

where k_s; ν, E and R are the conductivity, Poisson's ratio, Young's modulus, and radius of the particles, respectively, and the subscript i and j indicates the corresponding particle number. F_n describes the normal contact force caused by the normal overlap dn. E_e and R_e are the equivalent modulus and radius defined by Hertzian contact theory. The solid heat conductivity k_s in Eq. (4.279) is replaced by $2k_{si}k_{sj} = (k_{si} + k_{sj})$ to account for particles with different thermal properties. Equation (4.279) shows the relationship between the conductance and the normal contact force, which allows the exact model to account for the effects of an external mechanical load on the effective thermal conductivity.

Solid–fluid-Solid Contact

In most liquid-filled packed beds, the solid phase has a greater thermal conductivity than the liquid phase, so that the temperature distribution over the solid phase is more uniform, affecting in particular the core part of the particle (grey background in Fig. 4.47b). Therefore, it can be assumed that each particle has an isothermal core part with radius R_c with a representative temperature. For bulk beds of the same particle size, the gaps are small, while for beds with different particle sizes, the gaps are larger. As an approximation to the far-field boundary condition, one assumes a fluid bed at rest in a large zone with a radius 30-times the maximum particle radius. The heat flux by solid–fluid-solid heat conduction from the particle centroid of higher temperature to the particle centroid of lower temperature passes through a point P, as shown in Fig. 4.47b, which is equidistant from both particle surfaces. Based on the above assumptions of an isothermal particle core and heat flow path, the heat flow $d\dot{Q}_{ij}^{c}$ through the infinitesimal volume inside the smaller sphere i shown in Fig. 4.47b)is given by

$$d\dot{Q}_{ij}^{c} = 2\pi r^2 sin\theta_i d\theta_i \left(-k_s \frac{dT}{dr}\right), \tag{4.281}$$

θ_i is the angle between the central axis and the investigated heat flux path and r is the distance from the corresponding particle center. By integrating from the edge of the core part ($r = R_{ci}$; $T = T_i$) to the outer boundary of the smaller particle ($r = R_i$; $T = T_{\theta,\,i}$), we obtain the heat flux to particle j; and doing this in a similar way for the opposite direction, we finally obtain the fraction of differential heat flux dQ_{ij}^{c} between particles i and j

$$d\dot{Q}_{ij}^{c} = \frac{2\pi k_g (T_i - T_j)}{\frac{1/R_{ci} - 1/R_i}{K_{si} sin\theta_i d\theta_i} + \frac{1/R_i - 1/r_i}{sin\theta_i d\theta_i} + \frac{1/R_j - 1/r_j}{sin\theta_j d\theta_j} + \frac{1/R_{cj} - 1/R_j}{K_{sj} sin\theta_j d\theta_j}}, \tag{4.282}$$

with the details of the calculation can be found in the article of (Chen et al., 2019). where $K_{si} = k_{si}/k_g$ and $K_{sj} = k_{sj}/k_g$ denote the dimensionless solid conductivities of particles i and j, respectively. Applying the cosine law for the triangle spanned by the points O_i, O_j and P from Fig. 4.47b yields the following equations.

$$cos\theta_i = \frac{r_i^2 + L_{ij}^2 - r_j^2}{2r_i L_{ij}} = \frac{1 - \Delta\alpha^2}{2\delta_i}; \; cos\theta_j = \frac{r_j^2 + L_{ij}^2 - r_i^2}{2r_j L_{ij}} = \frac{1 - \Delta\alpha^2}{2(\delta_i + \Delta\alpha)} + \Delta\alpha, \tag{4.283}$$

in which $\delta_i = r_i/L_{ij}$, $\delta_j = r_j/L_{ij}, \alpha_i = R_i/L_{ij}, \alpha_j = R_j/L_{ij}$ and $\Delta\alpha = (r_j - r_i)/L_{ij} = (R_j - R_i)/L_{ij}$ are the dimensionless length scales of the different radii and the particles radius difference. By this dimensionless parameter Eq. (4.282) mutates

$$dQ_{ij}^{c} = \frac{\pi k_g L_{ij} (T_i - T_j)(1 - \Delta\alpha^2) d\delta_i}{(A + B)\delta_i^2 - 2(1 - \Delta\alpha B)\delta_i - (1 - \Delta\alpha B)\Delta\alpha}, \tag{4.284}$$

in which

$$A = \frac{1}{\alpha_i K_{si}}\left(\frac{R_i}{R_{ci}} - 1\right) + \frac{1}{\alpha_i}; \; B = \frac{1}{\alpha_j K_{sj}}\left(\frac{R_j}{R_{cj}} - 1\right) + \frac{1}{\alpha_j}, \tag{4.285}$$

The integration of Eq. (4.282) yields an analytic solution in the following way:

$$H_{ij}^{c} = \frac{\dot{Q}_{ij}^{c}}{(T_i - T_j)} \begin{cases} \frac{\pi k_g L_{ij}(1-\Delta\alpha^2)}{2\sqrt{|\Lambda|}} ln\left|\frac{1-Y_1}{1+Y_1}\right| & \text{for } \Lambda > 0 \\ \frac{\pi k_g L_{ij}(1-\Delta\alpha^2)}{A+B}\left(\frac{1}{\delta_{imin}} - \frac{1}{\delta_{imax}}\right) & \text{for } \Lambda = 0\,, \\ \frac{\pi k_g L_{ij}(1-\Delta\alpha^2)}{\sqrt{|\Lambda|}} arctan Y_2 & \text{for } \Lambda < 0 \end{cases} \tag{4.286}$$

with

$$Y_1 = \frac{X_{max}-X_{min}}{1-X_{max}X_{min}};\ Y_2 = \frac{X_{max}-X_{min}}{1+X_{max}X_{min}},$$
$$X_{max} = \frac{\Delta\alpha B-1+(A+B)\delta_{imax}}{\sqrt{|\Lambda|}};\ X_{min} = \frac{\Delta\alpha B-1+(A+B)\delta_{imin}}{\sqrt{|\Lambda|}}, \tag{4.287}$$

where $\Lambda = ((1+\Delta\alpha A)/(1-\Delta\alpha B))$ represents the discriminant for the integration. The upper limit of the integral δ_{imax} is directly coupled with the nominal area A_n at the interface can be expressed using a geometric relationship as follows:

$$\delta_{imax} = 1/2\left(\sqrt{1+\frac{4A_n}{\pi L_{ij}^2(1-\Delta\alpha^2)}} - \Delta\alpha\right). \tag{4.288}$$

The lower integral value δ_{imin} is given by the direct particle contact so that δ_{imin} can be calculated as follows.

$$\delta_{imin} = 1/2\left(\sqrt{1+\frac{4r_t^2}{L_{ij}^2(1-\Delta\alpha^2)}} - \Delta\alpha\right). \tag{4.289}$$

where the contact radius r_t can be solved from Eq. (4.279). For spherical particles of the same size ($R_i = R_j$ and $\Delta\alpha = 0$), the solution of the Cheng-Yu -Zulli (Cheng et al., 1999) model is obtained directly.

Radiation Heat Transfer

In order to determine the radiative heat transfer through a packed bed, we must first make some assumptions to make the physical description computationally accessible. These are:

- All particles in the bed consist of opaque spheres whose size is larger than the wavelength of the radiation
- All particle surfaces should be grey radiating so that E is not a function of solid angle.
- The temperature gradient with respect to the absolute temperature should be much smaller than one in the packed bed, i.e. $\Delta T/T \ll 1$;
- Radiative heat transfer between two adjacent particles takes place in a cone spanned by an imaginary re-radiating surface spanned by the points $MP_mNN'P_mM'$ as shown in Fig. 4.47c.

Given these assumptions we can immediately calculate the radiative heat flux between i and j which reads to

$$\dot{Q}_{ij}^{rad} = \frac{\sigma\left(T_i^4 - T_j^4\right)}{\frac{1-\varepsilon_i}{\varepsilon_i A_i} + \frac{1}{A_i F_{ij} + [1/(A_i F_{iR}) + 1/(A_j F_{jR})]} + \frac{1-\varepsilon_j}{\varepsilon_j A_j}}, \quad (4.290)$$

in which ε_i and ε_j are the emissivty of the partcles i and j and $A_i = 2\pi R_i^2(1 - cos\theta_i)$ and $A_j = 2\pi R_j^2(1 - cos\theta_j)$ describe the areas through which the radiation heat exchange occurs, i.e. MM' and NN'. The variables F_{ij}, F_{iR}, and F_{jR} describe the view factors of the surface pairs MM' and NN', MM' and $MP_mNN'P_m'M'$, and NN' and $MP_mNN'P_m'M'$. Using the reciprocity rule of the view factor we get

$$A_i F_{iR} = A_i(1 - F_{ij}); A_j F_{jR} = A_j(1 - F_{ji}) = A_j - A_i F_{ij}. \quad (4.291)$$

Inserting Eqs. (4.291) in (4.290) we obtain

$$H_{ij}^{rad} = \frac{\dot{Q}_{ij}^{rad}}{(T_i - T_j)} = \frac{\sigma A_i\left(T_i^4 - T_j^4\right)/(T_i - T_j)}{\frac{1-\varepsilon_i}{\varepsilon_i} + \frac{1-\varepsilon_j}{\varepsilon_j A_j/A_i} + \frac{1+A_j/A_i-2F_{ij}}{A_j/A_i - F_{ij}^2}}$$
$$\approx \frac{4\sigma A_i T_i^3}{\frac{1-\varepsilon_i}{\varepsilon_i} + \frac{1-\varepsilon_j}{\varepsilon_j A_j/A_i} + \frac{1+A_j/A_i-2F_{ij}'}{A_j/A_i - F_{ij}^2}} \quad (4.292)$$

while the approximation is justified only if the temperature difference is small.

By summing the heat fluxes over all particles at a given externally applied heat flux density $\dot{q}$ or a temperature difference ΔT, the total heat flux through the packed particle bed is obtained, which is characterized by the geometrical parameters such as packing density (porosity-μ), the coordination parameter and the particle size distribution. The method is indispensable if one has a complex particle bed and is dependent on an accurate determination of the effective thermal conductivity, since the temperature differences lead to a thermal expansion of the particles and thus, depending on the temperature, the heat conduction through point contacts but also the other heat conduction mechanisms change. Figure 4.48 shows a result of such a DEM calculation in a vertical cube with a heat flux $\dot{q}$ $(100W/m^2)$ applied at the top and a lower constant wall temperature $T_w = 421°C$, for a ceramic bed (Li_2TiO_3 particles of a size $d_p = 1.91mm$) and a packing density $\mu = 0.385$. A comparison with the measured data shows a deviation of max. 16%, which occurs especially near the wall.

The complete numerical determination is very time-consuming and only makes sense if the safety behaviour of the system is also connected with the bed behaviour, as for example in nuclear applications.

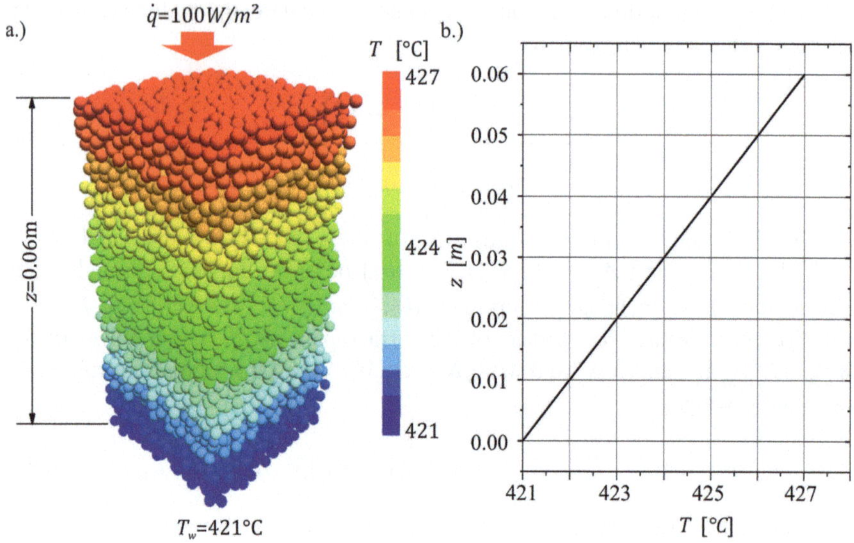

Fig. 4.48 Computed temperature distribution in a monosized spherical particle bed (Li_2TiO_3, $d_p = 1.91mm$, $\mu = 0.385$, $\varepsilon = 08$) exposed to a heat flux $\dot{q}$ ($100W/m^2$) with a constant bottom wall temperature $T_w = 421°C$ with helium as gas adapted from (Chen et al., 2019)

For pebble beds subjected to small mechanical loads, such as in thermal storage or insulation materials, however, the determination of the effective thermal conductivity (k_{eff}) can be substantially simplified by imagining the individual heat transfer mechanisms as analogous to an equivalent electrical circuit. This concept, shown in Fig. 4.49b, considers the entire heat transfer in the gas and solid phases as electrical resistors connected in parallel. The heat transfer in the solid consists of the heat conduction across the contact surface of the particles, the heat conduction through the stagnant gas layer, and the radiative transfer which is in series with the heat conduction through the spheres. In parallel to this is the heat conduction in the gas as well as the radiative component. This cell-based approach to the evaluation of the effective thermal conductivity (k_{eff}) leads to the following equation:

$$k_{eff} = \mu\left(k_g + h_{rad,v}\Delta L\right) + \frac{(1-\mu)}{\frac{1}{\frac{k_g}{l_v}+h_p+h_{rad,s}} + \frac{l_s}{k_s}}; \quad (4.293)$$

in which k_g and k_s describe the thermal conductivities of the gas and the solid; $h_{rad,v}$ and $h_{rad,s}$ reflect the heat transfer coefficients for the thermal radiation from void to void and between solid surfaces. ΔL denotes the effective length between the centres of two adjacent particles along the main direction of the heat flux, l_v specifies the thickness of a film of fluid which yields the same heat resistance as the stagnant fluid close to the contact point, l_s is the thickness of a slab of solid

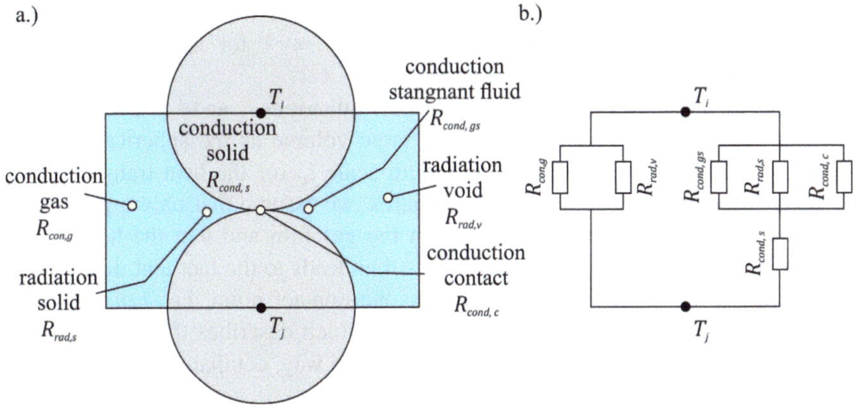

Fig. 4.49 Presentation of a unit cell with the heat transfer mechanisms involved (**a**) and transfer to an equivalent electrical circuit (**b**)

providing the same heat transfer resistance as the pebble and h_p is the heat transfer coefficient for the heat transfer through the contact surfaces between adjacent particles.

This Eq. (4.293) implies that one can estimate the effective thermal conductivity of the entire packed bed from the effective conductivity of a single contact within a uniform bed, which is reasonable because most thermal contacts in a bed of uniformly sized spherical particles with low mechanical loading have similar effective conductivities, and thus the entire bed has the same properties for all local contacts. This also explains why these cell-based theoretical models for predicting the effective thermal conductivity of pebble beds with the spherical sized particles, such as the model of (Kunii & Smith, 1960), are so successful. Ultimately, the problem can thus be reflected in form of a unit cell as illustrated in Fig. 4.49a.

The contribution of thermal radiation is taken into account by means of the heat transfer coefficients $h_{rad,v}$ and $h_{rad,s}$, which are defined as follows

$$h_{rad,v} = \left(4\sigma T^3 + \frac{\mu}{2(1\quad\mu)}\frac{(1-\varepsilon)}{\varepsilon}\right); h_{rad,s} = 4\sigma T^3\left(\frac{\varepsilon}{2-\varepsilon}\right), \tag{4.294}$$

in which ε denotes the emissivity of the particle surface, while σ is the Stefan-Boltzmann constant given by $5.67 \cdot 10^{-8} W/\left(m^2 K^4\right)$.

Using the electrical equivalent circuit (i.e., the notion of unit cell), one can represent the effective thermal conductivity in the following way,

$$\frac{k_{eff}}{k_g} = \mu\left(1 + \frac{h_{rad,v}}{k_g}\beta d_p\right) + (1-\mu)\frac{\beta}{\frac{1}{\frac{d_p}{l_v} + \frac{d_p}{k_g}(h_p + h_{rad,s})} + \frac{l_s}{d_p}\left(\frac{k_g}{k_s}\right)}, \tag{4.295}$$

where β is defined in the following way $\beta = \Delta L/d_p$, where β is in the range between 0.895 for dense sphere packing and $\beta = 1$ for loose packing, see (Trevisan et al., 2021).

What is unknown in Eq. 4.290 so far are the quantities l_s and l_v. Assuming that l_s is the length of a cylinder that has the same volume as the spherical particle, we get $l_s = 2/3d_p$. To determine the length scale l_v for the heat transfer of the gas phase near the contact point of the spheres, we assume that only a part of the total heat flux $(1/n)$ is transferred through the gas film and that the heat flux is parallel to the axis of the two particles. This then leads to the fact that this fraction corresponds to an angle θ associated with one contact point, i.e. $1/n = sin^2\theta$. Thus, a distribution function ϕ can be derived which describes the heat flux over the near-contact fuid region, given in the following way, compare (Kunii & Smith, 1960).

$$\phi = \frac{1}{2}\frac{\left(\frac{K-1}{K}\right)^2 sin^2\theta}{\ln(K-(K-1)cos\theta)-\left(\frac{K-1}{K}\right)(1-cos\theta)} - \frac{2}{3K}, \tag{4.296}$$

in wich K is $K = k_s/k_g$.

What is missing at last is an information about the packing state of the particles, which reflects the number n of contact points. For the densest possible sphere packing one obtains $n = 4\sqrt{3}$ while a loose packing has $n \sim 1.42$ only. This gives $\mu_1 = 0.476$ for a loose packing and $\mu_2 = 0.26$ for a dense ball packing. For real packed beds, it can be assumed that the actual one is a mixture of the densest and the loose ball packing. Since ϕ is proportional to the thermal resistance, the real value for such a bed should approximately correspond to an additive function of ϕ_1 (loose packing) and ϕ_2 (dense packing), so that for a spherical packed bed with a given void fraction μ, the following relation is obtained for the distribution function ϕ:

$$\phi = \phi_2 + (\phi_1 - \phi_2)\frac{\mu - \mu_2}{\mu_1 - \mu_2}. \tag{4.297}$$

The effective thermal conductivities (k_{eff}) determined using the unit cell model show a slightly larger deviation ($\pm 25\%$) than the exact DEM models when compared to experimental data but are conservative in an engineering sense. Figure 4.50 shows experimental results of a graphite pebble bed ($\varepsilon = 0.8$; $\mu = 0.385$) consisting of monodisperse spheres of a diameter of $d_p = 0.06m$ in helium ($k_s/k_g \approx 135$) with the model calculation.

This allows us to analyse qualitatively the influence of different bed parameters on the effective thermal conductivity, which is a function of the bed packing (ϕ, μ), the particle diameter (d_p), the emissivity (ε), the ratio of solid to gas thermal conductivity (k_s/k_g). Details on the influence of these parameters, as well as a set of correlations and their constraints for the determination of packed bed parameters, radiation exchange factors, mean coordination number as a function of porosity can be found in the review studies of van Antwerpen et al. (2010), Chen et al. (2019). Here is just some general information provided.

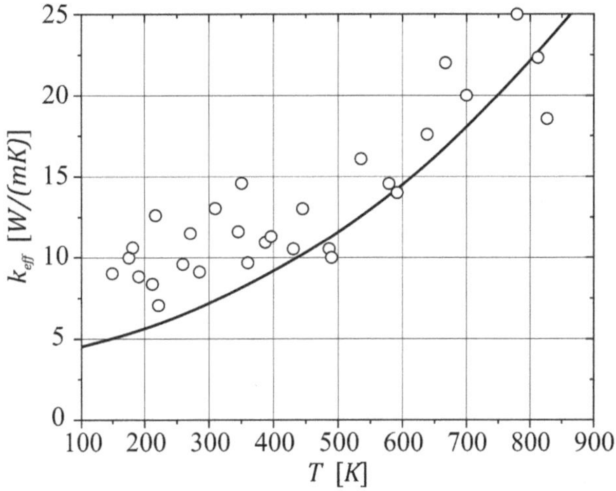

Fig. 4.50 Comparison of experimental (Niessen & Ball, 2000) and predicted (Kunii & Smith, 1960) effective thermal conductivity of a mono-sized spherical graphite particle bed ($\varepsilon = 0.8$; $\mu = 0.385, d_p = 0.06m$) for varying temperatures in helium ($k_s/k_g \approx 135$)

It is obvious that the radiation exchange increases with increasing emissivity (ε), so that a high emissivity is desirable for a high effective thermal conductivity (k_{eff}). What is not so obvious is that the interaction between thermal conduction and radiation can be neglected when the ratio between the conductivity of the solid and that of the gas k_s/k_g is less than 100. For $k_s/k_g > 100$, both the heat conduction and radiation fluxes are affected by the k_s/k_g ratio, and quantitatively in different ways.

The relative contributions of each heat transfer mechanism to the total heat transfer also depend on the particle diameter. For a small particle diameter, the contributions due to conduction are larger than the heat transfer due to radiation and are only slightly dependent on the bed temperature. Conversely, heat transfer by radiation increases rapidly with increasing particle diameter and overtakes heat transfer by conduction; for particles of $2cm$ (and $\varepsilon = 0.8$), a relevant heat radiation contribution starts at about $400^\circ C$. The effective thermal conductivity also increases with increasing bed temperature, mainly because of the increasing contribution of radiation. In packed beds consisting of two components with the same size distribution but different materials, the effective thermal conductivity increases with increasing volume fraction of the particles with the higher thermal conductivity, because they contribute more to the k_{eff} compared to the particles with the lower thermal conductivity.

References

Alados, I., Foyo-Moreno, I., & Alados-Arboledas, L. (2012). Estimation of downwelling longwave irradiance under all-sky conditions. *International Journal of Climatology, 32*(5), 781–793.

Alduchow, O., & Eskridge, R. (1996). Imporved Magnus form approaximation of saturation vapr pressure. *Journal of Applied Meteorology, 35*(4), 601–609.

Baehr, H. D., & Stephan, K. (2008). *Wärme- und Stoffübertragung* (6th ed.). Springer.

Berdahl, P, Fromberg, R. (1982). The thermal radiance of clear skies. *Solar energy, 29*, 299–314.

Bird, G. A. (1994). *Molecular gas dynamics and the direct simulation of gas flows.* Oxford University Press.

Bliss, R. (1961). Atmospheric radiation near the surface of the ground: A summary for engineers. *Solar Energy, 5*, 103–120.

Carslaw, H., & Jaeger, J. C. (1986). *Conduction of Heat in Solids* (2nd ed.). Oxford University Press.

Chen, l., Wang, C., Moscardini, M., Kamlah, M., & Songlin, L. (2019). A DEM-based heat transfer model for the evaluation of effective thermal conductivity of packed beds filled with stagnant fluid: Thermal contact theory and numerical simulation. *International Journal of Heat and Mass Transfer, 132*, 331–346. https://doi.org/10.1016/j.ijheatmasstransfer.2018.12.005

Cheng, G. J., Yu, A. B., & Zulli, P. (1999). Evaluation of e!ective thermal conductivity from the structure of a packed bed. *Chemical Engineering Science, 54*(19), 4199–4209. https://doi.org/10.1016/S0009-2509(99)00125-6

Dombrovsky, Leonid A. (2022): Specialty Grand Challenge for Heat Transfer and Thermal Power. In: Front. Therm. Eng. 2, Artikel 862070. https://doi.org/10.3389/fther.2022.862070

Eismann, R. (2015). Accurate analytical modeling of flat plate solar collectors: Extended correlation for convective heat loss across the air gap between absorber and cover plate. *Solar Energy, 122*, 1214–1224. https://doi.org/10.1016/j.solener.2015.10.037

ElSherbiny, S. (1996). Free convection in inclined air layers heated from above. *International Journal of Heat and Mass Transfer, 39*(18), 3925–3930. https://doi.org/10.1016/0017-9310(96)00047-6

EN 673. (2011). *Glass in building—Determination of thermal transmittance (U-value)-calculation method.* ICS 81.040.20, European Committee for Standardization, Avenue Marnix 17, B-1000 Brussels.

Filippov, L. P. (1968). Liquid thermal conductivity research at Moscow University. *International Journal of Heat and Mass Transfer, 11*(2), 331–345. https://doi.org/10.1016/0017-9310(68)90161-0

Grifoni, R. C., & Passerini, G. (2006). *Transport properties of organic liquids -advances in fluid mechanics* (5th ed.). WIT Press.

Hollands, K., Unny, T., Raithby, G., & Konicek, L. (1976). Free convective heat transfer across inclined air layers. *Journal of Heat Transfer, 98*(2), 189–193. https://doi.org/10.1115/1.3450517

Howell, J. T., Mengüc, M., & Siegel, R. (2020a). *Thermal radiation heat trasnfer* (7th ed.). CRC PRess, Taylor & Francis Group, LLC. https://doi.org/10.1201/9780429327308

Howell, J., Mengüc, M. P., Daun, K., & Siegel, R. (2020b). *Thermal radiation heat transfer* (7 ed.). CRC Press, Taylor & Francis Group, LLC. https://doi.org/10.1201/9780429327308

IfB GmbH Wolfgang Sorge. (2009). Presentation "Vakuumdämmung im Bauwesen" (in German). Nürnberg. Retrieved September 20, 2022, from http://vip-bau.de/

Inaba, H. (1984). Experimental study of natural convection in an inclined air layer. *International Journal of Heat and Mass Transfer, 27*(8), 1127–1139. https://doi.org/10.1016/0017-9310(84)90040-1

Kaganer, M. G. (1969). Thermal insulation in cryogenic engineering. *Israel Program for Scientific Translations*, 212–220.

Kalogirou, S. (2013). *Solar energy engineering* (2nd ed.). Academic Press. https://doi.org/10.1016/B978-0-12-5

Kunii, D., & Smith, J. M. (1960). Heat transfer characteristics of porous rocks. *AIChE Journal, 6*(1), 71–78. https://doi.org/10.1002/aic.690060115

Li, M., & Yiang, Y. C. (2017). On the determination of atmospheric longwave irradiance under all-skyconditions. *Solar Energy, 144*, 40–48.

Lienhard, J. (2001). *A heat transfer textbook* (3rd ed.). Massachusetts Institute of Technology.

Matuska, T. Z. (2009). *A mathematical model and design tool "Kolektor 2.2"*. Czech Technical University in Prague Faculty of Mechanical Engineering, Department of Environmental Engineering.

Modest, M. (2014). *Radiative heat transfer*. Academic Press.

Niessen, H., & Ball, S. (2000). *Heat transport and afterheat removal for gas cooled reactors under accident conditions*. International Atomic Energy Agency.

Ozoe, H., Sayama, H., & Churchill, S. (1975). Natural convection in an inclined rectangular channel at various aspect ratios and angles—experimental measurements. *International Journal of Heat and Mass Transfer, 18*, 1425–1431. https://doi.org/10.1016/0017-9310(75)90256-2

Pflüger, A., & Wittwer, V. (1984). Exact Analysis Of Radiative And Conductive Heat Transfer Through Radiative Grey Films. In: *Optical Materials Technology for Energy Efficiency and Solar Energy Conversion III*. SPIE.

Platzer, W. J. (1988). *Solare transmission und wärmetransportmechanismen bei transparente wärmedämmmaterialien*. Dissertation. Albert-Ludwigs-Universität, Freiburg. Fakultät für Physik.

Platzer, W. J. (1992). Calculation procedure for collectors with a honeycomb cover of rectangular cross section. In: *Solar Energy, 48*(6), S. 381–393.

Randall, K., Mitchell, J., & El-Wakil, M. (1979). Natural convection heat transfer characteristics of flat plate enclosures. *Journal of Heat Transfer, 101*(1), 120–125. https://doi.org/10.1115/1.3450901

Schinkel, W. M. (1980). *Natural convection in inclined air-filled enclosures*. Pijnacker.

Stephan, P., Kabelac, S., & Kind, M., Mewes, D. (2019). *VDI-Wärmeatlas* (12th ed.). Springer Vieweg.

Summ, T., Abdalsalam, M., Ehrenwirth, M., Trinkl, C., & Zörner, W. (2021). A detailed ivestigation on the cnvective heat transfer inside the enclosed cavity of insulated glass solar thermal flat-plate collectors. *Proceedings of the international solar world congress 2021* (p. 11). International Solar Energy Society. https://doi.org/10.18086/swc.2021.22.09

Trevisan, S., Wang, W., & Laumert, B. (2021). Coatings utilization to modify the effective properties of high temperature packed bed thermal energy storage. *Applied Thermal Engineering, 185*, 116414–116512. https://doi.org/10.1016/j.applthermaleng.2020.116414

van Antwerpen, W., du Toit, C., & Rousseau, P. (2010). A review of correlations to model the packing structure and effective thermal conductivity in packed beds of mono-sized spherical particles. *Nuclear Engineering and Design, 240*, 1803–1818. https://doi.org/10.1016/j.nucengdes.2010.03.009

von Boeckh, P., & Wetzel, T. (2011). *Heat transfer -basics and practise*. Springer.

Wakao, N., & Vortmeyer, D. (1971). Pressure dependency of effective thermal conductivity of packed beds. *Chemical Engineering Science, 26(10)*, 1753–1765.

Momentum and Energy Transport 5

Abstract

When solar radiation is converted into heat, thermal losses occur mainly due to passive heat transport modes as there are conduction through gases and solids, heat radiation and natural convection. In the latter one, already macroscopic mass transport driven by buoyancy effects can be observed. Convective heat transport is connected to momentum transport and movement of mass. Driven by mechanical pumps or ventilators the forced convective heat transport is supporting the thermal system by transporting heat with the help of a heat transfer medium from the place of optical conversion to the consumer. Effective heat and mass transport thus plays a key role in the design of a solar thermal system. Convective momentum and heat transport covers many aspects and is a-priori a nonlinear phenomenon. For example, flows can differ in the type of momentum transport, which can be forced, mixed or buoyancy-driven. In addition, the mass flow can have different flow patterns, for example laminar or turbulent. In the engineering treatment, a stability parameter plays a role in predicting the transition between the flow regimes. These regimes are even more complex in two-phase flows. Furthermore, the mass flow can be confined by walls, e.g. in tubes or between plates, often called *bounded flows.* On the other hand mass flow may fully include bodies as *external flow* for instance when an air flow occurs around a thermal storage tank. The interaction of mass flow and momentum transport within the media and with solid bodies not only leads to heat transfer but also to friction, which determines the hydraulic performance. External drives like pumps have to be dimensioned in a way that the pressure loss due to these hydraulic forces can be overcome. Last but not least, the effects of cold/heat can lead to phase changes, such as boiling, evaporation and condensation, which do not alter only the momentum characteristics but also the heat transfer characteristics of the transport. There is an extensive literature on convective momentum and energy transport in its various forms. The objective of this section is to elaborate the basic ideas of single-phase and multiphase momentum and energy transport, to show methods for classifying

R. Stieglitz and W. Platzer, *Solar Thermal Energy Systems*,
https://doi.org/10.1007/978-3-031-43173-9_5

the flow type, and based on this, to present engineering approaches as well as relevant correlations for quantitative calculations, which can be used to carry out assessments.

Keywords

Laminar • Turbulent momentum transport • Boiling • Two-phase flow • Friction • Hydraulic resistance • Natural • Mixed • And forced convection

5.1 Single Phase Convective Momentum and Heat Transport

The aim of this basic first section is to formulate the conservation equations of mass and heat transport in fluids, which will enable us to determine characteristic parameters which can be used to classify different flow forms. For example we may distinguish between laminar and turbulent flows. Since the latter is analytically scarcely accessible, approaches and methods are discussed which allow a numerical calculation. These will be outlined with some examples. In the calculation it is crucial whether the model considers the physical effect(s) occurring in the engineering problem in an appropriate way, which will be highlighted in the last section.

In order to get a more transparent approach to convective transport, which in contrast to passive transfer mechanisms is associated with mass transport, we start with a heuristic consideration.

In a macroscopic mass flow of fluid volumes we find a superposition in the microscopic scale of random motions of individual atoms and molecules. Both scales contribute to the heat transport. At low densities and/or low flow velocities, and when the fluid has a sufficiently high thermal conductivity, heat conduction created by microscopic stochastic kinetics dominates the macroscopic convective heat transport. As the heat capacity or the velocity of the flowing fluid increases, the heat transport due to the macroscopic fluid motion exceeds the molecular processes of heat conduction until these can finally be completely neglected.

Not only the heat transport of a fluid, but also the heat transfer between a flowing fluid and a solid wall is of interest if both have different temperatures. The interaction of the fluid flowing close to the wall creates a thin layer with large velocity and temperature gradients in the fluid, the so-called boundary layer, in which the velocity increases from zero at the wall, henceforth noted as $y = 0$, to almost the value of the undisturbed flow. For flows in pipes, the velocity in the core flow u_0 plays a similar role. The thickness of this "*viscous*" boundary layer is denoted by δ_v. Similarly, the fluid temperature changes across the near-wall zone, called the "*thermal*" boundary layer, starting from the wall temperature $T = T_w$ at $y = 0$ to almost the temperature $T = T_\infty$ at a greater distance from the wall. The thickness of this thermal boundary layer is denoted by δ_{th}. For pipe flows, the mean fluid temperature T_f averaged over the pipe cross-section is used instead of T_∞. The hydrodynamic and thermal boundary layers can differ considerably in

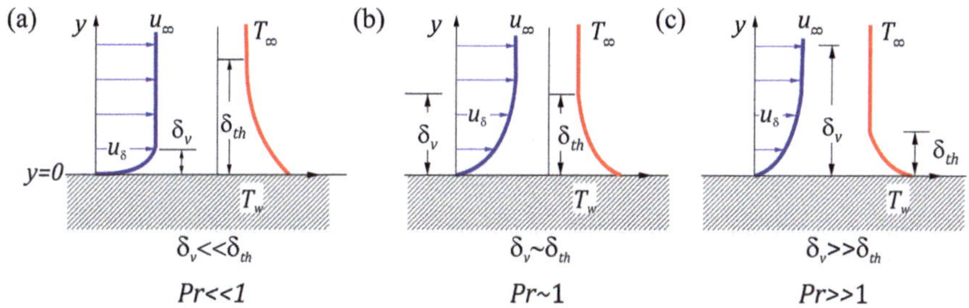

Fig. 5.1 Influence of the Prandtl number on the length scale of the thermal and viscous boundary layer in a two-dimensional flow at a constant wall temperature T_w

thickness from each other and depend largely on the thermophysical properties of the fluid, as indicated in Fig. 5.1.

The origin of the fluid motion can be a forced flow driven by an external drive such as a pump, wind, etc. Moreover, also density or concentration gradients may give rise to convection termed as free or natural convection. Here the buoyancy forces caused by density differences as a result of temperature differences, different phase distributions or different concentrations, e.g. in solutions, result in fluid motion. In practice, superpositions of several forms of convection may occur. If the buoyant flow moves the fluid perpendicular to the direction of the main forced flow, free convection can account for a considerable fraction of the total convective heat transfer. Convective heat transfer need not be limited to the transport of sensible heat. In the case of vapor formation or condensation, the transport of latent heat is added. The formation of vapor bubbles on a wall also causes a change of fluid movements and even may enhance or restrict the heat transfer. Very high heat transfers therefore occur during the condensation of steam on cold walls, for example.

Regardless of the type of convection, the relationship of the heat flux density at the fluid wall interface ($\dot{q}_w$) with the driving temperature difference at the wall and the fluid ($T_w - T_\infty$) is linked with a heat transfer coefficient h in $\left(\mathrm{W}/\left(\mathrm{m}^2\mathrm{K}\right)\right)$ in the following form as already developed in Chap. 4:

$$\dot{q}_w(y=0) = h \cdot \left(T_w - T_f\right) \ or \ \dot{q}_w(y=0) = h \cdot (T_w - T_\infty). \tag{5.1}$$

The treatment of heat transfer at the interface is transformed into the functional determination of the heat transfer coefficient by Eq. (5.1). Directly at the wall, the velocity of the fluid approaches zero (non-slip condition). Thus, the heat transfer there takes place only by heat conduction and Fourier's law of heat conduction applies:

$$\dot{q} = -k\left(\frac{dT}{dy}\right)_w, \tag{5.2}$$

in which k denotes the thermal conductivity. This yields for the heat transfer coefficient:

$$h = -\frac{k}{\left(T_w - T_f\right)} \cdot \left(\frac{dT}{dy}\right)_w, \tag{5.3}$$

h is thus determined by the slope of the temperature profile at the wall and the nominal difference between the wall and fluid temperatures. The calculation of the heat transfer coefficient thus requires a knowledge of the temperature field in the fluid, which is strongly influenced by the local velocity field.

The heat transfer coefficient h has been introduced and defined in Chap. 4. The concept of boundary layers has to be filled in the following section in order to understand methods with may treat convective heat transfer quantitatively. Similarly to the heat transfer, ad- and desorption, evaporation or condensation at a wall surface will also depend on the exact properties of the boundary layer.

First, a few order-of-magnitude considerations should be made. The different fluid types, such as air, oil, water, or liquid metals, have completely different thermal conductivities. Liquid metals exhibit a high thermal conductivity k (W/(mK)) and a comparatively low specific heat capacity c_p (J/(kgK)). Oils, on the other hand, have a high kinematic viscosity ν (m^2/s) compared with water or air. The thermal conductivity can be combined with the density ρ (kg/m^3) and the heat capacity c_p to give the thermal diffusivity a $\left(m^2/s\right)$. This thermal diffusivity has the same SI unit as the viscosity ν. The ratio ν/a yields a dimensionless ratio called the Prandtl number Pr. The physical meaning of the Prandtl number is that it weights the transport coefficient of the velocity field versus that of the temperature field, and thus the Prandtl number describes the ratio of momentum transport versus that of diffusive thermal transport. It is defined as follows:

$$Pr = \frac{\rho \nu c_p}{k} = \frac{\nu}{a}. \tag{5.4}$$

While oils typically show Prandtl numbers of the order of $O\left(10^2 - 10^6\right)$ and "conventional fluids" such as air or water show a value around unity, however, depending on the temperature, the Prandtl number of liquid metals is $Pr = 10^{-3} - 10^{-2}$. Table 5.1 shows the Prandtl number of some Newtonian media at different temperatures. It shows that there exists no Newtonian fluid in the Prandtl number range between $0.05 < Pr < 0.5$. A Newtonian fluid is one where there is a linear relationship between stress and strain-rate. The ratio of stress to strain-rate is the viscosity of the fluid.

With this in mind, consider the flow of a Newtonian fluid of temperature T_∞ over a semi-infinite plate of constant temperature T_w, as plotted in the Fig. 5.1 for various fluids of different Prandtl numbers such as $Pr \ll 1$, $Pr \sim 1$ and $Pr \gg 1$. Case (a) corresponds to liquid metals, which have good molecular thermal conductivity and low viscosity, hence $Pr \ll 1$. The thickness of the viscous boundary layer δ_v is negligible compared to the thickness of the thermal boundary layer δ_{th}.

Table 5.1 Typical temperature dependent Prandtl numbers of some fluids

Temperature (°C)	Mercury	Air	Water	Engine oil
0	0.0288	0.72	13.6	4.7×10^4
20	0.0249	0.71	7.02	1×10^4
100	0.0162	0.70	1.74	2.6×10^2

For water or air, both have approximately the same order of magnitude, while for oils the viscous boundary layer is significantly thicker than the thermal one.

The following sections are structured as follows. First, the basic differential equations of energy and momentum transfer in a continuum as well as the boundary and initial conditions are explained. Based on this, the basic ideas of the solution are shown, where the focus is on the identification of the essential dimensionless quantities or scaling dependencies.

A balance in methodical approach is needed in the engineering applications, between a detailed calculation of local quantities, with the help of finite difference/volume methods, so called computational fluid dynamics (CFD), and an empirical integrated description of the system behaviour or the system yield. The latter make use of scale separation methods and considers mainly the central physical effects quasi one-dimensionally. For example, CFD is required for the design of thermally highly loaded receivers, while a simplified integral description is relevant for the control of mass flows, the dynamic overall system behaviour or also economic performance quantities. In order to take both aspects into account, the basic ideas for the treatment of both local and system level problem solving approaches are presented and correlations or challenges are listed, and reference is made to relevant literature.

5.1.1 Conservation Equations of Mass, Momentum, and Energy

The conservation equations are balances of mass, momentum, and energy at a control volume, in which all fluxes of these control quantities are considered over all inlet and outlet surfaces of the three spatial directions.

This basic idea shall be demonstrated by a simple two-dimensional example of an incompressible ($\rho = const.$), single-phase, single-component flow in a closed control volume to illustrate the logic. In the volume element of dimensions dx aand dy shown in the figure, the mass flow leaving the control volume must be equal to that entering it. The fluid mass flows into the element and leaves it again exclusively as a result of the flow of the fluid. A transport of this type is called convective or advective. For mass flows that go beyond the boundaries of the element, conservation of mass must apply. This states that the sum of the mass flows across the volume boundaries is equal to the change in density in the volume element, or zero for incompressible fluids. Using the mass flows given in Fig. 5.2,

Fig. 5.2 Sketch for the derivation of the mass conservation equation

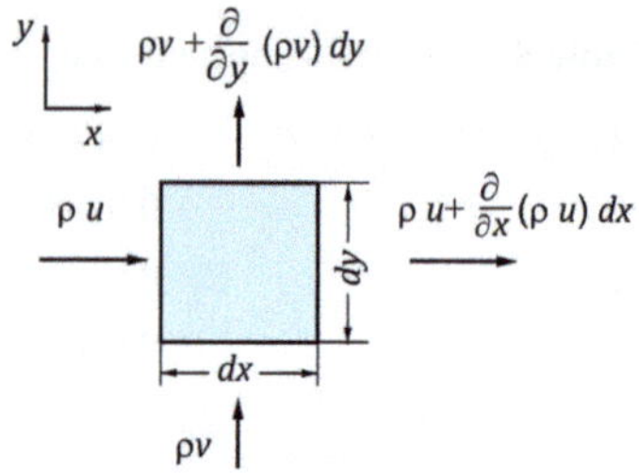

a mass balance can be established for the simplified, two-dimensional case in the following manner:

$$(\rho u)dy + (\rho v)dx - \left[\rho u + \frac{\partial(\rho u)}{\partial x}dx\right]dy - \left[\rho v + \frac{\partial(\rho v)}{\partial y}dy\right]dx = 0. \quad (5.5)$$

After rearranging the equation, some terms cancel out and after dividing by $dxdy$, we finally get:

$$\frac{\partial(\rho u)}{\partial x} + \frac{\partial(\rho v)}{\partial y} = 0. \quad (5.6)$$

For an incompressible medium, the mass conservation or continuity equation reads to:

$$\nabla \cdot \vec{u} = 0 \text{ or } \frac{\partial u}{\partial x} + \frac{\partial v}{\partial y} + \frac{\partial w}{\partial z} = 0. \quad (5.7)$$

where $\vec{u}$ is the velocity vector composed of the velocity components (u, v, w) in the x, y and z-direction.

The dynamic behaviour of a fluid motion is described by the momentum conservation equations. In analogy to the conservation of mass, a balancing of the momentum changes $(\rho \cdot u \cdot dxdydz)$ and forces acting on the infinitesimal volume element is carried out and as a result a vector equation is obtained for each of the spatial directions in the manner:

$$\rho\left(\frac{\partial u}{\partial t} + u\frac{\partial u}{\partial x} + v\frac{\partial u}{\partial y} + w\frac{\partial u}{\partial z}\right) = \rho\, f_x + \frac{\partial \sigma_{xx}}{\partial x} + \frac{\partial \tau_{yx}}{\partial y} + \frac{\partial \tau_{zx}}{\partial z},$$
$$\rho\left(\frac{\partial v}{\partial t} + u\frac{\partial v}{\partial x} + v\frac{\partial v}{\partial y} + w\frac{\partial v}{\partial z}\right) = \rho\, f_y + \frac{\partial \tau_{xy}}{\partial x} + \frac{\partial \sigma_{yy}}{\partial y} + \frac{\partial \tau_{zy}}{\partial z},$$
$$\rho\left(\frac{\partial w}{\partial t} + u\frac{\partial w}{\partial x} + v\frac{\partial w}{\partial y} + w\frac{\partial w}{\partial z}\right) = \rho f_z + \frac{\partial \tau_{xz}}{\partial x} + \frac{\partial \tau_{yz}}{\partial y} + \frac{\partial \sigma_{zz}}{\partial z}, \quad (5.8)$$

where $\vec{f} = (f_x, f_y, f_z)$ represent the volume forces occurring e.g. by gravity or imposed electric or magnetic fields. Herein, σ describes the stress normal to the surface and τ the shear stresses tangential to the sides of any control volume, with a linear relationship between stress and strain for a Newtonian medium (Lamb, 1993). On this basis, one can derive the following equations for Newtonian fluids:

$$\sigma_{xx} = -p + 2\mu\frac{\partial u}{\partial x} - \frac{2}{3}\mu\nabla\cdot\vec{u}, \quad \sigma_{yy} = -p + 2\mu\frac{\partial v}{\partial y} - \frac{2}{3}\mu\nabla\cdot\vec{u}\,,$$
$$\sigma_{zz} = -p + 2\mu\frac{\partial w}{\partial z} - \frac{2}{3}\mu\nabla\cdot\vec{u},$$
$$\tau_{xy} = \tau_{yx} = \mu\left(\frac{\partial v}{\partial x} + \frac{\partial u}{\partial y}\right), \quad \tau_{xz} = \tau_{zx} = \mu\left(\frac{\partial w}{\partial x} + \frac{\partial u}{\partial z}\right),$$
$$\tau_{yz} = \tau_{zy} = \mu\left(\frac{\partial w}{\partial y} + \frac{\partial v}{\partial z}\right), \tag{5.9}$$

where $\mu = (\rho\ \nu)$ represents the specific dynamic viscosity (Pa s) of the fluid and p the pressure.

Substituting Eq. (5.9) into (5.8), we obtain the complete momentum conservation equation, known in the literature as the Navier–Stokes equations, which is a set of partial non-linear differential equations of elliptic character. Almost all analytical studies of viscous fluid flows are based on them. They are general in the sense that they also apply to compressible Newtonian fluids of variable viscosity, such as gases. For constant density and viscosity (incompressible fluid and small temperature differences) the Navier–Stokes equations simplify to:

$$\rho\left(\frac{\partial u}{\partial t} + u\frac{\partial u}{\partial x} + v\frac{\partial u}{\partial y} + w\frac{\partial u}{\partial z}\right) = \rho\ f_x - \frac{\partial p}{\partial x} + \mu\left(\frac{\partial^2 u}{\partial x^2} + \frac{\partial^2 u}{\partial y^2} + \frac{\partial^2 u}{\partial z^2}\right),$$
$$\rho\left(\frac{\partial v}{\partial t} + u\frac{\partial v}{\partial x} + v\frac{\partial v}{\partial y} + w\frac{\partial v}{\partial z}\right) = \rho\ f_y - \frac{\partial p}{\partial y} + \mu\left(\frac{\partial^2 v}{\partial x^2} + \frac{\partial^2 v}{\partial y^2} + \frac{\partial^2 v}{\partial z^2}\right),$$
$$\rho\left(\frac{\partial w}{\partial t} + u\frac{\partial w}{\partial x} + v\frac{\partial w}{\partial y} + w\frac{\partial w}{\partial z}\right) = \rho\ f_z - \frac{\partial p}{\partial z} + \mu\left(\frac{\partial^2 w}{\partial x^2} + \frac{\partial^2 w}{\partial y^2} + \frac{\partial^2 w}{\partial z^2}\right),$$
$$\text{or}\quad \underbrace{\frac{d\vec{u}}{dt}}_{\text{inertial term}} = \underbrace{f}_{\substack{\text{gravity,}\\ \text{buoyancy,}\\ \text{capillary forces,}\ldots}} - \underbrace{\frac{1}{\rho}\nabla p}_{\substack{\text{pressure}\\ \text{force}}} + \underbrace{\nu\,\nabla^2\vec{u}}_{\substack{\text{viscous friction}\\ \text{forces}}}\,, \tag{5.10}$$

in which $\nabla^2 = \nabla\cdot\nabla$ is the Laplace operator, which is given by the relation $\nabla^2 = \left(\partial^2/\partial x^2 + \partial^2/\partial y^2 + \partial^2/\partial z^2\right)$.

Any fluid dynamic problem involving the determination of velocity and pressure distribution as a function of spatial coordinates and time requires the simultaneous solution of the continuity equation and the Navier–Stokes equations, considering the special initial and boundary conditions. Although in most cases this system of equations is not accessible to an analytical solution, in many cases it can be solved numerically. Nevertheless, there are a few specific cases where the nature of the flow is such that it can be simplified considerably and thus be submitted to an analytical solution.

A similar procedure can be applied to the derivation of the energy conservation equation. The derivation includes the first law of thermodynamics, which links the thermal energy with the work done on the system and the total internal energy within the control volume, which finally reads to:

$$\begin{aligned}\rho\frac{dU}{dt} &= \nabla\cdot(k\nabla T) + \sigma_{xx}\frac{\partial u}{\partial x} + \sigma_{yy}\frac{\partial v}{\partial y} + \sigma_{zz}\frac{\partial w}{\partial z}\\ &+ \tau_{xy}\left(\frac{\partial v}{\partial x} + \frac{\partial u}{\partial y}\right) + \tau_{yz}\left(\frac{\partial w}{\partial y} + \frac{\partial v}{\partial z}\right) + \tau_{xz}\left(\frac{\partial u}{\partial z} + \frac{\partial w}{\partial x}\right),\end{aligned} \tag{5.11}$$

where k is the thermal conductivity of the fluid, U is the internal energy and T is the temperature.

The internal energy U is related to the fluid enthalpy h in the following way: $h = U + p/\rho$. The operator d/dt describes the total derivative with respect to time. If the relations for the stress and strain on a fluid element (Eq. (5.9)) are substituted into Eq. (5.11), the simplified energy conservation equation is obtained:

$$\begin{aligned}\rho\frac{dU}{dt} &= \nabla\cdot(k\nabla T) - p\nabla\cdot\vec{u} + \mu\Phi \text{ with}\\ \Phi &= 2\left[\left(\frac{\partial u}{\partial x}\right)^2 + \left(\frac{\partial v}{\partial y}\right)^2 + \left(\frac{\partial w}{\partial z}\right)^2\right] + \left(\frac{\partial v}{\partial x} + \frac{\partial u}{\partial y}\right)^2 + \left(\frac{\partial w}{\partial y} + \frac{\partial v}{\partial z}\right)^2\\ &+ \left(\frac{\partial u}{\partial z} + \frac{\partial w}{\partial x}\right)^2 - \frac{2}{3}(\nabla\cdot\vec{u})^2.\end{aligned} \tag{5.12}$$

In this, Φ represents the dissipation function and describes the rate of dissipation of mechanical energy per unit mass due to viscosity. The first term on the right describes the net rate of heat conduction in a fluid element per control volume, the second the reversible work done on the control volume, and the last term gives the amount at which viscous forces do irreversible work in the form of viscous dissipation or viscous heating on the control volume. Assuming an incompressible fluid where the change in internal energy is described by $dU = c_p \cdot dT$, the following formulation is obtained:

$$\begin{aligned}\rho c_p\frac{dT}{dt} &= \nabla(k\nabla T) + \mu\Phi \text{ with}\\ \Phi &= 2\left[\left(\frac{\partial u}{\partial x}\right)^2 + \left(\frac{\partial v}{\partial y}\right)^2 + \left(\frac{\partial w}{\partial z}\right)^2\right] + \left(\frac{\partial v}{\partial x} + \frac{\partial u}{\partial y}\right)^2 + \left(\frac{\partial w}{\partial y} + \frac{\partial v}{\partial z}\right)^2\\ &+ \left(\frac{\partial u}{\partial z} + \frac{\partial w}{\partial x}\right)^2.\end{aligned} \tag{5.13}$$

If the thermal conductivity λ in the considered temperature regime is constant, the equation simplifies further, and we obtain:

$$\underbrace{\frac{dT}{dt}}_{\substack{\text{convective}\\ \text{heat transport}}} = \underbrace{\frac{k}{\rho c_p}\nabla^2 T}_{\text{heat conduction}} + \underbrace{\frac{\mu}{\rho c_p}\Phi}_{\substack{\text{heat generation by}\\ \text{viscous dissipation}}} \quad . \tag{5.14}$$

The continuity equation, the Navier–Stokes equations, and the conservation of energy equation together provide a complete set for describing any flow within a considered control volume. But, they also pose an almost insurmountable difficulty in the mathematical sense, given the large number of simultaneous equations to be solved and the presence of non-linear terms such as $(u\partial u/\partial x)$.

Because of the non-linearities, the superposition principle of engineering mechanics is not applicable and therefore complex flows cannot be composed of several simple flows as for example by the theory of potential flows, which are irrotational and incompressible and thus the viscous term vanishes, see e.g. Kreith and Goswami (2005). Exact solutions exist for some simple cases where nonlinear effects are small (approximate solutions) or zero. This class of solutions occurs in creeping flows and is important in lubricant film theory. In most engineering applications, however, the nonlinear terms are often larger than the other of the Navier–Stokes equations. An important parameter in this is the hydraulic Reynolds number (Re), a dimensionless quantity that weights the ratio of inertial forces versus viscous forces in a problem and is shown in Eq. (1.16). Finally, the product of the Prandtl number Pr with the Reynolds number Re gives the Peclet number Pe, which describes the ratio of convective heat transport to that by heat conduction:

$$Re = \frac{u_0\, d}{\nu} \text{ and } Pe = RePr = \frac{u_0 d}{a}, \tag{5.15}$$

in which d denotes a characteristic dimension of the hydraulic problem considered.

According to the definition of Reynolds number, creeping flows are characterized by small Reynolds numbers, while in most engineering problems Reynolds numbers well above unity occur.

At last, two observations should be noted. First, velocity and temperature fields are always coupled when the fluid has a temperature dependent density and/or viscosity. Second, temperature fields are similar to velocity fields provided certain conditions are met. If the pressure gradient is zero ($\nabla p = 0$) and $\Phi = 0$ and $f = 0$ and if additionally, the Prandtl number is one ($Pr = 1$), then this similarity is fulfilled (however, this also requires a similarity of the boundary conditions).

5.1.2 Boundary Conditions and Similarity Laws

The system of partial differential equations requires for the solution the specification of boundary and initial conditions either at the in-/outlet and the boundaries of the considered domain. The type of the initial and boundary conditions corresponds mathematically to that specified for the passive heat transfer mechanisms to be of Dirichlet, Neumann or Robin type, however, they scope a wider set of physical quantities such as scalars like flow rate or pressure as well as vectoral or tensorial quantities like e.g. velocity fields, shear stress at interfaces at walls or gas/liquid boundaries and thus impact the flow motion type and the fluid state. Especially, the Navier–Stokes equations can show large variations in mathematical behaviour, ranging from almost completely elliptic to almost completely hyperbolic, which has implications when it comes to prescribing admissible boundary conditions. There is also a discrepancy between mathematically valid boundary conditions and practically useful boundary conditions.

It is, therefore, of paramount importance that physically realistic, well-posed boundary conditions are set either in computational fluid dynamic (CFD) computations or piping network simulations, otherwise severe difficulties are encountered in obtaining solutions. The single, most common cause of rapid divergence of thermal-hydraulics simulations is the inappropriate selection of boundary conditions. Due to the multitude of potentially applicable boundary conditions we limit here to the most common ones as representatives to cover most applications.

5.1.2.1 Hydraulic Boundary Conditions

The hydraulic boundary conditions to be applied at the boundary of a fluid in contact with another medium depends on the nature of this other medium either it is solid, liquid, or gas. It has to be distinguished between in-/outlet and domain confining boundary conditions.

A common pressure boundary condition (p) at in-/outlet of incompressible flows assuming there are no external forces is of Dirichlet type reading to:

$$p = \rho \cdot \vec{g}\left(\vec{r} - \vec{r_0}\right) + p_0 \text{ or } p = p_0, \tag{5.16}$$

in which $\vec{g}$ denotes the gravity vector, $\vec{r}$ the location vector, ρ the density of the fluid and the index “0” indicates the reference conditions. The first term provides the hydrostatic pressure, while the second one determines the system reference pressure. Whenever possible, it is recommended to locate the pressure boundary in a region where the approximate definition of the hydrostatic pressure is applicable, or to define a boundary that is perpendicular to the gravity vector.

Often also a Neumann type boundary condition is applied imposing a constant pressure gradient, which imposes intrinsically a fully developed flow at the outlet, meaning that the velocity gradient in flow direction does not change. It reads to:

$$\frac{\partial p}{\partial \vec{n}} = const. \text{ or } \vec{u} - (\vec{u} \cdot \vec{n})\vec{n} = 0, \tag{5.17}$$

where $\vec{n}$ denotes the outward unity vector on the surface.

At a *solid–fluid interface* the velocity and temperature are continuous. Contained in this boundary condition is the “no-slip” condition: namely, that the tangential velocity of the fluid in contact with the boundary of the solid is equal to that of the boundary. If the wall is at rest, the condition posed along the wall boundary $\partial\Omega$ of the domain Ω is given by:

$$\vec{v}|_{\partial\Omega} = 0. \tag{5.18}$$

For an inviscid fluid, the no-slip condition does not apply, and only the normal component of velocity is continuous. If the wall is permeable, the tangential velocity is continuous and the normal velocity is arbitrary; the temperature boundary condition for this case depends on the nature of the injection or suction at the wall. Assuming an inviscid fluid no boundary layer develops, and the condition reads to:

$$(\vec{v} \cdot \vec{n})|_{\partial\Omega} = 0; \tag{5.19}$$

expressing that there is no flow across the boundary.

At a *liquid–liquid interface* the velocity components are continuous across the interface and the same holds for the normal and tangential stress components.

At a *liquid–gas interface*s as they occur e.g. in two-phase flows the interaction of the liquid with the gas is of relevance. In most cases the tangential stresses in the gas are generally negligible and the boundary conditions imposed are, first, that the normal velocity of the liquid at the interface equals to the normal velocity of the interface; and secondly the pressure of the liquid at the interface is the same as that of the gas. If surface tension (σ) is important across the interface a jump of the pressure occurs according to

$$\nabla p = \left(p_l - p_g\right) = \sigma \cdot \left(\frac{1}{R_1} + \frac{1}{R_2}\right); \tag{5.20}$$

where R_1 and R_2 are the radii of curvature of the intercepts of the interface by two orthogonal planes containing the normal to the interface.

For hydraulic piping networks instead of a pressure often a mass flux boundary condition is imposed. It specifies the mass flux ($\dot{m}$) flowing into the domain across an inlet boundary $\partial\Omega$ and can be specified in a number of ways, while most often the following formulation is used:

$$-\int_{\partial\Omega} \rho \cdot (\vec{u} \cdot \vec{n}) \partial\Omega = \dot{m}. \tag{5.21}$$

In case of a CFD computation this condition must be complemented by a further constraint that the tangential velocity components are set to zero on the boundary $\partial\Omega$:

$$(\vec{u} \times \vec{n})|_{\partial\Omega} = 0. \tag{5.22}$$

Although many combinations of constraints are conceivable, only a few of these combinations allow a solution without over constraining the problem. Here, some permissible combinations are stated:

- walls only;
- walls and inlet and at least one outlet;
- walls and inlet and at least one constant pressure boundary;
- walls and constant pressure boundaries.

Particular care must be taken in applying the outlet boundary condition. It can only be used if all flows entering the calculation domain are given by means of inlet boundary conditions (i.e. velocity and scalars fixed at inlet) and is only recommended for flow domains with a single exit. From a hydraulic point of view the exit pressures determines the flow split between multiple outlets so it is better to specify this quantity at outlets than setting there (zero-gradient) outlet conditions. It is not permitted to combine an outlet condition with one or more constant pressure boundaries, because the zero-gradient outlet condition specifies neither the flow rate nor the pressure at the exit, thus leaving the problem underspecified.

5.1.2.2 Thermal Boundary Conditions

In addition to hydraulics, the description of energy transport requires a definition of thermal wall conditions; a systematic description can be found in Shah and London (1978). Only the most common conditions will be addressed here.

The simplest Dirichlet-type thermal boundary condition along the flow boundary $\partial\Omega$ is that of a constant wall temperature (T_w) circumferentially and axially (i.e. in the flow direction x). This boundary condition occurs in solar thermal applications in condensers and evaporators and is as follows:

$$T_w|_{\partial\Omega} = const. \tag{5.23}$$

The second boundary condition to be assigned to the Neumann type is that of constant axial and circumferential heat flux.

$$\dot{q}_w|_{\partial\Omega} = const. \tag{5.24}$$

Technically, this characterizes a case in which the wall material has low thermal conductivity, and the wall thickness is constant.

Both of the above-named boundary conditions are somehow academic or artificial boundary conditions since they are technically barely feasible. Nevertheless, they are relevant for rough estimations because they reflect the upper and lower limits of the temperatures at the fluid/wall interfaces, respectively.

A condition typical for electric resistance heaters, but also for the one in heat exchangers, is the assumption of a constant heat flow with circumferentially constant wall temperature and axially constant wall heat flux.

$$\dot{q}_w|_{\partial\Omega} = \dot{q}_w(y, z), \quad T_w = T_w(x), \tag{5.25}$$

in which x denotes the axial direction and y, z the circumferential coordinates. It must be emphasized that this condition may only be used if the wall materials exhibit a high heat conductivity compared to that of the fluid, i.e. $\lambda_w \gg \lambda$, or the wall is much thinner than the characteristic dimension of the problem.

A mixed boundary condition of the Robin type is the assumption of an axially constant wall temperature and a finite thermal wall normal resistance.

$$T_{w0} = T_{w0}(y, z)\ ;\ T_w = T_w(x, y, z); \lambda\left(\frac{\partial T}{\partial n}\right)_w = h_0(T_{w0} - T_w);$$
$$\frac{1}{h_0} = \frac{t_w}{\lambda_w} + \frac{1}{h_i}. \tag{5.26}$$

Here T_{w0} is the temperature on the outside of the heater/cooler, T_w the temperature at the fluid/wall interface, h_0 the heat transfer coefficient at the fluid/wall interface, λ_w represents the thermal conductivity of the wall, t_w is its thickness, n is the wall normal unit vector, and h_i is the heat transfer coefficient at the inlet where heat transfer begins. Physically, the finite thermal wall resistance leads to an upstream heating of the fluid.

Convective boundary condition with axially constant heat flux and finite wall-normal thermal resistance. In addition to Eq. (5.26), it is assumed here that the heat conduction at the channel periphery (lateral direction) is negligibly small, which is particularly applicable to selective absorbers. This results in:

$$\dot{q}_w = \dot{q}_w(y, z)\ ;\ k\left(\frac{\partial T}{\partial n}\right)_w = h_0(T_{w0} - T_w). \tag{5.27}$$

Heat conduction boundary condition with axially constant heat flux and finite heat conduction at the channel wall circumference, in which s represents a circumferential coordinate. This boundary condition is an extension of Eq. (5.24) and reads:

$$q_w = q_w(y, z)\ ;\ \frac{q_w}{k} - \left(\frac{\partial T}{\partial n}\right)_w + \frac{k_w t_w}{\lambda}\left(\frac{\partial^2 T}{\partial s^2}\right) = 0. \tag{5.28}$$

This list of boundary conditions can be applied to both laminar and turbulent flows and illustrates the need for a detailed description when recalculating experiments and designing a thermal system. Regardless of the correct choice of boundary conditions in the simulation, this says nothing about the quality of the experiments, which can also be influenced by buoyancy effects in the laminar case, for example.

5.1.3 Similarity Laws in Fluid Mechanics

In the context of the conservation equations at the beginning of the chapter, the Reynolds, Peclet and Prandtl numbers were introduced as dimensionless parameters. We found that the magnitude of these dimensionless ratios separates different transport phenomena of momentum and energy transport. Thus, the equality of dimensionless ratios ensures the conversion of geometrically similar problems into each other. Therein lies the most important significance of the dimensionless quantities for the application in convective problems. It is the basis of every model experiment, in which with the help of measurements on reduced/enlarged models, even with different fluids, statements can be made about the subsequent application execution on a large or small scale. Already in the context of passive heat transfer in Chap. 4 we have found the Fourier number (Fo) and the Biot number (Bi) as important dimensionless quantities to characterize time dependent transient heat conduction problems. But how does one systematically arrive at the derivation of characteristics and similarity laws?

In principle, there are four different methods, which are briefly outlined in the following. As further literature we refer to e.g. to Kreith et al. (2011).

- π-Theorem of Buckingham

In every technical system force, energy, time and one or more characteristic lengths occur. So, there is a number of n different quantities Q_k, for which there is a relation $f(Q_1, \ldots, Q_n) = 0$. Each of these quantities Q_i is dimensionally related, i.e. its physical quantity can be represented in SI units in the form of kg, m, s and degree K. These basic quantities $A_1, \ldots, A_m$ cannot be represented by power products of the others. Thus, each of the quantities Q_i can be determined as a power product of the basic quantities A_i. The question arises: how many dimensionless quantities π of the form $\pi = Q_1^{i_1} \cdot Q_2^{i_2}, \ldots\ldots, Q_n^{i_n}$ are there? Comparing the exponents of the base variables yields a homogeneous linear system of equations of m equations for n unknown i_1, i_2 to i_n. The solvability of the system of equations is thus reduced to the rank r of a dimension matrix. Buckingham's π-theorem now states: If there are n quantities Q_1, ..., Q_n and there is a relation between them, then there are exactly $(n - r)$ dimensionless ratios, where $r \leq m \leq n$ is the rank of the dimension matrix. There is then also a relation $f(\pi_1, \pi_2, \ldots, \pi_{n-r}) = 0$ which solves the problem. This is a relatively mathematical consideration of the derivation of similarity parameters, and the question often arises for the user: how are the free unknowns to be chosen so that one gets meaningful π-ratios? There is no rule here and only experience helps to filter out the physically meaningful parameters.

- Ratios from conservation equations

Ratio formation on the basis of conservation equations is a heuristic procedure in principle. The starting point is the respective conservation equations of the considered problem. For example, the conservation of momentum Eq. (5.10) provides a balance of forces at a control volume of the fluid. As shown in the vectoral notation of the equation, the individual terms describe the contribution of a force to the balance. This differential representation can be expressed in terms of characteristic quantities of the flow, geometry, and fluid by selecting characteristic quantities of the thermal–hydraulic problem as e.g. the mean flow velocity u_0, a characteristic dimension l, and a refe-rence temperature T. This results in:

$$F_i \sim u\frac{\partial u}{\partial x} \sim \frac{u_0^2}{l}\ ;\ F_p \sim \frac{1}{\rho}\frac{\partial p}{\partial x} \sim \frac{p}{\rho \cdot l}\ ;$$
$$F_v \sim \nu\frac{\partial^2 u}{\partial x^2} \sim \nu\frac{u}{l^2};\ F_g \sim g, \qquad (5.29)$$

in which F_i represents the inertia force, F_p the pressure force, F_v the friction force and F_g the gravity force. The ratio of inertia force to friction force, i.e. F_i/F_v, thus gives the Reynolds number. A similar procedure can be followed for the energy conservation Eq. (5.15). Table 5.2 shows a selection of the ratios (π-quantities) to be obtained by this procedure.

Table 5.2 Dimensionless numbers determined from force and energy ratios

Force ratio		Energy ratio	
Euler number	$Eu = \frac{\text{pressure}}{\text{inertia}} = \frac{p}{\rho u_0^2}$	Peclet number	$Pe = \frac{\text{convective transp.}}{\text{heat conduction}} = \frac{u_0 l}{a}$
Reynolds number	$Re = \frac{\text{inertia}}{\text{viscosity}} = \frac{u_0 l}{\nu}$	Eckert number	$Ec = \frac{\text{viscosity}}{\text{convective transp.}} = \frac{u_0^2}{c_p \Delta T}$
Froude number	$Fr = \frac{\text{inertia}}{\text{gravity}} = \frac{u_0^2}{gl}$	Damköhler number	$Da = \frac{\text{reaction enthalpy}}{\text{convective transp.}} = \frac{\Delta H}{c_p \Delta T}$
Weber number	$Wb = \frac{\text{inertia}}{\text{capillary}} = \frac{\rho u_0^2 l}{\sigma}$	Fourier number	$Fo = \frac{l^2}{at}$
Grashof number	$Gr = \frac{\text{buoynancy}\cdot\text{inertia}}{(\text{viscosity})^2} = \frac{gl^3\beta\Delta T}{\nu^2}$		
Strouhal number	$Str = \frac{l}{u_0 t}$		
Composite numbers			
Stokes number	$Sto = Eu \cdot Re = \frac{p}{\mu}\frac{l}{u_0}$	Prandtl number	$Pr = \frac{Pe}{Re} = \frac{\nu}{a}$
		Rayleigh number	$Ra = Pr \cdot Gr = \frac{g\beta l^3 \Delta T}{\nu a}$

In the relationships of Table 5.2, p indicates pressure, ρ density, ρ kinematic viscosity, g gravity, σ surface tension, β volume expansion coefficient, μ dynamic viscosity, and t time. l and u_0 denote a characteristic length and a velocity, respectively; ΔT is a characteristic temperature difference, a is the temperature conductivity, c_p is the heat capacity, and Q is a heat source or sink.

- Method of differential equations

This method provides the best possible similarity statement, since one starts directly from the differential equations describing the corresponding problem. The scaling laws for the individual influencing variables are obtained directly. The method consists of several successive steps:

- Formulation of the differential equations and boundary conditions.
- De-dimensioning of all occurring quantities with the help of adjusted parameters, for example a characteristic length l, velocity u_0, ...
- Summarization of the incoming parameters to dimensionless parameter groups yields the ratios and as a result the similarity laws.
- Size order approximation of the dimensionless differential equation leads to the reduction of the problem, since only terms of the leading order are considered, and supplies thereby the solution approach for the treatment of the problem.

In the presentation of the boundary layer equations, this method is explicitly applied so that it will not be further discussed here.

- Similarities by transformation of variables

The convective transport of momentum and energy is usually described by a set of partial differential equations and the associated boundary and initial conditions. With the help of the method of the transformations of the independent and dependent variables variable combinations are searched in such a way that the number of the independent variables is reduced by one. By the resulting summary of the variables a similarity statement is obtained in such a way that different fields for example temperature or flow velocities can be converted into each other. The applicability of the method of the transformation of the variables is given if the solution function is continuous thus none of the independent variables takes an excellent value.

This method was already used unconsciously in Sect. 4.2.3 (transient heat conduction) and allows the conversion of different temperature fields at different heat conductivity and times into each other. Other examples of application of the method of transformation of variables known in fluid mechanics are the 1st Stokes problem for the plate, similar solutions of the boundary layer equations, heat transfer at the heated vertical plate, and many others.

5.1.4 Vorticity, Stream Function and Total Head in Incompressible Fluids

The rotation of a fluid element at a given point in the flow domain is related to the vector cross product of the velocity, which reads to:

$$\nabla \times \vec{u} = \left(\frac{\partial w}{\partial y} - \frac{\partial v}{\partial z}\right) \cdot \vec{e}_x + \left(\frac{\partial u}{\partial z} - \frac{\partial w}{\partial x}\right) \cdot \vec{e}_y + \left(\frac{\partial v}{\partial x} - \frac{\partial u}{\partial y}\right) \cdot \vec{e}_z, \tag{5.30}$$

in which $\vec{u} = (u, v, w)$ is the velocity field and $\vec{e}_i$ the unit vector in each direction of a 3D-space. This vector product $\nabla \times \vec{u}$ is called vorticity or $\vec{\omega}$. The Navier–Stokes equation for incompressible ($\rho = const$) flow, Eq. (5.10), may also be written in the form:

$$\frac{d\vec{u}}{dt} = -\frac{1}{\rho}\nabla \mathrm{p} + \vec{g} - \mu(\nabla \times (\nabla \times \vec{u})), \tag{5.31}$$

from which can observe that if viscous forces act on a fluid, the flow intrinsically contains vorticity. Thus, if $\nabla \times \vec{u} \neq 0$ the flow is called rotational, reversely a flow in which the vorticity vanishes is termed as irrotational. Further applying the vector identity,

$$(\vec{u} \cdot \nabla)\vec{u} = \nabla\left(\frac{\vec{u}^2}{2}\right) - \vec{u} \times (\nabla \times \vec{u}) \rightarrow \Delta\vec{u} = \nabla(\nabla \cdot \vec{u}) - \nabla \times (\nabla \times \vec{u}), \tag{5.32}$$

together with the definition of the vorticity $\vec{\omega}$ and using mass conservation ($\nabla \cdot \vec{u} = 0$) the Navier–Stokes equation can be re-written:

$$\rho\frac{\partial \vec{u}}{\partial t} + \nabla\left(p + \frac{\rho}{2}u^2 + \rho g z\right) = \rho\, \vec{u} \times \vec{\omega} - \mu(\nabla \times \vec{\omega}). \tag{5.33}$$

If the flow is said to be *irrotational*, i.e. $\vec{\omega} = \nabla \times \vec{u} = 0$, a velocity potential $\Phi = \nabla\vec{u}$ can be defined and by integration with respect to all spatial variables the unsteady Bernoulli-equation for irrotational, incompressible flows is obtained.

$$\rho\frac{\partial \Phi}{\partial t} + \left(p + \frac{\rho}{2}u^2 + \rho g z\right) = F(t). \tag{5.34}$$

The arbitrary function $F(t)$ introduced by the integration can be summarized in the potential function Φ or is determined by the boundary conditions.

Hence, irrotational flows are always potential flows, even if the flow is compressible. Because the viscous term in Eq. (5.33) vanishes identically for $\vec{\omega} = 0$, it would appear that the preceding Bernoulli equation is valid even for viscous flow. Potential solutions of hydrodynamics are in fact exact solutions of the full Navier–Stokes equations. Such solutions, however, are not valid near solid boundaries or bodies because the no-slip condition generates vorticity and causes non-zero $\vec{\omega}$.

The potential flow solution often used in external flows such as in air foil theory for the determination of lift is invalid in all those parts of the flow field that have been "*contaminated*" by the spread of the vorticity by convection and diffusion.

If we take the curl $(\nabla\times)$ of Eq. (5.33), being aware that the curl of any gradient is zero, we obtain that for the vorticity an equation which is independent of the pressure p:

$$\frac{d\vec{\omega}}{dt} = (\vec{\omega}\cdot\nabla)\vec{u} + \nu\Delta\vec{\omega}, \tag{5.35}$$

where $\nu = \mu/\rho$ is the kinematic viscosity in (m^2/s). Therein, first inviscid term on the right describes the rate of change of vorticity by stretching and tilting of vortex lines while the second one represents the rate of change of $\vec{\omega}$ due to diffusion of vorticity, which may be generated, for example, at solid boundaries due to the no-slip condition or by rotation such as Coriolis forces. In inviscid, two-dimensional flow both terms on the right-hand side vanish, and the equation reduces to $d\vec{\omega}/dt = 0$. Physically, this means the vorticity of a fluid particle remains constant as it moves, which is called Helmholtz's theorem.

For a two-dimensional planar, incompressible flow the mass conservation (continuity) reduces to $\partial u/\partial x + \partial v/\partial y = 0$, is identically satisfied by introducing a stream function Ψ defined by:

$$u = \frac{d\Psi}{dy}, v = -\frac{d\Psi}{dx}. \tag{5.36}$$

Physically, the stream function Ψ expresses the volume flow rate between streamlines. Stream functions can be similarly defined to satisfy identically the continuity equations for incompressible cylindrical and spherical axisymmetric flows and for these flows, as well as the preceding planar flow, as long as they are steady.

In inviscid flows the physical meaning of irrotational flow can be illustrated by the relation between the rotation or vorticity $\vec{\omega}$, and the total head, as given by $\left(p/\rho + v^2/2 + gz\right)$. If we assume a steady flow $(\partial/\partial t = 0)$ the Bernoulli Eq. (5.33) shows that that the gradient of the total head depends upon the vorticity. Since the vorticity is by definition perpendicular to the velocity vector $\vec{u}$, the gradient of the total head has no component along a streamline. This in turn means that along a streamline in a compressible, inviscid, steady flow the following relation must be satisfied:

$$\frac{p}{\rho} + \frac{v^2}{2} + g\,z = const. \tag{5.37}$$

Thus, an irrotational and steady flow of an incompressible fluid has a constant total head throughout the flow field and Bernoulli's equation is valid throughout the flow field.

Although this section may seem somewhat theoretical at first, the result of considering irrotational flows is that one can introduce a superposition principle of forces along a streamline, which is actually denied by the nature of the Navier–Stokes equations. Thus, the Bernoulli approach prepares the basic pre-requisite for the stream tube (or stream filament) theory considered in a later section, which also allows the inclusion of viscous effects such as pressure loss along a current filament in an additive manner, provided certain conditions are met.

5.1.5 Newtonian and Non-Newtonian Fluid Behaviour

The knowledge of fluid rheology is important in the design of process equipment such as pipelines, pumps, heat exchangers, mixers and lots of other thermal components and the rheological behaviour of fluids can be divided into Newtonian and non-Newtonian categories. Unfortunately, most fluids show non-Newtonian characteristics over at least part of the shear rate range that can be achieved in practice. There are many different forms of non-Newtonian behaviour, and each of these forms has to be considered separately in the design process. Although fluids, organic and inorganic liquids, nearly all gases, inorganic salts, molten metals and salts, and gases exhibit Newtonian flow characteristics this mostly holds when the temperature is well above the melting point. Nonetheless, many liquids industrial significance, especially of multi-phase nature as mixtures, emulsions such as nanofluids, dispersions and suspensions, solutions (both natural and manmade) do not conform to the Newtonian postulate. For most liquids, the viscosity decreases with temperature and increases with pressure, while for gases it increases with both temperature and pressure. Broadly, higher is the viscosity of a substance, more resistance it presents to flow and thus more challenging to pump. Table 5.3 provides typical values of viscosity for scores of common fluids and we immediately recognize that viscosity spreads of several orders of magnitude. One can argue that a solid can be treated as a fluid whose viscosity tends towards infinity, $\mu \rightarrow \infty$.

A Newtonian fluid is defined in such a way that at constant temperature and pressure the response of the fluid is characterized by a linear relationship

Table 5.3 Dynamic viscosity μ in (Pa s) of some liquids at room temperature

Fluid	μ (Pa s)	Fluid	μ (Pa s)
Air	10^{-5}	100% Glycerol	1.5
Water	10^{-3}	Honey	10
Ethyl alcohol	1.2×10^{-3}	Syrup	10^{2}
Mercury	1.5×10^{-3}	Bitumen	10^{8}
Ethylene glycol	2×10^{-2}	Molten glass	10^{12}
Olive oil	0.1		

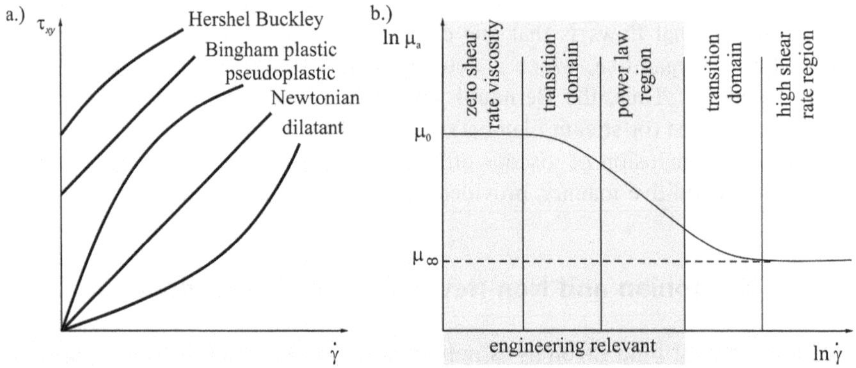

Fig. 5.3 **a** Flow curves characteristics of different viscous fluid types. **b** Flow curve of a pseudoplastic fluid with the different domains of apparent flow viscosites μ_a as function of the strain rate $\dot{\gamma}$

between the applied shear stress (τ) and the rate of shear ($\dot{\gamma}$) and the constant of proportionality is the dynamic viscosity (μ):

$$\tau_{xy} = \mu \frac{du}{dy} = \mu\, \dot{\gamma}, \tag{5.38}$$

where x refers to the direction of the shear stress y the direction of the velocity gradient. It is important to note that in a Newtonian fluid the dynamic viscosity is independent of the shear rate and thus Eq. (5.38) represents a constitutive equation specific to each fluid. Plotting a so-called flow curve, indicating the relation of shear stress as a function of strain rate as schematically shown in Fig. 5.3a is the usual procedure to visualize the viscous properties of a fluid. In such a figure, a Newtonian fluid represents a straight line of origin.

The dynamic viscosity μ has to be distinguish from the frequently used kinematic viscosity ν to which it is related by quantity $\mu = (\nu\ \rho)$. For μ the term shear viscosity is also in common use, and it varies first from substance to substance, but beyond that it is also a function of temperature. Strictly, μ even depends on the pressure (in contrast to the kinematic viscosity), but the pressure dependence can almost always be neglected. The kinematic viscosity ν can also be interpreted as a mass-specific viscosity.

Figure 5.4 contains dynamic viscosities of a larger number of substances as a function of reciprocal temperature. The plot shows that the numerical values of the gas viscosities are all within about one power of ten. In contrast, those of liquids differ by more than five orders of magnitude. Basically, viscosity is the internal resistance to motion due to molecular motion. This leads to collisions of molecules of different average velocities and consequently to an exchange of momentum. Macroscopically, this can be recognized as shear stress. Due to the low density of gases, the number of such molecular collisions is low and thus their viscosity is small. Because of the high density of liquids, the number of molecular collisions and therefore their dynamic viscosity is relatively large. While gases become more viscous with increasing temperature, the viscosity of liquids generally decreases with T. This is due to two different mechanisms. Two different mechanisms are responsible for this, one of which predominates in liquids, the other in gases. In liquids, intermolecular attractive forces must be overcome during shear. These forces generally weaken with increasing temperature because the liquid expands and the mean distance increases. In addition, the mean kinetic energy of the molecules increases, and with it the frequency of space changes. Both lead to reduced toughness with increasing temperature. Since the liquid density hardly changes with pressure, the pressure dependence of the viscosity of liquids can almost always be neglected. The average molecular distance for gases is about one order of magnitude higher than that for liquids. Therefore, intermolecular forces between them can often be neglected (ideal gases; the viscosity is therefore almost pressure-independent for ideal gases up to pressures of 10 MPa, although the molecular distances decrease sharply at this pressure change and thus the forces acting between the molecules increase by powers of ten). With increasing temperature, the molecular motion and thus the number of molecular collisions and, as a consequence, the toughness increase. The kinetic theory of gases leads to an increase in viscosity with $\sqrt{T}$. The application form chosen in Fig. 5.4 shows that for many fluids there is a proportionality $\ln \mu \sim 1/T$, at least to a first approximation, since an approximately linear progression is obtained in the diagram. The real behaviour, however, shows more or less strong deviations from the simple proportionality.

The non-Newtonian fluids are categorized in pseudoplastic (or shear thinning) and dilatant (or shear thickening). Other types as those exhibiting first an initial yield stress after which it acts as a Newtonian fluid, called Buckingham plastic, as e.g. toothpaste and those named Hershel-Buckley, which show a yield stress after which it becomes pseudoplastic. Although non-Newtonian fluids do not have a constant viscosity, it is useful to assign them an apparent viscosity (μ_a) for description using the fluid dynamic equations. This is defined as the ratio of the local shear stress to the shear rate at this point, i.e. $\mu_a = \tau_{xy}/\dot{\gamma}$.

In a next step the constitutive equations need to be defined, which relate the shear stress or apparent viscosity and the shear rate through the rheological properties of the fluid. For engineering purposes, simpler equations are mostly sufficient, and the general idea is described here only.

Since many of the non-Newtonian fluids considered in engineering applications are pseudoplastic, we use them to illustrate typical flow curves and constitutive equations. Figure 5.3b shows qualitative flow curve for a typical pseudoplastic fluid using logarithmic coordinates. The graph exhibits that at low shear rates the fluid shows a Newtonian behaviour with a constant apparent viscosity of μ_a (so-called the *zero-shear rate viscosity*). Increasing the shear rates entering the apparent viscosity begins to decrease until it becomes a straight line. This domain is called the power law region and is an important region in fluid mechanics and heat transfer. Increasing the shear rates beyond the power law region, another transition region appears until again the fluid gets Newtonian again. This behaviour is explained in

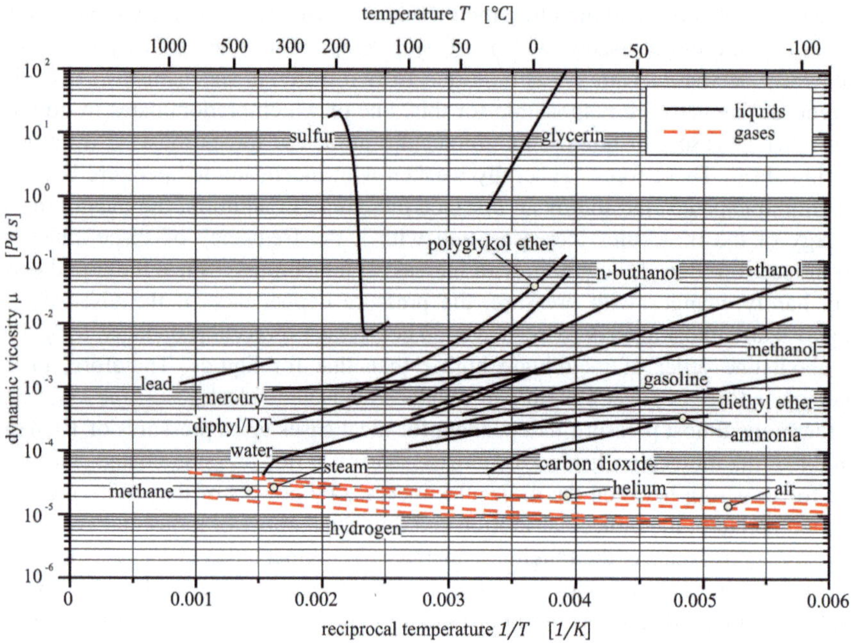

Fig. 5.4 Dynamic viscosity μ (Pa s) of some gases and liquids as a function of temperature T

polymer solutions or melts by the stretching of the molecular chains as a result of shear. The molecules become more mobile, and the viscosity decreases to a certain final value. Low shear rates do not lead to any significant change in the molecular clusters, and the viscosity initially remains unchanged. The grey shaded region indicates the domain where most of the engineering applications occur. A power law region as indicated in Fig. 5.3b can be in the simplest case defined by a constitutive equation as

$$\tau_{xy} = K \cdot \dot{\gamma}^n \text{ or } \mu_a = K \cdot \dot{\gamma}^{n-1}. \tag{5.39}$$

Here, K is termed fluid consistency and n the flow index. If $n = 1$, the fluid becomes Newtonian and K turns into the dynamic viscosity. Because of its simplicity, the power law constitutive equation is most often used in rheological studies, but at times it is inappropriate because it has several inherent flaws and anomalies. For example, if one considers the flow of a pseudoplastic fluid ($n < 1$) through a circular duct, because of symmetry at the centre of the duct the shear rate (velocity gradient) becomes zero and thus the apparent viscosity from Eq. (5.39) tends to infinity, for more details see Boger and Yeow (1988).

In order to avoid misunderstandings, it must be emphasized that a non-Newtonian media behaviour has nothing to do with a strongly variable viscosity change of a fluid, as it can be observed, for example, as a temperature-variable behaviour in water–glycol mixtures and must be considered accordingly in the pump design. Especially for fluid mixture for which the properties are not available in handbooks or other reference materials a measurement of the rheological properties is recommended and a discussion of the various instruments for rheological instruments can be found e.g. in Whorlow (1979).

5.1.6 Laminar Momentum Exchange

The laminar momentum exchange of a steady two-dimensional flow of an incompressible fluid of constant kinematic viscosity ν is described by the following equations:

$$\begin{aligned} &\frac{\partial u}{\partial x} + \frac{\partial v}{\partial y} = 0; \\ &u\frac{\partial u}{\partial x} + v\frac{\partial u}{\partial y} = -\frac{1}{\rho}\frac{\partial p}{\partial x} + \nu\left(\frac{\partial^2 u}{\partial x^2} + \frac{\partial^2 u}{\partial y^2}\right); \\ &u\frac{\partial v}{\partial x} + v\frac{\partial v}{\partial y} = -\frac{1}{\rho}\frac{\partial p}{\partial y} + \mu\left(\frac{\partial^2 v}{\partial x^2} + \frac{\partial^2 v}{\partial y^2}\right). \end{aligned} \tag{5.40}$$

This is a system of nonlinear partial differential equations of elliptic type. A uniform solution does not exist and there are few exact solutions under rather restrictive boundary conditions. Fortunately, most laminar exchange processes can be simplified considerably. At large Reynolds numbers, almost all exchange processes occur in a thin layer called the viscous boundary layer. Boundary layer simplifications are not limited to flows over solid walls; free jets or wakes behind a body can also have boundary layer character.

First, let us consider a stationary, planar and a fully developed flow in a channel as shown in Fig. 5.5. In this context, fully developed means that the velocity component u does not change in the x-direction; thus $\partial u/\partial x = 0$. For continuity reasons, $\partial v/\partial y = 0$. This ultimately means that the v-component of the velocity is constant. Excluding suction or blowing across the walls, one immediately obtains that the transverse velocity is $v = 0$. Thus, the convective terms in Eq. (5.40) on the left side disappear and in the force balance only the expression $dp/dy = 0$ remains, which means that the pressure in wall-normal direction is constant and depends only on the flow direction x.

$$\frac{dp}{dx} = \mu\frac{d^2 u}{dy^2} = \frac{d\tau}{dy}. \tag{5.41}$$

Since u is only a function of the wall-normal direction y, the partial derivative ∂ can be replaced by the simple derivative d. Only $\tau = \mu\partial u/\partial y$ remains from the stress tensor. An integration leads directly to a linear relation for the shear stress because of the symmetry which requires $\tau(y = H) = 0$ and we obtain:

$$\tau_w = -\frac{dp}{dx}H. \tag{5.42}$$

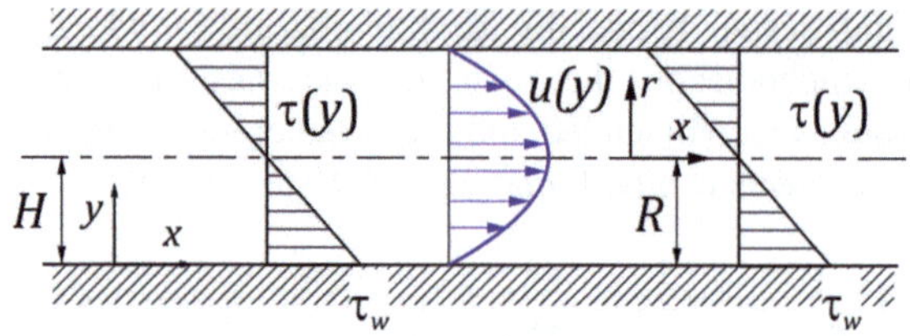

Fig. 5.5 Shear stress and velocity distribution of a fully developed flow in a planar duct or pipe

This equation links the wall shear stress with the pressure gradient and results in dimensionless form in the relationship

$$\frac{\tau(y)}{\tau_w} = 1 - \frac{y}{H}. \tag{5.43}$$

With $\tau = \mu \partial u/\partial y$ and a further integration the velocity distribution is obtained. If we set $u(y = H) = u_{max}$ we get:

$$u_{max} = \frac{\tau_w H}{2\mu} = -\frac{dp}{dx} \cdot \frac{H^2}{2\mu} = \frac{\Delta p H^2}{2\mu L}, \tag{5.44}$$

where $\Delta p/L = -dp/dx$ is the pressure drop along the channel of length L. Finally, for the dimensionless velocity distribution, we obtain:

$$\frac{u}{u_{max}} = \frac{y}{H}\left(2 - \frac{y}{H}\right), \tag{5.45}$$

which corresponds to the classical parabolic Hagen-Poiseuille profile shown in Fig. 5.5. The same result is obtained for a laminar, fully developed pipe flow leading to:

$$\begin{gathered}\frac{\tau(r)}{\tau_w} = \frac{r}{R};\ u(r) = \frac{\Delta p}{4\mu\ L}\left(R^2 - r^2\right); \\ u_{max} = \frac{\Delta p\ R^2}{4\mu\ L};\ \frac{u}{u_{max}} = 1 - \left(\frac{r}{R}\right)^2,\end{gathered} \tag{5.46}$$

where R in this context represents the pipe radius. The factor ½ is explained by the fact that in a pipe flow, compared to a plate flow, the ratio of the channel cross-sectional areas on which the pressure acts are twice as large as the surface on which the viscous forces act to compensate for the pressure. The volumetric flow $\dot{V}$ in a pipe flow thus calculates to:

$$\dot{V} = \int_{r=0}^{r=R} 2\pi r\ u(r) dr = \frac{\pi\ \Delta p\ R^4}{8\ \mu\ L}. \tag{5.47}$$

It is important to emphasize the scaling of the flow rate $\dot{V}$ with $\dot{V} \sim \Delta p/L$ and $\dot{V} \sim R^4$. For the same applied pressure gradient, a 10% increase in the tube diameter leads to a 46% increase in flow rate. Assuming a mean velocity u_0 is defined using $u_0 = \dot{V}/A$ with A as cross-sectional area, $u_0 = 1/2u_{max}$ is obtained. For engineering purposes, the pressure drop is a crucial parameter. Using $\dot{V} = \left(u_0\, \pi\, R^2\right)$ and for the diameter $D = 2R$, we get for the pressure loss over a duct length L:

$$\Delta p = \frac{\rho}{2} u_0^2 \frac{L}{D} \frac{64}{\mathrm{Re}}. \tag{5.48}$$

Defining a dimensionless quantity in form of a friction factor f_D in the way $\Delta p = (\rho/2)(L/D)u_0^2$ results in the Darcy formulation of the friction law of a pipe flow:

$$f_D = \frac{64}{Re}. \tag{5.49}$$

In it, Re is the hydraulic Reynolds number defined as $Re = u_0 D/\nu$. Thus, the pressure drop in a laminar pipe flow depends linearly on the velocity and the Reynolds number is the only occurring variable in all momentum exchange processes. One has to be careful about the normalization of the friction factor, if e.g. the Fanning friction factor f_F is mentioned, the friction factor is defined as the ratio of the wall shear stress τ_w to the dynamic pressure $(u_0^2 \cdot \rho/2)$. Then the Fanning friction factor for the tube reads instead of Eq. (5.49) to $f_F = 16/Re$. It should be noted that the Fanning friction factor is also used in the technical literature. The Fanning friction factor is 1/4 of the Darcy friction factor.

Forming an expression for frictional head loss and normalizing instead of the dynamic pressure with the gravitational pressure ($\rho\ g$), we get the so-called frictional head loss h_L in (m):

$$h_L = \frac{\Delta p}{\rho\, g} = 32 \frac{\nu\, u_0}{g} \frac{L}{D^2}. \tag{5.50}$$

If we use the Darcy formulation for the friction factor f_D, we obtain:

$$h_L = f_D \cdot \frac{L}{D} \frac{u_0^2}{2g}. \tag{5.51}$$

5.1.7 Boundary Layer Approximation

As we have already seen, due to the wall non-slip condition, a boundary layer is formed as a result of the viscosity of the fluid, within which the velocity changes from zero to the value of the external flow, as schematically illustrated in Fig. 5.6.

According to Prandtl's hypothesis developed in 1904, the effects of fluid friction at high Reynolds numbers are limited to a thin layer near the boundary of a body, while the rest of the domain behaves as inviscid. The thickness of the boundary layer is unknown a priori. Nevertheless, an estimate of its wall-normal extension can be obtained by means of order of magnitude estimations. If one introduces the following dimensionless variables,

$$x' = \frac{x}{L} \; ; y' = \frac{y}{\delta(x)} \; ; u' = \frac{u}{u_0} \; ; p' = \frac{p}{\rho_0 u_0^2}, \tag{5.52}$$

choosing the scaling in such a way that the ratios are of $O(1)$, and in a further step takes the length L and the mean flow velocity u_0 of the external flow as a reference variable in such a way, that

$$v' = \frac{v}{u_0} \; ; \rho' = \frac{\rho}{\rho_0} = 1 \quad (\text{since } \rho = const.), \tag{5.53}$$

and inserts all into the continuity equation the following result is obtained,

$$\frac{v}{u_0} \sim \frac{\delta}{L}, \tag{5.54}$$

since u', x' and y' are all of order $O(1)$. Inserting the chosen scaling into the Navier–Stokes equation (5.40) for the x-direction, the result is that the Reynolds number appears as a determinant parameter characterizing a viscous flow. Returning to the estimation and postulating that the viscous terms are of the same order of magnitude as the inertia terms and the pressure forces, it immediately follows that the thickness of the boundary layer scales as follows:

$$\frac{\delta}{L} \sim \frac{1}{\sqrt{Re}}. \tag{5.55}$$

A similar insertion of the dimensionless scaling in momentum equation for the y-direction shows that for $Re \gg 1$ the derivative of the pressure in wall-normal direction disappears. This means that the pressure perpendicular to the wall is constant within the boundary layer and is thus imposed by the inviscid external

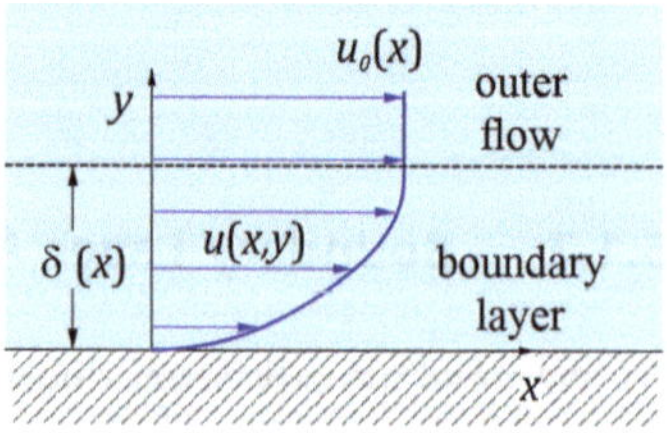

Fig. 5.6 Flow profile of a boundary layer flow

flow. Thus, the pressure p in the boundary layer is not an unknown quantity but a boundary condition. The boundary layer equations are thus:

$$\frac{\partial u}{\partial x} + \frac{\partial v}{\partial y} = 0 \text{ and } u\frac{\partial u}{\partial x} + v\frac{\partial u}{\partial y} = -\frac{1}{\rho}\frac{dp(x)}{dx} + \nu\frac{\partial^2 u}{\partial y^2}. \tag{5.56}$$

There are two equations to determine the unknowns u, $v = f(x, y)$. The pressure and the velocity gradient of the external flow are coupled via the Bernoulli equation (5.34). The boundary conditions to the system of equations are:

$$\begin{aligned} y = 0: \quad & u = v = 0, \\ y = \delta: u = & u_0(x). \end{aligned} \tag{5.57}$$

The system of equations of the boundary layer equations is thus parabolic and is a typical initial boundary value problem. Because of the simplification of the viscous terms, viscous dissipation is actively restricted to the y-direction only. There is no upstream information, and all flow quantities can be calculated straight forward, i.e. downstream. This is a considerable simplification compared to the elliptic Navier–Stokes equations, where a simultaneous calculation in all flow regions is required.

The main objective of laminar momentum exchange is to determine the local friction loss coefficient f_D and/or the global friction coefficient f_F. The reason is that such solutions include both friction factor–Reynolds number relations and a shear rate parameter, which is of relevance if non-Newtonian liquids are considered. The latter allows the designer to determine the shear rate region in which the system considered is operating and thus to choose the appropriate solution. Thus, for a laminar fully developed flow or a Newtonian or a modified power law fluid in a circular duct, the product of the friction factor and a certain Reynolds number is a constant depending on the flow index, n, and the shear rate parameter β.

$$f_D \cdot Re_h = const.(n, \beta), \tag{5.58}$$

where f_D is the Darcy friction factor and Re_h the Reynolds number built with the hydraulic diameter, which is defined as:

$$d_h = A/P, \tag{5.59}$$

with A the cross-sectional area of the conduit and P the wetted perimeter. For Non-Newtonian fluids the modified power law Reynolds number must be used, which contains a modified dynamic viscosity μ^* given by:

$$\mu^* = \frac{\mu_0}{1+\beta} \text{ with } \beta = \frac{\mu_0}{K}\left(\frac{u}{d_h}\right)^{1-n}. \tag{5.60}$$

Therein, β is the shear rate parameter and K the fluid consistency. When the $\log_{10}\beta$ is less than approximately −2, the duct is operating in a Newtonian region as indicated in Fig. 5.3b and therefore classical Newtonian solutions can be used. Considering the laminar momentum exchange of Newtonian fluids, the Darcy factor f_D and the Faning friction factor f_f can be expressed as,

$$f_f(x) = f\left(Re_x^m, e/d_h\right) \quad \text{and} \quad f_D = f\left(Re_L^n, e/d_h\right), \tag{5.61}$$

where Re_x is the local Reynolds number and Re_L is the Reynolds number formed with the characteristic dimension of the channel or along the body contour. Herein, the ratio e/d_h accounts for the surface roughness of the channel or on the body.

For a channel flow $n = -1$ holds and there the friction factor does not change in flow direction for fully developed flows. This differs from the boundary layer flow along a plate since there the boundary layer thickness along the plate increases in flow direction x while the local friction coefficient decreases with x. In this case the exponent $m = -1$ results for the local friction coefficient. In this case, the local friction coefficient has the exponent $m = -1/2$ and the total friction coefficient has the exponent $n = -1/2$. It is crucial in this context that the only two dimensionless quantities occurring are the Reynolds number and the surface property scaled with a characteristic dimension D or L. For fully developed flows, the inertial force can be replaced by the driving pressure gradient.

The following solution approaches for the calculation of laminar flows can be distinguished:

- *Analytical solutions* of the Navier–Stokes equations existing only for a few special cases.
- *Numerical solutions* of the Navier–Stokes equations are extremely time-consuming, due to the elliptic character of the nonlinear partial differential equations (PDE). They constitute a boundary value problem and no consecutive methods as for parabolic partial differential equations can be applied. In engineering practice, however, these are of little importance. For the limiting case $Re \to \infty$, the Navier–Stokes equations mutate into the boundary layer equations. In the practical cases, the Reynolds number is usually large enough to apply the boundary layer equations. An exception is the plate leading-edge problem for which, as a pre-requisite, $\partial\delta/\partial x \ll 1$ and $\partial^2/\partial x^2 \ll \partial^2/\partial y^2$ are not satisfied.
- *Exact solutions* of the boundary layer equations exist only for a few cases and denote here an arbitrarily exact numerical solution of an ordinary differential equation. For the class of self-similar solutions, the shape of the individual velocity profile does not change in the direction of flow, and it can be regressed to a similar profile by a suitable coordinate transformation.

For arbitrary external flows, the velocity profile changes in the flow direction and there are no self-similar solutions. In such cases, the boundary layer equations must be solved numerically. This can be done in three ways:

- *Numerical solutions* of the boundary layer equations exist for a variety of problems—using difference methods, the parabolic boundary layer equations can be numerically integrated as an initial value problem. For stability reasons, implicit procedures are preferable to explicit ones; several procedures are described in detail in Cebeci and Bradshaw (1977).
- *Integral methods* for solving the boundary layer equations such as Pohlhausen and V. Karman schemes, demanding low computational and memory requirements. Integral methods are still used, especially for the calculation of turbulent boundary layers. An extensive discussion of the various numerical solution approaches can be found in Walz (1966). Despite many indisputable advantages of integral methods, they have the central disadvantage; that the quality of the results depends on the adequate assumptions of the chosen initial profile.
- *Empirical solutions* are mainly used where complex geometries do not allow a direct numerical solution.

Finally, for the sake of completeness, two problems should be mentioned.

- These are higher orders of boundary layer theory, such as perturbation approaches, the so-called coupled asymptotic expansions, which are a further development of Prandtl's boundary layer concept. An overview of the different models is given in the books of van Dyke (1975) or Cole (1968).
- Another not considered aspect are three-dimensional boundary layers. Solution approaches for this are described (Schlichting & Gersten, 2000 or White, 1991).

A summary of the pressure losses or friction factors occurring in laminar flows is given together with the representation of the laminar heat transfer.

5.1.8 Laminar Energy Exchange

In a flowing fluid, energy and momentum are exchanged if

- thermal energy is supplied to or dissipated from the system;
- the kinetic energy of the flow is of the order of magnitude of the internal energy, so that dissipation (internal friction) causes the fluid temperature to rise.

In the first case a distinction is made between forced, mixed, and buoyant convection. In forced convective systems, the flow is driven by an external source, for example, a pressure gradient generated by a pump, while in free or natural convection, density differences (resulting from temperature differences) are the driving forces of the flow.

Considering stationary, two-dimensional, and single-phase flow of a Newtonian fluid, we obtain the following system as the set of equations to be solved for laminar flows with energy transport:

$$\frac{\partial u}{\partial x} + \frac{\partial v}{\partial y} = 0;\ \rho\left(u\frac{\partial u}{\partial x} + v\frac{\partial u}{\partial y}\right) = -\frac{\partial p}{\partial x} + \mu\left(\frac{\partial^2 u}{\partial x^2} + \frac{\partial^2 u}{\partial y^2}\right) + \rho g_x;$$
$$\rho\left(u\frac{\partial v}{\partial x} + v\frac{\partial v}{\partial y}\right) = -\frac{\partial p}{\partial y} + \mu\left(\frac{\partial^2 v}{\partial x^2} + \frac{\partial^2 v}{\partial y^2}\right) + \rho g_y;$$
$$c_p\left(u\frac{\partial T}{\partial x} + v\frac{\partial T}{\partial y}\right) = k\left(\frac{\partial^2 T}{\partial x^2} + \frac{\partial^2 T}{\partial y^2}\right) + u\frac{\partial p}{\partial x} + v\frac{\partial p}{\partial y} + \Phi. \tag{5.62}$$

where k is assumed to be the specific heat conductivity constant and $\vec{g}$ is the gravity field vector in the form $\vec{g} = (g_x, g_y)$. Φ is the dissipation function formulated in Eq. (5.12).

In this section, only steady-state, incompressible laminar flows of a Newtonian fluid with constant material properties are considered. All forms of bulk forces such as phase change, mass transport and chemical reactions are omitted here. A very complete presentation of laminar heat transport correlations and solution approaches can be found in Shah and London (1978).

Until now we had assumed that the hydrodynamic and thermal quantities do not change in the direction of flow, i.e. $d/dx = 0$, which is an idealized idea, since for both the viscous and thermal boundary layers are already fully formed. However, when entering a channel, both must develop along the channel walls, so that actually four types of laminar channel flow domains can be distinguished:

- fully developed,
- hydrodynamically evolving,
- thermally developing and
- simultaneously developing

flows. The latter form states that the flow is hydrodynamically and thermally evolving simultaneously.

Figure 5.7a sketches the distinguishing features of the individual flow forms and regions. It represents a flow of a fluid with constant velocity u_0 and temperature T_0 entering a duct of an arbitrary cross-section at $x = 0$.

If we first assume an isothermal flow and postulate that there is no heat generation or dissipation in the fluid, the fluid experiences no temperature change and a viscous boundary layer is formed by the viscosity of the fluid starting at $x = 0$. The viscous boundary layer grows in wall-normal thickness downstream according to Prandtl's boundary layer theory proportionally with the duct length corresponding to $1/\sqrt{Re_x}$. The boundary layer divides the flow field into two domains; a viscous region at the walls and a frictionless region around the centre of the channel. At $x = l_{hy}$, the viscous boundary layers have merged and cover the entire channel cross-section. The length $0 \leq x \leq l_{hy}$ is called the hydrodynamic inlet

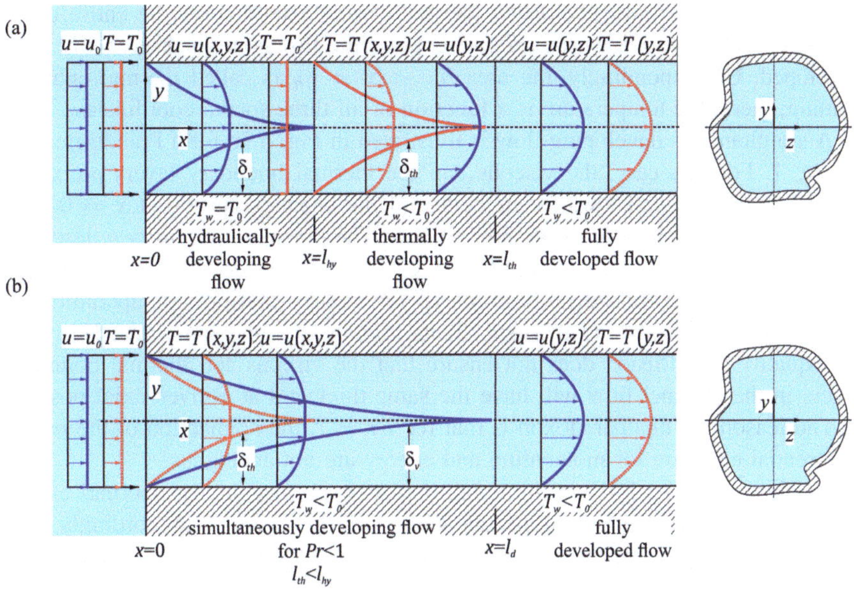

Fig. 5.7 Classification of a laminar channel flow at a boundary condition with constant wall temperature T_w into different regions. **a** Hydrodynamically developing flow followed by thermally evolving (and hydraulically developed) flow; **b** simultaneously developing flow in liquid metals ($Pr \ll 1$). Solid lines indicate velocity profiles, while dotted lines outline temperature profiles

region. In the inlet region, the velocity depends on all three spatial coordinates, while for $x > l_{hy}$ the velocity distribution is independent of the axial coordinate x.

After reaching the entrance length we assume that the channel wall temperature is lowered to a constant value with respect to the inlet temperature of the fluid, i.e. $T_w < T_0$. In this case, the thermal effects gradually diffuse from the channel wall towards the core flow starting at $x = l_{hy}$. The magnitude of the thermal diffusion is determined by the thickness of the thermal boundary layer δ_{th}, which increases along the axial coordinate x.

A similarity analysis based on the boundary layer approximation approach shows that the thermal boundary layer becomes thicker according to $\delta_{th}(x) \sim 1/\sqrt{Re \cdot Pr}$. The application of the boundary layer approximation to the thermal case also reveals that the thermal boundary layer separates the flow field into a zone near the wall influenced by the thermal effects and an uninfluenced region in the flow core. One result of the dimensional analysis is the scaling of the ratio of the thicknesses of the thermal and viscous boundary layers:

$$\frac{\delta_{th}}{\delta_v} \sim \frac{1}{\sqrt{Pr}}. \tag{5.63}$$

At $x = l_{th}$, the influence of the thermal effects extends over the entire channel cross-section and downstream of this point the flow is termed thermally fully developed. Correspondingly, the area $l_{hy} < x < l_{th}$ is called thermal entrance domain; there, the temperature is a function of all three spatial coordinates.

A simultaneous developing flow is illustrated in Fig. 5.7b for a Prandtl number $Pr \ll 1$. For this case, the viscous and the thermal boundary layers propagate from the channel wall towards the centre of the channel starting at $x = 0$. The decisive parameter here is the Prandtl number. The kinematic viscosity ν describes the propagation of the momentum flow (velocity) in the same way as the thermal diffusivity a describes the propagation of the energy transport (temperature). If $Pr = 1$, then viscous and thermal effects propagate at the same rate. However, the equality of diffusion does not ensure that the viscous and thermal boundary layers in the channel flow will have the same thickness at a given axial position x. The reason for this paradoxon is that for $Pr = 1$, the application of the partial differential equation for momentum and energy are not analogous.

As shown in Fig. 5.7b, in the range $0 < x < l_d$, viscous and thermal effects simultaneously propagate toward the center of the channel. Accordingly, this region is also called the inlet (entrance) region in this case. Obviously, the length l_d also depends here on the Prandtl number and velocity as well as temperature depend on all three spatial coordinates. Only for $x > l_d$, i.e. in the entrance region, both become axially invariant and depend only on y and z, i.e. the following applies: $u = u(x, y)$ and $T = T(x, y)$.

5.1.9 Flow and Heat Transfer Parameters

The flow characteristics of all channel flows are specified in terms of hydraulic parameters, which are universally specified as dimensionless quantities whenever possible. For a hydraulic inflow, the axial inlet length x^+ is defined as follows:

$$x^+ = \frac{1}{Re}\frac{x}{d_h}, \tag{5.64}$$

where d_h is the hydraulic diameter. The hydraulic entrance length l_{hy} is described as the length in flow direction required by the flow to reach 99% of the magnitude of the fully developed flow velocity. The dimensionless hydraulic entrance length, as it is mostly given in textbooks, is thus given by the expression $l_{hy}^+ = l_{hy}/(d_h \cdot Re)$.

The mean fluid temperature often also called bulk (-sometimes mixing) temperature T_m is defined by the following expression:

$$T_m = \frac{1}{A\, u_0}\int\limits_A uT\, dA, \tag{5.65}$$

and thus, the circumferentially averaged but local axial heat transfer coefficient h_x is given by:

$$\dot{q}_{w,x} = h_x \left(T_{w,m} - T_m \right), \tag{5.66}$$

where $T_{w,m}$ is the mean wall temperature and T_m is the fluid bulk temperature (from Eq. (5.65)).

The heat flux $\dot{q}_w$ and the temperature difference $\left(T_{w,m} - T_m \right)$ are directional and thus vectorial quantities. In the notation of Eq. (5.66), the direction of heat flow points from the wall toward the fluid and, consistent with this, the temperature falls from the wall toward the fluid. In contrast, if $\dot{q}_w$ points the heat flow from fluid in the direction of the wall, the difference is $\left(T_m - T_{w,m} \right)$. The heat transfer coefficient hm averaged over the path of the flow x is obtained by integration from $x = 0$ to x in the following form:

$$h_m = \frac{1}{x} \int\limits_{x=0}^{x=x} h_x \, dx. \tag{5.67}$$

The ratio of the convective heat transfer h to the heat transfer by molecular heat conduction k/d_0 is described as the Nusselt number Nu. The circumferentially averaged but local axial Nusselt number Nu_x is defined by:

$$Nu_x = \frac{h_x d_h}{k} = \frac{\dot{q}_{w,x} d_h}{k \left(T_{w,m} - T_m \right)}. \tag{5.68}$$

Thus, the Nusselt number describes nothing other than the dimensionless temperature gradient at the wall. However, the Nusselt number can also be understood as the ratio of two lengths, namely that of the characteristic channel dimension to the local thickness of the thermal boundary layer. The average Nusselt number Nu_m based on h_m in the thermal inlet (entrance) region is thus given by:

$$Nu_m = \frac{1}{x} \int\limits_0^x Nu_x dx = \frac{h_m d_h}{k} = \frac{\dot{q}_{w,m} d_h}{k (\Delta T)_m}. \tag{5.69}$$

The expression for $(\Delta T)_m$ could be problematic to determine since it depends on the thermal boundary conditions. The dimensionless axial thermal inlet length x^* is defined by:

$$x^* = \frac{1}{Pe} \frac{x}{d_h} = \frac{1}{Re \; Pr} \frac{x}{d_h}. \tag{5.70}$$

The thermal entrance length l_{th} is defined as the axial length required to attain a local Nusselt number Nu_x corresponding to 1.05 times the Nusselt number of a thermal developed flow. The dimensionless thermal inlet length is thus given by the following expression $l^*_{th} = l_{th}/(d_h\ Pe)$. The Stanton number St is also often used. It describes the ratio of the heat flux dissipated by the wall to the enthalpy difference of the external flow. Accordingly, the Stanton number is defined as follows:

$$St = \frac{h}{\rho\ c_p\ u_0} = \frac{\dot{q}_w}{\rho\ c_p u_0\ \Delta T} = \frac{Nu}{Re\ Pr}. \tag{5.71}$$

5.1.10 Laminar Heat Transfer in Tubes

The qualitative behaviour of the laminar momentum exchange as a function of the individual initial and boundary conditions can be clearly illustrated in the technically most relevant case of a pipe flow. It is useful to distinguish the four different flow domains from Fig. 5.7, considering the respective thermal boundary conditions.

- Fully developed flow

Assuming a constant wall temperature ($T_w = const.$), the temperature distribution in circular tubes can be specified exactly, provided the flow is non-dissipative, has no energy sources or axial heat conduction. The expression for the mean fluid bulk temperature is given by the following asymptotic formula for $x^* > 0.0335$, see Bhatti (1985):

$$\frac{T_w - T_m}{T_w - T_{in}} = 0.819048\ exp\left(-2\lambda_0^2 x^*\right) \text{ with } \lambda_0 = 2.70436442, \tag{5.72}$$

where T_{in} denotes the inlet temperature. It should be noted that although the local temperature T is simultaneously a function of the radial and axial coordinates, the mean fluid temperature T_m depends only on the axial coordinate (the dimensionless temperature $(T_w - T)/(T_w - T_m)$ is a function of the radius only). The mean Nusselt number of a fully developed flow is thus:

$$Nu_m = \frac{\lambda_0^2}{2} =^* 3.6568. \tag{5.73}$$

For developed flows with constant wall temperature, the Nusselt number is thus a constant and independent of the Prandtl number. However, if the Peclet number Pe is less than 10, axial heat conduction cannot be neglected and the following asymptotic expression is obtained, cf. Michelsen and Villadsen (1974):

$$Nu_m = \begin{cases} 4.180654 - 0.18346 \cdot Pe & \text{for } Pe < 1.5 \\ 3.656794 + \frac{4.4870}{Pe^2} & \text{for } Pe > 5.0 \end{cases}. \tag{5.74}$$

For the local Nusselt number Nu_x at a location x, calculated from the beginning of heating or cooling, the asymptotes are as follows:

$$Nu_x = \begin{cases} 3.6568 & \text{for } \; Pe \cdot (d_h/x) < 1 \\ 1.077 \cdot (Pe \cdot d_h/x)^{1/3} & \text{for } \; Pe \cdot (d_h/x) > 1 \end{cases}, \tag{5.75}$$

where Pe is the Peclet number defined as $Pe = Re \cdot Pr$.

For a constant wall heat flux ($\dot{q}_w = const.$), the temperature distribution in circular tubes for non-viscous dissipative systems, without energy sources and axial heat conduction can also be derived analytically and it is obtained:

$$\left(\frac{T_w - T}{T_w - T_m}\right) = 6\left(1 - \left(\frac{r}{R}\right)^2\right)\left(\frac{3 - (r/R)^2}{11}\right);$$
$$k\left(\frac{T_w - T}{\dot{q}_w d_h}\right) = \frac{11}{48} = \frac{1}{Nu} = \frac{\lambda}{h d_h}. \tag{5.76}$$

The Nusselt number for the case with constant heat flow is thus about 20% larger than that at constant wall temperature. The qualitative form of the mean fluid temperature distribution for both boundary conditions as a function of the axial coordinate x is shown in Fig. 5.8.

Regarding the boundary condition of a convectively heated/cooled wall, it is assumed that the temperature along the channel is constant in flow direction, but a finite wall-normal thermal resistance exists. This thermal resistance can be embedded in an external convective heat transfer coefficient h_e, which is expressed by a dimensionless Biot number. The Biot number Bi is defined by $Bi = (h_e d_h)/k$ (see Sect. 4.2).

Via the Biot number the influence of the thermal wall resistance can be considered. In this case, the Biot number is inversely proportional to the total thermal resistance R_w, i.e. $Bi = 1/R_w$. The total thermal resistance is given by

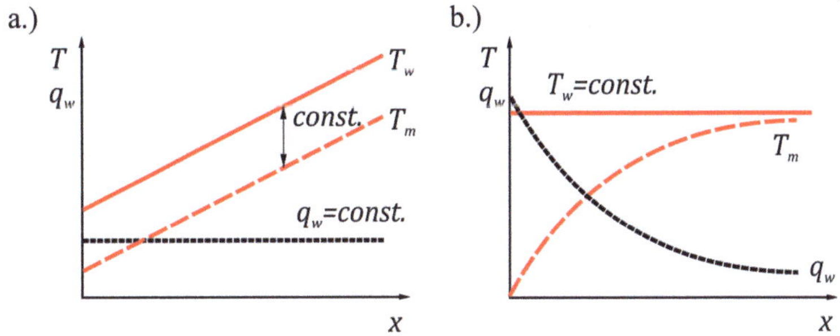

Fig. 5.8 Qualitative distribution of the mean fluid temperature T_m, the wall temperature T_w and the wall heat flux $\dot{q}_w$ as a function of the axial coordinate x for **a** constant wall heat flux $\dot{q}_w$ and **b** const. wall temperature T_w

$R_w = (kt_w)/(k_w d_h) + k/(h_e d_h)$. Thus, the limiting case of constant wall temperature corresponds to $Bi \to \infty$ and the other limiting case of constant wall heat flux implies $Bi = 0$. Therefore, the Biot number can be taken as the ratio of the wall thermal resistance to that of the fluid. For a constant h_e (Hickmann, 1974), an asymptotic solution for the Nusselt number is obtained and the global mean Nusselt number $Nu_{o,m}$ is:

$$Nu = \frac{\frac{48}{11} + Bi}{1 + \frac{59}{220} Bi}, \quad \frac{1}{Nu_{o,m}} = \frac{1}{Nu} + \frac{1}{Bi} \quad \text{with} \quad Nu_{o,m} = \frac{h_e d_h}{k} = \frac{q_w d_h}{k(T - T_m)}. \tag{5.77}$$

$Nu_{o,m}$ is insensitive with respect to a circumferential change in the Biot number after Sparrow et al. (1978).

- Hydraulically developing flow

The hydraulic inflow problem has been studied in many ways and depending on the Reynolds number, different solutions exist, which can be categorized as follows:

- Solutions using the boundary layer equations for $Re > 400$;
- Solutions using the Navier–Stokes equations for $Re < 400$;
- Solutions for creeping flows for $Re \to 0$.
- Boundary layer type solutions are described in detail in Shah and London (1978). Technically, the solution of Horbeck is mostly used. According to this, the dimensionless hydraulic inlet length l_{hy}^+ + and the dimensionless pressure gradient Δp^* are obtained:

$$l_{hy}^+ = 0.0565, \quad \Delta p^* = 13.74\sqrt{x^+} + \frac{1.25 + 64x^+ - 13.74\sqrt{x^+}}{1 + 2.1 \times 10^{-4}\left(x^+\right)^{-2}}. \tag{5.78}$$

Near the inlet of the flow into the channel, both the axial diffusion of the momentum and the radial pressure distribution are relevant. Adequate consideration of these aspects inherently leads to the inclusion of the Reynolds number and requires accurate knowledge of the velocity profile at the inlet. With the help of computational fluid dynamic codes (CFD) an arbitrarily exact solution can be obtained. Assuming a constant velocity far from the inlet, Chen (1973) obtains a relation for the hydraulic inlet length.

$$l_{hy}^+ = 0.056 + \frac{0.6}{Re(1 + 0.035Re)}. \tag{5.79}$$

- Thermally developing flows

At a constant wall temperature ($T_w = const.$) and neglecting viscous dissipation, axial fluid heat conduction, and other energy sources/sinks, the solution for the local Nusselt number Nu_x is

$$Nu_x = \begin{cases} 1.077 \cdot (x^*)^{-\frac{1}{3}} - 0.7 & \text{for } x^* \leq 0.1 \\ 3.657 + 6.874 \cdot \left(10^3 x^*\right)^{-0.488} exp(-57.2x^*) & \text{for } x^* \geq 0.1 \end{cases}, \tag{5.80}$$

and the relation for the mean Nusselt number Nu_m in the whole range of x^* is

$$Nu_m = 3.66 + \frac{0.0668}{x^{*1/3}\left(0.04 + x^{*2/3}\right)} \tag{5.81}$$

The error of the analytical relation is of the order of 2% compared to a full numerical solution. The influence of axial heat conduction is negligible for $Pe > 50$, but the thermal entrance length l_{th}^* increases considerably, from 0.033 for $Pe \to \infty$ to 0.5 for $Pe = 1$. The dimensionless temperature distribution T as a function of r/R and x^* is shown Fig. 5.9. The mean local Nusselt number Nu_x as well as the mean Nusselt number Nu_m are shown in Fig. 5.10.

Analogously, the thermal inlet length can be solved for a constant wall heat flux ($\dot{q}_w = const.$). The mean local Nusselt number Nu_x and the mean Nusselt number Nu_m can be derived by means of approximations:

$$Nu_x = \begin{cases} 1.302 \cdot \left(x^*\right)^{-1/3} - 1.0 & \text{for } x^* \leq 5 \times 10^{-5} \\ 1.302 \cdot \left(x^*\right)^{-1/3} - 0.5 & \text{for } 5 \times 10^{-5} \leq x^* \leq 1.5 \times 10^{-3} \\ 4.364 + 8.68 \cdot \left(10^3 x^*\right)^{-0.56} \exp(-41x^*) & \text{for } x^* \geq 1.5 \times 10^{-3} \end{cases}; \tag{5.82}$$

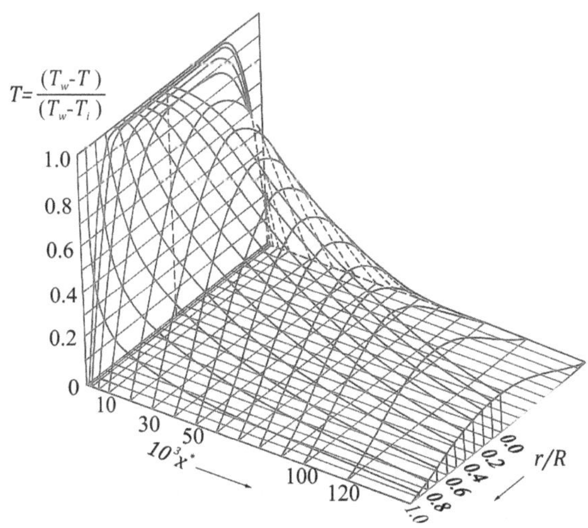

Fig. 5.9 Dimensionless temperature distribution of a laminar thermal pipe inlet flow as a function of the dimensionless length x^* for $T_w = const.$ from Grigull and Tratz (1965)

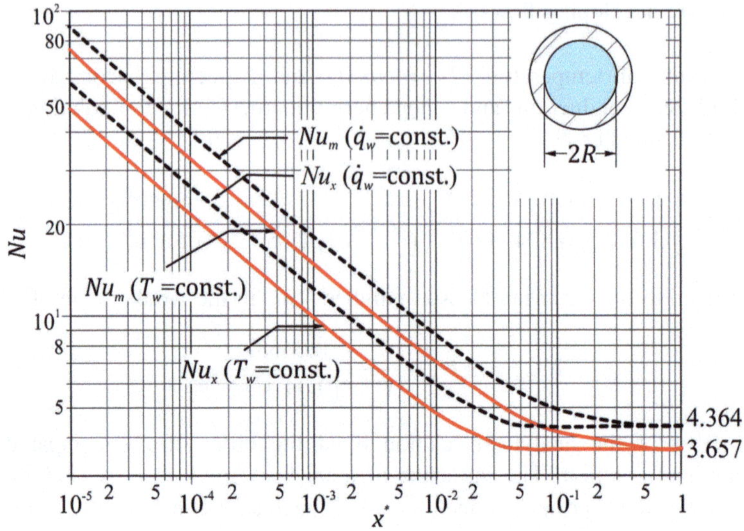

Fig. 5.10 Local Nu_x and mean Nu_m as a function of dimensionless inlet length x^* for a hydraulically inlet and thermally inlet pipe flow for $T_w = const.$ and $\dot{q}_w = const.$

$$Nu_m = \begin{cases} 1.953 \cdot (x^*)^{-1/3} & \text{for} x^* \le 0.03 \\ 4.364 + 0.0722/x^* & \text{for} x^* \ge 0.03 \end{cases}. \tag{5.83}$$

The approximation error is about $\pm 2\%$ compared to the exact solution. Again, the axial heat conduction is negligible for $Pe > 10$, provided $x^* > 0.005$. This is less than for the constant wall temperature case. The thermal inlet length l_{th}^* is 0.0430527. The temperature distribution is given in Fig. 5.11 and the Nusselt numbers can be taken from Fig. 5.10.

For the convective heated/cooled wall boundary condition, a solution involving the Biot number can also be determined using a series expansion (Hsu, 1968). The local Nusselt numbers as a function of the dimensionless distance x^* for different Biot numbers are shown in Fig. 5.12.

- Simultaneous developing flow

If the flow is not hydraulically fully developed, the heat transfer solution depends directly on the Prandtl number.

Constant wall temperature ($T_w = const.$)

For $Pr \to \infty$, the flow is already fully developed hydraulically before a temperature profile form. In the case of a well molecular conducting fluid, assuming the idealized boundary case $Pr = 0$, the temperature profile forms much faster than the velocity profile, so that in this case during the formation of the temperature field the velocity field changes only marginal downstream. The unmodified

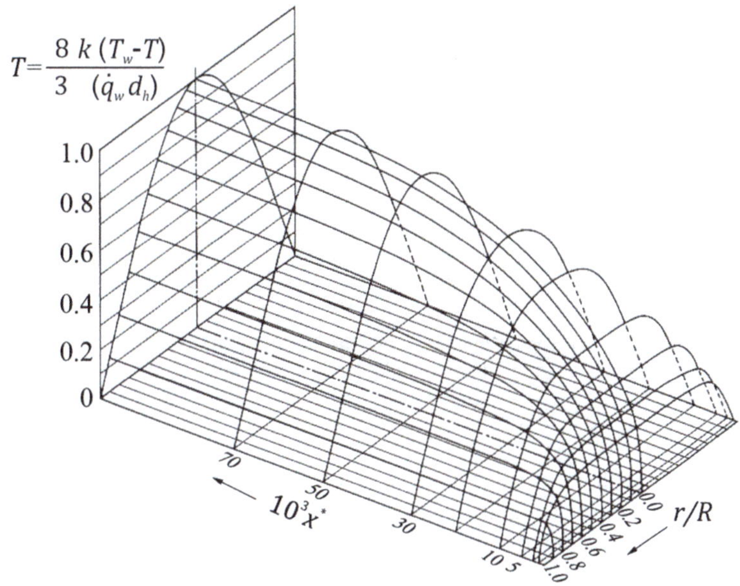

Fig. 5.11 Dimensionless temperature distribution of a laminar thermal pipe inlet flow as a function of the dimensionless length x^* for $\dot{q}_w = const.$ from Grigull and Tratz (1965)

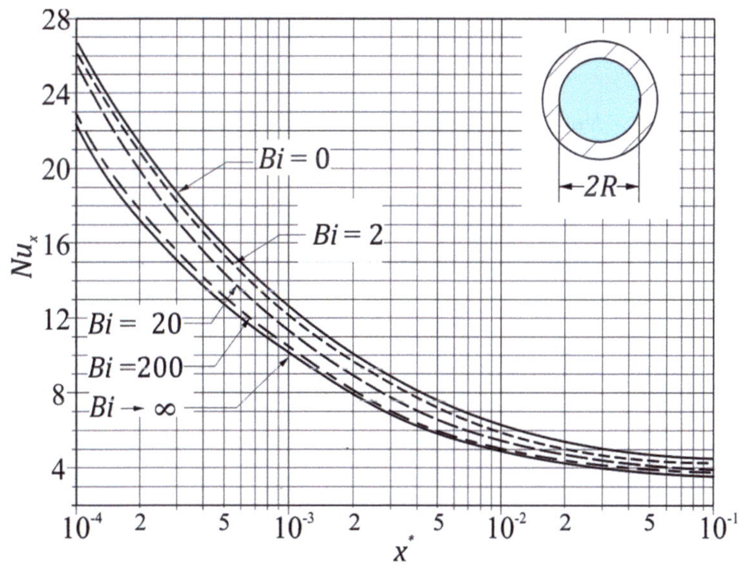

Fig. 5.12 Local Nu_x as a function of the dimensionless inlet length x^* of a hydraulically developed and thermally evolving laminar pipe flow for mixed convective thermal boundary conditions expressed by means of the Biot number Bi

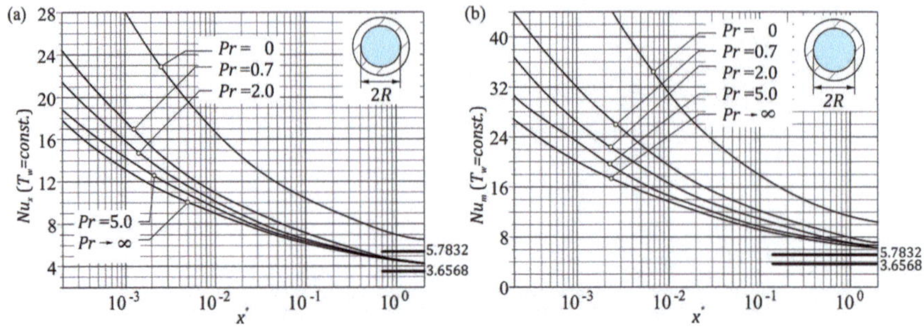

Fig. 5.13 Local and mean Nusselt numbers Nu_x (**a**) and Nu_m (**b**) as a function of the dimensionless entrance length x^* for simultaneous inlet laminar pipe flows with $T_w = const.$ and for different fluids (Pr)

velocity profile with its small boundary layer has a high velocity near the wall, which leads directly to a significantly increased heat transport and thus to larger Nusselt numbers than for the developed flow. Analytical solutions can be found for the idealized cases $Pr = 0$ and $Pr \to \infty$, while series expansions involving Bessel functions are required for the other Prandtl numbers. The mean and local Nusselt numbers Nu_x and Nu_m as a function of the dimensionless inlet length x^* for different Prandtl numbers are shown in Fig. 5.13a, b.

The dependence of the local heat transfer coefficient Nu_x is well described by the following approximate equation for media with different Prandtl numbers.

$$Nu_x = \frac{1}{2}\left(\frac{2}{1 + 22\,Pr}\right)^{\frac{1}{6}} \cdot \sqrt{Re\,\mathrm{Pr}(d_h/x)}. \tag{5.84}$$

Note that for $Pr = 0$, the local and mean Nusselt numbers have the value 5.7832 as the lowest limit, while the others have the value 3.6568 as the lower limit. The thermal entrance length for this type of flow is:

$$l^*_{th} = \begin{cases} 0.028 \text{ for } Pr = 0 \\ 0.037 \text{ for } Pr = 0.7 \\ 0.033 \text{ for } Pr \to \infty \end{cases}. \tag{5.85}$$

Constant wall heat flux ($\dot{q}_w = const.$)

Using a similar approach shows the limit value for the local mean Nusselt number and the idealized fluid with $Pr = 0$ for large x^* of 8, which is about 40% more than for constant wall temperature. The dependence of the local Nusselt number as a function of the inlet length is shown in Fig. 5.14. Not only is the heat transfer greater at constant wall heat flux, but the thermal entrance length also increases.

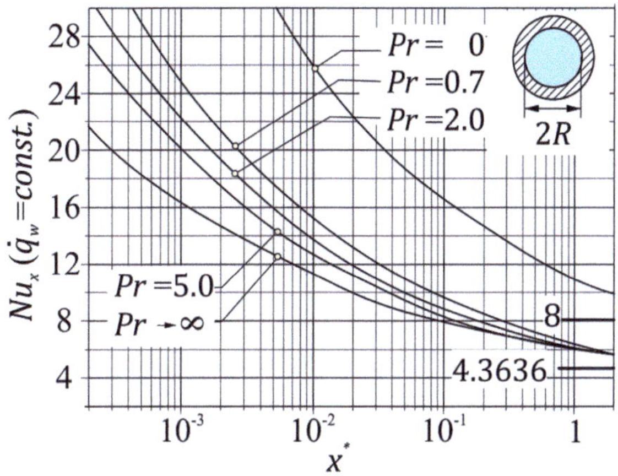

Fig. 5.14 Local and mean Nusselt numbers Nu_x (**a**) and Nu_m (**b**) as a function of the dimensionless entrance length x^* for simultaneous evolving pipe flows with $\dot{q}_w = const.$ and for different fluids (Pr)

It is about 50% more for well molecular conducting fluids than at constant wall temperature. Shah and London (1978) suggest the following values:

$$l_{th}^* = \begin{cases} 0.042 \text{ for } & Pr = 0 \\ 0.053 \text{ for } & Pr = 0.7 \\ 0.043 \text{ for } & Pr \to \infty \end{cases} . \tag{5.86}$$

The local Nusselt number Nu_x for a pipe flow reads to:

$$Nu_x = \begin{cases} 0.459 \cdot Pr^{1/3} \cdot \sqrt{(Re \cdot d_h/x)} \text{ for } & Pr \approx 1 \\ 0.464 \cdot Pr^{1/3} \cdot \sqrt{(Re \cdot d_h/x)} \text{ for } & Pr \to \infty \end{cases} . \tag{5.87}$$

- Influence of the temperature dependence of the thermo-physical properties

The dimensionless parameters relevant for laminar heat transfer, such as Nu and Re, depend on thermophysical properties of the fluid, and the solutions were derived assuming constant material values. For the practical use of the equations, one can use for the material values those at the mean temperature of the fluid flow $T_m = (T_e + T_o)/2$, where T_e is the temperature at the entrance and T_o that at the outlet, however, this refers only if the temperature differences are not too large. The direction of the heat flow (heating or cooling) influences the heat transfer for temperature-dependent material data. Laminar flow experiments of gases, such as air, nitrogen, and helium in the range $0.5 < T/T_w < 2.0$ have shown that the heat transfer coefficient changes less than 10%, where T is the mean Kelvin temperature of the gas and T_w is the Kelvin temperature of the tube wall. For liquids, a correction factor must be considered, which is obtained from the ratio of the Prandtl numbers of the liquid at the mean bulk temperature T_m and

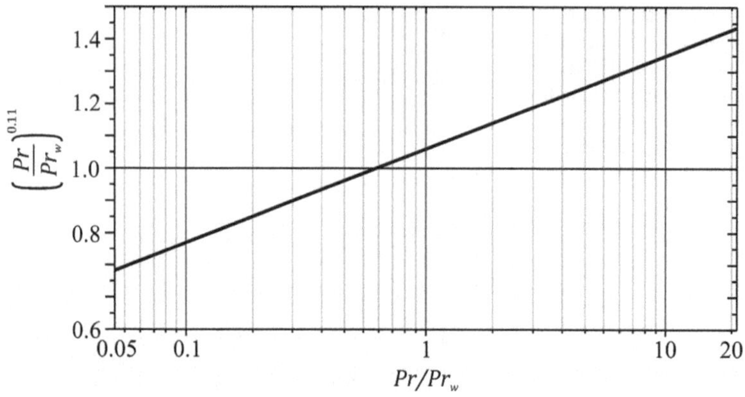

Fig. 5.15 Consideration of the temperature dependence of fluids during heat transfer in the tube

that at the wall, respectively. The effective Nusselt number Nu is then obtained multiplicatively from, see Fig. 5.15:

$$Nu = Nu_m \cdot \left(\frac{Pr}{Pr_w}\right)^{0.11} . \tag{5.88}$$

Note, that for turbulent flows also for gases a Prandtl number ratio of bulk fluid temperature to fluid wall interface temperature is required as well.

The main subject of laminar heat transfer was the different thermal boundary conditions, which lead to considerably different Nusselt numbers depending on the respective hydrodynamic inlet conditions. The influence of the fluid properties in dimensionless form thus plays a decisive role in inlet flows. The main point in laminar heat transfer is directed to the heat that can be dissipated from the wall as a function of geometry and boundary conditions. The wall heat flux, on the other hand, which is given by Fourier's law ($\dot{q}_w = -k(\partial T/\partial n)_w$, depends only on the molecular heat conduction of the fluid and the temperature gradient at the wall.

Although laminar heat transport is a purely molecular transport process, the flow field plays a central role. The flow is responsible for the convective transport of internal energy or enthalpy; it influences the temperature gradient. Obviously, heat transfer can be improved by increasing the flow velocity. The higher kinetic energy allows a larger amount of energy to be removed convectively from the fluid wall interface. For incompressible flows, this process is described only by one characteristic number. In forced convective systems this number is the Reynolds number, while in buoyant flows of natural convection it is given by the Grashof number. In both characteristic numbers, the molecular viscosity appears, which is responsible for momentum transport. In the heat exchange in a fluid, the ratio of the molecular viscosity to the molecular thermal conductivity $a = k/(\rho\, c_p)$, and thus the Prandtl number (Pr), plays the decisive role. In contrast to the Reynolds or Grashof number, Pr is a pure material constant and depends only on the thermophysical fluid properties.

The normalization of the wall heat flux with a suitable temperature difference is called the heat transfer coefficient $h = \dot{q}_w / \Delta T$. The heat transfer coefficient can be de-dimensioned in different ways. The Stanton number is the ratio of the wall heat flux to the enthalpy difference and does not include heat conduction. More common is the use of the Nusselt number, in which the heat transfer coefficient is multiplied by a characteristic length and weighted with respect to molecular heat conduction.

For pure forced convection, the results can all be presented in universal form:

$$\frac{Nu_x}{\sqrt{Re_x}} = f\left(Pr^n\right) \text{ with } 0 < n < 1. \tag{5.89}$$

- The expression $Nu_x/\sqrt{Re_x}$ grows with increasing Prandtl number. This does not mean that the heat transfer also increases with increasing Prandtl number, because the Nusselt number itself includes the molecular heat conduction.
- For fully developed flows, the Nusselt number is constant. The constant depends only on the thermal boundary conditions.

The Prandtl number only influences the heat transfer for developing flows. There are several ways to derive the heat transfer relations:

- Analytical solutions of the conservation equations (only possible in simple cases with drastic restrictions).
- Numerical solution of the conservation equations (very computationally intensive).
- Exact solutions of the boundary layer equations. Numerical solutions of the boundary layer equations (require mathematical effort but can be easily applied with commercial CFD programs).
- Approximate solutions using integral methods with lower accuracy than numerical methods.
- Analogies between heat and momentum transport are only applicable for $Pr = 1$ and result in an exact solution.
- Empirical correlations must be used for complex geometries.

5.1.11 Turbulent Momentum Exchange

In most engineering applications, however, the flow is turbulent, which is illustrated in Fig. 5.16 by the famous colour thread experiment first conducted by Osborne Reynolds 1883. If the flow velocity is continuously increased expressed by an increasing Reynolds number a transition from laminar flow characterized by a stratified ink layer as illustrated in Fig. 5.16a change first into transitional flow and finally into a turbulent flow if a critical velocity is exceeded. He determined a

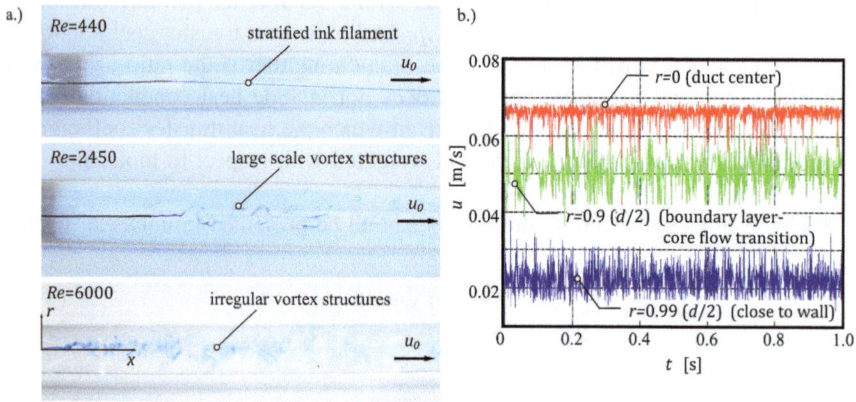

Fig. 5.16 **a** Reproduction of the ink filament Osborne Reynolds experiment by a water flow in a horizontal duct at three different Reynolds numbers; **b** measured velocities art $Re = 6000$ at three radial positions of the experiment in (**a**)

value for this laminar turbulent transition to occurs in a circular duct at a critical Reynolds number of $Re_{crit} \approx 2.300$ and assumed that this transition is a stability problem. From this simple experiment two main questions arise:

- How does turbulence evolve from laminar flow? This is indeed a stability problem, first solved by Tollmien in 1929 using a perturbation approach. There is a myriad of literature on this for example Schlichting and Gersten (2000), Betchov and Criminale (1967) and White (1991).
- What are the specific characteristics of a turbulence flow? Why does a turbulent flow remain turbulent? Here, the key to answering this question lies in the nature of turbulence production. As shown in the velocity recordings in the water ink flow in Fig. 5.16b obviously viscous shear of the boundary towards the core impacts the time dependent behaviour of fluid motion.

Turbulent flows are always unsteady, three-dimensional, and stochastic. From experiments we know that the transition from laminar to turbulent is accompanied by the following phenomena:

- increase of the pressure drop;
- increase of the boundary layer thickness.

Both are caused by turbulent diffusion. Due to the intense macroscopic fluctuations, the turbulent velocity profile becomes "fuller", and the turbulent boundary layer becomes thicker compared to the laminar flow. All variables in the turbulent flow field depend on the spatial coordinates x, y, and z and time t. The unknowns of a turbulent isothermal flow are thus all the velocity components of the velocity vector $\vec{u}$ namely u, v, w and the pressure p, which are all functions of the three spatial coordinates and the time.

There are several ways to treat turbulent flow problems and the user has to specify, which type of solution approach is required. In simplified terms, a distinction can be made between two classes of application and analysis questions.

The design and analysis of thermally highly stressed components requires detailed local two- or three-dimensional, often transient, computationally validated thermal–hydraulic modelling, which is solved by means of computational fluid dynamics (CFD). In contrast to this are system analyses, as they are relevant for regulation and control or for techno-economic analyses (yield, performance,). These are based on scale-separation methods and are therefore often one- or one-and-a-half-dimensional in nature. Therefore, the system analysis tools used there are mainly based on simplified validated correlations, which require a qualified user to adequately apply flow domain decomposition approaches and to select suitable models.

The focus of a textbook cannot be a detailed presentation of the model-technical subtleties of a three-dimensional turbulent CFD simulation as well as the details of a system-analytical calculation, so that the following presentation concentrates on the description of the essential methodical thought processes in the solution for both solution classes. The next two sections are dedicated to elaborate the modelling used for the detailed computation of turbulent flows by means of CFD approaches.

Detailed thermal hydraulic simulations can be roughly categorized into three larger classes.

- Direct numerical simulation (DNS) solve the Navier–Stokes equations are numerically without applying any turbulence model requiring that the entire range of spatial and temporal scales of the turbulence must be resolved in the computational grid from the smallest dissipative scales (Kolmogorov scales η) where the fluctuations are dissipated, up to the integral scale L, associated with the motions containing most of the kinetic energy. The Kolmogorov scale is $\eta = \left(\nu^3/\varepsilon\right)^{1/4}$, in which ν is the kinematic viscosity and ε the rate of kinetic energy dissipation, while the integral scale L is of order of the boundary layer dimensions. The challenge of a DNS is that the integration of the integral scale L requires a spatial resolution of N of points along a given mesh direction with increments h so that, $(N \cdot h) > L$, while simultaneously $h < \eta$ needs to be matched. Since the rate of the kinetic energy dissipation scales $\varepsilon \approx u'^3/L$, with u' the root mean square (RMS) of the velocity fluctuation the spatial resolution requirement of a 3D DNS scales with $N^3 \geq \left(Re'\right)^{9/4}$, where Re' is the Reynolds number of the turbulent fluctuation $Re' = \left(u' \cdot L\right)/\nu$. Additionally, the smallest vortices have to be resolved in time requiring in order to be accurate a time step (Δt) small enough such that a fluid particle moves only a fraction of the mesh spacing h in each step, i.e. $C = \left(u' \cdot \Delta t\right)/h < 1$, which is called Courant number condition and the ratio correspondingly C is the Courant number. In order to get a representative solution, the total time interval simulated must be at least proportional to the turbulence time scale $\tau = L/u'$. Merging all definitions, the number of time steps required by a DNS scales with $L/\eta \sim Re'^{3/4}$ and together with required spatial resolution required about Re'^3 operation is required. This disqualifies DNS for practical engineering applications since memory requirements exceed the capacity of the most powerful computer currently available. Nonetheless, the DNS is a powerful and indispensable complementary tool to get access to the fundamental physics of turbulence, e.g. by identifying transfer processes of eddy cascades. Moreover, DNS provide a quasi-accurate solution to develop and validate turbulence models for practical applications, such as sub-grid scale models for Large eddy simulation (LES) and to improve models for methods that solve the Reynolds-averaged Navier–Stokes equations (RANS), since the DNS serves data experimentally not accessible even by future sensing technologies.

- Large eddy simulation (LES) relies on the theory of self-similarity, which assume that the large eddies of the flow are dependent on the geometry while the smaller scales are quasi-"universal". By this approach the large eddy scales are explicitly solved in a simulation and implicitly also the smaller eddies by means of a subgrid-scale model (SGS model). Formally, one may think of filtering as the convolution of a function with a filtering kernel G, which yields $\overline{u}_i = \int G\left(\vec{x} - \vec{\xi}\right) u\left(\vec{\xi}\right) \, \mathrm{d}\,\vec{\xi}$. By this filtering the velocity is decomposed in a mean and a fluctuating part, $u_i = \overline{u_i} + u_i'$ similar as in a RANS simulation. Substituting in this decomposition and a similar one for the pressure and then filtering provides the filtered Navier–Stokes equations for the resolved "large scale" field.

$$\frac{\partial \overline{u_i}}{\partial t} + \rho \overline{u_j} \frac{\partial \overline{u_i}}{\partial x_j} = -\frac{1}{\rho} \frac{\partial \overline{p}}{\partial x_i} + \frac{\partial}{\partial x_j} \left(\nu \frac{\partial \overline{u_i}}{\partial x_j} \right) + \frac{1}{\rho} \frac{\partial \tau_{ij}}{\partial x_j}. \tag{5.90}$$

Thereby it is assumed that the filtering operation and the differentiation operation commute and the errors associated with this assumption are small. The additional term $\partial \tau_{ij} / \partial x_j$ originates from the non-linear advection terms and the turbulent shear stress tensor termed as Reynolds stress tensor is $\tau_{ij} = \overline{u_i} \cdot \overline{u_j} - \overline{u_i u_j}$. For the sub-grid-scale or residual field a similar equation is developed. Therein, the stress tensor is related proportionally proportional to the trace-less mean strain rate tensor S_{ij}, which is called Boussinesq approximation. Thereby, the the SGS is correlated with mean flow quantities:

$$\tau_{ij} - \frac{1}{3} \tau_{kk} \delta_{ij} = -2 \; \rho \; \nu_t S_{ij} = -2 \; \rho \; \nu_t \left(\frac{\partial \overline{u_i}}{\partial x_j} + \frac{\partial \overline{u_j}}{\partial x_i} \right), \tag{5.91}$$

in which ν_t is the subgrid-scale turbulent viscosity. The subscale information is the inserted into the filtered Navier–Stokes equations leading to:

$$\frac{\partial \overline{u_i}}{\partial t} + \rho \overline{u_j} \frac{\partial \overline{u_i}}{\partial x_j} = -\frac{1}{\rho} \frac{\partial \overline{p}}{\partial x_i} + \frac{\partial}{\partial x_j} \left(\nu + |\nu_t| \frac{\partial \overline{u_i}}{\partial x_j} \right). \tag{5.92}$$

In subgrid-scale models, there are a variety of approaches, ranging from isotropic dissipation of small eddy structures (Smagorinski), to algebraic models, to dynamic models that consider wall distance (Wale, RNG), or consider local stress tensors (structural modelling). These are:

- Smagorinsky model (Smagorinsky, 1963),
- Algebraic Dynamic model (Germano et al., 1991),
- Dynamic Global-Coefficient model (You & Moin, 2007),
- Localized Dynamic model (Kim & Menon, 1995),
- WALE (Wall-Adapting Local Eddy-viscosity) model (Nicoud & Duos, 1999),
- RNG-LES model,
- Structural modelling (Lu et al., 2008).

It must be mentioned that also the spatial resolution is substantially reduced compared to a DNS, the spatial resolution is substantially reduced, the Courant number requirements still applies for the temporal resolution. Also, for the spatial resolution the mesh—Reynolds number, in which as characteristic length scale the mesh dimension h is used, must be below 2. This limits the application of LES simulations considerable, but as computing power substantially increases is gaining more and more importance also for technical flows. However, the drawbacks potentially impacting the solution quality is that the dissipation and its commuting properties with the main flow field is imposed a-priori and requires a pre-emptive knowledge of the solution by the user to choose adequate SGS-models and additionally the frequency information of the subgrid scale is lost.

- Reynolds averaged Navier-Stokes based turbulence models (RANS). For engineering applications most relevant are the RANS models, since they are robust and fast posing limited computational efforts. Hence, some space is devoted to them subsequently.

Fundamental ideas of turbulence modelling

If we consider a statistically stationary flow, the flow variables can be divided into a time-averaged value and an additional fluctuation value around this average or mean value. This kind of consideration, which is called Reynolds decomposition, provides the correlation for the velocity in flow direction (x-direction):

$$u_i(x_i, t) = \overline{u_i(x_i)} + u_i'(t, x_i). \tag{5.93}$$

The time-averaged velocity is described by the following equation:

$$\overline{u_i(x_i)} = \frac{1}{\Delta t} \int_t^{t+\Delta t} u_i(x_i, t)\, dt, \tag{5.94}$$

in which the time interval Δt over which integration is performed must be large enough to maintain the ability to capture possible time variations of the flow in the control volume. By definition, the time average of the fluctuations is zero, but the mean of their squares is different from zero. The turbulence intensity (ϕ) or Root Mean Square (RMS) is given in the following manner.

$$\phi = \sqrt{\overline{u_i'^2}} = \sqrt{u'^2 + v'^2 + w'^2}. \tag{5.95}$$

The relative intensity of the velocity fluctuations, defined as the degree of turbulence Tu, is therefore:

$$Tu = \frac{\sqrt{\overline{u_i'^2}}}{|\overline{u_i}|}, \tag{5.96}$$

To get a sense of the orders of magnitude of turbulence intensity, it is about 1% downstream of a fine screen, about 10% near a wall, and more than 10% in a free jet. In general, turbulent flows can be divided into three classes:

- *Isotropic turbulence*: The statistical data of the flow field in the whole flow field are independent of the direction ($\overline{u'^2} = \overline{v'^2} = \overline{w'^2}$). They are invariant with respect to the translation and rotation of the coordinate system.
- *Homogeneous turbulence*: All statistical properties depend only on the direction and not on the location. They are invariant with respect to translational motion.
- *Anisotropic or shear turbulence*: This type is of practical interest in engineering applications, as it occurs in boundary layers, free jets, and channel flows, etc.

Reynolds equations

The steady state Navier–Stokes equations are valid for all turbulent flows of a continuum, but the complexity as well as the random nature of the turbulent exchange processes prevents for most technical configurations a direct numerical access to obtain an exact solution of these equations. Thus, most current methods for describing turbulent flows are based on the time-averaged Navier–Stokes equations. These equations are referred to as the Reynolds equations. Time averaging of the equations of motion generates new terms that can be interpreted as apparent stress gradients or heat flux quantities associated with turbulence. These new quantities must be related to the main flow using appropriate turbulence models. This modelling approach involves further assumptions and approximations. Therefore, the solution of a turbulent flow problem with the help of the Reynolds equations does not completely follow a self-contained basis based on basic physics, since the additional assumptions made must be made close to the system of equations on the one hand and on the other hand establishes well-founded but simplified links. The following equations based on Reynolds averaging are valid for a flow of an incompressible Newtonian fluid:

The application of Reynolds decomposition yields for the continuity equation:

$$\frac{\partial \overline{u_i}}{\partial x_i} = 0 = \frac{\partial \overline{u}}{\partial x} + \frac{\partial \overline{v}}{\partial y} + \frac{\partial \overline{w}}{\partial z}, \tag{5.97}$$

and accordingly, for the Reynolds averaged momentum equation:

$$\rho \overline{u_j} \frac{\partial \overline{u_i}}{\partial x_j} = -\frac{\partial \overline{p}}{\partial x_i} + \frac{\partial}{\partial x_j}\left(\mu \frac{\partial \overline{u_i}}{\partial x_j} - \rho \overline{u'_i u'_j}\right). \tag{5.98}$$

In Eq. (5.95), the apparent or virtual Reynolds shear stress tensor appears $\tau_{tur} = -\rho \overline{u'_i \cdot u'_j}$, which is significantly different from the molecular stress tensor given by $\tau_{ij} = \mu \cdot \left(\partial \overline{u_j}/\partial x_i + \partial \overline{u_i}/\partial x_j\right)$, as shown in Eq. (5.9). It is important to note that the additional Reynolds stress tensor appears only with respect to the time-averaged flow. The equation of motion for the instantaneous quantities does not include any additional turbulence term. The transition from the time-dependent equations of motion to the Reynolds equations formulates the challenge of turbulence modelling. The "price" of the time averaging is the introduction of the unknown Reynolds stress tensor arising from the nonlinear convective terms of the Navier–Stokes equations. The main task of solving the turbulence problem is to relate this Reynolds stress tensor to the time-averaged quantities of the velocity field using appropriate turbulence models. This can be done by three main categories of RANS-based turbulence models:

- Linear eddy viscosity models,
  - algebraic models (Cebeci-Smith model, Baldwin-Lomax model, ...),
  - one equation models (Prandtl's one equation model, Baldwin Lomax),
  - two equation models ($k - \varepsilon$ types, $k - \omega$ types).
- Nonlincar cddy viscosity models
  - Explicit algcbraic Reynolds stress models (EARSM),
  - V2f models.
- Reynolds stress model (RSM)

The procedure to evaluate the unknown shear stress tensor with the help of balance equations. Like any other transport quantity, these must satisfy a general form. The types of balance equations, often referred to as transport equations, are the basis of almost all recent closure approaches regardless of whether they are algebraic or tensorial in nature. The starting point for the formulation of the transport equations is the derivation of the Navier–Stokes equation for the instantaneous values. Following a strict mathematical derivation, the transport equations for the Reynolds stress tensor are:

$$\underbrace{\overline{v_k}\frac{\partial}{\partial x_k}\left(\overline{v_i'v_j'}\right)}_{C}+\underbrace{\overline{v_i'v_k'}\frac{\partial\overline{v_j'}}{\partial x_k}+\overline{v_j'v_k'}\frac{\partial\overline{v_i'}}{\partial x_k}}_{P}+\underbrace{2\nu\frac{\partial\overline{v_i'}}{\partial x_k}\frac{\partial\overline{v_j'}}{\partial x_k}}_{DS}$$
$$\underbrace{-\overline{\frac{p'}{\rho}\left(\frac{\partial v_i'}{\partial x_j}+\frac{\partial v_j'}{\partial x_i}\right)}}_{PSC}+\underbrace{\frac{\partial}{\partial x_k}\left[\overline{v_i'v_j'v_k'}-\nu\frac{\partial}{\partial x_k}\left(\overline{v_i'v_j'}\right)+\overline{\frac{p'}{\rho}\left(\delta_{kj}v_i'+\delta_{ki}v_j'\right)}\right]}_{D}=0. \tag{5.99}$$

In this equation C describes the convective change and P the turbulence production, which is the product of the Reynolds stress tensor with the time-averaged velocity gradients. The term DS represents turbulent dissipation, which is essentially a negative production term, and PSC is the pressure-shear correlation, which contributes to flow reorganization in a similar way to diffusion D. The production term is one of the most important, as it describes that only in flows with a velocity gradient (shear flow) can preserve turbulence.

The velocity gradient of the time-averaged velocity is thus the turbulence generator and the energy required for this is extracted from the main flow. The tensor equation consists of nine components. An important case of the transport equation is obtained by the formal reduction $i = j$, for which the transport equation of the turbulent kinetic energy k is obtained. Defining k as $k = 1/2 \cdot \overline{v_i'^2} = 1/2 \cdot \left(\overline{u'^2}+\overline{v'^2}+\overline{w'^2}\right)$, the transport equation of the turbulent kinetic energy has the following appearance:

$$\underbrace{\overline{v_k}\frac{\partial k}{\partial x_k}}_{C}+\underbrace{\overline{v_j'v_k'}\frac{\partial\overline{v_j'}}{\partial x_k}}_{P}+\varepsilon+\underbrace{\frac{\partial}{\partial x_k}\left[\overline{\left(k+\frac{p'}{\rho}\right)v_k'}-\nu\frac{\partial k}{\partial x_k}-\nu\frac{\partial}{\partial x_j}\left(\overline{v_j'v_k'}\right)\right]}_{D}=0, \tag{5.100}$$

where ε is the dissipation of turbulence. The turbulent energy is not an averaged quantity and the turbulent dissipation ε is described by the following equation:

$$\varepsilon=\nu\overline{\frac{\partial v_j'}{\partial x_k}\left(\frac{\partial v_j'}{\partial x_k}+\frac{\partial v_k'}{\partial x_j}\right)}. \tag{5.101}$$

The transport equations play an essential role in the development of modern turbulent flow modelling and methods. Although the Reynolds stress tensor can be formulated using the transport equations, these equations do not solve the closure problem because they themselves again contain unknown correlations. For these correlations, new transport equations must be formulated, in which, in turn, unknown correlations appear. The order of the obtained tensors increases with each step by a factor of one.

To summarize, the closure problem cannot be solved directly. There are always more unknowns than available equations. The only possible solution is to use closure assumptions, which in the simplest case can be applied directly to the Reynolds equations, such as Prandtl's mixing length approach. Further closure concepts consist in the use of numerically determined correlations, the application of physical model concepts and last but not least semi-empirical methods. Here, only the main ideas shall be communicated.

Fundamental ideas of linear eddy viscosity RANS turbulence modelling

Even simple algebraic models show reasonable results for plane two-dimensional flows but are hardly relevant for technical applications where three-dimensional flows occur, since higher orders of turbulence models are needed. These use one or more partial differential equations consisting of the modified Navier–Stokes equations for the turbulent kinetic energy k, the dissipation of the energy ε and the components of the turbulent stress tensor τ_{ij}.

A simple turbulence model in which a partial differential equation for the turbulent kinetic energy is related to an algebraic expression for the turbulent length scale l is the Prandtl mixing length approach. This model type is called a one-equation model or $k - l$ model.

An extension is represented by the likewise linear two-equation models, in which partial differential equations for the kinetic energy and its dissipation are solved. The best-known model is the $k - \varepsilon$ model, introduced by Harlow and Nakayama (1968)and later modified by Jones and Launder (1972) and Launder and Spalding (1974). It is the most widely used model in engineering applications and is implemented in almost all commercial fluid dynamic computational programs (CFD codes). Many applications of the $k - \varepsilon$ model use analytical wall functions to handle near-wall flow. In the alternative case, additional terms to the $k-$ and $\varepsilon-$equations must be introduced to extend the applicability to the treatment of the viscous sublayer (Chien, 1982). In this context, the viscous sublayer is often referred to as the low turbulent Reynolds number region $\left(\sqrt{k}l/\nu\right)$. This inner model is essential for flows in complex geometries, flow separation, or abrupt transverse cross-sectional area changes. The uncertainties in modelling such inner region models in complex flows with large anisotropy significantly limits the applicability framework of the $k - \varepsilon$ model.

The $k - \varepsilon$ models have been modified to include ta description of viscous sublayer. The necessary extensions are called "low Reynolds number" $k - \varepsilon$ models (Rodi, 1993). These additional extensions to the $k - \varepsilon$ model to capture lift effects or streamline changes require two-equation models. The most widely used models are those of Ng and Spalding (1972) and Wilcox and Traci (1976). All such models involve a formulation of turbulent kinetic energy and differ substantially in their treatment of the gradient of diffusion. The most striking difference, however, is the choice of dependent variable for the second transport model equation from which the characteristic length scale is determined. Two-equation models often coexist, as in the "shear stress turbulence model" (SST). The SST model combines the advantages of the $k - \varepsilon$ model with those of the $k - \omega$ model, where ω is the vorticity. The main advantage of the SST model is that in the near-wall region in the $k - \omega$ model, an analytical solution for the viscous sublayer is known (Wilcox, 1986). The linkage of the near-wall $k - \omega$ model to the $k - \varepsilon$ model treating the bulk flow is done by means of blending functions. Complicated multi-equation methods require the solution of partial differential equations of all components of the turbulent shear stress tensor. A summary of the description of the turbulence models with a large number of details can be found in the works of Reynolds (1976) and Arpaci and Larsen (1984).

The non-linear eddy viscosity models use a similar concept as the linear eddy viscosity concepts by relating the mean turbulence field to the mean velocity field. The main difference, however, is to introduce a nonlinear function not only dependent on the mean strain but additionally also the vorticity fields or even other turbulence variables. Thus, they do not require blending functions; prominent representatives are the EARSM and the V2f model widely used in computations due to their robustness.

Basic thoughts on RSM models in RANS turbulence modelling

Here, only the essential ideas of the Reynolds Stress Model (RSM) are described which a-priori do not assume that the turbulent shear stress is proportional to the strain rate, which leads for a two-dimensional, incompressible flow to:

$$-\rho\overline{u'v'} \neq \mu_t\left(\frac{\partial u}{\partial y} + \frac{\partial v}{\partial x}\right). \tag{5.102}$$

μ_t is the turbulent viscosity given by $\mu_t = \left(k^2/\varepsilon\right)$. The most widely known Reynolds stress models are those of Hanjalic and Launder (1972), Launder et al. (1975) and Donaldson (1972). RSM models require the solution of at least three additional partial differential equations (PDE) compared to the $k - \varepsilon$ model. Moreover, even RSM models still require approximations and assumptions for the double and triple correlation, being barely accessible by experimental techniques. Fortunately, the progress in DNS Simulation opens here a gate by providing a data basis for a closed-form confirmation. A simplification of RSM modelling known as the algebraic stress or flow model ASM is described in detail in Rodi (1981). In the ASM model, the transport of Reynolds stresses is assumed to be proportional to the transport of kinetic energy. For boundary layer flows without buoyancy effects, the ASM approach results in the following relationship for the turbulent shear stress.

$$\overline{u'v'} = C_\mu \frac{k^2}{\varepsilon}\frac{\partial u}{\partial y}, \tag{5.103}$$

which at first sight looks identical to the result of the $k-\varepsilon$ model, but in the ASM C_μ is a function of the ratio of turbulence production to dissipation of turbulent kinetic energy and thus no longer a constant.

Turbulent boundary layer equations

Many technical flows have boundary layer character. The laminar boundary layer equations cannot be applied to the turbulent case because the fluctuation velocities v'_k are fast in space and time and thus their derivatives cannot be neglected with respect to the other terms, even though their absolute value is an order of magnitude smaller than that of the mean values. Therefore, the boundary layer equations must be derived directly within the Reynolds equations and the transport equations. However, the procedure is the same as in the laminar case; details can be found in the books by Burmeister (1993) or Anderson et al. (1984). In two-dimensional case, the turbulent boundary layer equations are:

$$\frac{\partial \overline{u}}{\partial x}+\frac{\partial \overline{v}}{\partial y}=0,\ \rho\left(\overline{u}\frac{\partial \overline{u}}{\partial x}+\overline{v}\frac{\partial \overline{u}}{\partial y}\right)=-\frac{dp_\delta}{dx}+\frac{\partial}{\partial y}\left(\mu\frac{\partial \overline{u}}{\partial y}-\rho\overline{u'v'}\right), \tag{5.104}$$

where p_δ is the pressure at the edge of the boundary layer towards the bulk flow.

A similar procedure can be used for the transport equations and the transport equations for the shear stress and the turbulent kinetic energy are obtained, which for a two-dimensional flow are as follows:

$$\underbrace{\overline{u}\frac{\partial\left(\overline{u'v'}\right)}{\partial x}+\overline{v}\frac{\partial\left(\overline{u'v'}\right)}{\partial y}}_{C}+\underbrace{\overline{v'^2}\frac{\partial \overline{u}}{\partial y}}_{P}+\underbrace{2\nu\left(\overline{\frac{\partial u'}{\partial x}\frac{\partial v'}{\partial x}+\frac{\partial u'}{\partial y}\frac{\partial v'}{\partial y}+\frac{\partial u'}{\partial z}\frac{\partial v'}{\partial z}}\right)}_{DS}$$

$$+\underbrace{\overline{\frac{p'}{\rho}\left(\frac{\partial u'}{\partial y}+\frac{\partial v'}{\partial x}\right)}}_{PSC}+\underbrace{\frac{\partial}{\partial y}\left(\overline{u'v'^2}-\nu\frac{\partial}{\partial y}\left(\overline{u'v'}\right)+\overline{\frac{p'u'}{\rho}}\right)}_{D}=0;$$

$$\underbrace{\overline{u}\frac{\partial k}{\partial x}+\overline{v}\frac{\partial k}{\partial y}}_{C}+\underbrace{\overline{u'v'}\frac{\partial \overline{u}}{\partial y}}_{P}+\varepsilon+\underbrace{\frac{\partial}{\partial y}\left(\overline{\left(k+\frac{p'}{\rho}\right)v'}-\nu\frac{\partial k}{\partial y}-\nu\frac{\partial \overline{v'^2}}{\partial y}\right)}_{D}=0. \tag{5.105}$$

Most terms of the two transport equations except for the local pressure fluctuations can be determined experimentally. Fortunately, pressure fluctuations do not appear in the kinetic energy equation. If channel walls limit the flow, the velocity fluctuations disappear as the wall is approached, and so do the Reynolds shear stresses; the result is

$$lim_{y\to 0}\mu\frac{\partial \overline{u}}{\partial y}=\tau_w. \tag{5.106}$$

If one assumes a fully developed, two-dimensional, turbulent flow, then in the vicinity of the wall all statistical quantities depend only on the lateral (y-) coordinate. Since the derivative of the velocity in the axial direction is zero, the continuity equation leads to the result that the mean velocity in the y-direction is also zero. In this particular case, the convective terms and the pressure in the Reynolds equation are zero, and there remains

$$\frac{\partial}{\partial y}\left(\mu\frac{\partial \overline{u}}{\partial y} - \rho\overline{u'v'}\right) = 0. \tag{5.107}$$

With the boundary condition on the wall we get:

$$\frac{\tau_w}{\rho} = \nu\frac{\partial \overline{u}}{\partial y} - \overline{u'v'}. \tag{5.108}$$

Obviously, only the parameters ν and τ_w/ρ occur and one can define a shear stress velocity u_τ which is as follows:

$$u_\tau = \sqrt{\frac{\tau_w}{\rho}} = u_0\sqrt{\frac{f}{2}}, \tag{5.109}$$

in which f is the dimensionless friction factor (normalized by with $f = 2\tau_w/\left(\rho u_0^2\right)$. If one defines the dimensionless coordinates u^+ and y^+ in the way:

$$u^+ = {}^{\overline{u}}/_{u_\tau}, \; y^+ = {}^{(yu_\tau)}/_{\nu}, \tag{5.110}$$

the velocity distribution near the wall can be given in a universal representation of the form $u^+ = f\left(y^+\right)$. The solution for f in the immediate vicinity of the wall is thus immediately given, since the velocity fluctuations near the wall are zero and the adhesion condition holds at the wall, so that $u^+ = y^+$. This describes the so-called viscous sublayer, in which the molecular shear stresses are considerably higher than the Reynolds stresses. With increasing wall distance, the Reynolds stresses increase and at the same time the molecular fraction becomes negligible. This leads to a logarithmic expression for the dimensionless velocity in the form:

$$u^+ = \frac{1}{C_1} lny^+ + C_2. \tag{5.111}$$

Between the viscous sublayer and the fully turbulent boundary layer there is a transition layer, for which different but mostly logarithmic functions are used. Van Kármán suggests for the velocity distribution in the boundary layer:

- Viscous sub-layer: $u^+ = y^+$ for $y^+ < 5$;
- Transition layer: $u^+ = 5lny^+ - 3.05$ for $5 < y^+ < 30$;
- Fully turbulent boundary layer: $u^+ = 2.5lny^+ + 5.5$ for $y^+ > 30$.

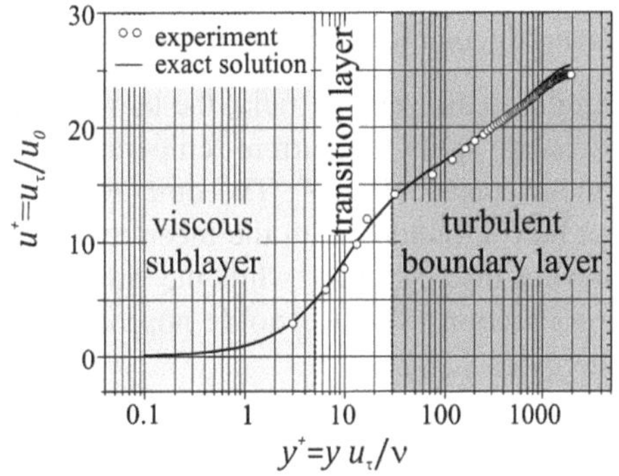

Fig. 5.17 Measured dimensionless wall velocity u^+ of a pipe flow as a function of the dimensionless wall distance y^+ in lead bismuth at $Re = 8 \times 10^4$ from Schulenberg and Stieglitz (2010) in comparison with the exact solution (Reichhardt, 1951)

At the transition from the anisotropic turbulent boundary layer to the region of isotropic turbulence of the bulk flow, the logarithmic wall law no longer applies. It changes into a wake law, which is also called the centre law in pipe or channel flows.

An overview of the different definitions between the viscous sublayers and the developed boundary layer is given by Schlichting and Gersten (2000). The existence and form of the laminar sublayer, the transition layer, and their attachment to the logarithmically scaling turbulent boundary layer has been experimentally proven even at high Reynolds numbers, as illustrated in Fig. 5.17.

The transition of the anisotropic turbulent boundary layer into an isotropic core flow causes the logarithmic functions to lose their applicability. Within this range, wake laws are introduced. There is a myriad of articles on this, most of which differ only in the choice of the constants of the wake profile. A good overview is given here by the works of Zagarola and Smits (1998), Barenblatt and Chorin (1998) and Guo and Julien (2003).

For the modelling of turbulence in a broad set of literature is available. The work of Rodi (2005) provides a detailed overview of the entire range, but also references to best practice guidelines, as well as the cross-references listed therein. The technical implementation within the framework of finite difference and volume methods is well described in Versteeg and Malalasekera (2007), among others.

5.1.12 Turbulent Energy Exchange

A similar formalism to that used for momentum exchange can also be used to derive the turbulent equations of energy exchange, which are for steady incompressible flow:

$$\rho c_p\left(v_k\frac{\partial T}{\partial x_k}\right) = \lambda\frac{\partial^2 T}{\partial x_k^2} \text{ or } \rho c_p\left(u\frac{\partial T}{\partial x} + v\frac{\partial T}{\partial y} + w\frac{\partial T}{\partial z}\right)$$
$$= \lambda\left(\frac{\partial^2 T}{\partial x^2} + \frac{\partial^2 T}{\partial y^2} + \frac{\partial^2 T}{\partial z^2}\right). \tag{5.112}$$

Reynolds equations of turbulent energy exchange

Similar to the velocity field, the temperature field in the Reynolds turbulent energy exchange splits the current temperature T into a fluctuating component T' and a time-averaged value $\overline{T}$. The time-averaged component of the temperature should not be confounded with the mean temperature T_m. The former is a local temperature at a discrete point, while the latter is a flow-averaged quantity over the channel cross-section. The Reynolds approach yields the following result for turbulent energy exchange:

$$\rho c_p\left(\overline{v_k}\frac{\partial \overline{T}}{\partial x_k}\right) = \lambda\frac{\partial^2 \overline{T}}{\partial x_k^2} - \underbrace{\rho c_p \frac{\partial}{\partial x_k}\left(\overline{v_k' T'}\right)}_{RH}, \tag{5.113}$$

in which the term RH represents a virtual turbulent heat flux vector arising from the Reynolds averaging approach. It thus describes a macroscopic energy exchange arising from the fluctuation components. In contrast to the Reynolds shear stress tensor, the turbulent heat flux is a vector quantity, since the transport quantity of the energy is a scalar (temperature), while for the momentum a vector quantity is already present. Similarly, transport equations must be derived for the turbulent heat flux. In contrast to momentum transport, turbulent energy exchange is in most cases a passive transport process; exceptions are buoyant flows. As in the case of momentum transport, boundary layer equations can also be derived for energy transport, which read for a two-dimensional, stationary, incompressible flow to:

$$\rho c_p\left(\overline{u}\frac{\partial \overline{T}}{\partial x} + \overline{v}\frac{\partial \overline{T}}{\partial y}\right) = -\frac{\partial}{\partial y}\left(-\lambda\frac{\partial \overline{T}}{\partial y} + \rho c_p \overline{v'T'}\right). \tag{5.114}$$

5.1.13 Buoyancy and Mixed Convection in Closed Cavities

Until now, we had considered energy transport only as a forced convective transport, but the energy thermal supply also changes the density of a fluid regardless of whether it is a gas, a liquid, or a mixture. With an appropriate energy input, the buoyancy velocity induced by the density gradient can be significantly greater than the mean flow velocity u_0 produced by an external pressure gradient, e.g. by body forces or a pump. If both types of flow drive appear the flow regime is called mixed convection. In the extreme case, for example, a solar thermal system can be driven only by means of buoyancy, such as a thermosiphon as shown in Fig. 5.18.

In buoyant flows, the mean velocity can thus no longer play the excellent role, since the density differences due to the temperature difference between the wall T_w and the bulk T_m generate the flow itself. These density differences, which give the buoyancy to the flow, originate from a temperature coefficient of volume expansion

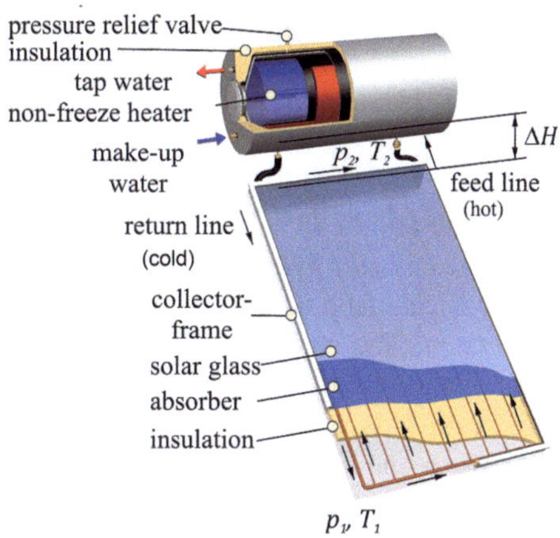

Fig. 5.18 Example of open thermosiphon system for drinking water preparation

(or thermal expansion) β, which is defined as follows:

$$\beta = \rho - \rho_m = -\frac{1}{\rho}\left(\frac{\partial \rho}{\partial T}\right), \tag{5.115}$$

where ρ_m is the density of the bulk flow and ρ the local fluid density.

In many engineering applications as e.g. the heat losses at solar receiver panels, both mechanisms occur at the same time. Therefore, heat transfer by mixed convection is the result of a superposition of buoyancy-driven natural and forced convection, which due to the non-linearity of the convective terms cannot simply be added. Thus, to account for buoyant flows, a new dimensionless quantity arises, the so-called Grashof number Gr, which can be conceived as the ratio of the buoyancy force acting on a fluid element to the counteracting viscous force. The Grashof number corresponds entirely to the Archimedes number used in civil engineering if the variation of the density is caused by a change of temperature T only as in wall bounded/internal flows. Since in most cases (heat flow by radiation on the absorber) not the wall temperature but the heat flux is provided, the wall temperature T_w can be converted into a relation containing the wall heat flux $\dot{q}_w$ by means of Fourier's law.

$$Gr = \frac{g \cdot \beta \cdot (T_W - T_m) \cdot L^3}{\nu^2} = \frac{g \cdot \beta \cdot \dot{q}_w \cdot L^4}{\lambda\ \nu^2}, \tag{5.116}$$

with ν the kinematic viscosity and g the acceleration due to gravity. Thus, for free convection, the Nusselt number no longer depends only on the Reynolds and Prandtl numbers, but also on the Grashof number, i.e. $Nu = Nu(Gr, Pr)$. With the Grashof number describing the contribution of the natural convection flow (buoyant force), and the Reynolds number Re_L describing the imposed flow (inertia), the relative effect of buoyancy on mixed convection can also be expressed though the Richardson number:

$$Ri = \frac{Gr}{Re_L^2}, \tag{5.117}$$

where the length L denotes the scale where the heat is applied to. For varying length scales and hydraulic circuits, i.e. containing a horizontal L and a vertical H, the Richardson number is considered to be more suitable to characterize mixed convection. The extrema $Ri \to 0$ and $Ri \to \infty$ represent pure forced and pure natural convection, respectively. A rough separation of the transition between the flow regimes is suggested by Gebhart (1988) for $Ri \approx 0.3$ from forced to mixed convection and $Ri \approx 16$ from mixed to pure natural convection. New, numerical based studies supported by an unfortunately rare experimental data bases (Garbrecht, 2017), however, show that the especially the onset of the transition from forced to mixed convection starts already at $Ri \approx 0.1$.

A first estimation based on analytical considerations of the transition from forced convection to mixed convection provides the following relationship suggested by Müller and Ehrhard (1999):

$$\frac{\sqrt{Gr}}{Re} < 1. \tag{5.118}$$

This ratio considers the natural convective lift influence on the flow via the Grashof number. The influence of the forced flow is characterized by the Reynolds number Re. If Re dominates in the ratio and it becomes smaller than one, buoyancy can be neglected. If the Grashof number Gr increases due to a higher heat flow or due to the given dimensions (see power of L!), the buoyancy effects caused by density differences must also be considered. According to this criterion, however, it is not possible to estimate whether mixed convection or pure natural convection is present. In order to distinguish more precisely between natural, mixed, and forced convection, the following criterion determined on the basis of experiments has been extended by Müller and Ehrhard (1999):

$$\frac{Ra^{1/3}}{Re^{4/5} \cdot Pr^{2/5}} = c^*. \tag{5.119}$$

The Rayleigh number $Ra = Gr \cdot Pr$ contained therein is formed from the product of the Grashof number Gr and the Prandtl number Pr and also represents a characteristic dimensionless parameter for describing natural convective effects. The Rayleigh number is an extension of the Grashof number and includes molecular viscous fluid properties as well as molecular thermal properties. The exponent of the Rayleigh number in the condition for c^* results from a boundary layer consideration of the thermal boundary layer of a buoyant flow whose thickness scales with $Ra^{-1/3}$. For pure forced convection, the thickness of the turbulent thermal boundary layer scales with $Re^{-4/5}/Pr^{-2/5}$ (Aicher & Martin, 1997). Thus, for given Ra, Re, and Pr, the parameter c^* provides information about the mechanism of heat transfer from the wall across the boundary layer into the flow core. Aicher and Martin (1997) define the following ranges; for $c^* < 0.2$ forced convection dominates, while for $c^* > 0.4$ heat transfer by natural convective effects is dominant. By plotting the ratio of the measured Nusselt number Nu to the Nusselt number Nu_{FT} resulting from pure forced convective heat transfer under the same boundary conditions, it is possible to classify measurement results into natural, mixed, and forced convection.

A wide range of literature exists to estimate the magnitude of buoyancy velocities caused by density differences versus forced convective mean flow velocity generated by external pressure forces (pumps). Before performing a CFD-based simulation of the flow field, it is recommended to use analytical relationships to estimate the buoyancy-induced velocities and to determine the Grashof and Rayleigh numbers, respectively. Although collector flow geometries can have complex geometric configurations, they can often be limited to the geometry of a point or line source or a vertical and horizontal plate. Analytical relationships for determining the lift velocities achievable in this context can be found, for example, in Kakac et al. (1985) or in the section on engineering relationships.

Another aspect of the methodology of analytical penetration should not be forgotten. Buoyant flows exhibit a multitude of flow patterns. At low heating powers, thermal conduction exists and only after a certain temperature difference is exceeded, a density difference occurs in which the buoyancy overcomes the viscous resistance, laminar convection sets in, which is characterized by an ordered cellular structure. As the simplest case, consider the Rayleigh–Benard cell shown in Fig. 5.19, in which a fluid is heated from below and cooled from above. If the fluid height (d) in the direction of the gravity vector g is smaller than the dimensions occurring in the $x - y$ plane ($x \gg d$, $y \gg d$), a critical Rayleigh number Ra_{crit} can be determined for the transition from the heat conduction state to laminar convection. The critical Rayleigh number is $Ra_{crit} = 1708$ for very good thermal conductive walls (Biot number $Bi \gg 1$), which usually occur in the copper absorber tubes in low temperature systems. Laminar convection is essentially two-dimensional. In high-temperature systems, where stainless steel tubes with a smaller molecular thermal conductivity are used, the Biot number effect must be considered. In this case, higher wall temperatures occur and thus the critical Rayleigh number decreases, in the adiabatic case ($Bi \rightarrow 0$) down to $Ra_{crit} = 1296$.

The laminar convection with its mainly two-dimensional cellular pattern is followed by the region of three-dimensional convection. This transition depends now in addition to the Rayleigh number also on the Prandtl number. As the Rayleigh number continues to grow, the still structured three-dimensional convection is accompanied by a stochastic three-dimensional convection in which distinct flow patterns are no longer discernible.

First flow pattern maps, in which the flow domain boundaries are illustrated as a function of the Rayleigh number and the molecular Prandtl number, were developed for the Rayleigh–Benard cell. Corresponding to these maps, Nusselt number

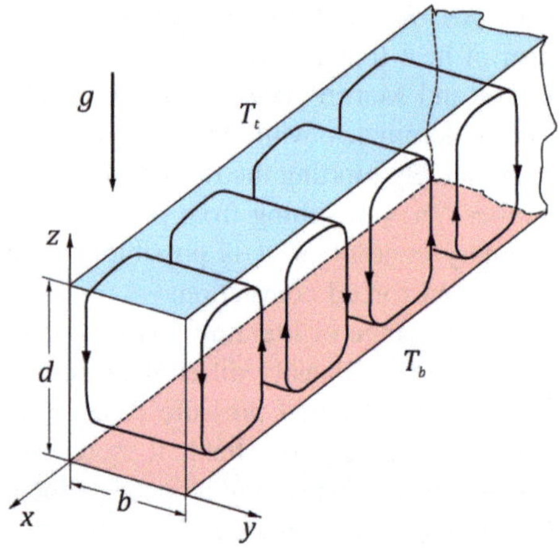

Fig. 5.19 Structure of a Rayleigh–Benard cell, schematic flow pattern of a quasi-two-dimensional buoyant flow

correlations as a function of Prandtl number and Rayleigh number were filed, based on the work of O'Toole and Silveston (1961), Castaing et al. (1989), and Kek and Müller (1993). Table 5.4 provides common Nusselt number relationships and their Rayleigh number range of validity for different fluids (Prandtl number) and Fig. 5.20 shows flow pattern map of a Rayleigh–Benard cell as a function of Pr and Ra.

For the vertical gap of height h and lateral thickness d, in which one vertical side is heated ($T = T_h$) and the other respectively cooled ($T = T_c$), a different flow dynamic results than in the case of the Rayleigh Benard cell (compare Fig. 5.21a) because even the smallest temperature differences lead to the formation of a large laminar convection cell, which has boundary layer character. In the boundary layer, there is an upward movement on the hot side and a corresponding downward movement on the cold side, while the core is almost at rest. This quiescence in the core leads to a linear temperature drop across the core characterized

Table 5.4 Ranges of flow patterns as a function of the Rayleigh number Ra and the Nusselt numbers Nu that can be calculated with it as a function of the molecular Prandtl number Pr for a Rayleigh–Benard cell from Müller and Ehrhard (1999)

$0 < Ra < 1.7 \times 10^3$	$Nu = 1$	
$Pr < 0.05$	$0.1 \leq Pr \leq 100$	$0.65 < Pr \leq 1.5$
$1.7 \times 10^3 < Ra < 7 \times 10^3$	$1.7 \times 10^3 < Ra < 3.5 \times 10^3$	
$Nu = 0.59 \cdot Ra^{0.072}$	$Nu = 0.0024 \cdot Ra^{0.82}$	
$10^4 < Ra < 4 \times 10^4$	$3.5 \times 10^3 < Ra < 10^5$	
$Nu = 0.62 \cdot Ra^{0.31}$	$Nu = 0.23 \cdot Ra^{1/4}$	
$4 \times 10^4 < Ra < 2.5 \times 10^5$	$10^5 < Ra < 10^9$	$10^7 < Ra < 10^{14}$
$Nu = 0.2 \cdot Ra^{1/5}$	$(\delta_{th}/d) \sim Ra^{-1/3}$	$(\delta_{th}/d) \sim Ra^{-2/7}$
	$Nu = 0.1 \cdot Ra^{0.31} \cdot Pr^{0.84}$	$Nu = 0.23 \cdot Ra^{0.282}$

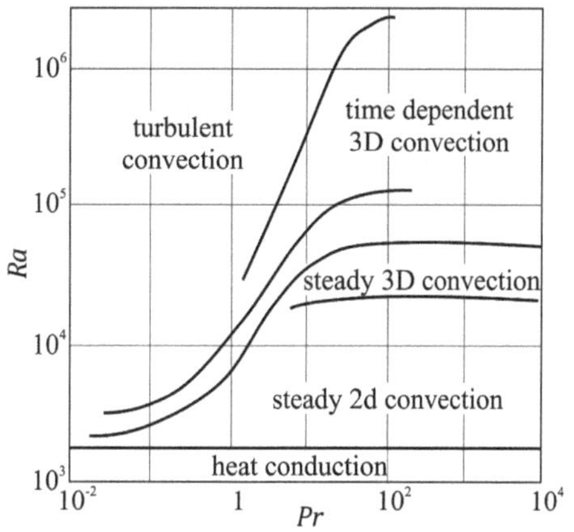

Fig. 5.20 Flow map of convection patterns establishing in a Rayleigh–Benard cell as a function of Rayleigh number Ra and molecular Prandtl number Pr according to Müller and Ehrhard (1999)

by molecular heat conduction. This region occurs between $10^5 < Ra < 10^8$. Further temperature increase also results in a cellular structure in the core. These are co-rotating vortices whose axes are horizontal in the gap and have a fixed spacing. Depending on the fluid properties, the cellular structure can be stationary or time dependent. For $Pr \ll 1$ they are mostly stationary while $Pr \gg 1$ leads to transient waves. A further increase in the temperature difference $\Delta T = (T_h - T_c)$ subsequently leads to counter-rotating secondary vortices, which are formed at the points between the large vortices where the maximum shear occurs. If the temperature difference is increased even further, a condition is created which is determined by the formation of turbulent boundary layers on both sides of the plate. The interaction between the boundary layers of both sides is only slight. In the core region of the flow, the flow is almost ideally mixed and thus almost isothermal.

In contrast to the Benard cell, two length scales occur inherently in the vertical gap, characterized by the aspect ratio $\epsilon = (h/d)$. Analogous to the Benard cell, flow maps can be generated which depend on the Rayleigh number Ra, the Prandtl number Pr, and the aspect ratio ϵ, and are thus three-dimensional. One such plot is illustrated in Fig. 5.21b, which shows the results of the transitions for two different assumptions. The solid boundary lines show the limits for the assumption that buoyancy is considered only in momentum conservation (so-called Boussinesq approximation). If the buoyancy effects are also considered in the energy equation, the other boundary lines result, which, however, require considerably more far-reaching modelling approaches. Table 5.5 gives some Nusselt number relationships for the vertical gap given in the literature.

From the consideration of flow in closed cavities, it is clear that an analytical consideration of the flow problem is mandatory prior to the creation of a numerical calculation model, so that with their help the selection of correct turbulence models and closure approaches can be made.

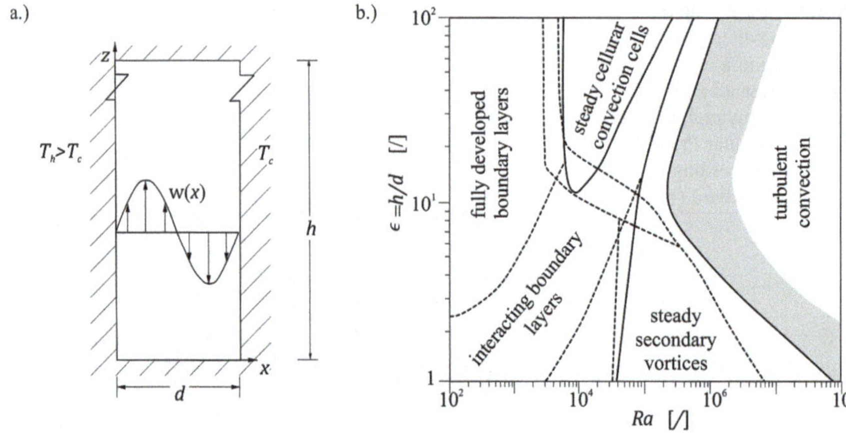

Fig. 5.21 **a** Schematic sketch of a vertical gap heated with $T = T_h$ on the left and cooled with $T = T_c$ on the right; **b** flow map of buoyant flow in a vertical gap for air ($Pr = 0.71$) from Chenoweth and Paolucci (1986), showing aspect ratio e as a function of Rayleigh number Ra. (—) ideal Boussineq approximation; (- - -) without Boussinesq approximation

Table 5.5 Nusselt number relations in heat transfer in a vertical gap as a function of Rayleigh number Ra, aspect ratio ϵ, and Prandtl number Pr from Müller and Ehrhard (1999)

$Ra < 7 \times 10^3$	$10^3 < Ra < 10^5$	$10^7 < Ra \cdot \epsilon^{-3} < 5 \times 10^9$
$Nu = 1$	$Nu = \epsilon^{1/4} \cdot Ra^{1/4}$	$Nu_m = 0.36 \cdot Pr^{0.051} \epsilon^{0.36} \cdot Ra^{1/4}$
$Nu = 1$	$10^3 Ra 5 \cdot 10^5$, $Pr = 0.7$ $Nu_1 = 0.0605 \cdot Ra^{1/3}$ $Nu_2 = \left[1 + \left(\frac{0.104 \cdot Ra^{0.239}}{1 + \left(\frac{6310}{Ra} \right)^{1.36}} \right)^3 \right]^{\frac{1}{3}}$ $Nu_3 = 0.242 \cdot (Ra \cdot \epsilon)^{0.272}$ $Nu_m = \max(Nu_1, Nu_2, Nu_3)$	$5 \times 10^9 \leq a \cdot \epsilon^{-3} \leq 10^{11}$ $Nu_m = 0.096 Pr^{0.051} \cdot \epsilon^{0.1} \cdot Ra^{0.3}$

5.1.14 Analogy of Heat and Momentum Transport, Solution Approaches for Local Turbulent Simulations

In analogy between Newton's friction law for the laminar shear stress τ, a law for the virtual turbulent shear stress τ_{tur} can be derived:

$$\tau = -\mu \frac{\partial \overline{u}}{\partial y} \quad \text{and} \quad \tau_{tur} = -\mu_t \frac{\partial \overline{u}}{\partial y}, \tag{5.120}$$

where μ_t is a virtual viscosity caused by the turbulence, which is called eddy viscosity. It should be noted here that the eddy viscosity is not a property of the fluid like the dynamic viscosity μ, but depends on the time-averaged flow variables. In a similar way as the molecular momentum diffusion ν, the molecular eddy diffusion ε_M is often used to characterize a turbulent flow, ν and ε_M being defined as:

$$\nu = \frac{\mu}{\rho} \quad \text{and} \quad \varepsilon_M = \frac{\mu_t}{\rho}. \tag{5.121}$$

By analogy with Fourier's law of heat conduction, a law for turbulent heat conduction $\dot{q}_{tur}$ can also be formulated:

$$\dot{q} = -k\frac{\partial \overline{T}}{\partial y} \quad \text{and} \quad \dot{q}_{tur} = k_{tur}\frac{\partial \overline{T}}{\partial y}, \tag{5.122}$$

in which k_{tur} represents an apparent (eddy) thermal conductivity, which is also not a fluid property. It also depends on the time-averaged flow variables and the mean temperature. By analogy with the molecular thermal diffusivity a, an eddy diffusivity of heat ε_H is defined:

$$a = \frac{k}{\rho c_p} \quad \text{and} \quad \varepsilon_H = \frac{k_{tur}}{\rho c_p}. \tag{5.123}$$

Thus, a turbulent Prandtl number Pr_t can be introduced complementary to the molecular Prandtl number Pr:

$$Pr = \frac{\nu}{a} \quad \text{and} \quad Pr_t = \frac{\varepsilon_M}{\varepsilon_H}. \tag{5.124}$$

Experiments to measure the turbulent Prandtl number Pr_t in tubes showed that the turbulent Prandtl number is a function of the Reynolds number Re, the molecular Prandtl number Pr and the wall distance (y/R), and exactly the determination definition of the turbulent Prandtl number Pr_t is thus:

$$Pr_t = f(Re, Pr, y/R) = \frac{\overline{u'v'}}{\overline{v'T'}} \cdot \frac{\partial T/\partial y}{\partial u/\partial y}. \tag{5.125}$$

Eckert and Drake (1972) collected measurement results of turbulent Prandtl number for different fluids including liquid metals. They write, "The experimental difficulties for detecting Pr_t are apparently so great that there is no satisfactory coherence of results for small Prandtl numbers", which holds still today. Assuming a two-dimensional turbulent pipe inlet flow, the resulting wall shear stress τ and heat flux $\dot{q}$ can be expressed using the turbulent exchange quantities in the following way:

$$\frac{\tau}{\rho} = (\nu + \varepsilon_M)\frac{d\bar{u}}{dy} \quad \text{and} \quad \frac{\dot{q}}{\rho c_p} = -(a + \varepsilon_H)\frac{d\overline{T}}{dy}. \tag{5.126}$$

The Reynolds analogy is used in most technical applications. In this context, it must be explicitly pointed out that in flows with recirculation, separation, shear flows, free jets, there is a strong anisotropy of the turbulence field. There, the validity of the $k-\varepsilon$ model and the applicability of the Reynolds analogy presented here is problematic. This cannot be improved by the user by choosing a higher spatial resolution of the computational grid. Rather, the applicability of the chosen viscous and thermal wall functions and boundary conditions must be verified by empirical values and with the help of analytical considerations before starting the calculation.

In fully developed flows, $\nu \ll \varepsilon_M$ and $a \ll \varepsilon_H$ are valid except in the immediate vicinity of the wall. If one assumes that the ratio of turbulent heat transport to turbulent momentum transport is constant, in the extreme case $\varepsilon_M = \varepsilon_H$ this leads directly to the result $Pr_t = 1$, which is called Reynolds analogy:

$$c_p d\overline{T} = -\frac{\dot{q}}{\tau} d\overline{u}. \tag{5.127}$$

The equation can be easily integrated, provided that a no-slip condition at the wall, the wall temperature T_w and $\overline{T}(y=0)$ are known. If the following assumptions are made for the connection of the flow and temperature field to the external flow, $\overline{u} = u_0$ and $\overline{T} = T_m$, which are defined in the following way:

$$c_p \int_{T_w}^{T_m} d\overline{T} = -\frac{\dot{q}}{\tau} \int_0^{u_0} d\overline{u} \quad \text{or} \quad c_p(T_w - T_m) = \frac{\dot{q}_w}{\tau_w} u_0, \tag{5.128}$$

the Reynolds analogy in a pipe flow yields the following analytical results for the friction factor f_λ and the Stanton number St:

$$f_\lambda = 8\frac{\tau_w}{\rho u_0^2} \quad \text{and} \quad St = \frac{Nu}{Re\ Pr} = \frac{f_\lambda}{8}. \tag{5.129}$$

The Stanton number measures the ratio of the heat transferred into a fluid to the thermal capacity of the fluid. By the above relation the Reynolds analogy relates heat transfer to the friction. Generally, this approach leads to relatively good results for molecular Prandtl numbers with $Pr > 0.71$, provided the anisotropy of the flow field is small. Therefore, it is usually implemented as a default setting in commercial computational codes that include the $k-\varepsilon$ model. Nevertheless, it must be emphasized that a $k-\varepsilon$ model reduces the turbulent dissipation from a tensor to a constant (question of anisotropy) and the Reynolds analogy is an additional constraint with the assumption $\varepsilon_M \sim \varepsilon_H$.

This problem of anisotropy of the momentum field as well as the use of the Reynolds analogy in turbulence heat transfer problem is known for a long time and critically discussed by several authors (Kays, 1994 or Carteciano & Grötzbach, 2003). To solve the closure problem of the turbulent heat fluxes RH in Eq. (5.113), different approaches can be taken, filling several textbooks. In the following, only the basic ideas of the solution approaches of the closure problem for the turbulent heat flows as well as the order of the methods shall be touched upon.

- *Semiempirical or empirical solution methods of zero and first order.* All such models do not allow insight into turbulent exchange mechanisms between temperature and velocity fluctuations and are only applicable after careful examination of the valid parameter range of the dimensionless numbers (Re, Pr, Ra) and for Pr_t around unity.
- *Turbulent Prandtl numbers from analytical solutions* are difficult to treat and are essentially limited to simple geometries but provide important information about the near-wall behaviour and the thermal wall functions. Unlike transport equation models, they include the temporal behaviour of the fluctuating variables and thus their spectral behaviour, which allows comparison with results from a direct numerical simulation. Unfortunately, for engineering applications, such methods require considerable effort and intervention in commercial CFD design tools.
- *Modelling of turbulent heat flux using transport equations* require an experienced CFD user to analyse the posed flow problem in terms of the potentially occurring phenomena and to choose appropriate thermal and viscous wall functions for this purpose. In highly anisotropic flows, heat transfer calculations using the $k - \varepsilon$ turbulence model and a turbulent Prandtl number $Pr_t = 0.9$ (Reynolds analogy) can lead to unphysical results such as a negative wall heat flux. In such cases, modelling the turbulent heat fluxes using transport equations of, for example, an algebraic heat flux model (AHFM) (Shams et al., 2019) or the Reynolds stress turbulence model (RSM) (Khan & Joshi, 2015) is essential. Nevertheless, such a procedure always leads to triple correlations, which are not known "a priori" and have to be determined either by experimental benchmarks or by "Direct Numerical Simulation" (DNS). Increasing computational power provided substantial progress in computing local turbulent heat transfer problems by means of AHFM, RSM and also large eddy simulations (LES); nonetheless, pro-found expert knowledge is still required to tune the computations to experimental observations. Nonetheless, the degree of uncertainty is continuously decreasing.
- *Direct numerical simulation (DNS)* of the time dependent Navier–Stokes equations is in principle an exact solution to the problem. Although memory resources and computational speeds have increased considerably, their application has so far remained limited to a few mostly academic cases. Recent DNS studies have focused on the influence of the Prandtl number in the case of isotropic turbulence (Kasagi & Ohtsubo, 1992), near-wall flows (Kawamura et al., 1999 or Piller et al., 2002), and buoyant flows (Otic et al., 2005). All of them include results for the time-dependent as well as the mean values of the temperature and velocity field depending on the boundary conditions. Therefore, DNS is one of the few tools that allows to validate the approaches and constants chosen in the transport sliding methods. This is especially important because the experimental determination of these quantities is hardly possible

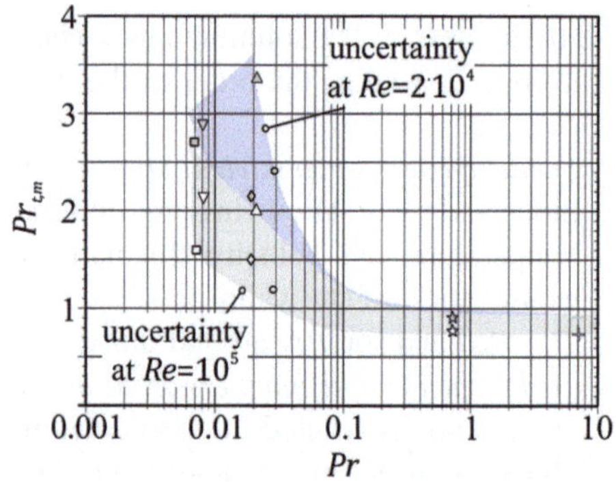

Fig. 5.22 Measured mean turbulent Prandtl number $Pr_{t,m}$ of a pipe flow as a function of molecular Prandtl number Pr (multiple fluids) for two different Reynolds numbers (Stieglitz, 2007)

in the absence of suitable measurement methods. Today's measurement methods are not capable of simultaneously measuring several field quantities with sufficiently precise time and spatial resolution.

Apart from the turbulent Prandtl number, which is difficult to determine experimentally, the assumption of a constant Pr_t value is problematic for many flows, especially for liquid metals. Figure 5.22 shows the measured mean turbulent Prandtl number $Pr_{t,m}$ of a circular pipe flow as a function of the molecular Prandtl number (several fluids) for two different Reynolds numbers (Stieglitz, 2007). It is clear from the graph that for $Pr < 1$ the uncertainties are considerable, although with the help of improved calculation methods the confidence has steadily increased in recent years. All experimentally determined observations allow the following conclusions concerning the mean turbulent Prandtl number $Pr_{t,m}$:

- For $Pr \geq 1$ (air water and oil, etc.) the turbulent Prandtl Pr_t number varies little and is almost independent of the Reynolds number and the molecular Prandtl number.
- The influence of turbulent Prandtl number and Reynolds number increases simultaneously with falling molecular Prandtl number and falling Reynolds number.
- The turbulent Prandtl number increases with falling Prandtl number and falling Reynolds number.

5.1.15 External Flows

Flow losses or heat transfer in external flows are often neglected in solar thermal textbooks. However, they play an important role in heat transfer, e.g. at plate collectors or thermal heat storages, but also along the often kilometre-long absorber rows of linear-focusing solar power plants or at the absorbers of solar towers. The

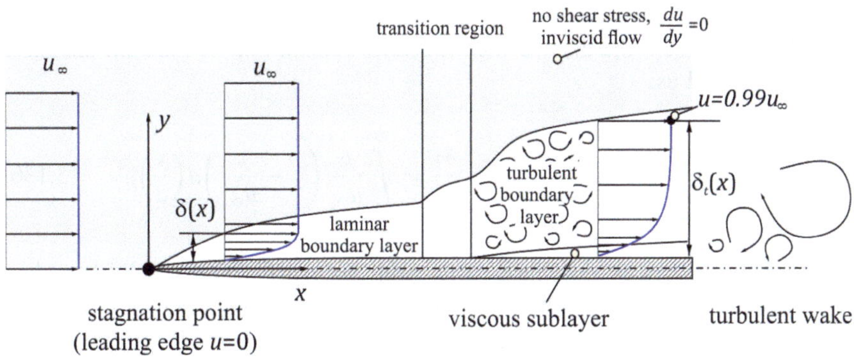

Fig. 5.23 Schematic evolution of a developing boundary layer flow over a flat plate

same applies to the flow resistance of the large mirrors, the structural integrity of solar updraft towers, which have to take environmental influences into account in an appropriate manner.

A number of phenomena that occur in external flow at high Reynolds number over a plate. Let's assume a plate as illustrated in Fig. 5.23. As the flow hits the object the free stream flow divides at the stagnation point, where due to the non-slip condition the velocity is zero. Due to the viscosity first, a laminar boundary layer develops which grows as the flow proceeds downstream in x-direction. The initially laminar boundary layer turns at a transition point into a turbulent boundary layer. Boundary layers thickening on the surfaces cause only a slight displacement of the streamlines of the external flow. Once the flow leaves the body separation may occur in the region of increasing pressure on the rear of the body, since the boundary layer fluid does not longer stay in contact with the wall surface. The fluid domain that formed the boundary layers generates a viscous wake behind the plate.

Because the boundary layer is thin, disturbance of the inviscid flow outside the boundary layer is negligible, and the pressure gradient along the surface is close to zero. The transition from laminar to turbulent boundary layer flow on a flat plate occurs if the Reynolds number based on x exceeds a critical Reynolds number of $Re_x = 5 \times 10^5$. This transition may occur at lower Re_x if the surface is rough, or the pressure increases downstream in flow direction, or separation occurs. For very low-turbulence inflow, considerably larger values of $Re_{x,crit}$ are obtained, and for high-turbulence inflow, considerably lower values. Downstream the transition, the turbulent boundary layer thickens more rapidly than the laminar boundary layer caused by increased shear stress at the body surface.

Boundary layer disturbance thickness δ is defined as the wall normal distance y, from the surface to a position at which the velocity within the boundary layer, $u(y)$ reaches $0.99u_\infty$ of the local freestream, and its velocity profile merges smoothly and asymptotically into the freestream, being challenging to measure. Thus, two integral measures of boundary layer thickness, which are easier to acquire experimentally are defined, the displacement thickness δ_1 and the momentum thickness δ_2, being defined as:

$$\frac{\delta_1}{\delta} = \int_0^\infty \left(1 - \frac{u}{u_\infty}\right) d\left(\frac{y}{\delta}\right) \quad \text{and} \quad \frac{\delta_2}{\delta} = \int_0^\infty \frac{u}{u_\infty}\left(1 - \frac{u}{u_\infty}\right) d\left(\frac{y}{\delta}\right). \tag{5.130}$$

Physically, δ_1 can be precepted as the wall normal distance a solid boundary would need to be shifted towards the free stream in a frictionless flow to generate a volumetric flow deficit caused by the boundary layer; while δ_2 corresponds to a thickness of a fluid layer with velocity u_∞, for which the momentum flux is the same as the deficit in momentum flux within the boundary layer. For a laminar flat-plate boundary layer according to Blasius similar self-similar velocity profiles in the boundary layer expressed in the form $u/u_\infty = f(y/\delta)$ can be assumed so that velocity profiles do not vary downstream with x. The result of the Blasius solution for the boundary layer thickness δ and the dimensionless wall shear stress C_f read to,

$$\frac{\delta}{x} = \frac{5}{\sqrt{Re_x}} \quad \text{and} \quad C_f = \frac{\tau_w}{(\rho/2)u_\infty^2} = \frac{0.664}{\sqrt{Re_x}}, \tag{5.131}$$

indicating that the laminar boundary layer thickness grows downstream with $\sqrt{x}$ and the shear stress accordingly with $1/\sqrt{x}$.

Regarding a turbulent boundary layer, the velocity profile can be expressed by a power law, $u/u_\infty = (y/\delta)^{1/n}$, in which n is an integer between 6 and 10, mostly 7 is taken. Both, boundary layer thickness and shear stress cannot be analytically calculated and for both empirical correlations are used:

$$\frac{\delta}{x} = \frac{0.382}{Re_x^{1/5}} \quad \text{and} \quad C_f = \frac{\tau_w}{(\rho/2)u_\infty^2} = \frac{0.0594}{Re_x^{1/5}}. \tag{5.132}$$

Hence, for a turbulent boundary layer the thickness increases with $x^{4/5}$ while the wall shear stress follows $x^{-1/5}$. A comparison of velocity distribution of laminar and turbulent flows in Fig. 5.23 exhibits that the turbulent velocity profile has much more momentum than the laminar profile.

Viscous shear always causes a net breaking force on any fluid particle within the boundary layer and even assuming zero pressure gradient ($dp/dx = 0$) shear forces only can never bring a fluid particle to rest, because the shear stress never vanishes for a finite downstream value of x (remind in laminar and boundary layers the wall shear stress scales with $x^{-1/2}$ and $x^{-1/5}$). Since the wall shear stress is defined by $\tau_w = \mu(\partial u/\partial y)_{y=0}$ the velocity gradient cannot diminish. In accelerated flows characterized by a negative pressure gradient ($dp/dx < 0$) as shown in Fig. 5.24 the pressure forces occur, conversely a positive pressure gradient ($dp/dx > 0$) yield pressure forces opposing the fluid motion and thus are a necessary condition for flow separation. Since turbulent boundary layers have a higher momentum than laminar ones, they are less fragile to flow separation in case of a positive pressure gradient.

Anybody subjected to viscous flow experiences shear stresses at the interfaces due to the viscosity of the fluid. At the same time, pressure differences result from the flow around the body. Both lead to a force acting on the body, whereby the force in the main flow direction is defined as drag and that perpendicular to it as lift. Both drag and lift forces are of importance for large-scale objects such as heliostats, solar upwind draft towers are well as parabolic troughs, which have to withstand environmental extreme environmental conditions without losing performance. For engineering purposes, the resulting drag and lift forces are calculated from experimentally derived coefficients (C_D, C_L) where both are normalized by the dynamic pressure.

$$C_D = \frac{F_D}{(\rho/2)u_\infty^2 \cdot A} \quad \text{and} \quad C_L = \frac{F_L}{(\rho/2)u_\infty^2 \cdot A}. \tag{5.133}$$

Note here that A is the area upon which the coefficient is based on. While A for drag coefficients is given by the projected frontal area in case of lift it is based on projected planform area. The total friction drag force acting on a plane plate aligned with the flow direction is obtained by integrating the shear stress distribution along the surface and the corresponding drag coefficient is defined as friction force divided by dynamic pressure and wetted area in contact with the fluid. As the shear stress is a function of the Reynolds number, so is the drag coefficient, which is depicted in Fig. 5.26a for a flat plate.

A thin flat surface exposed normal to a fluid flow has no area parallel to the flow direction, thus, there can be no friction force parallel to the flow; all drag is caused by pressure forces. Observations shows that drag coefficients C_D for objects with sharp edges are almost independent of the Reynolds number, if the

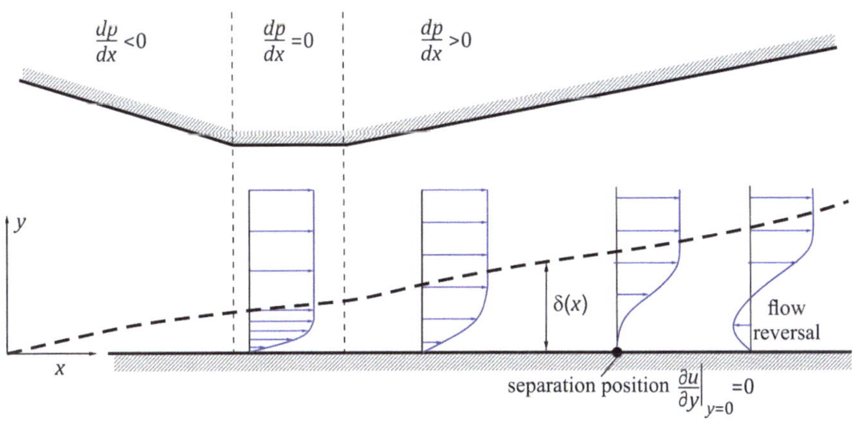

Fig. 5.24 Sketch of the boundary layer shape evolution as function of the pressure gradient

Reynolds number is large ($Re > 10^3$), because the flow separation points are pinned to the geometry discontinuity of the object. Smoothing the object shape reduces the drag considerably. A comparison of the drag coefficients for the hemisphere and the parabolic shapes as they appear for solar mirrors facing into and away from the flows is illustrated in Fig. 5.26b. Also note that the drag coefficient for a two-dimensional object (long square cylinder) is about twice that for the corresponding three-dimensional object (square cylinder with $b/h = 1$).

For smoothed objects exposed to wind loads the drag coefficients depend on the Reynolds number, the angle of attack of the wind α, and the height to diameter ratio. Moreover, for smoothed structures its own roughness but also the interaction with other adjacent structures yields Reynolds number dependent effects such as vortex shedding posing additional loads acting on the structures. Detailed drag and lift coefficient data and methods to evaluate the loads exposed to environmental effects may be taken from Veritas (2010). Exemplarily, here the values for a cylinder are shown since they are relevant for towers or chimneys of upwind power plants. For a smooth cylinder the flow domains are termed sub-critical flow if the drag-coefficient remains constant ($Re_D < 2 \times 10^5$), which is characterized by a laminar boundary layer. It is followed by a transcritical domain, at which the drag coefficients undergo at first a rapid drop reaching a local minimum before it increases again. At about $Re_D > 3 \times 10^6$ the supercritical flow domain with a fully turbulent boundary layer is reached where the drag coefficient recovers to an almost constant but lower value than in laminar flow. The dependence of the drag coefficient as function of the Reynolds number built with the diameter D of the cylinder is shown in Fig. 5.25.

If the total drag force on a slender body with the characteristic cross-sectional dimension d and the finite length l is estimated, the integrated sum of sectional force contributions is reduced due to the finite dimensions and the drag coefficient is multiplied by a reduction factor κ according to Table 5.6: Values of drag reduction factor κ for cylinders of finite length l and slenderness d. For cylinders with one end abutting on to another cylinder or a wall in such a way that free

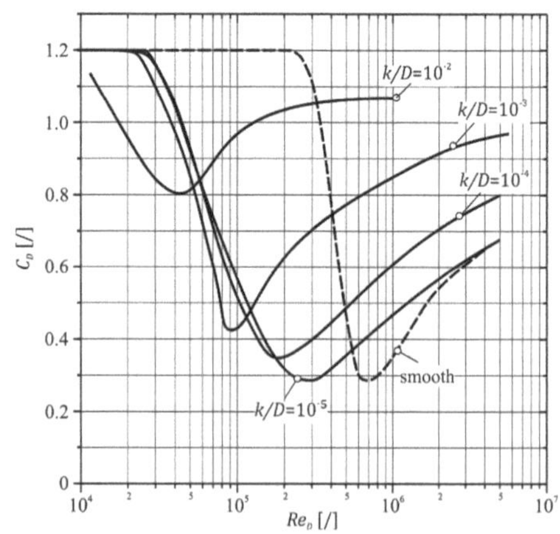

Fig. 5.25 Drag coefficient C_D for a fixed circular cylinder as a function of Re_D for steady flow in critical flow regime at various surface cylinder roughnesses

Table 5.6 Values of drag reduction factor κ for cylinders member of finite length l and slenderness d

l/d (/)	2	5	10	20	40	50	100
Cylinder sub-critical flow	0.58	0.62	0.68	0.74	0.82	0.87	0.98
Cylinder super-critical flow	0.80	0.80	0.82	0.90	0.98	0.99	1.00
Flat plate normal to flow	0.62	0.66	0.69	0.81	0.87	0.90	0.95

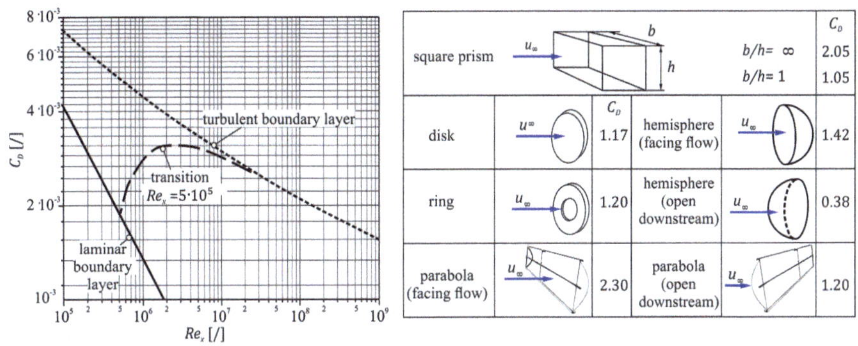

Fig. 5.26 Drag coefficient C_D as a function of the Reynolds number Re_x for a flat plate (left). Drag coefficient for selected geometries with sharp edges (for $Re > 10^3$, right)

flow around that end of the member is prevented, the ratio l/d should be doubled for the purpose of determining the drag reduction factor κ. When both ends are abutted as mentioned, the drag coefficient C_D should be taken equal to that for an infinitely long cylinder.

In the case of unbounded flow or flows around bodies, as in closed ducts, there is a distinction between forced and mixed convection and, accordingly, the heat transfer depends on the respective characteristic number and the thermal boundary conditions. As soon as the surface temperature of the body differs from that of the fluid, heat is transferred as forced convection. Conversely as soon as the bulk motion of the fluid results exclusively from the difference in temperature of the solid surface and the fluid, the mechanism is natural convection. For both type of problems, the velocity and temperature of the fluid far away from the solid surface are the free-stream velocity and free-stream temperature. Both, u_∞ and T_∞ are usually known or specified.

The task is usually to assess the heat flux from or to the surface, dependent on the given temperature or heat flux or to evaluate the temperature distribution, which requires the determination of the convective heat transfer coefficient h,

$$\dot{q} = h \cdot (T_w - T_\infty), \tag{5.134}$$

whereby in the following elaborations it is to be paid attention carefully whether in each case the local heat transfer expressed by the subscript "x" or the averaged heat transfer with the index "l" is addressed. It should also be pointed out, in the case of flows around body problems, that many correlations for determining the convective heat transfer coefficient are based on experimental data, which are subject to some uncertainty, even though the experiments are conducted under carefully controlled conditions. The resulting uncertainties are therefore often of the order of $\pm 10\%$.

Similar as in a duct flow the temperature of the fluid changes from the surface temperature (T_w) to the free-stream temperature (T_∞) at the edge of the temperature boundary layer. We have seen that laminar velocity boundary layer thickness δ depends on the Reynolds number scaling with $\delta/x = 5/\sqrt{Re_x}$. Hence, also the thermal boundary layer thickness δ_{th} depends both on Re_x and Pr. Based on a similitude boundary layer analysis the thickness scaling of velocity to thermal boundary layers is:

$$\begin{aligned} Re_x < Re_{crit}: & \ \frac{\delta}{\delta_{th}} = \mathrm{Pr}^{1/3} \text{ for } Pr > 0.7; \\ & \ \frac{\delta}{\delta_{th}} = \mathrm{Pr}^{1/2} \text{ for } Pr \ll 1; \quad . \\ Re_x > Re_{crit}: & \ \frac{\delta}{X} = \frac{0.37}{Re_x^{0.2}} \quad \delta \approx \delta_{th} \end{aligned} \tag{5.135}$$

Viscous dissipation and high-speed effects can be neglected if $Pr^{1/2} \cdot Ec/2 \ll 1$, where the Eckert number Ec is defined by $Ec = u_\infty^2 / \left(c_p \cdot (T_w - T_\infty)\right)$.

Flat plate

For a laminar flow over a plate, the local heat transfer (Nu_x) and the heat transfer averaged over the plate length l (Nu_l) are calculated at isothermal wall surface temperature ($T_w = const.$) as:

$$Nu_{x,lam} = 0.332 \cdot \sqrt{Re_x} \cdot Pr^{1/3}, \tag{5.136}$$

in which $Nu_{x,lam} = (h \cdot l)/\lambda$ and $Re_x = (u_\infty \cdot x)/\nu$. The mean Nusselt number $Nu_{l,lam}$ of a plate averaged over the length l is obtained by integrating the Eq. (5.136) to:

$$Nu_{l,lam} = 0.664 \cdot \sqrt{Re_l} \cdot Pr^{1/3}, \tag{5.137}$$

where for the Reynolds number the length l is to be used.

In case of a turbulent flow over plate the local Nusselt number $Nu_{x,turb}$ at constant wall temperature ($T_w = const.$) can be calculated with the following equation:

$$Nu_{x,turb} = \frac{0.0296 \cdot Re_x^{4/5} \cdot Pr}{1 + 2.185 \cdot Re_x^{-0.1} \cdot (Pr - 1)}, \tag{5.138}$$

for which the thermophysical material values for the mean temperature T_m of the fluid are to be used. For the length averaged turbulent Nusselt number $Nu_{l,turb}$, the average drag coefficient $C_D = 0.037 \cdot Re_l^{-0.2}$ of the turbulent boundary layer flow over the plate length l can be used, so that for $Nu_{l,turb}$ the following equation applies:

$$Nu_{l,turb} = \frac{0.037 \cdot Re_l^{0.8} \cdot Pr}{1 + 2.443 \cdot Re_l^{-0.1} \cdot \left(Pr^{2/3} - 1\right)}. \tag{5.139}$$

The range of validity of the equation is $5 \times 10^5 < Re_l < 10^7$ and $0.5 < Pr < 2 \times 10^3$.

With constant wall heat flux ($\dot{q}_w = const$), the local laminar Nusselt number at the location x away from the beginning of the plate can be calculated using the equation:

$$Nu_{x,lam} = 0.460 \cdot \sqrt{Re_x} \cdot Pr^{1/3}, \tag{5.140}$$

which is recommended for $Re_x < 10^5$ and $0.6 < Pr < 2 \times 10^3$. Churchill and Ozoe (1973) recommend the following single correlation for all Prandtl numbers:

$$Nu_{x,lam} = \frac{0.886 \cdot \sqrt{Re_x} \cdot Pr^{1/2}}{\left(1 + \left(\frac{Pr}{0.0207}\right)^{2/3}\right)^{1/4}}. \tag{5.141}$$

Note that for surfaces with uniform heat flux the local convective heat transfer coefficient is used to determine the local surface temperature. The total heat transfer rate being known, an average heat transfer coefficient is not needed and not defined.

For turbulent boundary layers with constant wall heat flux ($\dot{q}_w = const$) the following relations are recommended to evaluate the local turbulent Nusselt number $Nu_{x,turb}$:

$$\begin{aligned} &Re_{cr} < Re_x < 10^7: && Nu_{x,turb} = 0.0296 \cdot Re_x^{4/5} \cdot Pr^{1/3} \\ &Re_x > 10^7: && Nu_{x,turb} = 1.596 \cdot Re_x \cdot (\ln Re_x)^{-2.584} \cdot Pr^{1/3}. \end{aligned} \tag{5.142}$$

In laminar boundary layers, the convective heat transfer coefficient with uniform heat flux is approximately 36% higher than with uniform surface temperature. In case of turbulent boundary layers, the difference is negligible and the correlations for the local convective heat transfer coefficient can be used for both uniform surface temperature and uniform heat flux.

If the boundary layer is initially laminar followed by a turbulent boundary layer at $Re_x = Re_{cr}$, the for $0.7 < Pr < 60$ the following correlations are suggested for the length averaged turbulent Nusselt number $Nu_{l,turb}$:

$$Re_{cr} < Re_L < 10^7: \quad Nu_{l,turb} = \left(0.664\sqrt{Re_l} + 0.037\left(Re_l^{4/5} - Re_{cr}^{4/5}\right)\right)Pr^{1/3},$$

$$Re_x > 10^7: \quad Nu_{l,turb} = \left(1.963 Re_l \cdot (\ln Re_l)^{-2.584} - 871\right) Pr^{1/3} \tag{5.143}$$

In most practical cases, even at intermediate Reynolds numbers, a laminar boundary layer will not form along the entire length of the plate due to blunt plate edges and the turbulence of the inflowing medium. If measurement results are compiled, they can be represented by a "mean curve" which also includes the area of the turbulent boundary layer. As Gnielinski (1975) has shown, a mean curve for a large range of the Prandtl number can be represented by the relation:

$$Nu_{l,0} = \sqrt{Nu_{l,lam}^2 + Nu_{l,turb}^2}, \tag{5.144}$$

in which the relations for the laminar and turbulent averaged Nusselt numbers according to the thermal boundary conditions are inserted. The range of application applies to mean Nusselt numbers at $10^1 < Re < 10^7$ and $0.5 < Pr < 2 \times 10^3$ and for the fluid properties the values are to be used at the mean temperature $T_m = (T_{in} + T_{out})/2$.

Flow over cylinders

The heat transfer coefficients of cross-flow over tubes and profile cylinders play a role, for example, in tower receivers, the hot tanks of sensitive heat storage systems, but also in the often kilometre-long absorber tubes of solar farms. For the calculation of the mean Nusselt numbers of cross-flowed tubes or profiled cylinders, the same heat transfer relationships apply as for the flat plate; the so-called "overflow length" l is to be used as the characteristic length when calculating the corresponding Reynolds and Nusselt numbers. The overflow length is the distance a flowing particle can travel along the entire surface and is defined as the total surface area A of a body divided by the perimeter l_c of the shadow area of the body, i.e. $l = A/l_c$, which is sketched in Fig. 5.27. For a long pipe of diameter d and length L, the overflow length is thus obtained as:

$$l = \frac{\pi\, d\, L}{2\, L} = \frac{\pi}{2} d. \tag{5.145}$$

For the average Nusselt number of cross-flowed tubes and profile cylinders, the following relation is given as an approximation equation:

$$Nu_{l,0} = 0.3 + \sqrt{Nu_{l,lam}^2 + Nu_{l,turb}^2}, \tag{5.146}$$

for which the laminar and turbulent relations at the corresponding thermal boundary conditions are to be used. The minimum value for $Nu_{l,0}$ originates from the

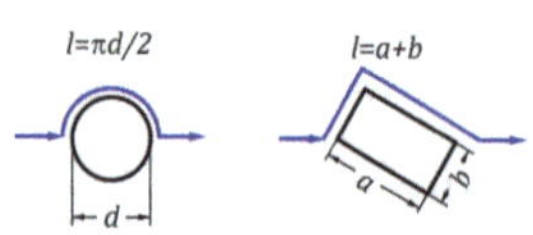

Fig. 5.27 Sketch for the definition of the overflow length l

Table 5.7 Exponents of heat transfer on a transversely flowed cylinder of constant wall temperature ($T_w = const.$) in different Reynolds number ranges

Re_d	Flow regime	C	m	n
$Re_d < 40$	Creeping	0.76	0.4	0.37
$40 < Re_d < 10^3$	Laminar	0.152	0.5	0.37
$10^3 < Re_d < 2 \times 10^5$	Transitional	0.26	0.6	0.37
$2 \times 10^5 < Re_d < 4 \times 10^7$	Turbulent	0.023	0.8	0.40

fact that cross-flowed cylinders always have a finite length, so that a minimum heat flow occurs even in a stationary environment. The mean Nusselt number of the cross-flowed cylinder therefore asymptotically approaches a minimum, which is approximately given by $Nu_{min} \approx 0.3$.

Exactly, the mean heat transfer (Nu_l) on transversely flowed cylinders with $T_w = const.$ is calculated according to the basic Eq. (5.147), where depending on the Reynolds number range Re_d related to the diameter, the exponents shown in Table 5.7 have to be used.

$$Nu_d = C \cdot Re_d^m \cdot Pr^n. \tag{5.147}$$

For the thermal boundary condition $\dot{q}_w = const.$ the following relation can be found in the literature, which is valid in the range $10^{-2} < Re_d < 10^5$:

$$Nu_d = 0.42 \cdot Pr^{0.2} + 0.57 \cdot Re_d^{0.5} \cdot Pr^{\frac{1}{3}}. \tag{5.148}$$

In the case of a cylinder with longitudinal flow, the following relations are obtained for the local Nusselt number Nu_x and the mean Nusselt number (Nu_l) in the range for $Re_{x,crit} < 2 \times 10^5$:

$$Nu_x = 0.332 \cdot Re_x^{1/2} \cdot Pr^{1/3} \cdot \left[1 + \frac{4.6}{Re_x^{1/2}}\left(\frac{x}{d}\right)\right];$$

$$Nu_l = 0.664 \cdot Re_L^{1/2} \cdot Pr^{1/3} \cdot \left[1 + \frac{2.3}{Re_x^{1/2}}\left(\frac{L}{d}\right)\right]. \tag{5.149}$$

For turbulent inflow, the correlation (5.149) expands multiplicatively by a geometry factor K_{cyl}:

$$K_{cyl} = \left[1 + 0.37 \cdot \frac{2 \cdot l}{d \cdot Re_d^{1/5}}\right]^{0.14}. \tag{5.150}$$

Free natural convection occurs in particular where temperature differences between the surfaces of two bodies occur in open cavities, for example in cavity receivers of solar thermal tower power plants, but also behind transparent heat insulation in thermal insulation. Again, the Nusselt number describes the heat transfer $\dot{Q}$ from the body to the environment. It is defined as follows and depends, as in closed systems, on the Rayleigh and Prandtl numbers, and furthermore on the geometry and the thermal boundary conditions

$$Nu^* = \frac{\overline{h} \cdot L}{k} = \frac{\dot{Q} \cdot L}{A_w \cdot k \cdot (T_w - T_\infty)} = f\left(Ra, Pr, \varepsilon_f\right), \tag{5.151}$$

In which $\overline{h}$ denotes the geometry averaged heat transfer coefficient, L a characteristic geometric length, k the heat conductivity of the fluid, A_w the area through which the heat is transferred and $\Delta T = (T_w - T_\infty)$ the temperature difference between the body wall and the ambient. The heat transfer correlation in the case of free convection can be calculated with the help of the general equation according to Michejew:

$$Nu = C_i^j(Pr) \cdot Ra^n \cdot \varepsilon_f, \tag{5.152}$$

where ε_f represents a form factor. The subscript i indactes in the following correlations, whether the flow is laminar (lam) or turbulent ($turb$), while the superscript indicates if the surface orientation is vertical (v) or horizontal (h) with respect to the gravity vector g. When calculating the Rayleigh number, it is necessary to pay attention to the correct choice of the characteristic length l. For vertical plane plates, for example receivers, the characteristic length l corresponds to the plate height. For example, the transition from laminar to turbulent heat transfer occurs for a vertical plate at $Ra_l = 10^{12}$. The fluid properties are taken as average of the wall and ambient temperatures, whereas the temperature coefficient of volume expansion β is evaluated at T_∞ for external natural convection.

For all external body flows, a similar dependence on the Prandtl number Pr is observed, provided it is larger than $Pr = 0.5$, so that according to Churchill and Usagi (1972) the following correlations average Nusselt-number correlations can be used for laminar and turbulent flows:

$$\begin{aligned} C_{lam} &= \frac{0.671}{\left(1 + (0.492/Pr)^{9/16}\right)^{4/9}}; \\ C_{turb}^{v} &= \frac{0.13\ Pr^{0.22}}{\left(1 + 0.61\ Pr^{0.81}\right)^{0.42}}; \\ C_{turb}^{h} &= 0.14\left(\frac{1 + 0.0107\ Pr}{1 + 0.01\ Pr}\right). \end{aligned} \tag{5.153}$$

Since there is often a coexistence of laminar and turbulent flows, the laminar and turbulent Nusselt numbers (Nu_{lam}, Nu_{turb}) are superimposed similar to Gnielinski's approach (compare Eq. (5.144)) but complemented by introducing a blending exponents accounting for body shape and orientation in order to compute an average Nusselt number according to.

$$Nu_l = \left((Nu_{lam})^m + (Nu_{turb})^m\right)^{1/m}. \tag{5.154}$$

The equation for Nu_{lam} in this section is usually expressed in terms of Nu_0, the Nusselt number that would be valid if the thermal boundary layer were thin. The difference between Nu_{lam} and Nu_0 accounts for the effect of the large boundary layer thicknesses encountered in natural convection.

Subsequently the correlations for the body shapes depicted in Fig. 5.28 are provided considering them to be heated ($T_w > T_\infty$) and immersed in a fluid at rest. In case $T_w < T_\infty$ the same correlations apply with good approximation, for a geometry that is rotated 180° relative to the gravitational vector; for example, the correlations for a horizontal heated upward facing flat plate applies to a cooled downward-facing flat plate of the same plane form, and for the driving temperature gradient the absolute value is taken, i.e. $|T_w - T_\infty|$.

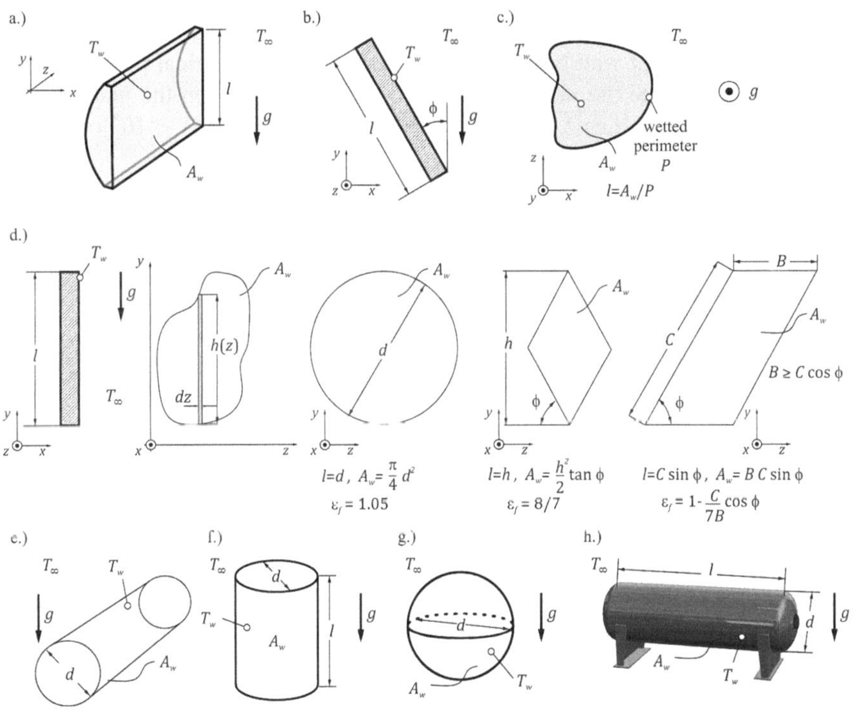

Fig. 5.28 Nomenclature for external heat transfer **a** assignment of general geometry and variables, **b** tilted plate, **c** horizontal plate, **d** planar vertical surfaces of various shape, **e** infinitely long horizontal cylinder, **f** vertical circular cylinder, **g** sphere and **h** horizontal tank

- The Nusselt number along a vertical plate ($\phi = 0°$) as shown in Fig. 5.28b for $1 < Ra < 10^{12}$ is given for a constant wall temperature ($T_w = const.$) by:

$$Nu_0 = C_{lam} \cdot Ra^{1/4};\ Nu_{lam} = \frac{2}{\ln(1 + 2/Nu_0)};$$
$$Nu_{turb} = C^v_{turb} \cdot \frac{Ra^{1/3}}{\left(1 + 1.4 \times 10^9 Pr/Ra\right)}, \tag{5.155}$$

In which C_{lam} and C^v_{turb} are to be taken from Eq. (5.153) and Nu_l is computed by inserting the expressions of Eq. (5.155) for Nu_{lam} and Nu_{turb} in relation (5.154) using as exponent $m = 6$. The average heat transfer at vertical surfaces with fully turbulent convection ($Ra > 10^{12}$) and constant wall temperature is

$$Nu_l = \left[0.825 + 0.387 \cdot (Ra_L \cdot K(Pr))^{\frac{1}{6}}\right]^2 \text{ with } K(Pr) = \left[1 + \left(\frac{0.492}{Pr}\right)^{\frac{9}{16}}\right]^{-\frac{16}{9}}. \tag{5.156}$$

For gases, $K(Pr)$ is constant and is $K(Pr) = 0.348$.

In case of constant wall heat flux ($\dot{q}_w = const$) along a vertical plate ($\phi = 0°$) the aim is to evaluate the mean temperature difference between the plate surface and the ambient fluid, i.e. $\overline{\Delta T}$. In the range of $15 < (Nu^* \cdot Ra) < 10^5$ one gets:

$$Nu_0 = G_{lam} \cdot \left(Nu^* \cdot Ra\right)^{1/5};\ Nu_{lam} = \frac{1.83}{\ln(1 + 1.83/6Nu_0)};$$
$$Nu_{turb} = C^v_{turb}{}^{3/4} \cdot \frac{(Nu^* \cdot Ra)^{1/4}}{\left(1 + 1.4 \times 10^9 Pr/Ra\right)} \text{ with}$$
$$G_{lam} = \frac{6}{5} \cdot \left(\frac{Pr}{4 + 9\sqrt{Pr} + 10Pr}\right)^{1/5}. \tag{5.157}$$

The Nusselt number Nu_l is again obtained by evaluating Nu_{lam} and Nu_{turb} (Eq. (5.153)) inserted in Eq. (5.154) and using $m = 6$.

- For tilted downward facing plates ($-90° < \phi < 0°$) at constant wall temperature the procedure is first to calculate the removed heat $\dot{Q}$ for a vertical plate using for the gravity vector instead of g the expression $g \cos\phi$. For the horizontal downward plate also, the removed heat $\dot{Q}$ is to be evaluated using $g \sin\phi$ instead of g. It is recommended to take the larger of both values.

- In case of inclined upward facing plates ($0° < \phi < 90°$) a similar procedure is applied and also the heat flux $\dot{Q}$ is calculated for a vertical plate with $g \cos\phi$ and correspondingly for a horizontal plate using $g \sin\phi$ and use the maximum of both values
- For horizontal upward isothermal surfaces ($\phi = 90°$) of various platforms, correlations are given in terms of a modified length scale l, scaled as $l = A_w/P$ according to Fig. 5.28c for $Ra > 1$:

$$Nu_0 = 0.835 C_{lam} \cdot Ra^{1/4};\ Nu_{lam} = \frac{1.4}{\ln(1 + 1.4/Nu_0)};$$
$$Nu_{turb} = C^h_{turb} \cdot Ra^{1/3}. \tag{5.158}$$

The procedure is similar as before to obtain, however the exponent is $m = 10$, while for constant heat flux ($\dot{q}_w = const$) ΔT is to be replaced by $\overline{\Delta T}$.
- For horizontal downward-facing plates ($\phi = -90°$) of various plane forms, the buoyant fluid motion is directed towards the plate so that only a very weak force drives the fluid along the plate and only laminar flows are reported in literature. For $Ra < 10^{10}$ and $Pr \geq 0.7$ the next equations apply:

$$Nu_0 = H_{lam} \cdot Ra^{1/5};\ H_{lam} = \frac{0.527}{\left(1 + (1.9/Pr)^{9/10}\right)^{2/9}};$$
$$Nu = \frac{2.45}{\ln(1 + 2.45/Nu_0)}. \tag{5.159}$$

- In case of vertical, tilted plates of various planform at constant temperature the line of constant z in in Fig. 5.28d represents the line of steepest gradient on the plate. Provided all such lines intersect the plate edges twice the thin-layer (Nu_0) heat transfer value can be found by subdividing the body into strips of width dz calculating the heat transfer from each strip and adding. For laminar flow from an isothermal vertical plate, this results in:

$$Nu_0 = \varepsilon_f \cdot C_{lam} \cdot Ra^{1/4};\ \varepsilon_f = \left(\frac{l^{1/4}}{A_w} \int_0^z h(z)^{3/4} dz\right). \tag{5.160}$$

The form factor of some plate shapes is provided in Fig. 5.28d. In case of a vertical plate the expression for Nu_0 in Eq. (5.155) is replaced by Eq. (5.160), while for inclined plates depending on the orientation (upwards/downwards) the procedure for an ideal plate is to be used.
- Horizontal cylinders, as they appear e.g. in parabolic trough collectors, can be considered as infinitely long cylinders and the expression for Nu_{lam} and Nu_{turb} read to

$$Nu_0 = 0.772\, C_{lam} \cdot Ra^{1/4};$$

Table 5.8 Coefficients C_{turb} for flow along a horizontal cylinder at different Prandtl numbers

Pr	0.01	0.022	0.1	0.71	2.0	6.0	50	100
Horizontal cylinder	0.077	0.81	0.90	0.103	0.108	0.109	0.100	0.097

$$Nu_{lam} = \frac{2f}{(1+2f/Nu_0)}; \; Nu_{turb} = C_{turb} \cdot Ra^{1/3}, \tag{5.161}$$

for which C_{turb} depends strongly on the Prandtl number Pr and values are provided in Table 5.8. For $Ra > 10^{-2}$, $f = 0.8$ can be used, while for $Ra < 10^{-2}$, the functional relationship $f = \left(1 - 0.13/Nu_0^{0.16}\right)$ is used. To find Nu, the values of Nu_{lam} and Nu_{turb} from Eq. (5.161) are substituted into Eq. (5.154) taking $m = 15$.

- Vertical cylinders are used as sensitive heat storage tanks and for high Ra values and large diameter d, the heat transfer from the wall of a vertical cylinder approaches that for a vertical flat plate. Thus, the equations for Nu_0 and Nu_{lam} for a vertical plate of length l (Eq. (5.155)) can be rewritten as $Nu_{0,p}$ and Nu_p. At smaller Ra and diameter d, transverse curvature yields lateral exchange which is has to be considered using

$$Nu_{lam} = \frac{0.9\,\xi\, Nu_p}{\ln(1+0.9\xi)}, \; \xi = \frac{2\,l}{d\, Nu_{0,p}}, \tag{5.162}$$

for which only in the laminar case correlations are reported in literature. In order to get Nu_l for Nu_{lam} the Eq. (5.162) has to be used and for Nu_{turb} Eq. (5.155), which have to inserted in Eq. (5.154) using the exponent $m = 10$.
- In case of spheres also Eq. (5.154) is taken with the exponent $m = 6$ and

$$Nu_{lam} = 2 + 0.878\, C_{lam} Ra^{1/4}, \; Nu_{turb} = C_{turb} \cdot Ra^{1/3}, \tag{5.163}$$

where for C_{turb} the data displayed in Table 5.8 can be used.
- Regarding combined shapes, such as the tank displayed in Fig. 5.28h with spherical end caps, calculate the heat transfer from the cylinder of length l, the heat transfer from a sphere of diameter d and add to obtain the total transfer. Other shapes can be treated in a similar manner.

High vertical components are found, for example, in pipelines of solar towers, as well as open gaps behind transparent thermal insulation. To prevent condensation from forming, which contributes to degradation, these are back-ventilated, and the open interfaces allow air circulation, which is associated with heat loss. In addition, ribbed structures are used as emergency capacitors to shorten stagnation periods.

- In vertical narrow cavities ($l/w << 1$), as shown in the figure Fig. 5.29a, with constant wall temperature (where wall temperatures T_1 and T_2 can be different but not between T_∞), the Nusselt number is given by:

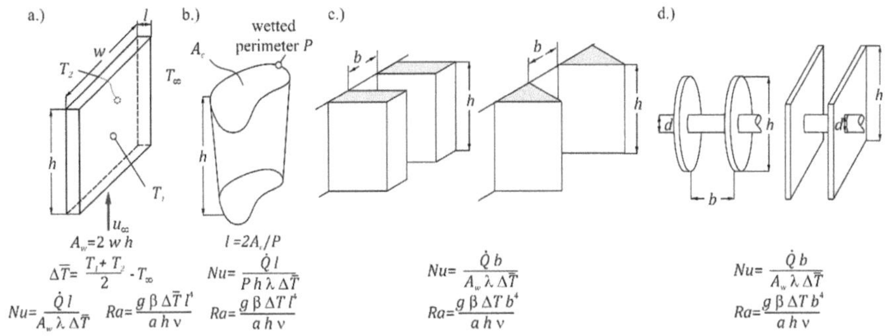

Fig. 5.29 Geometries for natural convective heat transfer in open cavities and corresponding nomenclature; **a** parallel plates, **b** arbitrary surface, **c** vertical fins and **d** horizontal fins

$$Nu = \left(\left(\frac{Ra}{f Re} \right)^m + \left(C_1 C_{lam} Ra^{1/4} \right)^m \right)^{1/m}, \tag{5.164}$$

for $Ra \leq 10^5$, where the product of friction factor and Reynolds number ($f Re$) is taken for fully developed plate flow at the corresponding aspect ratio and C_1 characterizes the increased heat transfer compared to a vertical plate according to the chimney effect. In Eq. (5.164) the Nusselt number refers to the total heat power transfer $\dot{Q}$ to the fluid form all cavity surfaces. For parallel arranged plates $f Re = 24$ with $m = -1.9$ using $C_1 \approx 1.2$ for gases applies. C_1 tends to unity for increasing Pr as well as for increasing l/w.

For cylindrically shaped gaps $f Re = 16$ holds with $m = -1.03$ and $C_1 \approx 1.17$ for gases. Other cross-sectional areas, as depicted in Fig. 5.29b, like concentric squares ($f Re = 14.23$), hexagonal shrouds ($f Re = 15.05$) Eq. (5.164) applies as well by inserting $m = -1.5$ and $C_1 \approx 1.2$ for gases (Kreith & Goswami, 2005). The heat transfer per unit cross-sectional area, $\dot{Q}/A_c$, for a given channel length h and temperature difference, passes through a maximum at approximately $Ra_{max} = \left(f Re C_1 C_{lam}/2^{1/m} \right)^{4/3}$ and provides the hydraulic radius $l = 2A_c/P$.

- Fins are frequently used as heat sinks, e.g. in condensers, and the heat transfer at low Ra is associated with a large uncertainty. Conservatively, for $Ra < 2 \times 10^2$ and U-shaped fins according to Fig. 5.29c, the following relation equation finds use,

$$Nu = \left(\left(\frac{24}{Ra} \right)^2 + \left(C_1 C_{lam} Ra \right)^{-2} \right)^{-1/2}; C_1 = \left(1 + \left(\frac{h}{b} \right), 1.16 \right)_{min}, \tag{5.165}$$

underestimating the heat transfer since heat losses at the fin edges and three-dimensional heat conduction within the structures are not considered. In case of triangular fins, the relation:

$$Nu = C_{lam} Ra^{1/4} \left(1 + \left(\frac{3.26}{Ra^{0.21}} \right)^3 \right)^{-1/3}, \tag{5.166}$$

can be used for $0.4 < Ra < 5 \times 10^6$.

- Horizontal fins, as illustrated in Fig. 5.29d, separate in circular or rectangular geometries for which the following relations may be taken for $5 < Ra < 10^4$ and air:

$$\text{circular fins: } Nu = \frac{1}{8} Ra^{0.55} \left(1 - \exp\left(-\frac{137}{Ra} \right) \right)^{-0.294},$$

$$\text{rectangular fins: } Nu = \left(\left(\frac{Ra^{0.89}}{18} \right)^{-2.7} + \left(0.62 Ra^{1/4} \right)^{-2.7} \right)^{-1/2.7} \tag{5.167}$$

5.1.16 Thermal-Hydraulics of Pipes and Engineering Correlations

Pipe system and pipe grids are central elements to transfer the heat generated within the solar absorber to a functional component, be it a steam generator for power generation, a thermal storage unit for intermediate storage of thermal energy, or direct heat application in processes or by the consumer. Pipelines or pipe networks play a central role here, and their optimization goals are the lowest possible pressure drop (to keep pumping work low), efficient heat transfer (to keep the components compact), and small heat losses to the environment.

The fundamentals and methods of one-dimensional stream filament theory often termed also as stream tube theory are the starting point for the design of piping networks or for system analyses. Here the flow is considered along a stream thread, which is a model concept, and which must be imagined as a thin stream tube with variable cross-sectional area formed by streamlines. The flow along the streamline is treated as a one-dimensional process. The state variables velocity, pressure, density, and temperature do not change over the cross-section of the flow filament, but only in the direction of flow. In the considerations of the streamline theory, among others, a one-dimensional form of the continuity equation is used, which considers the law of conservation of mass.

Since the streamlines are by definition the tangents of the velocity vectors, no fluid mass passes through the shell of the stream tube. This means that solid-walled channels with flow through them form stream tubes. If the changes in the flow variables across the cross-section of the flow tube are small compared to the changes along the flow tube, the approximately one-dimensional changes in the flow variables along the abstracted flow path can be calculated. The image of a flow field is represented more accurately by stream filaments the smaller

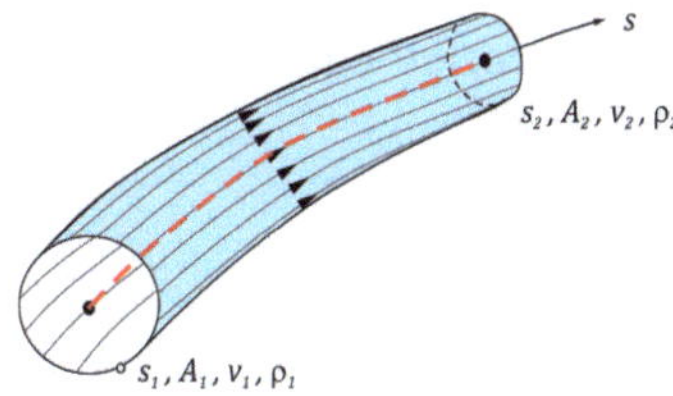

Fig. 5.30 Stream filament in a stream tube

their cross sections. Flows through pipes and in flumes can be treated as a single stream tube if values of the flow variables averaged over the cross section are used. This is the importance of streamline theory in engineering: It is often possible to treat two- or even three-dimensional flows approximately in one dimension. When applying relationships from the stream filament theory, it must be checked whether the idealizations made describe the processes correctly, e.g. whether the changes in the state variables are actually much smaller in the transverse direction than in the direction of flow.

Since the mass of a material volume remains constant, the following applies to such a material piece of a current filament, which is shown in Fig. 5.30:

$$\frac{d}{dt}\int\limits_{s_1(t)}^{s_2(t)} \rho(s,t)A(s,t)ds = 0. \tag{5.168}$$

For the derivation of an integral with respect to time, whose integrand and limits depend on time, Leibniz's rule applies and using $ds/dt = u_i$, the following is obtained:

$$\int\limits_{s_1(t)}^{s_2(t)} \frac{d(\rho A)}{dt} ds + \rho_2 A_2 u_2 - \rho_1 A_1 u_1 = 0. \tag{5.169}$$

For the steady-state flow of an incompressible fluid through a stream filament, it follows: $(u_1 A_1) = (u_2 A_2)$, $(u\ A) = const.$ or $((u\ dA) + (A\ du)) = 0$. In a stream filament, $(u\ A)$ just represents the volumetric flow ratc through a cross section A of the stream filament, it is $\dot{V} = (u\ A)$. Trivially, this mcans that for a steady flow of an incompressible fluid, the volumetric flow through a stream tube is the same in any cross section, i.e., $\dot{V}_1 = \dot{V}_2$ or $d\dot{V} = 0$.

For steady-state pipe flow, the momentum equation relates the sum of forces in a given direction acting on a stream filament in a closed conduit, to the net momentum flux through the control volume in the same direction. The most common forces F are pressure and friction

$$\sum_{i=1}^{i=n} F_i = (\rho_2 A_2 u_2) \cdot u_2 - (\rho_1 A_1 u_1) \cdot u_1. \tag{5.170}$$

The energy equation in a general form for an incompressible pipe flow reads then to:

$$\frac{p_1}{\rho g} + z_1 + \frac{u_1^2}{2g} = \frac{p_2}{\rho g} + z_2 + \frac{u_2^2}{2g} - h_p + h_t + h_f, \qquad (5.171)$$

in which the unit is a length scale (m), and p, ρ and g denote the pressure, the density, and the gravity constant. The first three terms of both sides denote the pressure head $(p/(\rho g))$, the elevation head with respect to a ground normal level, and the velocity head $(u_i^2/(2g))$. The last three terms on the right-hand side are the total dynamic head supplied by the pump (h_p), the head loss by a turbine (h_t) and the total friction losses from 1 to 2.

5.1.16.1 Friction Losses in Pipes and Pipe Elements

In this context pressure changes resulting from incompressible flow for pipes with circular geometry are considered, however, the results can be generalized for a pipe of noncircular geometry by replacing diameter D in any of the equations with the hydraulic diameter, D_h, defined as $D_h = A/P$, in which A is the cross-sectional area of the conduit and P denotes the wetted perimeter. To calculate the factors influencing the volumetric flow rate, which are particularly important in the design of pipelines, the minimum volume flow is usually given and the pressure difference between the inlet and outlet is sought. With the help of the energy equation of the stream tube theory, the Darcy-Weisbach relationship can be derived:

$$h_f = f \frac{L}{D} \frac{u^2}{2\,g} = f \frac{L}{g} \frac{8}{\pi^2} \frac{\dot{V}^2}{D^5}, \qquad (5.172)$$

in which L denotes the duct length, $\dot{V}$ the volumetric flow rate and f the tube friction factor depending on the material properties of the pipe and the flow velocity.

The friction factor f is a measure of pipe roughness and has been evaluated experimentally to create the Moody chart. According to that laminar flow exists if $Re < 2.300$, beyond that a critical zone exists up to $Re \sim 4 \times 10^3$, there friction loss calculations yield ambiguous results because it is impossible to determine a unique value of f. For $Re > 4 \times 10^3$, the flow becomes turbulent and f is a function of Re and relative pipe roughness (e/D), where e is the equivalent roughness height and D is pipe diameter. At $Re > 10^4$, f depends only on the relative roughness height, e/D and this domain is termed fully turbulent flow. The Reynolds number at which this transition occurs depends on the relative roughness of the pipe; a Moody chart for pipe flow is illustrated in Fig. 5.31.

According to Blasius, in the range $3 \times 10^3 < Re < 10^5$ the simple formula applies:

$$f = \frac{0.3164}{Re^{1/4}}, \qquad (5.173)$$

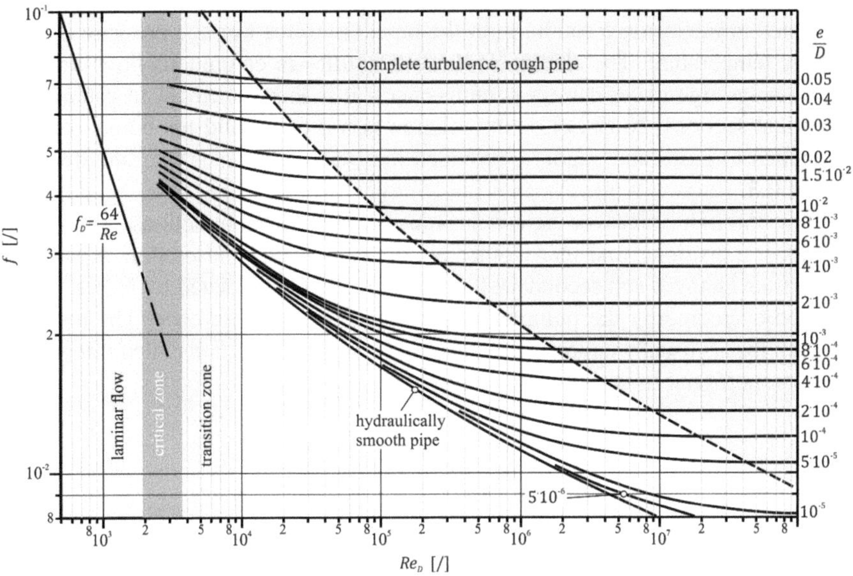

Fig. 5.31 Moody diagram for pressure losses in a pipe flow

and for higher ranges $10^4 < Re < 10^6$ the equation of Konakov is still convenient to handle:

$$f = (1.8 \cdot \lg Re - 1.5)^{-2}. \tag{5.174}$$

In the range $10^5 < Re < 5 \times 10^7$ the implicit formula of Filomenko is considered to be reliable with deviations of less than 0.5%:

$$\frac{1}{\sqrt{f}} = 1.819 \cdot \lg Re - 1.64. \tag{5.175}$$

Considering not ideally smooth tubes but also rough tubes, then, if the roughness flow is fully developed, the quadratic resistance law holds, and f is independent of Re. According to Prandtl and v. Kármán the following relation holds

$$\frac{1}{\sqrt{f}} = 2 \cdot \lg\left(\frac{D}{e}\right) + 1.14. \tag{5.176}$$

According to Nikuradze, the implicit equation of Colebrook and White applies to the turbulent flow in the entire range of smooth and rough surfaces:

$$\frac{1}{\sqrt{f}} = 1.74 - 0.869 \cdot \ln\left(\frac{2\,e}{D} + \frac{18.7}{Re\sqrt{f}}\right). \tag{5.177}$$

The material for the kinematic viscosity values ν as well as the velocity u are to be related to the average condition in the pipe, i.e. to the average pressure and the average temperature. The change of these quantities in continuous flow direction can be very large under certain circumstances; it is then expedient to calculate in individual sections. Nonetheless the relation is not only valid for liquids but can be applied to gases and vapours, provided the Mach number in the duct does not exceed 0.3.

Flow through valves or gates, throttles, expansions/contractions, elbows, T-junctions, transitions, and many other pipe grid elements causes flow separation, momentum changes and secondary flows which are all associated by losses, which must be overcome by the pump or the driving pressure difference. The resulting pressure loss is usually expressed as resulting head loss, h_m, and called minor loss. h_m is proportional to the velocity height

$$h_m = K_l \frac{1}{2g}\left(\frac{\dot{V}}{A}\right)^2 = K_l \frac{u^2}{2\,g}, \tag{5.178}$$

in which K_l is the minor loss coefficient being analogous to $f \cdot (L/D)$, g the gravity constant, A_m the cross-sectional area at the inlet of the minor loss element and u the corresponding velocity there. The summation of all friction and minor losses in a pipe system can be expressed by a head loss h_l:

$$h_l = h_f + h_m = \left(\sum\left(f\frac{L}{D}\frac{1}{2g\,A_p^2}\right) + \sum\left(K_l\frac{1}{2g\,A_m^2}\right)\right)\dot{V}^2. \tag{5.179}$$

It is mandatory to use the correct inner pipe diameter for each pipe section and minor loss. In order to simplify calculations, minor losses are often expressed also in form of an equivalent pipe length $l_{equ} = K_l(D/f)$. It represents the length of pipe that produces the same head loss as the local or minor loss. This is a simple, but potentially inaccurate, method of including minor losses. The problem is that the friction coefficient varies from pipe to pipe, so the equivalent length will not have a unique value. When minor losses are truly minor, this problem becomes academic. However, when local losses are an important part of the analysis, it is recommended that the minor loss coefficients K_l be used rather than an equivalent length.

The geometry of a pipe helix can be described, as shown in Fig. 5.32a, by the inner diameter D of the pipe wound into a helix, by the mean diameter D_h of the helix and by the pitch h of the helix. As a result of the curvature of the tube, centrifugal forces occur during the flow, which cause a secondary flow in the form of a double vortex termed as Dean vortices. The secondary flow causes an increase in pressure loss compared to the straight pipe. The pressure loss that a fluid suffers when flowing through a pipe helix is calculated analogous to a straight pipe

$$\Delta p = f\frac{L}{D}\rho\frac{u^2}{2}, \tag{5.180}$$

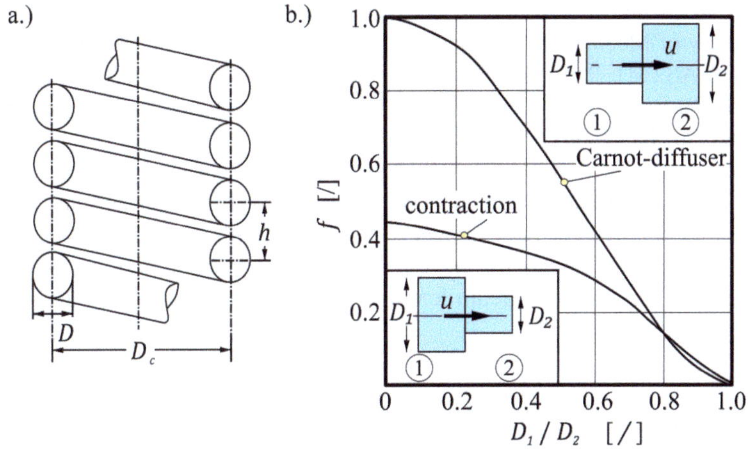

Fig. 5.32 **a** Geometry of a helical coil; **b** dimensionless pressure drop coefficients f for a discontinuous pipe cross-sectional constriction

where L is the length of the pipe wound into the helix. The drag coefficient f defined by this relation depends on the Reynolds number Re and on the ratio D/D_m, where D_m denotes the mean diameter of curvature of the pipe helix, which is determined according to

$$D_m = D_c\left(1 + \left(\frac{h}{\pi\ D_c}\right)^2\right). \tag{5.181}$$

Noticeable differences between D_m and D_c arise only for strongly curved pipes with a large pitch h of the helix, and for calculating f $D_m = D_c$ can be set if $h < D_c$. The drag coefficients for isothermal laminar and turbulent flows can be calculated using:

$$f = \frac{64}{Re}\left(1 + 0.33\left(\lg\left(Re \cdot \sqrt{\frac{D}{D_m}}\right)\right)^4\right) \text{ for } Re < \left(Re_{crit} \cdot \sqrt{D/D_m}\right)$$

$$f = \frac{0.3164}{Re^{1/4}}\left(1 + 0.095 \cdot \sqrt{\left(\frac{D}{D_m}\right) Re^{1/4}}\right) \text{ for } Re < 10^5;$$

$$\text{with } Re_{crit} = 2300\left(1 + 8.6\left(\frac{D}{D_m}\right)^{0.45}\right). \tag{5.182}$$

In solar thermal piping networks, many pipe connections are required due to the low power density, so the most common geometries are discussed in this context. A collection of formulas for the determination of head losses for pipe installations can be found in numerous standards handbooks available worldwide. Here, relations are provided from American Water Works Association (AWWA), American Society of Mechanical Engineers (ASME), American Society for Testing and Materials (ASTM), American National Standards Institute (ANSI) and Stephan et al. (2019), for details of the characteristics of pipe elements it is useful to use data sheets of the vendor/manufacturer.

Discontinuous duct cross-sectional changes occur, for example at manifolds and headers or at penetrations such as Carnot diffusers or pipe contractions, result in pressure loss coefficients f, as shown in Fig. 5.32b. Of central importance is that the pressure loss coefficients for this type of constriction are directional. Regarding contractions a certain downstream length l_d is required to equilibrate the velocity, which calculates to.

$$\begin{aligned} l_d &= 0.13 \cdot Re_2 \cdot D_2 \quad \text{(laminar flow);} \\ l_d &= 0.0575 \cdot Re_2 \cdot D_2 \text{ (tubulent flow),} \end{aligned} \tag{5.183}$$

where the subscript 1 denotes the position upstream the discontinuity and 2 downstream, respectively. If a fluid enters a larger space from a tube, e.g. into a tube extension or from the tube bundle of a heat exchanger, an expanding jet with strong turbulence is formed which reattaches to the tube wall after 8–10 times D_2. Compared to a change of the static pressure due to the velocity change, which can be calculated according to Bernoulli's law, a momentum or shock loss occurs in the so-called Carnot diffuser, which—for the case of constant flow velocity over the tube cross-section ($Re \to \infty$)—can be calculated exactly.

$$\Delta p = (1 - A_1/A_2)^2 \cdot \frac{\rho}{2} \cdot u_1^2 \tag{5.184}$$

In the case of turbulent flow, the deviation from Eq. (5.184) is small (<5%), in the case of laminar flow, due to the more pronounced velocity profile, the loss according to Eq. (5.184) is about 33% smaller.

With continuous contraction according to Fig. 5.33a, no flow separation occurs, provided the opening angle $\alpha < 40°$ remains. A special case of discontinuous cross-section narrowing is the pipe inlet, which also depends on the shape of the inlet. The corresponding pressure loss coefficients f_i are shown in Table 5.9. For continuous cross-sectional changes of the pipe inlet geometry at angle α (Fig. 5.33), the following relationship between pressure loss coefficient f and pipe diameter ratios D_1/D_2 is obtained:

$$f = k(\alpha) \cdot \left(1 - \frac{D_1}{D_2}\right)^2 \tag{5.185}$$

in which the factor $k(\alpha)$ is obtained from experimental data and assumes the values assumed in provided in Table 5.10.

Thus, in convergent tubes almost the entire pressure is converted into flow energy. With larger angles α, one approaches the case of sudden contraction, which causes much stronger losses and the data may be taken from Fig. 5.33b. To small

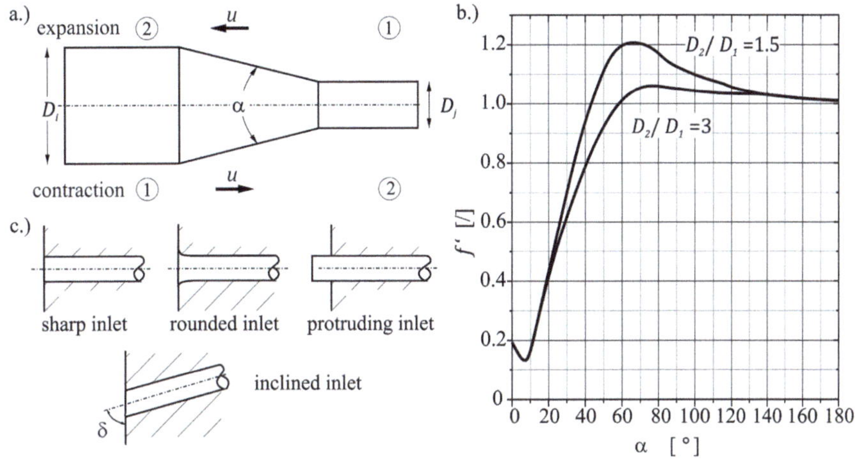

Fig. 5.33 **a** Nomenclature for continuous pipe expansion and contractions; **b** drag coefficient f' of expanded pipe sections as a function of the opening angle α and the ratio of pipe diameters D_2/D_1; **c** representation of common pipe inlet geometries

Table 5.9 Pressure loss coefficient f_i of different pipe inlet geometries

Pipe inlet type	f_i (/)
Sharp inlet	1.69 (90°) 0.45 (normal curved)
Rounded inlet	0.005 (very smooth) 0.05 (normal)
Protruding inlet	~ 3 (Borda mouth)
Inclined inlet	$= 0.5 + 0.3 \cos\delta + 0.2 \cos^2\delta$

Table 5.10 Experimentally determined loss factors $k(\alpha)$ of a continuous pipe cross-section contraction as a function of the opening angle α (from Fig. 5.33a)

α (°)	5	7.5	10	15	20	40	180
$k(\alpha)$	0.13	0.14	0.16	0.27	0.43	1.0	1.0

angles yield long conical pipe segment; then the additional pipe length must be calculated, typically using the mean cross-sectional area A_m and the mean velocity u_m.

Conversely, flow separation cannot be prevented in continuous pipe expansions due to wall friction and pressure build-up and there is no ideal conversion of flow energy into pressure energy (according to the Bernoulli equation). The losses are

Table 5.11 Reynolds number at which the minimum loss coefficient in an expansion is obtained at different divergence angles α (from Fig. 5.33a)

$Re\,(\times 10^5)$	0.5	1	1.5	2
α (°)	10	8.4	7.6	7.1

calculated as:

$$\Delta p = f' \cdot (1 - A_1/A_2)^2 \cdot \frac{\rho}{2} \cdot u_1^2, \tag{5.186}$$

with f' as the drag coefficient, whose dependence on the opening angle α is shown in Fig. 5.33b. The minimum of the losses occurs at low expansions and depends on the Reynolds number, as shown in Table 5.11. At even higher Re numbers, it is necessary to go down to $\alpha = 5\text{–}6°$. The roughness of the tube wall is of influence, so that, for example, $\alpha = 7°$ was found to be the preferred angle for a smooth surface ($f = 0.02$), and $\alpha = 9°$ for a rougher surface ($f = 0.04$).

When discharging from vessels, the area ratio in Eq. (5.186) $A_1/A_2 = 0$. In addition, there is an extremely strong dependence of the loss coefficient f_i on the design of the orifice, for which the values from Table 5.9 can be used. In order to calculate the outgoing volumetric flow rate $\dot{V} = (u_2 \cdot A_2)$ when the overpressure $(p_2 - p_1)$ prevails at the orifice, the acceleration work, and the resistance of a connecting pipe f (with the length L) as well as possibly further fittings must be added to the resistance coefficient of the orifice. The force balance around the outlet translates into a pressure difference:

$$\Delta p = (p_2 - p_1) = \left(1 + f_i + f \cdot \frac{L}{D_2}\right) \cdot \frac{\rho}{2} \cdot u_2^2, \tag{5.187}$$

which allows the outlet velocity and volume to be determined.

In circular pipe bends, the pressure loss coefficient f_b depends on the bend radius to diameter ratio (R/D), the angle of curvature and the surface roughness inside the pipe. In hydraulically smooth pipes, the following formula is obtained for f_b, which is shown in Fig. 5.34a for a bend angle of $\varphi = 90°$.

$$f\left(\frac{D}{R}\right) = 0.986859 - 0.658175 \cdot \left(\frac{D}{R}\right) + 0.0207726 \cdot \left(\frac{D}{R}\right)^2 - 0.027726 \cdot \left(\frac{D}{R}\right)^3 + 0.001274 \cdot \left(\frac{D}{R}\right)^4;$$
$$k(\varphi) = 0.06364 + 0.01149 \cdot \varphi - 1.3167 \times 10^{-5} \cdot \varphi^2;\ f_{Kr} = k(\varphi) \cdot f(D/R) \tag{5.188}$$

With rough tubes, the pressure drop increases significantly, as illustrated by the curve in Fig. 5.34a. Standard orifices, standard nozzles or Venturi tubes are often used for flow measurement, the shape of which is described in DIN EN ISO 5167

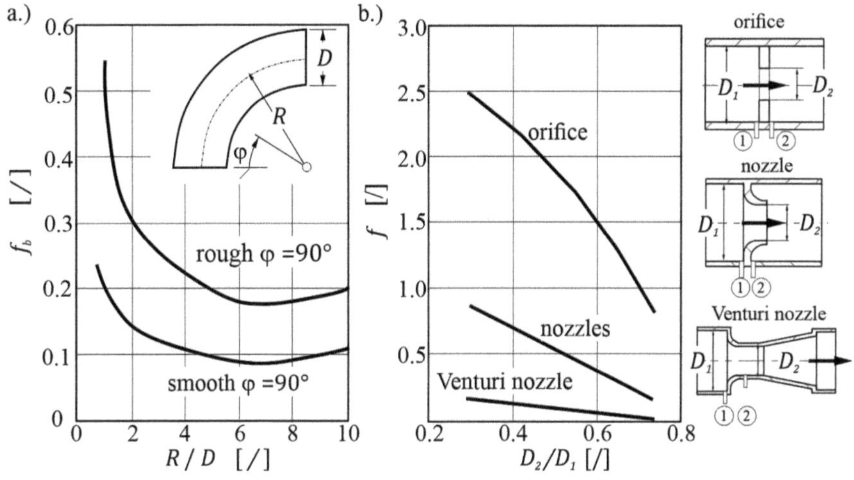

Fig. 5.34 **a** Pressure loss coefficients f_b of a circular tube elbow with smooth and rough surface as a function of the radius to diameter ratio (R/D); **b** pressure loss coefficients f of orifices, nozzles and Venturi tubes as a function of D_2/D_1

(2004). These also generate a pressure drop characterized by loss coefficient f, which is shown in Fig. 5.34b for the different designs.

For right-angled and oblique tees, bends, and elbows, the respective pressure loss coefficients can be found in the manufacturers' data sheets. The same applies to different valve types (gate or butterfly type), throttles, where the pressure loss coefficient is a function of the opening fraction.

5.1.16.2 Convective Heat Transfer in Pipes and Piping Elements

The engineering design of components often requires an estimation of the achievable heat transfer. In this section, some of the most important single-phase correlations for duct flows are compiled, similarly as before for external flows or flows through cavities. According to the previously discussed sections, the heat transfer relationships can be divided into different types of equations, which are shown in Table 5.12. The problem-specific Nusselt number correlations also include correction factors, exponents and constants resulting from the geometry, the temperature-dependent thermophysical properties of the fluid and boundary layer considerations.

Whenever heat transfer correlations are used, correction factors K_p must be considered to cover the following aspects:

- temperature influence on the material values (K_T),
- developing or fully developed flows (K_l),
- geometry influences (K_{dh}),
- convection influences (K_c),
- combined influences (K_p).

Table 5.12 Equation structure and incorporated correction factors/exponents for the different single-phase heat transfer modes

Heat transfer type	Dim.-less quantity equation type	Correction factors and exponents
Forced convection	$Nu = f(Re, Pr)$	$Nu = C \cdot Re^m \cdot Pr^n \cdot K_p$
Natural convection	$Nu = f(Gr, Pr)$	$Nu = C \cdot (Gr \cdot Pr)^n \cdot K_p$
Mixed convection	$Nu = f(Re, Gr, Pr)$	$Nu = C \cdot Re^m \cdot Pr^n \cdot Gr^k \cdot K_p$

The application of the adequate correction factors must be checked for each calculation based characteristic number equations. Additionally, care has to be taken of using the correct length scaling in the equations, the validity range of the respective correlations as well as their confidence interval must be considered in particular. Furthermore, the reference temperature T_{ref} for the material values must be considered, since this results in the correction factors K_T, which accommodates for the influence of viscosity due to temperature gradients. For gases and liquid metals, the following applies to T_{ref}:

$$T_{ref} = \frac{T_f + T_w}{2}, \tag{5.189}$$

in which T_f denotes the fluid temperature and T_w that of the wall, for all other liquids $T_f = T_w$ is assumed. Since the fluid temperatures change due to heat transfer, a nomenclature has been adopted in determining the fluid temperature value (T_f) for the individual flow problems, which Table 5.13 reflects. After calculating the fluid temperature and the reference temperature, the correction factors of the temperature effect (K_T) can be determined for the different heat transfer modes, as shown in Table 5.14. Here, the molecular Prandtl numbers Pr at the fluid temperature T_f and the value at the wall Pr_w at the wall temperature T_w are to be used. In case of missing data, a value of $K_T = 1$ is applied.

The hydraulic (l_{hy}) and thermal (l_{th}) inlet lengths of laminar flows depend on the geometry and the thermal boundary condition. Table 5.15 gives an estimate for this. The correction factor of developing or inflowing flows (K_l) results in $K_l = \left(1 + (d/l)^{2/3}\right)$ for laminar flows.

Technically more relevant are turbulent pipe flows, which constitute nearly all piping grids of thermal systems and for molecular Prandtl-numbers $Pr > 0.5$ the

Table 5.13 Fluid reference temperature T_f to be used when using heat transfer correlations

Flow configuration	Fluid reference temperature T_f
Bounded flows (pipes, ducts, bundles)	$T_f = \frac{T_{f,in}+T_{f,out}}{2}$
Boundary layers, external flows	$T_f = T_\infty$
Short pipes, inflows	$T_f = T_{in}$
Condensation, evaporation of fluids	$T_f = T_{sat}$ or $T_f = \frac{T_w+T_{sat}}{2}$

Table 5.14 Temperature correction factor calculation K_T for different heat transfer types and fluids

	Liquids	Gases
Duct lows	$K_T = (Pr/Pr_w)^{0.11}$	$K_T = 1$
Unbounded flows	$K_T = (Pr/Pr_w)^{0.25}$	$K_T = \left(T_f/T_w\right)^{0.12}$
Free convection	$K_T = (Pr/Pr_w)^{0.25}$	$K_T = 1$

Table 5.15 Dependence of hydraulic (l_{hy}) and thermal (l_{th}) entrance lengths on geometry and thermal boundary condition for laminar duct flows

Geometry	$l_{hy}/(d_h Re)$	$l_{hy}/(d_h Re Pr)$	
		$T_w = const.$	$\dot{q}_w = const.$
Circular duct	0.065	0.055	0.07
Plane channel	0.01	0.014	0.02
Annulus	0.01–0.015	0.05	0.06

Reynolds analogy can be used for technical assessments, which ties the friction factor f to the heat transfer characteristics expressed by the Nusselt number Nu, as e.g. the Colburn relation $\left(St \cdot Pr^{2/3}\right) \sim f/2$ for pipe flows, for details see Bejan (2013). Moreover, similar as for the laminar energy exchange (cf. Sect. 5.1.10) also for the turbulent pipe flow three different flow domains can be separated:

- fully developed flow (either hydraulically or thermally),
- hydraulically developing and
- thermally developing,

for which also the cases of $T_w = const.$ and $\dot{q}_w = const.$ can be separated, which fortunately do not differ as strong as in the laminar case.

- Fully developed pipe flow

For engineering assessment fully thermal and hydraulically developed pipe flow establishes due to the effective wall normal momentum transfer after a downstream entrance length of about $l/d_h \approx 10 - 40$ for $Pr > 0.5$. More exactly, Cengel and Cimbala (2006) provide a Reynolds number dependent relation $l/d_h - \left(4.4/Re^{1/6}\right)$. DNS calculations show that even after hundred hydraulic diameters upstream information still exists and, thus, regarding scientific measurements specific precautions have to be set to achieved fully developed flow. Fully developed turbulent flow is present at Reynolds numbers $Re \geq 10^4$. For both thermal boundary conditions $T_w = const.$ and $\dot{q}_w = const.$, practically the same mean Nusselt numbers result and the following equation for the heat transfer applies for $Pr > 0.5$:

$$Nu_m = \frac{(f/8)Re\ Pr}{1 + 12.7\sqrt{f/8}\left(Pr^{2/3} - 1\right)}\left(1 + \left(\frac{d}{l}\right)^{2/3}\right) \quad \text{with}$$

$$f = (1.8 \log Re - 1.5)^{-2}, \tag{5.190}$$

where f is the friction factor illustrating the tight connection of heat transfer and friction in higher Prandtl number fluids. This formula is valid up to $Re \leq 10^6$. For large length $l/d > 60$ the simpler Dittus-Bölter relation can be applied with an accuracy of about 20%

$$Nu_m = 0.23\ Re^{4/5} Pr^n, \tag{5.191}$$

in which the Prandtl number exponent is $n = 0.4$ when the fluid is being heated and $n = 0.3$ if the fluid is being cooled. The Dittus-Boelter relation is valid in the range $10^4 \leq Re \leq 1.2 \times 10^5$ for $Pr > 0.5$. For Reynolds number in the transition regime $2 \times 10^3 \leq Re \leq 10^4$ an intermittency factor γ is introduced blending the Nusselt number of the laminar flow ($Nu_{m,lam}$) to that of the turbulent flow ($Nu_{m,turb}$), e.g. from Eq. (5.190)

$$Nu_m = (1 - \gamma) \cdot Nu_{m,lam} + \gamma \cdot Nu_{m,turb} \text{ with } \gamma = \frac{Re - 2300}{10^4 - 2300}, \tag{5.192}$$

in which for the laminar Nusselt number the values at the corresponding thermal boundary are to be used. The reference temperature T_f is to be calculated according to Table 5.13 and to obtain the effective Nusselt number (Nu), Nu_m must be multiplied by a temperature correction factor K_T from Table 5.14. For liquid metals ($Pr < 0.1$) the accuracy of the correlations is in the range of the Dittus-Bölter relation at $Re \sim 10^4$ getting more accurate for higher Re. According to Kakac et al. (1987) the following relations exhibit high confidence

$$\begin{aligned} Nu_m &= 4.5 + 0.0156\ Re^{0.85} Pr^{0.86} \quad \text{for } T_w = const., \\ Nu_m &= 5.6 + 0.0165\ Re^{0.85} Pr^{0.86} \quad \text{for } \dot{q}_w = const., \end{aligned} \tag{5.193}$$

which are valid from $10^4 \leq Re \leq 5 \times 10^6$. Similar as indicated for high Prandtl number fluids a temperature correction factor must be introduced to account for change of the temperature dependence of the thermo-physical properties, which require a multiplication of Nu_m with a temperature ratio in the form, $Nu = Nu_m (T_m/T_i)^n$ to obtain the effective Nusselt-number (Nu), where T_m represents the bulk temperature and T_i is the inlet temperature. The exponent for the different thermal boundary conditions read to

$$\begin{aligned} T_w = const.: n &= 0.08 + 2.2 \times 10^{-4} T_m & T_m &> 600\,\text{K} \\ n &= 0.08 & T_m &< 600\,\text{K}, \\ \dot{q}_w = const.: n &= \exp\left(5.9 \times 10^{-3} T_{m-6.91}\right) & T_m &> 600\,\text{K}, \\ n &= 0 & T_m &< 600\,\text{K}. \end{aligned} \tag{5.194}$$

- hydraulically developing flow

Turbulent hydraulically developing flow exhibit with respect to the heat transfer a similar behaviour as fully developed flow and thus the correlations for turbulent flow can be used unaltered respecting the corresponding Re and Pr regimes.

- thermally developing flow

Turbulent thermally developing flows occur at the entrance e.g. of absorber pipes of solar farms and for Prandtl number fluids $Pr > 0.5$ also the correlations of fully developed turbulent can be used with high confidence. For liquid metals the thermal entrance length differs from those of higher Prandtl number fluids and depend not only on the Reynolds number but also on the thermal boundary condition. Figure 5.35a, b illustrate the thermal developing length l_{th}/d in turbulent pipe flows for various molecular Prandtl numbers as function of Re. The mean Nusselt number correlations for this condition are (Bejan, 2013):

$$Nu_m = 4.8 + 0.0156\ Re^{0.85} Pr^{0.93} \quad \text{for } T_w = const.,$$
$$Nu_m = 6.3 + 0.0167\ Re^{0.85} Pr^{0.93} \quad \text{for } \dot{q}_w = const., \tag{5.195}$$

which are close to that of fully developed flow and are reported to be valid for $10^4 \leq Re \leq 10^6$. Here, also the temperature correction applies as in fully developed turbulent flows.

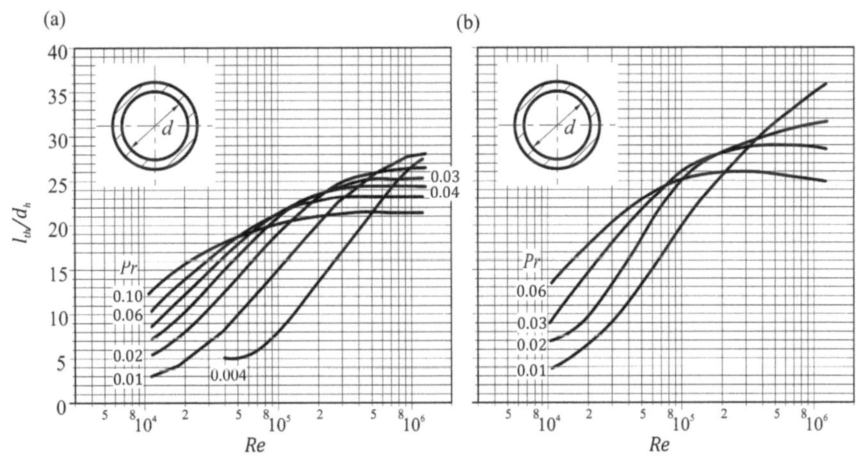

Fig. 5.35 Dimensionless thermal entrance length l_{th}/d as a function of the Reynolds number in a hydraulically fully developed flow of a smooth circular tube for different Prandtl numbers **a** $T_w = constant$ and **b** $\dot{q}_w = const$

5.2 Multiphase Momentum and Energy Exchange

If a collector exceeds the saturation temperature of the heat transfer fluid, which is observed, for example, in connection with the collector stagnation, evaporation of part of the cooling medium may occur. Due to the expansion during the phase transition, there may be a significant increase in pressure and temperature in the system. The existence of a two-phase flow substantially changes the thermal hydraulics discussed so far and therefore requires basic knowledge of two-phase flow.

The simplest and most familiar form of two-phase flow occurs during boiling. Starting from boiling, an attempt will be made to show some model ideas which are important for the further description, as they are also valid for the operating conditions which are encountered in solar collectors. Further details can be found in the books Collier and Thome (1996), Incropera and DeWitt (2001) as they are beyond the scope of this script. Empirical relations of pressure losses and heat transfer in two-phase flow can be found in the work of Streicher (1991) and the literature listed therein. Heinzel et al. (1994), in addition to providing the basics, also addresses the two-phase problem in the solar collector, focusing on the development of direct-evaporation boiling water collectors. When a collector or several collectors are at a stagnation, the same processes occur in principle, and in almost all cases the occurrence of flow instabilities is observed during the transition from single-phase to two-phase flows, which are addressed in the last section.

Two-phase flow effects must also be appropriately considered in the system analytical design and construction of collector systems in order to predict potential stall scenarios or their duration as a function of boundary conditions. The design of large-scale solar systems or even thermal solar power plants necessarily requires the consideration of potential shutdown cases since both their duration and the associated scenarios influence the performance of the system and thus its economic efficiency. For this purpose, it must be possible to predict the individual flow patterns that can occur in a collector or a segment. For example, if the collector field is connected incorrectly, reverse flows may even occur, resulting in no usable power being available. For this reason, an overview of the possible flow forms in an absorber tube will be given below.

5.2.1 Boiling Phenomenology-Homogeneous Boiling-Phase Diagrams

If a vapor bubble forms on a heated wall along which a fluid flows, the bubble will not only remove the latent heat (heat of vaporization) when it detaches from the wall, but its interaction with the fluid will cause mixing and thus agitation. This mixing disturbs the flow boundary layer but leads to an improvement in heat transfer. The formation of a bubble requires superheating of the fluid, the dynamics of which will be considered below.

Let us first consider homogeneous boiling. A vapor bubble of radius r in a liquid is subjected not only to the external pressure p_l imposed by the liquid column, but also to an additional pressure caused by the surface tension σ, so that the vapor bubble pressure p_v is given by:

$$p_v = p_l + \frac{2\,\sigma}{r}. \tag{5.196}$$

In addition, the vapor pressure in a bubble that has a concave curved surface is less than the pressure over a flat surface p_∞ of the same fluid at the same temperature T. Here one can derive the following relation:

$$p_v = p_\infty exp\left(\frac{-2\sigma\ v_l\ M}{r\ RT}\right) \approx p_\infty\left(1 - \frac{2\sigma\ v_l}{p_\infty r\ v_g}\right), \tag{5.197}$$

where R is the universal gas constant ($R = 8.314\,\mathrm{J/(mol\ K)}$), M is the molecular mass (g/mol), and v_g and v_l are the specific volumes of vapor and liquid, respectively. Substituting the Eq. (5.196) into the relation (5.197) yields:

$$(p_\infty - p_l) = \frac{2\sigma}{r}\left(1 + \frac{v_l}{v_g}\right). \tag{5.198}$$

To calculate the superheat due to the pressure, difference $(p_\infty - p_l)$ the Clausius–Clapeyron equation can be used. This is obtained by considering a closed adiabatic system in which there are vapor and liquid which are in thermodynamic equilibrium. According to the 2nd law of thermodynamics, the equilibrium state of a closed system is characterized by the maximum of its entropy. Under the constraints that the internal energy U and the volume V are constant:

$$dS = 0 \quad (U = const., V = const.) \tag{5.199}$$

The entropy (S) of the two-phase system is composed additively of the entropies of the two phases, whose state variables are marked with the subscript "v" and "l" for the gas phase (vapor) and the liquid phase, respectively. This results in the following for the entropy:

$$S = S_l + S_v = m_l s_l(u_l, v_l) + m_v s_v(u_v, v_v), \tag{5.200}$$

in which the small letters m, s, u and v represent the specific molar fractions, entropies, internal energies and volume fractions of the vapor phase and liquid phase, respectively. Inserting the relations for the thermodynamic quantities and forming the differential, we obtain for dS:

$$dS = m_l\left(\frac{1}{T_l} - \frac{1}{T_{v\,l}}\right)du_l + m_l\left(\frac{p_l}{T_l} - \frac{p_v}{T_v}\right)dv_l$$

$$+\frac{1}{T_v}(u_v - u_l + p_v(v_v - v_l) - T_v(s_v - s_l))dm. \tag{5.201}$$

In thermodynamic equilibrium, $dS = 0$. Since the differentials du_l, dv_l, and dm_l are independent of each other, the value of each of the three brackets must be zero, which yields the following condition for the two-phase equilibrium:

$$T_l = T_v = T, \quad p_l = p_v = p, \quad u_v + pv_v - Ts_v = u_l + pv_l - Ts_l. \tag{5.202}$$

As a result, two phases can only be in equilibrium if they have the same temperature and pressure. In addition, their specific free enthalpies h or Gibbs energies g must match, thus it applies:

$$g = u + pv - Ts = h - Ts. \tag{5.203}$$

A differentiation of the Gibbs function yields:

$$dg = -s\,dT + v\,dp, \tag{5.204}$$

to which the boundary condition $dg_l(T, p) = dg_v(T, p)$ is applied and this gives:

$$-s_l dT + v_l dp = -s_v dT + v_v dp. \tag{5.205}$$

The result is the Clapeyron relationship:

$$\frac{dp}{dT} = \frac{s_v - s_l}{v_v - v_l}. \tag{5.206}$$

If instead of the entropy difference $(s_v - s_l)$ one introduces an enthalpy difference $(h_v - h_l)$, then using $((h_v - h_l) = T(s_v - s_l)$ one obtains the Clausius-Clapeyron equation in the form:

$$T\frac{dp}{dT} = \frac{(h_v - h_l)}{(v_v - v_l)} \quad \text{or} \quad \frac{dp_{sat}}{dT} = \frac{\Delta h}{T_{sat}(v_v - v_l)}, \tag{5.207}$$

where p_{sat} is the saturated steam pressure, T_{sat} the saturated steam temperature and h the enthalpy of vaporization. The Clausius–Clapeyron equation can be used to calculate the phase boundary line between the liquid and gaseous phases, the so-called boiling point curve. Equation (5.207) can be further simplified under the following boundary conditions:

- the specific volume of the liquid is small, i.e. $v_l/v_v \ll 1$, and
- for small temperature changes the saturated vapor can be treated as an ideal gas, thus $(pv_v) = (R/M)T$ holds, and for
- the enthalpy of vaporization shall be $h(T) = const$.

Under these assumptions, the following result is initially obtained:

$$\frac{dp}{p} = \frac{M}{R\,T^2}\Delta h\, dT. \tag{5.208}$$

An integration of the pressure from p_l to p_∞ and from saturation temperature (T_{sat}) to vapor bubble temperature (T_b) leads to

$$\ln\left(\frac{p_\infty}{p_l}\right) = -\frac{M}{R}\left(\frac{1}{T_b} - \frac{1}{T_{sat}}\right)\Delta h = -\frac{M\Delta h}{RT_bT_{sat}}(T_b - T_{sat}). \tag{5.209}$$

and considering the surface tension, the following is obtained

$$(T_b - T_{sat}) = \frac{RT_bT_{sat}}{M\Delta h}\ln\left(1 + \frac{2\sigma}{rp_l}\cdot\left(1 + \frac{v'}{v''}\right)\right). \tag{5.210}$$

The assumptions $v_l/v_v \ll 1$ and $(2\sigma/(rp_l)) \ll 1$ further simplify Eq. (5.210) by using series expansion for $\ln(1+x)$ while neglecting higher order terms:

$$(T_b - T_{sat}) = \Delta T_{sat} = \frac{RT_{sat}^2}{M\Delta h}\left(\frac{2\sigma}{r\ p_l}\right). \tag{5.211}$$

$(\ln(1+x) = x - x^2/2 + x^3/3 - x^4/4 + \cdots$. The pressure difference between the bubble and the fluid must thus be compensated by an appropriate superheat temperature to maintain a bubble with a stable size. Because of the reciprocal dependence of the pressure p_l on the bubble radius ($\sim 1/r$), bubble formation in a homogeneous fluid starts only at the temperature significantly higher than the boiling temperature, i.e. the spontaneous or homogeneous nucleation temperature.

In this context, it is useful to first take a look at phase and state diagrams, as schematically shown in Fig. 5.36. Within such a space spanned by the specific volume v, the pressure p and the temperature T, there are curves, the so-called phase boundary lines, which delimit different aggregate states from one another. In the areas enclosed by them, several phases are in thermodynamic equilibrium. The individual phase boundary lines are called

- vaporization curve, which indicates the phase boundary liquid to vapor between the tripe line and the critical point.
- the sublimation curve, which separates the solid from the vapor state between the zero point and the triple line, and
- the dew point line, which is often referred to as the condensation curve.

The vaporization and sublimation curves can be combined to form the vapor pressure curve. The critical point has no degree of freedom, while all other curves exhibit one degree of freedom of either temperature or pressure. Within the phase spaces, two degrees of freedom exist, namely pressure and temperature.

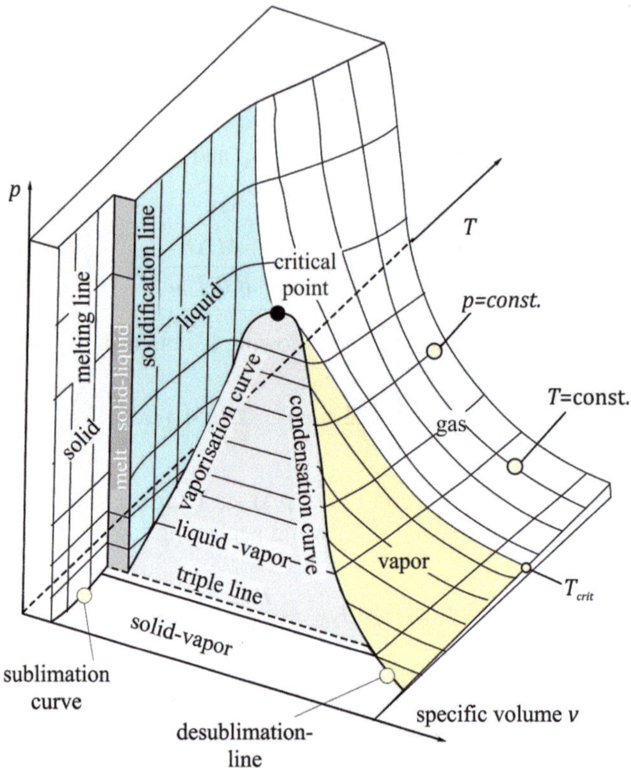

Fig. 5.36 Three-dimensional schematic representation of a phase diagram

Table 5.16 provides boiling temperatures at normal pressure, temperatures and pressures of the triple point and critical points of some selected fluids. It illustrates that as the molar mass number increases, the critical pressure and temperature increases as well. A similar increase of both quantities can be observed with increasing surface interaction of the molecules.

Figure 5.37 shows an example of a $p - v$ diagram of carbon dioxide (CO_2). To the left of the critical point, the vaporization separates the region of subcooled, saturated liquids from the two-phase domain. To the right of the critical point, the condensation (dew point) curve separates the two-phase region from the saturated vapor region. Above the critical point, there is no more phase transition. Points within the boundary line correspond to stable states with a mixture of saturated vapor and saturated liquid. The mass ratio of vapor and liquid of a state is described by the ratio of the distances on the line AE to the boundary lines of the saturated liquid and saturated vapor, respectively.

For an ideal gas, the isotherm is characterized by the stability condition $(\partial p/\partial v)_{T=const.} < 0$. In the $p - v$ diagram, the critical point denotes the inflection point of the critical isotherms. As the boundary line is approached, the state

Table 5.16 Boiling point, triple point, and critical point of some fluids

	Fluid	Boiling temp. @ 0.1 MPa (K)	Triple point T (°K)	p (MPa)	Critical T_{crit} (°K)	Point p_{crit} (MPa)
H_2	Hydrogen	20.28	13.80	7.04×10^{-3}	33.19	1.315
N_2	Nitrogen	77.34	63.14	1.252×10^{-2}	126.19	3.396
O_2	Oxygen	90.23	54.36	1.463×10^{-4}	154.60	5.043
CO_2	Carbon dioxide	194.69	216.59	0.518	304.13	7.377
NH_3	Ammonia	239.81	195.50	6.1×10^{-3}	405.40	11.359
H_2O	Water	373.15	273.16	6.117×10^{-4}	647.10	22.104
CH_4	Methane	111.55	90.67	1.169×10^{-2}	190.56	4.599
C_2H_6	Ethane	184.60	91.60	1.169×10^{-6}	305.30	4.89
C_3H_8	Propane	231.11	85.03	1.685×10^{-10}	369.82	4.248
C_4H_{10}	Iso-butane	262.12	113.55	1.948×10^{-8}	425.13	3.796
CH_3OH	Methanol	337.82	175.52	8.11	513.38	8.216
C_2H_5OH	Ethanol	351.52	150.00	6.34	489.00	6.591
$C_2H_6O_2$	Glycol	470.48	260.57	*n.a.*	720.00	8.23
He	Helium	4.222	2.177	5.04×10^{-3}	5.30	0.228
Ne	Neon	27.104	24.556	4.34×10^{-2}	44.00	2.720
Ar	Argon	87.302	83.806	6.89×10^{-2}	151.08	4.85
	Air	78.8	59.75	5.27×10^{-3}	133.00	3.95
$NaNO_3$	Sodium nitrate	653.15	*n.a.*	*n.a.*	*n.a.*	*n.a.*
KNO_3	Potassium nitrate	673.15	*n.a.*	*n.a.*	*n.a.*	*n.a.*
Na	Sodium	1156.90	370.97	*n.a.*	2503.70	35.00
Hg	Mercury	629.81	234.15	1.65×10^{-9}	234.29	*n.a.*
Pb	Lead	2016.65	*n.a.*	*n.a.*	4460.00	160.00

of a gas increasingly deviates from the ideal gas. This behaviour of the real gas is described by the van der Waals equation:

$$\left(p + \frac{a}{T^n v^2}\right)(v - b) = RT. \tag{5.212}$$

where a and b are constants and n is the Berthelot or van der Waals exponent. In addition to the external pressure p, there is the internal pressure described by $\left(a/\left(T^n v^2\right)\right)$, which is based on the attractive forces between the molecules. The co-volume b must be subtracted from the volume because the molecules with their finite radius require a certain volume which is no longer available for free movement. Within the two-phase transition region in Fig. 5.37, the van der

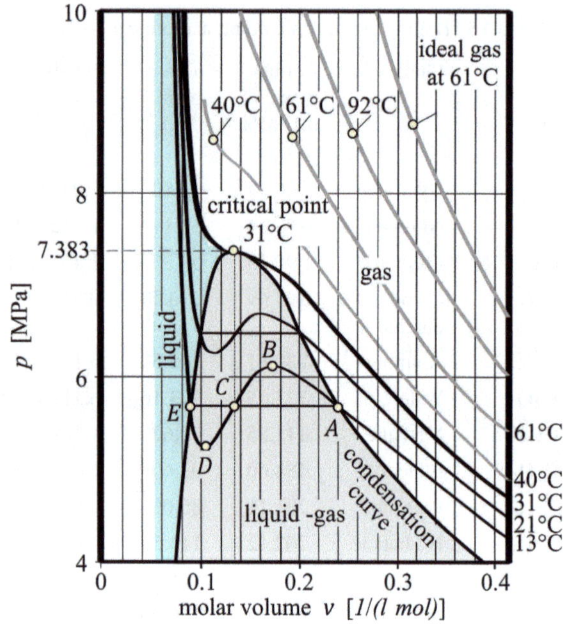

Fig. 5.37 Section of the $p - v$ diagram of carbon dioxide (CO_2)

Waals relationship assumes a curve corresponding to the line through the points $ABCDE$.

With $T < T_{crit}$, on the other hand, an initially unphysical behavior is observed. The pressure decreases with compression. This weakness of the van der Waals equation is corrected by the so-called Maxwell construction. Here, a horizontal straight line is placed in such a way that the areas enclosed by the straight line and the graph are equal. This horizontal line describes the phase transition. A further compression now leads to the transition of vapor particles into the liquid phase. The pressure remains constant and corresponds to the vapor pressure. Once the phase transition is complete, further compression leads to an extreme increase in pressure, which is also expected when liquids are compressed. The value of the intrinsic volume b, which cannot be undercut, can also be read off.

In technical systems, changes of state are mostly isobaric, e.g. imposed by a pump pressure, and therefore the $T - v$-diagram schematically shown in Fig. 5.38 shall be used for explanation. If the isobar p_1 is considered, the two-phase region is located between E and A, in which all states between E and A are additively composed of two phases. With the van der Waals equation there are now further points, or lines DE, BD and AB.

The state on line ED can also be assumed by water, although with respect to the stability condition $(\partial p/\partial v)_{T=const.} < 0$. It is called metastable and as a consequence fluids can be superheated. Coming from E, the fluid is slowly heated as a supercooled fluid and should actually boil at E. Instead, you get what is called a superheated fluid, between E and D. This is unstable. If any disturbance occurs,

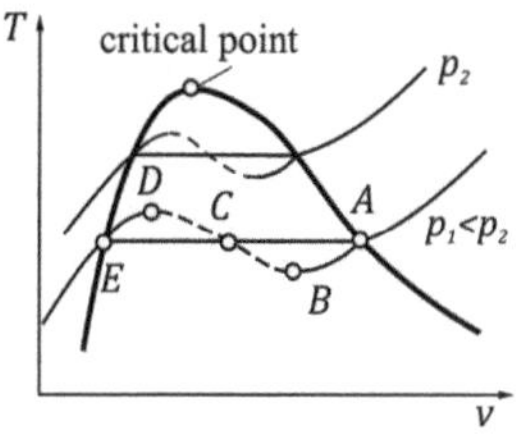

Fig. 5.38 Schematic sketch of a $T - v$ diagram to explain superheating and subcooling

the superheated liquid jumps from a point between ED to a point on the line AE. This means that the fluid suddenly changes into two phases.

The corresponding reverse phenomenon, called supercooling, is known from meteorology, and is depicted by the branch AB. Coming from the right from the superheated vapor region, the vapor should start to condense at A, e.g. in clouds. But it does not do this. It remains supercooled vapor. If there is a disturbance, for example by an airplane jet, the cloud suddenly forms droplets. This supercooling plays an important role in meteorology and can amount to several degrees Celsius.

At points B and D, the van der Waals curves exhibit a maximum and a minimum, respectively. There $(\partial p/\partial v)_{T=const.} = 0$. In between there is an unstable region where $(\partial p/\partial v)_{T=const.} > 0$. In this region, metastable states cannot exist and cannot occur. The latter can be seen again better in the $p-v$ diagram. Between B and D any arbitrary small expansion would lead to a pressure increase and thus to a further expansion.

In the $p - v$ diagram, the area, and the points CDE above the equilibrium line AE must equal the area and ABC below the equilibrium line.

With the understanding of the phase diagrams it is possible to define the limit against overheating, because at the minima the conditions must be fulfilled

$$(\partial p/\partial v)_{T=const.} = 0 \text{ and } \left(\partial^2 p/\partial v^2\right)_{T=const.} > 0. \tag{5.213}$$

With the conditions at the critical point $(\partial p/\partial v)_{T=crit.} = 0$ and $\left(\partial^2 p/\partial v^2\right)_{T-crit.} > 0$, the constants in the van der Waals equation can be determined. The result is $a = 3$ and $b = 3$. A normalization of the state variables with the values at the critical point in the manner $\pi = (p/p_{crit})$, $\theta = (T/T_{crit})$ and $\phi = (v/v_{crit})$ transforms the Eq. (5.212) into:

$$\left(\pi + \frac{3}{\theta^n \phi^2}\right)(\phi - 3) = \frac{8}{3}\,\mathrm{T}. \tag{5.214}$$

For the exponent n, van der Waals propose $n = 0$ and Berthelot $n = 1$. Along the superheat limit of a liquid, according to Eq. (5.212) $(\partial \pi/\partial \phi)_{\theta=const.} = 0$ must hold, yielding the following expression for the reduced temperature, the so-called spontaneous nucleation "sN":

$$\theta_{sN}^{n+1} = \frac{(3\phi - 1)}{4\phi^3}. \tag{5.215}$$

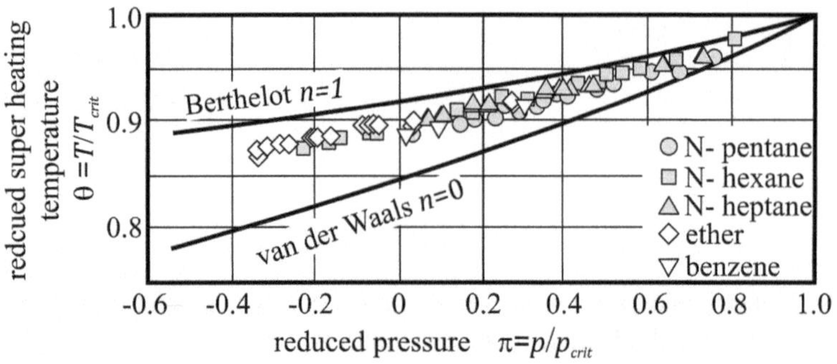

Fig. 5.39 Comparison of the superheat limit according to Eq. (5.215) with the exponent of van der Waals with $n = 0$ and Berthelot with $n = 1$ as well as measured values from Eberhart and Schnyders (1973)

Eberhart and Schnyders (1973) published a comparison of measured superheat limits with calculated values (compare Fig. 5.39). The best fit of the measured values is obtained for $n = 1/2$. Substituting in Eq. (5.215), a modified Berthelot equation is obtained. (The negative reduced pressure in Fig. 5.39 corresponds to conditions where the fluid is subjected to external forces.) The atmospheric pressure is taken to be 0.1 MPa small compared to the critical pressure of water at 22.5 MPa, so that the reduced pressure $\pi \to 0$. For $\pi = 0$, it follows from Eqs. (5.214) and (5.215) $\phi = 2/3$. Substituting this result into the relation (5.215), we obtain for the spontaneous nucleation temperature:

$$\theta_{sN} = \left(\frac{27}{32}\right)^{1/(1+n)}. \tag{5.216}$$

Thus, the reduced temperature of the superheat limit takes the value of 0.844, 0.893, and 0.919, depending on van der Waals, Berthelot, or the modified Berthelot equation.

Collier and Thome (1996) quote a value of $T_{sN} = 320.7\,°C$ for water under atmospheric conditions as the temperature for homogeneous nucleation. These values considerably exceed the temperature level aimed at for boiling water collectors during operation. Such stagnation temperatures can only be reached in case of malfunctions, when the heat sink fails, and full solar irradiation is present. However, in the presence of gases, the superheating temperature decreases rapidly because the partial pressure p_p of trapped gases stabilizes the steam bubble. In such a case, Eq. (5.196) modifies to:

$$p_v - p_p = p_l + \frac{2\,\sigma}{r}. \tag{5.217}$$

The superheat temperature T_{sh} thus results in:

$$(T_{sh} - T_{sat}) = \frac{R T_{sat}^2}{M \Delta h}\left(\frac{2\sigma}{r p_l}\right) \ln\left(1 + \left(\frac{2\sigma}{r p_l}\right)\left(1 + \frac{v_l}{v_v}\right) - \frac{p_p}{p_l}\right). \tag{5.218}$$

5.2.2 Boiling at Surfaces

In technical systems, especially in solar collectors for steam generation, which are aerated and deaerated at least once a day, gases dissolved in water must always be expected. In addition to these gas bubbles carried in the water, gas-filled pores in the pipe wall provide additional boiling nuclei. In the latter case, the wetting behaviour of the water on the pipe wall and thus the contact angle plays an excellent role. The free enthalpy of formation $\Delta G(r)$ of a boiling nucleus is reduced by the factor φ at a planar interface, which depends on the wetting angle θ in the following way (Bankoff, 1957):

$$\varphi = \frac{\left(2 + 2\cos\theta + \cos\theta\sin^2\theta\right)}{4}. \tag{5.217}$$

In the limiting case $\theta = 0°$, which means complete wetting, $\varphi = 1$ and the enthalpy of formation does not change. Without any wetting $\theta \to 180°$ and with $\varphi = 0$ no superheating is required. In wall pores opening to the liquid with an angle β, the effective contact angle θ increases, as shown in Fig. 5.40. In the case of a shallow cavity or pores, this effect is marginal. However, with more acutely tapered cavities, the effective contact angle can reach 180° and the required superheat approaches zero. In the case of poor wetting and small β, the curvature of the liquid surface in the notch can reverse and the enthalpy of formation becomes negative. Thus, supercooled boiling is possible. In the later sections describing the flow forms, attention must therefore be paid to the wetting effect because bubble boiling at the duct walls significantly impacts the flow.

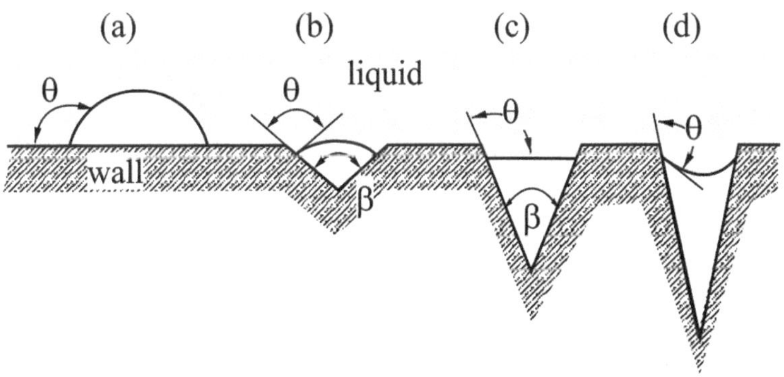

Fig. 5.40 Boundary angle and curvature of the liquid/vapor interface at flat surfaces (**a**) and differently shaped cavities (**b–d**)

5.2.3 Pool Boiling

Vaporization of liquid at the surface of a body immersed in a pool of liquid at rest is generally referred to as pool boiling. A technical distinction is made between pool boiling (boiling in free flow) and flow boiling in tubes, since the nature of the pool boiling process differs considerably depending on the conditions at which boiling occurs from that in tubes. Numerous parameters as the level of heat flux, the thermophysical properties of the liquid and vapor, the surface material and finish, and the physical shape of the heated surface affect the boiling process.

Pool boiling can be explained very well using the Nukiyama curve, which is depicted in Fig. 5.41 for water at 0.1 MPa in the form of a plot of the transferred heat flux ($\dot{q}$) versus the difference between wall (T_w) and saturated steam temperature (T_{sat}). Using the boiling curve, five different regions can be distinguished during boiling.

Since a liquid does not immediately change its phase when it reaches saturation temperature unless it is already in contact with vapor, no vapor bubbles form in region I. This is due to the difference in density between the wall and saturated vapor temperatures. Natural convection, driven only by density differences, sets in, continuously transporting colder fluid towards the wall. This transfer process leads to an increase of the heat transfer from $A - B$.

Once the fluid temperature is above the saturation temperature, small vapor bubbles form at surface pores or cavities of the wall, called vapor nuclei. The formation of the vapor bubbles indicates the transition to region II, that of nucleate

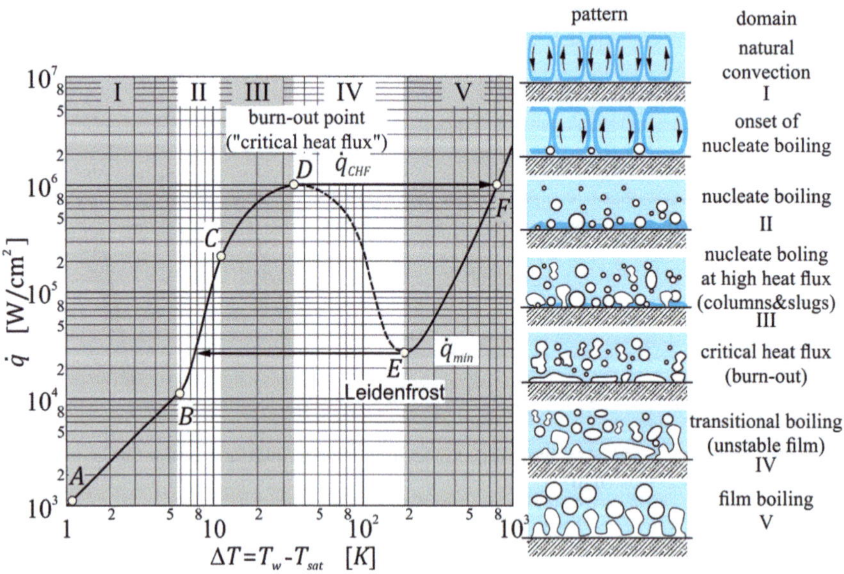

Fig. 5.41 Boiling curves during pool boiling for water at atmospheric pressure (left) and various stages of vessel boiling modified from Collier and Thome (1996)

boiling, at point B ($B = ONB$ "onset of nucleate boiling"). The onset of nucleate boiling is associated with a steep increase in the heat transfer coefficient; even a small increase in the wall heating temperature is associated with a significantly higher heat flux transferred. First, vapor bubbles appear, grow, and slowly leave the wall. There, is a sufficient distance between the bubbles, which is why region II is called individual or isolated bubble boiling.

If the temperature is increased further, the distance between the bubbles near the wall is shortened and one bubble has barely left the wall before the next one is formed. This continues until the bubbles coalesce into continuous columns or slugs of vapor departing from the wall. This marks the transition to Region III, which is called slug or column nucleate boiling ("high-heat-flux nucleate boiling regime"). This process cannot continue indefinitely.

As temperature difference rises further, vapor bubbles are generated so quickly that they prevent the fluid from entering the heated wall. The heat flux at which this boiling crisis occurs is called burn-out-point or critical heat flux (CHF). It marks the beginning of area IV, the transitional boiling. In this region, large parts of the wall are covered by a vapor film and liquid reaches the heated wall only intermittently. It is characterized by a rapid drop in the heat transfer coefficient as the wall temperature increases.

The transition boiling is extremely unstable and the dashed line in Fig. 5.41 can hardly be achieved technically, since in most cases the surface temperature is not controlled but rather the heating power, e.g. solar irradiation. Thus, in power-controlled configurations if the critical heat flux is exceeded, the temperature of the wall abruptly jumps to a higher value (F) and film boiling (area V) starts instantaneously. This is indicated by the arrow. This sudden rise in temperature often has considerable consequences for the heater or the interface material. During film boiling, there is a homogeneous vapor film on the wall and the wall temperature is so high that the heat conduction through the vapor film is large enough to vaporize liquid on the other side.

The lowest wall temperature at which a closed vapor film covers the wall is called the Leidenfrost temperature. If the wall temperature is continuously lowered after entering the film boiling region (region V), this hysteresis phenomenon becomes clearly visible. The wall superheat follows the line from E to D and the closed vapor film remains stable until the Leidenfrost temperature is reached. Then the vapor film collapses abruptly. This collapse leads to an abrupt transition from film boiling back to region II of the nucleate boiling along the arrow.

For the domain of bubble boiling (bubble boiling in free flow) in pool boiling the following relation applies:

$$\frac{\alpha}{\alpha_0} = c_w \cdot F\left(p^*\right) \cdot \left(\frac{\dot{q}}{\dot{q}_0}\right)^n, \tag{5.220}$$

α_0 is a reference value at $\dot{q} = 20\,\text{kW/m}^2$ and $p^* = p/p_{crit} = 0.1$, which is determined experimentally. The impact of roughness is considered by $c_w = (R_a/R_{a0})^{0.133}$, in which as reference roughness $R_{a0} = 0.4\,\mu m$ is used. For the pressure correction $F(p^*)$ from $p^* = 0.1$ to p^* the following relations can be used for liquids other than liquid metals, which exhibit a too high surface tension:

$$\text{fluids except water: } F\left(p^*\right) = 1.2 \cdot p^{*0.27} + \left[2.5 + \frac{1}{1-p^*}\right] \cdot p^*,$$
$$\text{water: } F\left(p^*\right) = 1.73 \cdot p^{*0.27} + \left[6.1 + \frac{0.68}{1-p^{*2}}\right] \cdot p^{*2}. \quad (5.221)$$

In solar collectors, the transition from nucleate boiling to film boiling will occur only when the liquid supply is limited and stagnant. A comparative consideration of the numerical values of the heat flux densities with surface powers of the solar irradiation below 1000 W/m^2 quickly shows that in a container, even without convective heat transfer, the advantages of the increased heat transfer during boiling are not utilized due to the lack of heat flux density. Only a significant concentration of solar irradiation with mirrors allows an improvement of the efficiency by boiling. This can be further increased by improved heat conduction in the absorber board to the heat dissipation tubes.

The following relationships then apply to the exponent n:

$$\text{fluids except water: } n\left(p^*\right) = 0.9 - 0.3 \cdot p^{*0.3},$$
$$\text{water: } n\left(p^*\right) = 0.9 - 0.3 \cdot p^{*0.15}. \quad (5.222)$$

If no experimental value is known for α_0, it can be estimated using relationships based on the bubble break-up diameter and the wetting angle. Details on this can be found in Collier and Thome (1996), among others.

The maximum attainable heat flux in the nucleate boiling regime termed as departure of nucleate boiling (DNB) or critical heat flux (CHF) is reached when a hydrodynamic instability destabilizes the vapor jets rising from the heater surface; this domain of bubble boiling ends at the critical heat flux density $\dot{q}_{CHF}$, which for water is given by the relation

$$\dot{q}_{CHF} = 0.149 \cdot \sqrt{\rho_v} \Delta h \cdot (\sigma \cdot g \cdot (\rho_l - \rho_v))^{1/4}, \quad (5.223)$$

where ρ_l represents the density of the liquid and ρ_v that of the vapor, σ is the surface tension and Δh the vaporization enthalpy (latent heat). This relation considers that Taylor and Helmholtz-instabilities govern the collapse of a liquid vapor interface if a certain velocity is exceeded, for details see Bejan and Kraus (2003). Equation (5.223) holds for strictly planar surfaces, other geometries require a correction factor which may be taken from Bejan and Kraus (2003) as well.

For an upward-facing heated surface, vapor generated at the interface during stable film boiling is released as bubbles at the nodes of a standing two-dimensional Taylor wave pattern. The Taylor instability of the interface assumes that the minimum is reached when the vapor generation rate becomes too small to sustain Taylor wave action on the free interface for stable generation of bubbles.

This formulates the relation for the minimum heat flux at the Leidenfrost-point reading to:

$$\dot{q}_{min} = 0.09 \cdot \rho_v \Delta h \cdot \left(\frac{\sigma \cdot (\rho_l - \rho_v)}{(\rho_l + \rho_v)^2} \right)^{1/4}. \tag{5.224}$$

Regarding the film boiling the heat transport across the vapor film from the wall to the liquid interface depends on convection, conduction, and radiation. The radiation depends on the wall surface shape. However, if radiation is negligible the film boiling heat transfer is independent of the material properties and surface quality. For buoyancy-driven laminar film boiling over a vertical flat isothermal surface in a pool of saturated liquid, the local heat transfer coefficient h from the surface is given by:

$$h = \left(\frac{k_v^3 g \, \rho_v (\rho_l - \rho_v) \Delta h}{4 \, \mu_v (T_w - T_{sat}) x} \right)^{1/4}, \tag{5.225}$$

where k_v is the heat conductivty of the vapor, x the void fraction and the μ_v dynamic viscosity of the vapor.

5.2.4 Flow Boiling

In a two-phase flow, different flow patterns are observed depending on the volume fraction of the vapor and liquid phases. Their knowledge is of fundamental importance, since in most cases the various processes in two-phase flow can only be calculated more precisely if the distribution of the two phases is adequately understood. Flow boiling in tubes is perhaps the most complex convective process encountered in applications, since the heat transfer includes both a convective contribution and a nucleate boiling contribution, whose relative importance depends on the specific conditions. The local flow boiling heat transfer coefficient is depends on vapor quality, mass velocity, heat flux, duct geometry and orientation, the fluid properties and as a result of the preceding mentioned quantities certain characteristic flow patterns establish. These two-phase flow patterns have a particular identifying characteristic and thus flow pattern maps are used to predict the transition from one type of two-phase flow pattern to another. Thereby not only the flow pattern occurring at the particular local conditions under consideration is identified but also the heat transfer characteristics delineating the individual flow patterns are separated and accessible via correlations. Because the flow boiling coefficient is a function of vapor quality, these calculations are typically done locally in thermal design methods.

For a better understanding of two-phase flows, flow patterns are defined which are characterized by certain features. For an upward directed two-phase flow in a vertical pipe, the flow patterns illustrated in Fig. 5.42a can be observed and qualitatively, the individual patterns are described as follows:

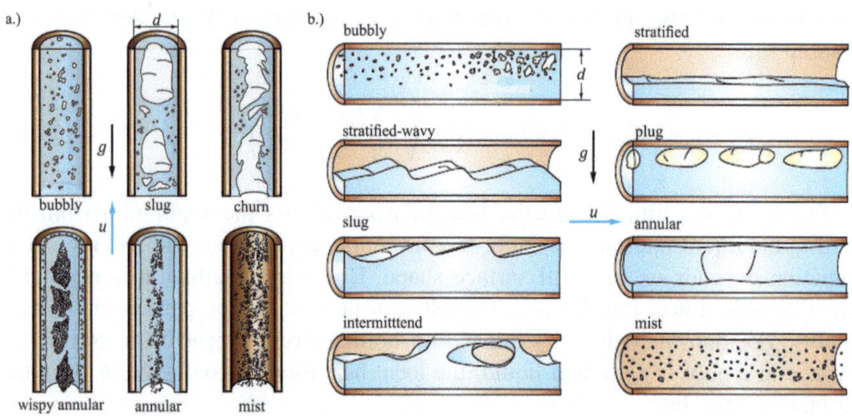

Fig. 5.42 Observed flow patterns in adiabatic tubes (increasing gas fraction from left to right, or from top to bottom, modified according to Thome and Cioncolini (2015)): **a** vertical tube, **b** horizontal (or low inclined) tube

- *Bubbly flow*: a large number of small bubbles being considerably smaller than the channel diameter d are dispersed in a continuous liquid flow. The bubbles vary in size and shape but are rather densely packed. Moreover, they interact with each other by collisions, coalescence, break-up. This flow form occurs at small gas volume fractions and is often observed in subcooled boiling before the fluid attains T_{sat}.
- *Slug flow*: The gas flows as plugs which fill almost the entire pipe cross-section and is separated by liquid slugs which may contain small bubbles. These bubbles exhibit an almost cylindrical shape with dimensions of the order of the tube diameter with a hemispherical nose and a blunt fluctuating tail and are often termed as *Taylor bubbles.*
- *Churn flow*: This transitional flow type occurs between stable slug flow to stable annular flow and features tails of the elongated bubbles beginning to break up and creating large voids in the liquid slugs, which in turn become unstable themselves. This flow type is to be avoided in technical applications since it is associated with pressure and flow rate oscillations. It is usually observed at larger pipe diameters and higher pressure.
- *Annular flow*: In annular flow, a liquid film flows on the inner wetted perimeter of the channel with a higher velocity of the gas (or vapor) core in the centre. The film gas interface is characterized by short wavelength waves, due to the shear stress by the velocity difference of liquid film and gas. Thereby a bidirectional transfer occurs, gas bubbles can be trapped by the film and vice versa liquid droplets are entrained in the gas. When the flow depicts entrained droplets traveling in periodic coherent structures as clouds or wisps of liquid in the central vapor core, some flow pattern maps refer to this as *wispy annular flow*, see Thome and Cioncolini (2015).

- *Mist flow*: In mist flow, the entire liquid is transported by the gas in the form of droplets. Mist flow is encountered only at high gas velocities and low liquid volume fractions if the annular film is thinned by the shear of the gas core on the interface until it becomes unstable. It is also observed in flow boiling if the critical heat flux is surpassed and thus appears often during evaporation under higher pressure.

In the flow of two-phase mixtures in a horizontal or only slightly inclined pipe, different flow patterns occur than in vertical pipes, since, due to gravity and density, the liquid flows predominantly in the lower half of the pipe and the gas for the most part in the upper half. The flow patterns in a horizontal pipe and in a vertical pipe differ from each other the more the influence of gravity is greater than that of frictional forces on the flow. Especially at low velocities, flow forms develop in the horizontal pipe that are not observed in the vertical pipe. The flow forms in the horizontal pipe can be classified as bubbly, stratified, slug, surge, annular, and splash flow. Figure 5.42b qualitatively outlines these flow patterns.

- Bubbly flow: Multiple small gas bubbles are fine dispersed in the continuous liquid-phase, depending on the liquid velocity, where at lower values an accumulation in the top of the duct occurs.
- Stratified flow: At low liquid and gas velocities, gas and liquid separate due to buoyancy. Since the shear stress cause by the velocity difference between the two phases is small it can be balanced by the surface tension and the interface is smooth. Typically, more than half of the perimeter of the tube is dry while the remainder is wet.
- Stratified wavy flow: If the gas velocity in the stratified flow increases the interfacial shear between the phases causes waves directed towards the top of the tube. A considerable fraction of the upper pipe part keeps dry.
- Intermittent flow. A further rise of the gas velocity, yields waves to wet the top perimeter, typically leaving behind a liquid film such that the entire perimeter of the tube is nearly always wet. The large amplitude waves are separated by smaller amplitude waves not wetting the top and entrained bubbles can be identified in the large waves, origination from being trapped by the rollover of the waves having been trapped by the rolling motion of the waves. „The liquid slugs separating such large, elongated bubbles could also be described as large periodic amplitude waves and hence gives the name of this regime": intermittent flow. Intermittent flow is sometimes subdivided into two other regimes: plug flow and slug flow, see Thome and Cioncolini (2015).
- Annular flow: Even higher gas flow rates lead to the formation of a continuous liquid annular ring like film around the perimeter of the tube, which at high gas flow rates has a uniform thickness. Rising the gas flow fraction further, the top of the tube with its thinner film becomes dries out first, so that the annular film covers only the lower tube perimeter.

- Mist flow: Finally, at very high gas velocities, the liquid is dispersed as small droplets in the almost continuous gas phase. From time to time small rivulets of liquid can be observed after the breakdown of the liquid film.

Flow patterns in non-adiabatic heated tubes
A typical example of complete evaporation of a liquid at not too high pressure in a vertical tube and the successive flow regions occurring in the process is shown in Fig. 5.43. Supercooled liquid enters the tube. When the liquid layer near the wall is sufficiently superheated as a result of the external heating, bubbles form on the wall which break off at a certain size and migrate into the flow core. This is initially subcooled, and the bubbles condense again (range of boiling in subcooled liquid). When the entire liquid has reached the boiling temperature T_{sat}, chains of small bubbles are entrained in the flow. As the number of bubbles increases, the bubbles combine, and larger bubbles are formed (so-called bubble flow). With further heat input, these can grow together to form long plugs (slug flow region) or form very large irregular bubbles, with many small bubbles also present (churn flow region). With increasing evaporation, the liquid walls between the plugs or large bubbles tear open and an annular flow develops. Droplets are torn out of the liquid film by the fast-flowing vapor and transported away from the vapor core. As a result of the evaporation, the liquid film becomes thinner and thinner. At very low film thickness, it begins to breakup and then disappears completely. A mist flow is present. When all the liquid droplets have evaporated, the vapor is superheated.

During complete condensation (heat removal from the tube), superheated or saturated steam enters the tube. The vapor condensation on the pipe wall forms a liquid film and an annular flow is created. As the film thickness increases, droplets are torn out of the film. With further condensation, slugs form on the film, which grow together and fill the entire cross-section of the tube. During the subsequent bubbly flow, the bubbles become smaller due to heat dissipation and finally disappear. If the vapor condenses in the horizontal tube, the wavy flow and stratified flow are the main phenomena that occur with larger liquid volume fractions.

In a simplified form, the transitions between the individual flow states in a horizontal pipe are shown in Fig. 5.44, which also indicates qualitatively. The heat transitions in the horizontal pipe are also shown. Correspondingly, the heat transitions are indicated qualitatively the distribution of the heat transfer coefficient along the flow path. The processes in the horizontal tube, as they occur for example in Fresnel collector or parabolic trough collector systems, differ from those in the vertical tube, as they typically occur in a low-temperature collector or a solar tower configuration.

Starting from pure single-phase convective heat transfer characterized by a heat transfer coefficient h, the heat transfer in the two-phase region increases significantly with the onset of boiling. The convective fraction also increases with the rising vapor content. In the domain of stratified and wavy flow, a considerable

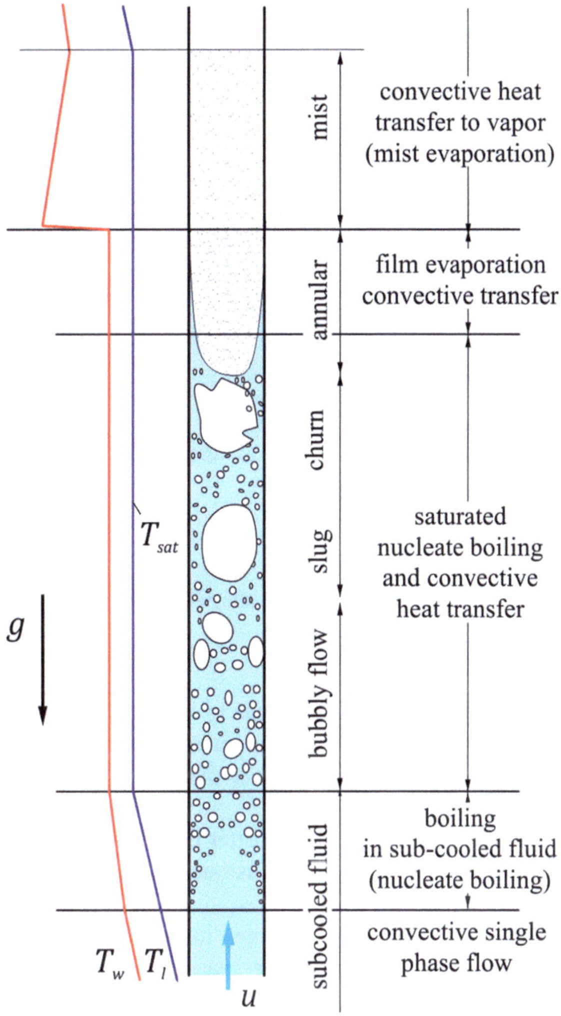

Fig. 5.43 Sketch of flow patterns, heat transfer regions and course of the wall temperature T_w and the fluid temperature T_l during evaporation in the vertical pipe

degradation of the heat transfer can occur with large differences at the tube perimeter. In slug or annular flows, the convective fraction of the heat transfer recovers again, since sufficient or sufficiently frequent wetting begins around the entire pipe circumference. In annular flow, the tube is wetted with fluid around the entire circumference and evaporation takes place at the film surface. If the water film dries out, the heat transfer changes to a state in which only steam exists. The transition from nucleate boiling dominated regime to evaporation dominated ones (slug, annular or intermittent), is termed a departure from nucleate boiling (DNB) or the CHF condition. This transition is associated with a decreasing total effective heat transfer performance causing a rapid rise of the wall temperatures. An uncontrolled

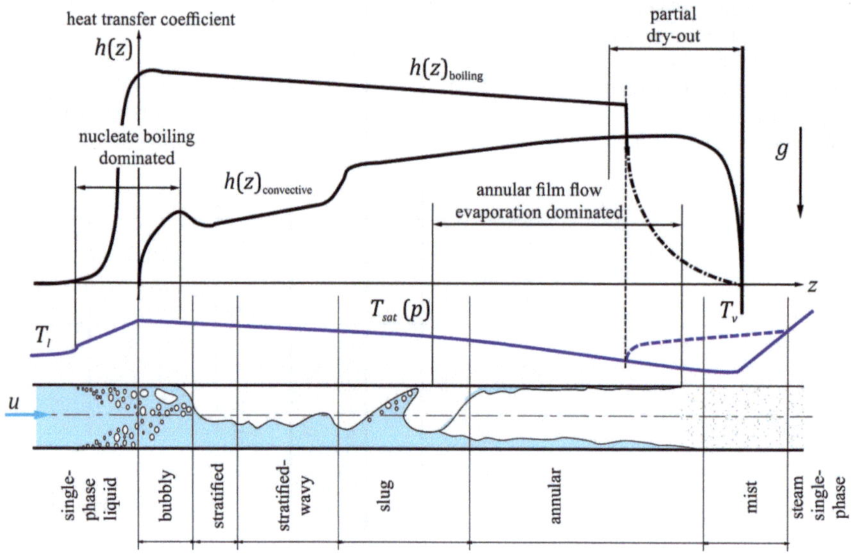

Fig. 5.44 Flow patterns and heat transfer regions with indication of the qualitative course of the fluid temperature T_l and the heat transfer coefficient (h) during evaporation in a horizontal pipe

operation in such a domain has to be prevented because the equipment is not usually designed to operate under such conditions. Another option as e.g. followed in direct evaporation linear power plants is the integration of steam dryers within a recirculation concept.

The corresponding flow patterns are of great importance for heat transfer and pressure drop, as Fig. 5.44 illustrates. In order to be able to predict their occurrence and the associated pressure loss and heat transfer, they are represented in the form of so-called flow pattern maps, in which the occurrence of the flow patterns is indicated as a function of flow variables and material parameters.

5.2.4.1 Definitions and Characteristic Numbers in Two-Phase Flows

Two-phase flows of gas and liquid are characterized by a temporally and spatially variable phase boundary and interaction forces between both phases. This results in difficult analytical accessibility for both momentum and heat transport. Reliable analytical models require a detailed investigation of the individual sub-processes that occur in two-phase flow and often depend on each other in many ways. Due to the increase in computing power in recent years, it is now possible to use numerical computational programs to solve the analytical models, which consist of complex systems of equations, some of which are multiply indeterminate. For a macroscopic description of the processes in the two-phase region, simplified part-empirical models are mostly still used today and state of the art in all system thermal–hydraulic codes. This requires the definition of the variables essential to characterize a two-phase flow. The most important of the parameters are:

- $\dot{M}_g, \dot{M}_l$ mass flow of the gas and liquid,
- ρ_g, ρ_l density of gas and liquid,
- μ_g, μ_l dynamic viscosity of gas and liquid,
- σ surface tension,
- d pipe diameter and β the inclination angle versus the horizontal plane and
- g the gravitational acceleration.

In addition, there are inlet effects, influences of the channel geometry, nature of the channel walls and many more. In the following derivations, the subscript "g" denotes the gas/vapor phase and "l" (liquid) the liquid phase, u_m is the mean velocity and ν the kinematic viscosity. As for single-phase flows, the main variables can be combined to form a set of independent, dimensionless numbers, which can be conceived force ratios, such as the Reynolds number Re, the Froude number Fr and the Weber number Wb, which are defined as follows:

$$Re = \frac{u_m \cdot d}{\nu}; \; Fr = \frac{u_m^2 \cdot d}{g \cdot d}; \; Wb = \frac{u_m^2 \cdot \rho_l \cdot d}{\sigma}, \tag{5.226}$$

Here, the Froude number can be understood as the ratio of inertial forces versus gravity, and the Weber number as the ratio of inertial forces versus surface tension forces. Ratios of thermo-physical properties, such as the density ratio of the two phases ρ and the viscosity ratio E of the two phases, form a further class of dimensionless ratios:

$$\rho = \frac{\rho_g}{\rho_l}; \; E = \frac{\mu_g}{\mu_l}. \tag{5.227}$$

Last, the fraction of the total mass flux $\dot{x}$ often termed as flow quality and the mean velocity u_m are defined by

$$\dot{x} = \frac{\dot{M}_g}{\dot{M}_l + \dot{M}_g} \quad \text{and} \quad u_m = \frac{\dot{M}_g}{\rho_g \cdot A_C} + \frac{\dot{M}_l}{\rho_l \cdot A_C}, \tag{5.228}$$

in which A_C represents the total flow cross-section intersected by both phases.

The formation of a kinematic viscosity from the values ν_g and ν_l of the two phases is difficult and will be discussed later. In the same way, another set of dimensionless combinations could be formed from the ten influencing variables mentioned. However, the number of variables is always reduced to seven according to Buckingham's π-theorem (cf. Sect. 5.1.3). If heat transfer processes are also considered, four further influencing variables are added. These are the specific heat capacities and the heat conduction coefficients of the two phases, from which the Prandtl numbers for the vapor and the liquid result as two further dimensionless parameters.

Additional definitions are required for further consideration of the models and the flow form maps, for which the following, all time-related quantities are indicated by a dot as superscript. In addition, specific quantities, i.e. quantities related to area or mass, are written in lower case. The balance equations known from the single-phase flow can be formed for the two-phase flow with the quantities of the individual phases as well as with average quantities of the two phases. The total flow cross-section A_C is occupied partly by the gas and partly by the liquid.

$$A_C = A_g + A_l. \tag{5.229}$$

The total mass flux or volume flux of the two-phase flow is the sum of the fluxes of the individual phases:

$$\dot{M} = \dot{M}_g + \dot{M}_l \quad \text{and} \quad \dot{V} = \dot{V}_g + \dot{V}_l. \tag{5.230}$$

The proportion of the gas mass flux is given in Eq. (5.228). Accordingly, the following applies to the share of the vapor volume flow $\dot{\alpha}$ in the total volume flow:

$$\dot{\alpha} = \frac{\dot{V}_g}{\dot{V}_g + \dot{V}_l}. \tag{5.231}$$

For the liquid phase, the fractions according to Eq. (5.230) are thus $(1 - \dot{x})$ and $(1 - \dot{\alpha})$. Analogous to Eqs. (5.230) and (5.231), the volumes and masses V, M and volume (α) or mass (x) specific vapor fractions are determined:

$$M = M_g + M_l;\ V = V_g + V_l;\ \alpha = \frac{V_g}{V_g + V} \ \text{and}\ x = \frac{M_g}{M_g + M_l}. \tag{5.232}$$

There is a functional dependence between the quantities of phase velocities of gas and liquid u_g, u_l, their mass flows $\dot{M}_g$, $\dot{M}_l$ and densities ρ_g, ρ_l. For two-phase flows, gas–liquid, the dependence can be given in the form

$$\frac{\dot{M}_g}{\dot{M}_l} = \frac{\rho_g}{\rho_l} \cdot S \cdot \frac{\alpha}{(1 - \alpha)}, \tag{5.233}$$

where for any kind of averaging and for reasons of mass conservation $\dot{x}_l = 1 - \dot{x}$, and $\alpha_l = (1 - \alpha)$ holds here. In particular, it is clear from this relationship that the volume fraction alpha depends quite generally on the velocity ratio S, the density ratio and the vapor content of a two-phase flow. In general, the values $\dot{x}$ and x or $\dot{\alpha}$ and α differ because the two phases move at different velocities in the channel. To illustrate this, imagine the limiting case where vapour bubbles rise upwards through a column of liquid at rest. There $\dot{x} = 1$ or $\dot{\alpha} = 1$ but $x < 1$ or $\alpha < 1$. The average density in a volume element is given by:

$$\rho_m = \frac{M}{V} = \alpha \cdot \rho_g + (1 - \alpha) \cdot \rho_l. \tag{5.234}$$

It is determined from the respective density of the individual phases multiplied by their volume fractions, and the average flow density can thus be determined.

$$\dot{\rho}_m = \frac{\dot{M}}{V} = \dot{\alpha} \cdot \rho_g + (1 - \dot{\alpha}) \cdot \rho_l. \tag{5.235}$$

The respective mean axial velocity u_j of phase j can be calculated from the volume flux $\dot{V}$ of the phase j considered and its occupied cross-sectional area A_j via:

$$u_v = \frac{\dot{V}_g}{A_g} \text{ and } u_l = \frac{\dot{V}_l}{A_l}. \tag{5.236}$$

The average velocity of the two-phase mixture u_m in the flow channel is thus obtained as follows

$$u_m = \frac{\dot{V}}{A_C} = \dot{\alpha} \cdot u_g + (1 - \dot{\alpha}) \cdot u_l. \tag{5.237}$$

If the mass flux or specific volume flux density of the mixture is related to the cross-sectional area, the corresponding density of mass flux $\dot{m}$ or volume flux is obtained:

$$\dot{m} = \frac{\dot{M}}{A_C} = (u_m \cdot \rho_m) \text{ and } \dot{v} = \frac{\dot{V}}{A_C} = u_m. \tag{5.238}$$

When forming the ratios of two-phase flow, a velocity v_j is often calculated, which would be obtained if phase j flows exclusively in the duct. This velocity v_j is termed superficial velocity.

$$v_g = \frac{\dot{V}_g}{A_C} = u_v \cdot \alpha \text{ and } v_l = \frac{\dot{V}_l}{A_C} = u_l \cdot (1 - \alpha). \tag{5.239}$$

Because two-phase flows are of particular importance in applications, additional terms for velocities have been introduced to describe the transport processes. Drift velocity $u_{j,u}$ is the deviation of the actual phase velocity u_j from the total related phase velocity ($u = u_g + u_l$) and these drift velocities are linked in a simple way to the relative velocity ($u_g - u_l$):

$$\begin{aligned} &u_{g,u} = (1 - \alpha)\left(u_g - u_l\right) \text{ and } u_{l,u} = -\alpha\left(u_g - u_l\right) \\ &\text{with } u_{g,u} = \left(u_g - u\right) \text{ and } u_{l,u} = (u_l - u). \end{aligned} \tag{5.240}$$

Another characteristic parameter for two-phase flows was introduced by Lockhart and Martinelli (1949). It is defined as the ratio of the frictional pressure losses for the case where gas and liquid each flow through the pipe alone. If $(dp/dx)_l$ and $(dp/dx)_g$ denote respectively the pressure loss of the liquid and gas volume flows averaged over the total cross-section, then the Martinelli parameter X^2 is as follows

$$X^2 = (dp/dx)_l/(dp/dx)_g. \tag{5.241}$$

This parameter is thus, among other things, a measure of the volume fraction of the flow, because for $X^2 \gg 1$ the two-phase flow consists predominantly of liquid while for $X^2 \ll 1$ predominantly gas occur.

The most essential parameters expressed by the j-th phase are the mass flow rate ($\dot{m}_j$), the velocity (u_j) of the phase and the volume fraction (α_j) of the phase, while subsequently $G_j = m_j/A_C$ is used as mass flux of the j-th phase. All quantities are interrelated via the following relations:

$$v_j = \frac{G_j}{\rho_j}; u_j = \frac{v_j}{\alpha_j}; x_j = \frac{\dot{m}_j}{\sum_j^n m_j} = \frac{G_j}{\sum_j^n G_j}; \alpha_j = \frac{\left(x_j/\left(\rho_j u_j\right)\right)}{\sum_j^n \left(x_j/\left(\rho_j u_j\right)\right)}. \tag{5.242}$$

In most engineering calculations, the above parameters are defined as average quantities across the whole flow area A_C. It should be noted, however, that details of the multiphase flow may incorporate local variations across the flow area. Then, G_j, v_j, and α_j are often defined on a local basis, varying with transverse position across the flow area.

5.2.4.2 Flow Maps

Flow maps have been developed with the help of experimental observations to delineate the different flow forms from each other. A considerably simplified representation of the states can be made for a given gas–liquid mixture, the most common being here experimentally studied the mixture of air/water, by means of the two control parameters related liquid and gas velocity $U_l = (1 - \alpha)u_l$ and $U_g = \alpha\, u_g$. Such a map was constructed from a large experimental database first by Mandhane et al. (1974) and it depicts the individual flow patterns observed as a function of liquid and gas volume flows for a horizontal/vertical pipe configuration, as e.g. illustrated in Fig. 5.45.

The boundaries between flow patterns are not sharp and the experimental observations show hysteresis depending on the transition type. In addition, for vertical pipe flows, the pipe length (l) and the shape of the flow inlet play a role. It significantly influences the transition between slug flow and dispersed/annular flow. By linearly stretching the referenced velocities for air–water mixtures with property parameters that consider density and surface tension, Mandhane et al. (1974) were able to generalize their flow maps to other gas–liquid mixtures. Since the more detailed investigation of the two-phase flow, therefore, a large number of different maps for the prediction of the flow pattern have been developed. Essentially, flow regimes are assigned to different flow patterns in a two-dimensional way. One flow pattern map used in petrochemistry and applied to different pairs of substances is the flow pattern map outlined in Fig. 5.46, which is based on the derivation according to Baker (1954). It is based on extensive experiments with air/water mixtures

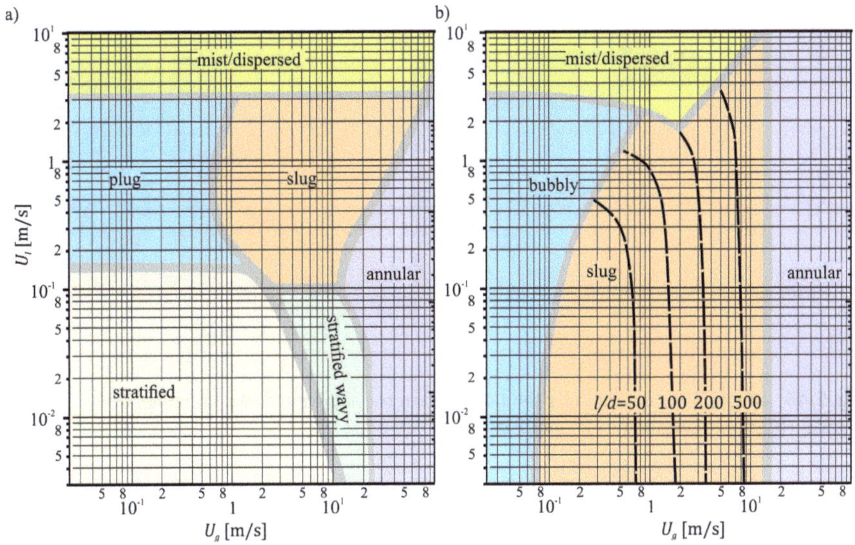

Fig. 5.45 Flow map after Mandhane et al. (1974) for air–water flows at 0.1 MPa for a horizontal tube (**a**, $d = 25$ mm) and a vertical tube (**b**, $d = 51$ mm). l/D denotes the ratio of pipe length to pipe diameter, modified according to Müller (2016)

at in pipes of diameter $25 < d < 100$ mm. For application to other gas/liquid systems and other pressure ranges the gas or liquid mass flow densities used as parameters are extended. Ratios are formed from the material properties of the fluid pair at hand with those of air and water, respectively, at atmospheric pressure and 20 °C:

$$\lambda_b = \left[\frac{\rho_g}{\rho_{Luft}} \cdot \frac{\rho_l}{\rho_{H2O}}\right]^2 \psi_b = \left(\frac{\sigma_{H2O}}{\sigma_l}\right) \cdot \left[\frac{\mu_l}{\mu_{H2O}} \cdot \left(\frac{\rho_{H2O}}{\rho_l}\right)^2\right]^{\frac{1}{3}} \tag{5.243}$$

With the flow maps, one usually obtains satisfactory results for rough calculations. At higher pressures, however, larger deviations result. Various researchers tried to improve the Baker diagram and to transfer it to a non-adiabatic flow, but so far, no satisfactory solution has been achieved. By this type of representation, one is limited to two independent variables associated with the x-axis and the y-axis. Most often, the superficial velocities of the fluid and gas, or generalized parameters containing them, are chosen for these. Hewitt and Roberts (1969) use as axis parameters the momentum current densities ($\rho_j u_j^2$) of the two phases. However, this map can only serve as a rough estimate of the flow shape, since the plot versus the momentum current density is not adequate to reflect the influence of the physical fluid properties and the channel diameter.

Therefore, the idea is to study the transitions from one flow form to another, and to describe them for a wide range of applications if possible. In this case,

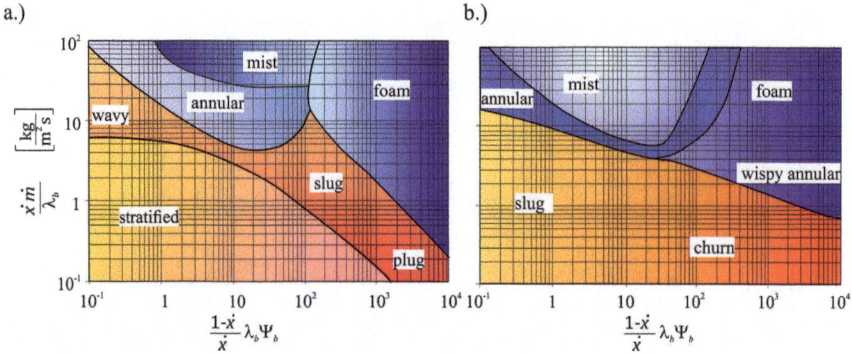

Fig. 5.46 Diagram for determining the flow pattern for adiabatic flows in the horizontal (**a**) and vertical (**b**) pipes according to Baker (1954)

the shape of the transition lines is described as a function of other important flow parameters such as the liquid and vapor mass flow densities, the fluid properties, the duct diameter, and the inclination of the duct. A flow shape map can then be generated for a set of these quantities. Taitel and Duckler (1976) developed such a model for predicting the flow shapes, which was used early with some modifications in programs for safety calculations of nuclear power plants. Here, the transition lines of the flow forms were extended by transition areas, as shown in Fig. 5.47 by using hatched areas. There exist in this formulation dependencies to the homogeneous velocity u_h, the gas volume fractions α and the steam superheat $(T_g - T_{sat})$. Furthermore, a distinction is made between stratified and unstratified flow (depending on the homogeneous velocity). Limits are the Taylor bubble velocity u_{Tb} and a minimum gas volume fraction in the considered node as well as in the nodes below and above it. If unstratified flow is present, this model distinguishes between regions of different heat flux densities (or temperature difference $\left(T_g - T_{sat}\right)$). Below the critical heat flux density (Pre-CHF), the flow patterns shown in Fig. 5.43 occur. Above the film evaporation limit (post-dry out), flow reversal can occur (e.g., inverted annular, inverted slug). The transition functions to the different flow forms are described in detail, for example, in the program manual (RELAP Code Development Team, 1995) or similar manuals of code packages. Decisive for the user of such simulation programs is the examination whether and to what extent the implemented models and in particular their limitations satisfy the respective application.

Since two-phase flows in particular play a role in solar farm concepts (parabolic trough or Fresnel power plants), this area is analysed again in more detail here. Three flow families can be observed in horizontal and slightly inclined pipes,

- stratified flows in smooth or wavy form,
- intermittent flows in the form of surge and plug flows, and
- dispersive flows in the form of bubble or ring-droplet flow.

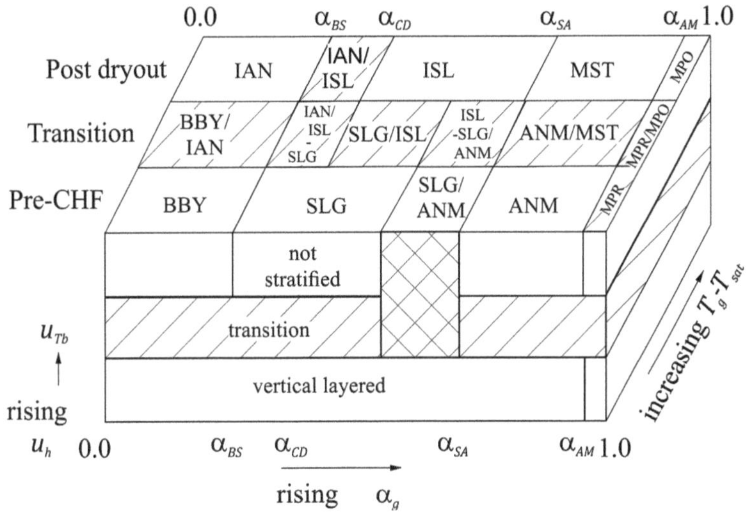

Fig. 5.47 Schematic of the 3D flow pattern map for a vertical pipe from the RE-LAP simulation program (RELAP Code Development Team, 1995) Hatched areas indicate transition regions. IAN-inverted annular flow, ISL = inverted plug flow, BBY = bubbly, SLG = slug, ANM = annular mist flow, MST = mist flow, MPR = mist flow pre-CHF, MPO = mist flow post-CHF

A condition for the transition from stratified to intermittent flow can be derived using an instability condition for a soliton wave. The boundary curve is given by a modified Froude number Fr:

$$Fr = \sqrt{\frac{\rho_g}{\rho_l - \rho_g}} \frac{U_g}{\sqrt{d\, g \cos \beta}}, \tag{5.244}$$

as a function of Martinelli parameter X (Eq. (5.241)). Here, β is the inclination angle of the tube, g is the gravity and d is the tube diameter. The transition from smooth to wavy stratified flow is determined by the Kelvin–Helmholtz instability condition

$$K = \sqrt{\frac{\rho_g}{\rho_l - \rho_g}} \frac{U_g}{\sqrt{dg \cos \beta}} \sqrt{\frac{U_l d}{\nu_l}} \geq 20\alpha\sqrt{(1-\alpha)}, \tag{5.245}$$

in which ν_l denotes the kinematic viscosity of the liquid phase. For the volume fractions, a clear relationship to the Martinelli parameter can be established, so that a curve in the form $K(X)$ can be formulated.

The transition from annular flow to intermittent flow in pipes which are not too steeply inclined is determined by the minimum possible liquid fraction in a slug interspersed with gas bubbles. Intermittent slug flow sets in according to experimental observations for $(1-\alpha) \geq 0.24$, which corresponds to a value of the

Martinelli parameter of $X \approx 1.6$. The state of dispersed bubble flow and intermittent flow is caused by turbulent agitation, by gravity, and by bubble breakup and coalescence. The turbulence intensity in the liquid phase is largely determined by the pressure drop in the liquid phase. For this purpose, Taitel and Duckler (1976) introduce the ratio of related pressure drop of the liquid phase and hydrostatic buoyancy of the bubbles as a parameter in the form:

$$T^2 = \left|\frac{dp}{dx}\right|_l \frac{1}{(\rho_l - \rho_g)g\cos\beta}. \tag{5.246}$$

The transition between the two regions can then be represented as a function T of the Martinelli parameter X. After evaluating the functional relationships, the flow map has the form shown in Fig. 5.48, where each of the boundary curves $K(X)$, $F(X)$, and $T(X)$ is individually assigned to an ordinate. Later this theory of flow pattern boundaries has been generalized such, that two-phase flows in pipes with any angle of inclination can be classified. The transition conditions are either represented graphically or can be determined point by point with the aid of a computer program. The various state variables and variables of the two-phase flow as well as their derivatives, such as the pressure gradient, the volume fraction, and the heat transfer coefficient, depend to a large extent on the flow form. Therefore, in many cases of one-dimensional and quasi-stationary processes, it is aimed to develop calculation methods for two-phase flows individually for the characteristic flow patterns, which is complex and extensive. The different calculation models are then to be linked with each other via flow maps or computer-aided transition conditions in order to determine the two-phase states sufficiently accurately for technical systems, for example for steam generators. The range-based calculation of two-phase flows has not been satisfactorily solved to date. Therefore, correlations derived on the basis of extensive experimental data are currently still used for certain technically relevant quantities such as pressure drop and heat flows. These two-phase laws are addressed in the following sections.

5.2.4.3 One-Dimensional Two-Fluid Model

The formal derivation of the one-dimensional two-fluid model is carried out in a simplified heuristic manner with the help of the notion of a layered flow in the pipe, as illustrated in Fig. 5.49. When deriving the one-dimensional two-fluid equations from the general basic equations for multiphase flows, some fundamental assumptions are made in connection with the spatial, in this case cross-sectional, and temporal averaging. In the case of spatial or temporal averaging of a product of state variables f and g, the following generally applies $\langle f \cdot g\rangle = c\langle f\rangle \cdot \langle g\rangle$. In the simplest representation of the one-dimensional two-fluid model, $c = 1$ is set.

The thermodynamic equations of state and the constitutive equations for local quantities also apply to the averaged quantities. The heat conduction and the change of shear stresses in the flow direction as well as the dissipation power of the frictional forces in the respective phases are neglected. Local thermodynamic equilibrium is assumed in each phase. Against each other, the phases can

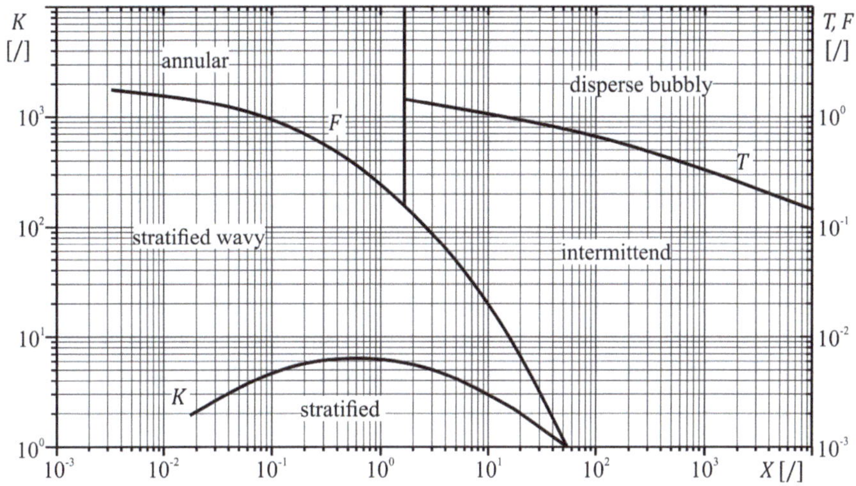

Fig. 5.48 Flow map for horizontal and slightly inclined pipes showing K, F and T as a function of the Martinelli parameter X according to Taitel and Duckler (1976)

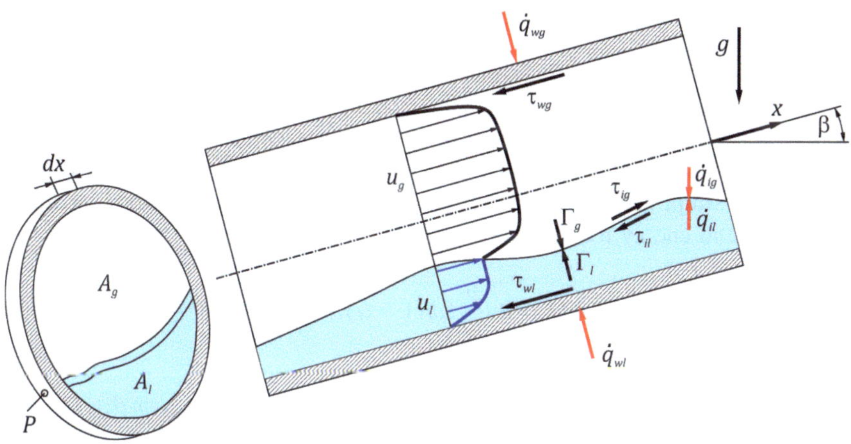

Fig. 5.49 Control volume for a separate two-phase treatment flow with designations

be in thermodynamic non-equilibrium. In vertical two-phase flow, the pressure is assumed to be constant over the cross-section of the pipe, which in many applications is also justified with good approximation for flows in horizontal pipes. The effect of interfacial tensions can be neglected for the time being. However, it reappears in a general formulation or when introducing constitutive equations to close the model equations.

To represent the one-dimensional conservation equations, the following terms are used for the averaged quantities.

- Geometry: cross-sectional area $A_g + A_l = A_C$, perimeter $P_g + P_l = P_C$, wetted perimeter of gas, liquid at position i P_{gi}, P_{li}, P_i, cross-sectional fraction of phase $\varepsilon_g = \alpha = A_g/A_c, \varepsilon_l = A_l/A_c, \alpha + \varepsilon_l = 1$.
- State variables: velocities u_g, u_l, pressures and densities p_g, p_l, ρ_g, ρ_l, mass flux densities $\dot{m}_g, \dot{m}_l$, specific inner energy and enthalpy e_g, e_l, h_g, h_l.
- Constitutive variables: wall shear stress τ_{wg}, τ_{wl}, wall heat flux $\dot{q}_{wg}, \dot{q}_{wl}$, interfacial shear stress at position i τ_{ig}, τ_{il}; interfacial heat flux densities $\dot{q}_{ig}, \dot{q}_{il}$, mass source densities at the phase interface Γ_g, Γ_l, velocity and pressure at interface at position i u_i, p_i.
- Transfer at phase transition: momentum source term due to mass exchange at interface area $M_g^{\Gamma_g}$, $M_g^{\Gamma_l}$, power source density due to wall shear stress at interfaces and walls $L_{\tau g}, L_{\tau l}$, power source density due to mass exchange at the interface $L_g^{\Gamma_g}$, $L_l^{\Gamma_l}$, power source density due to gravity and heat supply Lg, L_q.

Due to local equilibrium, the following relationships hold between the constitutive variables at the interface:

$$\Gamma_g - \Gamma_l = 0; \tau_{ig} - \tau_{il} = 0; \dot{q}_{ig} - \dot{q}_{il} = 0; M_g^{\Gamma_g} - M_g^{\Gamma_g} = 0; L_g^{\Gamma_g} - L_l^{\Gamma_l} = 0. \tag{5.247}$$

Those parts of the quantities that are based on mass transfer sum to zero, but if surface tensions play a role, other terms appear in the momentum and energy conservation equations for the respective phases that take surface tensions into account. For simplicity, surface tension effects will not be considered here. Under these simplified assumptions, the one-dimensional conservation of mass, momentum and energy for the two-fluid model are as follows:

$$\frac{\partial}{\partial t}(\rho_g \alpha A) + \frac{\partial}{\partial x}(\dot{m}_g \alpha A) = 0; \frac{\partial}{\partial t}(\rho_l (1-\alpha) A) + \frac{\partial}{\partial x}(\dot{m}_l (1-\alpha) A) = 0. \tag{5.248}$$

$$\begin{aligned}
&\frac{\partial}{\partial t}(\dot{m}_g \alpha A) + \frac{\partial}{\partial x}(\dot{m}_g u_g \alpha A) = -\alpha A \frac{\partial p_g}{\partial z} - \tau_{wg} P_g - \tau_{ig} P_i - \alpha A \rho_g g \sin\beta + M_g^{\Gamma_g};\\
&\frac{\partial}{\partial t}(\dot{m}_l (1-\alpha) A) + \frac{\partial}{\partial x}(\dot{m}_l (1-\alpha) u_l A)\\
&\quad = -(1-\alpha) A \frac{\partial p_l}{\partial z} - \tau_{wl} P_l - \tau_{il} P_i - (1-\alpha) A \rho_l g \sin\beta + M_l^{\Gamma_l};
\end{aligned} \tag{5.249}$$

$$\begin{aligned}
&\frac{\partial}{\partial t}(\rho_g E_g \alpha A) + \frac{\partial}{\partial x}\left(\dot{m}_g \left(h_g + \frac{u_g^2}{2}\right) \alpha A\right) = L_{\tau g} + L_{gg} + L_g^{\Gamma_g} + L_{qg};\\
&\frac{\partial}{\partial t}(\rho_l E_l (1-\alpha) A) + \frac{\partial}{\partial x}\left(\dot{m}_l \left(h_l + \frac{u_l^2}{2}\right)(1-\alpha) A\right) = L_{\tau l} + L_{gl} + L_l^{\Gamma_l} + L_{ql}
\end{aligned} \tag{5.250}$$

Herein $E_j = e_j + u_j^2/2$ represents the energy density, h_j is the specific enthalpy of the respective phase, and on the right-hand side of the equation are listed, without further specification, the power components of shear stress, gravity, momentum exchange by mass transfer between phases, and heat fluxes. Here one can also introduce the assumption $p_l = p_g = p$. For the simplest case of an incompressible flow, the equations contain the six state variables u_g, u_l, e_g, e_l, p and α. In addition, there are a large number of constitutive variables, which can be reduced to the following independent variables $\Gamma, (\tau_{wg} \cdot P_g), (\tau_{wl} \cdot P_l), (\tau_{ig} \cdot P_{wi}), (\dot{q}_{wg} \cdot P_g), (\dot{q}_{wl} \cdot P_l), (\dot{q}_{il} \cdot P_{wl})$ when the momentum and energy equations are derived in detail. For the constitutive variables, correlations to the state variables must be established on the basis of experimental findings or theoretical considerations in order to close the conservation equations and apply them to the solution of two-phase flow problems. In each case, the model correlations for the constitutive variables are to be developed for specific flow domains. Such correlations have been given in general form by Ishii and Hibiki (2011).

5.2.4.4 Homogeneous Model of Two-Phase Flow

In the homogeneous model, it is assumed that a homogeneous fluid flows in the flow channel, as indicated in Fig. 5.50. Average values are formed from the properties of the two phases. These are assumed to be constant over the entire flow cross-section.

The two-fluid model can be simplified if, for technical reasons, one is only interested in the integral states of the two-phase flow such as total mass flow, total pressure drop and total heat transfer. By adding two conservation equations for each of the individual phases, one then obtains three conservation equations for the total mass, total momentum, and total energy of the two-phase mixture. These equations can be represented in a form corresponding to the form of the basic one-dimensional fluid mechanics equations for compressible media if the densities, wall shear stress, and wall heat flux are introduced into the mixture equations as weighted quantities. Using the general definitions and with simple algebraic transformations, one obtains for the two-phase mixture the following

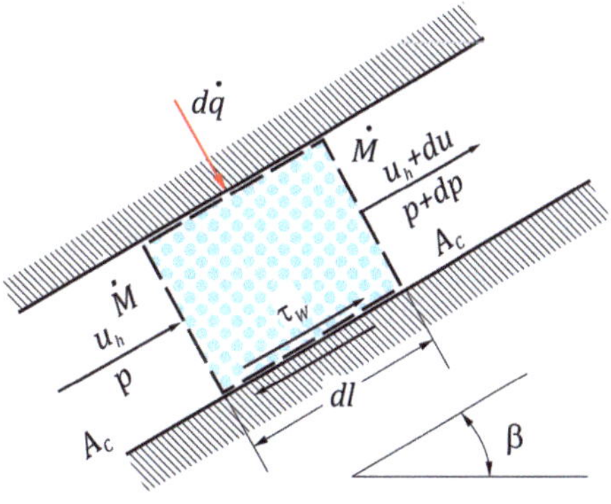

Fig. 5.50 Control volume of a non-adiabatic two-phase flow in the inclined channel assuming the homogeneous two-phase flow model

three conservation equations:

$$\begin{aligned}
&\frac{\partial \rho_h}{\partial t} + \frac{1}{A_c}\frac{\partial \dot{M}}{\partial x} = 0 \text{ with } \rho_h = \alpha\rho_g + (1-\alpha)\rho_l;\\
&\frac{\partial \dot{M}}{\partial t} + \frac{\partial}{\partial x}\left(\frac{1}{\rho_M}\cdot\frac{\dot{M}^2}{A_C}\right) = -A_C\frac{\partial p}{\partial x} - \tau_w \cdot P - A_C\rho_h g \sin\beta;\\
&\text{with } \dot{x}_g = \dot{x};\ \frac{1}{\rho_M} = \frac{\dot{x}^2}{\alpha\rho_g} + \frac{(1-\dot{x})}{(1-\alpha)\rho_l};\ \tau_w = \tau_{wg}P_g + \tau_{wl}P_l;\\
&\frac{\partial E}{\partial t} + \frac{1}{A_C}\frac{\partial}{\partial x}\left(\dot{M}\left(h + \frac{1}{\rho_E^2}\frac{\dot{M}^2}{2A_C}\right)\right) = L_{\tau w} + L_g + L_{qw};\\
&E = \rho_g E_g + \rho_l E_l;\ h = \dot{x}h_g + (1-\dot{x})h_l;\ \frac{1}{\rho_E^2} = \frac{\dot{x}^3}{\alpha^2\rho_g^2} + \frac{(1-\dot{x})^3}{(1-\alpha)^2\rho_l^2}.
\end{aligned} \tag{5.251}$$

It can be seen that the densities of the individual phases are weighted differently in the various conservation equations, so in addition to the homogeneous density of the mixed phase ρ_h, the momentum density ρ_M and the energy density ρ_E. They have different effects compared to single-phase flows. Although the model for the mixture flow has simplified by reducing the equations, the new problem now arises that the volume fraction α must be correlated with the vapor content $\dot{x}$ by an empirical relationship in order to use the simplified model for problem solving. In general, α is related to $\dot{x}$ by the velocity ratio $S = u_g/u_l$ of the phases (see Eq. (5.233)). Therefore, empirical links are sometimes also determined for the velocity ratio S and the relationship is inserted into the mixture equations.

By their nature, the mixture equations are suitable for application calculations when dispersed flow forms such as bubbly or droplet flows are present. They can be simplified even further if mechanical equilibrium between the phases is assumed, i.e. if the disperse phase has the same velocity as the homogeneous phase in each case. The mixture model simplified in this way is also called a "*homogeneous flow model*". It has special significance because of its simplicity and can be reliably applied to disperse flows with very small volume fractions of the disperse phase. With the assumption $S = 1$ there is an unambiguous relation between α and $\dot{x}$ according to Eq. (5.233):

$$\alpha_h = \frac{1}{1 + \frac{1-\dot{x}}{\dot{x}}\cdot\frac{\rho_g}{\rho_l}}, \tag{5.252}$$

where is the α_h is the homogeneous void fraction. With this relation, all other density definitions for momentum ρ_M and energy ρ_E reduce to the expression $\rho_h = \alpha\rho_g + (1-\alpha)\rho_l$ in the mixture balance, i.e., $\rho_h = \rho_M = \rho_E$.

There are a number of averaging methods for the homogeneous dynamic toughness μ_h. The following relationships for calculating the viscosity of the homogeneous fluid are described (Collier & Thome, 1996)

$$\frac{1}{\mu_h}=\frac{\dot{x}}{\mu_g}+\frac{(1-\dot{x})}{\mu_l};\ \mu_h=\dot{x}\mu_g+(1-\dot{x})\mu_l;\ \mu_h=\rho_h\left[\dot{x}u_g\mu_g+(1-\dot{x})u_l\mu_l\right]. \tag{5.253}$$

These averaging calculations can lead to very different results. Since the calculation of mean viscosities already encounters great difficulties for liquid mixtures, the formation of mean values for gas/liquid mixtures is very questionable. Thus, even results for pressure drop calculated with different definitions for μ_h show sometimes considerable deviations from each other. The homogeneous model is the simplest of all two-phase flow models. It can certainly be extended to thermally non-equilibrium disperse flows with evaporation and condensation processes. Mechanical equilibrium is not possible for dispersive two-phase flows with larger of the disperse phase is generally not given. In order to account for this fact and still use the idea of a good mixing of the phases for simplification the drift flow model has been developed, which is described next.

5.2.4.5 Drift Model

The drift flow model is based on the basic idea that both phases are well mixed but move relative to each other and generally have different thermodynamic states. The scope of the model is therefore primarily in dispersed flows, i.e., bubbly, foam, and droplet flows. However, there are also attempts to extend the model to plug and ring flows. The model is based on a density-weighted mixture velocity um

$$u_m=\frac{\alpha\,\rho_g u_g+(1-\alpha)\rho_l u_l}{\alpha\,\rho_g+(1-\alpha)\rho_l}. \tag{5.254}$$

and considers the relative velocities by so-called drift velocities, which are first introduced locally as follows:

$$u_{g,u}=u_g-U;\ u_{l,u}=u_l-U, \tag{5.255}$$

where U is the local total volumetric flow density, represented by the related local velocities $U=U_g+U_l$. From these definitions, the mean values of volume fraction alpha, related total volume flow U and a mean drift velocity yet to be defined can be linked by cross-sectional averaging. This is done by multiplying the relation Eq. (5.255) with the local volume fraction α and a cross-sectional averaging. After algebraic transformations one first obtains the relation:

$$\langle\alpha\,u_{g,u}\rangle=\langle\alpha\,u_g\rangle-C\,\langle\alpha\rangle U, \tag{5.256}$$

Using the definitions $\overline{u}_{g,u}=\langle\alpha u_{g,u}\rangle/\alpha$ and $U_g=\langle\alpha u_g\rangle$ as cross-section-weighted drift and averaged related gas velocity, the cross-section-averaged gas volume fraction is given by

$$\langle\alpha\rangle=U_g/\left(CU+\overline{u}_{g,u}\right). \tag{5.257}$$

To define a mean drift velocity, the volume-weighted quantities $\overline{u}_{g,u}$ and $\overline{u}_g = \langle \alpha u_g \rangle / \langle \alpha \rangle$ are now introduced. For the representation in the conservation equations the relations $\overline{u}_{g,u} = \overline{u}_g - U$ and $\overline{u}_{l,u} = \overline{u}_l - U$ are used. To simplify the notation, the averaging symbols are again omitted in the following. The drift velocity can thus be understood as the velocity of the phase relative to a surface moving at the mixing velocity U. With the relation for the mixture velocity $U = \alpha u_g + (1-\alpha)u_l$, the relation of drift velocity and relative velocity of the phases to each other can be established directly.

$$u_{g,u} = (1-\alpha)\left(u_g - u_l\right);\ u_{l,u} = -\alpha\left(u_g - u_l\right). \tag{5.258}$$

From this relationship and the defining Eq. (5.254) follows a relationship between u_g, u_m, and $u_{g,u}$ of the form:

$$u_g = u_m + \frac{\rho_l}{\rho_h} u_{g,u};\ u_l = u_m - \frac{\rho_l}{\rho_h}\frac{\alpha}{(1-\alpha)} u_{g,u}. \tag{5.259}$$

If these expressions are inserted into the conservation equation for the mixtures and if, furthermore, the conservation equation for the gas phase is retained for the description of phase transitions, then, after some algebraic transformations, four conservation equations result for the state variables mean velocity u_m, pressure p, mean enthalpy h_m and volume fraction α.

$$\begin{aligned}
&\frac{\partial \rho_h}{\partial t} + \frac{1}{A_c}\frac{\partial(\rho_h u_m A_c)}{\partial x} = 0;\\
&\frac{\partial(\alpha\rho_g)}{\partial t} + \frac{1}{A_c}\frac{\partial(\alpha A_c \rho_g u_m)}{\partial x} + \frac{1}{A_c}\frac{\partial}{\partial x}\left(\alpha A_c \frac{\rho_g \rho_l}{\rho_h} u_{g,u}\right) = \frac{\Gamma_g}{A_c};\\
&\frac{\partial(\rho_h u_m)}{\partial t} + \frac{1}{A_c}\frac{\partial\left(A_c \rho_h u_m^2\right)}{\partial x} + \frac{1}{A_c}\frac{\partial}{\partial x}\left(A_c \frac{\alpha}{1-\alpha}\frac{\rho_g \rho_l}{\rho_h} u_{g,u}^2\right)\\
&= -\frac{\partial p}{\partial x} - \tau_w \cdot \frac{P}{A_c} - \rho_h g \sin\beta;\\
&\frac{\partial(\rho_h h_m)}{\partial t} + \frac{1}{A_C}\frac{\partial}{\partial x}(A_C \rho_g u_m h_m)\\
&= \frac{1}{A_C}\left(\dot{q}_{wg} P_g + \dot{q}_{wl} P_l\right) + \frac{\partial p}{\partial t} + u_m \frac{\partial p}{\partial x} + u_{g,u}\frac{\rho_l - \rho_g}{\rho_h}\frac{\partial p}{\partial x}\\
&- \frac{1}{A_C}\left(A_C \alpha \frac{\rho_g \rho_l}{\rho_h} u_{g,u} \Delta h_{lg}\right) + \frac{L_{diss}}{A_C};
\end{aligned} \tag{5.260}$$

in which Δh_{lg} denotes the vaporization, L_{diss} the power released by dissipation enthalpy and h_m is the density averaged enthalpy given by:

$$h_m = \frac{\alpha\,\rho_g h_g + (1-\alpha)\rho_l h_l}{\alpha\,\rho_g + (1-\alpha)\rho_l}. \tag{5.261}$$

This model is used especially for ring and plug flow. Since in ring flow the liquid droplets in the gas core flow much faster than the liquid film on the wall, large errors occur, especially at high loadings, due to the formation of an average velocity for the liquid. Therefore, in this case, the liquid transport in the film and core are often considered separately. A successful application of the simple drift flow model depends essentially on whether constitutive relations can be developed for the drift velocity $u_{g,u}$.

5.2.4.6 Pressure Loss Calculations in Multi-phase Flows

Pressure losses in two-phase flows are of great importance in power plant and process engineering. Therefore, robust empirical pressure drop relationships similar to those in single-phase flow hydraulics have been developed on the basis of measurements and simple model notions that do not distinguish between specific flow forms. Nevertheless, the notion of a dispersed flow on the one hand and that of a separate two-phase flow on the other has led to two different variants for pressure drop relationships. In general, the total pressure drop in a pipe or line section is composed additively of the pressure drop due to viscous friction (f), acceleration (a) and gravity (g) as sketched in Fig. 5.51 in the following way.

$$\left(\frac{dp}{dx}\right)_{tot} = \left(\frac{dp}{dx}\right)_f + \left(\frac{dp}{dx}\right)_a + \left(\frac{dp}{dx}\right)_g . \tag{5.262}$$

While in horizontal straight pipe sections of constant cross-section usually only the friction component (laminar/turbulent) is relevant, in contraction or expansion elements, such as valves or elbows, the acceleration component often predominates, similar to single-phase flow, but far more pronounced. This is because in two-phase flows, heat supply or pressure change not only mean a change in density, which is reflected in variable gas/liquid velocities, but can also lead to evaporation and/or condensation. Also, for two-phase flows lots of correlations are available, see e.g. Friedel (1979), Collier and Thome (1996), however, great care has to be

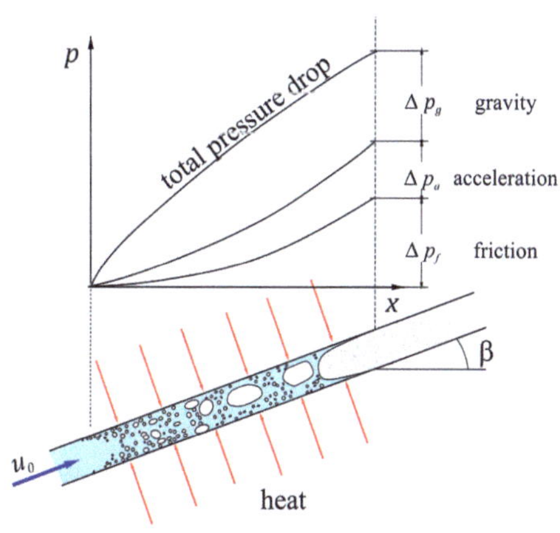

Fig. 5.51 Schematic sketch showing the components of the pressure drop in an inclined pipe with heat supply

taken, on which basis the correlations are formed. The most common approaches are the homogeneous model or the separate model:

Homogeneous model (friction)

The pressure gradient in a pipe is calculated similarly to a single-phase flow as follows:

$$\left(\frac{dp}{dx}\right)_f = \frac{4}{d}\tau_w \text{ with } \tau_w = \frac{f_{F,2ph}}{4}\frac{\rho_h}{2}u^2, \tag{5.263}$$

In which d is trhe hydraulic diameter, $f_{F,2ph}$ the Fanning friction factor of the two-phase flow, $f_{F,2ph}$ the homogeneous density. The coefficient of friction is considered as a function of a Reynolds number of the two-phase flow to be defined and for the respective velocity a corresponding coefficient of friction is to be chosen, for example for a annular-droplet flow a value of $f_{F,2ph}$ of $f_{F,2ph} \cong 0.02$ is valid for a rough pipe wall—if the volume fraction of the gas is large, $\alpha \to 1$ the f_F—value of the single-phase gas flow can be chosen, i.e. $f_F = f_{F,g}$, vice versa at $\alpha \to 0$ the value of the liquid $f_F = f_{F,l}$, where in the two-phase Reynolds number (Re_{2ph}) appears the viscosity of the gas or liquid. Often based on the classical implicit relation of Prandtl for single-phase fully turbulent flows, the following relation is also used

$$\frac{f_{F,2ph}}{4} = 0.0014 + \frac{1}{8}Re_{2ph}^{-0.32}. \tag{5.264}$$

To form the Reynolds number, the total mass flow density and a mixture viscosity are used, for example, in the following form

$$\mu_{2ph} = \frac{U_g}{U}\mu_g + \frac{U_l}{U}\mu_l = \dot{x}\mu_g + (1-\dot{x})\mu_l \text{ or } \frac{1}{\mu_{2ph}} = \frac{\dot{x}}{\mu_g} + \frac{1-\dot{x}}{\mu_l}. \tag{5.265}$$

For condensable mixtures such as water and water vapor, a so-called two-phase multiplier is also frequently used for practical calculations. This multiplier is defined as the ratio between the pressure drop in the current two-phase flow with mass flux $\dot{M}$ to the pressure drop of the condensed liquid flow with the same pipe cross-section and mass flux:

$$\Phi_{l0}^2 = \frac{(dp/dx)_{2ph}}{(dp/dx)_{l0}}, \tag{5.266}$$

where the suffix "$l0$" indicates that the pressure gradient of the pure liquid at the same mass flux form the reference. Textbooks give correlations that reflect values essentially as a function of gas–liquid properties and vapor content and take the following form $f = \left(\mu_g/\mu_l, \rho_g/\rho_l, \dot{x}\right)$.

Separate model (friction)

The separate model is based on the idea that both phases flow in two separate regions of the pipe, as in a layered or annular flow, but that pressure equilibrium nevertheless exists in both phases regardless of the flow pattern. The total pressure drop is then described by the momentum equation for the two-phase mixture given and e.g. for the horizontal pipe of constant cross-section, the following follows for steady-state flows:

$$\frac{dp}{dx} = \langle\tau_w\rangle\frac{P}{A_C} + \frac{\dot{M}^2}{A_C^2}\frac{d}{dx}\left(\frac{\dot{x}^2}{\alpha\rho_g} + \frac{(1-\dot{x})^2}{(1-\alpha)\rho_l}\right). \tag{5.267}$$

If no heat is applied across the edge of the tube, $\dot{x}$ and α do not change along the tube and the wall shear stresses are in equilibrium with the compressive force. If one introduces a Martinelli parameter analogous to Eq. (5.241) for the two-phase multiplier in such a way that they relate the pressure drop in the two-phase flow to the pressure drop of the gas or liquid phase flowing alone in the tube, one obtains and alpha do not change along the tube and the wall shear stresses are in equilibrium with the compressive force.

$$\Phi_g^2 = \frac{(dp/dx)_{2ph}}{(dp/dx)_g} \text{ and } \Phi_l^2 = \frac{(dp/dx)_{2ph}}{(dp/dx)_l}. \tag{5.268}$$

These quantities are related to the so-called Martinelli parameter, the pressure decay ratio, based on experimental data. The Martinelli can be calculated explicitly if the related velocities of the two-phase flow and its laminar or turbulent flow state are known. The turbulence states are determined by the Reynolds number of the gas or liquid flow and for $Re > 2300$ a turbulent state is assumed. Accordingly, there are four possible forms of the Martinelli parameter depending on whether the gas and liquid phases are laminar or turbulent flow.

The Martinelli pressure drop relationship is based on a relatively small in system pressure ($p < 0.4\,\text{MPa}$) and pipe diameter ($d < 3\,\text{cm}$) and have an uncertainty of up to 40%. For more accurate design calculations, empirical pressure drop correlations have been developed based on a large database and taking into account other specific dependencies on two-phase mass flow density (Reynolds number), surface tension (Weber number $We = \left(\dot{m}^2 d\right)/(\rho_h\sigma)$), and gravity (Froude number $Fr = \dot{m}^2/(gd\,\rho_h^2)$) all calculated using the homogeneous density ρ_h. They represent complex functional relationships between dimensionless ratios that capture the physical relationships. The Friedel correlation for a horizontal tube is given here as an example.

$$\Phi_{l0}^2 = E + \frac{3.24FH}{Fr^{0.045}We^{0.035}} \text{ with } E = (1-\dot{x})^2 + \dot{x}^2 \frac{\rho_l}{\rho_g}\frac{f_{F,g0}}{f_{F,l0}};$$
$$F = \dot{x}^{0.78}(1-\dot{x})^{0.24};\ H = \left(\frac{\rho_l}{\rho_g}\right)^{0.91}\left(\frac{\mu_g}{\mu_l}\right)^{0.19}\left(1-\frac{\mu_g}{\mu_l}\right)^{0.7}. \tag{5.269}$$

Acceleration pressure losses

Compared to single-phase flow, acceleration pressure losses occur more frequently in pipe extensions or contractions, in pipe bends and pipe junctions. For design purposes in apparatus engineering, the pressure loss relationships are given in analogy to corresponding relationships from single-phase flow, neglecting wall friction.

Since a deceleration of the flow generally occurs with cross-section expansion, a separation of the phases is to be expected with two-phase flows. Therefore, it is obvious to use the separate flow model in the form of the mixture balance equations for the calculations of a pressure change. From the conservation equations for the two-phase mixture of the separate model, a relationship for the pressure recovery in the Carnot diffuser (cf. Fig. 5.32b) can be easily given if wall friction is neglected.

$$p_2 - p_1 = \dot{m}_1^2 \frac{A_1}{A_2}\left(\frac{1}{\rho_{M,1}} - \frac{A_1}{A_2}\frac{1}{\rho_{M,2}}\right) \tag{5.270}$$

Here $\rho_{M,1}, \rho_{M,2}$ denote the momentum density (cf. Eq. (5.252)). If we switch to single-phase flow, i.e. choose $\dot{x} = 0$ or $\dot{x} = 1$, the momentum density changes to the density of single-phase flow and so does the expression for pressure recovery. By its nature, this expression can only be evaluated if volume fraction and vapor content can be linked between cross sections 1 and 2. At low system pressures, when $p \ll p_{crit}$ holds and no phase transitions occur due to evaporation, the volume fraction of the gas α and the vapor content $\dot{x}$ do not change and in this case the analogy to single-phase flow is evident, $\rho_{M,1} = \rho_{M,2}$. Observations show that downstream, however, equilibrium of two-phase flow is reached only after relatively long running distances, about 30–70 diameters. This fact requires a pressure-dependent correlation for the change of the vapor content in more accurate calculations. In the case of vapor–liquid flows with phase transitions, empirical correlations between volume fraction α and vapor content $\dot{x}$ are used.

For pipe contraction from 1 to 2, an acceleration of the two-phase flow occurs, which leads to better mixing, therefore a pressure loss calculation can be carried out here according to the homogeneous flow model with a good approximation. Since the law of single-phase flow applies to the homogeneous model, with the homogeneous density as the only characteristic feature, the known relationship of single-phase flow is obtained:

$$p_2 - p_1 = \frac{\dot{m}_2^2}{\rho_h}\left(1 - \left(\frac{A_2}{A_1}\right)^2 + \left(\frac{A_2}{A_1} - 1\right)^2\right), \tag{5.271}$$

in which the two fractions in the parenthesis can be identified as the reversible and irreversible fractions of the pressure drop.

In the pipe elbow, detachment also occurs on the inside at higher velocities, causing a phase separation due to centrifugal forces in two-phase flows. In this case, gas phase usually accumulates on the inside, while the fluid flows in the outer area of the elbow. Dispersed flows temporarily become layered flows. A new equilibrium between the phases corresponding to the inlet conditions is also established only after 30–70 pipe diameters. This means that fully developed two-phase flows can rarely be assumed for technical designs in plant engineering.

For arbitrary cross-sections, the acceleration pressure loss is determined integrally over a distance x along the duct. It results from the equation of inertia. The acceleration du/dt of the mass $dM = (A\Delta x\,\rho)$ corresponds to the force difference, i.e. the change in the product $(A\,p)$.

$$\frac{\partial(A \cdot p)}{\partial x}\Delta x = -A\Delta x \cdot \rho\frac{du}{dt} \text{ with } \frac{du}{dt} = \frac{\partial u}{\partial t} + \frac{\partial u}{\partial x} \cdot \frac{\partial x}{\partial t} = \frac{\partial u}{\partial t} + u \cdot \frac{\partial u}{\partial x}. \tag{5.272}$$

The insertion of the right side into the left provides:

$$p \cdot \frac{\partial A}{\partial x} + A \cdot \frac{\partial p}{\partial x} = -A \cdot \rho\frac{\partial u}{\partial t} - A \cdot \rho \cdot u \cdot \frac{\partial u}{\partial x}. \tag{5.273}$$

For a pipe with an invariant cross-sectional area, as is the case for an absorber tube, the term with $\partial A/\partial x$ cancels out and the acceleration pressure loss is composed of two components, the acceleration over distance and the acceleration over time.

$$\frac{\partial p}{\partial x} = \rho\frac{\partial u}{\partial t} + \rho \cdot u \cdot \frac{\partial u}{\partial x} = \Delta p_{a0} + \Delta p_{a1}. \tag{5.274}$$

The acceleration along the flow path is given by $\dot{m} = (\rho u)$ and after local discretization into the nodes i and $i+1$:

$$\Delta p_{a1} = \frac{\left(\dot{m}_i^2 - \dot{m}_{i+1}^2\right)}{2\rho}. \tag{5.275}$$

When accelerating over time, one obtains after time discretization:

$$\Delta p_{a0} = \frac{\Delta x}{\Delta t}\left(\dot{m}_{i,t} + \dot{m}_{i,(t-1)}\right). \tag{5.276}$$

Pressure change due to height

The change in pressure due to the geodetic change in elevation is given by the following equation, where β is the angle of slope with respect to the horizontal plane.

$$\Delta p_H = g \cdot sin\beta \cdot \rho \cdot \Delta x = \rho \cdot g \cdot \Delta H. \quad (5.277)$$

5.2.5 Density Waves, Critical Mass Flux and Cavitation

When a two-phase mixture flows out of a pressure reservoir through a narrow cross-section, a limitation of the mass flow occurs above a certain critical pressure difference of container, for example a dam drum, and the consumer's extraction pressure ($p_1 - p_0$). A further reduction of the pressure, for example by a further opening of a consumer valve, then leads to no further increase of the mass flux. An analogous phenomenon can be observed with compressible flow through a Laval nozzle. The physical cause is identical in both cases, because above the critical pressure ratio, wave-like pressure or density disturbances in the fluid can only propagate downstream because the flow velocity has become larger than the propagation speed of small disturbances. It is not possible to influence the flow region upstream by a change of state further downstream. In compressible flow, these are sonic or Mach waves, and in two-phase flow, they are small changes in the volume fraction of the gas phase. The propagation velocity a of these small disturbances is given in both cases by the isentropic change of density with pressure:

$$a^2 = \left(\frac{\partial p}{\partial \rho}\right)_s. \quad (5.278)$$

The calculation of the critical mass flow in a cross-sectional constriction is performed as in gas dynamics from the conservation equations for mass and momentum using certain equations of state for the gas and the liquid phase. The initial equations are those for the two-phase mixture Eq. (5.251) or (5.260). The critical mass flux $\dot{m}_{crit} = (A^* a^* \rho^*)$ is given by the propagation speed a^* of the density wave at the narrowest flow cross-section A^*.

The relation shows again the typical problem in a two-phase flow, since the critical mass flow depends on the definition of the two-phase density ρ_{2ph}, which has a different shape depending on the mixing model (separate or homogeneous model). Since the flow is always accelerated during outflow or overflow, it is usually assumed that there is good mixing of the phases and that the homogeneous density ρ_h describes the mixing well. Differentiation of the expression for this density at constant entropy, i.e., in the approximation that the changes of state occur adiabatically in each phase but with phase changes at the phase boundaries, then yields

$$\left(\frac{1}{a_{2ph}^2}\right)_h = \left(\frac{\partial \rho}{\partial p}\right)_s = \frac{1}{\rho_h^2}\left(\frac{1}{\rho_l^2 a_l^2} + \dot{x}\left(\frac{1}{\rho_g^2 a_g^2} - \frac{1}{\rho_l^2 a_l^2}\right) - \left(\frac{\partial \dot{x}}{\partial p}\right)_s \left(\frac{1}{\rho_g} - \frac{1}{\rho_l}\right)\right), \tag{5.279}$$

in which a_g, a_l are the sound speed of the gas and liquid, and the index "h" refers to the homogeneous model. Since the speed of sound in the liquid is much greater than that in the gas, one can further simplify the equation:

$$\left(a_{2ph}^2\right)_h = a_g^2\left(\dot{x} + \frac{\rho_g}{\rho_l}(1-\dot{x})\right)^{-2} \cdot \frac{1}{\left(\dot{x} - \left(\frac{\partial \dot{x}}{\partial p}\right)_s a_g^2 \frac{\rho_g}{\rho_l}(\rho_l - \rho_g)\right)}. \tag{5.280}$$

Here, it can be clearly seen that the propagation velocity of the density wave depends essentially on the vapor content and on the change of the vapor content in isentropic thermodynamic equilibrium. In many technically relevant flows, an evaporation process in thermodynamic equilibrium by pressure reduction does not take place in the narrowest flow cross-section because the pressure reduction occurs too quickly and is too small. This means that the relaxation time for evaporation is significantly greater than the transit time of the flow through the narrowest cross-section. This case is referred to as inhibited or frozen thermodynamic equilibrium, characterized by $(\partial \dot{x}/\partial p)_s = 0$. If the vapor content is divided by the volume fraction α, the frozen density-wave velocity in a homogeneous two-phase flow is given by the expression:

$$\left(a_{2ph}^2\right)_h = \frac{a_g^2}{\alpha\left(\alpha + \frac{\rho_l}{\rho_g}(1-\alpha)\right)^2} \approx \frac{\rho_l}{\rho_g}\frac{a_g^2}{\alpha^2(1-\alpha)}. \tag{5.281}$$

It is noteworthy here that the velocity of the density wave is much lower than the speed of sound in the gas and assumes the smallest value for $\alpha \approx 0.5$. This behavior has been well confirmed experimentally in air–water flows. Surprising here is the low value of the wave velocity compared to the sound velocity of the gas. It is in the minimum below 10% of the sound speed of the gas. This has the consequence that the mass flow limitation at geometrical constrictions of flow conduits already occurs at very low two-phase mass flux densities. This is technically extremely significant in connection with the outflow of gas–liquid mixtures from of gas–liquid mixtures from pressure vessels.

Analytical models for the calculation of critical mass flux can be roughly classified according to whether thermodynamic and mechanical equilibrium between the phases is assumed or not. Outflow experiments show that, in general, neither thermodynamic nor mechanical equilibrium is established locally in an outflow process. During flash evaporation, temperature differences develop between the phases, which are not balanced by heat transfer at the interfaces during the short overflow process. At the same time a differential velocity between the phases.

Long pipe connections between pressure reservoirs, on the other hand, lead to the reattachment of the free jet to the wall after a certain distance, and independently of this, after a certain relaxation run length, to the setting of the thermodynamic equilibrium with decisive steam formation rates. The noticeable increase of the vapor content in the flow then results in a significant two-phase friction and acceleration pressure drop.

The two situations shown can each be described in rough approximation by a one-dimensional homogeneous two-phase flow in thermodynamically fully inhibited or in full equilibrium. The two limiting cases of complete equilibrium and completely inhibited equilibrium are captured by simplifying Eqs. (5.279) and (5.280) for the critical propagation speed of a homogeneous two-phase flow. Assuming that $a_g^2 \ll a_l^2$ holds for the sound velocities in the vapor and liquid phases, the simplified expression is obtained for the critical and maximum mass flow densities, respectively:

$$\dot{m}^*_{crit} = \frac{1}{\sqrt{\frac{\dot{x}}{\rho_g a_g^2} - \left(\frac{1}{\rho_g} - \frac{1}{\rho_l}\right) \cdot \left(\frac{\partial \dot{x}}{\partial p}\right)_s}}, \tag{5.282}$$

in which the outflow at fully inhibited thermal equilibrium is characterized by $(\partial \dot{x}/\partial p)_s = 0$. The linkage of the states in the narrowest flow cross-section with the rest values in the overpressure reservoir is done via the equation of state of the corresponding gas–liquid mixture, for example water vapor state tables.

In summary, the thermodynamic and mechanical equilibrium assumptions made in the simple models are too broad for a quantitative comparison of experiment and model. In order to capture the actual conditions when two-phase mixtures flow through nozzles, orifices or pipe nozzles under strong pressure relief, the local and temporal deviation from thermodynamic equilibrium and the mechanical interaction between the phases must be considered in the model, here the devoted reader is referred to expert literature.

In fluid flows of high velocity, a reduction in pressure when flowing around bodies and corners can lead to localized vapor or gas formation. This phenomenon is called cavitation. It is localized flash evaporation followed by condensation, or the release of dissolved foreign gases from the fluid due to pressure relief. They sometimes occur in hydraulic fluid machinery such as pumps and turbines and other hydraulic components such as valves. Undesirable side effects such as deterioration of operating performance, noise generation, mechanical vibration and localized material wear can be observed. The prevention and control of cavitation is therefore of great importance for hydraulic machine and plant engineering. A detailed description of cavitation can be found in Franc and Michel (2006).

Cavitation can start in a single-component flow when the local static pressure in the flow reaches and falls below the thermodynamic saturation vapor pressure p_{sat} of the fluid. This necessary condition for cavitation onset is characterized in a frictionless incompressible flow by a dimensionless cavitation number. The cavitation number is similar in structure to the Euler number and gives a measure of when the fluid cavitates. It expresses the relationship between a local pressure difference caused by a restriction $(p - p_{sat})$ and the kinetic energy per volume of the flow $(\rho u^2/2)$ and is used to characterize energy losses in the flow, where a perfect frictionless flow corresponds to a Euler number of 0.

$$Ca = 2\frac{(p - p_{sat})}{\rho u^2}. \tag{5.283}$$

For values of $Ca \leq 0$, evaporation of the fluid may occur. Due to mechanical and thermodynamic real effects, the "ideal" critical value of the cavitation coefficient $Ca = 0$ for the onset of cavitation, however, shifts to positive or negative coefficients.

The observed manifestations of cavitation are similarly diverse as the flow forms of two-phase flows in pipes and always transient. They are classified according to increasing vapor content as cloud, bubble, layer, and super cavitation. Collectives of bubbles in the micron and submicron range, which can form in strong sheared boundary layers, are called cloud cavitation. Extended coherent bubble flows develop in the negative pressure regions of detached flows behind body edges, on the suction side of adjusted profiles in pumps, or in regions of detached flows. This is referred to as bubble cavitation. The formation of coherent vapor or gas films occurs in partial areas of the profile or body edge due to bubble coalescence when the vapor or gas volume fraction exceeds a critical level of $\alpha \sim 0.5$. This form is called layer cavitation. If the body is surrounded by the vapor film at very high flow velocities and correspondingly low local pressures on the whole or on the largest parts of its contour area, this is called super-cavitation.

In addition to the cavitation number, other parameters influence cavitation; these include geometry, fluid properties as viscosity, surface tension, parameters of the real gas equations, thermal conductivity, heat capacity, latent heat, and concentration of foreign substances in the fluid such as foreign gases or particles. Furthermore, for the onset and the dynamic behaviour of the cavitation process, the thermal micro-processes for the activation of nucleation nuclei and the degree of turbulence of the flow are also important.

5.2.6 Instabilities in Two-Phase Flows

Instabilities can occur in two-phase flows. The considerations in the following refer to heated channels—single or several parallel channels—with increasing steam content in the direction of flow. A pressure difference is imposed on the channels together from the outside. Transients may involve changes in mass flow rate, pressure or pressure drop within the duct, density or changes in flow patterns and consequently heat transfer.

From Fig. 5.44 it could be seen that two-phase flow—except for layered flow—provides better heat transfer than single-phase flow at the same mass flow rate. On the other hand, there is a higher pressure drop and the occurrence of flow instabilities. The instabilities can be aperiodic or periodic in nature. If parallel channels are operated with flow ratios near transitions to instabilities, periodic oscillation can arise from inherently aperiodic processes due to communication between the parallel channels and in interaction with the heat transfer and vapor formation rate.

Aperiodic instabilities arise primarily when several flow states are stable under the same external conditions. Instabilities are triggered by transitions from one flow form to another, resulting in, among other things, a different pressure drop. If such a transition in the boiling tube of a collector results in a reduced steam release rate and heat transfer, the steam utilization is impaired, or the solar efficiency deteriorates due to its temperature dependence. The boiling tube wall can be overheated to such an extent that, for example, the selective coating of the absorber is damaged.

Periodic oscillations can result in vibration excitations, against which the collector and its components must be protected. From the preceding sections and experiments presented later, it is clear that oscillations of the mass flow rate can lead to an improvement of the heat transfer and can be advantageous in the application. The pre-requisite for advantageous behaviour is that the oscillations do not grow undamped to unacceptably high amplitudes.

A general distinction is made between *static* and *dynamic instabilities*, depending on whether it is sufficient to describe the problem with static equations, or to predict the onset or occurrence of the instability.

The simplest flow instability at a phase interface is wave instability, which is caused by the effect of relative velocity between phases, surface tension, and acceleration forces, and is known as Kelvin–Helmholtz and Rayleigh–Taylor instability. These types of instabilities likewise play a central role in the deformation and decay of bubbles, drops, liquid lamellae, and liquid films.

Schematically, Fig. 5.52 depicts an interface in which a gas and a liquid move relative to each other with the phase velocity $u_g - u_l$. As soon as the velocity difference is greater than zero, small wave-like perturbations are fanned in time at a certain magnitude of surface tension and forces acting externally on the interface, e.g., gravity. Here, the fanning rate of the disturbance generally depends on the wavelength.

The perturbation with a wavelength m with the largest amplification rate, leads to the decay of the smooth interface and the formation of a new spatial phase distribution characterized by the intrinsic new length scale λ_m. This length scale also determines the size of droplets and bubbles after the remodelling process. From a linear stability analysis for a frictionless infinitely extended two-layer flow

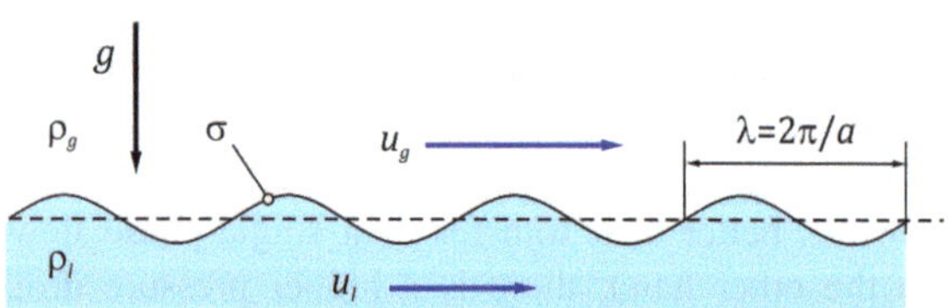

Fig. 5.52 Sketch of the stability of an interface of two stratified fluids with different velocities u_l, u_g

yields an amplification rate c_i of the form:

$$c_i = \frac{1}{a}\sqrt{\frac{\rho_l \rho_g (u_g - u_l)^2 a^2}{\rho_l + \rho_g} - \frac{\sigma\, a^3 - g(\rho_g - \rho_l)a}{\rho_l + \rho_g}}, \tag{5.284}$$

in which ρ_l, ρ_g are the densities, a is the wave number ($\lambda = 2\pi/a$), σ and the surface tension. The acceleration here is due to the gravity g. Any other acceleration with corresponding effect can take its place. The maximum of the amplification rate as a function of the wave number can be determined from $\partial c_i/\partial a = 0$. If there is no velocity difference between the phases ($u_l = u_g$), the maximum amplified perturbation has the wavelength λ_m:

$$\lambda_m = 2\pi\sqrt{\frac{3\,\sigma}{g(\rho_l - \rho_g)}}. \tag{5.285}$$

This assumes that the denser liquid is accelerated in the direction of the less dense one. Otherwise, only a damped oscillation occurs. This is referred to as the Rayleigh–Taylor instability. The cut off wavelength for amplified perturbations is given by a vanishing amplification rate c_i and is:

$$\lambda_c = 2\pi\sqrt{\frac{\sigma}{g(\rho_l - \rho_g)}}. \tag{5.286}$$

This length scale λ_c is called the Laplace length. It can be concluded that the interface is stable against small perturbations with smaller wavelengths and does not undergo any permanent change because the surface tension keeps the accelerating forces in equilibrium. For perturbations with longer wavelengths, the interface deforms permanently. An example of this type of interfacial instability is the prevention of leakage of liquids from containers with sufficiently small orifice openings in thc bottom or thc dccay-frcc risc of gas bubbles of diameter $d_b \leq \lambda_c$ in a liquid.

When significant velocity differences $(u_g - u_l)$ occur between phases and interfacial tensions are present, the amplification of interference is limited downward by the cutoff wavelength:

$$\lambda_c = 2\pi\left(\frac{\sigma}{\rho_g (u_g - u_l)^2}\right). \tag{5.287}$$

Here, it is assumed that the densities of the phases are very different as in gas and liquid under normal conditions, i.e. $\rho_g \ll \rho_l$. The perturbation with the strongest amplification has the wavelength $\lambda_m = 1.5\lambda_c$. For two-phase flows, the consequence of this capillary instability is that undamped capillary waves due to shear velocities at the phase interface cause the interfaces to break up. For example, large droplets break up into smaller components when the droplet dimension is larger than the cut-off wavelength λ_c. This instability of interfaces against shear motion is called Kelvin–Helmholtz instability. This fact also explains the Weber number criterion for the decay of liquid jets and liquid drops, which states that a decay of moving liquid volumes with free surfaces occurs when the Weber number We_L formed with the characteristic length L of the volume exceeds unity, i.e.

$$We_L = \frac{\rho_g \left(u_g - u_l\right) L}{\sigma} > 1. \tag{5.288}$$

The two instabilities presented are of great influence for bubble and droplet formation in two-phase flows and explain some important phenomena in boiling and condensation processes, for example, the collapse of bubble boiling and film boiling can be explained as Rayleigh–Taylor instabilities. The formation of wavy layer flows of homogeneous bubble and droplet flows is directly related to the Kelvin–Helmholtz instability, which can be described in terms of a Weber number criterion.

Dynamic instabilities—Ledinegg-instability

In addition to the small-scale interfacial instabilities that have an important function for the phase distribution, there are other fundamental large-scale instability mechanisms that determine the timing of two-phase flow in hydraulic systems with phase transitions, such as those involving boiling, condensation, and cavitation processes. Since such instabilities can lead to uncontrolled mechanical pressure and shock loads and further to thermally triggered stresses in channel walls of the system, the stability limit for such processes is practically a design and operational criterion for such systems as steam generators and fluid machinery. Typical two-phase instability can occur in a system consisting of two pressure reservoirs and two hydraulic components connected in series in between, for example a pump and an evaporator tube with constant heat supply, as shown in Fig. 5.53.

The stability behaviour of the system is determined by the different pressure-mass flow dependencies of the pump and the evaporator tube. While the pressure build-up (Δp) of a pump typically decreases monotonically with increasing mass flow ($\dot{m}$), the two-phase pressure drop in the evaporator tube shows a non-monotonic behaviour in an average mass flow range, which is mainly caused by the different contributions of friction and acceleration components to the total pressure drop of the two-phase flow in the evaporator tube (compare Sect. 5.2.4.6 Eq. (5.262)). The monotonic branches of the pressure loss curve are determined by the high vapor fraction at low mass fluxes and by the high liquid volume fraction at high mass fluxes. In general, therefore, the system can assume three steady-state operating conditions with a change in mass flow, given by the intersections A, B, C of the two pressure-mass curves for pump and evaporator tube. Assuming a small variation of the quantities $\dot{m}$ or p in the vicinity of the steady states, a simple consideration shows that states A and C are stable, while a pressure and mass flow change in the vicinity of B due to a power change of the pump causes a pressure loss change with exactly the opposite sign. This leads to a transition to one

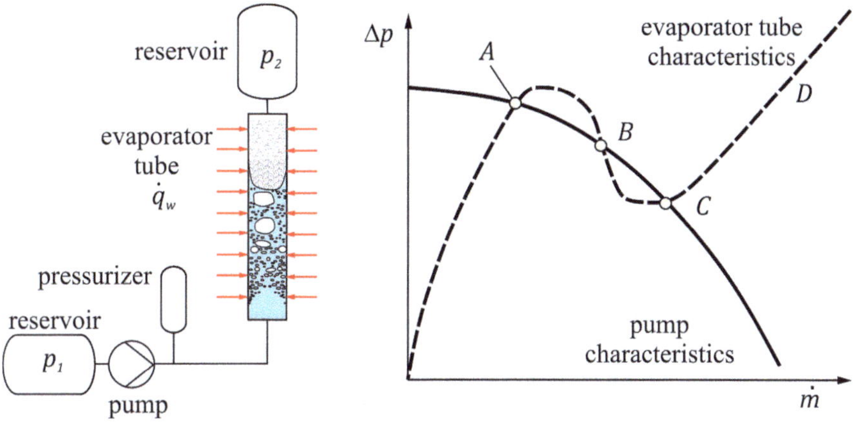

Fig. 5.53 Two-phase system with possible (Ledinegg) instability, left: Flow system, right: pressure-mass flow characteristics of the pump and the evaporator tube

of the two stable operating states A or C. The simple considerations show that for the stability of an operating point the condition $\partial \Delta p / \partial \dot{m} > 0$ is necessary, with Δp as the driving system pressure through the pump. Since this time behaviour triggered by the instability is given by a simple transient from an unstable to a stationary stable state, the system is said to be statically stable or unstable.

Ledinegg (1938) was the first to investigate this instability, which is named after him. According to Ledinegg, an aperiodic instability can occur in a flow channel with a constant heat flux ($\dot{q}_w$) when a subcooled fluid enters and is partially evaporated. In contrast to before, the occurrence of the instability requires that the static pressure drop increases with decreasing mass flow rate $\dot{m}$, that is, when $\partial \Delta p / \partial \dot{m} < 0$. The frictional pressure drop in turbulent flow is proportional to the square of the mass flow rate, i.e., $\Delta p \sim \dot{m}^2$. It dominates at large mass fluxes or small heat flux rates and accordingly affects the total pressure drop. In a certain range, the acceleration pressure drop can increase with increasing heat output or decreasing flow rate in such a way that a transition to a new stable condition with smaller total mass flux rate and large vapor volume fraction at the channel outlet occurs, resulting in the operating point A.

The influence of subcooling, given by the enthalpy difference $(h' - h_i)$, where h_i is the inlet enthalpy, is shown in Fig. 5.54. Instability arises with decreasing mass flow density and exceeding of a critical subcooling of the liquid at the channel inlet. In the case of several parallel channels and an externally imposed equal pressure drop for all channels, e.g. in steam boilers or in solar collectors, steam blockages with strongly reduced heat removal can develop.

A similar phenomenon, based on the same instability, can also be observed in nature geysers. Heat is supplied to a cavern filled with supercooled water and connected to a higher water reservoir via a slender channel. The water heats to saturation temperature and begins to boil. The steam that forms leaves the cavern in

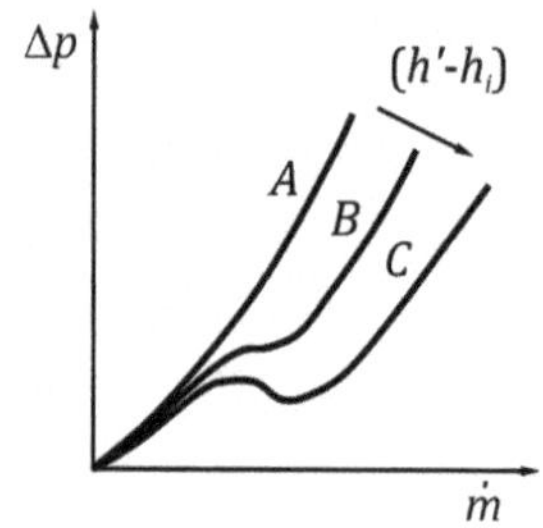

Fig. 5.54 Instability of the two-phase flow in the heated channel with increasing inlet subcooling

the form of a two-phase mixture via the connecting channel. Due to the increasing steam content, the hydrostatic pressure in the cavern is reduced. This leads to further increased evaporation and to further water discharge from the cavern. As the water deficit increases, steam generation is reduced and supercooled water can enter the cavern from the upper reservoir against a reduced steam flow and bring the boiling process to a complete standstill. As the heat source heats the cavern again the penetrated water to saturation temperature the process repeats itself and with constant external conditions (heat supply in the cavern, heat removal from the reservoir) a periodic process with typical relaxation character is established.

The Ledinegg criterion can be used to estimate whether aperiodic instabilities of the type described above occur during the evaporation of a liquid that enters the evaporator tube subcooled. The criterion considers the friction and acceleration pressure loss. The coefficient of friction $f_{F,2ph}$ is applied according to a homogeneous model of two-phase flow and heat flux density ($\dot{q}_w$) at the tube wall is assumed to be stationary as well as constant along the entire tube length. This provides the enthalpy difference $(\Delta h - h_i)$ by which the liquid may be at most subcooled at the pipe inlet:

$$(\Delta h - h_i) \leq \frac{K_1 \cdot K_2}{1 + K_1 - \sqrt{\frac{3}{2} K_1 \cdot \left(\frac{K_1}{2} + 1\right)}} \text{with } K_1 = \frac{f_{F,2ph} L}{2d} \text{ und } K_2 = h_r \frac{\rho_g}{\rho_l}, \tag{5.288}$$

where L is the tube length, d is the tube diameter, ρ_g and ρ_l are the vapor density and fluid density, respectively, and Δh is the enthalpy of vaporization. Then, instability is avoided by decreasing subcooling or with increasing friction pressure loss, for example, by using an orifice plate at the pipe inlet. Pederson and May (1982) describe a formula that treats the friction pressure loss divided by inlet or outlet zones as well as the pipe section. They assume the simplified relationship for total pressure loss. The studies were specifically set up for solar systems with parabolic trough collectors.

Dynamic instabilities—density wave instability

Another system-wide two-phase instability, which takes the form of intense oscillatory fluctuations in the gas volume fraction, is the so-called density-wave

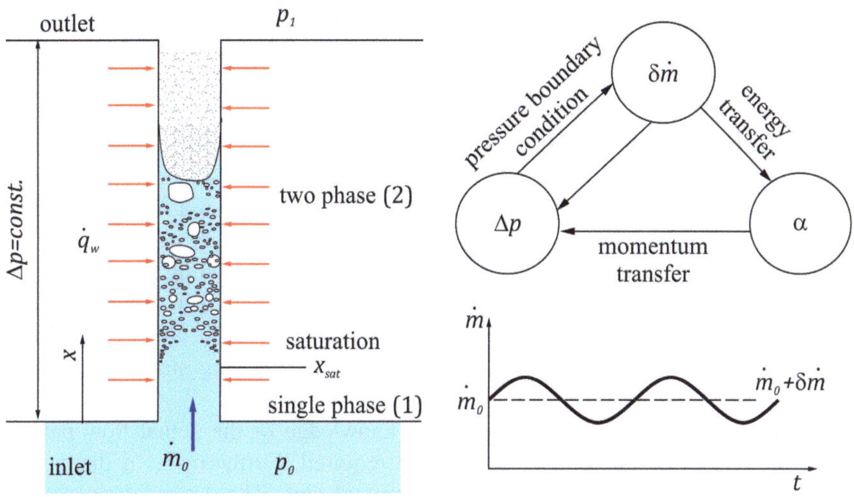

Fig. 5.55 Principle diagram for the formation of the density-wave instability (left) and feedback effects occurring there (top right) according to Müller (2016)

instability. The instability is common in systems with phase transitions and is due to a feedback of mass flow, vapor formation rate or condensation rate, and pressure drop in the boiling section or condensation section. This instability can be described analytically for a piping system with constant heat supply and removal using the one-dimensional two-fluid model and a linear stability analysis. Here, only the mechanism will be described using a simple evaporator tube tied between two pressure reservoirs with constant pressures p_0 and p_1, compare Fig. 5.55.

If the evaporator tube is subjected to a small periodic perturbation of the mass flow $\partial\dot{m}$ at the inlet, the position of the saturation temperature and thus the limit of the onset of evaporation inside the tube follows this fluctuation, because mass flow oscillations in the range of single-phase flows include enthalpy fluctuations. Mass flow variations and changes in the path of the single-phase flow in the pipe result in pressure variations $\partial\Delta p_1$ in the region of the single-phase flow. In the two-phase region, enthalpy disturbance also acts as a disturbance in the vapor volume fraction α, which propagates as a density wave in flow direction. The change in vapor content in conjunction with the mass flow and length perturbation leads to increased pressure perturbations $\partial\Delta p_2$ in the two-phase region. Since the total pressure difference is imposed on the evaporator tube by the system, the individual pressure fluctuations $\partial\Delta p_1$ and $\partial\Delta p_2$ must balance each other. This implies a feedback between the two-phase and single-phase flow domains and leads to a resonant amplification of the initial small disturbance if the disturbances are in phase. The result is massive fluctuations in the extension of the boiling domain and the vapor content of the evaporator section. This should be avoided for reasons of increased thermal wall loads in technical systems. The specification of limits for density wave instability is therefore important for the design of tubular steam generators. The tendency to unstable behaviour increases when several evaporator channels are connected in parallel.

5.2.7 Phase Change Heat Transfer

The calculation of heat transfer in evaporator tubes depends on a large number of influencing variables. These are mainly the mass flow density $\dot{m}$, the heat flux density $\dot{q}_w$, the fraction of the vapor mass flow $\dot{x}$, the boiling pressure p_{sat}, and the material properties of the liquid and gaseous phases of the heat transfer medium. Due to the occurrence of a two-phase flow, there are different flow pattern described earlier in this chapter with different wetting states of the tube wall.

To calculate the heat transfer, therefore, knowledge of the actual flow pattern in the pipe segment under consideration is required. Furthermore, a distinction must be made between a vertical and a horizontal pipe. The determination of the flow pattern by means of flow pattern maps is useful for manual calculations for estimations, but in the simulation the transitions of the flow forms must be brought into functional dependencies. In complex models, this leads to an accumulation of a large number of equations which is not very descriptive. For this reason, the description of the calculation of the heat transfer using the flow form maps is omitted here.

For heat transfer (except for subcooled boiling) in horizontal tubes, Streicher (1991) uses a subdivision of flow forms as shown in Table 5.17 to calculate heat transfer in heat pump circuits. The heat transfer coefficient for the different ranges is calculated as a mixture of the heat transfer for the saturated single-phase flow h_l and the heat transfer coefficient for bubble boiling h_b. For the bubble flow region (index I), the average heat transfer coefficient h_I, at $\dot{x} > 0$, the following relationship is assumed:

$$h_I = \sqrt{\left(h_l^2 + h_b^2\right)} \text{ with } Nu_l = \frac{h_l \cdot d_{hg}}{\lambda_l} = 0.023 \cdot \mathrm{Re}_l^{0.8} \cdot Pr_l^{0.4}, \tag{5.290}$$

where the Reynolds (Re_l) and Prandtl numbers (Pr_l) for the liquid are evaluated in the following way:

$$Re_l = \frac{u_l \cdot d_{hg} \cdot \rho_l}{\mu_l}; Pr_l = \frac{\mu_l \cdot c_{p,l}}{\lambda_l}, \tag{5.291}$$

where the actual mean flow velocity u_l and the modified hydraulic diameter d_h are determined in the following way.

$$u_l = \frac{\dot{m} \cdot (1 - \dot{x}_m)}{\rho_l \cdot (1 - \dot{\alpha}_m)}; d_{hg} = d_h \cdot \sqrt{1 - \dot{\alpha}_m}, \tag{5.292}$$

in which d_h the geometric hydraulic diameter.

Table 5.17 Limits of the fraction of the vapor volume flux $\dot{\alpha}_{lim}$ for the transition from one to another flow pattern according to Streicher (1991)

Froude number Fr	limiting value $\dot{\alpha}_{lim}$	Flow pattern transition type
Arbitrary	$\dot{\alpha}_{lim} = 0.38$	Bubbly→ stratified
$0.06 \leq Fr \leq 0.1$	$\dot{\alpha}_{lim} = 1.09 + 0.16 \ln Fr$	Stratified→ stratified wavy
$Fr \leq 0.06$	$\dot{\alpha}_{lim} = 0.38$	
$Fr \geq 0.1$	$\dot{\alpha}_{lim} = 0.52 + 0.085 \ln Fr$	Stratified→ plug-annular
$Fr \leq 0.1$	$\dot{\alpha}_{lim} = 0.60 + 0.087 \ln Fr$	Stratified wavy→ annular
$Fr \geq 0.1$	$\dot{\alpha}_{lim} = 0.80$	Stratified→ annular

The fraction of the steam volume flux $\dot{\alpha}$ is obtained by:

$$\dot{\alpha} = \beta - 2.55 \cdot \beta \cdot (1 - \beta)^{0.36} \cdot (Fr \cdot Ga)^{-0.11} \text{ with } \beta = \frac{1}{1 + \left(\frac{1-\dot{x}}{\dot{x}}\right) \cdot \frac{\rho_g}{\rho_l}}, \tag{5.293}$$

in which the Galilei number Ga appears, which is defined in the following way

$$Ga = \frac{g \cdot \lambda_c^3 \cdot \rho_l^2}{\mu_l^2}. \tag{5.294}$$

In the Galilei number appears the Laplace's constant λ_c (see Eq. (5.286)), which is a ratio of surface tension forces versus gravity and density differences. The heat transfer for bubble boiling h_b near ambient pressure is obtained by an empirical numerical equation of the type:

$$h_b = 1.95 \cdot \dot{q}_w^{0.72} \cdot p^{0.24}, \tag{5.295}$$

where the heat transfer coefficient h_b in $(\mathrm{W}/(\mathrm{m}^2\mathrm{K}))$, the heat flux $\dot{q}_w$ in $(\mathrm{W}/\mathrm{m}^2)$ and the pressure p in (bar) are to be used. With a range of validity of: $10^4 < \dot{q}_w\left(\mathrm{W}/\mathrm{m}^2\right) < 10^6$ and $0.5 < p\,(\mathrm{bar}) < 20$, the previous equation is well suited for transfer to a solar absorber tube, but other relationships are required for power plant applications. In the range of wave flow (index II) and layer flow (index III), for the calculation of the heat transfer coefficient $h_{II} = h_{II}I$ the relation is valid:

$$h_{II} = h_{III} = 0.58 \cdot \dot{x}_m^{-0.194} \cdot \sqrt{h_l^2 + h_b^2}, \tag{5.296}$$

where $\dot{x}_m$ represents the average vapor coefficient in the considered area. For the heat transfer coefficient in the area of the wavy annular flow (index IV) applies:

$$h_{IV} = \sqrt{h_l^2 + h_b^2}, \tag{5.297}$$

while in the case of annular flow (index V) the heat transfer coefficient h_V is calculated from

$$h_V = \sqrt{h_l^2 + \left(1 + 36.5 \times 10^{-9} \cdot K_D^{1.5}\right) \cdot h_b^2};\ K_D = \frac{\dot{m} \cdot \Delta h_g \cdot \dot{x}_m}{\dot{q}_w \cdot \dot{\alpha}_m \cdot \frac{\rho_g}{\rho_l}}. \qquad (5.298)$$

Δh_gin Eq. (5.298) denotes the evaporation enthalpy. With the equations it is possible to determine the heat transfer coefficient from the wall to the fluid with sufficient accuracy for the boundary conditions occurring in a thermal solar collector. However, intermittent flow forms such as slug flow or plug flow are not considered. For these, separate flow stability considerations may have to be made, the fundamentals of which can be found, for example, in Heinzel (1998).

5.2.8 Condensation in Tubes and Pipes

In a solar collector operated with water or water–glycol mixtures, or in direct evaporation solar farm concepts, steam generation intrinsically occurs in the collector absorber. The potential range of this steam in the piping network essentially determines the control of the system, its temporal availability as well as the dimensioning of the components of the solar circuit. In contrast to solar thermal power plants, the formation of steam in solar thermal low-temperature systems is undesirable.

In case of a collector stagnation, the generated steam is pressed from the collector into the connecting pipes. Since the temperature there is lower, it can condense there. The steam production capacity is also determined, among other things, by the thermal losses of the piping, so that it is necessary to consider the heat transport from the inside of the steam-filled piping towards the surroundings. This requires a consideration of the condensation processes in the piping. In addition, the amount of residual liquid in the collector, the fraction of steam and liquid in the individual pipe sections must be known. A detailed consideration of condensation in inclined pipes, as occurs in real solar thermal systems, is omitted here and only the basic physical processes in vertical or horizontal pipes are outlined below. Special effects such as condensate subcooling or steam superheating are not discussed, rather reference is made to the technical literature.

Although in low-temperature thermal solar systems mostly binary mixtures of water and propylene glycol are used as heat transfer medium, most technical considerations assume a pure water vapor flow in the pipes. Experimental investigations show that fractional evaporation of the fluid occurs in early stagnation phases (Rommel et al., 2007). The phase change that takes place in the collector initially allows a water vapor flow to develop in the piping, with the maximum vapor range being reached almost exclusively in this early stagnation phase. Only when the heat transfer medium is highly concentrated in the collector field and the steam production rates are comparatively low does the glycol fraction in the steam flow increase. Another advantage of the calculations of water vapor flows is that the necessary material values for water are known very precisely, in contrast to water-propylene glycol mixtures. The description of condensation within the tube requires knowledge of the flow state. Since flow and temperature fields are hardly accessible analytically even at low complexity of the problem, mostly dimensionless numbers are determined here as well by means of the similarity theory. An illustrative introduction of the indices used in this section can be found in Baehr and Stephan (2008).

5.2.8.1 Steam Velocities at Stagnation

First, the velocity of the steam flow in the stagnation case must be determined. The direct experimental measurement requires a considerable metrological effort. However, if the steam production rate is known, the inlet velocity of the steam from the collector ug into the piping can be estimated.

$$u_g = \frac{P_g}{\Delta h_V \ A_C} = \frac{P_g^* \ A_{col}}{\Delta h_V \ A_C}. \tag{5.298}$$

Therein, the flow cross-section A_C is given by the tube inner diameter d. Both the steam production rate P_g and d depend on the collector aperture area A_{col}. Table 5.18 shows typical required pipe sizes for different collector areas A_{col} and system types (low-flow or high-flow systems). If, for example, the values of a low-flow system $15\,\mathrm{l}/\left(\mathrm{m}^2\mathrm{h}\right)$ are taken as a basis, the steam flow velocities ug and steam mass flow densities as a function of the specific steam production rate P_g^* given in Table 5.19 result at the inlet to the pipe. Table 5.19 shows that the steam velocity at the inlet of the pipeline u_g depends only on the specific steam production power. The resulting steam mass flow, however, is additionally a function of the pipe cross-section d, which in turn depends on the collector area. In addition, it must be noted that in most cases, during stagnation, steam propagation occurs through both the supply and return lines, halving the flow values listed in Table 5.18. The largest steam production rates to date have been measured on an evacuated tube collector at Fraunhofer ISE and were approximately 400 W/m^2 (Rommel et al., 2007).

5.2.8.2 Condensation in a Vertical Pipe

When saturated steam comes into contact with a surface whose temperature is below the saturation temperature of the steam, the steam liquefies. The liquid precipitates on the surface and serves as a source of further condensate. This precipitation can occur in the form of individual drops or as a closed liquid film.

The occurrence of drop condensation is decisively accelerated by impurities on the cold surface, into which the wettability of the cooling surface and the surface tension of the condensate enter. However, droplet condensation plays only a minor role in solar thermal systems.

A simple configuration for phase change is the condensation of a vapor on a cold vertical surface, which is illustrated in Fig. 5.56. The condensate forms a film on the surface driven in a vertical configuration by gravity. At first the film

Table 5.18 Typical dimensioning of the solar circuit piping, in which the dimensions of the pipes were selected as a function of the volume flow so that a comparable flow velocity of the fluid u_g of about $u_g = 0.35$ m/s results (Remmers, 2001)

A_{col} (m²)	d (m) High flow $30\,\mathrm{l/(m^2h)}$	d (m) Low flow $15\,\mathrm{l/(m^2h)}$	d (m) High flow $15\,\mathrm{l/(m^2h)}$
20	0.025	0.016	0.016
30	0.032	0.020	0.020
50	0.039	0.025	0.025
80	0.050	0.032	0.032
130	0.072	0.039	0.032
160	0.072	0.050	0.039
200	0.085	0.050	0.050
250	0.085	0.072	0.050
300	0.103	0.072	0.072
400	0.103	0.085	0.072
u_g (m/s)	0.33	0.37	0.35
A_{col}/A_c	3.95×10^4	8.94×10^4	1.09×10^5

Table 5.19 Specific steam production rate P_g^* (reference area = collector aperture area) of a typical low-flow system $15\,\mathrm{l/(m^2h)}$ and resulting steam flow velocity and steam mass flow rate at the collector outlet

P_g^* (W/m²)	u_g (m/s)	$\dot{M}_g$ (kg/(m²s))
20	0.38	0.8
50	0.96	2.1
100	1.92	4.2
200	3.83	8.3
300	5.75	12.5
400	7.66	16.7

is laminar near the top and relatively thin. As the film proceeds downwards its thickness increases continuously vapor condenses at the vapor film interface. If the film exceeds a certain thickness first signs of a transition to a nonlaminar flow are observed referred to a *wavy flow region*, in which a regular sequence ripples are observed. This region is last followed by a turbulent film flow, provided that the cooled wall is long enough.

Figure 5.56b illustrates that even in a laminar film flow interacts with the vapor boundary layer. At the vapor film interface the temperature corresponds to the saturation temperature to the local pressure along the wall T_{sat}, while the saturation temperature is lower than that of the vapor far away the wall (T_∞) but higher than that of the wall T_w. Through the shear stress at the liquid–vapor interface, the falling jet of vapor aids the downward flow of the liquid film. The vapor in the falling jet is colder than the vapor reservoir and warmer than the liquid in the film attached to the wall. The complexity of the two-phase flow gets more

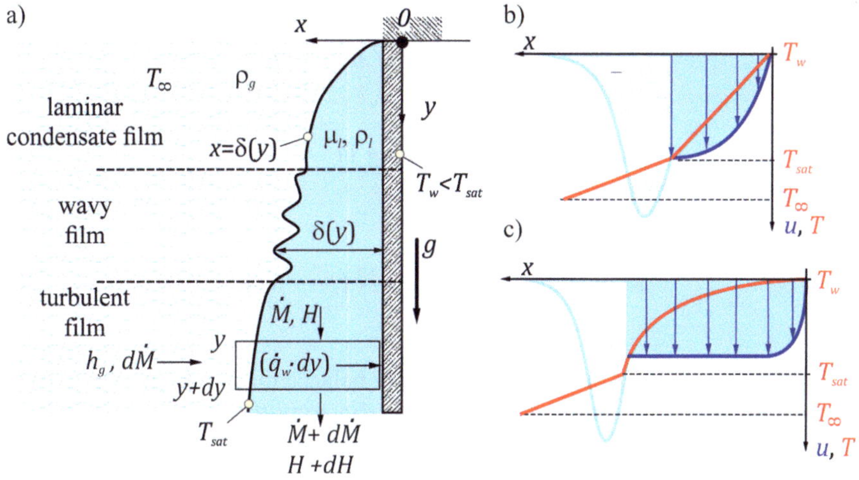

Fig. 5.56 (**a**) Flow regions of film condensation on a vertical cold surface and (**b**) qualitative wall-normal temperature and velocity distribution in laminar flow and (**c**) turbulent flow

sophisticated the wavy and turbulent flow domain (Fig. 5.56c). Moreover, in case of very large films covering all flow regimes, the overall heat transfer rate from the vapor reservoir to the wall is dominated by the contributions from the wavy and turbulent sections.

Laminar film condensation

If vapor condenses on a vertical cold surface, a liquid film forms which flows downward driven by gravity. If the vapor velocity in the vicinity of the condensate is low and the liquid film is thin, the flow in the condensate is laminar. The simplest case of laminar film flow when the wall surface is isothermal is described by Nusselt's water film theory. Heat removal is mainly by conduction from the condensate surface towards the wall and convective heat transfer within the film can be neglected. Assuming a Newtonian fluid, a constant wall temperature T_w, the saturated vapor temperature T_{sat} and a temperature-independent dynamic viscosity μ_l the motion in gravity direction reads to:

$$\underbrace{\rho_l\left(u\frac{\partial u}{\partial x}+v\frac{\partial v}{\partial y}\right)}_{\text{inertia}}=-\frac{dp}{dy}+\mu_l\frac{\partial^2 v}{\partial y^2}+\rho_l g=\underbrace{\mu_l\frac{\partial^2 v}{\partial x^2}}_{\text{friction}}+\underbrace{g(\rho_l-\rho_g)}_{\text{sinking}}, \tag{5.300}$$

where due to the slenderness of the film, the vertical pressure gradient in the liquid is the same as the hydrostatic pressure gradient in the external vapor, i.e. $(dp/dy)=\rho_g g$. Assuming inertia to be negligible compared the equation can be simply integrated twice in wall normal direction x assuming at the wall no-slip condition ($v(x = 0) = 0$) the vertical velocity profile is obtained:

$$v(x, y) = \frac{g}{\mu_l}(\rho_l - \rho_g)\delta^2\left(\frac{x}{\delta} - \frac{1}{2}\left(\frac{x}{\delta}\right)^2\right), \tag{5.301}$$

in which $\delta(y)$ is the film thickness. Integrating in wall normal direction the local mass flow ($\dot{M}$) is obtained.

$$\dot{M} = \int_0^\delta \rho_l v(x, y)dx = \frac{g\rho_l}{3\mu_l}(\rho_l - \rho_g)\delta^3. \tag{5.302}$$

The film thickness is obtained by an energy balance on an infinitesimal control volume ($dy \cdot \delta(y)$) as indicated in Fig. 5.56a. Entering this control volume from the right is the saturated vapor stream $d\dot{M}$, the enthalpy flow rate of which is $(hg \cdot d\dot{M})$. The vertical enthalpy inflow H is given by

$$H = \int_0^\delta \rho_l v\big(h_f - c_{p,l}(T_{sat} - T)\big)dx. \tag{5.303}$$

The specific enthalpy of the film is smaller than the specific enthalpy of saturated liquid (h_f) because the liquid is subcooled ($T < T_{sat}$). Considering that the temperature across the film from the vapor film towards the wall with the temperature T_w changes linearly due to heat conduction the enthalpy H can be related to the mass flow $\dot{M}$.

$$H = \dot{M}\left(h_f - \frac{3}{8}c_{p,l}(T_{sat} - T_w)\right). \tag{5.304}$$

The formation of a condensate mass flow requires the provision of a condensation heat dissipation at the wall ($\dot{q}_w$), which is given in a simple way by heat conduction ($\dot{q}_w = (h_l/\delta)(T_{sat} - T_w)$). Balancing the thermal power in the control volume and using the previous relation one obtains

$$\begin{aligned} 0 &= H - (H + dH) + h_g d\dot{M} - \dot{q}_w dy \\ &= \underbrace{\left(h_{fg} + \frac{3}{8}c_{p,l}(T_{sat} - T_w)\right)}_{h_{fg'}} d\dot{M} - \frac{h_l}{\delta}(T_{sat} - T_w), \end{aligned} \tag{5.305}$$

where the latent heat h_{fg} and a sensible-heat contribution accounting for the cooling of the fresh condensate to temperatures below forms a modified *augmented* latent heat of condensation h'_{fg}. Combining Eq. (5.302) with Eq. (5.305) and integrating from $x = 0$ to $x = \delta$ yields the laminar film thickness $\delta(y)$ as a function y:

$$\delta(y) = \left(\frac{4\, h_l \mu_l}{\rho_l(\rho_l - \rho_g) g} \frac{(T_{sat} - T_w)}{h'_{fg}} \cdot \right)^{\frac{1}{4}}. \tag{5.306}$$

Thus, the laminar film thickness increases with the power ¼ of the vertical length L. Since the local film heat transfer coefficient is via Fourier's law coupled to the film thickness and reads to:

$$h(y) = \frac{\dot{q}_w}{(T_{sat} - T_w)} = \frac{h_l}{\delta(y)} = \left(\frac{\rho_l(\rho_l - \rho_g) g\, h_l^3}{4\mu_l} \frac{h'_{fg}}{(T_{sat} - T_w)} \cdot \frac{1}{y} \right)^{\frac{1}{4}}. \tag{5.307}$$

This allows the average heat transfer coefficient $\overline{h}_L$ in the film of length L to be determined and also the average Nusselt number

$$\overline{h}_L = \frac{4}{3} h_{y=L};\ \overline{Nu_L} = \frac{\overline{h}_L L}{h_l} = 0.943 \left(\frac{\rho_l(\rho_l - \rho_g) g}{h\, \mu_l} \frac{h'_{fg} L^3}{(T_{sat} - T_w)} \right)^{\frac{1}{4}}. \tag{5.308}$$

For engineering evaluations, the liquid properties are best evaluated at the average film temperature $(T_{sat} + T_w)/2$. The condensation heat transfer coefficient may also be written in terms of the film Reynolds number ($Re_y = 4\dot{M}/\mu_l$), with $\dot{M}$ from Eq. (5.302) which is then

$$\frac{h_y}{h_l} \left(\frac{\mu_l^2}{\rho_l(\rho_l - \rho_g) g} \right) = 1.1 Re_y^{-1/3};\ \frac{\overline{h}_L}{h_l} \left(\frac{\mu_l^2}{\rho_l(\rho_l - \rho_g) g} \right) = 1.47 Re_L^{-1/3}. \tag{5.309}$$

Rohsenow (1998) showed that if the condensate temperature profile is allowed to be nonlinear to account for convection effects in the condensate film, the augmented latent heat expression requires a modification, for which the following expression has been shown to be successful $h'_{fg} = h_{fg} + 0.68 c_{p,l}(T_{sat} - T_w)$.

Laminar film condensation with wave formation

As the running length of the condensation film increases, its thickness and flow velocity increase. Even though laminar flow is still present within the film, waves form at the film-vapor interface which enhance the heat transfer, which is then about 10–25% higher than that of Nusselt's water skin theory. There are different models that take this wave formation into account when determining the heat transfer, which can be specified using a critical Reynolds number.

$$Re_{crit} = 0.392 \left[\left(\frac{\sigma}{\rho_l \cdot g} \right)^{\frac{1}{2}} \left(\frac{g}{\nu_l^2} \right)^{\frac{1}{3}} \right]^{\frac{3}{4}}, \tag{5.310}$$

where ν_l is the kinematic viscosity of the condensate and σ is its surface tension. According to Rohsenow, surface waves can be observed in the Reynolds number range of $256\,Pr^{0.44} \leq Re \leq 400$. Where Pr is the molecular Prandtl number. Within these run lengths, heat transfer is improved over laminar flow and textbooks suggest a multiplication factor of 1.15 over Eq. (5.308) for the average heat transfer coefficient in this range. Somewhat more accurately, the influence of film flow ripple can be accounted for by a Reynolds number dependent ripple correction factor f_{rib}. Here approaches of different authors exist, which do not differ very much; mostly the following relation is found

$$f_{rib} = \begin{cases} 1 & \text{for } Re < 1 \\ Re^{0.04} & \text{for } Re \geq 1 \end{cases} \tag{5.311}$$

Turbulent film condensation

As the film length increases, the flow changes into a turbulent condensate flow. There are different specifications as to the Reynolds number from which the transition occurs. In addition to the Reynolds number, the molecular Prandtl number is also included in the heat transfer relationships. The following relationship is commonly used:

$$\begin{aligned} Nu &= \frac{h\left(\frac{\nu_l^2}{g}\right)^{\frac{1}{3}}}{\lambda_l} = 0.0325\,Re^{1/4}\sqrt{Pr}\,; \\ Re &= \left(89 + 0.024\sqrt{Pr}\left(\frac{Pr}{Pr_w}\right)^{\frac{1}{4}}(f - 2300)\right); \\ f &= \frac{c_{p,l}(T_{sat} - T_w)}{h_{fg}}\,\frac{1}{Pr}\,\frac{y}{\left(\nu_l^2/g\right)^{\frac{1}{3}}} \end{aligned} \tag{5.312}$$

where $c_{p,l}$ is the heat capacity of the condensate. The approach is valid for $1 \leq Pr \leq 25$ and $400 \leq Re \leq 7 \times 10^5$. The material values are to be used at saturation temperature and Pr_w is to be formed with the wall temperature T_w. For practical purposes, the heat transfer in each region of condensate flow on a vertical surface is summarized as follows:

$$h = \sqrt[4]{(f_{rib} \cdot h_{lam})^4 + h_{turb}^4}, \tag{5.313}$$

f_{rib} is calculated according to Eq. (5.311), the heat transfer coefficient of laminar film flow according to (5.307) and that of turbulent film flow according to (5.312).

Influence of the steam flow velocity

The previous considerations include condensation of steam with low flow velocities or steam at rest. However, during the collector stagnation significant steam flow velocities occur, this also affects the condenser of a Rankine cycle. The influence of a steam flow on the condensate film at a vertical steam-water film interface can be neglected up to a steam velocity of approximately 5 m/s. The condensate film flow then depends on the steam velocity. The condensate film flow then depends only on gravity and the heat transfer relationships mentioned for steam at rest apply.

Only at very high steam production rates do considerable flow velocities occur, in which the flowing steam exerts a shear stress τ_δ on the condensate film, which depends on the drag coefficient f:

$$\tau_\delta = f \cdot \frac{\rho_g \cdot u_g^2}{4}, \tag{5.314}$$

For a laminar steam flow of steam ($Re < 2320$), the drag coefficient f_D is independent of the pipe wall roughness and $f_D = 64/Re$ applies (compare Eq. (5.49)). In the turbulent regime, the steam flow also behaves like a conventional Newtonian fluid, so that with respect to the pressure drop, the relationships worked out in Sect. 5.1.16 apply. The shear stress τ_δ affects the mean condensate flow velocity of the film uf and the condensate mass flow rate draining through a plane perpendicular to the wall. The following relationships are obtained for both:

$$\begin{aligned} u_0 &= \frac{(\rho_l - \rho_g)g}{3\,\mu_l}\,\delta^2 \pm \frac{\tau_\delta\,\delta}{2\,\eta_L}, \\ \dot{M}_l &= u_0\,\rho_l\,b\,\delta = \frac{(\rho_l - \rho_g)g\,b}{3\,\mu_l}\,\delta^3 \pm \frac{\rho_l\,\tau_\delta\,\delta^2\,b}{2\,\mu_l}, \end{aligned} \tag{5.315}$$

in which b is the wetted depth, i.e. for a tube $b = \pi d$. The positive sign applies to downward flowing steam, the negative sign accordingly to upward flowing steam. In the laminar film flow region, the heat transfer results directly from the film thickness via the relationship $h = \lambda_l/\delta$.

5.2.8.3 Condensation in the Horizontal Pipe

Whereas in vertical pipes the condensate flows off the wall as a film, in a horizontal pipe—depending on the flow velocity of the steam and the proportion of liquid in the pipe cross-section—various flow forms can occur which are similar to those during heating (compare Sect. 5.2.4.2 Horizontal pipes). At low flow velocities, the influence of gravity predominates and, depending on the fluid content, a stratified, slug or plug flow is formed. At high flow velocities, annular or even bubbly flow establishes, which can also occur at high steam velocities in vertical pipes if the steam shear stress is large compared to the drag effect of gravity.

A stratified flow is formed if

$$u_g = \frac{x\,\dot{M}}{A\left[g\,d\,\rho_g\,(\rho_l - \rho_g)\right]} \leq 1, \tag{5.316}$$

where x represents the vapor content. During stagnation, steam velocities of about 2 m/s typically occur at the inlet to the piping. Assuming a pipe cross-section of $d = 0.016$ m, Eq. (5.316) yields a value of $u_g = 0.233$ m/s and a Reynolds number of $Re = 4610$, so that mostly a stratified flow is obtained. In the calculations, it should be noted that the flow velocity decreases more and more with the rising flow length within the tube. Therefore, the following consideration is limited only to this type of two-phase flow in horizontal pipes.

In stratified flow, the condensate flows downward from the pipe wall and collects at the bottom of the pipe. Considering the pressure drop in the flow direction of the steam, the condensate fills the entire cross-section at the end of long tubes. The heat transfer coefficient h in this type of flow is largely dependent on the thickness of the bottom condensate and is calculated according to:

$$h = F(\gamma)\left(\frac{\rho_l\,(\rho_l - \rho_g) g\, h_{fg}\,\lambda_l^3}{d\,\mu_l\,(T_{sat} - T_w)}\right)^{\frac{1}{4}}, \tag{5.317}$$

in which h_{fg} is the latent heat. The thinner the soil condensate film, the greater the factor F and the heat transfer to the tube wall improves. Table values exist for F. The wetting angle γ can hardly be determined by measurement. Baehr and Stephan (2008) propose empirical equations for the calculation of the heat transfer coefficients, which consider the influence of the steam flow on the turbulent condensate film. For the local heat transfer, they give the following correlation:

$$Nu = 0.023\,Re^{0.8}\,Pr^{0.4}\left[(1-x)^{0.8} + \frac{3.8\,(1-x)^{0.04}\,x^{0.76}}{p^{0.38}}\right];$$

$$Nu = \frac{h\,d}{\lambda_l};\ Re = \frac{u_m\,d}{\nu_L};\ u_m = \frac{4\,\dot{M}}{\rho_L \pi\, d^2};\ Pr = \frac{\nu_l}{a_l}\ \text{and}\ p^+ = \frac{p}{p_{crit}}. \tag{5.318}$$

The critical pressure p_{crit} of water is 22.12 MPa. However, there is disagreement in the literature about the range of validity of this relation. Shah and Sekulic (2003) quote a validity of the calculation method for steam velocities from 3 to 300 m/s and mass flow densities between $10.8 \le \dot{M}\left(\text{kg}/\left(\text{m}^2\text{s}\right)\right) \le 210.6$, while Collier and Thome (1996) limit the validity range to mass flow densities above $\dot{M}\left(\text{kg}/\left(\text{m}^2\text{s}\right)\right) > 200$. For slow vapor flows, the latter recommend a different relationship:

$$Nu = C\,Re^n\,Pr^{1/3}\ \text{with}\ Re_e = \frac{\dot{M}_e\,d}{\mu_l}\ \text{and}\ \dot{M}_e = \left[(1-x) + x\left(\frac{\rho_l}{\rho_g}\right)^{1/2}\right]. \tag{5.319}$$

The quantities C and n are calculated depending on the flow pattern. For $Re_e < 5 \times 10^4$, $C = 0.0265$ and $n = 0.8$, and for $Re_e > 5 \times 10^4$, $C = 5.03$ and $n = 1/3$. Since the steam production rates escaping at stagnation lead to steam flow velocities in the range of 2 m/s and mass flow densities of $4\ (\mathrm{kg}/(\mathrm{m}^2\mathrm{s}))$ for most house plants, the determination method according to Eq. (5.319) is more suitable for these cases.

References

Aicher, T., & Martin, H. (1997, October). New correlations for mixed turbulent natural and forced convection heat transfer in vertical tubes (Elsevier, Ed.). *International Journal of Heat and Mass Transfer, 40*(15), 3617–3626.

Anderson, D. A., Tannehill, J. C., & Pletcher, R. H. (1984). *Computational fluid mechanics and heat transfer*. Hemisphere Publishing Co.

Arpaci, V. S., & Larsen, P. S. (1984). *Convection heat transfer*. Prentice Hall Inc.

Baehr, H. D., & Stephan, K. (2008). *Wärme- und Stoffübertragung* (6. Auflage ed.). Springer.

Baker, O. (1954, July). Design of pipe lines for simultaneous flow of oil and gas. *Oil and Gas Journal, 53*, 185–195.

Bankoff, S. G. (1957). Ebullition from solid surfaces in the presence of a pre-existing gaseous phase. *Transactions ASME, 79*, 735ff.

Barenblatt, G. I., & Chorin, A. J. (1998). Turbulence: An old challenge and new perspectives. *Meccanica, 33*(5), 445–468.

Bejan, A. (2013). *Convection heat transfer* (4th ed.). Wiley.

Bejan, A., & Kraus, A. (2003). *Heat transfer handbook*. Wiley.

Betchov, R., & Criminale, O. W. (1967). *Stability of parallel flows*. Academic Press.

Bhatti, M. S. (1985). *Fully developed temperature distribution in a circular duct tube with uniform wall temperature*. Owens-Corning Fiberglas Cooperation.

Boger, D., & Yeow, Y. (1988). *Ullmann's encyclopedia of industrial chemistry-chapter fluid mechanics* (Vol. 1). VCH Verlagsgesellschaft mbH.

Burmeister, L. C. (1993). *Convective heat transfer* (2nd ed.). Wiley-Interscience.

Carteciano, L. N., & Grötzbach, G. (2003). *Validation of turbulence models in the computer code FLUTAN for a free hot sodium jet in different buoyancy flow regimes*. Forschungszentrum Technik und Umwelt.

Castaing, B., Gunaratne, G., Heslot, F., Kadanoff, L., Libchaber, A., Thomae, S., & Zanetti, G. (1989). Scaling of hard thermal turbulence in Raleigh-Benard convection. *Journal of Fluid Mechanics, 204*, 1ff.

Cebeci, T., & Bradshaw, P. (1977). *Momentum transfer in boundary layers*. McGraw-Hill Inc.

Cengel, Y., & Cimbala, J. M. (2006). *Fluid mechanics fundamentals and applications*. McGraw-Hill Higher Education.

Chen, R. Y. (1973). Flow in the entrance region at low Reynolds number. *Journal of Fluids Engineering, 95*, 153–158.

Chenoweth, D. R., & Paolucci, S. (1986). Natural convection in an enclosed vertical air layer with large horizontal temperature differences. *Journal of Fluid Mechanics, 169*, 173–210.

Chien, K. Y. (1982). Prediction of channel and boundary-layer flows with a low-Reynolds-number turbulence mode. *AIAA Journal, 20*(1), 33–38.

Churchill, S., & Ozoe, H. (1973). Correlations for laminar forced convection with uniform heating in flow over a plate and in developing and fully developed flow in a tube. *Journal of Heat Transfer, 18*, 78ff.

Churchill, S., & Usagi, R. (1972). A general expression for the correlation of rates of transfer and other. *AIChE Journal, 18*, 1121–1128.

Cole, J. D. (1968). *Perturbation methods in applied mathematics*. Blaisdell Publishing Co. Ginn and Co.

Collier, J. G., & Thome, J. R. (1996). *Convective boiling and condensation* (3rd ed.). Oxford University Press.

DIN EN ISO 5167. (2004). *Durchflussmessung von Fluiden mit Drosselgeräten in voll durchströmten Leitungen mit Kreisquerschnitt.* Beuth Verlag GmbH.

Donaldson, C. (1972). Calculation of turbulent shear flows for atmospheric and vortex motions. *AIAA Journal, 10*, 4–12.

Eberhart, J. G., & Schnyders, H. (1973). Application of the mechanical stability condition to the prediction of the limit of superheat für normal alkanes, ether and water. *Journal of Physical Chemistry, 77*(23), 2730.

Eckert, E. R., & Drake, R. M. (1972). *Analysis of heat and mass transfer*. Mc Graw-Hill.

Franc, J.-P., & Michel, J.-M. (2006). *Fundamentals of cavitation.* Springer.

Friedel, L. (1979, June). Improved friction drop correlations for horizontal and vertical two-phase pipe flow. *European Two-phase Flow Group Meeting*, Paper E2.

Garbrecht, O. (2017). *Large eddy simulation of three-dimensional mixed convectionon a vertical plate.* Disseration RWTH Aachen, Fakultät Maschinenwesen.

Gebhart, B. (1988). *Buoyancy-induced flows and transport* (1st ed.). Springer.

Germano, M., Piomelli, U., Moin, P., & Cabot, W. H. (1991). A dynamic sub-grid scale eddy viscosity model. *Physics of Fluids A, 3*, 1760–1765.

Gnielinski, V. (1975). Berechnung mittlerer Wärme- und Stoffübergangskoeffizienten an laminar und turbulent überströmten Einzelkörpem mit Hilfe einer einheitlichen Gleichung. *Forsch-Ing Wes, 41*, 145–150.

Grigull, U., & Tratz, H. (1965). Thermischer Einlauf in ausgebildeter laminarer Rohrströmung. *International Journal of Heat and Mass Transfer, 8*, 669–678.

Guo, J., & Julien, P. Y. (2003). Modified log-wake law for turbulent flow in smooth pipes. *Journal of Hydraulic Research, 41*(5), 493–501.

Hanjalic, K., & Launder, B. E. (1972). A Reynolds stress model of turbulence and its application to asymmetric shear flows. *Journal of Fluid Mechanics, 52*, 609–638.

Harlow, F. H., & Nakayama, P. I. (1968). *Transport of turbulence energy decay rate.* Los Alamos Science Lab.

Heinzel, V. (1998). *Solare Prozeßwärmeerzeugung -Siedewasserkollektoren.* Universität.

Heinzel, V., Holzinger, J., & Simon, M. (1994, August). Flow patterns in horizontal and tilted tubes of direct evaporating solar (H. E. Soc., Ed.). In *Proceedings of the ISES Solar World Congress* (Vol. 4, pp. 135–140).

Hewitt, G. F., & Roberts, D. N. (1969). *Studies of two-phase flow patterns by simultaneous x-ray and flash photography.* HMSO. UKAEA.

Hickmann, H. J. (1974). An asymptotic study of the Nusselt-Graetz problem, Part 1: Large x behaviour. *International Journal of Heat and Mass Transfer, 96*, 354–358.

Hsu, C. J. (1968). Exact solution to entry region of Laminar heat transfer with axial conduction. *Chemical Engineering Science, 23*, 457–468.

Incropera, F. P., & DeWitt, D. P. (2001). *Fundamentals of heat and mass transfer* (5th ed.). Wiley.

Ishii, M., & Hibiki, T. (2011). *Thermo-fluid dynamics of two-phase flow* (2nd ed.). Springer.

Jones, W. P., & Launder, B. E. (1972, February). The prediction of laminarization with a two-equation model of turbulence. *International Journal of Heat and Mass Transfer, 15*(2), 301–314.

Kakac, S., Aung, W., & Viskanta, R. (1985). *Natural convection: Fundamentals and applications.* Hemisphere Publishing.

Kakac, S., Shah, R., & Aung, W. (1987). *Handbook of single phase convective heat transfer* (A. W. Publication, Ed.). Wiley.

Kasagi, N., & Ohtsubo, Y. (1992). Direct numerical simulation of low Prandtl number thermal field in a turbulent channel flow. In F. Durst et al. (Eds.), *Turbulent shear flows* (Vol. 8, pp. 97–119). Springer.

Kawamura, H., Abe, H., & Matsuo, Y. (1999, June). DNS of turbulent heat transfer in channel flow with respect to Reynolds and Prandtl number effects. *International Journal of Heat and Fluid Flow, 20*(3), 196–207.

Kays, W. M. (1994). Turbulent Prandtl number. Where are we? *Journal of Heat Transfer, Transactions ASME, 116*, 284–296.

Kek, V., & Müller, U. (1993, July). Low Prandtl number convection in layers heated from below. *International Journal of Heat and Mass Transfer, 36*(11), 2795–2804.

Khan, Z., & Joshi, J. (2015). Comparison of k–ε, RSM and LES models for the prediction of flow pattern in jet loop reactor. *Chemical Engineering Science, 127*, 323–333. https://doi.org/10.1016/j.ces.2015.01.054

Kim, W., & Menon, S. (1995). A new dynamic one-equation subgrid-scale model for large eddy simulation. In *33rd Aerospace Sciences Meeting and Exhibit, Reno NV-USA*.

Kreith, F., & Goswami, D. Y. (2005). *The CRC handbook of mechanical engineering* (2nd ed.). CRC-Press LLC.

Kreith, F., Manglik, R. M., & Bohn, M. S. (2011). *Principles of heat transfer* (7th ed.). Cengage Learning.

Kunii, D., & Levenspiel. (2013). *Fluidization engineering* (2nd ed.) (Brenner, Ed.). Butterworth-Heinemann-Elsevier.

Lamb, H. (1993). *Hydrodynamics* (6th ed.). Cambridge University Press.

Launder, B. E., Reece, G., & Rodi, W. (1975). Progress in the development of a Reynolds-stress turbulence closure. *Journal of Fluid Mechanics, 68*(3), 537–566.

Launder, B. E., & Spalding, D. B. (1974). The numerical computation of turbulent flows. *Computer Methods in Applied Mechanics and Engineering, 3*, 269–289.

Ledinegg, M. (1938). Instability of flow during natural and forced circulation. *Die Wärme, 61*, 8.

Lockhart, R.W., Martinelli, R.C. (1949). *Chemical engineering progress, 45*, pp. 39–48.

Lu, H., Rutland, C. J., & Smith, L. M. (2008). A posteriori tests of one-equation LES modeling of rotating turbulence. *International Journal of Modern Physics C, 19*(12), 1949–1964. https://doi.org/10.1142/S0129183108013394

Mandhane, J., Gregory, G., & Aziz, K. (1974). A flow pattern map for gas–liquid flow in horizontal. *International Journal of Multiphase Flow, 1*, 537–553.

Michelsen, M. L., & Villadsen, J. (1974). The Graetz problem with axial conduction. *International Journal of Heat and Mass Transfer, 17*, 1391–1402.

Müller, U. (2016). Strömung mit mehreren Phasen. In H. Oertel (Ed.), *Prandtl - Führer durch die Strömungslehre.* Springer Vieweg Fachmedien. https://doi.org/10.1007/978-3-658-08933-7_9

Müller, U., & Ehrhard, P. (1999). *Freie Konvektion und Wärmeübertragung.* C.F. Müller-Verlag.

Ng, K. H., & Spalding, D. B. (1972). Turbulence model for boundary layers near walls. *Physics of Fluids, 15*, 20–30.

Nicoud, F., & Duos, F. (1999). Subgrid-scale modelling based on the square of the velocity gradient tensor. *Flow, Turbulence and Combustion, 62*, 183–200.

Otic, I., Grötzbach, G., & Wörner, M. (2005). Analysis and modelling of the temperature variance equation in turbulent natural variance equation in turbulent natural (C. U. Press, Ed.). *Journal of Fluid Mechanics, 525*, 237–261.

O'Toole, J. L., & Silveston, P. L. (1961). Correlation of convective heat transfer in confined horizontal layers. *Chemical Engineering Progress Symposium Series, 67*, 81ff.

Pacio, J. W. (2013). Assessment of liquid metal technology status and research paths for their use as efficient heat transfer fluids in solar central receiver systems. *Solar Energy, 93*, 11–22. https://doi.org/10.1016/j.solener.2013.03.025

Pederson, R. J., & May, E. K. (1982). *Flow instability during direct steam generation in a line-focus solar collector system.* Solar Energy Research Institute Report.

Piller, M., Nobile, E., & Hanratty, T. J. (2002). DNS study of turbulent transport at low Prandtl numbers in a channel flow. *Journal of Fluid Mechanics, 458*, 419–441.

Reichhardt, H. (1951, Juli). Vollständige Darstellung der turbulenten Geschwindigkeitsverteilung in glatten Leitungen. *Zeitschrift für Angewandte Math Mech, 31*(7), 208ff.

RELAP Code Development Team. (1995). *RELAP5/MOD3 code manual volume I–IV: Code structure, system models, and solution methods.* Idaho National Engineering Laboratory: Lockheed Idaho Technologies Company.

Remmers, K.-H. (2001). *Große Solaranlagen: Einstieg in Planung und Praxis* (2. überarb. Auflage ed.). Solarpraxis AG.

Reynolds, W. C. (1976). Computation of turbulent flows. *Annual Review of Fluid Mechanics, 8*, 183–208.
Rodi, W. (1981). Progress in turbulence modelling for incompressible flows. *AIAA Journal, Paper 81-0045.*
Rodi, W. (1993). *Turbulence models and their application—a state of the art review.* Monograph Int.l Ass. for Hydraulic Res. (IAHR): A.A. Balkema.
Rodi, W. (2005). *Engineering turbulence modelling and experiments 6* (E. I. ETMM6, Ed.). Elsevier Science.
Rohsenow, W. M. (1998). *Handbook of heat transfer*. McGraw-Hill.
Rommel, M., Siems, T., Schüle, K., Mehnert, S., & Thoma, C. (2007). *Entwicklung von Techniken zur Beherrschung des Stillstandsbetriebs.* www.solarkombianlagen-xl.info
Schlichting, H., & Gersten, K. (2000). *Boundary-layer theory* (8th ed.). Springer.
Schulenberg, T., & Stieglitz, R. (2010). Flow measurement techniques in heavy liquid metals. *Flow Measurement Techniques in Heavy Liquid Metals, 240*, 2077–2087.
Shah, R. K., & London, A. L. (1978). *Laminar flow forced convection in ducts: A source book for compact heat exchanger analytical data* (Vols. Advances in heat transfer, Supplement 1). Academic Press.
Shah, R. K., & Sekulic, D. (2003). *Fundamentals of heat exchanger design.* Wiley.
Shams, A., De Santis, A., & Roelofs, F. (2019). An overview of the AHFM-NRG formulations for the accurate prediction of turbulent flow and heat transfer in low-Prandtl number flows. *Nuclear Engineering and Design, 355*, 110342. https://doi.org/10.1016/j.nucengdes.2019.110342
Smagorinsky, J. (1963). General circulation experiments with the primitive equations I the basic experiment. *Monthly Weather Review, 91*, 99–164.
Sparrow, E. M., Patankar, S. V., & Shahrestani, H. (1978). Laminar heat transfer in a pipe subjected to circumferentially varying external heat transfer coefficients. *Numerical Heat Transfer, 1*, 117–127.
Stephan, P., Kabelac, S., & Kind, M. M. (2019). *VDI-Wärmeatlas* (12th ed.). Springer Vieweg.
Stieglitz, R. (2007). Chapter 10: Low Prandtl number thermal hydraulics. In Nuclear Energy Agency NEA-OECD (Ed.), *Handbook on lead-bismuth eutectic alloy and lead properties, materials, compatibility, thermalhydraulics and technologies* (pp. 399–478). NEA-OECD.
Streicher, W. (1991). *Simulation des instationären Verhaltens von Kompressionswärmepumpen.* Universität Graz.
Taitel, Y., & Duckler, A. A. (1976). A model for predicting flow regime transitions in horizontal and near horizontal gas-liquid flow. *AIChE Journal, 22*(1), 47–55.
Thome, J. R., & Cioncolini, A. (2015). Chapter 2: Two-phase flow pattern maps for macrochannels. In T. J. R., *Encyclopedia of two-phase heat transfer and flow* (Vol. 3, p. 1112). World Scientific Publishing.
van Dyke, M. (1975). *Perturbation methods in fluid mechanics* (Annotated ed.). Parabolic Press.
Veritas, D. N. (2010). *Environmental conditions and environmental loads—Recommended practice.* DNV-RP-C205, October 2010, Bærum, Norway.
Versteeg, H. K., & Malalasekera, W. (2007). *An introduction to computational fluid dynamics* (2nd ed.). Pearson Education Limited.
Walz, A. (1966). *Strömungs- und Temperaturgrenzschichten.* G. Braun Verlag.
White, F. M. (1991). *Viscous fluid flow* (2nd ed.). McGraw-Hill Science/Engineering/Math.
Whorlow, R. W. (1979). *Rheological techniques.* Wiley & Sons.
Wilcox, D. C. (1986, January). Multiscale model for turbulent flows (A. I. Astronautics, Ed.). In *Proceedings of the 24th AIAA Aerospace Sciences Meeting* (pp. 1311–1320).
Wilcox, D. C., & Traci, R. M. (1976). *A complete model of turbulence.* NASA.
You, D., & Moin, P. (2007). A dynamic global-coefficient subgrid-scale eddy-viscosity model for large-eddy simulation in complex geometries. *Physics of Fluids, 19*(6), 065110.
Zagarola, M. V., & Smits, A. J. (1998). Mean flow scaling of turbulent pipe flow. *Journal of Fluid Mechanics, 373*, 33–79.

Stationary Solar Thermal Collectors

6

Abstract

Solar thermal collectors are the core components of solar thermal energy systems, converting the solar radiation into heat, which is transported to a demand location by active or passive means. Solar thermal collectors can use the whole solar spectrum, not only a specific spectral part. But converted solar radiation cannot be fully used as inevitable thermal losses from the hot absorbers to the colder environment cannot be completely avoided. Therefore, the design of solar collectors must compromise—depending on the specific application—between maximizing the optical gains and the minimization of thermal losses. Highly transparent covers and selective absorbers with small thermal emissivities are used to achieve this target. Concentration by mirror optics and consecutive reduction of receiver area is a powerful tool to obtain an even better energy balance also for very high temperatures. Stationary collector types without or with small concentration are described in their main characteristics and the internal processes described, as well as the concentrating collector types which need one- or two-axis tracking devices to be able to collect the direct solar radiation from the sun over the day. In a last part collector testing methods are described and discussed, as well as the stagnation of collectors at which the highest absorber temperatures are reached without any net energy output, as thermal losses and solar gains just cancel out.

Keywords

Flat-plate collector • Vacuum tube collector • Collector equation • Theory • Parabolic trough • Linear Fresnel reflector • Thermal performance testing

6.1 Introduction

Solar thermal collectors are special kinds of heat-exchangers which transform incoming radiation energy into heat, typically of a fluid which is used to transport the heat to a distant demand. They can be designed for a multitude of applications

R. Stieglitz and W. Platzer, *Solar Thermal Energy Systems*,
https://doi.org/10.1007/978-3-031-43173-9_6

and locations, each having specific requirements with respect to the solar radiation input and the temperatures of the heat output.

In Sect. 6.2 the energy balance of a collector and the physical characteristics will be described mathematically, mainly for the case of a simple flat-plate collector. However, the principles described, and the concepts developed are also valid for other collector times if applied accordingly.

For this reason, we have a huge variety of collector designs including non-concentrating collectors using natural global radiation with a maximum intensity of about 1100 W/m^2 and concentrating collectors which focus the incoming radiation onto small focal spots or linear focal lines with intensities of up to many kW/cm^2.

After dealing with some basic definitions and principles, in Sects. 6.3 and 6.4. different collector types in both categories will be described, including the functionality and specifications of the individual components used. Then the topic of performance testing and efficiency of solar collectors will be treated in Sect. 6.4. Section 6.5 is dealing with the specific problem of stagnation in collectors, which may occur if heat is not needed or cannot be transported away from the collectors.

6.1.1 Basic Structure and Operation of a Collector

Although designs of collectors differ substantially in geometry, choice of materials and mode of operation, very often one finds in some form or other the main components of a collector:

- absorber,
- transparent cover,
- heat transport system,
- thermal insulation.

The aperture is defined as the active optical part of the collector, which accepts solar radiation for conversion to heat. For concentrating collectors additionally, one finds a concentrator optic and an adapted aperture system.

The role of the absorber is the transformation of incoming solar radiation into thermal energy. This heat raises the temperature of the absorber above ambient. Resulting heat losses due to this temperature difference may be reduced by using a transparent cover. A heat transport system usually uses a fluid to transport the generated heat to a different place. However, it is also possible that the heat naturally is transported to a colder place by conduction, thermal radiation, or natural convection. Thermal insulation is used to suppress heat losses of the absorber parts which need not be irradiated or of the heat transport system.

We illustrate the components with a typical flat-plate collector (Fig. 6.1) as an example: A flat absorber sheet of metal is coated black on the side facing the sun. The glass cover at a distance of 2–3 cm from the absorber reduces convective and radiative heat losses of the absorber surface facing the environment. On the back side of the collector some mineral wool or insulating foam is used to reduce the

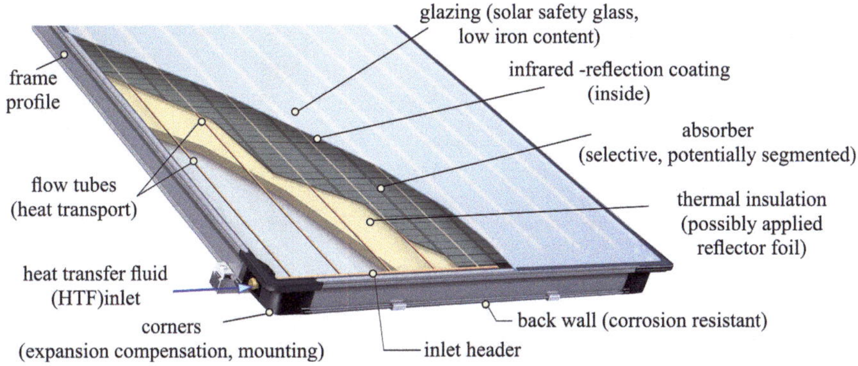

Fig. 6.1 Principal components of a solar flat-plate thermal collector

heat losses from the back side of the absorber sheet. And many pipes carrying water or a similar heat transfer fluid transport the resulting useful heat to a heat consumer or heat storage.

In a solar tower system with a central receiver and a heliostat field, as illustrated in Fig. 6.2, many mirrors of the heliostats—the optical concentrator—are oriented any time in a way to reflect the incoming direct radiation from the sun towards a limited receiver area on a tower. Here the absorber may consist of a multitude of pipes located in this area. A transparent glass cover does not exist, but for instance in the example shown in the picture an open cavity restricts the heat transport to the environment by protecting the absorber tubes from wind and reducing the view factor to sky and environment to reduce the thermal radiation exchange. Thus, this component is also "covering" the absorber but is 100% transparent for the incoming concentrated radiation from the heliostat field. The fluid in the absorber pipes must be suitable for the high temperatures of several hundred degrees reached. Therefore air, steam or molten salt are examples for such a fluid. Thermal insulation may be used in the cavity receiver to avoid peripheral heat losses.

Even the Earth, our blue planet, is a solar thermal collector. We have all the basic components present. Ground and water absorb solar radiation and convert it to heat. The atmosphere is partly withholding thermal radiation to escape to outer space and thus acts as a transparent cover during the day and as thermal insulation during the night. The air of the atmosphere transports heat to other shaded places, being responsible for compensating temperature differences on earth and having temperatures suitable for the existence of life everywhere.

So, in fact "solar collector" is a term which is very general and has many conceptual variations. Collectors may be mounted stationary or use a one-axis or two-axis tracking device to follow the sun's movement. The concentration of light varies from 1 (no concentration) to more than 10^3, the thermal insulation may be just the atmosphere (in an uncovered collector) or an evacuated space with low emissivity areas, so that convective-conductive as well as radiative heat transfer

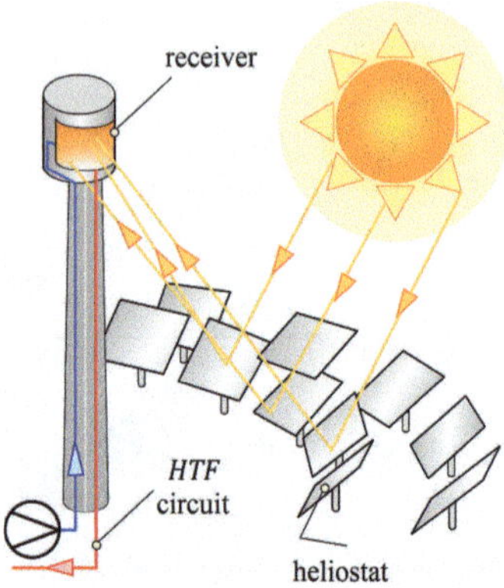

Fig. 6.2 Principal layout of a solar tower with heliostat field

is minimized. Figure 6.3 depicts selection of possible solar collectors using some fluid for transporting the heat to a user. But even this figure does not cover all variety. The heat may be transported away with different media, including solid walls for a passive transport. And in hybrid PVT collectors the absorber is combined with a photovoltaic module to produce heat and electricity simultaneously. But let us first talk about the distinction between passive and active solar.

6.1.2 Distinction Between Active and Passive Solar Thermal Energy

As mentioned already, we distinguish between passive and active solar thermal collectors depending on the existence of an active circulation of a heat transfer fluid. The system is called *active* if a mass transport induced by a mechanical system like a pump or ventilator is included. If the heat transport is taking place only based on a temperature gradient without any external means (e.g. by conduction through a massive absorber wall) the system is called *passive*. In the passive solar application, the heat converted from solar radiation should be used locally close to the absorber. Table 6.1 lists some general pros and cons of passive vs. active systems in a qualitative manner.

Popular examples for passive solar collectors are windows or transparently insulated solar walls. In fact, a complete building could be seen as a huge but inhomogeneous passive solar collector where the sun serves to partially cover the heat demand for comfort living during the heating season. With solar shading devices such a collector can be actively shut down to some extent to avoid overheating in summertime.

It is this aspect which led to the branding of "passive solar building design" or "passive-solar green architecture" which has a developed set of own design rules. Within these designs sometimes also heat transport by natural convection is considered. Hot air from a solar glass house or conservatory may be let into a house by openings or channels, thus providing passive solar heating even to rooms

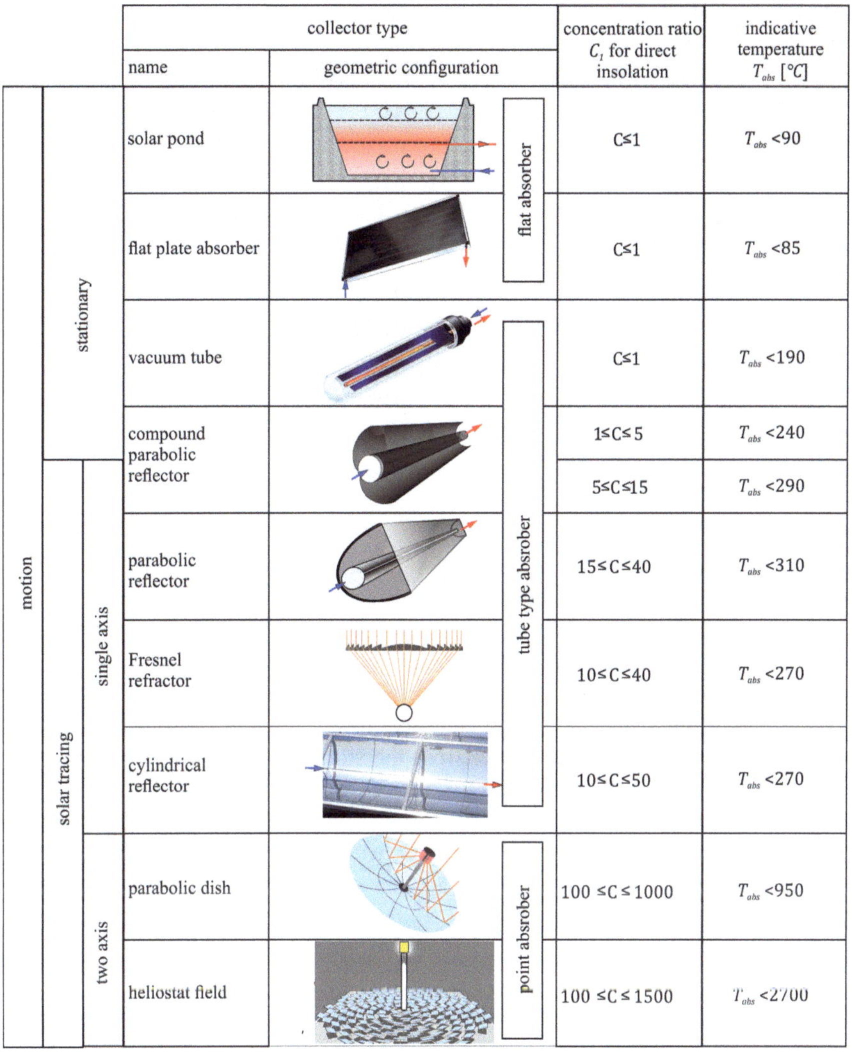

Fig. 6.3 Selection of different solar collector concepts

Table 6.1 Pros and Cons for active/passive solar systems

Area	Passive solar	Active solar
Efficiency	Low	High
Investment cost	Low	High
Operation cost	None	Noticeable
Control	Difficult	Good
Power use	Negligible	Noticeable
Maintenance	Little	Noticeable

Table 6.2 Normal solar transmittance (τ), reflectance (ρ) and absorptance (α) of selected transparent media

Material	Thickness t (mm)	Transmittance τ (–)	Reflectance ρ (–)	Absorptance α (–)
Float glass (soda lime)	4	0.836	0.075	0.089
Green glass (soda lime)	4	0.623	0.062	0.315
White glass (soda lime)	4	0.908	0.080	0.012
Borosilicate glass	3.3	0.926	0.048	0.026
Polyvinylchloride (PVC)	1	0.836	0.082	0.082
Polymethyl methacrylate (PMMA)	1	0.878	0.071	0.051
Polycarbonate (PC)	0.1	0.856	0.093	0.051
Polytetrafluoroethylene (PTFE)	0.1	0.924	0.046	0.030
Polyvinyl fluoride (PVF)	0.1	0.919	0.067	0.014
Polyethylene terephthalate (PETP)	0.1	0.882	0.097	0.022

further away from the actual collector. If ventilators would be used to increase the air transport volume, this system would be converted in an active system.

Another widespread use of a passive solar collector is the thermosyphon collector which works by a heat transport via natural convection. A solar liquid collector is connected directly to a storage tank directly above the collector. When the sun heats up the liquid inside, due to gravitational forces the less dense hot water rises to the storage and some colder water from the storage flows towards the collector inlet, thus circulating relatively slowly.

As consequence we can conclude that often components may be used both in a passive or in an active system. Nevertheless, we will see that an optimization of a collector for a passive or for an active application will result in differences of the design.

6.1.3 Impact of Concentration

Two fundamental categories of collectors are often called non-concentrating and concentrating collectors. Another way to label the different categories is whether the collectors require tracking mechanisms for their perfect operation or not. However, physics is not black and white and, therefore, one may find intermediate cases. The fundamental design parameter to classify the concentrating quality of a collector is the geometrical concentration ratio C_g. This parameter describes the relation between two areas of aperture (A_a) and receiver (A_r), as illustrated in Fig. 6.4. The receiver or absorber are receiving the concentrated radiation, where power is converted into heat and from which heat again can be lost by convection and thermal radiation (energy loss). Secondly, the aperture area of the collector which collects all the natural solar irradiation and redirects that as far as possible to the receiver (energy gain).

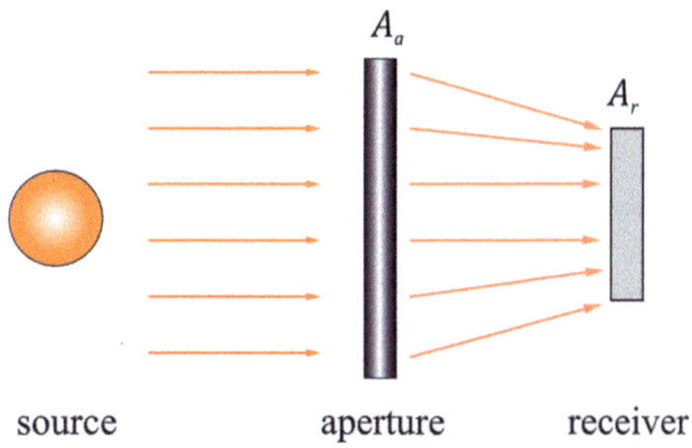

Fig. 6.4 Sketch aperture/receiver relation

For $C_g < 3$ we will find collectors using concentrators which are able to cover the whole sun-path over the day or even year without any adjustment, i.e. without tracking. This is a small concentration value. For values between 5 and 100 you have medium concentration where at least tracking along one axis is needed, and only the direct and circumsolar part of the solar irradiance can be used. Above 100 we speak about high concentration ratios, requiring two-dimensional tracking.

6.1.4 Typical Operating Temperatures and Applications

The relationship between concentration ratio and temperature is illustrated by the schematic diagram of different collector systems outlined in the Fig. 6.5, in which typical application temperatures and concentration ratios are given.

Wth increasing power density and temperature at the absorber, the demands on the collector designs increase. In case of high concentration ratios, the two-axis tracking is useful. Tracking not only increases the collection of direct solar radiation by better adjustment of the mirrors, but also ensures a constant level of power generation over a longer period.

When increasing typical operating temperatures for a collector, the heat losses usually increase more than linearly. Therefore, it is key to reduce the heat losses due to design when higher operating temperatures are to be achieved. Only then the heat losses will not reach the same magnitude as the solar gains, which would mean no useful heat can be transferred from the collector to the demand site. The stationary heat loss $\dot{Q}_{loss}$ of a collector is proportional to the heated (absorber) area, the temperature difference, and the so called U_L-value of the collector, the area-specific heat loss coefficient. One option to reduce of the value of U_L is to use low-emissive surfaces to lower thermal radiation emission, and to use convection barriers like additional glazings to reduce the convection heat transfer. Advanced methods are the utilization of noble gases or vacuum in gaps and cavities to reduce heat conduction. The other option is opened up by optics: If the solar radiation can be concentrated into a small absorber area compared to the collecting aperture area, the heat loss will be diminished accordingly for the same temperature difference—and hence it is possible to reach higher temperatures with higher concentration ratios.

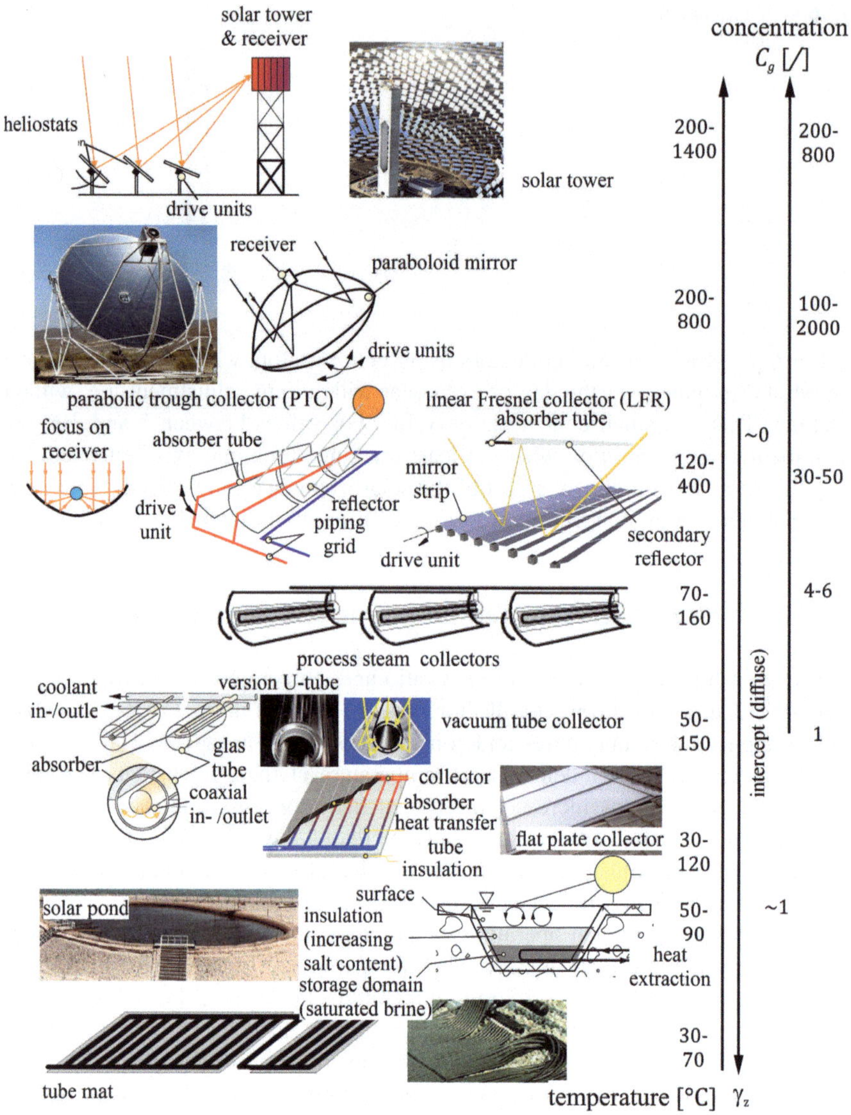

Fig. 6.5 Several collector types with indication of characteristic temperatures and geometric concentration factors (C_g)

The geometrical concentration ratio C_g is only one part of the story as the optical system has an efficiency—the optical quality of the materials used, and the perfect geometry determine the optical losses of a concentrator. Optical efficiency defined as the ratio of output of a real concentrator divided by the concentrator with ideal geometry and ideal surfaces (100% reflectivity) must be multiplied with C_g to get the real optical concentration factor C_{opt}.

The overview in Fig. 6.5 suggests that choosing a temperature level automatically leads to the choice of collector type. But that is not true in many cases. On the one hand side you have seasonal variations in the demand and the efficiency of collectors as well. When you need hot water with a certain temperature only for summertime where the outdoor temperature is high and solar irradiation is abundant, then a simpler collector may be the better and more economic situation compared to the case where the same hot water is needed in wintertime with possibly minus degrees outside. Then a collector with lower heat losses is needed.

On the other hand also, the climate and location are important for the collector choice, for example, when the production of 150 °C steam for an industrial site is perfectly possible with linearly concentrating collectors like Parabolic Trough or Linear Fresnel. The latter one might be the more economical solution for locations closer to the equator say up to $\Phi = 30°$ latitude. But the same steam is needed in a location (with similar *DNI*) further North, say at $\Phi = 40°$ latitude, due to the lower solar altitudes the Parabolic Trough might be the better solution. If in this location, however, due to meteorological conditions the *DNI* is relatively small and more diffuse irradiation is available, then a non-concentrating Vacuum Tube Collector (VTC) on an adequately tilted roof could perform better. Therefore, in many situations we have not to adhere to fixed rules and prejudices, but we should use our knowledge and adequate design rules to choose the best collector—whether that is the one producing more energy, or energy for less cost, or energy for the least investment.

6.2 Theoretical Description of Solar Collectors

6.2.1 Energy Balance

The theoretical description of a collector is based on the (quasi-)stationary heat energy balance between thermal losses and solar gains of a collector. For dynamical considerations this balance is later enhanced by thermal capacity terms for the collector materials and the heat transfer fluid (if that applies). The useful power $\dot{Q}_u$ which may cover an existing heat demand and thus the corresponding efficiency η of a solar collector is described in this so-called collector equation which in its first form has been introduced by Whillier and Hottel (Whillier, 1964), modified by and thus is called Hottel-Whillier(-Bliss) equation. Graphically, this energy balance is schematically shown in Fig. 6.6.

The area-related useful power $\dot{q}_u = \dot{Q}_u / A_c$ from a collector (in the picture it is a flat plate collector—but, the principle applies also to other collector types) is thus calculated from the balance of the solar gain, given by global irradiance G (in W/m^2)) times an optical efficiency η_{opt}, and the thermal losses from the absorber which in the stationary case may be described by a heat loss factor U_L of the collector times the temperature difference between mean absorber T_{abs} and ambient temperature T_{amb}. The optical efficiency often is also called the effective transmittance-absorptance-product $(\tau\alpha)_e$, because the main optical losses are

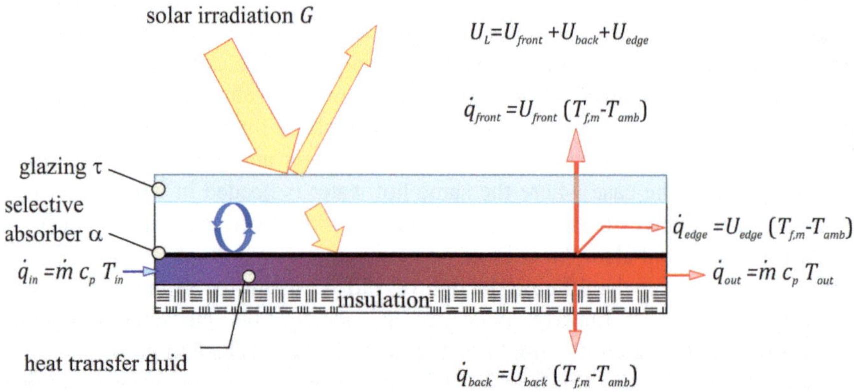

Fig. 6.6 Principal design of a solar thermal collector and energy balance

characterised by the angular-dependent transmission (τ) of the cover glass and the absorptivity of the thermal absorber (α). It is called the effective product, because multiple reflections between absorber and cover glass enhance the gain, as well as so-called secondary thermal gains induced by absorption in the cover glass(es). Of course, all optical quantities have a spectral dependence and are varying with the incidence angle of the incoming solar light. Similarly, the U_L-value is not a constant but in fact its value changes appreciably with temperatures involved, with wind velocities and even IR-radiation of the sky. The balance shown in Eq. (6.1) therefore shows a relationship for the specific net power where all quantities characterizing heat loss and optical efficiency are effective averages for the situation considered:

$$\dot{Q}_u = A_c \cdot \left[G \cdot (\tau\alpha)_e - U_\mathrm{L} \cdot (T_{abs} - T_{amb})\right]. \tag{6.1}$$

Here, the absorber temperature (T_{abs}) for which the useful power becomes zero is called the stagnation temperature. There are equivalent variants of this collector efficiency equation. There main difference is the use of different collector temperatures.

$$\dot{Q}_u = A_c \cdot F' \cdot \left[G \cdot (\tau\alpha)_e - U_\mathrm{L} \cdot \left(T_{f,m} - T_{amb}\right)\right]; \tag{6.2}$$

$$\dot{Q}_u = A_c \cdot F_R \cdot \left[G \cdot (\tau\alpha)_e - U_\mathrm{L} \cdot \left(T_{f,in} - T_{amb}\right)\right]. \tag{6.3}$$

$T_{f,m}$ is the mean fluid temperature in the collector, and $T_{f,in}$ is the inlet temperature of the fluid. F' is called the collector efficiency factor and F_R is called the heat removal factor. Both factors for real collectors are smaller than 100% but should be ideally very close to that limiting value. In Hottel and Whillier (1955) could show with this collector equation Formula (6.3) that efficiency η increases with increasing solar irradiation.

In the equations $(\tau\alpha)_e$ represents the optical conversion of transmitted short wave solar radiation at the absorber into long wave heat. The factor U_{L} can be conceptually split up in the case of a flat plate collector, being the sum of the heat losses through the front cover U_{front}, the rear insulation layer U_{back} and the side walls U_{edge} of the collector box:

$$U_{\mathrm{L}} = U_{back} + U_{edge} + U_{front}. \tag{6.4}$$

The first two factors in Eq. (6.4) can be kept small by using insulation layers and making a collector slim and large. The front side, on the other hand, requires optimization regarding high optical transmission of the incident radiation and, on the other hand, high thermal insulation in order to keep convective losses to the environment low.

Similar to an electrical circuit, the front cover heat loss factor U_{front} adds up reciprocally according to the Eq. (6.5) from the heat transfer coefficients between the absorber and the glass cover $h_{abs-\mathrm{gl}}$, the heat conduction through the glass pane (with the thickness δ_{gl} and its thermal conductivity k_{gl}) and the heat transfer coefficient $h_{\mathrm{gl}-amb}$ from the glass to the environment:

$$\frac{1}{U_{front}} = \frac{1}{h_{abs-\mathrm{gl}}} + \frac{\delta_{gl}}{k_{gl}} + \frac{1}{h_{\mathrm{gl}-amb}}. \tag{6.5}$$

In Eq. (6.5) the total heat transfer coefficient $h_{abs-\mathrm{gl}}$ usually dominates the front side heat loss. It is composed of heat transfer by convection and conduction ($h_{conv,abs-\mathrm{gl}}$) as well as heat radiation ($h_{rad,abs-\mathrm{gl}}$) in the air gap between absorber and glass:

$$h_{abs-gl} = h_{\mathrm{conv},abs-gl} + h_{\mathrm{rad},abs-gl}. \tag{6.6}$$

Conductive and convective heat transfer can be described $\dot{q} = h_{conv} \cdot \Delta T = h_{conv} \cdot (T_{abs} - \mathrm{T}_{amb})$ where h_{conv} again is a complex function depending on the temperature difference between glass and absorber, the tilt angle β oft he collector, the dimensions of the gap and the fill gas involved. The physics is described in detail in Chaps. 4 and 5. Similarly, idealized as the thermal radiation exchange between two infinitely large planes, from the absorber to the glass, the radiative thermal flux is:

$$\dot{q}_{\mathrm{rad},abs-\mathrm{Gl}} = \frac{\sigma}{\frac{1}{\varepsilon_{abs}} + \frac{1}{\varepsilon_{\mathrm{gl}}} - 1} \cdot \left(T_{abs}^4 - T_{\mathrm{gl}}^4\right). \tag{6.7}$$

wherein T_{abs} represents the absorber temperature and T_{gl} represents that of the glass plate; ε_{abs} and $\varepsilon_{\mathrm{gl}}$ represent the emission coefficients and σ is the Stefan-Boltzmann constant, the value of which is $\sigma = 5.67 \cdot 10^{-8}\mathrm{W}/(\mathrm{m}^2\mathrm{K}^4)$. Using the transformation $\left(T_{abs}^4 - T_{\mathrm{gl}}^4\right) = \left(T_{abs}^2 + T_{\mathrm{gl}}^2\right)\left(T_{abs} + T_{\mathrm{gl}}\right) \cdot \left(T_{abs} - T_{\mathrm{gl}}\right)$ one gets the result for the radiative heat transfer coefficient:

$$h_{\mathrm{rad},abs-\mathrm{gl}} = \frac{\dot{q}_{\mathrm{rad}}}{\left(T_{abs} - T_{\mathrm{gl}}\right)} = \frac{\sigma}{\frac{1}{\varepsilon_{abs}} + \frac{1}{\varepsilon_{\mathrm{gl}}} - 1} \cdot \left(T_{abs}^2 + T_{\mathrm{gl}}^2\right)\left(T_{abs} + T_{\mathrm{gl}}\right). \quad (6.8)$$

In all these considerations we must understand, that, the quantities given in the formulae like temperatures, irradiation, heat fluxes when investigated in detail are variable fields having complex distributions which may be modelled in so called Computational Fluid Dynamics programmes (CFD). However, for engineering purposes the simple description in Eq. (6.1) and following is extremely useful and powerful to describe the collector in a system and to determine characterizing parameters in testing procedures.

In a limited temperature range of several 10 °K both, convective and radiative heat transfer functions can be linearized based on average temperature or temperature difference. And for a very narrow temperature range, of 10–20 °K they can even be approximated by a constant value with good accuracy.

For pure heat conduction the gas conductivity k_{gas} determines the heat transfer coefficient $h_{cond,abs-gl} = \left(k_{gas}/d_{abs-gl}\right)$ through the gap of thickness d_{abs-gl} between the glass and absorber plate. Air is usually used as the gas filling here. If the temperature difference is large enough so convection sets on, the heat transfer coefficient for conduction must be multiplied by the Nusselt number Nu to get the convection coefficient $h_{conv,abs-gl}$ (in which $Nu = 1$ assumes pure heat conduction). Further details on the calculation of conductive and convective heat transport are discussed in Chaps. 4 and 5, respectively.

The heat exchange or heat loss due to radiation between the absorber and the glass plate decreases with the emission coefficients ε_i (see Eqs. (6.7) and (6.8)). For standard uncoated glass, however, $\varepsilon_{\mathrm{gl}} \approx 0.87$ applies, so the emission coefficient of the absorber surface in the range of long-wave thermal radiation must be kept small. The same surface absorbs the solar radiation. It should therefore have in the optimal case a high solar absorptivity and a low thermal emissivity. Such a surface is called a "selective absorber surface", as the optical properties reflectivity or absorptivity change drastically when going from the short wavelength solar range to the long wavelength thermal region, as illustrated in Fig. 6.7.

When we specify the collector area, different area is used depending on the purpose of the description or sometimes preference of the author. Collector efficiency might be defined relative to absorber or to aperture area. Gross collector area (total area) A_G should be used to specify the space needed for installation. It is the area between the outer boundaries of the collector, generally the outer boundaries of the collector casing.

The aperture area A_a is the transmission aperture of the collector for vertical or angular solar radiation incidence, generally the visible glass area within the pane seal. In the case of vacuum tube collectors without reflectors, it is the product of the inner diameter, the unshaded cylindrical inner length, and the number of tubes. In the case of reflectors attached to the reverse side or in the collector, it is

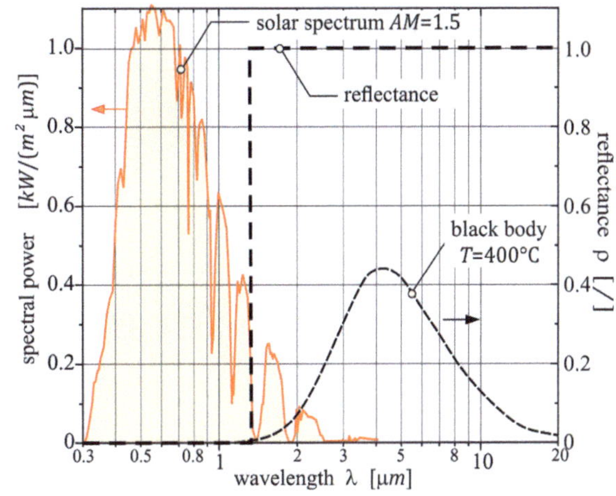

Fig. 6.7 Spectral reflectivity/absorptivity for an ideal selective absorber

determined using the projection of the area formed by the absorber and reflector in the collector level. The smallest area is the absorber area A_{abs} which equals the flat or curved absorber surface which is relevant for the thermal loss of the collector. For even more detailed explanations of these area definitions see the ISO 9806 standard specifying collector testing (ISO 9806, 2017) or the ISO 9488 on solar terminology (ISO 9488, 2021). The active absorber area generally corresponds to the absorber area A_a. Some collectors, however, are constructed so that the absorber area below the collector frame is larger than the aperture area. The active absorber area is then the absorber area not affected by shading from the frame. In this case it is equal to the aperture area A_a.

6.2.2 Efficiency Equation

By dividing the net thermal power $\dot{Q}_u$ in Eq. (6.1) with the collector area (A_a) and solar irradiance (G), the efficiency of the collector (η) is obtained directly:

$$\eta = \frac{\dot{Q}_u}{A_c \cdot G} = (\tau\alpha)_e - U_L \cdot \frac{(T_{abs} - T_{amb})}{G}. \tag{6.9}$$

The idealised Eq. (6.9) can first be analysed in more detail with the help of a diagram, which is shown in Fig. 6.8. For a temperature difference of $(T_{abs} - T_{amb}) = 0$, the collector characteristic curve corresponds exclusively to the conversion ratio $(\tau\alpha)_e$ and thus indicates which proportion of the solar radiation is converted into heat. The area above the horizontal of the conversion ratio reflects the proportion of optical losses of a collector. The area below the collector curve indicates the useful heat, while the area between the useful power and the optical losses quantifies the heat losses.

The graph can be further analysed. For small temperature differences, the efficiency is essentially determined by the conversion ratio $(\tau\alpha)_e$. This means that

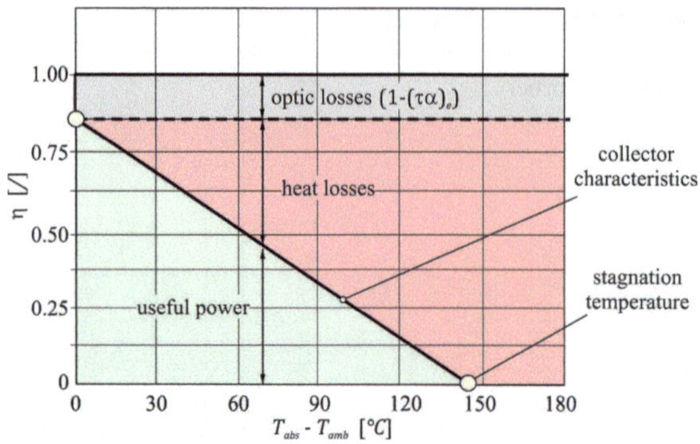

Fig. 6.8 Graphic analysis of the collector efficiency from Eq. (6.9)

even in the case of poorly thermally insulating collectors such as tube mats, a high degree of efficiency is still achieved with small temperature differences, provided that the conversion ratio is sufficiently high. However, as the temperature difference between the absorber surface and the environment ($T_{abs} - T_{amb}$) increases, all heat losses increase, and especially convective but also radiation heat losses then determine the total heat loss. For the latter, we must therefore introduce an addition that considers the non-linear character of the convective or radiant heat losses. The collector characteristic curve also shows another point, the stagnation or standstill temperature. This is the point at which the collector efficiency becomes zero, i.e. the collector does not produce any useful output. This occurs when, for example, the coolant circulation pump fails or the thermal storage tank of the collector circuit is full, so that the collector inlet temperature increases with time and thus also the heat losses to the environment. The occurrence of a stagnation temperature is a characteristic of all solar thermal collectors, regardless of whether they are concentrating or not. The stagnation temperature increases the lower the U_{L} value, i.e. the lower its heat transfer loss to the environment. A closer look at Eq. (6.9) also shows that the efficiency depends on the irradiation. The greater the irradiation, the lower the relative importance of the heat losses. For example, for a flat-plate collector for domestic hot water heating, this means that in winter, when irradiation is low (G), the importance of good thermal insulation is considerably more important than in summer, since G is in the denominator.

If the assumption of a narrow temperature difference with constant heat transfer coefficients and heat loss factor U_{L} does not apply the linearization leads to a quadratic loss term in the efficiency equation, as shown in Eq. (6.10), see c.f. (Frei, Häuselmann, Flückinger, & Frey, 1995).

$$\eta = (\tau\alpha)_e - \frac{1}{G}\left[u_0 \cdot (T_{abs} - \mathrm{T}_{amb}) + u_1 \cdot (T_{abs} - \mathrm{T}_{amb})^2\right]. \tag{6.10}$$

This is the typical approximative parametrization of collector testing for non-concentrating collectors. Other factors are also introduced in the empirical loss equation describing the dependence for example on wind speed. For concentrating collectors due to the high temperatures which often leads to a dominating radiative

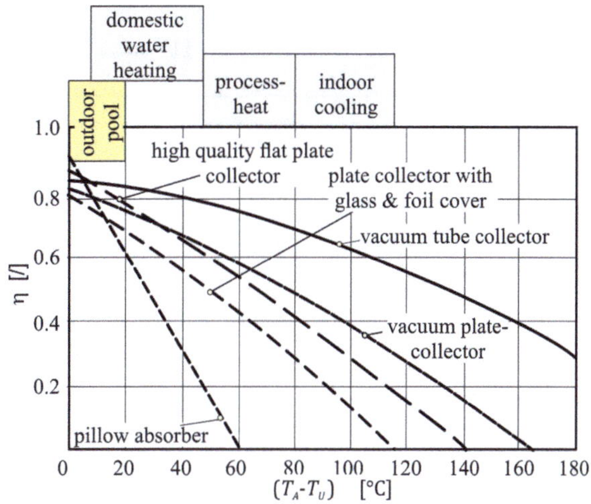

Fig. 6.9 Examples of different low temperature collector types with linear and quadratic efficiency curves with indication of a potential area of use

heat transport the polynomial expansion up to the 4th power in temperature difference is preferable. Figure 6.9 illustrates an example of a linear and a quadratic efficiency curve of different collector types and their potential utilization domain, which will be discussed subsequently.

6.2.3 Internal Flow and Temperature Distribution

To understand the different variations (6.1)–(6.3) of the collector equation, we now discuss the internal heat flows within the collector. At first, we will have a look at the heat transfer between absorber and heat transfer fluid. For simplicity we use the example of a fin-type absorber sheet of a flat-plate or vacuum tube absorber.

The thermal coupling of the absorber strip to the absorber tube influences the absorber efficiency factor F'. The thickness of the absorber plate, its thermal conductivity and the distance from the pipe is described by the so-called *fin factor* F. The absorber efficiency factor $F\prime$, however, relates the thermal resistance from the absorber fin to the ambient to the resistance from the fluid to the ambient. Therefore, also the thermal resistance of the bonding connection pipe-plate, the resistance of the pipe walls and the convective heat transfer in the pipe fluid must be considered as well. The decisive factor here is the effective heat transfer surface and thus the transfer to the fluid, which in turn depends on the flow rate. The fin factor F for a classical ideal fin absorber, as shown in a cross-sectional view in Fig. 6.10 neglecting the two-dimensional heat flux within the very thin absorber sheet can be modelled quite well discretizing the fin and solving the one-dimensional conduction heat transport within the fin for constant flux (absorbed solar power) from the fin end to the pipe section.

A simple thermal resistor model as sketched in Fig. 6.11 then is required to calculate the absorber efficiency factor F' for that situation.

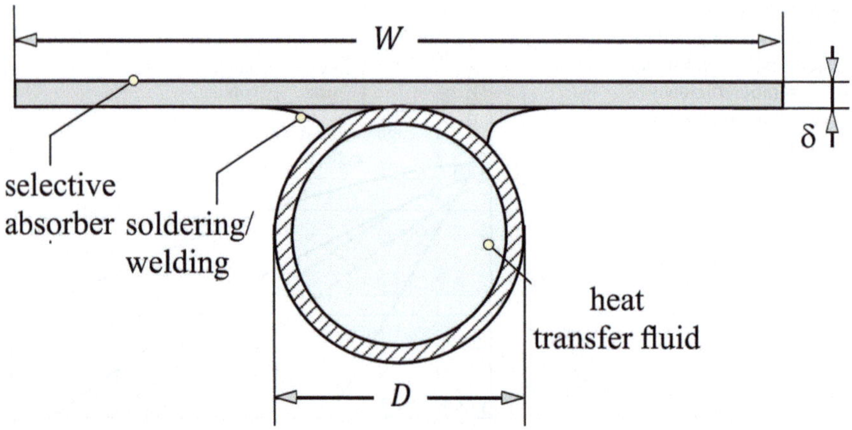

Fig. 6.10 Idealized absorber fin geometry

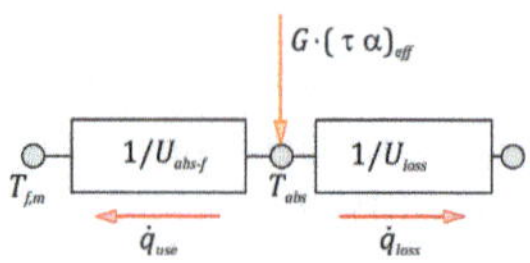

Fig. 6.11 Electrical equivalent diagram for determining the absorber efficiency factor F' of a flate plate collector

The temperature distribution in a thin absorber fin between tube and absorber sheet may be approximated with a one-dimensional heat transfer problem. The fin shall be discretized in infinitesimal elements receiving solar radiation and conducting the absorbed heat horizontally along the fin according to the temperature distribution and the conductance of the fin. Figure 6.12 shows schematically the situation with the infinitesimal discrete element of length Δx of the fin.

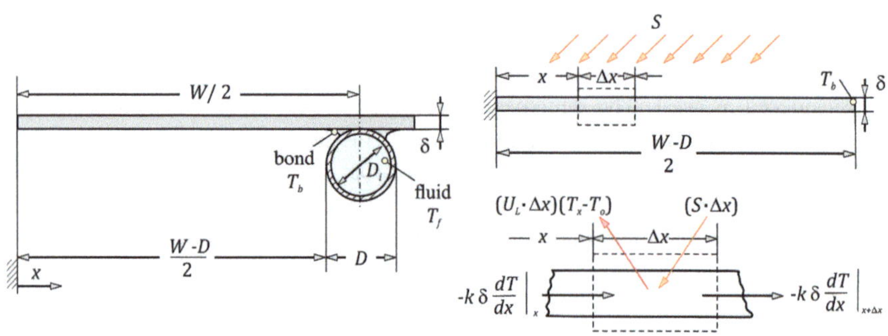

Fig. 6.12 Schematical drawing of absorber fin with tube after Duffie and Beckmann (2020)

The energy balance for the discrete element is

$$S \cdot \Delta x - U_L \cdot \Delta x \cdot (T_x - T_a) - k \cdot d \cdot \left(\left. \frac{dT}{dx} \right|_{x+\Delta x} - \left. \frac{dT}{dx} \right|_{x} \right) = 0. \quad (6.11)$$

Using $S = \left((\tau\alpha)_e \cdot G\right)$ and $x =$ position on fin. This leads to a differential equation of 2nd order with adiabatic boundary condition on one side and a center tube temperature on the other side of the symmetrical half fin. If we define the two variables:

$$m = \sqrt{\frac{U_L}{k \cdot d}}; \psi = (T - T_a) - \frac{S}{U_L}. \quad (6.12)$$

This leads to the simple differential equation

$$\frac{d^2\psi}{dx^2} - m^2 \cdot \psi = 0, \quad (6.13)$$

which can be solved taking into account the two boundary conditions. The solution for the temperature distribution as a function of x along the fin is then

$$\frac{T_x - T_{amb} - S/U_L}{T_0 - T_{amb} - S/U_L} = \frac{\cosh(m \cdot x)}{\cosh(m \cdot (W - D)/2)}, \quad (6.14)$$

where W denotes the fin width, D the tube diameter, T_0 is the temperature of the absorber right at the center of the tube position. Summing up the heat flux per unit length perpendicular to the fluid flow direction from the fin sheet to the central region and the part directly above the tube one gets the useful gain per unit length is

$$\dot{q}'_u = [(W - D) \cdot F + D] \cdot [S - U_L \cdot (T_0 - T_{amb})]. \quad (6.15)$$

The solution for the simple model of the absorber fin yields the following result

$$F = \frac{\tanh[m \cdot (W - D)/2]}{m \cdot (W - D)/2}, \quad (6.16)$$

where m is given by Eq. (6.12), U_L in $(\mathrm{W}/(\mathrm{m}^2\mathrm{K}))$ is absorber heat loss factor, k in (W/(mK)) the thermal conductivity of absorber plate, W the fin width in (m) and D the diameter of pipe also in (m). The fin factor F is shown as a function of fin width W, thermal conductivity k and thickness d of the absorber sheet in Fig. 6.13.

Now the heat collected in the fin must be transferred to the fluid circulating in the absorber tube. When the bonding resistance R_b and the heat transfer coefficient h_{fi} to the fluid inside the tube are considered, the resulting equation is

$$\dot{q}'_u = W \cdot F' \cdot \left[S - U_L \cdot \left(T_f - T_{amb}\right)\right]. \quad (6.17)$$

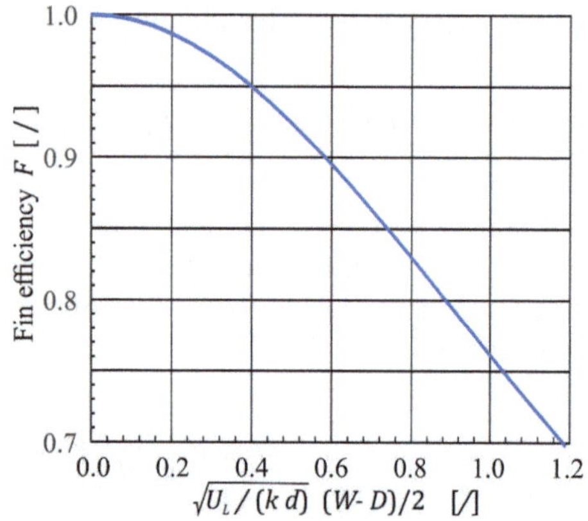

Fig. 6.13 Fin efficiency F for tube and sheet solar collectors

The fin factor F then is used to calculate F' according to Fig. 6.10 as

$$F' = \frac{1/U_L}{W \cdot \left[\frac{1}{U_L \cdot [(W-D) \cdot F + D]} + R_b + \frac{1}{\pi \cdot D_i \cdot h_{fi}}\right]}; C_b = \frac{k_b \cdot w_b}{\overline{\delta}_b}, \tag{6.18}$$

in which C_b in (W/(mK)) is bond conductance, k_b in (W/(mK)) the thermal conductivity of bond, $\overline{\delta}_b$ in (m) the average thickness of bond and w_b in (m) the width of bond, respectively.

A physical interpretation for F', the collector efficiency factor, results from examining Eq. (6.17). F' presents the ratio of actual useful energy gain to the useful gain if the collector fluid temperature would be identical to the absorber temperature. Another interpretation becomes clear when one recognises that the denominator of Eq. (6.18) is the heat resistance $1/U_0$ from fluid to ambient air. The numerator is the resistance from the absorber plate to ambient air. Therefore,

$$F\prime = \frac{U_0}{U_L}. \tag{6.19}$$

An example calculation again for typical parameters of a flat-plate liquid solar thermal collector is shown in Fig. 6.14. As can be seen there the overall heat loss to the ambient characterized by U_L is a decisive factor. Therefore, the curves are shown for simple collectors with black absorber coating and for state-of-the art collectors with selective absorbers and low emissivity values.

Fig. 6.14 Computed temperature distribution in an absorber produced from high-performance concrete (Hermann et al., 2015)

The exact form of the function F' may differ for appreciably for other collector geometries, e.g. for parabolic trough collectors. However, the fundamental principle and the considerations of designing a good collector by making the thermal resistance between absorber solar radiation and transporting fluid as small as possible is a general design law. While the approximate considerations above can be used for designing highly efficient fin-tube collectors, varying sheet thickness and tube spacing dependent mainly on the thermal conductivity of the sheet material, other more complex geometries may be simulated by 2D heat conduction programs, but the concept of F' is valid even then. When we integrate now Eq. (6.17) over the collector length L and use the average mean fluid temperature we arrive at the collector equation introduced already earlier, which reads to:

$$\dot{Q}_u = A_c \cdot F' \cdot \left[G \cdot (\tau\alpha)_e - U_\mathrm{L} \cdot \left(T_{f,m} - T_{amb}\right)\right], \tag{6.20}$$

with $T_{f,m}$ being the mean fluid temperature. The Fig. 6.15 illustrates the collector efficiency factor F' as a function of the tube spacing for different heat loss factors U_L and absorber fin conductance $k \cdot \delta$, where k is the absorber conductivity and δ the absorber fin thickness. Other assumptions taken for these calculations are: linear bond conductance $C_b = 500$ W/(m·K) and internal convection coefficient of absorber pipe $h_{fi} = 2000$ W/(m^2 · K).

It is obvious that the mass flow through a solar collector for given solar irradiation has a strong impact on the fluid temperature distribution along flow direction and hence also on the mean fluid temperature. A low mass flow would allow a much higher temperature rise during pass-through compared to a high mass flow, because a higher solar energy gain is accumulated during the longer passage time. While in reality the temperature distribution on an absorber area surface even with homogeneous irradiation has a two-dimensional distribution due to the

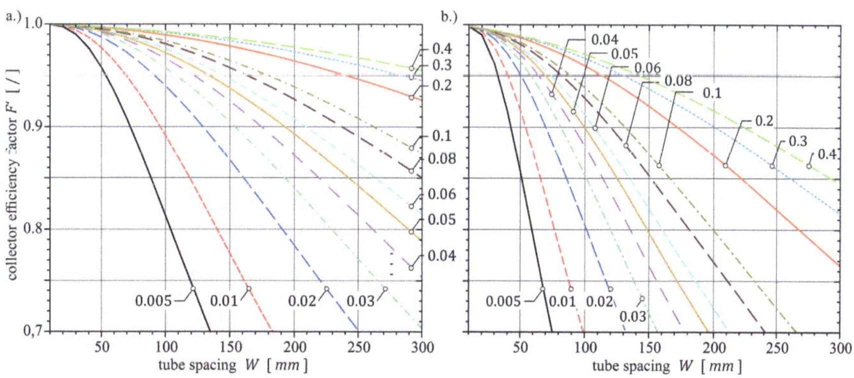

Fig. 6.15 Calculated collector efficiency η factor F' as a function of the tube spacing for different heat loss factors U_L and fin conductance $(k \cdot \delta)$ in (W/K) (curve indicator at right hand side of graphs); **a** $U_\mathrm{L} = 2\mathrm{W}/(\mathrm{m}^2\mathrm{K})$, **b** $U_\mathrm{L} = 4\mathrm{W}/(\mathrm{m}^2\mathrm{K})$

slightly inhomogeneous hydraulic flow of fluid through individual tubes, a one-dimensional temperature distribution along the main flow direction gives generally a good approximation for most geometries. Assume therefore a fin collector with identical fin elements. Here the fluid temperature along each tube is determined by a constant solar gain, but an increasing heat loss due to the rising fluid and absorber temperature. The energy balance of an infinitesimal tube element with length dy, and assuming that F' and U_L are constant over a small temperature range, one obtains a 1st order differential equation:

$$\dot{m} \cdot c_p \cdot \frac{dT_f}{dy} = n \cdot W \cdot F' \cdot \left[S - U_L \cdot \left(T_f - T_{amb}\right)\right], \tag{6.21}$$

where $\dot{m}$ is the total collector mass flow and c_p the fluid specific heat capacity. Integrating the equation over the tube length and assuming that all tubes inlet and outlet temperatures are the same, using $A_c = (n \cdot W \cdot L)$, the result is:

$$\frac{T_{f,out} - T_{amb} - S/U_L}{T_{f,in} - T_{amb} - S/U_L} = \exp\left(-\frac{A_c \cdot U_L \cdot F'}{\dot{m} \cdot c_p}\right). \tag{6.22}$$

It is convenient now to define a quantity that relates the actual useful energy gain of a collector to the useful energy gain if the absorber temperature were at fluid inlet temperature. This quantity is called the collector heat removal factor F_R:

$$F_R = \frac{\dot{m} \cdot c_p \cdot \left(T_{f,out} - T_{f,in}\right)}{A_c \cdot \left[S - U_L \cdot \left(T_{f,in} - T_{amb}\right)\right]}. \tag{6.23}$$

With some rearrangement of the terms and using Eq. (6.22) the factor F_R can be expressed as

$$F_R = F' \cdot F'' = F' \cdot \frac{\dot{m} \cdot c_p}{A_c \cdot U_L \cdot F'} \cdot \left[1 - \exp\left(-\frac{A_c \cdot U_L \cdot F'}{\dot{m} \cdot c_p}\right)\right], \tag{6.24}$$

where F'' is called the collector flow factor. Using the dimensionless quantity γ as the collector capacitance rate, being defined as

$$\gamma = \frac{\dot{m} \cdot c_p}{A_c \cdot U_L \cdot F'}, \tag{6.25}$$

we can see the importance of flow rate for the maximisation of the collector flow factor in Fig. 6.16. In solar thermal system design in fact different flow rate regimes are being used. There are so called low-flow systems where γ is in the range of 1–3 with a low F'', and high-flow systems with γ larger than 10 with a F'' close to 1. Although the transfer of heat to the fluid is reduced in low-flow systems there are other advantages which may counteract these disadavantages. We will discuss this further in the chapter on low temperature systems in Chap. 9.

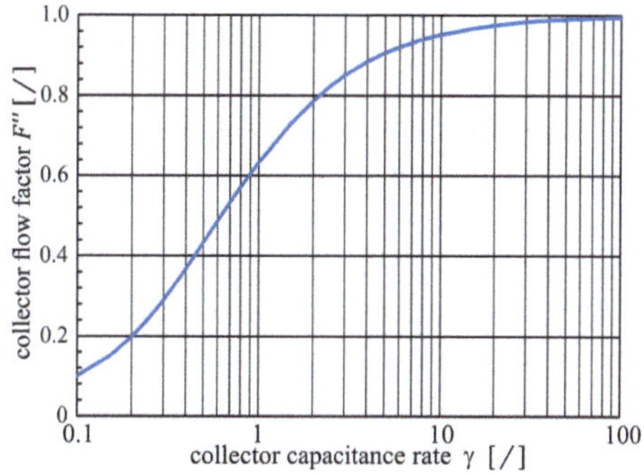

Fig. 6.16 Collector flow factor F'' versus dimensionless collector capacitance rate γ.

The quantity F_R is equivalent to the effectiveness of a conventional heat exchanger, which is defined as the ratio of the actual heat transfer to the maximum heat transfer rate. The maximum possible useful energy gain of a collector is when the whole collector is at the inlet fluid temperature—then heat losses to the ambient are minimum.

$$\dot{Q}_u = A_c \cdot F_R \cdot \left[G \cdot (\tau\alpha)_e - U_L \cdot \left(T_{f,in} - T_{amb}\right)\right], \tag{6.26}$$

with $T_{f,in}$ being inlet fluid temperature. This is the third form of the collector energy balance equation, the so-called Hottel-Whillier equation (Whillier, 1964). Whereas Eq. (6.20) serves as the basis for standardized collector testing in European standards (ISO 9806, 2017), the equation in the form (6.26) is the basis for the ASME collector test procedure (ANSI/ASHRAE Standard 93-2003, 2003).

6.3 Non-concentrating Collectors

According to the basic collector structure and geometry traditionally we distinguish three main collector groups which have established themselves in the market of solar low temperature collector systems:

- uncovered absorbers,
- flat plate collectors,
- vacuum collector,
- special collector types like air collectors, PV-T collectors.

In the following, the characteristics and performance of these three types are described. Certainly, we see numerous subtypes of collector designs that are specially adapted to specific applications and use. For example, the material choice is influenced by the targeted temperature range (polymeric or metallic absorbers), and also the heat transfer fluid used does vary (e.g. water, water–glycol, air).

6.3.1 Uncovered Collectors

The uncovered absorber design is certainly the simplest approach to produce hot water. No collector cover is used to lower heat losses to the environment, and thus the performance is very much dependent on the environmental variables like wind, ambient air and sky temperature. In the simplest case the absorbers consist of black UV-resistant flexible plastic tubes, which are placed on a suitable mostly horizontal surface. In more complex systems, the pipes are combined into absorber mats (see Fig. 6.17). Other uncovered collectors consist of metallic absorber sheets which may be used as complete facade or roof elements (Gombert, Reineke-Koch, Fenske, & Hofmann, 2007). An interesting variation is the transpired air collector, where fresh air for a building is sucked through tiny holes of an absorber sheet, heating up during the passage and in the rear cavity formed with a wall or roof.

The optical efficiency is high, not reduced by a cover glass, due to the high absorptance in the range of $0.8-0.95$. However, the thermal loss increases dramatically with fluid temperature rising above ambient temperature, as the uncovered surface due to convection and thermal radiation has on the average a heat transfer coefficient of around 20 $\mathrm{W/(m^2K)}$. This allows to raise the water temperature only few degrees above the ambient temperature.

In short, the characteristic features of uncovered collectors are:

- very simple construction,
- low price,
- easy installation. They can easily be laid on flat surfaces.
- Low efficiency due to high heat losses,
- low temperature applications, e.g. preheating of water and solar swimming pool heating,
- no winter operation possible, and hence no frost protection measures necessary.

Figure 6.17 shows a simple absorber mat using many tubes made of flexible, black coloured plastic with intermediate bars. The flexibility of the individual elements allows easy installation and decommissioning on any surface. Low system pressures ensure the use of conventional pipe connection systems for installation technology. The blackened water-filled pipes are made of polyethylene (PE),

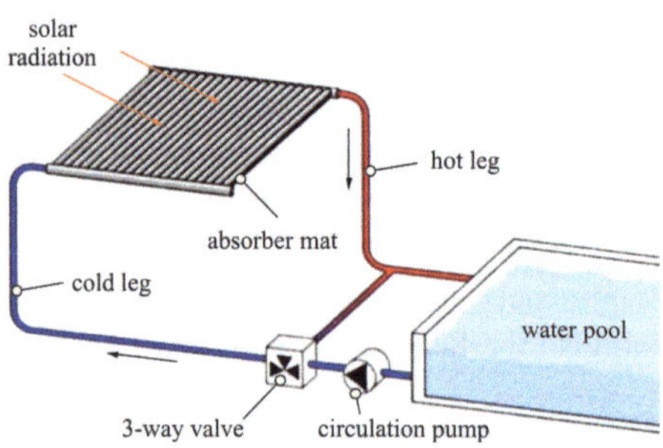

Fig. 6.17 Set-up of mat absorber collector circuit

polypropylene (PP) or more stable ethylene-propylene-diene monomer rubber (EPDM). For swimming pool applications, the absorber surface is about 50–80% of the pool surface. The simple design requires only a two-point control for efficient operation. Above a certain temperature difference between the pool and the absorber temperature, a pump is switched on, which sends the water back into the pool through the absorber.

6.3.2 Liquid Flat-Plate Collectors (FPC)

If a second basic element of collector design is added, a transparent cover for the absorber, then we can increase the working temperatures and efficiency of the solar collector substantially. In this case also a back-side insulation for the flat absorber is added thus both front and back side heat losses are reduced. Depending on the number and type of transparent collector covers and the absorber type this flat-plate collector as sketched in Fig. 6.18 may reach absorber temperatures well above 200 °C. Hence metal sheet absorbers from copper, aluminium or steel are an appropriate material choice. The absorber coating itself may be a simple black absorber e.g. using a paint which has a high thermal emissivity, or a selective absorber, which combines high absorptance α_s in the solar wavelength range with low thermal emissivity ε_{th}. The heat produced in the solar conversion process is transported in tubes connected to the backside of these metal absorbers. If only a simple absorber strip is connected to a single pipe, we call that an absorber fin. Nowadays however most collectors use full area metal sheet absorbers with a complete manifold of pipes attached. We distinguish between flat-plate liquid collectors and air collectors. Whereas in the first type the fluid is water or a water–glycol mixture and is pumped through pipes connected to the absorber sheet, in the case of air more design alternatives are common (cf. Fig. 6.33). Air may flow above, through or behind the absorber or in combinations of that. In the following paragraphs the design and effect on the collector performance of different components will be discussed.

6.3.2.1 Transparent Cover

The most popular solution for a transparent cover of a flat-plate collector is a sheet of glass. It has several functions. First of all, the cover shall reduce the thermal

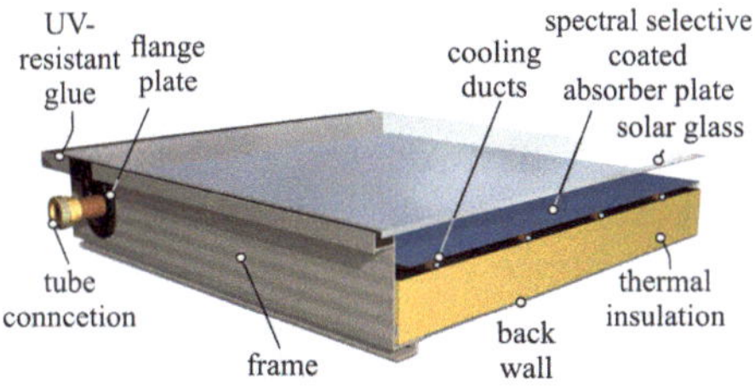

Fig. 6.18 Components of a liquid flat plate collector

front-side losses of the absorber due to convection and thermal radiation. For thermal radiation is glass a good absorber, if it has a sufficient thickness of 3 mm or more. It also creates a tight and narrow air gap between absorber and absorber, thus introducing a thermal resistance characterized by the natural convection of air. High convective heat losses due to wind are only possible from the outermost, much colder glass surface. A second requirements is, of course, that a very large amount of solar incoming radiation will be transmitted to the absorber. The solar transmittance of the glazing therefore shall be maximised. The third important task of the cover is to keep rainwater, hail, snow, ultraviolet radiation, birds remaining, organic growth of fungi or atmospheric pollution from the sensitive materials in the collector, i.e. the absorber coating and the thermal insulation. In days, when the absorber sheets had a less favourable appearance, also structured glass with micro-pyramids or other patterns obscured the view for the observers.

Depending on the glass quality and the treatment of the glazing different optical efficiencies of flat-plat collectors may be realized. Borosilicate glass and sodium lime glass are used for commercial collector covers. Soda lime glass is produced in huge thermal ovens on a fluid mercury bath with an extremely flat surface as float glass. Very often the glass is toughened in a thermal treatment process to avoid large sharp flakes when breakage occurs. As toughened glass breaks into very small fragments this type is often also called "security glass". Polymeric glazings are sometimes used as outer or as second cover materials, as they have less weight and may have higher transmittance. Very thin polymer films, however, are too sensitive for the outer surfaces, as they might degrade fast due to environmental stresses or even, they tear when birds or cats are using them as sitting area.

Table compares the solar transmittance for normal incidence of typical solar cover glass materials. The different reflectance is due to the index of refraction of the materials which is given as averaged over the solar range in Table.

The difference between green, normal float and white glass is that they are produced from sands which contains silicate oxide, but also different contaminations of iron oxide Fe_2O_3. White sands with very small iron oxide content produce little absorption in the near infrared. Green glass is produced from sand with high iron-oxide content. Therefore, a large part of the transmitted red light is absorbed, and the light emitted from the edges appears as green, which is the complementary colour.

If further increase of transmittance is needed, then surface treatment of the pure white glass may reduce the index of refraction at the glass-air interface, resulting in a lower reflectivity. This antireflection (AR) treatment of glass can be applied on one or on both sides of the glass sheet. One possibility is the etching of the glass surface with fluoric acid, resulting in a nano porous surface layer with intermediate refraction index. Alternatively, and a more environmentally friendly production is the coating with a thick solgel dip coating process. A silicate network structure is forming on the glass surface, and when the fluid solvent evaporates then similarly an open nano porous surface layer is created. A further production method could be a PVD coating of oxides on the glass. This AR-coatings reduce the reflectivity not

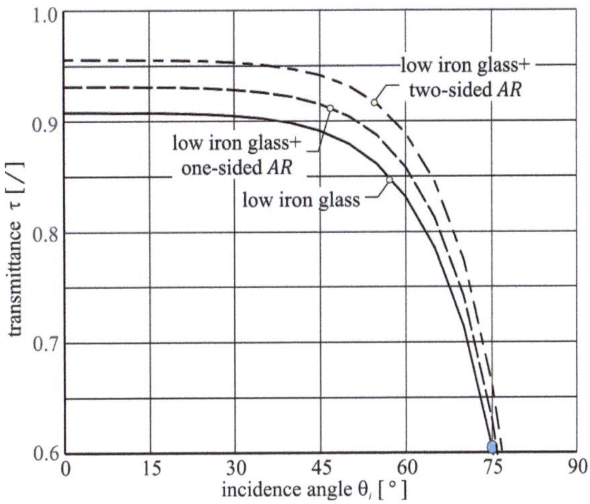

Fig. 6.19 Solar light transmittance as function of incidence angle θ_i of low iron-oxide content glass without Anti-reflection layer (AR) with one-sided and two-sided AR-coating

only for normal incidence θ_i, but have even a high effect for off-normal incidence, as illustrated in Fig. 6.19.

As a second collector cover—even with AR-coatings—reduces the solar gains and increases the weight of the collector, the benefit due to additional reduction of convective and radiative heat loss must be meticulously taken into account. The Fig. 6.20 shows the thermal network of a two-cover collector in terms of convective (c) and radiative (r) resistances. From that can be concluded that for a collector with low absorber emissivity ε_{abs} the radiative resistance between absorber and first cover glass is the dominating one. A second cover glass then may only be reasonable if the absorber temperatures are high, which implies that large temperature differences between absorber and glazings leading to a low convective resistance in the air gap between absorber and first glazing, which is "short-circuiting" the heat flux. In this case a second glazing might be appropriate.

To investigate in detail whether a second cover may be useful for increasing the useful energy output of the collector we have to investigate the physical models of the collector heat loss according to the presented network, not only the empirical parameterized heat loss of Eq. (6.10). Within the gaps between absorber, 1st glazing and 2nd glazing we must quantify radiative and convective heat transfer

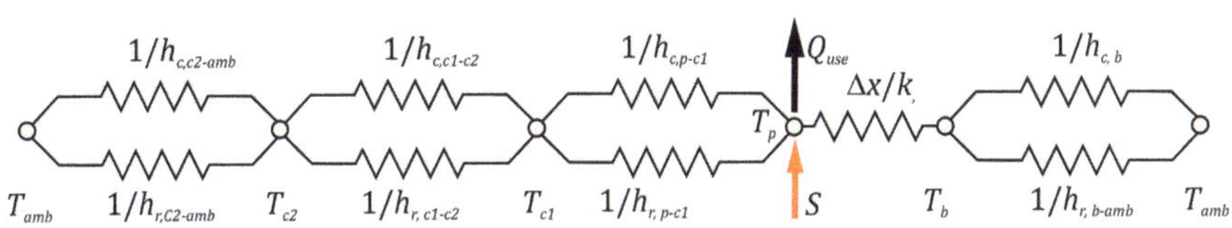

Fig. 6.20 Electrical equivalent circuit for a double-glazed collector consisting of convective and radiative resistors connected in parallel

coefficients, which in the case of convection also depend on the resulting temperatures. With some temperatures assumed we will determine the resulting heat fluxes, which in turn may change the temperatures in stationary equilibrium. Therefore, the solution of the network equations is iterative.

The radiative heat transfer in the gaps can be approximated by shallow cavities with the formula for infinite parallel plates:

$$h_{r,p-c1} = \frac{\sigma \cdot \left(T_{c1}^2 + T_p^2\right) \cdot (T_{c1} + T_p)}{\frac{1}{\varepsilon_{c1}} + \frac{1}{\varepsilon_p} - 1}, \tag{6.27}$$

$$h_{r,c1-c2} = \frac{\sigma \cdot \left(T_{c2}^2 + T_{c1}^2\right) \cdot (T_{c2} + T_{c1})}{\frac{1}{\varepsilon_{c1}} + \frac{1}{\varepsilon_{c2}} - 1}. \tag{6.28}$$

The radiative heat transfer to the environment from the outer cover "$c2$" is dominated by the radiation exchange with the sky having an equivalent sky temperature T_{sky} which for clear skies may be drastically below the ambient temperature T_{amb}. While for cloudy skies the equivalent radiative temperature of the sky dome is close to the air temperature, for clear skies due to low humidity content of the atmosphere the collector "sees" partially the cold background of outer space, resulting in equivalent temperatures 40–50° Kelvin below air temperature.

$$\begin{aligned} h_{r,c2-sky} &= \varepsilon_{c2} \cdot \sigma \cdot \left(T_{sky}^2 + T_{c2}^2\right) \cdot (T_{sky} + T_{c2}) \cdot (1 + \cos\beta)/2 \\ &+ \varepsilon_{c2} \cdot \sigma \cdot \left(T_{amb}^2 + T_{c2}^2\right) \cdot (T_{amb} + T_{c2}) \cdot (1 - \cos\beta)/2. \end{aligned} \tag{6.29}$$

Convective heat transfer is characterized by Nusselt numbers depending mainly on the temperature differences of adjacent walls, the gap width d and the inclination β of the collector. We use for illustration the following correlation for convective heat transfer in a closed and inclined shallow cavity:

$$\begin{aligned} Nu = 1 + 1.44 \cdot &\left[1 - \frac{1708 \cdot (\sin 1.8\beta)^{1.6}}{Ra \cdot \cos\beta}\right] \cdot \left[1 - \frac{1708}{Ra \cdot \cos\beta}\right]^+ \\ &+ \left[\left(\frac{Ra \cdot \cos\beta}{5830}\right)^{1/3} - 1\right]^+, \end{aligned} \tag{6.30}$$

where Nu ist he Nusselt number, Ra the Rayleigh number and β in (rad) the tilt angle of collector with respect to the horizontal plane. The Rayleigh number is defined by:

$$Ra = \frac{g \cdot T_m^{-1} \cdot \Delta T \cdot d^3}{\nu \cdot \alpha}, \tag{6.31}$$

in which g in (m/s^2) denotes gravitational constant, T_m in (°K) mean absolute gas temperature in the gap, d in (m) gap width, ΔT in (K) temperature difference over gap, ν in (m^2/s) the kinematic viscosity of gas, α in (m^2/s) the thermal diffusivity of gas.

The convective heat transfer coefficients in the gap then are:

$$h_{c,p-c1} = Nu_1 \cdot \frac{k}{d_1}, \tag{6.32}$$

$$h_{c,c1-c2} = Nu_2 \cdot \frac{k}{d_2}, \tag{6.33}$$

where k (W/(mK)) gas conductivity, d_1 in (m) is gap width between absorber and 1st cover and d_2 in (m) the gap width between 1st and 2nd cover.

For the convective heat transfer from outer cover to the air we take a simple correlation, e.g.

$$h_{c,c2-amb} = \left(2.8 + 3.0 \cdot \frac{u_0}{\mathrm{m/s}}\right) \mathrm{in} \left(\frac{\mathrm{W}}{\mathrm{m^2K}}\right), \tag{6.34}$$

where u_0 in (m/s) represents the wind speed, and d_1 in (m) is gap width between absorber and 1st cover and d_2 in (m) the gap width between 1st and 2nd cover. More sophisticated correlations for flat areas with different inclination angles or bodies can be found in literature for wind driven situations with forced convection (using the Reynolds number) or for wind-still situations with natural convection (using the Rayleigh number), for more details see e.g. Rohsenow (1998), Stephan et al., 2019) or Chap. 5 of this book.

From this a solution for the thermal network is obtained in terms of the different heat loss coefficients. Figure 6.21 illustrate the calculated total front heat loss coefficient (U_t) as function of absorber temeprature for different ambient temperatures for single glazed collectors (top) and for double glazed collectors (bottom).

A different possibility of suppressing convection is the use of so-called transparent insulation materials like honeycombs or capillary structures. Especially useful for the collector cover are combinations of a protecting glass-sheet with an open absorber-vertical cover behind. The narrow cavities or slits between the transparent walls of the structure are subduing convection efficiently. It is important that the structures do not touch the absorber surface. In direct contact this might on the one hand side enhance infrared emission from the absorber, on the other hand polymeric materials will not be stable enough the take the high absorber temperatures. Figure 6.22 shows a commercial collector using such a honeycomb structure (TIGI Solar, 2011) and a close up on the material.

Convection suppression by choosing gas fillings with a lower conductivity than air is a common approach and utilized e.g. in noble-gas filled window glazings. Similarly, solar thermal collectors can be built based on such fillings. However, one critical design criterion is the collector gas volume. As the collector shall be gas-tight, a warming up of the collector results due to the gas expansion in a higher pressure level on the glass pane, where a breakage must be avoided. Evacuation of the air gap between absorber and cover glass or even of the complete collector

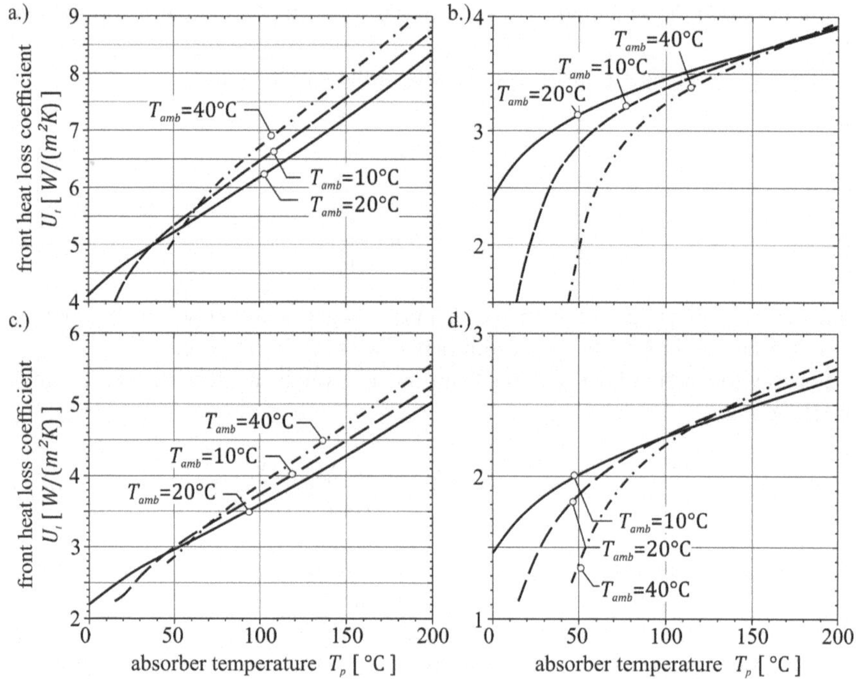

Fig. 6.21 Calculated heat loss coefficient U_t for different ambient temperatures as function of absorber temperature T_p and for a single glazed cover (**a**, **b**) and double glazed version (**c**, **d**). Left graphs black absorber $\varepsilon_p = 0.95$; right graphs selective absorber $\varepsilon_p = 0.05$

Fig. 6.22 Commecially available Honeycomb Collector from TIGI Solar (2011)

theoretically is the best suppression method of convection. The Rayleigh number Ra scales with the square of the air density and a temperature rise yields thus a pressure increase and thereby to an enhanced heat loss to the ambient. One may see the in Fig. 23 the resulting Nu-number of an inclined single-glazed collector of tilt $\beta = 45°$ and air gap 20 mm for a temperature difference of 40 °K as a function of density. At a pressure of $p = 0.2$ bar in the collector convection is suppressed effectively which is expressed by $Nu = 1$. If the conductive heat transport through the air gap shall be reduced further, even lower pressures are needed challenging the solar glass integrity. Such a low-vacuum collector thus is able to suppress air

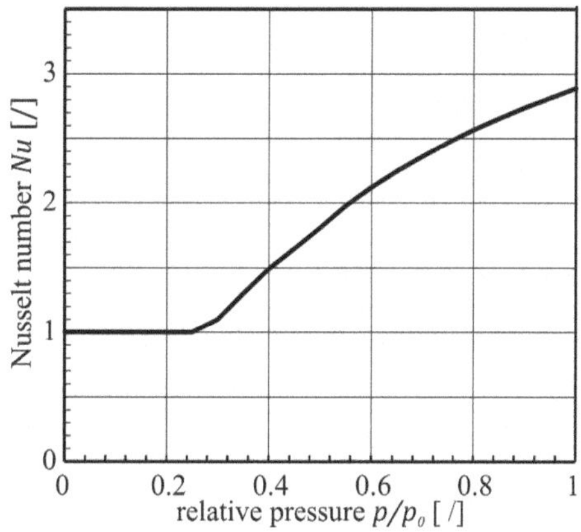

Fig. 6.23 Nusselt number for an inclined flat plate collector ($\beta = 45°$) as a function of the relative pressure (~ air density) for a collector with an air gap of 20 mm and a temperature difference of $\Delta T = 60$ °K

convection; however, the gas conductivity is still present and contributing to the heat loss.

Suppressing gas conduction needs a further evacuation to a high-vacuum collector. When the free mean path length of air molecules between mutual collisions is reaching the dimensions of the air gap (Knudsen regime), then the thermal gas conductivity is determined by the equation:

$$\frac{k_{gas}(p)}{k_{gas}(p = 1atm)} = \frac{1}{1 + 3.75 \cdot Kn} \text{ with } Kn = \frac{\delta}{L} \approx \frac{0.008313}{\left(1 + \frac{116}{T}\right)L \cdot p}, \qquad (6.35)$$

in which Kn is the dimensionless Knudsen number, k_{gas} in (W/(mK)) the thermal conductivity of gas, δ in (m) the mean free path of gas molecules, L in (m) a characteristic dimension of container (pore, air gap) and p in (Pa) the gas pressure.

At about $p = 10^{-5}$ Pa the final limit of conductive heat transport through the air gap in the collector is reached, if that gap dimension is $L = 1$ mm, as illustrated in Fig. 6.24, for more details see Chap. 4.

However, the practical problem is that with a pressure difference of close to an atmosphere from inside to outside, the forces on the cover glass are enormous, and in order not to break even special toughened glass must be supported about every 50–100 mm. This in turn creates a problem of heat conduction through the support pillars, which have to be extremely slim, but mechanically stable.

Figure 6.25 shows a solution of a low-vacuum collector commercially produced in the 2000 years, where holes have been punched into the absorber plates so that the stainless-steel support pillars only tough the cold back and front surfaces. The convection was well suppressed in both air gaps on the front and back side of the absorber. The collector U-value was therefore reduced considerably compared to an unevacuated selective flat-plate collector. The low pressure within this collector leads to a slow infiltration of air, as the sealings were not completely air-tight. For this reason, a possibility to reduce the pressure in the collector in a routine maintenance step using a simple pump (e.g. a waterjet pump) was foreseen by the manufacturer (see Fig. 6.25).

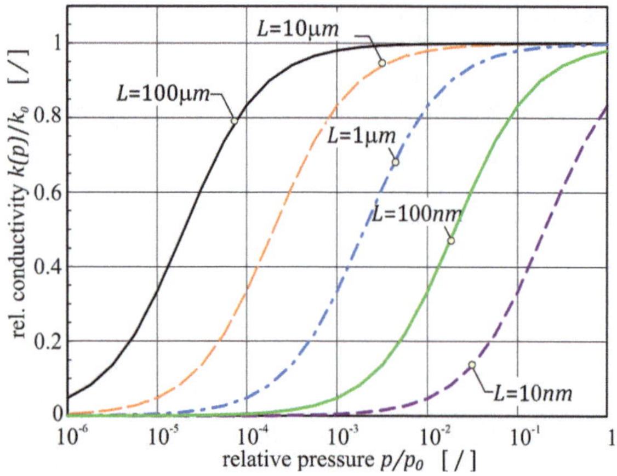

Fig. 6.24 Relative conductivity of the air gap to normal conditions as a function of the relative pressure (p/p_0) for different dimensions of the container L; free molecular regime (Knudsen number $Kn > 1$) requires small dimensions L or low relative pressures (p/p_0)

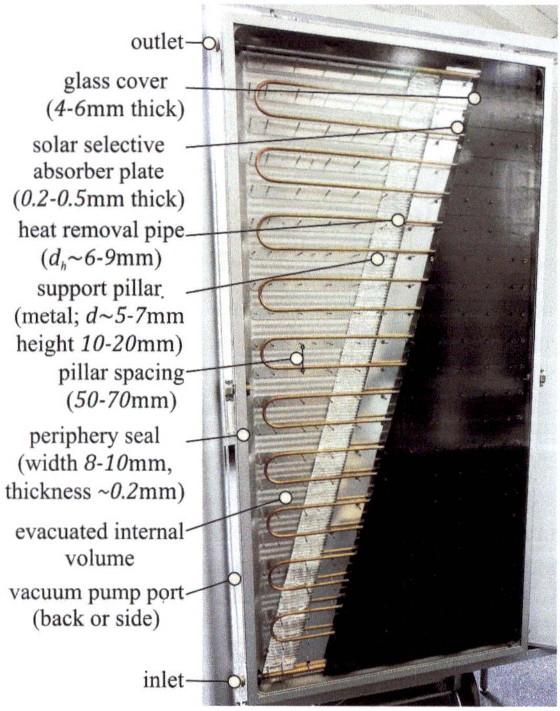

Fig. 6.25 Low-vacuum flat-plate collector width some typical component dimensions

For a real high-vacuum flat-plate collector evacuation not only requires a vacuum pump (e.g. a turbomolecular pump), it also would be very difficult to evacuate a thin layer of say 1 mm air through one valve, as this would take a long time. Therefore, such high-vacuum collectors need a much better sealing technology, so they do not need to be serviced because of vacuum loss. High-vacuum collectors have been developed some years ago utilizing glass-to-metal sealings and a getter

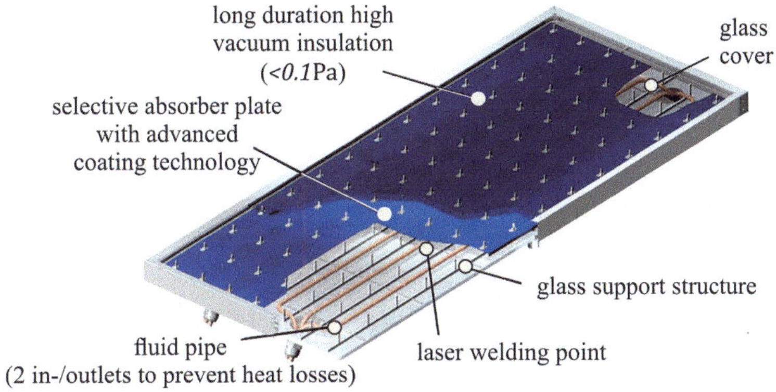

Fig. 6.26 Cross-sectional cut through a high-vacuum flat-plate collector

Fig. 6.27 Collector efficiency based on gross area for three different flat-plate collectors (unevacuated, low and high vacuum) vs temperature difference between collector and ambient $G = 850\ \text{W/m}^2$

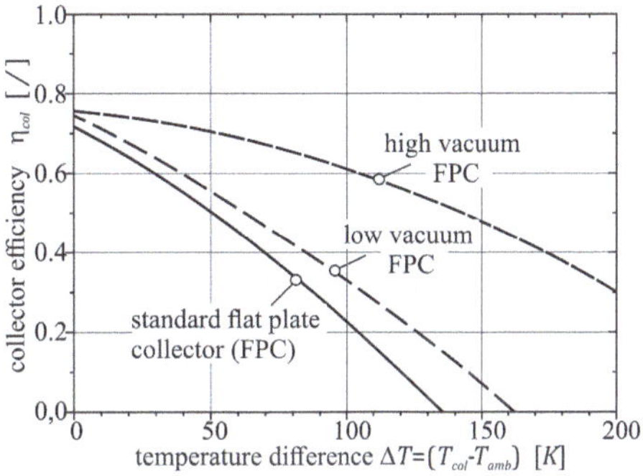

solution to keep the ultra-high vacuum (UHV) inside. No organic materials can be used inside these extremely slim collectors (see as an example Fig. 6.26). The collector efficiency of a low-vacuum and a high vacuum flat plate collector is shown in the next Fig. 6.27. Here it can be seen the convection suppression increases the efficiency noticeably in the temperature range 60–100 °C (here the comparison is with a non-evacuated collector of the same manufacturer). High vacuum, however, enables the flat-plate collector to be used also in up to 180 °C for medium temperature applications (solar cooling or solar process heat).

6.3.2.2 Absorber Plate

The connection of the absorber plate with the absorber tubes carrying the heat transfer fluid plays an important role. Several different joining techniques are available here, as the Fig. 6.28 shows. The absorber strips can either be connected in a fixed way with the absorber by welding, soldering, or gluing (left 4 options). Other options are attachment by mechanical means (roll-bond option pressing two sheets together or clamping) or extrusion as a whole structure, while the latter is mainly conducted mainly with polymer absorbers.

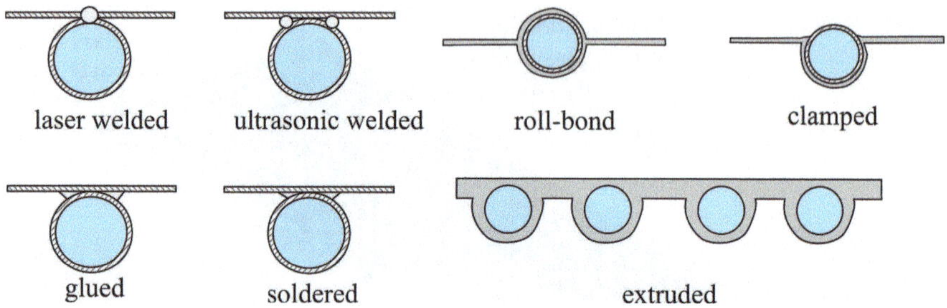

Fig. 6.28 Different possible way of connector flow channel (pipes) with the absorber sheet material

Other possibilities include pressing entire structures including cooling channels or inflating pressed sheets using gas pressure. The cooling channel cross-section can also be adapted to the individual applications, although this requires computer-aided simulation during the design phase. Figure 6.29 shows some examples.

Table 6.3 lists typical absorber efficiency factors F' as a function of the flow rate for the three joints shown. Other joining options include pressing or blowing processes in which parts are left out of the sheet metal and treated with release agents to prevent them from joining together. Afterwards these areas are inflated, so fine canals can be built there.

Another important aspect of the collector design is the internal piping. There are potentially several possibilities. The arrangement of the flow channels in the absorber can be serial (Fig. 6.30a)—serpentine collector—for low-flow systems

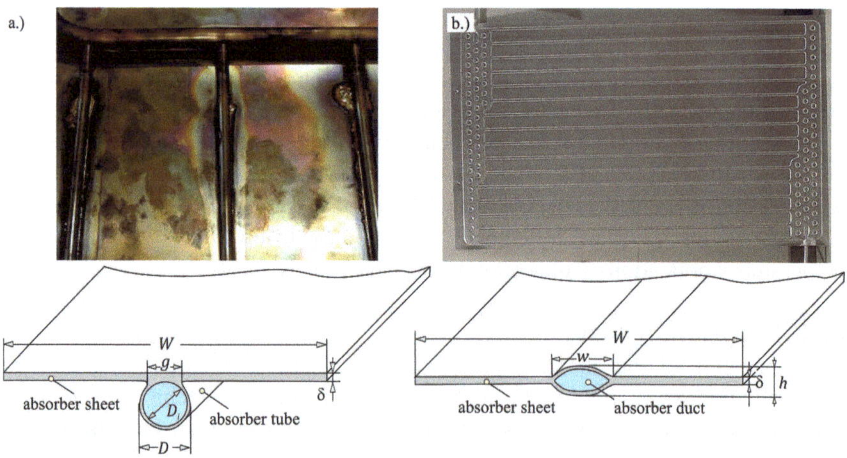

Fig. 6.29 Tube-fin absorber (**a**) and roll bond absorber **b**) in view (top row) and in cross-section (bottom row); © Fraunhofer ISE (left) © CGA Technologies S.p.A.(right)

Table 6.3 Absorber efficiency factor F' for three connection methods of absorber plate with absorber pipe for different mass flows

Joining method absorber and tube	F' at (20 L/min)	F' at (40L/min)	F' at (80 L/min)
Roll-in procedure	0.89	0.91	0.92
Soldering/welding	0.87	0.91	0.93
Rollers	0.88	0.89	0.92

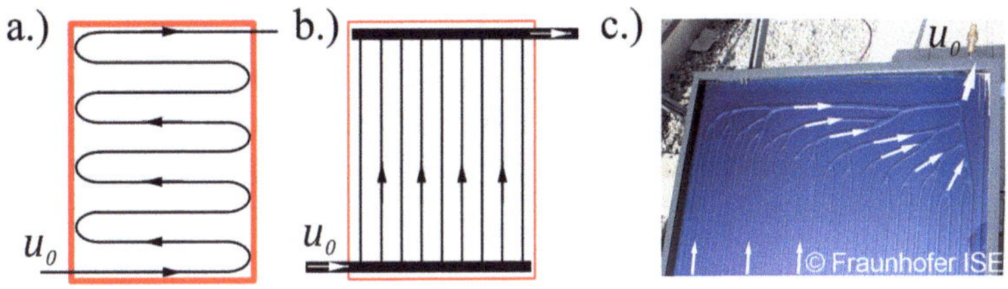

Fig. 6.30 Piping and distribution arrangement in flat plate collectors in a serial circuit (**a**), parallel (**b**) and a type taken from biological (**c**) models (Hermann, 2011)

and collectors working on natural convection, parallel (Fig. 6.30b)—harp collector—which is the standard design for high-flow systems, or in a branched structure (Fig. 6.30c)–bionic collector which uses the principles developed in nature to approach a very even temperature distribution over the collector width combined with a minimum hydraulic resistance. With serial flow control in a meander, high flow rates through the long flow path result in high pressure losses, which require a powerful pump and lead to high electricity consumption. In addition, there are large temperature differences in the collector box and between the fluid inlet and outlet. Therefore, this kind of collector is suitable for low mass flow systems, e.g. in thermosyphon systems. When the flow channels are connected in parallel, the temperature distribution in the collector is more uniform. It must be ensured that the flow rates in the subchannels are approximately equal, so that local evaporation of the heat transfer fluid does not cause the collector to come to a standstill. Also, uneven temperature distribution between different riser pipes would result in a lower resulting efficiency and exergy. In the case of flat plate collectors with their relatively small differences in useful temperature, this is ensured by suitable pressure loss ratios between the distributor—or header—and absorber tubes. For vacuum tube collectors with higher useful heat, more complex numerical calculations are required. This also applies to the complex distribution of flows and temperatures in the bionic branching systems.

6.3.2.3 Thermal Insulation

Thermal insulation is necessary in flat plate collectors to reduce heat losses. Generally, the back wall and the sides of the collector casing are lined with these materials. As well as rock and glass wool, PUR foams and melamine resin are used. In rare cases, smaller manufacturers offer collectors with sheep's wool, flax or other»natural« insulation. The insulation materials must withstand the high stagnation temperatures of the collectors. In particular, the rear side insulation, on which the absorber mostly lies, must have high thermal stability. The insulation should not evolve gas, even at very high temperatures, otherwise inside deposition on the cover pane is to be expected. Some typical values of thermal conductivity for insulation materials are shown in Fig. 6.31. Important for the protection of insulation against wetness are seals, at the casing, at the glass cover and at the tube ducts. They must prevent the penetration of rainwater over the entire operating period, however due to pressure and humidity transport (micro-climate) considerations, the collector should be able to exchange air with the environment (slight ventilation). Above all, the seals at the tube ducts must withstand very high temperatures in the case of stagnation and mechanical loading (movements of tubes due to temperature expansion).

Thermal Insulation Durability

Peuser, Remmers, and Schnauss (2011) investigated materials degradation of collectors, having served already 10–20 years of operation. In these collectors the insulation consisted almost exclusively of polyurethane (PUR) rigid expanded

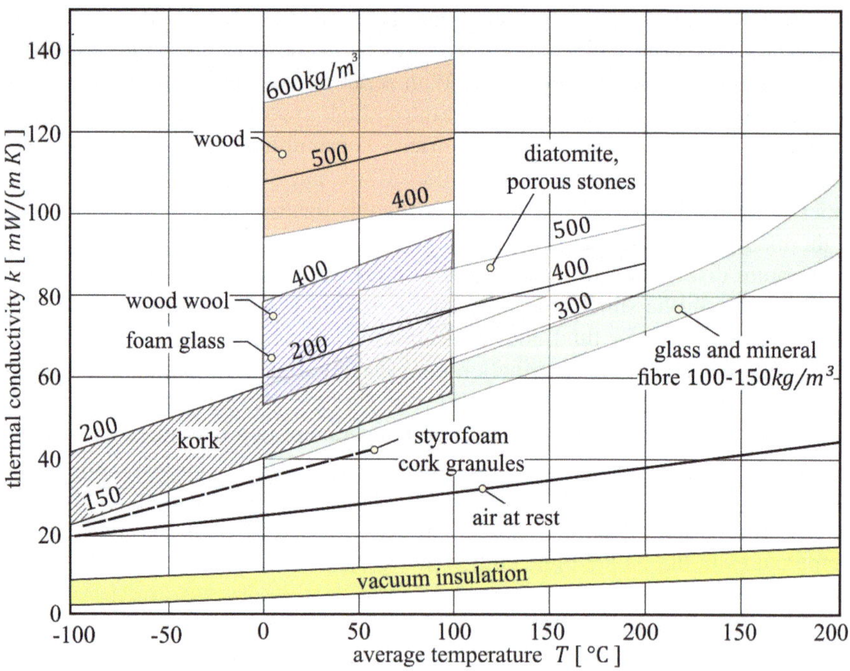

Fig. 6.31 Thermal conductivity k (W/(mk)) of selected insulating materials as function of the temperature in (°C)

Fig. 6.32 Right: rock wool insulation of a selective flat plate collector

Fig. 6.33 Seal between collector casing and collector tubes, (new-right and after more than one decade of use-left)

sheets or mineral wool. Often the insulation was having an aluminium lamination (laminated on one or both sides). Mineral wool is still used frequently today for which an examples is illustrated in Fig. 6.32. PUR sheets (not in combination with other materials) are rare. When opening the collectors, the mineral wool insulation was undamaged and seemed to be in good condition, regardless of whether it was used in selective or non-selective collectors or whether aluminium lamination was present or not. Also, the thermal conductance of insulation from a 16-year old collector resulted in a value of $U(10\ °C) = 0.034\ W/(m^2K)$, which is close to the value for new material.

The results of the investigations proved that except for some minor optical impact, the mineral wool materials had still good performance after many years of operation while the PUR mats were discoloured brown on the surface at the absorber and hardened at the top. The intensity of the discolouration follows the temperature development in the absorber. Even when the insulation mat was protected from the absorber by the mineral wool, damage to the PUR foam could be observed. In selective collectors where PUR foam was directly touching the absorber back, the polymer foam material partly crumbled away. In contrast, none of the PUR material in non-selective collectors showed discolouration which is a clear indication of lower temperature load. As the cell structures of the PUR foam were partially destroyed due to high temperatures and gas could escape, the thermal conductance of the PUR increased, however still being lower in comparison to mineral wool (PUR foam manufacturers state thermal conductivity values of around $k = 0.030\ W/(mK)$ for insulation mats without propellant). Laboratory tests showed that thermal conductivities are still low even after partial gas evolution of the propellant, but have deteriorated by approx. 20% and by 25–45% for the degraded 15 year old material depending on the colouration—compared with the new material $k(10\ °C) = 0.020\ W/(mK)$; after one week in an oven at 150 °C: $U(10\ °C) = 0.024\ W/(m^2K)$ The continuing good insulating properties of

PUR rigid expanded foams even after many years are confirmed by measurements on this material from two 13–14 year old collectors. The thermal conduction was maintained. For a non-discoloured insulation from a non-selective collector still good insulating properties of PUR rigid expanded foams can be seen after 15 years with about $k(10\ °C) = 0.025\ W/(m^2K)$, whereas it is comparable with that of new mineral wool for selective absorbers. Therefore, even if PUR rigid foam is not optimal for use in selective collectors due to their low temperature resistance the damage which occurs is not critical to the function of the collector. Problematic possibly are very efficient collectors used in large-scale installations which are frequently exposed to high stagnation temperatures. Also, the seals at the tube ducts are to be viewed more critically than the insulation. Figure 6.33 shows two of the seals made of an elastomer, after several years of operation with some stagnation periods in new and aged conditions. The seals at the collector inlet are typically facing lower temperatures, while those at the outlet get deformed and highly crumbled and therefore more likely to fail than those at the collector inlet.

Current Thermal Insulation Materials

The thickness of the rear insulation is often between 40 and 70 mm. Increasing the thickness of the insulation beyond this leads only to a small improvement in the U-value. In collectors with frames made of aluminium or stainless steel, an edge insulation can prevent a thermal bridge between the absorber and frame. This markedly improves the U-value. Furthermore, some thermal bridges to the tube ducts should be avoided by using suitable insulation. Careful construction also considering the vibration during transport, has proven to be more effective than further increases in insulation thicknesses. Panels with deposits on the inside of the cover should be avoided as interaction with other materials in the collector, such as glass and absorber coatings. Therefore, stability and compatibility should be tested before their use. It should be carried out with every material before its use since the high temperatures in the collector cause gas evolution in almost all materials. Sheep´s wool, flax, hemp, or similar natural materials are used in some collector designs, but some of these materials have proven to be unsatisfactory in practice. This may be noticed by condensation on the inside of the collector cover or by decomposition.

Requirements for Sealing Materials

Materials based on ethylene-propylene-dien-monomer (EPDM) have proven themselves for the seal profiles used between the frame and the pane of glass. The adhesives used are mostly silicon-based. In this case, care should be taken since many of the products on offer have completely different compositions despite having the same name. When using the adhesives, adhesive power, material softening, and gas evolution must be tested as well as their long-term stability. The seals in the tube ducts should be temperature resistant to the stagnation temperature (larger than 180 °C) to guarantee permanent sealing against leaks. Some silicon-based materials are suitable for this purpose.

6.3.3 Air Collectors (Absorber Design, Ventilation Energy)

Air collectors use air as a heat transfer medium, and therefore need channels with large cross-section for the air transport. Otherwise their basic construction corresponds to that of a flat plate collector, as Fig. 6.1 sketches. The surface-structured absorber is placed in a metal tray and transfers the heat energy provided by the sun to the air flow below. However, it is also possible to use other materials with large surface area, like woven textiles or matrix absorber, even pebble heaps consisting, for example, of lava stones, which are heated by solar radiation and transfer the heat to the air. Modern air collectors achieve efficiencies $\eta = 50–70\%$ in terms of heat transfer. Since the air ducts have a large cross-section, a solar thermal system with air collectors is more voluminous than liquid collectors. Depending on the basic air heater designs the transfer of solar heat to the air has a different collector efficiency factor F'. These are shown for a wider variety of geometries in Fig. 6.34.

Since air itself can store almost no energy (over a long period of time), there are two alternatives for using the heat:

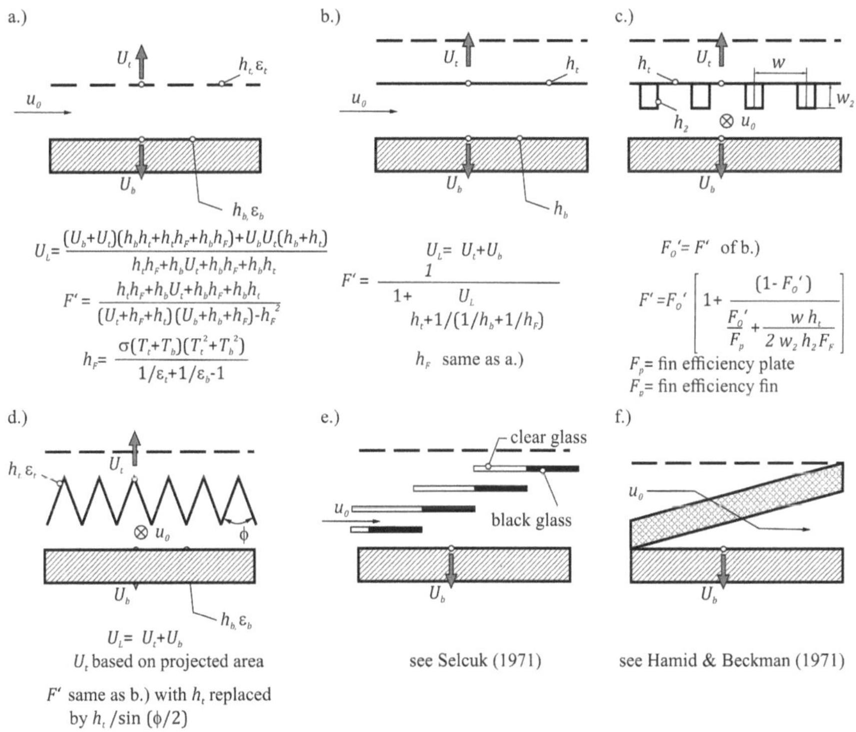

Fig. 6.34 Typical design of absorber geometries for air collectors and corresponding F' factors; Selcuk (1971) and Hamid and Beckman (1971)

Table 6.4 Pros and Cons of air-cooled collectors for active/passive solar systems

Advantages	Disadvantages
No freezing	Large duct cross-section
No steam formation	High ventilation energy
No damage of HTF by stagnation	Lower efficiency
Leaks are less problematic	Collector thickness
Lower critical start-up level	

- direct air heating with moderately heated warm air ventilated directly into a building with high fresh air demand (holiday homes, apartment building, swim halls, factory halls etc.). The combination with an air-heating system is most favourable.
- Air is heated to high temperatures and heat is transferred to a storage by an air–liquid heat exchanger (water storage) or solid storage materials (rocks, pebbles, bricks, or hollow walls etc.). Only after charging the storage, the cooler air may be used for direct heating of rooms.

In agriculture and industry, air collectors have been successfully used for several years to dry hay, grain, herbs, or other biomass. Very favourable applications are solar dryers and swim halls, as they have a high ventilation heat requirement even in summer. But also the heating of buildings with lower comfort temperatures and high ventilation requirement is a popular application. Some pros and cons of air cooled collectors are listed in Table 6.4.

A variant which is especially suitable for large industrial buildings and halls is the transpired uncovered metal sheet collector (Wang, Shukla, & Liu, 2017), which is depicted in Fig. 6.35. In this product the outside air is sucked through a perforated absorber sheet and heated up in the passage. The air flow must be adjusted to reach the desired temperature. Behind the absorber fixed to a vertical façade the air is guided through the air-channel between absorber and wall to the building interior. With this technology preheating is of fresh air is the target. The coloured and stable absorber coatings provide an architecturally interesting option of a wall cladding.

6.3.4 Vacuum Tube Collectors

A vacuum tube collector (VTC) consists of an evacuated glass tube, inside which there are either concentric cylindrical or fin-type absorbers. A cost-efficient way to produce vacuum tubes is by using two concentric glass tubes which are containing an evacuated volume (the Dewar-type Sydney collector). The required vacuum—as the characteristic distance L between absorber and glass tube is in the order of magnitude 1 cm—must be as low as 10^{-7}–10^{-8} atmospheres. In this case the thermal conduction of gas can be suppressed and only the radiative component contributes to the heat loss. We can see some cross sections of VTCs in Fig. 6.36.

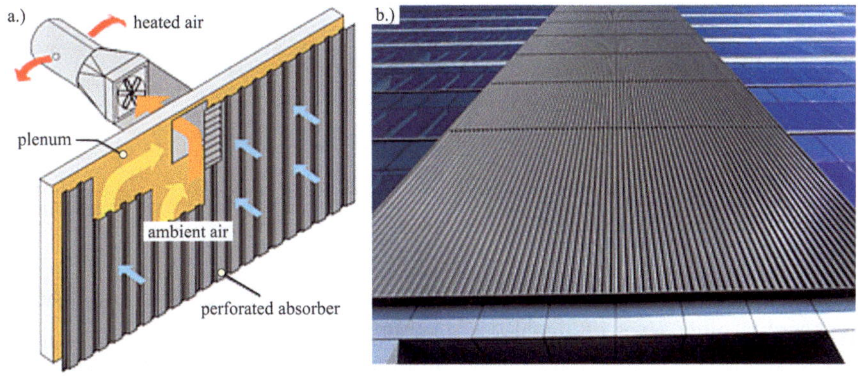

Fig. 6.35 Transpired façade air-collector—functional principle (**a**) and example building façade using SolarWall air collector (**b**) (SolarWall Europe Sarl., 2023)

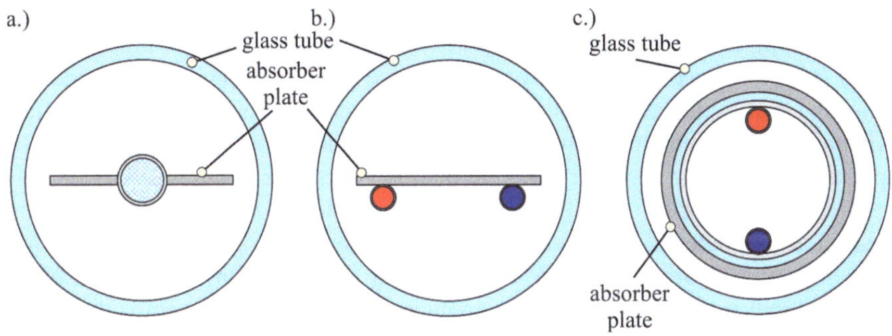

Fig. 6.36 Cross sectional cut of different vacuum tube collector (VTC) types; **a** heat pipe, **b** direct flow and **c** Dewar type

Vacuum collectors differ from flat plate collectors by an almost optimal thermal insulation within the collector box. The absorber strips are located inside an evacuated glass flask. This practically eliminates convection losses completely. Only the heat radiation occurs as a loss factor. This is minimized by a selective coating of the absorber strips. The heat transfer to the plant circuit is either direct or by means of the so-called heat-pipe system. We will discuss the three absorber basic VTC types, heat-pipe VTC, direct-flow VTC with fins and Sydney collector in the next subsections individually.

6.3.4.1 Heat-Pipe Vacuum Tube Collector

Vacuum collectors based on the heat-pipe principle use no circulation of fluid through the vacuum tube as other types of vacuum collectors with direct flow, and their mode of operation is shown in the Fig. 6.37. The heat pipe inside the glass tube is coated black and absorbs the solar radiation. It is closed on both sides

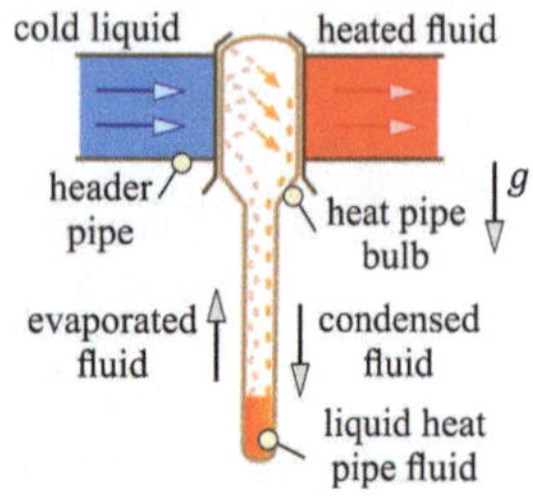

Fig. 6.37 Operation principle of a heat pipe

and is hermetically sealed itself, containing a rapidly evaporating liquid, usually methanol or other types of alcohol, at an appropriate pressure.

Due to the heat transferred from the absorber to the liquid in the heat pipe, this liquid evaporates and rises to the top due to its lower density. Therefore, usually a heat-pipe requires an installation inclined at a certain angle as it is dependent on gravity. A heat exchanger is located at the upper end. Here, the steam condenses and transfers the heat of evaporation to the secondary circuit. This secondary circuit is connected to the heat circulation system of the solar circuit. The advantages of the heat pipe system are greater ease of maintenance, as defective pipes in a collector can be replaced individually without having to drain the transport medium. Since the heat exchanger is connected to the secondary circuit "dry", a better south orientation of the individual absorber fins can be made possible in larger plants by specially rotating the absorber tubes. In addition, this type requires a considerably smaller pump capacity, since there are no pressure losses within individual tubes. Heat Pipes are inherent safe and protect the solar circuit from overheating—when the temperatures are too high no condensation can occur anymore and transfer heat. A disadvantage of the heat-pipe compared to the coaxial system is the higher price, which is mainly due to the more complex design. Such a dry connection is illustrated in Fig. 6.38.

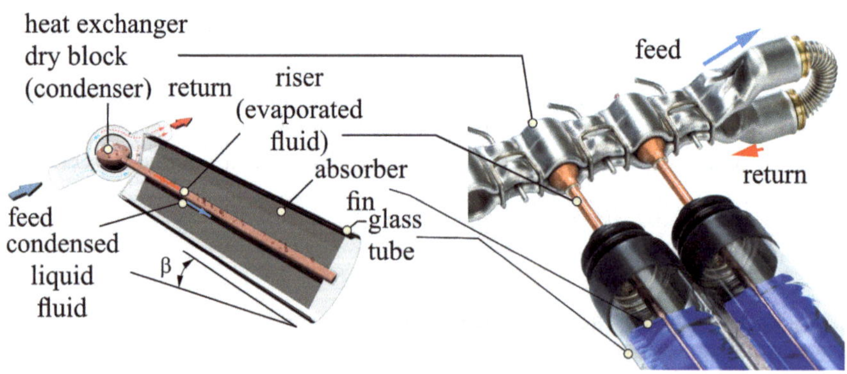

Fig. 6.38 Vacuum tube according to the heat pipe principle with dry connection

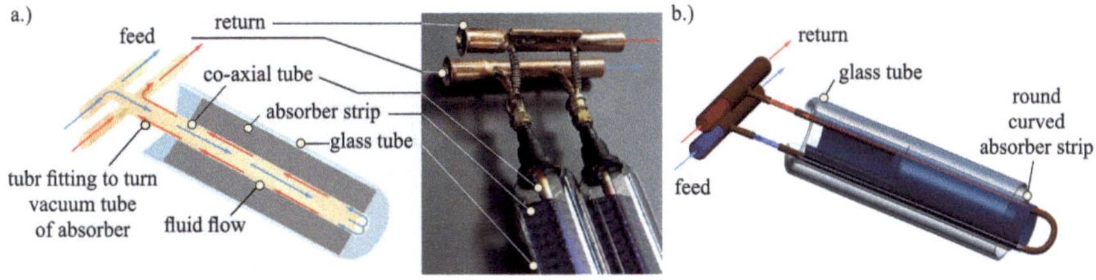

Fig. 6.39 Vacuum tube collectors (VTC) with fins; coaxial tube design **a** with a typical head design and **b** U-type tube design

The inside of the glass tube can be evacuated with a vacuum-tight connection between the metal tubes and the surrounding glass tube. The vacuum thermally insulates the absorber against the glass tube. One of the central challenges is to ensure a vacuum-tight connection between the glass tube and the metallic absorber over many years. Occasionally the glass tubes contain a getter. This can be heated inductively from the outside to restore the vacuum during the service life.

6.3.4.2 Direct-Flow Vacuum Tube Collector with Fins

There are two different types of direct flow vacuum tubes, the coaxial vacuum tube collector, and the U-tube collector. Both are collector tubes connected on one side to the circulation. However, in principle the fin-type collectors exist also as two-sided flow through collectors. In the coaxial version, the absorber fin is directly connected to a single coaxial heat exchanger tube through which the heat transfer medium flows directly. The heat transfer medium absorbs the heat from the absorber via the heat exchanger tube. The heat exchanger tube opens into the collector pipe. The fin collectors have the advantage that solar radiation may hit the absorber fin vertical when the sun is at an optimal position. Deviations from the south direction can be compensated for by axial rotation (maximum approx.25°) of the vacuum tubes. One problem for the usual oneside collectors is the temperature difference between inlet and outlet being very close to each other and connected to the same fin. The critical upper side, the so-called head, of such a collector tube is shown in the Fig. 6.39. There, relative strains occur because of the different temperatures, which must exclude failure in every potentially conceivable operating condition by suitable design measures. Typical design measures to minimize stress are metal bellows integrated.

6.3.4.3 Dewar-Type Tube Collector "Sydney Collector"

Another solution is a double glass tube with the vacuum in the intermediate area similar to the principle of a thermos flask or Dewar. The absorber is coated onto the inner glass tube. The absorber cross-section is smaller than the outer glass tube, so that the usable absorbing surface is smaller than the collector surface. This can be compensated by rear side reflector sheets.

The two glass tubes lying one inside the other are welded together and evacuated. The absorber is vapour-deposited on the inner glass tube and the absorbed

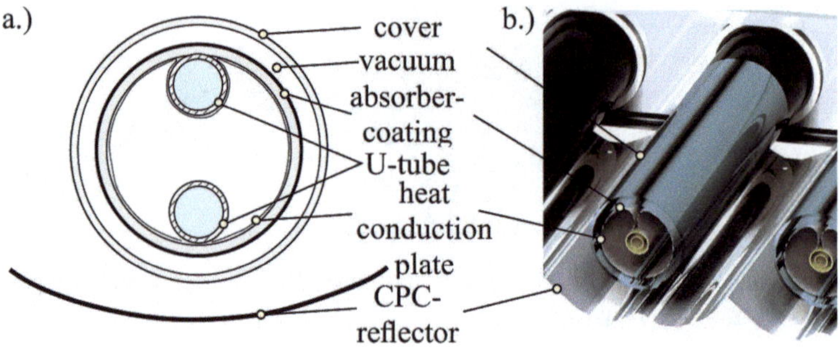

Fig. 6.40 **a** Schematic diagram of a direct flow U-tube collector. **b** Improvement of the heat generation of vacuum collectors by rearward mounting of CPC mirrors

heat is transferred to the heat transfer medium via a heat conduction sheets and absorber tubes inserted in them. This is sketched in Fig. 6.40a. The cold fluid enters at the bottom for preheating and returns in the top to reach temperatures. CPC mirrors are used at the rear of the absorber to ensure that the incident radiation can also be utilized. As an alternative also here, a coaxial tubing is possible (Fig. 6.40b), which exhibits better standstill behaviour than a U-tube.

During standstill the absorber tube and the pipes reach high temperatures. If, after a standstill, cooler fluid first flows back into one of the two U-tube legs on one side—which can hardly be avoided—then this leg is cooled down considerably within a very short time. Cooling can be more than 100 °C within a few seconds. The pipe is stiffened by fixing it to the absorber plate. The only possibility to reduce resulting thermal stresses remains in the pipe section between the absorber plate and the penetration of the vacuum pipe. In many cases, only a few such situations are sufficient until the already critical vacuum-tight connection glass tube/ metal tube fails. In addition, the U-tube type has a lower square metre capacity.

The optical losses of vacuum collectors are comparable or slightly higher than those of flat collectors (single-glazed, curved surface). The conversion factor η_0 is in the range between 0.65 and 0.85. Due to the excellent thermal insulation of the vacuum in the order of 10^{-4}–10^{-5} mbar at a collector temperature of 150 °C and an ambient temperature of 20 °C, vacuum collectors achieve high efficiency even at very high operating temperatures or low irradiation. The heat loss factor U_L is in the range 0.5–1.2 W(m^{2}K) which is due to the thermal radiation loss.

Due to these values the efficiency curve of a typical vacuum tube collector as shown in Fig. 6.41 indicates efficiencies of 50% for high irradiation even at higher temperatures above 100 °C. VTC are preferably used for heating support in older buildings (high-temperature heating) and for process heat applications. They are also an option if space is limited. A disadvantage of vacuum tube collectors compared to flat plate collectors is their significantly higher price for high quality products in Europe. In China VTCs of a simpler quality with shorter lifetime, however, are very cheap being a mass product.

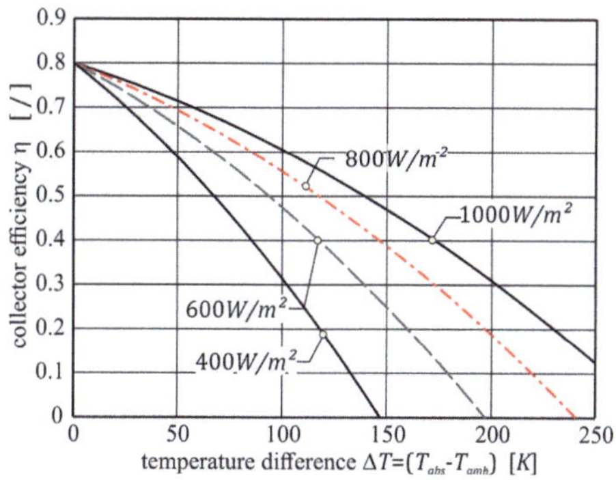

Fig. 6.41 Efficiency curve η of a vacuum tube collector as a function temperature difference for different solar irradiation intensities G in (W/m^2) for a collector with $u_0 = 1.45$ W/(m^{2}K) and $u_1 = 0.005$ W/(m^{2}K) according to Eq. (6.10)

The quality of the vacuum in the glass tube can be easily checked. At a sufficiently low internal pressure, a "good vacuum", there is neither heat conduction nor convection in the free space around the absorber. The absorber is selectively coated and behaves approximately like a Lambertian spreader in terms of thermal radiation. The temperature is the same almost everywhere around the glass tube circumference. If the pressure increases, heat conduction occurs, and the glass tube temperature is highest at the point with the smallest distance to the absorber. With a temperature peak at the zenith of the glass tube, i.e. at the top, the internal gas pressure is sufficiently high that natural convection can form. Additionally, getters are built into vacuum tubes as indicators. If the internal pressure increases due to diffusion or a leak in the glass tube, the getter absorbs the acid and nitrogen. This can be done at operating temperature, or the getter is heated by induction heating. If the leakage rate exceeds the absorption capacity of the getter, this can be seen from its colour change—usually white.

6.4 Stationary Collectors with Low Concentration

The use of solar thermal energy to supply industrial process heat is one of the most promising areas for solar market growth. Whereas low temperature ($T < 100$ °C) applications and the higher end of medium temperature applications (200 °C $< T <$ 400 °C) are well suited for stationary collectors and line-focusing tracking collectors, respectively, for the temperatures gap in between there are few developments available. Non-imaging collectors of the type flat-plate collectors with integrated CPC (Compound Parabolic Concentrator, see Chap. 3), if carefully designed, could possibly serve this temperature range. Another option is to use flat-plate collectors with reduced heat loss for higher temperatures and/or a radiation enhancement method using external mirrors redirecting part of the global radiation towards the collector.

6.4.1 CPC-Collectors

CPC reflectors may be integrated into a flat-plate collector. This leads to a larger thickness of the collector casing, and a CPC volume with an internal tubular or sheet absorber, in which due to the depth of the structure air usually is convecting. Therefore, the heat loss in such collectors is a challenge. Commercial first developments were not very successful, as described in IEA TASK 33 (Weiss & Rommel, 2008), see e.g. Fig. 6.42.

Due to the laws of optics it can be shown that collectors with $C_{max} < 2$ has a resulting half acceptance angle of $\theta_a > 30°$. As the declination of earth's orbit determining the seasonal path variation of the sun is $\delta_{max} = \pm 22.5°$, in practice, even lower values of C are preferred.

Non-imaging CPC collectors can be designed for quasi-stationary (or seasonally tilt adjusted) operation, making use of the advantages of a low concentration $C < 2$ while not depending on tracking systems. In this way a low-cost production could benefit from the dismissal of tracking devices. Many factors, from the geometrical design to materials, production, installation conditions, O&M, operation temperature and resource availability have an impact on the performance and cost, and must be accounted for to find the optimum collector design (Osório et al., 2019).

Stationary collectors with appropriate tilt and CPC trough orientation East–West could cover then the solar path for the whole year. At equinox the solar path should be just in front of the aperture of the collector, such that the normal on the aperture would be in the orbit of the sun. And the tilt of the CPC collector would be equal to the latitude:

$$\beta = \Phi. \tag{6.36}$$

For the whole year the projected transversal incidence angle on the aperture is within $\Phi \pm \delta_{max}$, and provided, the acceptance angle is large enough, the direct beam radiation from the sun should be accepted throughout the year by the optical concentrator. If a smaller acceptance angle is chosen, a seasonal adjustment is sufficient.

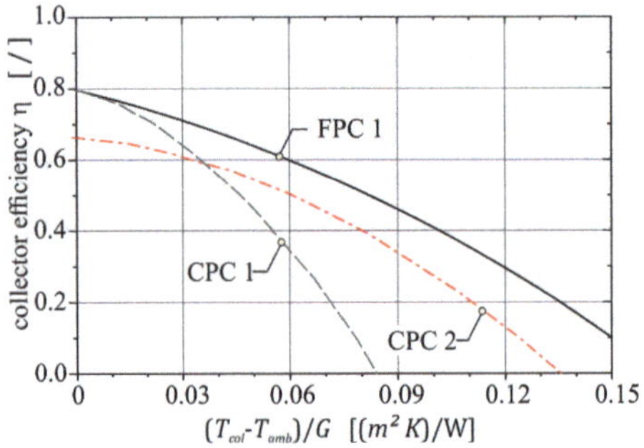

Fig. 6.42 Efficiency curve η of a two CPC collectors as a function reduced temperature difference for $G = 800$ W/m^2

One would think that for a CPC trough with an acceptance angle $\theta_a > \delta_{max}$ the condition should be always fulfilled. However, the analysis of the incoming solar radiation with direction cosines μ_E, μ_N, μ_H with the normalized radiation vector $\overrightarrow{k_s} = (\mu_E, \mu_N, \mu_H)$ shows a different situation (Spirkl et al., 1998). Rays are only accepted if and only if

$$\left|\frac{\mu_N}{\mu_H}\right| \leq \tan\theta_a. \tag{6.37}$$

Using the fact that the vector is normalized the direction cosine along the aperture normal μ_H can be eliminated, which leads to the equation

$$\mu_N^2 \leq \left(1 - \mu_N^2 - \mu_E^2\right)\tan^2\theta_a. \tag{6.38}$$

Putting this into the normalization equation where the sum of cosines squared equals unity, one obtains

$$\frac{\mu_N^2}{\sin^2\theta_a} + \mu_E^2 \leq 1. \tag{6.39}$$

The Fig. 6.43 shows the critical acceptance angles for the incidence direction cosines μ_N and μ_E for a CPC trough with critical acceptance angle 35° and 23.5°.

The solid collector just touches the sun path at the solstice and as a result the collector achieves no operational hours on the summer and winter solstice days. The dashed collector has a large enough acceptance angle (θ_a) to still capture a significant fraction of solar energy even during the solstices. The unit circle in the plot represents the "horizon" of the collector (as for a horizontal collector at the equator)—if the collector would be at another latitude and then would be tilted, a part of the ground would be included there. The horizon to be seen by the collector in this case then would appear also as a half ellipse in the lower half of the map.

The combination of a CPC type reflector with an evacuated tube collector is a commercial solution already on the market, as already shown in Sect. 6.3.4. However, here the concentration factor C is close to 1, and the advantage of the reflector is mainly to collect also global direct and diffuse radiation falling in between the tube absorbers. However, with a different shape also higher concentration can be reached—considering that then only a part of the diffuse radiation can be redirected to the absorber.

If higher temperatures shall be reached by a CPC type collector, the design must be more sophisticated. Absorber shape and concentrator shape must match each other. As already shown in Chap. 3 the CPC shape is different for tubular and flat absorbers. There are also geometries where two sides of a flat absorber are irradiated. A critical issue with the optical design is always that the hot absorber on the one hand side should not touch the cold reflector, and that rays falling into a gap between absorber and reflector shall not escape again.

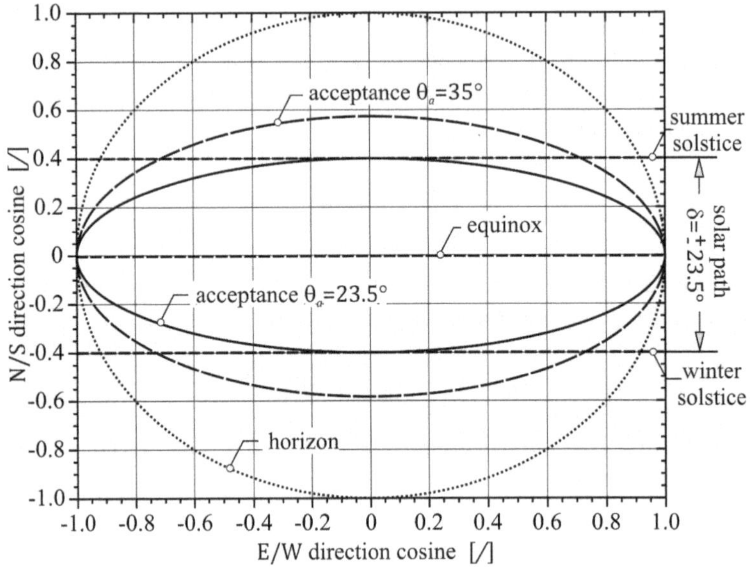

Fig. 6.43 Direction Cosine Map. The full hemisphere is shown as the unit circle. The sun travels along horizontal paths each day, bound by the upper and lower dashed black lines representing the summer and winter solstices. The sun path during the equinox is the horizontal dashed line ($y = 0$). The parts of the sky which are "seen" by a solar collector are drawn for acceptance angles $\theta_a = 23.5°$ (solid) and $\theta_a = 35°$ (dashed)

The design and development of an experimental medium-temperature solar thermal collector, the external compound parabolic concentrator (XCPC), is presented in Widyolar et al., (2018). Here, a non-imaging reflector is paired with an evacuated tube absorber for efficient and low-cost heat collection between 100 and 250 °C. The absorber geometry is optimized under the constraint of being assembled with an ultrasonic welding machine, with a final pentagon-shaped absorber selected. This approximates a circular absorber. The external CPC trough is designed with an acceptance angle of 40°. The modified absorber shape, gap loss, and truncated reflector result in a geometric efficiency of 93% compared to an ideal CPC. The simulations for the manufactured prototype predicted an optical efficiency of 71% and thermal efficiency of 50% at 200 °C. Experimental test results (optical, thermal, stagnation) have shown an optical efficiency of 62% and a thermal efficiency near 50% at 200 °C with a final stagnation temperature of $333°C$. This example shows that in principle with such a design high temperature can be produced with a stationary collector. Whether that collector is cost efficient and practical enough, must be shown, but the theoretical potential is demonstrated.

6.4.2 Radiation Enhancement with External Reflectors

Another option to reach higher temperatures, this time with flat-plate collectors, is the possibility to enhance the radiation on the collector aperture by external reflectors. When between two rows of tilted flat-plate collectors the gap can be filled with mirrors, then a part of the diffuse radiation from the sky, and for most of the time direct beam radiation may be reflected on the collector aperture. Due to reflectivity losses and spillage due to incomplete reflection matching with the collector aperture, of course not a doubling of the incident radiation can be achieved, but a factor between 1.3 and 1.6. In the German Reflec Project (Hess et al., 2010) such a collector assembly has been designed and optimized. A prototype of such a collector has been installed to provide heat for a laundry at elevated temperatures. The flat-plate collectors used were double-glazed selective collectors with low heat losses.

The concept of a collector field with reflector segments in between the rows is shown in the next picture as a CAD impression (Fig. 6.44).

After optimization of the reflector concept, where an expensive CPC shape reflector has been replaced by an assembly of three flat reflector segments, prototypes have been built and tested, which is illustrated in Fig. 6.45. The incidence angle modifier (IAM) of such a collector-reflector unit is biaxial and asymmetric for incidence of beam radiation in the vertical plane. The incidence angle modifier in the longitudinal plane is just the same as of an ordinary flat-plate collector without reflector.

The transversal incidence angle modifier for the collector in the vertical plane is depicted in Fig. 6.45. It shows a maximum enhancement of radiation at about 30° incidence. This is the angle where the direct sun is fully projected onto the flat-plate collector. The reflector itself shades in continuously radiation from the ground in the range $-10°$ until $-40°$. After that some radiation enters again through the gap between reflector and collector.

However, this part is shielding ground reflected diffuse radiation, which is much lower typically than the diffuse sky irradiance.

The performance at higher temperatures compared to other flat-plate collectors can be deduced from the efficiency curves as shown in Fig. 6.46b.

Taking into account the complex optical performance of the collector, the benefit over a year can be seen in the annual solar gains. For the location of Würzburg,

Fig. 6.44 CAD impression of RefleC collectors 4 on a flat roof. Three flat segments of 900 mm approximate the shape of a one-sided CPC with $\theta_c = 35°$ with a trough slope of $\beta = 55°$

Fig. 6.45 Prototype of the RefleC collector on the outdoor test rig of Fraunhofer ISE; the collector is tilted β = 55°

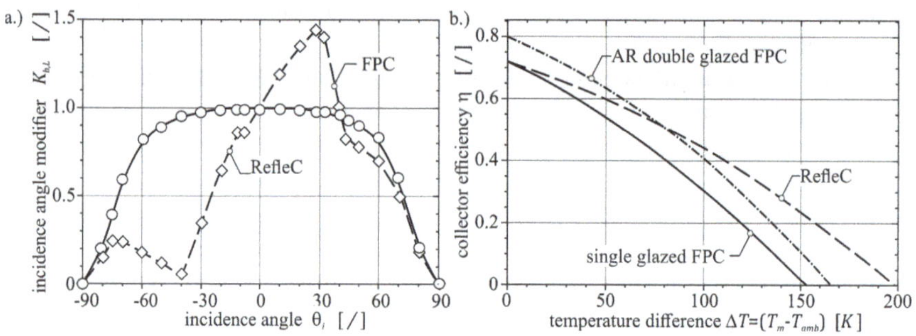

Fig. 6.46 **a** Transversal incidence angle modifier of the RefleC collector (beam irradiance in the vertical plane); the negative values between −35° and −90° correspond the view angle of the ground for a collector til of 55° (after Hess et al., 2010); **b** Efficiency curves of an improved FPC with booster reflector (Reflec), a single glazed standard FPC and an improved AR double-glazed FPC without booster as function of temperature difference for G = 1000 W/m²

Germany, a comparison between the RefleC collector using a double-glazed improved flat-plate collector (cover glass is using antireflexive (AR) coatings on both side; second glazing is a highly transmitting polymeric film) and the same collector without a reflector shows an impressive larger yield especially for higher operational temperatures (Table 6.5).

Table 6.5 Annual gain of RefleC and AR-double glazed flat-plate at constant inlet temperatures T_{in}. The RefleC gains are related to 1 m^2 flat-plate aperture to illustrate the additional gain by the booster reflectors (location Würzburg/Germany). Extrapolated from Hess (2015), Hess, et al. (2010)

System	Annual yield of collector w/o booster reflectors $[kWh/(m^2a)]$	Annual yield of collector with booster reflectors $[kWh/(m^2a)]$
Optimal tilt β	40°	55°
Inlet $T_{in} = 40°C$	645	771
Inlet $T_{in} = 80°C$	350	427
Inlet $T_{in} = 120°C$	145	271

6.5 Advanced Collector Topics

Although solar thermal collector technologies for hot-water and heating applications are mature and highly efficient, innovative approaches exist, which improve efficiency, reduce costs and materials use, integrate new functions in a collector or improve the aesthetical appearance of the technology. In the next sections some of the more recent developments are shortly characterized.

6.5.1 Bionic Collectors

Solar flat-plate collectors with harp absorber are a state-of-the-art technology with very good performance. However, the hydraulic resistance of the different riser channels is slightly different, leading to different flow and temperature distribution over the complete absorber surface. The aim of a new development of improved absorber sheets was to develop solar collectors with improved absorbers based on bionic principles (Hermann, 2005a, 2005b). The bionic FracTherm® channel structures as shown in Fig. 6.47 are multiply branched in a fractal way to obtain a uniform flow distribution, a low pressure drops as well as a high thermal efficiency. The pattern resembles the veins of leaves or in the human body. The aluminium absorbers are produced by a roll-bond process. In the roll-bond process, the channel structure is printed with a separating ink on an aluminium sheet. A second sheet is then rolled over the first sheet, and they are cladded together except for the areas printed with the separating ink. In a further step, pressurized air is applied to inflate the flow channels through which the solar fluid is to flow. When the channels produced in this way are visible on both sides of the absorber, they are called Double-Side-Inflated panels (DSI), as opposed to One-Side-Flat panels (OSF).

Not only rectangular shaped absorbers can be design but also other shapes, for example triangles. This allows in principle a good building integration. The collectors have been experimentally and numerically investigated for a wide temperature range to cover various applications, as illustrated in Fig. 6.48. A specific FracTherm® algorithm for designing the flow channels has been developed. A

Fig. 6.47 Partial view of selectively coated FracTherm® roll-bond absorber

challenge of the development has been the coating of the absorber after the production of the channel plate. To prevent corrosion in Aluminium, it is necessary to work with appropriate heat transfer fluids. The collector has been tested and developed in a European project Bionicol (Hermann, 2011).

Through a partnership in 2012 between CGA Technologies and the Finnish company Savosolar Oy, spectrally selective roll-bond solar absorbers can meanwhile be offered commercially. In the usual production process, absorber tubes are connected to pre-coated metal sheets. In the roll bonding process, however, the coating is applied to the finished absorber.

The FracTherm® principle is not limited to the design of solar absorber but can be applied also to other applications where homogeneous flows with minimum hydraulic resistance are useful, for example in cooling structures of semiconductor elements or for heating elements in buildings.

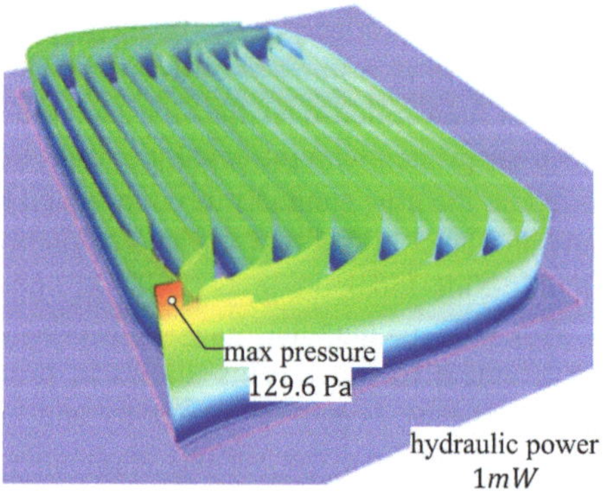

Fig. 6.48 Numerically simulated pressure distribution in a FracTherm® absorber plate (Hermann, 2005a, 2005b)

6.5.2 Collectors Using High-Performance Concrete

Usually absorbers are fabricated from metals with high thermal conductivity, because the thickness of the absorber sheet and the conductivity determine the F'-factor to a large extent (see Sect. 6.2.3). An attempt to design similarly efficient absorbers, but produced from a cheaper building material, is realized in the TABSOLAR® concept (https://www.tabsolar.de/). A special ultra-high performance concrete (UHPC) can be used to manufacture complex structures as needed. The material combines mechanical robustness with good thermal properties. The target of producing large area collectors for integration in a façade used the approach of developing prefabricated flow-through façade elements made of for different façade systems (ventilated curtain wall (VHF), external thermal insulation composite system (ETICS) and sandwich wall). A solar fluid flowing in channels through the absorber element made of UHPC transports the extracted heat to the use (Fig. 6.49). Inside the building, corresponding elements can also be used as so-called thermoactive building systems (TABS), possibly in addition to a classic concrete core activation system. Hence the name TABSOLAR®. The channels are formed directly from UHPC using the so-called membrane vacuum thermoforming process. The forming of the shape is shown in Fig. 6.50.

In the development of the concept several technological challenges had to be met. The first and basic challenge was the production of a stable UHPC absorber, which would withstand internal fluid pressure and temperature changes with leakage. In order to reach higher temperatures, the development of a selective coating on the UHPC absorber is necessary. As the substrate concrete has a high emissivity as such, a low emitting coating for the microscopically rough surface of concrete had to be developed. A prototype is depicted in Fig. 6.51.

The third challenge using the UHPC material with rather low thermal conductivity for an absorber was to reach a high F'-factor of collector (Fig. 6.52).

Three different product families have been conceived. The TABSOLAR® Premium design may achieve yields and temperature levels comparable to commercially available solar flat-plate collectors with a spectrally selective coating

Fig. 6.49 Bionic UHPC absorber cut into two pieces (modified from Hermann et al., 2020)

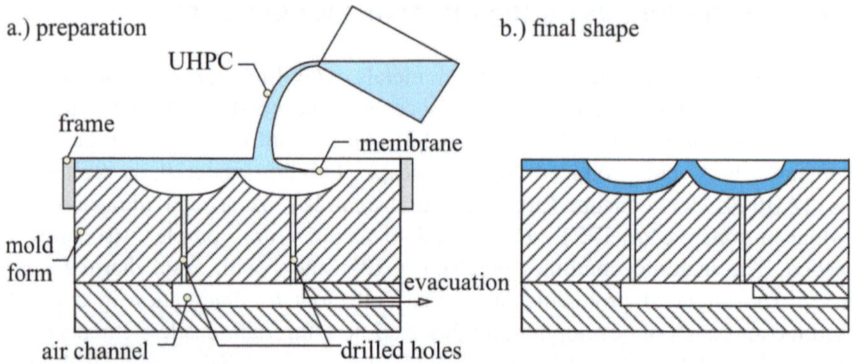

Fig. 6.50 First steps in membrane vacuum deep-drawing process **a** preparation of UHPC paste, **b** forming in mould by evacuation

Fig. 6.51 Spectrally selective coating on UHPC absorber (https://tabsolar.de/spektralselektive-beschichtung)

Fig. 6.52 Numerical simulation of temperature distribution and heat flux in UHPC-absorber resulting in $F' = 95.6\%$ from Hermann et al. (2015)

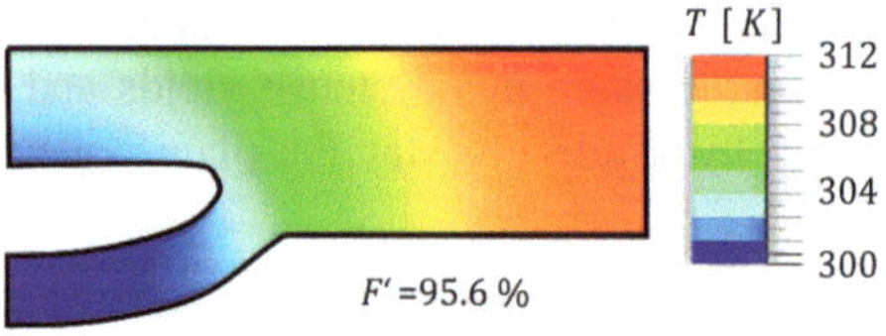

and anti-reflective glazing. TABSOLAR® Economy stands for painted or integrally coloured UHPC collectors with low-E glazing and slightly lower yields and temperature levels. Finally, in the TABSOLAR® Design product family, the collectors are unglazed and can be provided with a wide variety of structures and colours—see also Fig. 6.53. The focus here is on the design aspect; yields and temperature levels are even lower here (Hermann et al., 2020).

The first two product families are designed for direct solar applications for domestic hot water preparation or for use in solar combi systems. The third product family is characterized by a wide range of architectural design options. Since it

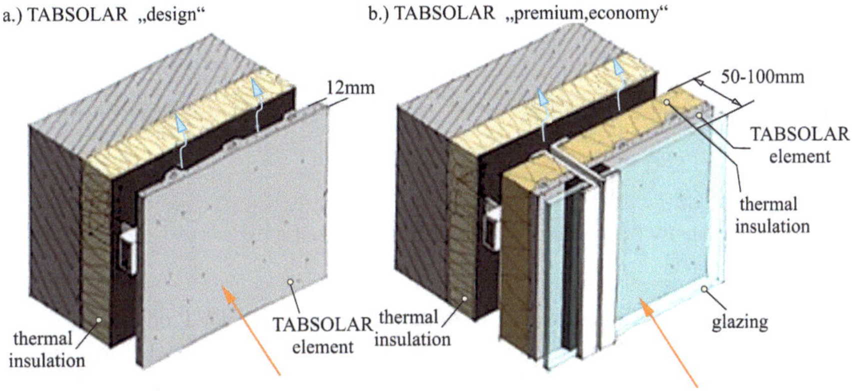

Fig. 6.53 TABSOLAR® rear-ventilated elements; **a** product family "Design" as uncovered absorber and **b** product family "Premium/economy" as glazed collector with low-ε glazing or selective absorber (modified from Hermann et al., 2020)

achieves significantly lower temperature levels and efficiencies, it is suitable, for example, as a low-temperature heat source for heat pumps, for domestic hot water preheating or for a swimming pool heating system.

6.5.3 Photovoltaic-Thermal Collectors (PVT)

The drop in photovoltaic (PV) module prices and the resulting area competition on the building roofs has stimulated interest in the development of photovoltaic/ thermal (PVT) collectors which convert solar energy into both electricity and heat. At present modern PV modules typically convert about 20% of the incident solar radiation into electricity, depending on the employed cell technology. This implies that a major part of the solar spectrum remains unused and is transformed into heat. Hybrid PVT collectors on the other hand can make use of this excess heat by cooling the solar cells using a heat transfer fluid. The electrical and the thermal performance of PVT collectors is lower than that of separate PV panels and conventional thermal collectors. However, they emphasized that two PV/T collectors together produce more energy per unit surface area than one PV panel and one thermal collector next to each other. This is particularly important in applications where surface area availability is very significant. In this way, solar energy can be harvested on cascading exergy levels: first high-grade exergy is produced in the form of electricity, then low-grade exergy in the form of heat. In this way, PVT collectors make optimum use of the solar resource with high utilization of available area. PVT collectors can be categorized according to their heat transfer medium (air and liquid) and their design (unglazed, glazed, and concentrating). Each collector design has its specific temperature levels and accordingly suitable applications.

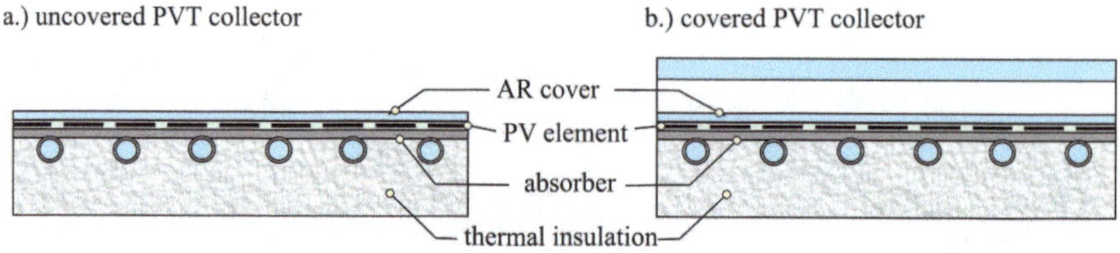

Fig. 6.54 Flat-plate PVT collectors **a** uncovered and **b** covered

At present the European market is dominated by unglazed PVT collectors delivering heat at low temperatures, however, at operating temperatures below ambient they can achieve some heat gains to be used by heat pumps. Glazed PVT collectors reach higher temperature levels. However, these collector types have often relatively high thermal losses, as they do not use selective coatings. In the next sections we will discuss the main types in practice, uncovered and covered flat-plate collectors and discuss their design and performance (Fig. 6.54).

Extensive reviews have been written on the large variety of PVT collectors (which we will not cover here), for example by Bandaru et al. (2021) who included a survey of different reviews, or Chow (2010) by looking into design variations and the importance of parameters like the impact of temperature on PV-cells, of the bonding quality of PV cells to liquid heat exchanger, of efficient cooling of the cells and thermal design options using different cover glasses with and without air gap (Zondag, 2008). Single sheet and tube absorbers were used very often in designs, but channelled plates are also interesting because of the more homogeneous temperature distribution. Uncovered collectors are widespread, as it is not easy to reduce the heat losses from a single cover collector. A selective absorber is neither possible on top nor below the PV layer. Early products have heat losses like non-selective black flat-plate collectors.

To achieve a sustainable PVT market development, performance and reliability of products is a key factor, and it can only be characterized by establishing functioning testing standards. The testing of thermal and electrical performance is complex in the case of PVT, and different situations must be considered.

Especially for uncovered PVT collectors the thermal performance has not only be determined with collector temperatures above ambient, as covered in (ISO 9806, 2017). The quasi-dynamic test method seems to be able to cover the wide range of PVT products. But as the uncovered PVT collectors are connected to a heat pump and used as an ambient energy source, also operating temperatures below ambient must be considered. For extracting heat from the ambient, the topics of condensation, rainfall and infrared radiation must be considered in the testing equation. An overview over existing approaches is given in Kramer et al. (2020).

The electrical condition of the PV cells is even important for thermal testing. The standard testing in hybrid operation where the PV module is operated in the maximum power point (MPP), yields lower thermal gains than a testing in open circuit mode (OC, disconnected with no current flowing). The optical efficiency

η_0 is higher in OC-mode because the photons excited in the PV-cells eventually thermalize and produce thermal power, whereas in the MPP-mode the electrical power is extracted. The electrical power extracted is. On the other hand, dependent on the cell temperatures. In most designs the absorber with the heat transfer fluid is coupled to the cells and should cool the PV as compared to the standard PV module operation. Especially for covered PVT-collectors, this contradicts the objective of reaching higher temperatures in the solar fluid. As a consequence, the electrical and thermal output ideally should be parameterized in a testing method simultaneously. Here a promising method has been developed (Kramer & Helmers, 2013) but not yet evaluated for all types of PVT collectors.

The reliability and durability of PVT modules are especially challenged at elevated temperatures and higher humidity loads. The test methods available from the IEC and ISO standards are covering the specifics of PV and ST module's, which often are similar for PVT modules. But the PVT specific loads and the operating conditions arising from the system integration of PVT are only partially covered in the set of existing standards (e.g. stagnation temperature approval, mechanical load resistance). Therefore, further work is needed to close this gap.

6.5.3.1 Uncovered PVT Collectors

Unglazed PVT collectors are beneficial for operation near ambient temperatures particularly because of high heat transfer to ambient air even in times of low radiation. Among PVT collector types, unglazed PVT collectors are widely popular in the northern European market mainly as an additional heat source for heat pumps. Such a collector is based on a PV module on the back of which an absorber plate (a heat extraction device) is attached. The purpose of the absorber plate is twofold. Firstly, to cool the PV module and thus improve its electrical performance and secondly to collect the thermal energy produced, which would have otherwise been lost as heat to the environment.

The thermal performance of a coverless PVT collector is reduced especially at high temperatures due to heat losses from the top. However, the uncovered PVT collectors have a better electrical performance than covered ones as already shown in a design comparison by Zondag et al., (2003).

The most common connection to couple the photovoltaic panel to the thermal absorber is the gluing technique. Here the entire PV laminate is bonded to the flat plate, by means of thermo conductive glues or adhesives, which shall be resistant to high temperatures. However, this technique presents several problems. With the gluing technique, air bubbles may be trapped between the PV laminate and the plate decreasing the heat transfer coefficient and creating inhomogeneity in the cell's temperature distribution. Additionally, a risk of condensation between the two parts occurs (Aste et al., 2014).

Concerning the manufacturing of uncovered, liquid-based photovoltaic-thermal (PVT) collectors, design requirements or low expenses have certainly high priority, not only the overall optimization of the thermal efficiency. The specific impact of inevitable thermal resistances caused by wavy surfaces and imperfect contacts between PV modules and heat exchangers is still a topic of investigation and optimization. In their paper (Brötje et al., 2018) focus on the performances of uncovered PVT collectors with detachable mounting of PV modules on heat exchangers and the effect of macroscopic air gaps. Two prototypes with thin film modules were built without using silicon glue, other adhesives or heat conductive paste. They measured the thermal zero-loss efficiencies for two adhesive-free collectors at $\eta_{0,M} = 66.8\%$ (copper pipe meander laser-welded on aluminium sheet) and $\eta_{0,M} = 69.2\%$ (aluminium roll-bond) based on gross areas. These results are similar or even better than the efficiencies of commercially available collectors. Due to the thermal resistance of air gaps, they found temperature differences of 12 K and higher between absorbing sheets and heat carrier, which have to be considered while estimating the electrical yield by means of system simulations. As comparison they also built a prototype with a very thin adhesive layer between PV module and a roll-bond heat exchanger. This optimized heat transport led to a higher optical efficiency of $\eta_{0,M} = 76.8\%$. Simulations using numerical, two-dimensional models show good agreement with experimental results and proved to importance of the thermal coupling (Brötje et al., 2018).

A second important parameter seems to be the backside heat exchanger efficiency, especially for the use as thermal source for a heat pump system. The solar gain itself is only a fraction of the ambient energy to be provided. The heat exchange with ambient air is key. With an PVT collector optimized for heat exchange with ambient air, front and back side should be used to increase heat exchanger area Chugani (2020) investigated the performance of a PVT collector which is thermally coupled with a fin-tube heat exchanger on the backside of PV. The fin surface area is 10 times larger than the PV area in each module, which specifically increases the use of ambient energy from ambient air and hence the PVT collector works as a good environmental heat exchanger. In addition to convective heat gains in times of low irradiation particularly in winters or nights, condensation heat gains, as well as heat gains through the phase change to frost occur on larger fins surfaces, which improves the thermal collector output and makes it suitable for working as a sole heat source of a heat pump. The authors show that especially when the PVT collector is the only heat source for a heat pump, the improved heat exchange allows to use smaller collector areas compared to a standard unglazed PVT collector as a reference for reaching the same seasonal performance factor.

Using the PVT collector as an ambient heat exchanger naturally leads to a large value of the thermal loss coefficients in the collector testing (see (ISO 9806, 2017) testing of WISC collectors, WISC = wind and infrared sensitive collectors). As a solar collector therefore, the efficiency is only good for very low temperatures. The solar gain seems to be less important than the gains from ambient air.

6.5.3.2 Covered PVT Collectors

A different development route than the one just presented is followed in the design of covered PVT collectors. The cover glass certainly leads to a lower optical solar gain, due to the additional reflectance. On the other hand, its purpose is to decrease the thermal loss coefficient to the environment. A collector like that may produce heat at higher temperatures with good efficiency as long as the sun is shining, however, the heat gain through ambient air at low temperatures is also drastically reduced.

Hybrid PVT collectors generate solar electricity and heat simultaneously in one component. PV modules typically have efficiencies between 15 and 20%. So, it makes sense to dissipate the unused part of the solar radiation as heat to a fluid and use it for domestic hot water or space heating. This makes better use of limited solar surfaces and generates more electricity and heat. However, covered PVT collectors often have a low thermal efficiency for operating temperatures 40–80 °C. A selective absorber cannot be applied on top of the PV-cells in the airgap of a covered collector such as Fig. 6.54b as all solar radiation would be absorbed. On the other hand, below in direct contact with the PV cells and the protecting glass and possibly lamination film the low emissivity surface of the absorber is blocked. One idea to design a good selective PVT collector is to have a partially transparent PV front cover, as also used in building integrated PV (BIPV), combined with a selective absorber. Here, the rear contacts of the PV cells must be provided by a grid. However, this option seems not be investigated often, probably because the transparency of the solar cells is limited (Zondag, 2008). Another option is the use of transparent selective coatings, as in principle used for windows, for the PVT collector as cover glass or as IR-reflective coating on the PV cover glass, e.g. on either side of the air gap. Both approaches have their challenges. Several design aspects of such a covered PVT-collector using a low-e coating have been tackled successfully.

The combination of photovoltaics and solar thermal energy in one component represents an exciting technological challenge. Good thermal coupling of the PV cell to the fluid is of central importance to ensure the lowest possible cell temperatures. Low-E coatings must be optimized to simultaneously have a high transmittance for high PV efficiency and a low emissivity for high thermal efficiency. The high temperatures in stagnation can be addressed either by overheating protection or adapted materials (glues, laminates). Finally, the PVT collector, solar thermal application, and building technology must be coordinated to exploit the full potential of this technology. A suitable low-e coating has been demonstrated by Lämmle et al., (2016). The highly transparent low-e coating is based on silver, specifically optimized for the application in PVT collectors. The coating combines a thin silver layer with a multi-layer system of low-refractive SiO_2 and high-refractive TiO_2. It reaches a low emissivity of ε (100 °C) = 0.13 while maintaining a high solar transmittance of $\tau_{AM1.5} = 0.79$ (see Fig. 6.55).

The densely packed monocrystalline PV cells are directly laminated on an aluminium sheet absorber. Thus, providing a good contact between cells and fluid. Two layers of EVA serve as adhesive, encapsulant and vapour barrier for the PV cells. A thin electrical insulation layer is spray-coated on the aluminium sheet before lamination. The cross-section of the collector is shown in Fig. 6.56.

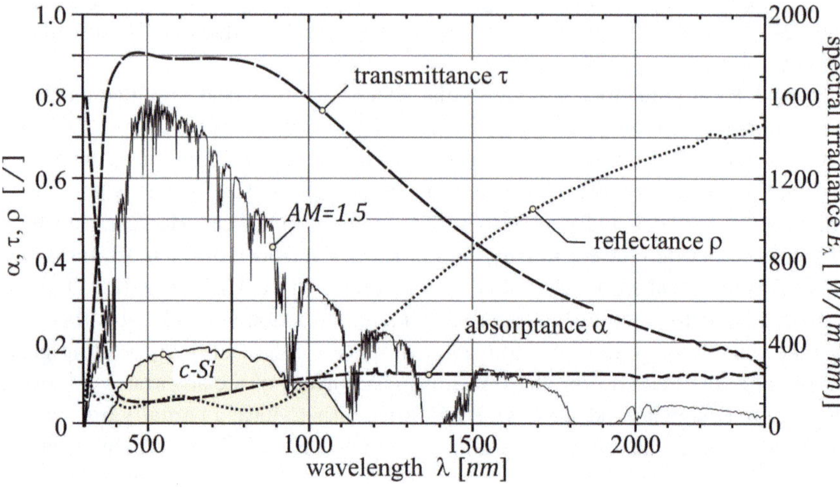

Fig. 6.55 Measured transmittance, reflectance and absorptance spectra for a PVT low-e glass cover, modified from Lämmle et al., (2016)

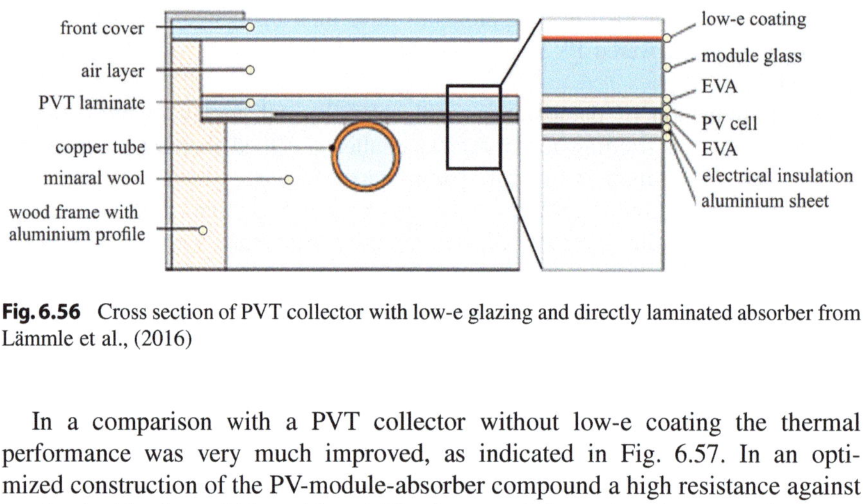

Fig. 6.56 Cross section of PVT collector with low-e glazing and directly laminated absorber from Lämmle et al., (2016)

In a comparison with a PVT collector without low-e coating the thermal performance was very much improved, as indicated in Fig. 6.57. In an optimized construction of the PV-module-absorber compound a high resistance against mechanical loads could be achieved. The improved thermal performance, however, leads to possible stagnation temperatures of about 160 °C, which either requires better temperature-resistant materials for glueing or an overheating prevention. The economic potential of such a development is large. When roof area is available without limitation, the individual thermal and electrical efficiencies of separate solar thermal collectors and PV-modules are higher than the ones of a PVT system. If, however, the area is limited, the best combined output can be produced with a PVT system. Several annual performance simulations have been performed comparing both alternatives in different applications.

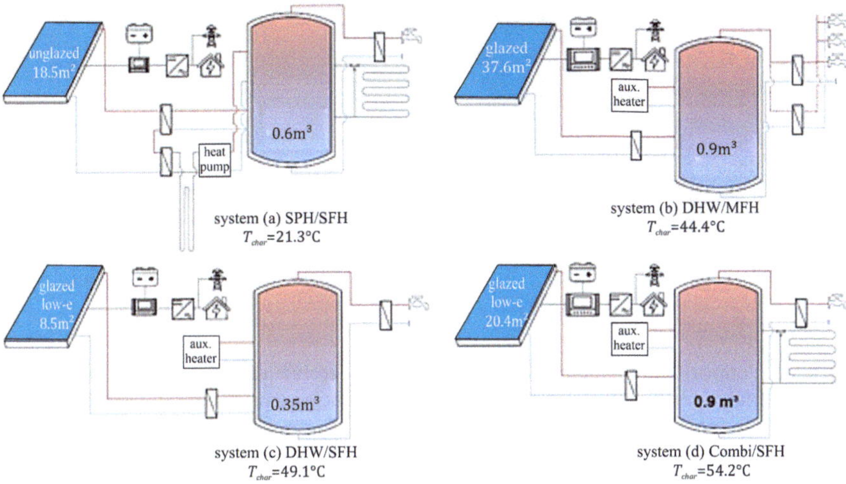

Fig. 6.57 Electrical and hydraulic layout of PVT systems **a–d** including PVT technology and required collector area (Lämmle, 2018)

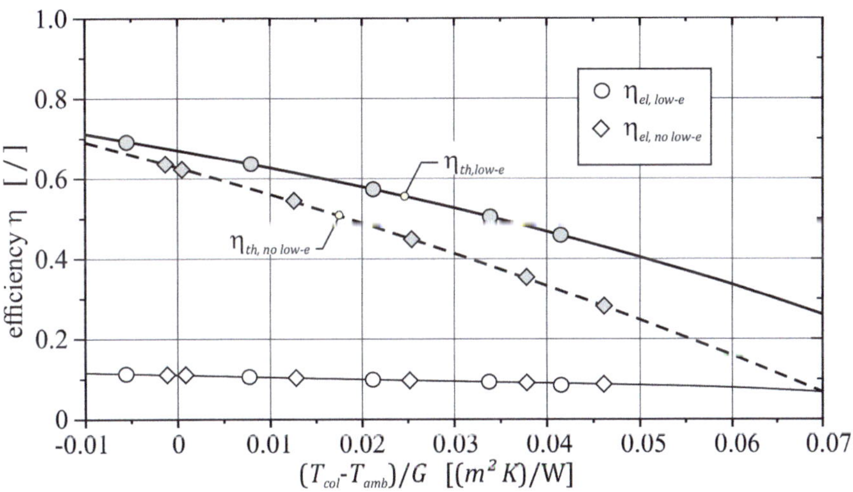

Fig. 6.58 Thermal and electrical efficiency curves (lines) with test points (markers) measured in MPP mode of PVT collectors with and without low-e from Lämmle et al., (2016)

(a) Solar-assisted heat pump system with ground heat exchanger for DHW and room heating,
(b) DHW in a multi-family house MFH,
(c) DHW in a single family house SFH,
(d) Combi system in a SFH.

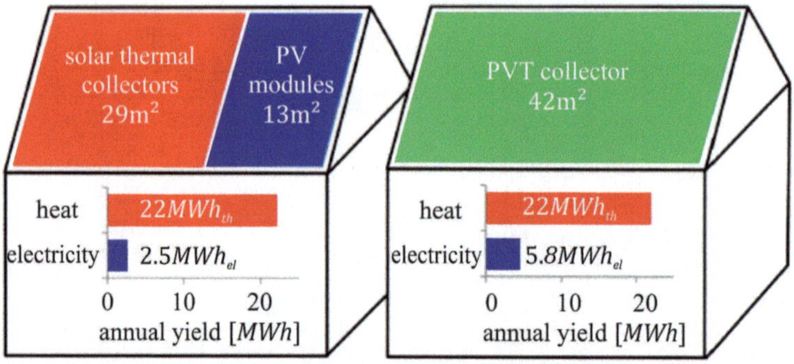

Fig. 6.59 Comparison of building with roof covered by separate solar thermal and photovoltaic panels vs. roof covered with covered PVT collectors producing the same solar thermal yield per year (Lämmle, 2018)

Table 6.6 Electricity surplus generation compared to a side-by-side PV + ST system with the same thermal yield and using the same roof area PV (Lämmle et al., 2017)

System	Covered PVT collector (%)	Covered PVT collector/low-e (%)
System a—SHP/SFH	187	318
System b—DHW/MFH	42	134
System c—DHW/SFH	16	90
System d—Combi/SFH	15	85

Figure 6.57 compares the thermal and electric efficiency of PVT collectors with and without low-e coating. Model calculations for a demonstration building show that on a given available roof area, 15–318% more electricity is generated with the same thermal yield compared to a separate installation of solar thermal and PV (Lämmle et al., 2017). The lower the solar thermal system temperature (lowest in case (a) and increasing towards (d)) the higher the electricity surplus. Using low-e coatings, covered PVT collectors can be suitable for conventional solar thermal applications such as domestic hot water and space heating. Table 6.6 shows that the collector using the low-e coating has a much higher improvement due to its better thermal performance.

6.5.4 Building Integration of Solar Thermal Collectors

In the building sector, the combination of energy efficiency measures and the use of renewable energy sources reduces the consumption of fuels and electricity and thereby reduces emissions of CO_2 and other pollutants (Aelenei et al., 2016).

The optimal use of the solar components in buildings has an influence on architectural design and building concepts. The complex architectural integration of solar components into the building envelope is of critical importance to the appearance, form, aesthetics, investment, and operating cost as well as the whole environmental impact within the life cycle of the buildings. From an architectural perspective, additional specifications for components may be required such as mounting position, appearance, dimensions, geometrical shape, physical functions, light permeability, colour, and construction details. Building integrated solar thermal systems (BISTS) can be used as well in the construction of new buildings as in the refurbishment of buildings. With true integration, not just add-on design, additional functions are fulfilled, foremost to be mentioned is the thermal performance which is strongly influenced by materials, constructional integration, and detailing. Thus, BISTS may also improve the quality of life for inhabitants (e.g. thermal comfort, daylight use). The increasing implementation of BISTS may also stimulate a transition of the building product industry towards a more open attitude with respect to innovative building designs, new functionalities, and competitiveness of conventional building claddings. The influence of BISTS on the appearance of single buildings and building groups can be significant. However, the correct application of BISTS is not without its challenges for architects and building industry.

The main function of BISTS is to produce thermal energy. In the case of hybrid systems BIPV/T electricity will also be produced. A whole range of additional functions related to building physics and constructional requirements can be addressed by BISTS (Platzer, Smyth, & Kennedy, 2014):

- Thermal insulation
- Acoustic insulation
- Humidity regulation
- Rain and wind tightness
- Solar protection
- Daylighting
- Structural functions
- Fire resistance
- Security protection

There are different levels of building integration depending on the number of functions being delivered by BISTS. While partially integrated solar thermal systems have a lower level of functionality, fully integrated systems are characterized by higher complexity. The external layers of the building envelope and of the solar thermal collectors has an influence on the aesthetic potential and design options (Krstic-Furundzic, 2007). Solar thermal components can be used to replace normal building components possibly showing a multifunctional potential in the external skin.

Clearly the multi-functionality of a collector makes it applicable to integration and can provide the advantage for the designer to use less building elements. A prominent example is the application of building integrated solar thermal façade collectors which may remove the need for conventional cladding materials. That again is reflected in investment costs.

In terms of functions, light permeability requires a new type of semi-transparent collectors (Giovannetti, 2014; Ritter, 2014). Various light transmission grades and interesting lighting effects can be produced inside a building. For example, the variation of profiles, partial use of absorber and transparent areas in the aperture, redirection of light by slats, different arrangements and distances between vacuum tubes can result in different effects achieved by the shadows and light (Fig. 6.60).

Achieving functional requirements must be accompanied by fulfilling aesthetic requirements.

This requires that the functional and aesthetic aspects are considered simultaneously, considering the various building aesthetics, building physics and solar thermal collector mounting criteria categories. The aesthetical possibilities like dimensions, form, jointing, colors, texture etc. of a solar thermal collector can all be influenced by the materials and constructional elements of the collector that are used. Even the type of fluid used (liquid for solar water heating systems or air for solar air heating systems) may affect the eventual appearance (Table 6.7), see Platzer, Smyth, and Kennedy (2014). Passive solar collectors like the Trombe wall or Translucent/transparent insulation use the thermal mass of the wall directly

Fig. 6.60 Visualization showing building integration of a horizontal strip collector in an opaque façade (left) and of a solar thermal collectors (opaque glazed absorber semi-transparent solar thermal blinds) in a glazed transparent façade (right) (Maurer, 2018)

Table 6.7 Connection between functionality and aesthetic criteria

Aspects	Individual criteria
Building aesthetics	Dimensions
	Color
	Materials (texture, surface structure, heat capacity)
	Optical reflectivity and glossiness
	Sealing, joints, and frames
	Design harmony
	Structural integration
	Use context
Building physics	Mechanical characteristics
	Moisture and water permeability
	Thermal insulation performance
	Acoustic characteristics
Practical operation	Mounting and installation ease
	Construction stability
	Maintenance

as absorbing element and thus some aspects are already dictated by the building structure (Chan, 2010; Platzer, 2000).

6.6 Collector Performance Testing and Certification

In performance testing we must distinguish the different approaches which all have their benefits and shortcomings. First of all, there is indoor and outdoor testing.

Indoor and Outdoor Testing

In both cases we need a good control of the environmental variables which affect the thermal performance of a collector:

- Solar irradiance level and direction.
- Ambient air and radiative temperatures.
- Wind speed and direction.

Whereas in outdoor testing we need to find representative situations which are sufficiently stable and characteristic for the range of operational conditions in indoor testing the ambient conditions can be controlled by using artificial solar simulators (lamps with a spectral distribution close to a standard solar irradiation spectrum) and by controlling temperatures by thermostats, heating and cooling of surfaces and rooms. In this way very constant and reproducible conditions can be created, which lead to very exact and reproducible measurements. On the other hand, it is not easy to be close to realistic conditions, especially for the solar simulation. Not only the spectral distribution of the best class of simulators is not exactly a solar spectrum, as illustrated in Fig. 6.61a, also it is extremely difficult to design simulators with sufficient homogeneity for large areas in the collector aperture plane and with similar beam divergence. The latter problem can only be solved for measuring relatively small concentrator devices indoors by using high precision paraboloids with a good point source. Flat-plate collectors and solar photovoltaic panels due to their flat angular response between 0° and 50° incidence on the other hand have little sensitivity to the divergence of simulator light as long as the main part comes from a source in front of the sample. Therefore, for such collectors, large simulator lamp arrays have been designed which provide a reliable and quite homogeneous irradiance in the collector plane, which are exemplarily shown in Fig. 6.61b.

Stationary and Quasi-Dynamic Testing

Solar collectors have a complex dynamic response to changes in environmental conditions. Material thermal capacities as well as the properties of the heat transfer fluid circulating through the interior of the collector influences the reaction to changes. Therefore, it is much easier to use a stationary approach using a collector energy balance equation as given by Eq. (6.1). The model has been used very successfully over many years, and a physical modelling of the dynamic thermal response is not required, when the conditions for testing are kept nearly constant within a certain small range. This is the approach of stationary testing as specified for example in ISO 9806 (2017), DIN EN 12975-1 (2011). Indoor testing lends itself naturally to this approach. For outdoor testing the conditions on constant conditions have to be relaxed to some extent, and small changes in the collector response are taken into account by adding a single-knot effective capacity to the time-dependent temperature change. This quasi-dynamic approach is not able to cope with all variable situations but in many labs outdoor periods can be found where solar irradiation (around noon) and temperatures are quite stable. The

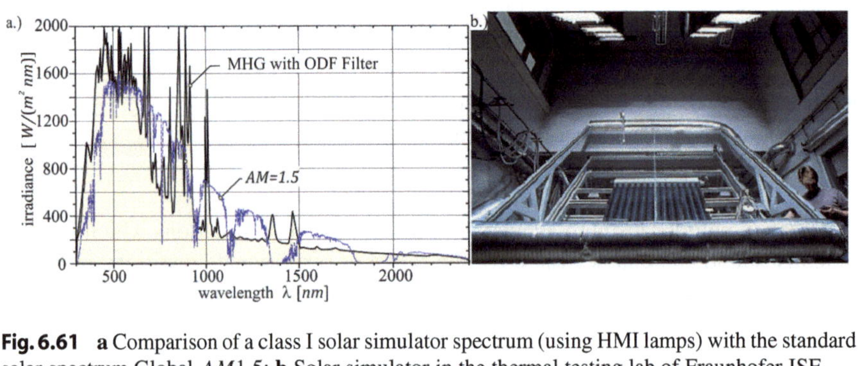

Fig. 6.61 **a** Comparison of a class I solar simulator spectrum (using HMI lamps) with the standard solar spectrum Global $AM1.5$; **b** Solar simulator in the thermal testing lab of Fraunhofer ISE

change in incidence angle can be compensated by tracking the sun with the collector at a constant incidence angle; a test stand for this type of testing is illustrated in Fig. 6.62.

Collector and Solar Field Testing

Although similar in theory the testing of large solar fields and individual collectors has different challenges. Small collectors can be tested indoors and outdoors, and the test equipment itself can be a standardized closed loop circuit where the returning fluid of the collector is cooled down by a chiller to the required and desired inlet temperature of the collector. High quality sensors can be used and included in the testing circuit, as the collector unit usually is installed *into* the existing circuit. A testing of a collector is in most cases used for a product certification that is to guarantee the product specifications of a certain type of collector which can be bought as commodity. The production of the manufacturer then must be certified to be able to guarantee the uniformity of the production.

However, for large solar fields the situation is different: First, these fields shall be tested as installed, they are unique entities and no mass product. Therefore, they should be tested at the site and in the condition as they were installed by a technology provider. A deinstallation and reinstallation at a test centre could give different results. Therefore, a temporary installation of the testing infrastructure including heat exchangers, pumps, fluid conditioning is required. Therefore, movable testing labs may be designed for smaller installations where all equipment of the circuit, sensors and data acquisition are including in a transportable container.

On the other hand, the thermal power produced by large solar fields is huge and dependent on the installation. Building a test circuit specifically laid out and designed for exactly the case considered is economically not feasible. A scientifically improved twin of the balance of plant already installed in the application (pumps, heat exchangers, expansion vessel, control, heat rejection and sensors) would drive up the project cost enormously. Therefore, field testing is for practicality relying on existing infrastructure, on the existing balance of plant. It may be possible to control mass flows and temperature levels to some extent, in order to cover a certain range of temperatures in the testing. But it is usually not possible to reach the qualification for a quasi-dynamic testing as put forward in the

Fig. 6.62 Outdoor tracker for testing thermal collectors at Fraunhofer ISE

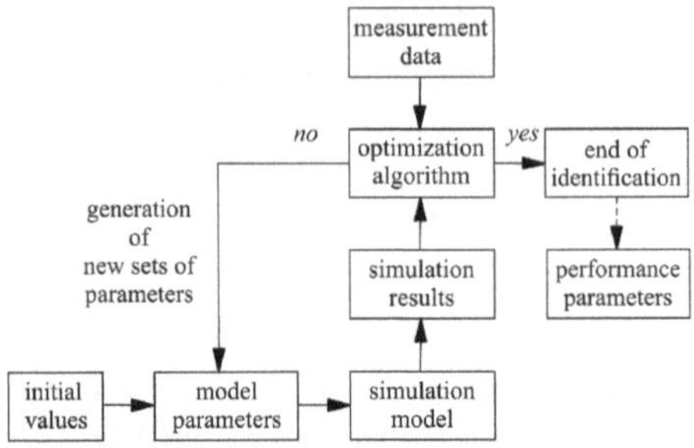

Fig. 6.63 Schematical procedure of parameter identification procedure for large scaled solar fields

standards. For these situations scientists have developed dynamic testing methods which allow to evaluate even completely dynamic environmental conditions – solar irradiation with moving clouds, temperature drops and other realistic changes. The basic idea of dynamic testing is that the collector field can be simulated with a dynamic simulation program. The underlying equations in the collector model must be parametrized (similar to the quasi-dynamic model), but the output at each time-step is dependent on the previous time step (contrary to the quasi-dynamic evaluation). It is still an area of current research how to reach reproducible and reliable detailed results with such a method, especially when industrial sensor qualities must be used which are not equivalent to scientific sensors. The principle of the dynamic testing with parameter identification is schematically illustrated in Fig. 6.63. One drawback is, that the simulation program using the identified parameters of the solar collector field may give realistic results and may predict the performance correctly, however, one may not transfer the parameters into a similar but different simulation program which might use different algorithms.

6.6.1 Standardization (ISO, ASHRAE)

As we have distinguished now the different types of performance testing, let us look into the standardization of testing. Here in international or national standards the exact conditions to be used during the testing as well as the prescribed instrumentation, the precision of sensors and the evaluation of the tests is described in detail.

As we discussed different fundamental approaches for performance testing, we did not yet focus on the kind of collector equation used for the evaluation of stationary or quasi-dynamic testing. However, although the theoretical basis is identical different standards use either the mean fluid temperature as variable or the collector inlet temperature. The European and international testing standards (ISO 9806, 2017) use the first approach while the American ASHRAE test (ANSI/ ASHRAE Standard 93–2003, 2003) uses the second one. Both approaches require the mass flows to be in a representative range of collector operation and for both the focus is small collectors with a few square meters collector area which experience only a small temperature rise within the collector under normal operation.

The collector heat loss and the heat loss factor U_L is considered representative as an average for the whole collector.

The advantage of the European way to test collectors is that the parameters are independent (within the limits said before) of flow rates using in testing, whereas the advantage of the ASHRAE approach is that with these parameters it is easy to calculate the performance of collectors put into a series. In the following discussion we focus now on the international standard ISO 9806 which describes both stationary and quasi-dynamic testing. In addition, we only present the case of liquid heating collectors, although in the standard also other collector types, e.g. air heating collectors, are being treated in a similar way.

Stationary testing is based on the stationary collector equation without a heat capacity. The heat loss is parametrized using possibly a few coefficients which describe the heat loss related to wind speed, to thermal radiation and

$$\dot{Q}_u = A_c \cdot \left[\eta_{0,hem} G_{hem} - a_1(T_m - T_{amb}) - a_2(T_m - T_{amb})^2 - a_3 u'(T_m - T_{amb}) + a_4\left(E_L - \sigma T_{amb}^4\right) - a_6 u' G_{hem} - a_7 u'\left(E_L - \sigma T_{amb}^4\right) - a_8(T_m - T_{amb})^4\right], \tag{6.40}$$

with $a_4 = \eta_{0,hem}(\varepsilon/\alpha)$, $a_6 = \eta_{0,hem} \cdot b_u$ and $a_7 = \eta_{0,hem} \cdot b_u \cdot (\varepsilon/\alpha)$. In this equation u' is a reduced wind speed, E_L is the thermal irradiance in the collector plane and b_u is a coefficient describing wind dependence. Depending on the collector, some coefficients can be safely set to zero. For details refer to the standard document.

In this case a hemispherical optical efficiency and irradiance is being used, because with solar simulators it is not practically possible to separate a direct beam radiation having a similarly small divergence as the sun's radiation from the diffuse irradiation.

For the *quasi-dynamic testing* outdoors, the testing conditions cannot be stationary due to the changing weather conditions. However only testing periods satisfying a partially stable measurement are specified in the standard. Due to the change of temperature with time and irradiation levels an additional term is being introduced in the testing equation for the extracted useful power, which is an effective heat capacity of the collector a_5:

$$\dot{Q}_u = A_c \cdot \left[\eta_{0,b} K_b G_b + \eta_{0,d} K_d G_d - a_1(T_m - T_{amb}) - a_2(T_m - T_{amb})^2 - a_3 u'(T_m - T_{amb}) - a_5 \frac{dT_m}{dt} + a_4\left(E_L - \sigma T_{amb}^4\right) - a_6 u'^{G_{hem}} - a_7 u'^{\left(E_L - \sigma T_{amb}^4\right)} - a_8(T_m - T_{amb})^4\right]. \tag{6.41}$$

The terms K_b and K_d described the incidence angle modifier functions for beam and diffuse irradiation. K_b can be a function depending one incidence angle (e.g. a flat plate collector) or on two incidence angles (e.g. for vacuum tubes or linearly concentrating collectors).

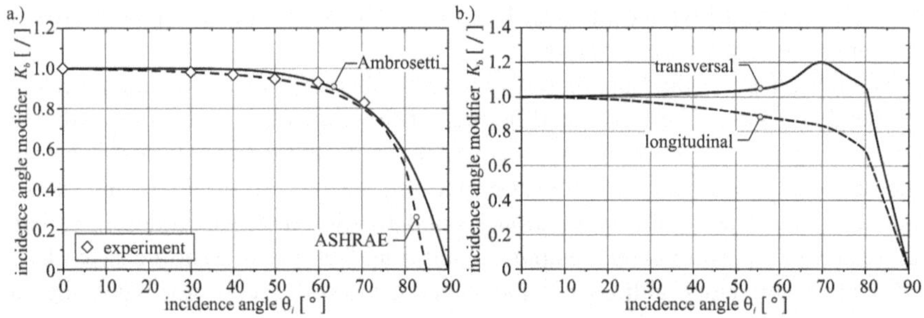

Fig. 6.64 **a** Incidence angle modifier K_b for a flat-plat collector (line: Ambrosetti function $r = 4.8$; dashed: ASHRAE function $b_0 = -0.1$); **b** Incidence angle modifiers $K_{b,T}$ and $K_{b,L}$ vacuum tube collector with interpolation (line: transversal data; dashed: longitudinal data)

Depending on the collector type the incidence angle modifier function could be one- or two-dimensional. For flat-plate collectors a radially symmetric one-dimensional incidence angle modifier is assumed which is usually described by a single fit parameter, see Fig. 6.64a. The function describing very well the optical performance of a collector covered with glazing(s) is the Ambrosetti function with the parameter r to be fitted to experimental results according to (ISO 9806, 2017):

$$K_b(\theta) = 1 - \left[\tan\frac{\theta}{2}\right]^r. \tag{6.42}$$

This equation improves the simple function specified by ASHRAE 93-77:

$$K_b(\theta) = 1 + b_0 \cdot \left(\frac{1}{\cos\theta} - 1\right), \tag{6.43}$$

where θ is the incidence angle, and r respectively b_0 are parameters for fitting.

For a collector without rotational symmetry like a vacuum tube collector, we have a longitudinal incidence angle modifier K_L for incidence angles θ_L in the plane along the absorber tube which defines the optical axis, and a transversal incidence angle modifier K_T for incidence angles θ_T in the plane perpendicular to the optical axis (ISO 9806, 2017), as e.g. illustrated in Fig. 6.64b. The incidence angles are given by

$$\theta_T = \tan^{-1}(\tan\theta \cdot \sin\gamma), \tag{6.44}$$

$$\theta_L = tan^{-1}(tan\theta \cdot \cos\gamma), \tag{6.45}$$

where θ is the incidence angle on the collector plane, and γ is the angle between the optical axis and the incidence plane of the solar radiation in the spherical coordinate system.

The bidirectional incidence angle modifier for such a collector and arbitrary incidence angles can be approximated by (ISO 9806, 2017):

$$K_b(\theta_L, \theta_T) = K_b(\theta_L, 0) \cdot K_b(0, \theta_T) = K_{b,L}(\theta_L) \cdot K_{b,T}(\theta_T). \tag{6.46}$$

When doing stationary testing in the lab or outdoors, the radiation should always come from the same direction, and a test can be performed at a certain incidence angle $\theta_{testing}$, which should be kept constant during the test. For outdoor testing this implies that collectors must be mounted on solar tracker racks, which move in a way that the incidence angle is constant during the test period. This implies small inaccuracies as collector inclination (with an impact on the convection heat transfer) and the view angle to the sky (with an impact on radiation heat transfer and on diffuse irradiance level on the collector) changes. In the lab it is often very difficult to produce directional and homogenous irradiation on a collector plane using a rack of lamps.

This is much easier in quasi-dynamical testing, because here the conditions of irradiation may change during the testing period. The collector is fixed and is irradiated by the moving sun. In this way the incidence angle modifier function can be parameterized, and fitting parameters can be extracted using the experimental time series data.

The standards mentioned up to now have mainly been developed to test individual collector products used for heating or hot-water production. However, it is also possible to test complete collector fields. Collectors like the concentrating collectors for solar thermal power already as individual modules have an aperture area which is exceeding several hundred square meters. In addition, very often the target is not the quality of an individual component but the quality of an installation. In both cases it would be costly to transport the collector to a test site, install and test it there, and then tear it down and move it to the intended installation site. Moreover, the new installation could lead to different performance. All these arguments lead to the requirement of in-situ testing at the intended installation site. Test equipment and experts must be available now at this site, which is challenging for the logistics and cost. From this background the requirement of shortening the test period as much as possible has led to the development of *dynamic test procedures* (DT), which are able to digest all different dynamically changing test conditions, where quasi-dynamic testing (QDT) would exclude these periods. In these procedures for the evaluation of the dynamic performance of a collector, a numerical simulation model is required that shall reproduce the dynamic behaviour of the collector. To this end, any dynamic simulation model can be used, mostly incorporating a discretization of the (simplified) Navier–Stokes equations. Along the fluid flow the collector is segmented therefore in several N nodes. A numerical difference between a QDT method and a DT method is that for QDT a parameterized collector equation is used for each individual time-step to minimize the difference between modelled and measured heat output of the collector. A multi-linear regression analysis is performed, which is a non-iterative, fast matrix method minimizing the error of the measured output power of the collector versus the approximated/modelled output power. Although there is a single heat capacity term with a parameter characterizing the thermal capacity of the collector and connected to the temperature derivative with time, essential it is not a dynamic method. The DT methods, on the other hand, can model a collector performance using a detailed model, in which thermal capacities of fluid and collector materials, as well as time-lags of in the fluid flow can be included. There is a real thermal simulation of the collector. The minimization may not only be done with collector output power, but also with collector temperatures. A flow scheme of the identification process comparing simulation and experimental time series of data is shown in Fig. 6.63. A validation of such a DT method, which can be used for a large collector field as well as individual collectors, has been done in Fahr, Gumbel, Zirkel-Hofer, and Kramer (2018).

Also, these methods use the comparison of the power output measured with the power output simulated in the numerical tool. The simulation model therefore needs parameters to quantify optical performance and thermal losses. Then the parameter identification method as sketched in Fig. 6.63 will be used, where an optimization method is used to find the parameters for which the agreement between the two power output time series is best (least-squares minimization). Depending on the collector, the testing period, and the location sometimes not the complete range of incidence angles can be realized (e.g. it is difficult to realize normal incidence for a Linear Fresnel collector for latitudes larger than 22.5°). If certain angles however are needed for a full simulation, interpolation or extrapolation using an optical raytracing tool is a possibility. A comparison of quasi dynamic testing (QDT) and dynamic testing (DT) for concentrating collectors including the complex incidence angle modifiers has been described by Hofer et al., (2015) and an overview over procedures and standards for testing concentrating collectors has been given in Hofer et al., (2016). Since then the IEA SolarPACES published guideline for dynamic collector testing, and a first standard is referring to this method (IEC6 2862-5-2, 2022). In the standards for collector testing also other tests are described, which are needed also for the quality assessment and certification. Apart from thermal performance testing gives several additional functional tests to be performed in a full cycle according to ISO 9806 which are addressed in Table 6.8.

6.6.2 Certification

A product certification is an action or process of providing something with an official document attesting to a status or level of achievement. With the certification document a certain quality can be attested. This helps the product in the market, as buyers/customers may trust that this product useful and has a guaranteed performance. For collectors and collector systems several certification schemes exist. They all look in the different performance criteria, from thermal performance to durability and resistance against external stresses. In Europe, the Solar Keymark scheme is implemented. The Solar Keymark was developed by the Solar Heat

Table 6.8 Tests according to ISO 9806 (2017)

Sub clause	Testing type
6	Internal pressure test for fluid channels
7	Leakage test
8	Rupture and collapse test
9	High-temperature resistance
11	Exposure test
12	External thermal shock test
13	Internal thermal shock test
14	Rain penetration test
15	Freeze resistance test
16	Mechanical load test
17	Impact resistance test
20	Thermal performance test
28	Pressure drop measurement
18	Final Inspection

Europe/ESTIF and CEN (European Committee for Standardisation) in close co-operation with leading European test labs and with the support of the European Commission. It is the main quality label for solar thermal products and is widely spread across the European market and beyond. In the USA the Solar Rating and Certification Council (SRCC) has a similar role, In the China the Golden Sun certification and in Australia the Standard Mark™ program exists.

Certification often is a requirement in many countries to receive subsidies in a support program. The certification is issued by a certification body, an independent organization watching and guaranteeing that the rules are followed. Solar Keymark has a network of such certification bodies and accredited test laboratories as well as technical inspectors. The test laboratories must perform tests according to international standards. When the products pass the physical tests, the certification can be issued. However, a regular inspection of production facilities is another part of the scheme. In this way the production uniformity and constancy can be guaranteed.

Products holding a certificate can be found in a publicly accessible data base, see for example (Solar Keymark, no date). Tools provided by Solar Keymark allow also to calculate in a simplified way the annual yield of different collectors for various European locations.

Accredited laboratories often do not only work for one certification scheme but are able to support the different global schemes. This also means that they must be able to follow different testing standards. In the area of thermal performance of collectors, the approaches are quite different. But with the following section at least an approximate conversion of results is possible between International ISO and American ASHRAE standards.

6.6.3 Conversion of Efficiency Equations

Generally, the results and parameter from the two approaches of ISO and ASHRAE can be converted into each other, using a procedure we will describe in the next paragraph. However, it should be stressed that ideally the test data should be determined at flowrates corresponding to those found in reality. The hydraulic turbulence is changing with Reynolds number (Re) and may give rise to small changes in the heat transfer between collector piping and fluid. For the conversion we have to assume in order to make analytical corrections for flow rates that the only noticeable effect of changing flow rate is to change the temperature gradient in the flow direction, and that we can neglect changes in F' due to different internal heat transfer coefficients.

The results of a collector thermal performance according to ISO 9806 is an efficiency curve as presented in Fig. 6.65. The parameters determined from the experimental data to fit the test equation are listed in Table 6.9.

The mass flow during testing in ISO 9806 is chosen in a way, that the collector is operating in a typical condition with respect to turbulent flow and heat transfer within the tubes of the collector. Now assuming that within a certain range the impact of a change of flow is negligible on the value of F', these results are valid for a range of mass flows. The mean fluid temperature in Eq. (6.2) is approximated by the arithmetic average of inlet and outlet temperature, which is equivalent in the limit of high mass flow, where the temperature profile within the collector is

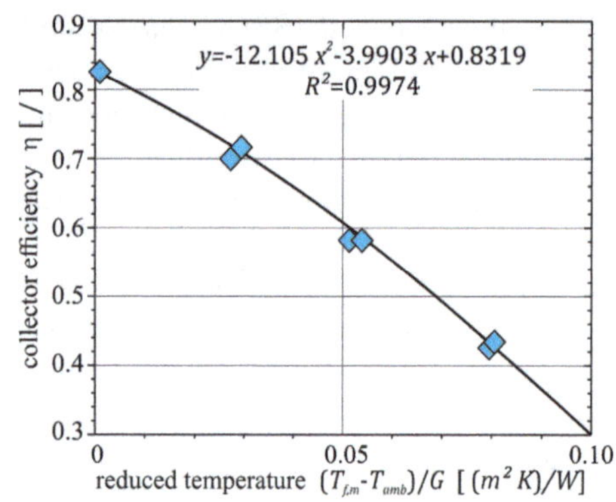

Fig. 6.65 Efficiency curve for a flat plate liquid collector tested according to ISO9806 for irradiation $G = 1000$ W/m^2

Table 6.9 Test parameter according to ISO 9806 for the collector curve of Fig. 6.65

Test parameter	Symbol	Value	Unit
Optical efficiency	$F_{av}(\tau\alpha)$	0.832	–
Thermal loss coefficient	$F_{av}(u_0)$	3.990	W/(m^{2}K)
Thermal loss coefficient (2nd order)	$F_{av}(u_1)$	0.0121	W/(m^{2}K)

linear. Therefore, the testing equation is correctly written (to distinguish it from Eq. (6.2)) as:

$$\dot{Q}_u = A_c \cdot F_{av} \cdot \left[G \cdot (\tau\alpha)_e - U_{\mathrm{L}} \cdot \left(T_{f,av} - T_{amb}\right)\right]. \tag{6.47}$$

In (ANSI/ASHRAE Standard 93-2003, 2003) the results on the contrary are presented in the form of collector Eq. (6.3) using the collector inlet temperature. Here, the mass flow is extremely important for the results—a low mass flow will result in much higher outlet temperatures than a high mass flow, and therefore the overall heat loss will be higher. Given a certain mass flow within the range of validity of test results in Table 6.9 one may convert these results to parameters used by ASHRAE. The conversion formulae are:

$$F_R = F_{av} \cdot \left[1 + \frac{A_c \cdot F_{av} \cdot U_L}{2\dot{m} \cdot c_p}\right]^{-1} = F_{av} \cdot \left[1 + \frac{1}{2\gamma_{av}}\right]^{-1}, \tag{6.48}$$

where F_{av} is approximately the same as F'. A flow rate correction for F_R can be done with the ratio r:

$$r = \frac{F_R|_{use}}{F_R|_{test}} = \frac{\frac{\dot{m} \cdot c_p}{A_c \cdot F' \cdot U_L} \cdot \exp\left(-\frac{A_c \cdot F' \cdot U_L}{\dot{m} \cdot c_p}\right)\Big|_{use}}{\frac{\dot{m} \cdot c_p}{A_c \cdot F' \cdot U_L} \cdot \exp\left(-\frac{A_c \cdot F' \cdot U_L}{\dot{m} \cdot c_p}\right)\Big|_{test}} = \frac{\gamma \cdot \exp\left(-\frac{1}{\gamma}\right)\Big|_{use}}{\gamma \cdot \exp\left(-\frac{1}{\gamma}\right)\Big|_{test}}. \tag{6.49}$$

Again, here F' may be approximated by F_{av} from a test according to ISO 9806—or from F_R under test conditions directly using the equation:

$$(F' \cdot U_L) = -\frac{\dot{m} \cdot c_p}{A_c \cdot} \cdot \ln\left(1 - \frac{A_c \cdot F_R \cdot U_L}{\dot{m} \cdot c_p}\right)\Big|_{test}. \tag{6.50}$$

These equations are straightforward if U_L can be approximated by a constant. However, very often a second order approximation for the temperature dependence of U_L is used, and hence U_L must be evaluated for the temperature conditions specified in the test. Then the heat loss U_L is expressed as a linear function of temperature difference as in the collector equation, which changes the dimensionless flow rate γ, the flow factor F' and consequently also the flow factor relation r. It is advisable to check whether for the mass flow conditions considered the linear approximation of temperature profile can be utilized.

6.7 Collector Stagnation

The *stagnation of* a solar system is defined as the state in which there is no more forced circulation in the solar circuit and thus no heat can be extracted from the absorber. A stagnation occurs if, on the one hand, heat can no longer be dissipated because the storage tank has reached its maximum temperature and the control system must therefore switch off the pump, or, on the other hand, the pump or the control system fail.

The first case can be prevented by additional heat dissipation or by optimized control. In the event of pump or controller failure (e.g. due to power failure or defective sensor), stagnation cannot be avoided in forced flow systems. In general, shutdown in today's efficient collectors is associated with evaporation of the fluid in the absorber.

When the system is stopped, the temperature of the collector rises until the losses are as high as the gains, i.e. the efficiency drops to zero. In the simplest case of a linear efficiency relation with a constant heat loss factor U_L, the absorber stagnation temperature $T_{abs,stag}$ can be calculated in a simple way:

$$T_{abs,stag} = \frac{\eta_0}{U_L} \cdot \mathrm{G} + T_a. \tag{6.51}$$

In the case of a linear temperature dependence of U_L a quadratic equation must be solved instead. In Fig. 6.66 the absorber stagnation temperature for the collector efficiency curve as shown in Fig. 6.65 is plotted as a function of the radiation intensity G for the constant and the linear approximation of the heat loss factor for an ambient temperature of 20 °C.

As can be seen in Fig. 6.66, the stagnation temperatures of modern flat plate or vacuum tube solar collectors with U_L values in the range 1–4 $(\mathrm{W}/(\mathrm{m}^2\mathrm{K}))$ are well above the evaporation temperature of water at atmospheric pressure. Although the evaporation temperature of the collector fluid rises with pressure in the collector loop (typically around 145 °C for 0.6 MPa pressure), a collector shutdown during sunny days usually is associated with an evaporation of the heat transfer medium. The processes during a stagnation situation with evaporation can be divided according to Lustig (2002) into five successive phases. Since the shutdown scenario affects the entire solar thermal system, Fig. 6.67 shows a typical system diagram of a simple solar circuit system with pressurized collector loop and forced circulation.

Due to the evaporation of the heat transfer fluid caused by solar radiation, the system pressure also changes over time, if no safety valve is opened and releases

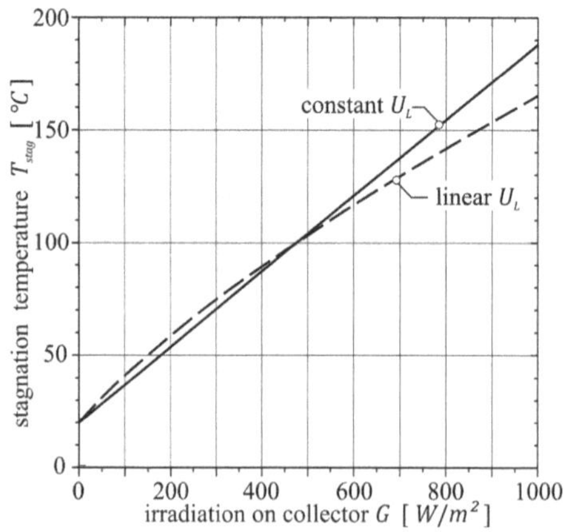

Fig. 6.66 Stagnation temperature T_{stag} in (°C) as a function of the irradiance G in $(\mathrm{W/m^2})$ on the absorber surface for constant and linear heat transfer factors U_L for ambient temperature of 20 °C for collector efficiency as in Fig. 6.65

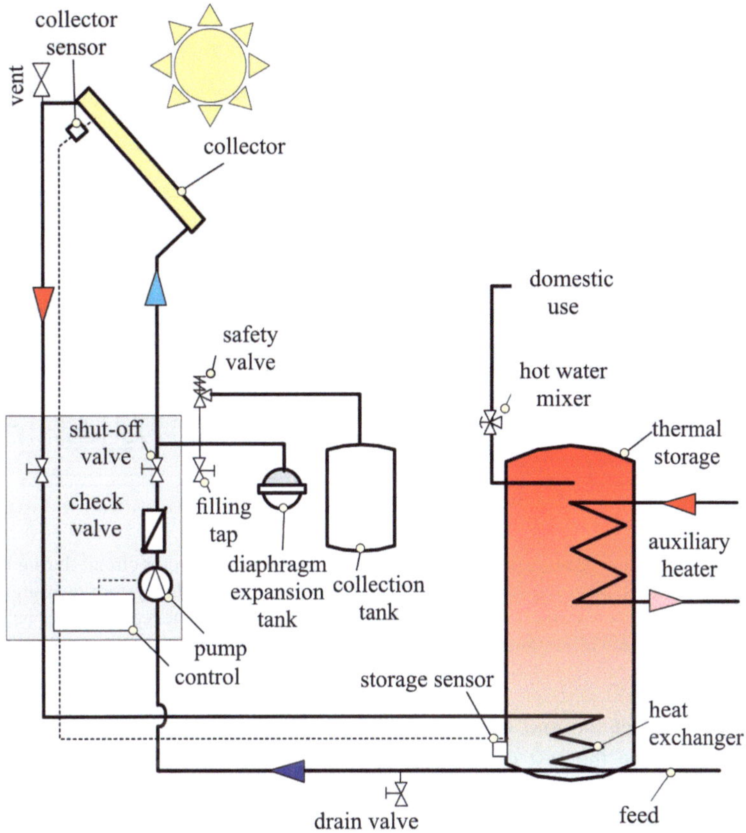

Fig. 6.67 Simplified schematic structure of a domestic solar thermal collector system

steam. This pressure change in turn affects the boiling temperature and the steam-filled pipe space. Figure 6.68 shows qualitatively the development of the system pressure and the steam/fluid fraction of the five different phases of a collector stagnation; details are described in Dimitrova (2006) or Scheuren (2008).

Phase 1: Liquid expansion

After switching off the collector circuit pump, the collector temperatures rise rapidly and evenly until the evaporation process begins at a point on the absorber material in the upper part of the collector. Up to this point, the increase in system pressure is only slight and corresponds to the temperature-related volume expansion of the fluid in the pipeline network.

Phase 2: Emptying the collector

Evaporation begins at a boiling temperature which is determined by the local pressure at this point of the collector. As the system pressure is still low at this point

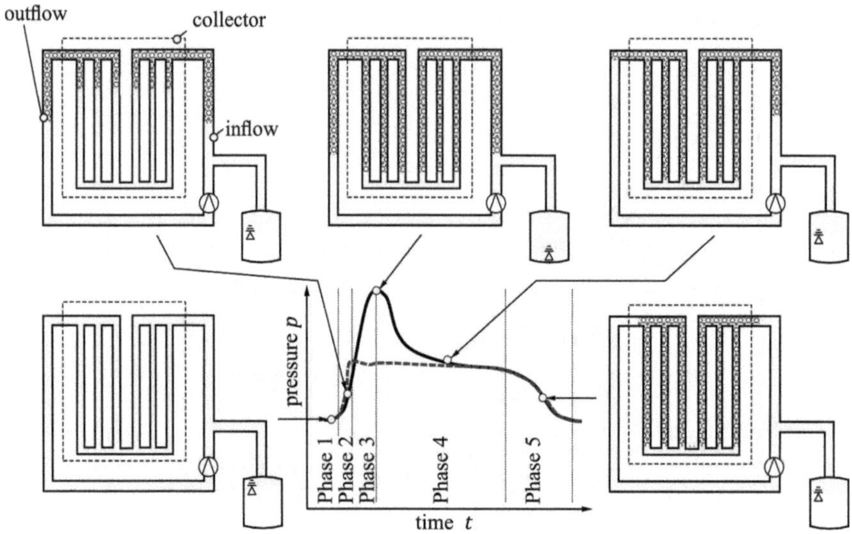

Fig. 6.68 Schematic representation of the steam distribution in a collector circuit during the five different phases of a collector shutdown. Bottom middle: Course of the system pressure p as a function of time t in the different collector stagnation phases; the solid line represents the behaviour of a system that is difficult to drain, the dashed line that of a system that is easy to drain

(immediately after the end of normal operation), the boiling temperature is also lower than during the rest of the system shutdown. A small amount of medium evaporates and pushes a larger part of the liquid collector contents out of the absorber tubes until finally the supply and return lines are also filled with saturated steam. This results in a relatively rapid increase in pressure, as large quantities of liquid are pushed into the expansion tank. During this phase, the system pressure increases rapidly and thus the boiling temperature in the system also increases. The duration of this phase is only a few minutes. The temperatures of the supply and return lines in the plant room and the supply line to the expansion tank are initially low (cooled down after pump stagnation) and later rise when the hot collector contents are inserted. However, they do not reach the values of the collector temperatures due to the cooling losses in the pipes. If the lines are short and have low cooling losses, high temperature loads can occur on the expansion tank and other system components, especially if evaporation only starts at higher temperatures at high system pressure. Long and uninsulated lines to the expansion tank significantly reduce this temperature load. Likewise, a hanging attachment of the membrane expansion vessel can prevent the vessel membrane from overheating.

Phase 3: Boiling in the collector—Phase with saturated steam

This is a phase with a slowing pressure increase. A smaller, but not insignificant part of the collector content is still present in liquid form in the absorber and collection pipes. In this phase, either liquid medium or saturated steam in thermal equilibrium with the liquid is observed at all points on the absorber. The supply and return lines are increasingly pressed downwards towards the membrane expansion vessel with approximately the same liquid level (communicating vessels) and the heat exchanger can be reached by steam. This can also lead to steam formation on the secondary side of the heat exchanger. The system pressure reaches its maximum value, which also results in the highest boiling temperature. The dissipated power is determined by the collector efficiency at high temperatures, taking into account the energy stored in the tubes and the absorber. According to the principle of the heat pipe, energy is transported very effectively at almost constant temperature from the heat source to all heat sinks, whereby the vapour condenses again and runs off downwards as a liquid. The temperature differences between heat source (collector) and heat sink (e.g. heat exchanger) are low (of the order of a few degrees Kelvin), as there are only slight pressure differences (flow pressure losses of the steam). However, concentration differences in the water–glycol system must also be considered (fractional distillation). The maximum temperature load of the plant components reached by the steam thus depends on the maximum plant pressure via the saturated steam temperature. At the beginning of this phase, the temperatures of the supply and return lines in the plant room and the supply line to the expansion tank can drop, as the steam area now increases much more slowly, and the lines filled with liquid cool down. A sudden rise in temperature up to boiling point only occurs when steam appears at these points.

Phase 4: Boiling in the collector—Phase with saturated steam and superheated steam

In the upper collector area the liquid begins to evaporate completely. These areas overheat and the absorber temperature increases. As a result, collector efficiency drops further and the amount of energy to be transported away with steam decreases, because less energy is transported into the system areas filled with steam, so that the "loss area" of the system can also decrease. The volume of steam decreases rapidly even with constant solar radiation. The system pressure decreases (and with it the saturated steam temperature) and liquid is slowly pushed from the expansion tank into the return line. The non-return valve prevents the latter on the flow side. As the superheating areas in the collector increase, the system pressure decreases further, and the liquid level of the return flow reaches the collector inlet. The flow line also fills slowly through condensate. The overheating phase can last several hours on cloudless days and only ends when the solar radiation decreases.

Phase 5: Refilling the collector

The collector is refilled via the return line, whereby the temperatures here drop rapidly. The expansion vessel presses the liquid back into the collector. Depending on the position of the non-return valve, this is done via the supply and return lines, or only via the return line. In the latter case, the flow line is filled via the collector with a time delay.

Measurements and numerical observations show that the collector stagnation depends essentially on the arrangement of the piping system and the internal piping of the collector. It is important that heat transfer fluid is driven out of the collector by the formed steam, and thus the maximum steam load is reduced when the fluid is pressed into cooler parts of the loop. If however the fluid is captured by steam pockets due to gravity in the hot collector, stagnation leads to the evaporation of nearly all fluid content of a collector.

References

Aelenei, L., Smyth, M., Platzer, W., Norton, B., Kennedy, D., Kalogirou, S., & Maurer, C. (2016). Solar thermal systems—Towards a systematic characterization of building integration. *Energy Procedia, 91*, 897–906.

ANSI/ASHRAE Standard 93-2003. (2003). *Methods of testing to determine thermal performance of solar collectors*. 1791 Tullie Circle, Ne, Atlanta, GA30329: ASHRAE, Inc.

Aste, N., Del Pero, C., & Leonforte, F. (2014). Water flat plate PV–thermal collectors: A review. *Solar Energy, 102*, 98–115.

Bandaru, S. H., Becerra, V., Khanna, S., Radulovic, J., Hutchinson, D., & Khusainov, R. (2021). A review of photovoltaic thermal (PVT) technology for residential applications: Performance indicators, progress, and opportunities. *Energies, 14*(13), 3853.

Brötje, S., Kirchner, M., & Giovannetti, F. (2018). Performance and heat transfer analysis of uncovered photovoltaic-thermal collectors with detachable compound. *Solar Energy, 170*, 406–418.

Chan, H.-Y. R. (2010). Review of passive solar heating and cooling technologies. *Renewable and Sustainable Energy Reviews, 14*(2), 781–789. https://doi.org/10.1016/j.rser.2009.10.030

Chow, T. T. (2010). A review on photovoltaic/thermal hybrid solar technology. *Applied Energy, 87*(2), 365–379.

Chugani, B. (2020). Model validation and performance assessment of unglazed photovoltaic-thermal collectors with heat pump systems. In A. G. Charalambides. International Solar Energy Society.

Dimitrova, K. G. (2006). *Thermische Effekte im Kollektorfeld solarthermischer Anlagen*. Dissertation, Techn. Universität, Fak. Maschinenbau, Chemnitz.

DIN EN 12975-1. (2011). *Thermal solar systems and components—Solar collectors—Part 1: General requirements*. Beuth Verlag GmbH. https://doi.org/10.31030/1728174

Duffie, J. A., & Beckmann, W. A. (2020). *Solar engineering of thermal process, photovoltaics and wind*. Wiley.

Fahr, S., Gumbel, U., Zirkel-Hofer, A., & Kramer, K. (2018). In situ characterization of thermal collectors in field installations. In *Conference EuroSun 2018* (pp. 1–10). International Solar Energy Society. Retrieved September 2, 2022, from http://proceedings.ises.org/paper/eurosun2018/eurosun2018-0167-Fahr.pdf

Frei, U., Häuselmann, T., Flückinger, F., & Frey, R. (1995). *Leistungsdaten Thermischer Sonnenkollektoren*. Solarenergie Prüf- und Forschungsstelle, Technikum Rapperswil. Infoenergie Brugg, Postfach 310, CH-5200 Brugg.

Giovannetti, F. K. (2014). Development of an insulated glass solar thermal collector. *Energy Procedia, 48*, 58–66. https://doi.org/10.1016/j.egypro.2014.02.008

Gombert, A., Reineke-Koch, R., Fenske, K., & Hofmann, T. (2007). Optische Beschichtungen für Solarkollektoren-Technologien und Qualitätssicherung. In F. Sonnenenergie (Ed.), *FVS-Jahrestagung 2007 zu Produktionstechnologien für Solaranlagen* (p. 23). Leibniz Universität Hannover.

Hamid, Y., & Beckman, W. A. (1971). Performance of air-cooled radiatively Heated screen matrices. *Journal of Engineering Power, 93*(2), 221–224. https://doi.org/10.1115/1.3445553

Hermann, M. (2005a). *Bionische Ansätze zur Entwicklung energieeffizienter Fluidsysteme für den Wärmetransport.* Dissertation, Universität Karlsruhe, Faculty for Mechanical Engineering, Karlsruhe.

Hermann, M. (2005b, April). Entwicklung des FracTherm-Absorbers—Simulationen und Experimente. In *Proceedings of 15. Symposium Thermische Solarenergie OTTI* (pp. 94–99).

Hermann, M. (2011). *Development of a bionic solar collector with aluminium roll-bond absorber.* BIONICOL Project Final Report, European Commission, 7th Framework Programme.

Hermann, M., et al. (2015). *TABSOLAR—Solarabsorber und andere thermisch aktive Bauteile aus Ultrahochleistungsbeton (UHPC)—Schlussbericht Förderkennzeichen: 03ET1117A/C/D.* Retrieved June 12, 2022, from https://tabsolar.de/externer-bericht/schlussbericht_TABSOLAR/at_download/file

Hermann, M., et al. (2020). TABSOLAR II – Solarabsorber und andere thermisch aktive Bauteile aus Ultrahochleistungsbeton (UHPC). *Schlussbericht Förderkennzeichen: 03ET1310A/B/C/D.*, 115.

Hess, S. (2015). *Low-concentrating, stationary solar thermal collectors for process heat generation.* Shaker Verlag.

Hess, S., Oliva, A., Di Lauro, P., Klemke, M., Hermann, M., Eisenmann, W., … Kramp, G. (2010). Solar heat for industrial processes: RefleC collector development and system design. In *Conference Proceedings/EuroSun 2010, International Conference on Solar Heating, Cooling and Buildings* (pp. 1–8). International Solar Energy Society.

Hofer, A., Büchner, D., Kramer, K., Fahr, S., Heimsath, A., Platzer, W., & Scholl, S. (2015). Comparison of two different (quasi-) dynamic testing methods for the performance evaluation of a Linear Fresnel Process Heat Collector. *Energy Procedia, 69*, 84–95. https://doi.org/10.1016/j.egypro.2015.03.011

Hofer, A., Valenzuela, L., Janotte, N., Burgaleta, J., Arraiza, J., Montecchi, M., … Scholl, S. (2016). State of the art of performance evaluation methods for concentrating solar collectors. In *AIP Conference Proceedings 1734* (pp. 020010, 9 p.). https://doi.org/10.1063/1.4949034

Hottel, H., & Whillier, A. (1955). Evaluation of flat-plate solar collector performance. *Transactions of the Conference on the Use of Solar Energy II, 3 (Part 2 Thermal Processes)*, 74–104.

IEC 62862-5-2. (2022). *Solar thermal electric plants Part 5-2. Systems and components. General requirements and test methods for large-size linear Fresnel collectors.*

ISO 9488. (2021). *Solar energy—Vocabulary.* ISO/TC 180 Solar energy.

ISO 9806. (2017). *Solar energy—Solar thermal collectors—Test methods.* ICS 27.160 Solar energy enigneering, ISO/TC 180 Solar Energy.

Kramer, K., et al. (2020). *Status Quo of PVT characterization.* Report B1, IEA SHC Task 60.

Kramer, K., & Helmers, H. (2013). The interaction of standards and innovation: Hybrid photovoltaic–thermal collectors. *Solar Energy, 98*, 434–439.

Krstic-Furundzic, A. (2007). PV integration in design of new and refurbishment of existing buildings: educational aspect. *Journal of the Association of Arab Universities for Basic and Applied Sciences, 4*(Supplement), 135–146.

Lämmle, M. (2018). *Thermal management of PVT collectors.* Technische Fakultät. Freiburg im Breisgau: Albert-Ludwigs-Universität.

Lämmle, M., Hermann, M., Kroyer, T., Mülhöfer, G., Oliva, A., Thoma, C., … Jäger, H. (2017). PVTgen2: Entwicklung und Pilotfertigung eines abgedeckten photovoltaisch-thermischen Hybrid-Kollektors der zweiten Generation mit verbesserter thermischer Leistungsfähigkeit. 113. Freiburg.

Lämmle, M., Kroyer, T., Fortuin, S., Wiese, M., & Hermann, M. (2016). Development and modelling of highly-efficient PVT collectors with low-emissivity coatings. *Solar Energy, 130*, 161–173.

Lustig, K. (2002). *Experimentelle Untersuchungen zum Stillstandsverhalten thermischer Solaranlagen.* Dissertation, Universität Karlsruhe, Fak. Maschinenbau, Karlsruhe.

Maurer, C. (2018). *Worldwide innovation: A venetian blind that supplies solar thermal energy.* Retrieved from https://blog.innovation4e.de/en/2018/09/19/worldwide-innovation-a-venetian-blind-that-supplies-solar-thermal-energy/

Osório, T., Horta, P., & Collares-Pereira, M. (2019). Method for customized design of a quasi-stationary CPC-type solar collector to minimize the energy cost. *Renewable Energy, 133*, 1086–1098.

Peuser, F. A., Remmers, K.-H., & Schnauss, M. (2011). *Solar thermal systems: Successful planning and construction* (1st ed.). Taylor & Francis Ltd.

Platzer, W. (2000). Transparent insulation materials and products. In Y. Goswami (Ed.), *Advances in solar energy* (Vol. 14, pp. 33–66). ASES.

Platzer, W., Smyth, M., & Kennedy, D. (2014). *State of the art on building integrated solar thermal systems*. TU1205—Building Integration of Solar Thermal Systems (BISTS), Deliverable.

Ritter, X. L. (2014). *Solarthermie in der Fassade*. Retrieved from http://ritter-xl-solar.com/uploads/media/Booklet_Fassadenkollektor.pdf

Rohsenow, W. M. (1998). *Handbook of heat transfer*. McGraw-Hill.

Scheuren, J. (2008). *Untersuchungen zum Stagnationsverhalten solarthermischer Kollektorfelder*. Dissertation, Universität, Fachbereich Maschinenbau, Kassel.

Selcuk, K. (1971). Thermal and economic analysis of the overlapped-glass. *Solar Energy, 13*, 165–191. https://doi.org/10.1016/0038-092X(71)90002-8

Solar Keymark. (n.d.). https://solarkeymark.eu/. Retrieved from The Origin and Composition of the Solar Keymark. https://solarkeymark.eu/

SolarWall Europe Sarl. (2023). SolarWall air heating systems-product information. 66 Avenue des Champs Elysees, 75008 Paris, France. Retrieved February 27, 2023, from https://www.solarwall.com/technology

Spirkl, W., Ries, H., Muschaweck, J., & Winston, R. (1998). Nontracking solar concentrators. *Solar Energy, 62*(2), 113–120.

Stephan, P., Kabelac, S., & Kind, M., Mewes, D. (2019). *VDI-Wärmeatlas* (12th ed.). Springer Vieweg.

TIGI Solar. (2011). Product whitepaper honeycomb collector. 3 Hamechonai St., Hod Hasharon, Israel. Retrieved from http://www.tigisolar.com/assets/files/Honeycomb%20Collector%20whitepaper%20Oct11(1).pdf

Wang, Y., Shukla, A., & Liu, S. (2017). A state of art review on methodologies for heat transfer and energy flow characteristics of the active building envelopes. *Renewable and Sustainable Energy Reviews*, 1102–1116.

Weiss, W., & Rommel, M. (2008). *Process heat collectors-state of the art*. IEA SHC-Task 33 and SolarPACES-Task IV: Solar Heat for Industrial Processes. IEA SHC. Retrieved September 20, 2022, from https://task33.iea-shc.org/publications

Whillier, A. (1964). Performance of black-painted solar air heaters of conventional design. *Solar Energy, 8*(1), 31–37. https://doi.org/10.1016/0038-092X(64)90008-8

Widyolar, B., Jiang, L., Ferry, J., & Winston, R. (2018). Non-tracking East-West XCPC solar thermal collector for 200 celsius applications. *Applied Energy, 216*, 521–533.

Zondag, H. A. (2008). Flat-plate PV-Thermal collectors and systems: A review. *Renewable and Sustainable Energy Reviews, 12*(4), 891–959.

Zondag, H. A., Vries, D. W., van Helden, W., van Zolingen, R., & van Steenhoven, A. A. (2003). The yield of different combined PV-thermal collector designs. *Solar Energy, 74*(3), 253–269.

Concentrating Collectors

7

Abstract

High-temperature collectors using optical concentration and tracking, are described in this chapter. The determination and optimization of the optical performance of line-focusing and point-focusing technologies is key for reaching high temperatures. We classify therefore the collectors by their optical concentrator principle. Although also the selected heat transfer fluid has an impact on the temperature level which can be achieved, due to fluid stability and receiver material durability issues, this aspect is only treated in Chap. 10 on high temperature collector systems and fields. Here, the focus is on the main components and designs of individual collectors. As solar radiation is transformed into heat in the receiver, the design of this element is mainly responsible for minimizing the thermal losses. However, thermal losses play a minor role, the higher the optical concentration. We present receiver designs, determine the selection of materials, and define the performance. Finally, some line- and point-focusing collector types are compared with respect to their optical performance characteristics.

Keywords

Classification of high temperature collectors • Line-focusing collectors • Parabolic trough and Fresnel collector • Point focusing collectors • Solar tower • Heliostats • Parabolic dish and Scheffler dish

7.1 Classification of High-Temperature Concentrating Collectors

We may classify concentrating tracking collectors by their optical concentration principle (Fig. 7.1). Line and point-focusing collectors have a very different range of achievable and real optical concentration factors. The receiver element converting the concentrated solar intensity into high temperature heat is a second key element. Its design is a consequence of the concentrator, and the thermal losses of

R. Stieglitz and W. Platzer, *Solar Thermal Energy Systems*,
https://doi.org/10.1007/978-3-031-43173-9_7

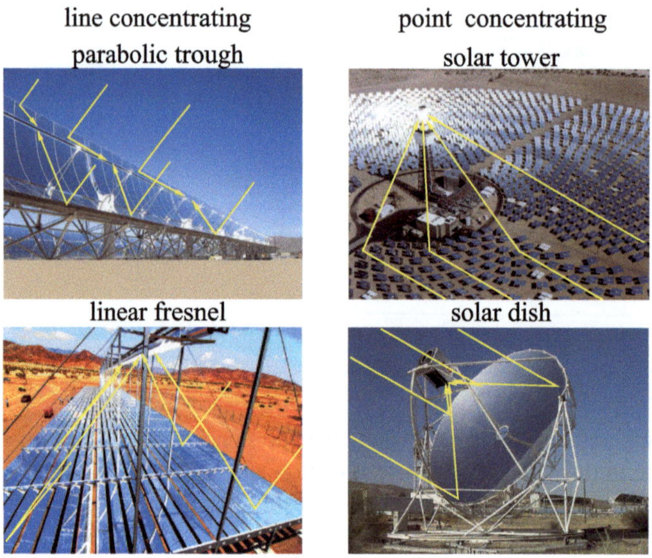

Fig. 7.1 Classification of solar thermal collector types according to the type of the solar concentration

the complete collector are mainly influenced by this design. The critical choice of materials which are able to withstand the high temperatures, is a part of the collector design. The geometrical concentration determined by design, and the optical quality which is dependent on the selection of optical materials like mirrors, their production and mounting, within the two main groups of line- and point-focusing systems determines to a large extent the maximum temperatures which can be achieved. Also, the land consumption of a collector assembly is associated with the concentration type.

However, another option of classification, is often useful with respect to the selection of the heat transfer fluid. In Chap. 6 on stationary collectors we found for example the classification air and liquid collector. In the case of temperatures well above 100 °C the choice of heat transfer fluid determines for example the maximum allowable fluid temperature, which has an impact on the capability to drive connected thermal processes, whether that is a thermodynamical cycle for producing electricity, desalination, or an industrial process. Materials in the open or closed heat transfer fluid (HTF) loop must be selected accordingly. As we will see in Chap. 10, the maximum achievable fluid temperature also has an influence on the size of the solar field (via the pressure and heat losses in the solar receiver piping system), the choice of thermal storage medium, and ultimately the type of thermodynamical cycle which needs the heat.

We focus solely on line-focusing and point-focusing collectors which track the sun and hence work only with direct solar irradiation, in contrast to stationary non-tracking collectors without concertation which also use the diffuse part of the global irradiance.

The concentrating collectors are able to reach high temperature levels exceeding by far 150–200 °C. The practical limits will be discussed. In a first step the optical concentration process of solar radiation from the source (sun) via the imaging system (mirrors) and the conversion process (receiver) is discussed. Based on these considerations, conversion efficiencies or cumulative yields can be obtained, describing the characteristic properties of the individual collector technologies.

Due to the different imaging technologies and geometric arrangements of both line- and point-concentrating mirror/receiver systems, there is also a different utilization of the beam radiation in the individual receiver technologies.

The angular distribution of the solar radiation intensity (solar disc and circumsolar radiation as discussed in Chap. 2), and the optical imaging of the concentrator determine the spatial optical conversion and distribution of heat generated in the receiver, which determines optical efficiency and heat losses of the receiver. The incoming beam radiation is subject to atmospheric influences and varying position of the sun during the day and the seasons. The conversion process therefore is dependent on the solar incidence angles and the geometrical arrangement of the mirror concentrator in relation to the solar source (e.g. for line-focusing collectors the axis might be oriented East–West, North–South or in between). Depending on the concentrator design and the tracking, the utilization of solar radiation is reduced by geometric restrictions leading to effects such as mutual shading of mirrors, or shading by structures, for example a tower construction, a linear receiver or support pylons.

7.1.1 Concentration, Exergy, and Operating Temperatures

In concentrating collectors, the solar irradiation power is directed towards the absorber with the help of mirrors and/or lenses. Mirrors are favored in solar thermal technology, as the lenses show strong achromatism due to the large wavelength range. In both cases the focal spot shall be much smaller than the collecting aperture. We distinguish between imaging concentrators where an image of the sun is formed on the absorber, or non-imaging concentrators without any image formation.

However, redirection of light towards a specific target, also imaging, is only possible with beam radiation. An immediate consequence is that focusing systems are only able to use the direct radiation component, but not the diffuse radiation component, which accounts for about 40% of the incident radiation in Germany. In arid regions around the equator, however, the proportion of direct radiation is between 70 and 80%.

Several concentrator types can be realized. In mirror geometries with rotational symmetry the focus may be a single small spot (ideally a point-focus), whereas in mirror shapes with translational symmetry the focal area will be a linear strip (ideally a line-focus). The concentration respectively the radiation intensity on the absorber with point-focus systems can be much higher than with line-focus concentrators.

The basic idea of concentrating light has been shown already in Chap. 3. Light from a source is incident on a concentrator aperture are A_a from a certain direction. Due to finite sizes of light source the light always has a small divergence and is not completely parallel. In the concentrator the directional light is redirect towards a smaller receiver area A_r. The geometrical concentration factor C_g then is defined as:

$$C_g = \frac{A_a}{A_r}. \tag{7.1}$$

An ideal concentrator is one with the maximum possible concentration. The maximum concentrator is determined by the θ divergence angle θ_s of the incoming light. From optical theory using the conservation of étendue the upper limit for concentration can be derived. We will find the same result considering the second law of thermodynamics. For the radiation coming from the sun the divergence angle due to the finite disk size of the sun is 0.27° (half-angle). The maximum concentration for line-focus systems using concentration in one dimension is:

$$C_{max,1D} \leq \frac{1}{\sin\theta_s} = 212. \tag{7.2}$$

The maximum concentration for point-focus systems focusing in two dimensions is

$$C_{max,2D} \leq \frac{1}{\sin^2\theta_s} \approx 4.5 \cdot 10^4. \tag{7.3}$$

Maximizing overall system efficiency in terms of energy is an appropriate goal. However, the restriction to purely energy terms may miss some information needed for optimization. The second law of thermodynamics addresses the inherent of irreversibility of processes and leads to the limiting conversion efficiency of a heat engine between isothermal reservoirs, given by the well-known Carnot limit on the efficiency, η_C:

$$\eta_C = 1 - \frac{T_2}{T_1}. \tag{7.4}$$

In this equation, temperatures T_1 and T_2 are absolute temperatures to be provided in Kelvin (K), where T_2 is the lower temperature of the Carnot cycle, at which heat is released, usually close to the ambient temperature. Clearly, operating at low driving temperatures T_1 to achieve high receiver efficiencies due to small heat losses does not result in the highest Carnot efficiency. The efficiency of an ideal Stirling engine is approximated by the Carnot cycle quite well. When mounted in the focus of a receiver, at steady state the Stirling engine operates between a single high-temperature source and a fixed low-temperature environment. However, for systems where a heat transfer fluid (HTF) is used between receiver and power cycle, the power cycle is no longer operating between effective isothermal reservoirs. The thermodynamic cycles approximating these situations can be the Rankine cycle for a steam turbine or the Brayton cycle for a gas turbine. Nevertheless, the Carnot efficiency give the upper limit of work that can be extracted between two isothermal reservoirs. It is the basis for the useful thermodynamic concept called exergy.

The exergy of a system is defined as the maximum amount of work that could usefully be performed when bringing the (thermodynamic) system to equilibrium with its low-temperature environment. In this sense, also a certain mass of a substance such as a heat transfer fluid has an exergy content. Thus, every energy flow can be associated with a corresponding exergy flow, being the ideal amount of work that could be extracted from it. Exergy analysis can be a powerful tool in understanding thermal systems which are designed to produce power or chemical actions, and it can be used to design systems in low-exergy applications to analyze which low temperature sources might be best used to provide a certain energy service as for example heating a building.

Table 7.1 illustrates the temperature dependence of the Carnot efficiency together with the exergy-to-energy ratio calculated for some potential HTFs at the same temperature. Considering realistically achievable temperatures yields a reduced potential to extract work. Also, when heat is transferred from an isothermal source to an HTF heated up to the same temperature the exergy content is reduced. It is an irreversible process destroying exergy. For a given temperature, the Carnot efficiency assuming an infinite isothermal source is higher than the exergy-to-energy ratio of a working fluid having the same temperature as this is cooling down when work is performed. Energy in a finite amount of a substance at a certain temperature can be considered as having a work potential according to the full range from its starting temperature down to ambient temperature according to temperature decrement. Steam on the other hand has a temperature profile that is strongly affected by the phase change from steam to water. Depending on where the condensation/evaporation temperature is relative to the maximum steam temperature, its exergy to- energy ratio is different.

The concept of regenerative feedwater heating in a steam cycle using bled steam can be understood using the exergy concept. Partially expanded steam has already given up some of its exergy to power generation, so using it to pre-heat feedwater destroys less exergy than using the highest temperature source to heat the feedwater from ambient temperature. Thinking about a concentrated solar power (CSP) system, preheated HTF in the receivers utilizes the high exergy potential of the solar radiation better than cold HTF. On the other hand, the consequence is a higher receiver average temperature, and the thermal losses are increased. Also, higher flow rates must be maintained as the energy increment per unit mass flow is smaller.

A simplified analysis of the performance of a solar thermal energy system is useful in demonstrating the competing design demands. From the theory of

Table 7.1 Thermodynamic efficiency metrics (relative to environment at 298 °K, 0.1 Mpa)

Temperature (K)	Carnot-efficiency η_C (/)	Constant c_p	Exergy-energy-ratio steam 15 Mpa	Steam 5 MPa	Typ. steam cycle efficiency (/)
5800	94.9	84.1			N.a.
1273	77.0	56.1			N.a.
973	69.9	48.3	0.469	0.435	0.4 – 0.45
673	56.5	35.9	0.421	0.388	0.35 – 0.4
573	48.9	29.8	0.294	0.376	0.25 – 0.3
473	38.1	22.0			

reversible heat engines, power cycles running from high-temperature thermal reservoirs are most efficient, but thermal losses increase as receiver temperatures increase. Using simplified expressions for the various loss mechanisms, and some typical parameter values, gives the results for receiver efficiency as a function of concentration ratio and temperature shown in Fig. 7.2. It can be seen that at low temperatures, all efficiencies reduce to the average absorptivity value. As temperature is increased, efficiencies increase with higher concentration systems at the same temperature difference. In a real system, operating temperature can be controlled through variation of HTF flow rate, for example having higher receiver efficiencies at any given temperature. The various curves intersect the x-axis at the corresponding stagnation temperature. To simulate an idealized operating regime for maximizing power production, the receiver efficiency must be multiplied by a power cycle conversion efficiency to yield an overall ideal system efficiency η_{sys}.

To a large extent, power cycle efficiency dependence on operating conditions is complex and must be established from empirical performance curves. The difference in exergy value of the HTF before and after transfer of heat to the working fluid at input to the power cycle provides a good indicator, with real power cycles extracting something over 70% of the available exergy as work. The Carnot expression provides a good qualitative illustration of the principles, however. If the receiver efficiency results in Fig. 7.2 are multiplied by the Carnot efficiency (according to Eq. (7.4)) pertaining to the receiver temperature at each point to give a semi-ideal system efficiency, the results are as shown in Fig. 7.3. There is now a clear peak efficiency at a particular temperature for each concentration ratio. Higher concentration ratios have higher peak efficiencies, and these occur at higher receiver temperatures.

Without going into the different technology options for point-focusing systems, it is useful to recall the basic idea for the development of high-concentration collectors. The development of central receivers is essentially targeting at high collector outlet temperatures. This is essentially driven by the increase of the solar-to-thermal receiver efficiency $\eta_{rec,th}$, which balances the effective useful heat extracted by the receiver to the solar radiation impinging on the receiver surface. The useful heat contains the optical conversion minus the heat losses. By doing

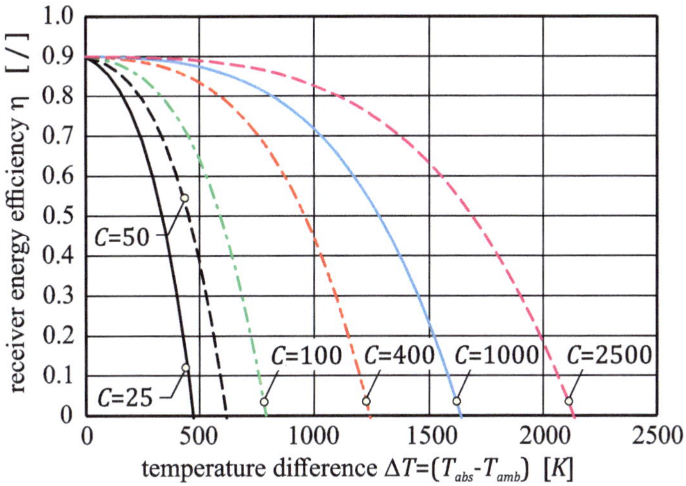

Fig. 7.2 Ideal receiver energy efficiency (η_{rec}) for $DNI = 800$ W/m^2 as function of the temperature difference between absorber and ambient for different concentration ratios C_g; optical efficiency (solar absorptance) 0.9 and emissivity 0.9, assuming $h_{conv} = 20$ W/(m^{2}K) (dashed lines)

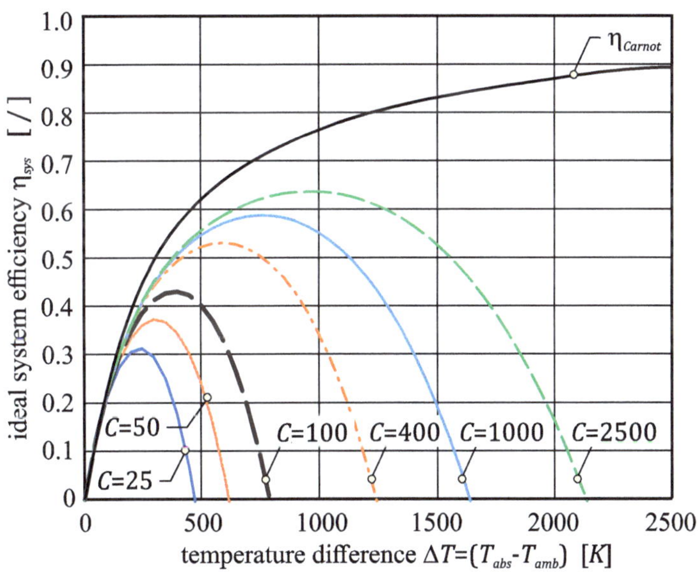

Fig. 7.3 Ideal system efficiency $\eta_{sys} = (\eta_{rec} \cdot \eta_C)$ given for $DNI = 800$ W/m^2 as function of the temperature difference between absorber and ambient for different concentration ratios C_g; optical efficiency (solar absorptance) 0.9 and emissivity 0.9, assuming no convective losses and $h_{conv} =$ 20 W/(m^{2}K)

so $\eta_{rec,th}$ can be derived in a simplified manner from a thermal balance in the following way:

$$\eta_{rec,th} = \frac{\alpha \dot{q}_{rec} - \dot{q}_{loss}}{\dot{q}_{rec}} = \alpha - \frac{\varepsilon\sigma\left(T_{rec}^4 - T_{amb}^4\right) \mp h_{conv}(T_{rec} - T_{amb})}{\eta_{opt,conc} \cdot DNI \cdot C_g}, \quad (7.5)$$

in which is $\dot{q}_{rec}$ the area specifc irradiance on the receiver and $\dot{q}_{loss}$ describes the sum of the radiative and convective losses from the receiver to the ambient, both in (W/m^2). α is the effective solar absorptance of the receiver and ε its thermal emittance, while T_{rec} and T_{amb} denote the receiver surface temperature and the ambient temperature, respectively. σ is the the Stefan-Boltzmann constant and h is the heat transfer coefficient from the receiver to the ambient. The optical efficiency of the concentrator system is described by $\eta_{opt,conc}$, while DNI is the specific irradiance on the aperture area in (W/m^2) and C_g is the concentration ratio, given as collector aperture area divided by the receiver area.

In principle, $\eta_{rec,th}$ for a given temperature difference increases the higher the optical concentration ratio $\eta_{opt,conc} C_g$ gets. In this sense point-focusing collectors from a thermodynamic point of view are more efficient than line-focusing systems. On the other hand, as receiver outlet temperature increases, the thermal receiver efficiency decreases, because the area specific heat losses increase, and, as shown in a simplified example in Fig. 7.2 finally a stagnation temperature is reached where the receiver efficiency is zero. A dilemma of the efficiency relation is not immediately apparent. Both, the absorptivity α and the total emissivity ε are spectral quantities, but moreover also strongly temperature dependent. While the solar irradiance spectrum does not change substantially, and α can be considered constant, this is not true for the thermal emissivity ε which is a strongly temperature-dependent quantity and increases dramatically with absorber temperature. The spectral emissivity is constant in the considered temperature range, but the thermal emission described by a blackbody spectrum inevitably increases with increasing absorber temperature and shifts its maximum closer to the solar wavelengths according to Wien's displacement law. It even extends into the visible spectrum at very high temperatures (c.f. the red glow of hot iron). The consequence of this behavior is that, to develop appropriate high-temperature absorbers, the focus must be on maintaining high absorptivity at high temperatures, while heat emission-related losses caused by an increasing emissivity at high temperatures are increasingly difficult to be avoided.

However, if the relative temperature difference to ambient temperature is used to drive a thermodynamic cycle for a working machine, a higher receiver fluid outlet temperature causes the Carnot efficiency η_C of the working machine (compare Eq. (7.4)) to grow monotonically. As a result the system efficiency η_{sys} being the product of the thermal collector efficiency and the Carnot efficiency, i.e.$\eta_{sys} = \eta_{rec,th} \cdot \eta_C$, exhibits a local maximum, which exceeds 300–400 °C for line focusing systems as depicted in Fig. 7.3. Thus, the potential capability of point-concentrating systems to achieve exergy efficiencies than line-focusing systems originates from two causes. At high concentration ratios the area-specific heat losses decrease since the concentration factor is in the denominator. Secondly, the Carnot efficiency increases with increasing temperature which, according to the first law of thermodynamics, allows more work to be extracted. It must be emphasized that the development of vacuum receivers with much smaller thermal losses than assumed in the Fig. 7.3 lead also to higher optimum working temperatures.

7.1.2 Radiation Density Distribution of the Sun and the Circum-Solar Radiation

The direct normal irradiance (DNI) describes the irradiance on a surface perpendicular to the vector from the observer to the center of the sun. The DNI is the essential input responsible for the yield of concentrating collectors. The angular distribution of the direct and circumsolar radiances, *i.e.* the sunshape, as detailed in Chap. 2, has a significant effect on the performance of concentrating collectors especially for high geometric concentration factors C_g. Depending on atmospheric conditions, a significant fraction of solar radiation coming from the sun is forward scattered by atmospheric constituents and forms the circumsolar region.

Concentrating collectors do not only collect the solar radiation directly originating from the sun's disk, but also a fraction of the circumsolar radiation. The absolute fraction depends strongly on the collector type and the concentrator design and quality.

7.2 Line Focusing Systems

Line-focusing collector systems have the largest market penetration of all high-temperature collector systems, such that more than 80% of the world's operating CSP power plants are based on line-focusing collector technologies, of which more than 90% are parabolic trough power plant types. In the following sections, we focus first on a full description of the parabolic trough collector's (PTC) for high-temperature applications, with the aim of describing the collector performance. Here, optical efficiency, incident angle modifier, thermal losses as key factors are influenced by design and material choices. Similarly, we will describe later the more variable designs of linear Fresnel collectors (LFC). Although both, parabolic trough, and linear Fresnel, are line concentrating technologies, the optical paths in the concentrators and the spatial distribution of optical conversion and heat transfer in the receivers differ.

7.2.1 Parabolic Trough Collectors

7.2.1.1 Optics of a Parabolic Trough

A parabolic trough solar collector (PTC) consists of mirrors of parabolic cross-section and a receiver element, which is composed of a tubular receiver containing

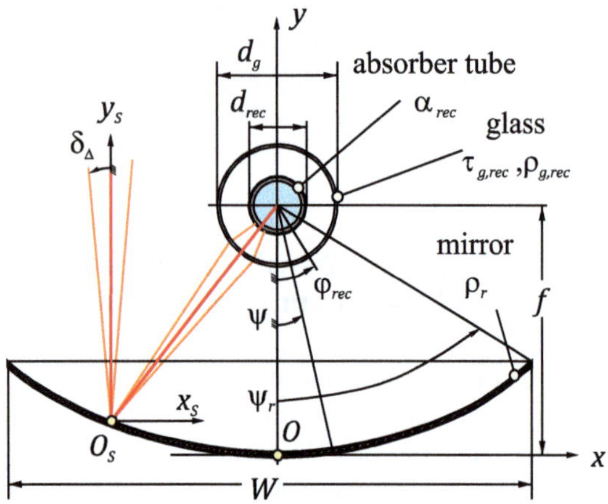

Fig. 7.4 Cross section of a parabolic trough collector (PTC) and definition of the geometry

the heat transfer fluid, which is concentrically enveloped by a glass tube. The cross-section of such a generic PTC is sketched in Fig. 7.4. In this simple case of a concentrator of a parabola in two-dimensional space, the light rays incident parallel to the y-axis are focused by the parabola contour to the focal point F of the parabola, as shown in the Eq. (7.6). According to geometric algebra, the following relationship for the contour is obtained

$$y + f = \sqrt{x^2 + (y - f)^2} \rightarrow y = \frac{1}{4f}x^2. \tag{7.6}$$

The incident beam of the solar radiation given by the direct normal irradiation including the circumsolar irradiation with an angular width δ_Δ first strikes the reflector of aperture width W, rim angle ψ_r, and focal length f. There, the incident beam is redirected towards the receiver with reflectivity ρ_r of the mirror, which is also a function of the incident angle θ. The change in direction of the beam at the reflector changes the distribution of the radiance in the direction of the receiver. At the glass envelope of the receiver, a part of the beam, ($\rho_{g,rec}$), is reflected dependent on the incidence on the glass tube into the environment, and the other part of the radiation continues in the direction of the receiver with a transmittance ($\tau_{g,rec}$) of the glass cover. A spatially distributed radiation density—the fraction which hits the receiver tube—then impinges on the absorber surface and is converted into heat by the absorptivity (α_{rec}). The process of the photothermal conversion is complex, and the distribution of the solar energy flux on the absorber is the boundary condition for heat transfer calculation. The heat is transferred through the absorber tube wall to the heat transfer fluid (HTF) circulating through the absorber tube.

For the determination of the thermal collector yield in the context of a power plant or process heat facility, the use of simple models calculating the energy balances is sufficiently accurate, since it aims at the determination of integral quantities, such as heat yields or losses per meter collector length. However, the design

Fig. 7.5 Schematical design of a parabolic trough collector

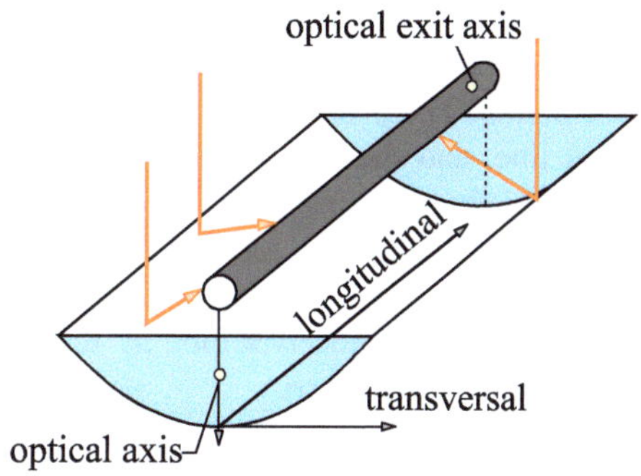

of a combination of reflector and receiver, or the detailed modelling of heat transfer to a two-phase flow pattern requires a detailed determination of local temperatures, heat fluxes and heat losses, for example to identify thermally highly loaded spots and to adjust or optimize the component design accordingly.

When the parameter focal length f of the parabola increases, a flatter parabola is the consequence. In practice, this means a smaller contour length and thus less material requirement for the same mirror width (i.e. constant concentration ratio). On the other hand, the distance to the focal point increases, so that possible mirror errors have a larger effect. Also, the irradiated section of the absorber tube becomes smaller.

Based on this optical element collectors are designed using a long reflecting mirror ideally with the cross section of this parabola. Figure 7.5 shows this design schematically defining the optical axis along the absorber tube and the longitudinal and transversal directions.

Looking at Fig. 7.4 this shows the parameters of a cross section. All incident rays with a projection into the transversal plane which is parallel to the optical axis are reflected by the mirror to the focal point where the absorber is located. However, depending on the longitudinal incidence projection the rays end on a different position along the tube. In fact, for large longitudinal incidence angles rays might be reflected to a focus outside the finite length of a parabolic trough. This is considered as end loss.

The end loss formula for a collector trough of finite length L and focal length f is

$$\eta_{endloss}(\theta) = 1 - \frac{f}{L} \cdot \left(1 + \frac{W}{48 \cdot f^2}\right) \cdot \tan\theta. \qquad (7.7)$$

where θ is the incidence angle on the tracked aperture, which is the angle in the longitudinal optical plane.

Basic Parameters of a Parabolic-Trough Collector

As solar concentrating devices, PTCs require solar tracking systems to modify their position with the changing apparent sun position in the sky from sunrise to sunset. Movement of this type of solar collector needs only one degree of freedom, an axial rotation with the axis ideally in North–South (*NS*–) or East–West (*EW*–) orientation. The position of the PTC must be such that the sun vector, the collector focal line, and the normal to the collector aperture plane are in the same plane. For PTC the resulting incidence angle is the angle between surface normal and solar beam direction. It affects the energy reaching the receiver pipe as mainly the cosine effect reduces the incident solar flux on the reflector. Diffuse solar radiation reflected onto the receiver tube by the concentrator is negligible compared to the concentrated direct part.

The most important PTC parameters are the geometric concentration ratio C_g, acceptance angle (δ_a), rim angle ψ_r, and peak optical efficiency at normal incidence $\eta_{opt,0}$. These parameters are explained in the following paragraphs. The geometric concentration ratio, C_g, is the ratio between the collector aperture area and the total absorber tube area (see Eq. (7.8)). This concentration ratio is usually about 25, although theoretically, the maximum is in the order of 70. High concentration ratios are associated with higher working temperatures. The geometric concentration ratio is given by Eq. (7.8):

$$C_g = \frac{W \cdot L}{\pi \cdot d_{rec} \cdot L} = \frac{W}{\pi \cdot d_{rec}}, \tag{7.8}$$

where d_{rec} is the outer diameter of the receiver steel pipe, L is collector length, and W is the parabola width. The acceptance angle, δ_a, is the maximum angle that can be formed by two rays on a plane transverse to the collector aperture in such a way that, when they are reflected by the parabolic mirrors, they intercept the absorber pipe. The wider the collector acceptance angle is, the less accurate the sun tracking system has to be. Small acceptance angles as explained in the optics Chap. 3 are associated with high concentration ratios and require the installation of very accurate solar-tracking systems leading often to higher costs. To capture all direct beam solar radiation the minimum acceptance half-angle should be at least the solar divergence half-angle $\delta_s = 0.27$ which contains the solar disk as seen from the Earth. It must be chosen by the designer which part of the circumsolar radiation shall be captured as well. Recommended acceptance angles for commercial PTCs are in the range 1–2°. Higher concentration with smaller acceptance angles demands very accurate and more expensive solar-tracking systems, while wider angles with lower concentration lead to lower working temperatures of the heat transfer fluid.

The relevant parameters identified for optical gain are the dimensions of the trough given by the rim angle ψ_r, and the geometric concentration ratio C_g, which are defined by:

$$\frac{f}{W} = \frac{1 + \cos\psi_r}{4\sin\psi_r};\ C_g = \frac{W \cdot L_{rec}}{\pi \cdot d_{rec} \cdot L_{rec}} = \frac{\sin\psi_r}{\pi \sin\theta}, \tag{7.9}$$

in which θ is the incidence angle on the aperture. The acceptance angle δ_a describes the maximum angle that can be formed by two rays in a plane normal to the collector aperture such that, when reflected from the parabolic mirrors, they just hit the absorber tube. Small acceptance angles thus imply large concentration factors that require the installation of accurate sun tracking systems. The minimum acceptance angle is ($2 \cdot \delta_S = 0.532°$), which is the average solid angle at which the solar disk is seen from Earth. Therefore, any parabolic trough collector (PTC) with an acceptance angle less than ($2 \cdot \delta_S$) always looses some direct solar irradiance. Real acceptance angles for commercial PTCs are in the range of 1–2°. The outer diameter of the absorber (d_{rec}) must be larger than the diameter of the light cone to prevent rays missing the absorber tube. When we only consider the sun disk as source (excluding circum-solar), the required minimum outer diameter of the absorber can be calculated by

$$d_{rec} = 2\left(\frac{W^2}{16f} + f\right)\sin\theta_s. \tag{7.10}$$

The rim angle (ψ_r) is the angle between the reflected rays from the reflector edge and y-axis, wich is given by

$$\psi_r = \begin{cases} \arcsin\left(\frac{8(W/f)}{(W/f)^2} + 16\right) & \text{for } 0° < \psi_r \leq 90° \\ 180 - \arcsin\left(\frac{8(W/f)}{(W/f)^2} + 16\right) & \text{for } 90° < \psi_r < 180° \end{cases}. \tag{7.11}$$

The upper half of the receiver tube absorbs solar radiation directly, while at the bottom side the concentrated solar beam focused by the reflector mirrors is optically converted by the receiver.

A further consequence is that the permissible acceptance angle δ_a also depends on the distance between absorber and reflector. In the case of the parabolic contour, this is largest at the edge or rim. The average value over the entire contour segment of a parabolic trough reaches a minimum if the rim angle (ψ_r) of the parabola is $\psi_r = 120°$. However, parabolas with this rim angle have a strong curvature at the vertex, which can lead to problems during production. Furthermore, the distance at the vertex can become too narrow for the installation of the glass envelope tube or mirror cleaning. Therefore, parabolas with smaller rim angles are often chosen.

For cylindrical tube receivers, usual rim angles are in the range 70–110°. Smaller rim angles are not advisable because they reduce the aperture surface and lead to comparatively large focal lengths. Rim angles over 110° need a lot of reflector material because the surface is enlarged without significantly increasing the area of the aperture plane.

Optical and thermal losses in a parabolic-trough collector. Optical losses characterized by the peak optical efficiency are very important in a PTC because they are about 25% of the total solar flux incident on the PTC aperture plane. Optical losses are associated with the following four parameters, illustrated in Fig. 7.6.

Reflectance $\rho_r < 1$ of the collector reflective surface. Typical solar spectral reflectance values of the best clean silver-based low iron glass mirrors are around 0.93–0.94. Reflective films have usually a specular reflectance $\rho_r < 0.9$.

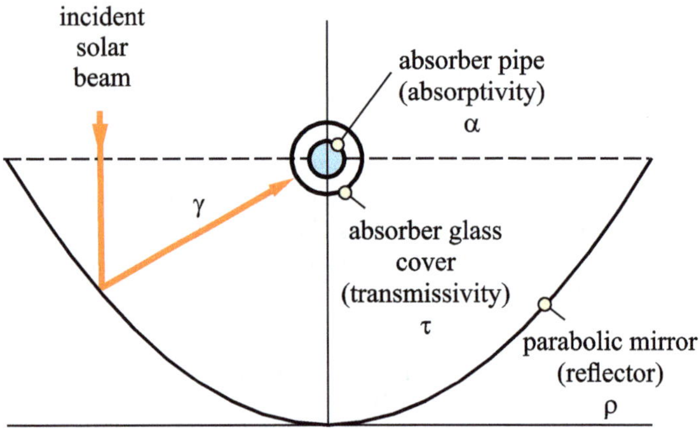

Fig. 7.6 Parameters impacting the optical efficiency ($\eta_{opt,0}$) of a PTC

Intercept factor γ. This is the fraction of the direct solar radiation reflected by the mirrors that finally reach the active surface of the receiver tube. The other part reflected by the mirrors does not reach the receiver pipe due to either microscopic imperfections of the reflectors, macroscopic errors in the parabolic-trough concentrator shape (e.g. inaccuracies during assembly), mechanical deformation of the PTC, flexible bellows, or shadowing by the receiver tube supports. Therefore the intercept factor for 0° incidence angle of the solar beam radiation can be calculated as the product of two parameters: a factor γ_1 related to optical imperfections (micro-roughness, errors in the parabolic shape, mechanical deformation (i.e. bending plus torsion) of the support structure) and a factor γ_2 related to shadowing by the receiver tube supports, the active length factor of the receiver tube, which accounts for the shadowing effect of the flexible bellows and their protective shield onto the steel absorber tube. Typical values are $\gamma_1 < 0.97$ and $\gamma_2 < 0.97$, giving an overall intercept factor γ in the range of 0.91–0.93 for high-quality PTCs.

Transmittance of the glass cover $\tau_{g,rec}$. The steel receiver tube is often covered by a glass cylinder to reduce thermal losses. A fraction of the direct solar radiation reflected by the mirrors onto the glass cover of the receiver pipe is unable to penetrate it. The transmittance $\tau_{g,rec}$ of the cylindrical glass for the relevant incidence angles on the glass surface is typically very close to the normal transmittance $\tau_{g,rec} = 0.93$ and can be increased up to 0.96 by antireflective coatings applied on both sides of the glass cover.

Absorptance of the receiver selective coating α_{rec}. This parameter quantifies the amount of energy absorbed by the steel receiver pipe divided by radiation reaching it. This parameter is typically 0.95 for receiver pipes with a cermet selective coating, and slightly lower (0.92) for pipes coated with black nickel or chrome.

Multiplication of these four parameters (intercept factor, mirror reflectance, glass transmittance and receiver absorptance) for an incidence angle $\theta_i = 0°$ gives the peak optical efficiency of the PTC $\eta_{opt,0}$:

$$\eta_{opt,0} = \rho_r \tau_{g,rec} \alpha_{rec} \gamma. \quad (7.12)$$

$\eta_{opt,0}$ is usually in the range of 0.74–0.79 for clean, good-quality PTCs. Taking the four optical parameters included in the peak optical efficiency into consideration, it represents the percentage of beam solar radiation converted to heat for normal incidence.

The effect of non-normal incidence angles is quantified by the incidence angle modifier $K(\theta)$, abbreviated *IAM*. This accounts for the fact that the change in the angle of incidence of direct solar radiation (θ_i) affects the four optical parameters mentioned previously and the collector's useful aperture area. The incidence angle modifier (*IAM*) corrects for these additional reflection and absorption losses. This parameter is usually provided by the manufacturer and results either from extensive calculations supported by validation experiments or as an empirical correlation to experimental data on a parabolic trough collector. The *IAM* $K(\theta)$ includes all optical and geometric losses in a PTC generated by an incidence angle $\theta_i = 0°$, without considering the $\cos\theta$ effect, and without accounting for the end-loss factor (7.7). Thus, we get for the incidence angle dependent optical efficiency $\eta_{opt}(\theta_i)$:

$$\eta_{opt}(\theta_i) = \eta_{opt,0} \cdot K(\theta_i)\eta_{endloss}(\theta_i). \quad (7.13)$$

This formulation contains by its factorization some simplifications, e.g. that the absorption or transmission of incident radiation on the absorber tube or on the cover glass is constant with circumferential position. However, the simple structure of Eq. (7.13) and the comparatively small deviations from the exact result make it ideally suited to determine different collector field or solar circuit arrangements with regard to the thermal yield for any location quickly and with good accuracy.

The incidence angle modifier is usually given by a polynomial equation such that it is equal to 0 for high θ_i (>85°), and to 1 for $\theta_i = 0°$. A representative incidence angle modifier for a PTC similar to a EuroTrough collector is given by:

$$K(\theta_i) = \max\left[0, 1 - \frac{7 \cdot 10^{-4} \cdot \theta_i + 36 \cdot 10^{-6} \cdot \theta_i^2}{\cos\theta_i}\right], \quad (7.14)$$

where θ_i is given in degrees.

However, this simple optical efficiency approach has some drawbacks. Although thermal collector yield calculations consider daily and annual variations of a typical meteorological year (TMY) and the associated direct normal irradiation (DNI), they do not include the influence of variable circumsolar radiation (CSR) on the intercept factor. This has to be modelled in more detail. What is more important from a technical point of view, especially for the constructive design of a collector and of the receiver, is a determination of local circumferential and longitudinal temperature distribution on the receiver (absorber) surface. This is the basic pre-requisite for the design of the receiver, the selection of suitable structural and functional materials, for example the selective absorber, and is dependent on the dimensioning and positioning of spacers, brackets, pylons carrying the support structure etc. and thus the structural design of the receiver relative to the parabolic reflector. The formula Eq. (7.13) does not produce any location-dependent power values of the radiation density at the thermally loaded components. This requires local modelling for example using ray tracing techniques as described in Chap. 3.

7.2.1.2 Thermal Performance

Regarding the thermal performance the integral approach models characterize the heat gain of the fluid along the flow path through the receiver tube using as heat source the optically converted heat minus the radiative and conductive-convective heat losses through via the glass envelope and the bracket and spacer elements. An energy balance model of heat loss in a vacuum receiver has been described for example in Burkholder & Kutscher (2008, 2009). In a similar way, even air-filled receivers many be treated, via engineering correlations outlined in Chaps. 4 and 5. Otherwise measured parameter determined in a receiver test can be used. Typically, the empirical thermal loss function is provided by the collector manufacturer in form of a polynomial function of a temperature difference between mean fluid or absorber surface temperature in the receiver and the ambient temperature. It contains linear terms representing conductive heat losses, quadratic and/or fourth order terms approximating convective and radiation heat losses. For a real local determination of heat losses to the ambient the temperatures at a discrete position z of the collector is obtained as a function of the flow rate, the flow regime and the imposed solar irradiation using computational fluid dynamic (CFD) models.

Thermal losses are very important in PTCs because they significantly affect the overall collector efficiency. The total thermal loss of a PTC $\dot{Q}_{loss}$ is due to radiative, convective, and conductive processes from the receiver tube to its surrounding. Although they all are well-known mechanisms and can be modelled explicitly, often—not knowing the precise design data—in testing they are experimentally determined under real solar conditions at several temperatures. These experimental results are then processed to determine the best-fit coefficients b_1 and b_2 *or* b_1 and b_4 depending on which fit is better. Usually, evacuated tube receivers are characterized by b_1 and b_4:

$$\frac{\dot{Q}_{loss}}{A_{ap}} = \dot{q}_{loss} = b_1 \cdot \Delta T + b_2 . \Delta T^2 + b_4 \cdot \Delta T^4, \tag{7.15}$$

where b_1,b_2 and b_4 are coefficients obtained from the experimental values, and ΔT in (°K) is calculated from the difference of the average temperature of the working fluid in the collector (T_m) and the ambient temperature (T_{amb}).

Although now an international standard exists for specifying this testing previously to this the manufacturers delivered their receivers with an equation different from (7.15) to calculate the thermal losses. Today's high-temperature PTCs are provided with evacuated receiver tubes, thus avoiding convection losses between the steel pipe and its glass cover.

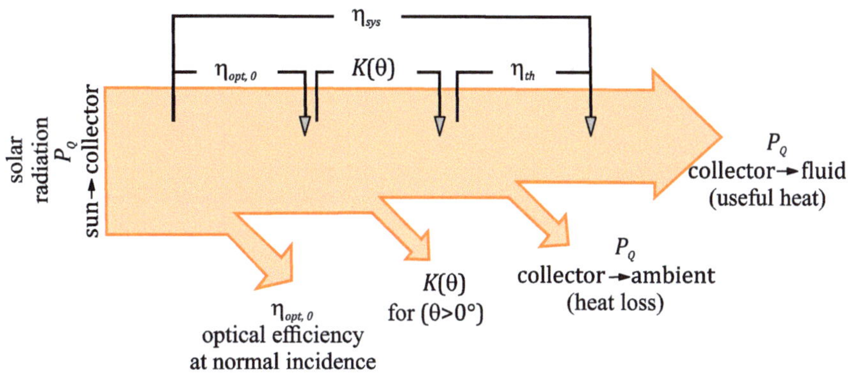

Fig. 7.7 Energy flows for PTC

With the determination of the heat losses, the entire loss chain of a collector is thus mapped, which results from the optical efficiency at perpendicular irradiation ($\eta_{opt,0}$), the consideration of the IAM ($K(\theta)$) and the inclusion of the heat losses, which is depicted in Fig. 7.7.

This allows the specific collector yield to be calculated at any time of the day depending on the collector orientation, for example north–south (NS) or east–west (EW). And from the integration over the collector area and over the day, the total energy yield results. Figure 7.8 shows the calculated specific power as a function of solar time for a Eurotrough-100 collector at the Seville site on a summer and winter day for NS and EW orientations, respectively. The yield in the NS orientation is greater in the EW direction over the year, with this being an advantage especially in the summer. Only in winter is the EW orientation superior to the NS orientation in terms of yield, but the irradiance values are significantly lower in winter.

7.2.1.3 Optical Errors and Effects on Collector Efficiency

For small concentration factors, the optical errors of the concentrator and the mirror material have only a minor effect. For C_g below 4–5 there is still a usable portion of the diffuse radiation which decreases with increasing concentration. On the other hand, the higher the concentration, the more important is the precise manufacturing of the concentrator shape and the selection of mirror materials with small optical spread. The quantification of the optical quality is given by the width σ of a Gaussian beam. The larger the value of σ the wider the beam spread. And diffusing surface or deformed surface may widen up the beam spread. Good concentrators usually achieve a total optical error σ_{opt} of a few mrad only. In the following sections we describe the methodology of coming to a total error and show some representative results of optical simulations.

Apart from optical quality of the mirrors which has a permanent effect, there may be also temporary effects due to soiling of mirrors. Soiling is the accumulation

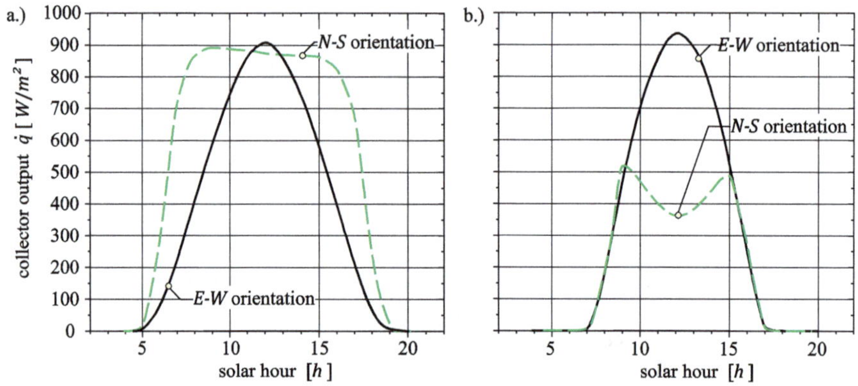

Fig. 7.8 Simulated daily thermal output of a Eurotrough-100 parabolic trough collector located in Seville/Spain ($\Phi = 37°14'$), on a summer (**a**) and winter (**b**) clear day and with two different PTC orientations

of dirt (dust particles, aerosols, salt particles, biological growth) on a mirror and may be at least partially removed by washing/cleaning processes. Especially in dry areas with little vegetation the concentration of fine dust particles in the air is high which come from wind erosion.

The accumulation of particles sticking to the mirror surface depends on the particle size and chemistry, the adhesive forces of the surface (characterized by the contact angle) and environmental parameters (humidity, wind velocities). Soiling leads to light scattering and absorption and hence to a drastical reduction of specular reflectance of a mirror, which in turn is characterized by an additional beam spread σ. From geometry considerations the impact of dirt particles on the reflectance varies with incidence angle. Different soiling processes can be parametrized very well with a model developed by Heimsath and Nitz (2019) even for different wavelengths as illustrated in Fig. 7.9.

Imperfections in material and construction, beam spread of the sun, soiling all need to a lowering of the optical efficiency. However, the sensitivity is different for large and small parabolic trough collectors and also differs to Linear Fresnel collectors as indicated in Fig. 7.10. The reason is that these collectors have different focal lengths, and the displacement of a beam is the product of an angular error and a length dimension, which is in the order of the focal length.

Optical Errors Affecting PTC Collector Performance

As introduced in Sect. 3.4.6 any deviation of from the ideal mirror receiver geometrical configuration strongly impacts the intercept factor γ. Theses geometrical inaccuracies are expressed by the term optical errors and its standard deviation. To achieve a low overall optical error σ_{opt} the entire chain from manufacturing to installation requires a stringent quality assurance posing a challenging task to ensure a high precision equipment. To the standard deviation of the overall optical

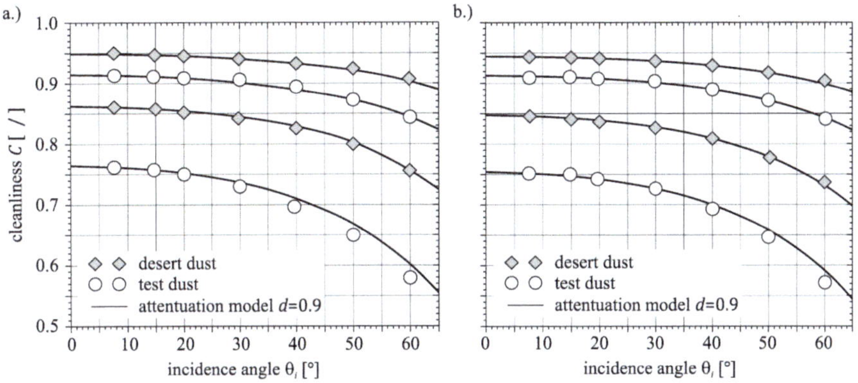

Fig. 7.9 Comparison of model (lines) and measured values (symbols) at different wavelengths as a function of incidence angle (θ_i) (left:$\lambda = 455$ nm, right $\lambda = 533$ nm). Crosses correspond to mirrors soiled with desert dust, circles correspond to samples soiled with test dust modified from Heimsath and Nitz (2019)

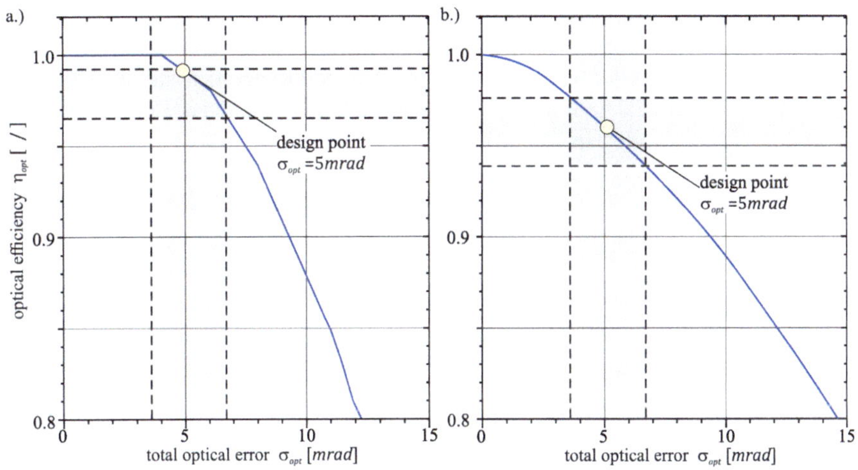

Fig. 7.10 Impact of total optical error on the optical efficiency (Heimsath, Bern, van Rooyen, & Nitz, 2014a, 2014b, 2014c); Heimsath, Cuevas, Hofer, Nitz, & Platzer, 2014a, 2014b, 2014c). **a** Large scale PTC and **b** a small scale LFC. Both collectors were designed for 5 mrad optical efficiency (Design Point). The effect of an increase or decrease of 1 mrad shape accuracy is indicated by grey lines

error σ_{opt} several factors contribute such as the standard deviations of the specularity error σ_{spec}, the slope error σ_{slope}, the misalignment or displacement error σ_{dis} and the tracking error σ_{trac}, which are depicted in Fig. 7.11. The specularity error σ_{spec} describes the glass quality or exactly to the degree of which the collected rays respect to the law of reflection. Macroscopic surface deviations in

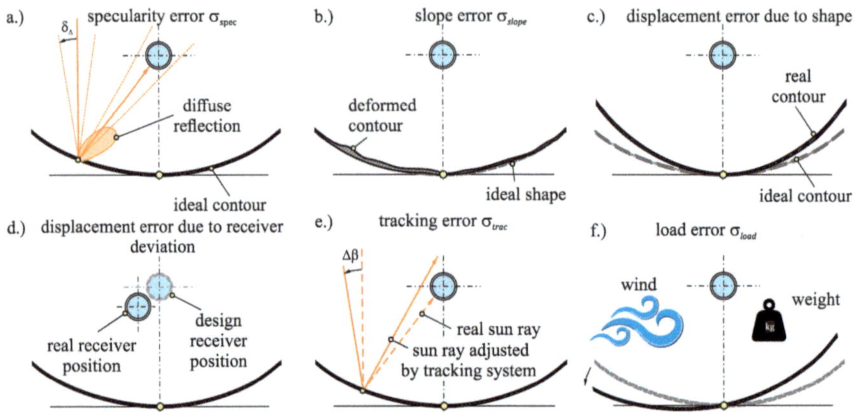

Fig. 7.11 Geometrical error types appearing in a PTC collector

position and slope are summarized as slope errors σ_{slope} are directly linked to the manufacturing quality and to the local ripples or the aberrations on the mirrors. Displacement errors σ_{dis} are caused either by a misalignment of the facets that compose the collector or a miscellaneous arrangement of the receiver with respect to the mirror. These displacement errors are attributed to geometrical deformations with respect to the ideal geometry as indicated in Fig. 7.11c and d. Finally, tracking errors σ_{trac} as illustrated in Fig. 7.11e occur either due to incorrect measurement or aging of the drives or their gearboxes.

To quantify and, moreover, to separate operational effects from quality assurance arguments mandatory to design, characterize and assess the collector performance all arguments associated with fabrication, erection and mounting are separated from operational impacts. This allows to analyse already computationally the robustness of a PTC design by executing sensitivity studies for all parameters affecting the intercept factor. Shape errors originating from operational loads such as bending or torsion of the structures due to wind or weight loads occur especially in periods with large ambient gradients (temperature, wind cloud, fields) are termed loading errors.

Thereby, averaged over time and over the entire collector array of collectors, all these errors can be assumed to be approximately linear independent and exhibiting a Gaussian distribution according to Rabl (1985). Additionally, all standard deviations of the optical errors can be summarized in the form of a sum of the contributing individual terms,

$$\sigma_{opt}^2 = \sum_i^n (a_i \sigma_i)^2 = \sigma_{spec}^2 + \left(2\sigma_{slope}\right)^2 + \sigma_{dis}^2 + \sigma_{trac}^2, \tag{7.16}$$

in which a_i represents a weighting factor. The beam spread of the sunrays, when interacting with the imperfect mirror (concentrator), is represented by a weighting factor of two and thus twice all other the standard deviations. This refers to all mirror components such as mirror shape, mirror support position and mirror support angle. Since the sun shape represents also a statistical random error independent of the collector design a so-called standard deviation of the total error σ_{tot} is used

$$\sigma_{tot}^2 = \sigma_{opt}^2 + \sigma_{sun}^2. \quad (7.17)$$

The most commonly used, straight forward and accurate method to quantify the standard deviation of the total optical error is statistical ray tracing (MCRT), in which normal distribution functions of the angular deviations of each contributing ray are evaluated by means of using perfect geometrical optics. Once the local deviations fall below a predefined threshold the solution is converged, see e.g. Pottler et al. (2014). The central advantage of this Monte-Carlo concept is that all statistical deviations of the contributing effects can be condensed in one single parameter, the so-called intercept factor γ (see Eq. (7.12)). This intercept factor γ is then a function of the parabolic trough rim angle ψ_r, the longitudinal incidence angle θ_i and the total beam width σ_{tot}, as qualitatively shown in Fig. 7.12a. An increasing beam width yields a monotonic decrease of the intercept factor. Bendt Rabl, Gaul, and Reed (1979) interpret the influence of the incidence angle on the beam pattern as the apparent broadening of the solar disk with increasing incidence angle, resulting in an increase in the path of the reflected beam from the mirror to the absorber. Larger total beam widths, have a particular effect at greater incidence angles, which becomes relevant in the winter period when the incidence angle at noon for the north–south oriented trough collector exceeds values $\theta_i >$ 30° at widths $\delta >$ 23.5°, which are additionally associated with a higher CSR. Figure 7.12b depicts that the intercept drastically declines for increasing incident angles.

Robustly designed and professionally installed parabolic trough collectors today achieve an overall quality of σ_{tot} = 6–6.5 mrad. This requires extensive quality assurance steps and control measures, which already start with the single part production in series production. Proof is provided by deflectometric measurements for mirrors and modules and 3D point measurements for metal frame structures. This process continues during installation on the solar field where the focus is to achieve a balanced quality across all components and assembly steps involved. At the end, after the assembly is completed, the achieved geometric quality is measured in the field by means of photogrammetry, deflectometry, tachymetry, inclinometers providing real data. The real geometry is then analysed again using mostly by means of statistical ray tracing models applying standard deviation parameters σ_i, which is shown in Fig. 7.12c broken down by the individual components of a collector according to specifications of the manufacturer or operational experience. The Fig. 7.12c illustrates that for a real PTC collector slope, displacement and sun-shape statistical standard deviations are of the same order of magnitude, while tracking deviations contribute usually to a much smaller fraction. Finally, Fig. 7.12d depicts a computed ray-tracing result based on measurement of a real PTC surface highlighting reflected rays in a 1 m width section with strong

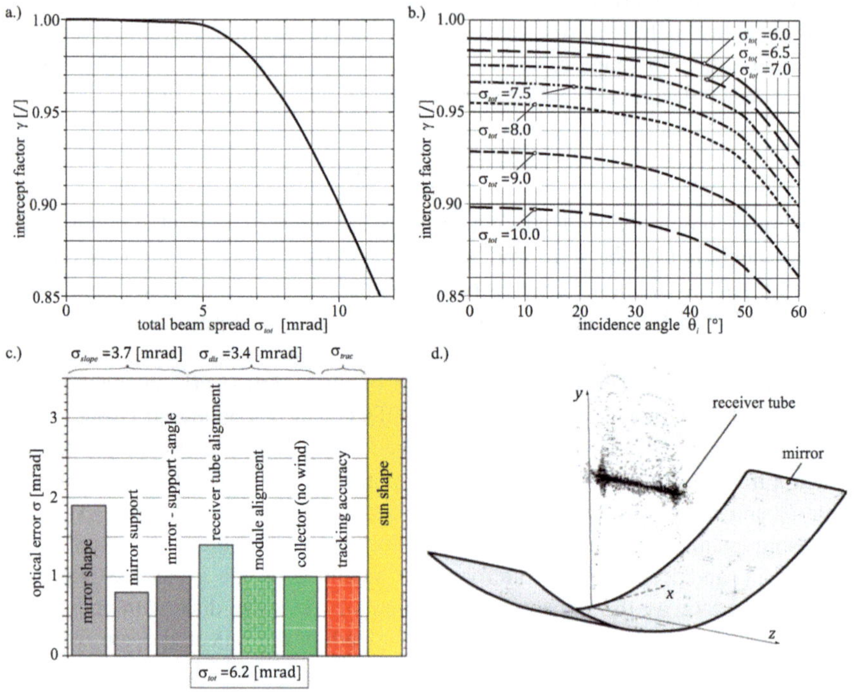

Fig. 7.12 **a** Computed intercept factor γ against total beam spread σ_{tot}; **b** Calculated intercept for different total beam spreads σ_{tot} as function of the incident angle θ_i on intercept factor; **c** share of the various optical error sources on the total beam spread σ_{tot} at $\theta_i = 0°$, **a–c** modified from Pottler et al. (2014); **d** simulated behaviour of the reflected rays in a 1 m width section with strong deformation modified according to García-Cortés et al. (2012). All graphs refer to the EuroTrough collector geometry

deformation. The graph in coincidence with operational experience shows that the positions of the greatest deviations depend strongly on the installation type of the mirror facets; many designs exhibit larger deviations occurring at the outer borders of the collector mirror (large x-values). Also, operational data for several plants reveal that larger deviations compared to an ideal optics is obtained for a collector in a horizontal position rather than in a vertical position. Stochastic effects are particularly relevant for the design of the collector, but during operation or installation of systems, systematic deviations occur, which in some cases have a more significant negative influence on the collection factor. Systematic deviations mainly include asymmetric shape deviations such as too shallow supports, deviations of the centre of gravity from the axis of rotation and misalignment of the tracking sensor. Symmetric systematic effects can also reduce the pickup factor, such as mirror fields that are too open or too closed and/or concentrator shapes.

Each collector design must be analysed with respect to the sensitivity of manufacturing, measurement deviations of the drive or inclination sensors, actuator deviations or bending of the collector support structure, which are reflected in the collection factor. Finite element methods (FEM) are typically used to analyse the strength of the support structures with the associated deformations in terms of bending and/or torsion of the structures, for more details see (Meiser, Schneider, Lüpfert, Schiricke, & Pitz-Paal, 2017). With the help of numerical calculations, it is possible to compensate for systematic alignment deviations, for example, by an offset in the tracking angle.

Simulated Optical PTC Collector Characteristics

PTC performance at Normal Incidence ($\theta_i = 0°$).

Figure 7.13a shows a simulated circumferential distribution of local concentration factors (C_L) at normal irradiance ($\theta_i = 0°$) for a PTC collector of realistic dimensions at different circumsolar irradiance fractions. First, the local concentration factor C_L shows four different ranges. A so called shelter zone (1) with a high but reduced C_L caused by the shadowing of the receiver tube, then a segment (2) with a significant increase of the concentration ratio up to a maximum, followed by a Sect. 3 characterized by a reduction of the local irradiance and finally a receiver Sect. 4, which only sees the sun ($C_L = 1$) and is called the isolation zone. If we first consider a $CSR \approx 0$, we find that significant local power density gradients over a few degrees (here for $\varphi_{rec} > 62°$) occur in circumferential direction. Furthermore, there is a significant decrease of the local concentration factor C_L in the shelter domain compared to the peak power position. For the considered receiver at $\varphi_{rec} \approx 89°$ no radiation reaches the receiver ($C_L = 0$), before with increasing φ_{rec} a value of $C_L = 1$ establishes. Analyzing the impact of a variable CSR, the local concentration factor in regions 1 and 2 falls strictly monotonically with increasing CSR. Thus, the optical efficiency of the PTC collector $\eta_{opt,0}$ decreases steadily with increasing CSR, which is immediately apparent, because the angular angle variation of the sun θ_s increases and thus the path of solar irradiation through the atmosphere (SOD). The reduction of $\eta_{opt,0}$ in the example is about 22% at an incidence angle $\theta_i = 0°$ from $CSR = 0$ (no atmospheric scattering) to a $CSR = 0.8$ (extremely hazy sky). At the same time, however, the power gradients expressed by C_L also flatten out in domains 1 and 2, and a shelter effect is hardly observed at high CSR. Conversely, the increase of CSR in region 3 leads to a minor increase of the local concentration ratio compared to $CSR \approx 0$.

The graph also shows that for $CSR > 0$ the optical efficiency $\eta_{opt,0}$ increases with increasing receiver diameter. This is because the larger the diameter, the larger the radial intercept angle (θ_γ), and thus more reflected rays are received by the absorber. It can also be seen from the figure that the larger the CSR value becomes, the smaller the optical efficiency is. The difference of optical efficiency between two CSR s decreases with the increase of receiver diameter, indicating that the optical efficiency is less sensitive to the CSR for a PTC with larger receiver diameter.

With the help of local determination of the radiation incident on the receiver, the geometry of the PTC collector (receiver diameter-d_{rec}, focal length-f, aperture

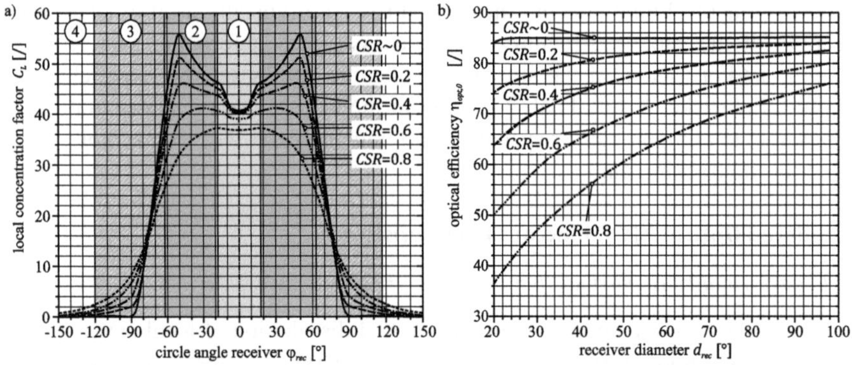

Fig. 7.13 Computed optical properties of a PTC collector for different CSR modified from Zou et al. (2017) (SEGS-LS-2PTC collector: $W = 5$ m, $f = 1.84$ m, $L_{rec} = 7.8$ m, $d_{rec} = 0.07$ m, $d_g = 0.115$ m, $\alpha_{rec} = 0.96$, $\tau_g = 0.95$, $\rho_r = 0.93$); **a** local concentration factor C_L as function of azimuthal receiver angle φ_{rec}; **b** optical efficiency $\eta_{opt,0}$ for different receiver diameters d_{rec}

width-W) can now be optimized to achieve an optimum optical efficiency. Analyzing the optical efficiency as a function of the receiver diameter, as shown in Fig. 7.13b, we see that there is a critical minimum diameter at $CSR \approx 0$. This is because at $CSR \approx 0$ all rays come from the solar disk (with negligible aureole) whose radial angle is 4.65 mrad ($\delta_s = 4.65$ mrad) and using Eq. (7.10) we can calculate this analytically. A diameter larger than the critical minimum diameter does not lead to any increase in the optical efficiency when $CSR \approx 0$. Conversely, for smaller values of the receiver diameter, some of the reflected rays escape around the absorber tube, which reduces the optical efficiency.

Figure 7.14a shows the distribution of the radial intercept angle (θ_γ) as a function with absolute abscissa value x_A. The intercept angle decreases monotonically with increasing the abscissa up to W, which means that the larger the aperture width W is, the smaller the average θ_γ becomes, whereby more rays do not reach the receiver and as consequence the optical efficiency $\eta_{opt,0}$ decreases with increasing aperture width for a given focal length f. This behavior requires a larger PTC receiver diameter d_{rec} to maintain a large optical efficiency. The effect of focal length f on the optical efficiency for $CSR = 0.4$ in Fig. 7.14b shows that the optical efficiency initially increases and reaches a local maximum and then decreases monotonically with further increase of the focal length. The cause is the dependence of the radial intercept angle (θ_γ) on the focal length f at different reflection points A shown in Fig. 7.14b. For $x_A \geq 1$, the radial intercept angle first increases with increasing focal length before it reaches a local maximum value. Beyond this local maximum θ_γ decreases monotonically with increasing f. Further, Fig. 7.14b shows that the intercept angle exhibits a peak at an increasing focal length fp as x_A rises. Therefore, f for a given PTC is determined by the combined effects of all reflection points A on the parabolic mirror. This indicates

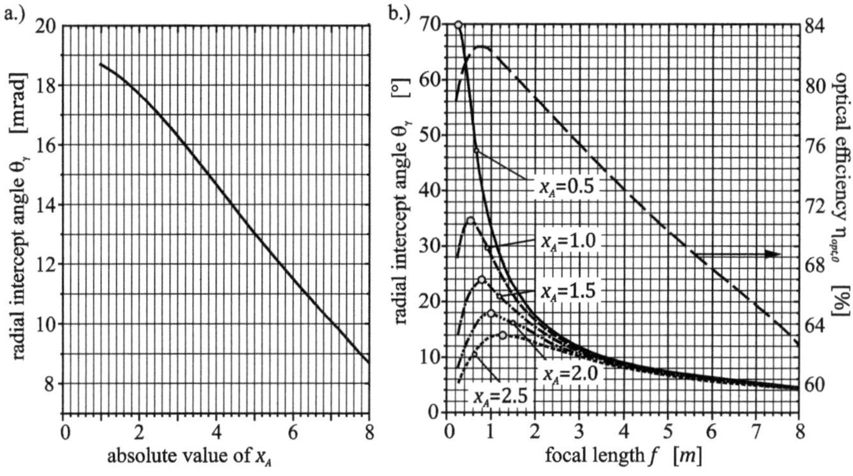

Fig. 7.14 Calculated optical properties of a PTC collector modified from Zou et al. (2017); (SEGS-LS-2PTC collector: $W = 5$ m, $L_{rec} = 7.8$ m, $d_{rec} = 0.07$ m, $d_g = 0.115$ m, $\alpha_{rec} = 0.96$, $\tau_g = 0.95$, $\rho_r = 0.93$; $CSR = 0.4$); **a** radial intercept angle θ_γ as function of x_A ($f = 1.84$ m) **b** radial intercept angle θ_γ for different x_A and optical efficiency $\eta_{opt,0}$ as function of focal length f (circles indicating maximum at lateral reflector width)

that the larger the aperture width W is, the larger the optimum f and the receiver diameter should be.

Figure 7.15a shows the results of the distribution of C_L for different aperture widths W. All C_L distribution curves as a function φ_{rec} exhibit the four regions as before. As the aperture width increases the maximum of C_L increases continuously, too, while the C_L value of the Shelter region for $\varphi_{rec} = 0°$ remains constant. Overall, the energy flux in regions 2 and 3 increases with increasing aperture width W, making region 4, which receives the direct irradiance, smaller. When the aperture width falls below a threshold value (here $W < 5$ m), the receiver area $\varphi_{rec} > 90°$ receives no concentrated radiation, while exceeding a certain aperture width (here $W > 12$ m) the entire receiver tube circumference is reached by concentrated radiation. Furthermore, it can be observed that the sheltering area becomes smaller and smaller when W falls below a threshold value (here 2 m) and disappears completely at a certain aperture width (here $W < 0.92$ m). The distribution of the intercept factor (γ), the dependence of the rim angle and the minium required receiver diameter as a function of the aperture width W are depicted in Fig. 7.15b. The intercept factor drops significantly above a critical aperture width W, since as the aperture width increases, the required absorber diameter also increases according to Eq. (7.10) and at some point, exceeds the size of the real absorber. Thus, part of the concentrating radiation is lost, which reduces the intercept factor. Increasing aperture width are associated with a rising rim angle (ψ_r) and require larger minimum receiver diameters (d_{min}). In other words, an increase of the aperture width

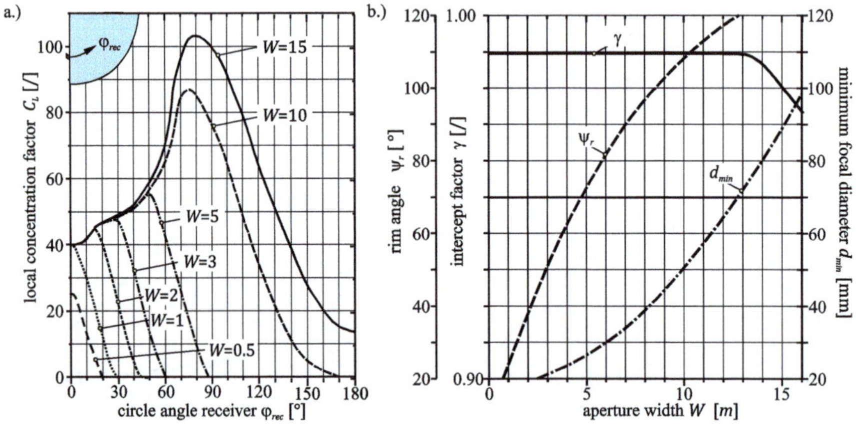

Fig. 7.15 Calculated optical properties of a PTC collector modified from Zou et al. (2017); (SEGS-LS-2PTC collector: $f = 1.84$ m, $L_{rec} = 7.8$ m, $d_{rec} = 0.07$ m, $d_g = 0.115$ m, $\alpha_{rec} = 0.96$, $\tau_g = 0.95$, $\rho_r = 0.93$); **a** circumferential local concentration factor C_L for different aperture width W; **b** intercept factor (γ), rim angle (ψ_r) and minimum receiver diameters (d_{min}) as a function of the aperture width W

W always means an increase of the rim angle (ψ_r) and an increase of the minimum receiver diameter (d_{min}). More radiation can be absorbed above the receiver surface, but with increasing aperture width the utilization of the radiation on the receiver decreases, so that above a certain aperture width, even with perpendicular irradiation, the intercept and thus the optical efficiency decreases.

PTC Performance at Non-normal Incidence Angle θ_i.

As illustrated in Chap. 6, for the integral collector modelling the impact of the incidence angle on the optical efficiency is depicted by the product of the peak optical efficiency $\eta_{opt,0}$ and the incidence angle modifiers (*IAM*) considering the temporal variations of the sun position with respect to the collector. Because PTCs are equipped with sun-tracking systems so that the transversal incidence angle θ_T of solar radiation is kept very small, ideally zero or very close to zero (<0.5°), so that the *IAM* due to transversal incidence angles does not play any significant role.

An increasing incidence angle θ_i causes part of the rays reflected by the mirror to exit at one end of the PTC and not reach the receiver, which naturally leads to end losses and thus to a reduction in optical efficiency. If the incidence angle θ_i exceeds a certain limit, the optical efficiency η_{opt} tends to zero because no beam reaches the receiver. Another effect of an increasing angle of incidence is that the image of the sun on the receiver becomes larger, because the distance of the reflected rays from the reflector to the receiver increases with growing incidence angle. This leads to a decrease in the power density and thus in the local concentration ratio C_L. Moreover, this effect smears the circumferential power distribution of regions 1 to 3 shown in Fig. 7.13a and results in a circumferential more uniform but smaller power distribution at large incident angles, similar to that at large

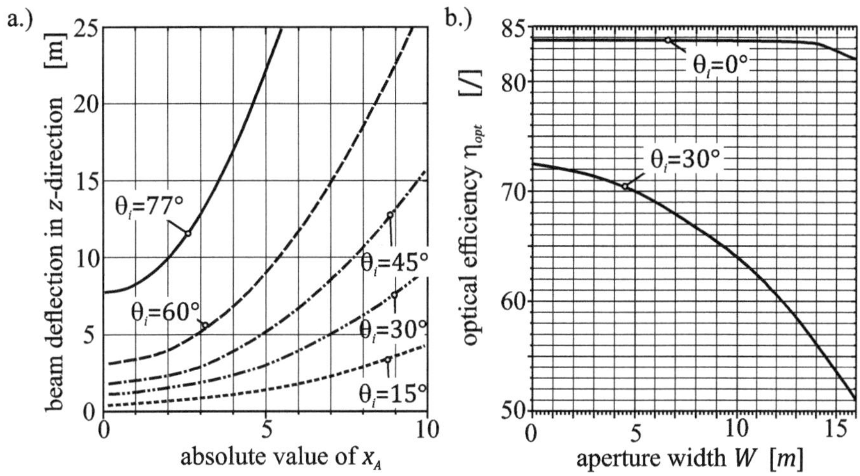

Fig. 7.16 Calculated optical properties of a PTC collector modified from Zou et al. (2017); (SEGS-LS-2PTC collector: $W = 5$ m, $f = 1.84$ m, $L_{rec} = 7.8$ m, $d_{rec} = 0.07$ m, $d_g = 0.115$ m, $\alpha_{rec} = 0.96$, $\tau_g = 0.95$, $\rho_r = 0.93$); **a** axial beam deflection (z-direction) for different incident angles θ_i; **b** optical efficiency η_{opt} as function of the aperture width W for $\theta_i = 0°$ and 30°

CSR, and characterized by a large-scale absence of a shelter effect at $\varphi_{rec} = 0°$. Considering the beam deflection in the longitudinal z-direction along the absorber tube for different incidence angles θ_i at a given point A (with the absolute value of the abscissa x_A, see Fig. 7.4 (A is the point where the incoming beam hits the mirror)), as shown in Fig. 7.16a, it is evident that the larger the aperture width (W), the larger the end losses become with increasing incidence angle θ_i. The position with a larger circumferential angle φ_{rec} can only receive reflected beams of larger values of x_A, which in turn leads to larger beam deflections in the axial z-direction. Therefore, the length of the section with very small local concentration factor C_L will be larger for larger φ_{rec} than for small φ_{rec}. For the SEGS-PTC considered, it is seen that at an angle of incidence of $\theta_i = 77°$, that the beam deflection by the reflector is larger than the length of the receiver ($L_{rec} = 7.8$ m), and no beam reflected on the mirror at any point of x_A reaches the receiver. Finally, Fig. 7.16b shows the optical efficiency η_{opt} as a function of the aperture width W at an angle of incidence of $\theta_i = 0°$ and 30°. As the incidence angle increases the optical efficiency decreases moreover; for increasing the aperture width this effect increases rapidly.

7.2.1.4 Technical Design Components of a Parabolic Trough Unit

A functional unit of a PTC receiver consists mainly of four components.

- a support structure,
- the receiver with the selective absorber enclosed in a glass envelope and the heat carrier fluid circulating inside,

- the curved mirrors mounted on the support structure, and
- a tracking system that tracks the receiver and mirror according to the position of the sun.

Each component must meet specific design criteria and ensure both functionality and availability. Any deviation due to faulty installation, poor design or tracking quality or great sensitivity to environmental influences such as wind, temperature and or humidity will result in a reduction of the optical yield. In Sect. 7.2.1.1, the basic design was discussed, while here the technical design of some realized collectors is considered.

Commercial Designs

Many commercial designs are based on a precommercial development from a European project, which is the Eurotrough design. In the years after 2000 mainly Spanish and US companies developed their own PTC design with little variations, as the aperture width of 5.8 m had been optimized for the 70 mm absorber tubes on the market. Some of them used prefabricated and preformed parts. Most structures were made of galvanized steel, but also aluminum and metal-polymer composites were used.

The implementation of the parabolic trough power plants during the first decade of the 21st century made evident that cost degression was slow using the same design. New developments investigated the possibility to achieve a larger solar field cost reduction if the size of the PTC was increased. The reduction in the number of specific components (e.g. ball joints, pylon foundations, drive units, local controllers, and loop specific piping) for solar fields with the same nominal power and the design of steel structures requiring less manpower cost for transport and on-site assembly could achieve significant cost reductions. Therefore, in 2010 the development of the Ultimate Trough design was started by the engineers of Schlaich Bergermann und Partner (SBP) and the Fraunhofer Institute for Material Flow and Logistics (IML) under the coordination of Flabeg Holding GmbH (Flabeg) (Riffelmann et al., 2014). The new aperture width of 7.5 m and receivers with 90 mm diameter were leading to an aperture area of 1689 m^2 and a claimed 25% solar field cost reduction. The main contribution to this was a 30% cost reduction of the solar field assembly cost. Another competitor design of large aperture PTC was the HelioTrough design with an aperture area of 1263 m^2 and an aperture width of 7.5 m (Janotte et al., 2014). Other industrial groups were following using optimized aperture widths from 6.9 to 8.2 m. The size difference between standard and large collector is shown in Table 7.2.

Although most of the R&D effort in the late 20th century and early in the 21st century was devoted to large aperture PTCs for large solar thermal power plants, new smaller troughs have also been developed for process heat applications with temperatures below 300 °C. These collectors can provide process heat to a wide range of applications replacing natural gas, such as crop drying and food processing. Industrial processes such as biofuel production, water purification, desalination, and absorption-chiller air conditioning for commercial and

Table 7.2 Typical technical design parameters of PTC for power plants

Parameter	ET-150 Standard	large PTC
Parabola width (m)	5.77	7.51
Collector length SCA (m)	147.8	242
Module length SCE (m)	12.27	24
Number of modules SCE	12	10
Outer diameter of steel receiver pipe (m)	0.07	0.09
Inner diameter steel receiver pipe (m)	0.065	0.085
Aperture area (m^2)	822.5	1689
Peak optical efficiency $\eta_{opt,o}(/)$	0.75	N.a

industrial buildings are similarly feasible with these small PTCs. These applications are grouped under the acronym SHIP (Solar Heat for Industrial Processes). The main difference to so-called large CSP applications is often, that the solar field sizes typically are in the range of a few hundred to a few thousand square meters, whereas for solar power plants in the 50–200 MW range the field sizes were several hundred thousand or even millions of square meters aperture area. Consequently, local production on-site is not cost-effective for those SHIP collectors and a high degree of pre-fabrication combined with standardized shipping and fast mounting is desired. This is reflected in the geometrical designs of the collectors, for which typical data are collected in Table 7.3. One of the first companies developing PTC for industrial uses was IST (USA) which installed several commercial plants in the United States. The technology was purchased by the Spanish group Abengoa, and their collectors (PT1 and RMT) were then marketed by Abengoa Solar (www.abengoasolar.com). Both collectors had non-evacuated receiver pipes with a black chrome coating and glass envelope, and maximum working temperatures of 288 and 204 °C, respectively. Another small PTC for process heat applications is the PTC-1800 collector developed by the German-Turkish company, SOLITEM (http://solitemgroup.com). The maximum working temperature of this collector is 200 °C and the collector aperture area is 36.65 m^2. The Australian company, NEP Solar Pty Ltd. (www.nep-solar.com), also developed small PTCs, the PolyTrough-1200 and PolyTrough-1800, suitable for temperatures up to 230 °C.

The 24 m long, 1.2 m wide PolyTrough-1200 collector (or 1800 mm wide PolyTrough-1800) consists of composite reflector panels with aluminum sheets which can be mounted either on roofs or at ground level. A more recent development is the PT950 solar collector from the company PROTARGET (www.protarget-ag.com) which has an aperture width of 3 m and due to the use of silicone-based oil as HTF may achieve a maximum working temperature of 425 °C. The Mexican company Inventive Power (https://inventivepower.com.mx) has been very successful economically and installed more than 100 SHIP projects in Mexico up

Table 7.3 Technical parameters of small PTC process heat collectors

Parameter	PowerTrough 110	PowerTrough 250	PTC-1100	PTC950	PolyTrough-1800
Manufacturer	Inventive power (MX)	Inventive power (MX)	Solitem (D)	Protarget (D)	NEP Pty (AUS)
Structure	Galvanized steel	Galvanized steel	Aluminum	Galvanized steel	Aluminum
Reflectors	Aluminum	Aluminum	Aluminum	Tempered glass	Composite
Module size (m)	1.21 × 3.06	2.5 × 4.16	2.5 × 1.1	3 × 12	1.8 × 24
Receiver tube	Non-evacuated, selective coating, borosilicate glass	Evacuated, selective coating, borosilicate glass	Non-evacuated, selective coating, borosilicate glass	Evacuated, selective coating, borosilicate glass	Evacuated, selective coating, borosilicate glass
Max. operating temp. T(°C)	130	260	200	425	230
HTF	Water			Silicone oil	

to 2019 using their own PTC designs. The success of Inventive Power demonstrates that there is a good market niche for SHIP applications in countries with a high level of direct solar irradiance.

Support Structure

The main functions of the support structure are to fix the collector components in a defined position and to provide structural stiffness and stability against varying environmental conditions. Steel or aluminum are mostly used as structural materials. From a structural-mechanical point of view, the PTC support structure consists of three parts: the main support (columns, piles, and box), the frame and the receiver brackets as depicted in Fig. 7.17.

- The main support serves as the structural anchor of the collector and holds it in place. The main support is structurally capable of resisting wind loads without exhibiting large torsion or bending which substantially impair the optical conversion and thus the yield.
- The frame provides rigidity to the mirrors, allowing them to always maintain their parabolic shape. In addition, it must serve torque transmission for tracking the mirrors.
- The brackets are attached to the mount of the mirrors, and the receiver is fitted via receiver supports to the frame so that it is in the focal line of the parabola.

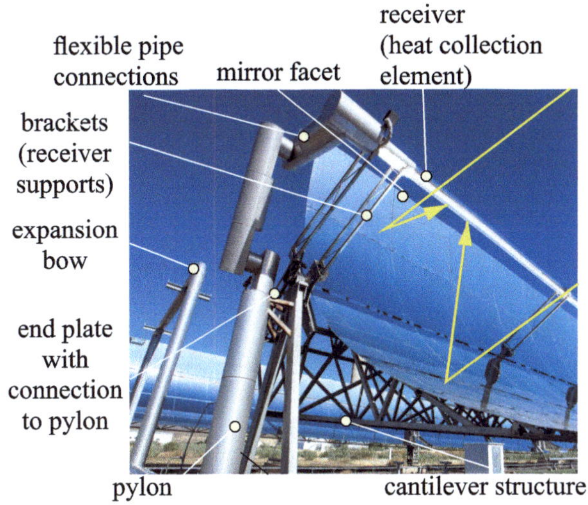

Fig. 7.17 Structural composition of a parabolic trough collector from DLR (2008)

The structural design of the connection between the frame and the receiver unit is such that there is no heat loss in the direction of the frame structure. Almost all designs use thermal insulation materials of various types.

The central task of the supporting structure of a parabolic trough is the correct mirror positioning and therefore primarily requires a high stiffness, since any deviation from the ideal contour leads to losses in the optical efficiency of the system. The stiffness refers to both the static loads, which are mainly due to the weight of the receiver elements, some of which are more than 100 m long, and the dynamic loads, which are mainly due to the aperture surfaces exposed to the wind. The entire collector structure, consisting of support structure and frame, must be designed to withstand these loads with only very slight geometric deviations. In addition, high stiffness allows for longer troughs and thus fewer poles and tracking units, and thus reduces solar field costs. High stiffness is realized with consequent use of lightweight construction while keeping material and manufacturing costs low, which not only allows for weaker foundations and fewer number of tracking mechanisms with lower power consumption but also reduces deformations due to dead weight. In particular, the power consumption of the actuators contributes not insignificantly to the parasitic energy consumption of a parabolic trough power plant, which is relatively high compared to other solar thermal power plants. It should be considered that the solar field cost contributes to more than 30–40% of the total cost of a PTC plant (IRENA—International Renewable Energy Agency, 2013). This means that reducing the cost of the solar field has a large impact on the total cost of the power plant and its economics. Usually, the supporting structure of a collector consists of a main body, which in most cases is a space frame

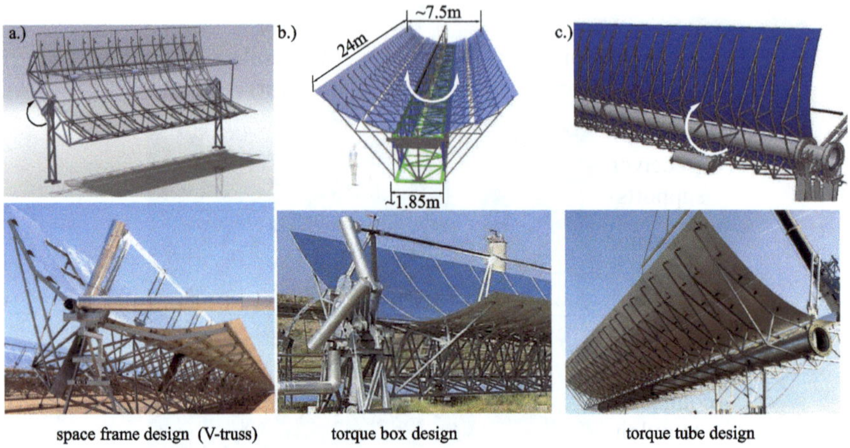

Fig. 7.18 Sketches of the support structure of PTC collectors (top) and realization in different power plants or prototypes (bottom); **a** space frame design—Solana 2 Power Station; **b** torque box design—Eurotrough—150 DLR; **c** torque tube design—Noor 2-SENER

or a tubular structure made of steel or aluminum. Other elements of the support structure are:

- mirror support points on the space frame structure or on special cantilevers,
- receiver support, also called heat collector element (*HCE*) support,
- structure for attachment to the pylons,
- pylons, including drive pylons, and foundations.

Modern structures target at a high degree of pre-fabrication rate off-site at the manufacturer and final assembling on site, where the focus is to minimize the work hour for assembling a solar collector assembly (SCA). Typical structural frame designs are frame space designs, torque box designs for large, scaled module of power stations of up to 200 m in length per single collector assembly composed of several collector modules which in length range of up to 20 m. Such designs do not only allow to use aluminum to reduce weigth but also allow for designs with a center of weight below the mirror and thereby minimizing stress for static loads. Torque tube designs as sketched in Fig. 7.18a mainly use steel as constructional material and are characterized by a simple and cost-effective design. Figure 7.18 show the different structural design types with support structure, cantilevers, and fixation of the *HCE*.

Table 7.4 shows that there is no uniform design of both the structural form as well as the structural material chosen; nevertheless, the trend in power stations is towards longer modules and thus also longer overall units to reduce the number of drives. At the same time, larger aperture widths are aimed at, with up to 7.5 m being achieved commercially. Nevertheless, the upward size growth is limited due to the strength of the available structural materials for the support structures, as the collectors must be in structural integrity even at wind speeds of up to 50 m/s. The verification for this is done by coupled fluid dynamic and structural mechanics calculations, which include the resistance and deflection as well as critical positions.

Substructure

A precise assembly of the steel structural profiles, overall structural rigidity and low cost is required to achieve a good parabolic trough concentrator. Therefore, most PTC designs for large collectors require assembly jigs, as depicted in Fig. 7.19 left, in order to meet the design tolerances, which are compared to the assembly size very small, typically of the order of ± 1 mm. The collector design shown in Fig. 7.19 right has a so-called central space frame, called torque box, which provides good rigidity and prevents torsion, which is important for a sufficient optical performance.

Another PTC design approach is the replacement of the torque box by a central steel tube (called the torque tube, see Fig. 7.20). The main advantage of PTC torque box designs is their usually better performance and rigidity under wind loads, while their main disadvantage is higher assembly cost due to the number of steel profiles that is required by high-precision specifications. Torque tube designs, on the other hand, are usually somewhat cheaper but may be more prone to deformation from gravity (bending) and wind loads (torsion) than those using a torque box. In both cases, rigidity in wind load situations of both designs is good enough to guarantee high optical performance.

Concentrating Mirror Surface

The most common parabolic mirrors of almost all commercially realized parabolic trough power plants consist of metal-coated glass mirrors, with the coating consisting almost exclusively of silver. The mirrors have proven to be durable, showing little decrease in mirror reflectance even after more than ten years of operation considering regular cleaning. Figure 7.22a shows the spectral reflectance of some potential mirror materials in comparison with the solar spectrum containing direct irradiance and circumsolar radiation at an air mass factor of $AM = 1.5$. The figure exhibits that for silver (Ag) the reflectance collapses in the UV spectral domain (≈ 320 nm) containg already a certain fraction of the spectral solar energy. This UV transparency of silver can cause a degradation of epoxy-based glues attaching the silver layer to the steel liner forming the mirror frame. This degradation is mainly expressed by blistering. To avoid blistering the silver layer backside is usually covered by copper. Except for this small spectral reflectance gap silver exhibits an excellent reflectance along the rest of the solar spectrum being close to unity, while all other metals except for aluminum such as gold, copper or chromium exhibit a smaller reflectance. Thus, integrated over the entire spectrum silver represents the prime candidate as mirror material.

The glass for the mirror facets is usually produced using the float glass process, in which molten glass is continuously applied to a liquid tin bath where it gradually solidifies. This process enables high planarity of the glass surface. The

Table 7.4 Structural design, dimensions and optical performance of selected parabolic trough concentrators (PTC) for solar thermal electric plants, data from Tagle-Salazar and Nigam (2020) complemented by manufacturer data sheets

Concentrator (developer/ name)	support structure design type/ material	Aperture width W/ module length L_{rec} (m)	Assembly length (m)	conc. ratio C^* (/) focal length f (m)	Optical efficiency $\eta_{opt,0}$ (/)	Mirror reflectance ρ_r (/)
Luz International						
LS -2	torque tube/ steel	5/7.8	47	71/1.49	0.737	0.94
LS 3	V-truss/ steel	5.76/	99	82/1.71	0.8	0.94
Eurotrough						
ET-100	Torque box/ steel	5.76/12	99.5	82/1.71	0.75	0.94
ET-150	Torque box/ steel	5.76/12	148.5	82/1.71	0.75	0.94
ENEA	Torque tube/ steel	5.76/12.5	100	77	0.78	0.94
Acciona SGX-2	Space frame/ alu	5.77/12	100–150	82/1.72	0.77	N.a
Sener Trough	Torque tube + stamped cantilever/ steel	5.76/12	150	80/1.70	0.76	0.935
Flagsol						
SKAL -ET 150	Torque box/ steel	5.77/12	148.5	82/1.71	0.8	N.a.
Helio Trough	Torque tube	6.77	191	82/n.a	0.82	0.95
Ultimate Trough	Torque box	7.51	246	76/n.a.	0.8	0.94
Skyfuel						
Sky trough	Space frame /alu	5.7/12	115	72/1.71	0.76	0.94
Sky trough DSP	Space frame /alu	7.6	150	75/n.a.	0.76	0.935
Abengoa						
Solucar PT-2	N.a	4.4/12	148.5	63/1.73	0.75	0.93
Astr0	Torque box/ steel	5.76	150	80/n.a	0.77	N.a
Phoenix	Space frame/ steel	5.76	150	81/n.a	0.79	0.95

(continued)

Table 7.4 (continued)

Concentrator (developer/name)	support structure design type/material	Aperture width W/ module length L_{rec} (m)	Assembly length (m)	conc. ratio C^* (/) focal length f (m)	Optical efficiency $\eta_{opt,0}$ (/)	Mirror reflectance ρ_r (/)
Solarlite						
SL 4600	Torque tube/ steel/alu	4.6/12	Up 120	66/1.2	0.75	0.94
SL 5770	N.a	5.77/12	Up 120	82/1.71	N.a	>0.94
Gossamer space frame						
LAT 73	V-struts/steel	7.3/12	192	103/2.56	0.77	0.95

NB: C^* is not the correct thermodynamically defined maximum concentration factor using the circular aperture $A_r = \pi.d_{rec}.L_{rec}$, but the commercially used factor based on the projection of the receiver: $A_r^* = d_{rec}.L_{rec}$.

Fig. 7.19 Assembly of Eurotrough Collector in a jig (left) and parabolic trough design using space frame (right) (Zarza Moya, 2020)

Fig. 7.20 PTC mirror support structure based on torque tube design (Zarza Moya, 2020)

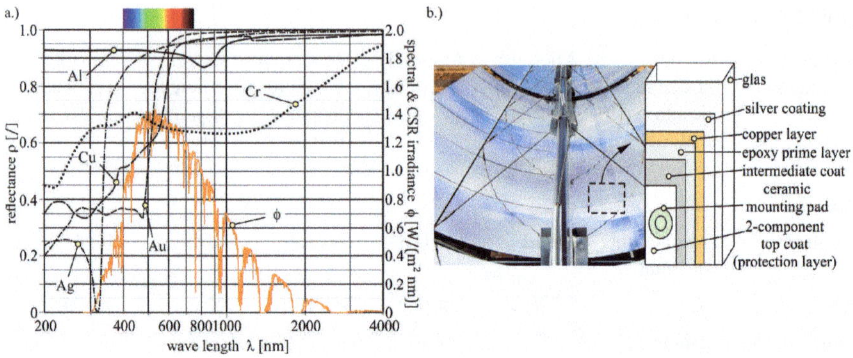

Fig. 7.21 **a** spectral reflectance of several mirror metals in comparison with the spectral irradiance of direct and circumsolar irradiation at air mass factor $AM = 1.5$ according to ISO 9845-1 (1992), ASTM G173-03 (2020); **b** Cross-section through the layered structure of a parabolic reflector mirror of a PTC

glass coating itself consists of metal oxide free glass to minimize light absorption, especially by iron oxide, which keeps transmissivity high. After the glass is cut to the right size, it is formed into the parabolic shape in a furnace. One option is to form the glass in a mold by its own weight; in other manufacturing routes, the glass is formed to the desired parabolic shape using a pressing process. Industrially, geometric mirror accuracies are achieved in which the tilt error is in the order of 0.132° from the ideal design contour shape. For Eurotrough applications, an intercept factor of 99.9% was obtained for a 70 mm diameter absorber tube (the common absorber tube diameter of most commercial receivers) and 95.5% for a 40 mm diameter absorber tube.

The mirrors have a multilayer structure as illustrated in Fig. 7.21b. The first layer below the glass is the reflective layer, is usually made of silver deposited by spray coating. Due to the transparency of silver in the UV range, there is a copper layer below the silver layer. Below this is a base layer, an intermediate layer, and a protective two component topcoat, which are epoxy coatings that encapsulate the metal–glass bond. In most solar mirrors applied to date, the base and intermediate layers contain for durability reasons some percentage of lead, but layers without copper or lead are under development. The thickness of the entire mirror assembly is about 4–5 mm. The following figure shows the multilayer structure. Below the topcoat ceramic pads are used to mount on the substrate to achieve good focusing. Typical average solar weighted direct reflectivity of the PTC mirrors is in the range of $\rho_r(SW, \theta_i, \theta_c) = 0.95$.

The solar field costs of a parabolic trough plants can reach up to 50% of the total investment cost and therein the mirror cost constitute a significant, for more details see Sect. 10.2. So, the search for alternative materials with high potential to reduce costs is a high priority. Research here is focused on mirror composites, which is directed either from silver-coated polymer films or from front-side aluminum mirrors.

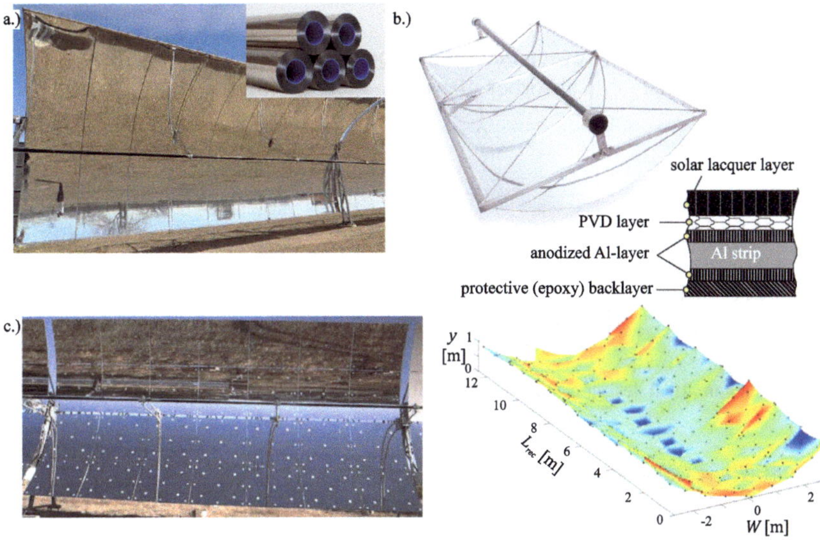

Fig. 7.22 **a** Advanced parabolic trough solar collector from SkyFuel covered with a silver coated polymer film by ReflecTech™ (Kurup & Turchi, 2015); **b** PTC collector with aluminized mirror from ALANOD GmbH & Co. KG (2021); **c** Photogrammetric 3D-analysis of Eurotrough PTC collector from measurements modified from Lüpfert et al. (2003) and reconstructed mirror shape right (red indicating 7 mrad deviation from ideal value denoted in blue)

Silver-coated polymer films as reflector materials are designed as roll-up reflective films, as shown in Fig. 7.22a (top right), which can be applied to any smooth, nonporous material and consists of multiple polymer layers with a reflective silver layer. Mirrors under the proprietary brand name ReflecTech were tested in the SEGS facilities for over seven years and showed no decrease in reflectance, although the reflectance of about 0.94 was slightly lower than that of the glass versions.

The other variant is front aluminized mirrors (Fig. 7.22b), which consist of an aluminum reflective layer, an anodized layer below and above it, and other protective layers. Due to the lower reflectance of the solar spectrum of aluminum compared to silver. The reflectance of the mirrors marketed under the brand name Alanod MIRO-SUN^R is lower than that of the silver-based mirror types. A value of $\rho \approx 0.9$ is given here. The mirror performance of a PTC is measured either during manufacturing or after installation on-site. In both cases the mirror is scanned spatially at discrete positions and by means of a photogrammetric analysis to gather a spatial picture of the area distribution in terms of surface quality at clean conditions as indicated in Fig. 7.22c, while during operation deflection measurements are executed at discrete mirror positions in the solar field.

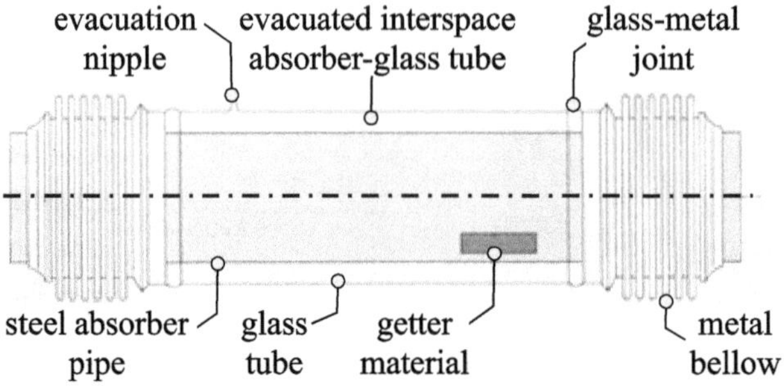

Fig. 7.23 Schematical design of an evacuated tube receiver

Receiver (Heat Collecting Element-HCE) Design

A receiver tube for PTC is composed by an inner steel pipe containing the heat transfer fluid and an outer glass tube surrounding the steel pipe. The glass tube is produced from a robust low iron borosilicate glass. The steel pipe is coated with a selective absorber surface with a high solar absorptance and low thermal emittance, for more details see in Chap. 3. The glass tube is usually coated with an antireflective layer to achieve a higher solar transmittance.

Receivers may be either evacuated or non-evacuated. The evacuated ones are still having small thermal losses even for temperatures above 300 °C because they have a high vacuum (i.e., 10^{-5} mbar) between the steel pipe and the glass cover, thus prohibiting thermal conduction and convection losses. Figure 7.23 shows the schematical design of an evacuated receiver. The glass cover of has to be connected to the steel pipe in a flexible way to avoid thermal stress, e.g. by using stainless steel expansion bellows which compensate the different thermal expansion of glass and steel and keep the annular gap hermetically sealed. The exact shape of the bellows is dependent on the manufacturer. A second means to keep the vacuum are chemical 'getters' placed in the gap. They shall absorb hydrogen molecules which might form in the HTF and diffuse through the steel tube. The diameter of the PTC receivers depends on the aperture width of the parabola. The standard outer diameter of the steel absorber tube is 70 mm for a parabola with up to 6 m. For bigger parabolas, steel tubes of 80 and 90 mm are used. For the small process heat collectors with aperture widths of 1–25 mm absorber tube of 1–2 in. are often used. Further typical technical data may be taken from Table 7.5.

The core of parabolic trough collectors are the receivers also named "heat collection elements"-HCE, which are located at the focus of the parabolic trough mirrors. They consist of a steel tube through which the heat transfer medium circulates enveloped by a glass cover tube, which is evacuated to reduce heat losses. Of crucial importance is the selective coating of the steel tube, which ensures high absorption of the incident concentrated solar radiation in the solar spectrum with

Table 7.5 Typical technical parameters of commercial vacuum tube receivers for PTC

Property	Typical value	Range of commercial products
Solar absorptance α	0.95	0.94–0.96
Thermal emittance ε (400 °C)	0.09	0.07–0.11
Steel pipe diameters (mm)	70	70–90
Pipe material	Stainless steel	–
Glass cover diameter (mm)	125	115–145
Glass cover material	Borosilicate glass + AR	
Transmittance cover glass τ	0.96	0.95–0.97
Heat transfer fluid (HTF)	Thermal oil	Thermal or silicone oil, molten salt, direct (water/steam)
Maximum temperature T(°C)	400	300–550

simultaneously high reflectance in the far IR spectrum to minimize radiation losses to the environment. The residual gas pressure in the gap is of order of 10–100 Pa.

Receiver Glass Tube

The receiver glass must exhibit a spectral selective behavior expressed by a high transmittance in the solar spectrum and a high reflectance in the heat emission spectrum, as outlined in Chap. 3 (optical conversion). The principal physics and the corresponding means to reduce either reflectance in the solar spectrum by means of anti-reflection coatings as well as absorptance by choosing low iron content glass, usually borosilicate glass, has been explained there. With the help of both measures, PTC glass tubes today achieve transmittance values of more than $\tau \approx 0.96$. Nonetheless, we focus here on specific PTC receiver collector challenges.

Since the glass envelope almost completely encloses the heat source, convective, conductive, and radiant heat losses emanating from the absorber tube must be minimized as far as possible. Convective losses can be substantially reduced by evacuating the gap between hull and receiver; typical residual gas pressures in the gap are about 1–100 Pa. This requires a minimum glass thickness, typically in the order of 3–4 mm to achieve structural hull integrity at all operations conditions. Conductive heat losses are reduced by appropriate choice of structural materials. A special challenge is the mismatch of the specific linear expansion between glass and the structural material. Therefore, materials with small specific thermal conductivities towards the glass envelope are used, for example, stainless steel. Moreover, wall thicknesses are adapted in such a manner to keep thermal stresses low and simultaneously maintain a high thermal insulation to ambient.

In PTC collectors, thermal radiation is the largest heat loss mechanism. For example, according to Wien's displacement law, the maximum of heat radiation at 600 °C is about 3.32 μm, of which already a small part is in the visible light range, irretrievably leaving the glass envelope. Fortunately, glasses exhibit a spectrally selective behavior (see Chap. 3), characterized by a low transmittance in

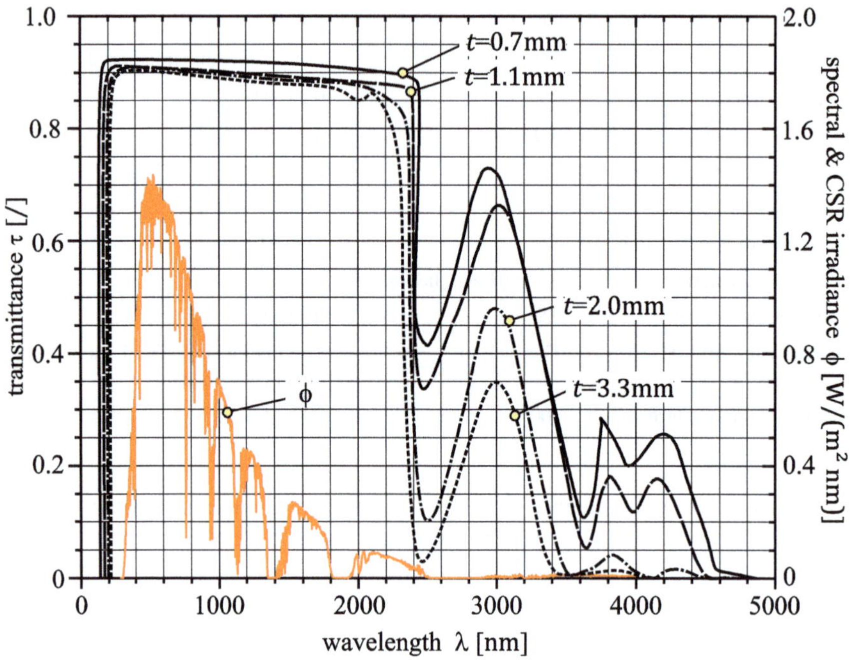

Fig. 7.24 Spectral transmittance of uncoated commercially available borosilicate glass of different thickness t (SCHOTT Technical Glass Solutions GmbH, 2018) in comparison with the spectral irradiance of direct and circumsolar irradiation at air mass factor $AM = 1.5$

the infrared region. Thus, the glass envelope acts as a thermal radiation reflector and contributes to thermal insulation, since the thermal radiation emitted by the absorber tube remains within the glass envelope. Nevertheless, the glass envelope is heated and thus generates additional thermal radiation back onto the absorber tube. Thus, glasses to be used must exhibit a high thermal resistance. To provide sufficiently high resistance to thermal shock and temperature gradients high boron content glasses are taken, compared to soda-lime glasses used in the past. These borosilicate glasses are usually coated with an anti-reflective layer enhancing transmittance to values exceeding 0.96 and allow for temperatures of about 450–500 °C.

Despite this favorable behavior the addition of boron reduces the thermal expansion coefficient of the glass considerably, posing significant design challenges at each end of the receiver assembly where a glass-metal bond is required. Care is taken to ensure that the glasses contain no iron or other mainly oxide metal impurities which cause high absorptance especially in the visible light spectrum. Even untreated glasses exhibit a high transmittance down to 300 nm UV with a value of $\tau \geq 0.92$ in the visible and near IR wavelength range, as well as low intrinsic fluorescence and low solarization, as shown in Fig. 7.24 for an uncoated Borosilicate glass in comparison with the solar spectrum.

Selective Absorber on Receiver Tube

In the selective absorber on the receiver tube the optical conversion takes places according to the principles mentioned in Chap. 3 based spectral selective behavior characterized by spectral high absorption of the incident concentrated solar radiation in the solar spectrum (250–2500 nm) over a long lifetime, typically 25 years are considered. Simultanously, the emissivity in the far infred ($\lambda > 3\ \mu m$) should be as small as possible. Since the glass envelope of the receiver is evacuated and thus losses due to thermal conduction or convection are reduced, heat losses due to radiation dominate. With increasing temperatures, the spectra of the solar radiation and that of a blackbody radiator increasingly overlap, as illustrated in Fig. 7.25. For an average collector temperature of 350 °C, high absorption is only possible at the expense of higher emission values. Silver has one of the highest reflectances of all materials in the infrared spectrum and, therefore, it is primarily used as an IR mirror. High absorber temperatures place high demands on the stability of the absorber layers due to the mobility and solubility of metals in other metals or composites. Even when used in a vacuum, in which no degradation occurs due to oxidation processes, the development of absorbers with high temperature stability is a major challenge in terms of materials technology. Thus, the details of the design of commercial absorbers as well as their fabrication route are the intellectual property of the manufacturers. Scientific developments of selective absorbers such as spectral analyses show that most of the used absorber types in PTC collectors consist of multilayer cermets ("*cer*" = ceramic and "*met*" metal) as described in Chap. 3. This cermet layer is applied on the IR-mirror layer and consists of materials such as Cr_2O_3, Al_2O_3, AlN, SiO_2 in which metal nanoparticles are integrated. The top layer is a pure ceramic anti-reflective coating mostly made of Al_2O_3 and SiO_2, more details can be found in (Cao, McEnaney, Gang, & Ren, 2014).

Depending on the material pairing, additional adhesive layers are required to ensure the adhesion of the functional layers to the structural tube material. In case of oil as heat transfer medium, additional gas diffusion barrier layers are applied to minimize gas diffusion through the absorber tube and to maintain the insulation vacuum. If molten salt is used as the heat transfer medium, coolant temperatures of about 550 °C are possible. Such temperatures require additional diffusion barriers to prevent the migration of metal from the IR mirror into the tube substrate. The principal spectral reflectance characteristics of a selective absorber is illustrated by Fig. 7.25.

Technical Design

High power plant efficiency requires high steam turbine inlet temperatures and thus high heat transfer fluid (HTF) temperatures. Above about 390 °C, most thermal oils preferably used in PTC troughs decompose and develop hydrogen in a dissociation reaction, which may be released into the vacuum gap, diffusing through the steel, and deteriorating the vacuum there. Getter material in the vacuum tube, improved diffusion barriers for the steel tube and a hydrogen removal in the circulating liquid can reduce the problem largely. In direct steam generation PTC concepts, in which the thermal oil as HTF is replaced by water, higher steam temperatures

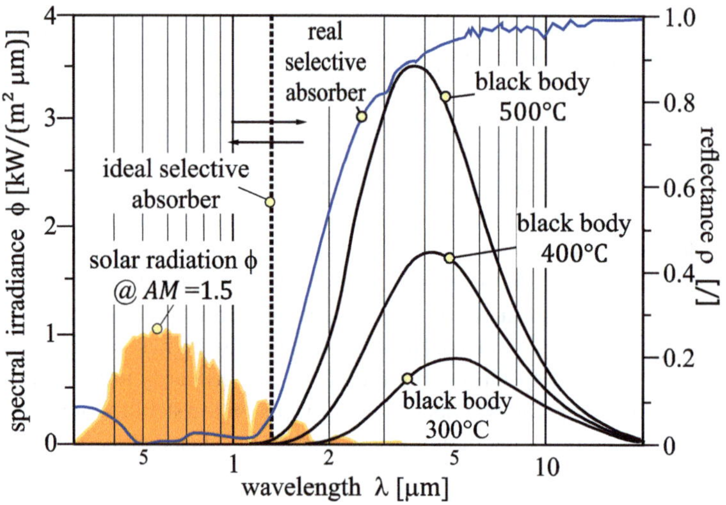

Fig. 7.25 Increasing overlap of the solar spectrum ($AM = 1.5$ direct, non-concentrated) with the spectra of a blackbody radiator for temperatures between 300 and 500 °C; a typical absorber spectrum and an optimal selective absorber are also shown for illustration modified from Hildebrandt (2009)

are achieved by superheating the steam in the collector directly. However, the use of water as HTF requires high pressures increasing the thickness of receiver tube. In alternative concepts using molten salts or liquid metals as HTF lower pressures are needed, and the degradation of the vacuum due to hydrogen diffusion through the tube can be avoided. But, due to their capability to attain substantially higher HTF temperatures, other technical measures are required to limit the enhanced heat losses through heat radiation via the glass to the ambient or heat conduction through the brackets connecting the HCE to the support structure. Additionally, using molten salts or metals as HTF demands sophisticated control and operating strategies to avoid solidification (or freezing) of the heat transfer fluid during the night.

A key technical aspect of all line-focusing systems is the piping connection of a collector module string to the manifold leading to the power plant unit. Due to the temperature differences during operation of the receivers, there is a differential elongation of the metallic absorber pipe compared to the mirror units and the support as well as the entire collector line. Since the lengths of a string can be several hundred meters, a single compensator in an HCE string would be very costly and hardly able to absorb the entire elongation. Therefore, compensation for the different thermal elongation of the collector and the structural-mechanical superstructure is achieved by means of bellows on each absorber element. From the effects described above, there are three immediate critical issues for the design of a long-life HCE:

Table 7.6 Coefficient of (linear) thermal expansion α_{th} of some metals and glasses (Schöttl et al., 2019)

Metals	$\alpha_{th}(\cdot 10^{-6}/K)$	Ceramics, glasses	$\alpha_{th}(\cdot 10^{-6}/K)$
Aluminum	25	Al_2O_3	6.5–8.8
stainless steel	17	Si	2.6
ASTM Nickel alloy	12	Soda-lime silicate glass	9.2
Titanium (Ti)	8.8	Borosilicate glass	4.6
Tungsten (W)	5	Silica (96% pure)	0.8
Kovar alloy (Fe–Ni–Co)	5	Silica (99.9% pure)	0.55

- breakage of the glass-to-metal seal,
- thermal oil degradation,
- weathering resistance of the glass tube.

Breakage of the glass-to-metal joint is the main cause of receiver module damage in power plants, mainly due to different thermal expansion coefficients of the thermal fluid tube and glass envelope. Table 7.6 shows the thermal expansion coefficients (CTE, α_{th}) of some metals and glasses. To compensate for the different thermal expansion coefficients between steel and glass, in the past the metal–glass joint was fused into the glass body by means of a continuously decreasing metal cross-section. All new designs use either tungsten or a Kovar between glass and steel, circumventing the problem of differential elongation; the details of the bond, however, are the intellectual property of the manufacturer. Figure 7.26a illustrates the principle of metal–glass bonding by the two approaches.

When thermal oils are used, but also when steam is superheated in direct evaporating collectors, hydrogen can be generated, which diffuses atomically through the steel and recombines at the interface to the vacuum space. This leads to a deterioration of the vacuum and increases heat losses by a factor of two to three. The diffusion of the hydrogen can be slowed down, but not prevented, by permeation barriers made of aluminum oxide, for example, which are arranged between the steel tube and the selective absorber layer. To maintain good vacuum over a long period of operation, a gettering or getter capsule (on the opposite side of the focal line) is preferably integrated at the coldest point of the vacuum chamber, which absorbs the hydrogen. Some products include vacuum level indicators to allow visual inspection of the receivers during operation. This is outlined in Fig. 7.26b and c.

All receivers are prefabricated at the manufacturer and only installed on site. As the size of the power plant has increased, so have the internal diameters of the tubes, and commercial systems have sizes of 70–90 mm with lengths of 2–4 m, as shown in Table 7.7 for some commercial products. To reduce the number of circuits in large solar arrays and allow for long arrays, coolant pressures of 6 MPa are not uncommon. The design of the connection between the receiver tubes aims at a compact construction to ensure a maximum active absorber length on the one

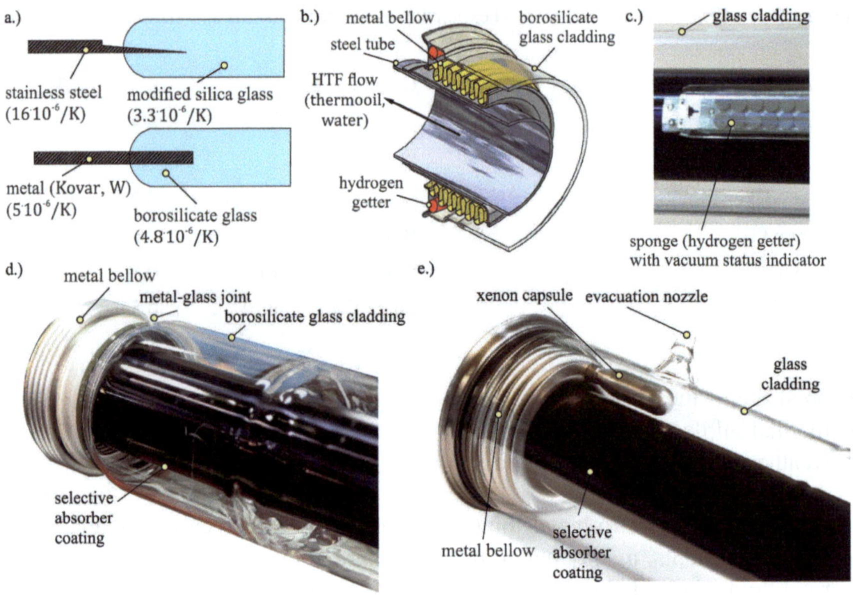

Fig. 7.26 **a** principle of metal–glass bonding to compensate different CTE by geometry and use of adapted material; **b** hydrogen getter options to maintain vacuum and allowing visual receiver integrity inspection (**c**), from Nagarajan and Barshilia (2010). Commercial receiver design external bellow (Siemens Concentrated Solar Power, Ltd., 2011 now Rioglass) and external bellow (SCHOTT Solar CSP GmbH, 2013)

hand and to keep the heat loss via the metallic tube connection elements small on the other hand, at the same time the compensation of the relative elongation of the tubes during operation must be ensured. Two development lines have become established on the market, a design variant with an external bellows and one with an internal compensation element, as shown in Fig. 7.26d and e, respectively.

Solar Tracking System

Parabolic troughs as every concentrating system have to track the sun to reach a continuous concentration of the direct solar radiation to provide maximum efficiency at the given site and environmental conditions. In contrast to linear Fresnel in a parabolic trough the whole assembly (mirror and absorber) is simultaneously single axis tracking the sun's position.

Two types of collector tracking directions are used in PTCs, namely north–south ($N - S$) and east–west ($E - W$) tracking, according to the direction of rotation of the collector aperture plane as sketched in Fig. 7.27. In north–south tracking, the receiver axis and the rotation axis are aligned in the east–west direction, which means that the collector aperture is rotated from north to south. Conversely, in east–west tracking, the aperture plane and the rotation axis are aligned in the

Table 7.7 Characteristic data of some PTC receivers from Tagle-Salazar and Nigam (2020) and manufacturer data sheets

Manufacturer/ model ID	Rioglass UVAC 90 7G	Rioglass PTR 70 4G	Archimede HCEMS11 (molten salt)	Archimede HCESHS12 (direct evap.)	Solar lite SL 4600	Sunda Seido 6-3
HTF tube						
Length (m)	4.061	4.06	4.06	4.06	4.06	4.06
Diameter d (mm)	88.9	70	70	70	70	70
Thickness t_w (mm)	2.5	2	3	N.a.	2–3	2.5
Material	Stainless steel	Stainless steel	Stainless steel	Stainless steel	Stainless steel	Stainless steel
Glass tube length (m)						
Length (m)	3.91	3.77	3.9	3.9	3.9	3.9
Diameter d (mm)	135	125	125	125	115–125	115
Thickness t_w (mm)	3	2.5	3	3	2	N.a.
Transmittance τ(/)	0.964	0.97	0.966	0.966	0.96	0.95
Selective coating						
Absorptance α (/)	0.962	0.94	0.95	0.95	0.095	0.94
Emittance ε@ 400 °C (/)	0.095	0.095	0.073	0.073	0.14	0.12
Solar field						
Max. pressure (MPa)	4	4.1	3	10.4	<4	4
Max. temp. T (°C)	450	400	580	550	393	450
Design						
Lifetime (years)	25	25	25	25	25	N.a.
Vacuum pressure (Pa)	<10	<10	<10	<10	N.a.	N.a.

north–south direction and the aperture plane rotates from east to west. The north–south tracking has the advantage of lower energy consumption during tracking, but is associated with higher end losses, while with east–west tracking systems it is the other way around: lower end effect but higher energy consumption. Without prejudice to solar field design, north–south tracking is almost exclusively used to achieve maximum thermal yield.

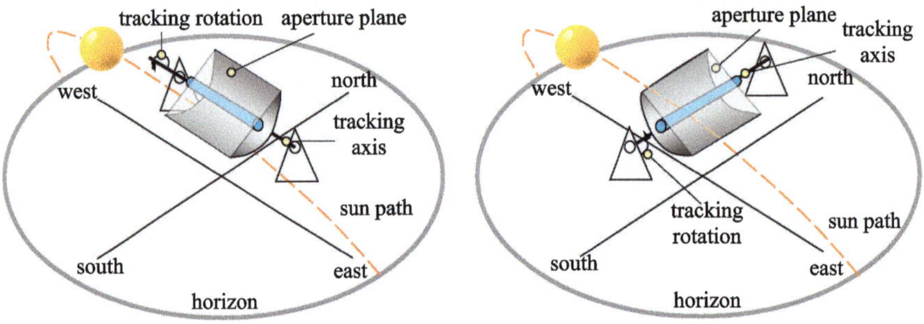

Fig. 7.27 Definition of east–west mounted collector with north–south tracking (left) and north–south mounted collector with east–west tracking (right)

The tracking angle φ_{track} as deviation from the normal of the horizontal plane is for any horizontal collector with its transversal plane azimuth angle γ_c relative to the south direction is defined by solar azimuth γ_s and altitude α_s by

$$\varphi_{track} = \sin^{-1}\left(\frac{\sin(\gamma_s - \gamma_c)}{\tan\alpha_s}\right), \tag{7.18}$$

$$\alpha_s = \sin^{-1}(\sin\lambda \cdot \sin\delta + \cos\lambda \cdot \cos\delta \cdot \cos\omega), \tag{7.19}$$

$$\gamma_s = \sin^{-1}\left(\frac{cos\delta \cdot \sin\omega}{\cos\alpha_s}\right), \tag{7.20}$$

where λ denotes the latitude, δ the declination and ω the hour angle. The incidence angles on the tracked collector aperture with tilt φ_{track} are given in the following equations—θ_{NS} for the north–south mounted collector ($\gamma_c = 0°$) and θ_{EW} for the east–west mounted collector ($\gamma_c = 90°$

$$\cos\theta_{NS} = \cos\delta\sqrt{\sin^2\omega + [\cos(\beta - \lambda)\cos\omega - \tan\delta\sin(\beta - \lambda)]^2} \tag{7.21}$$

$$\theta_{NS} = \cos^{-1}\left[cos\delta\sqrt{sin^2\omega + [cos(\beta - \lambda)cos\omega - tan\delta sin(\beta - \lambda)]^2}\right]; \tag{7.22}$$

$$\cos\theta_{EW} = \cos\delta\sqrt{\cos^2\omega + \tan^2\delta} \tag{7.23}$$

$$\theta_{EW} = \cos^{-1}\left[cos\delta\sqrt{\cos^2\omega + \tan^2\delta}\right]. \tag{7.24}$$

Here, β denotes a possible tilt of a parabolic trough along the rotation axis. This is not a realistic option for large power plant collector of several meter length, but possibly for short roof-integrated collectors or collectors mounted on a slope of a hill. When the tilt is equal to the latitude, then we speak of a polar-mounted collector, where the rotational axis is parallel to the earth's axis. In this case the incidence angle is constant over the day and equal to the declination of the sun's orbit.

Knowing exactly position and solar time of a collector, in principle the tracking angle can be calculated. However, to generally realize a tracking of the receiver, a solar field sensor, which detects the irradiation, an actuator consisting of a motor and a gearbox for the rotation of the collector, and a position sensor, which detects the angular position of the collector, are required. All these elements are part of a regulation and control chain and are themselves subject to manufacturing tolerances or inaccuracies and errors that must be considered in the regulation, regardless of whether they are sensors or drives or gears.

Active tracking systems are standard in concentrating solar power plants, with a further subdivision into closed-loop and open-loop systems, more information on details may be taken from Moga et al. (2015), Lee et al. (2009). The main difference between these types of active tracking systems is the way of signal conversion and feedback with the actuator.

Closed-loop active tracking systems consist of three components, a light sensor, a control system, and a drive system. The sensor sends a signal to the control system when the collector is not properly aligned with the sun, prompting the control system, where a comparison is made. If it is misaligned, the controller sends a signal to the actuator, which applies torque to the structure until the vector value falls below a threshold. Thus, closed loop tracking systems are pure feedback controls. Light receiving diodes or PV cells are the most commonly used sensors for closed loop tracking. The main advantage of this tracking method is the high precision tracking, but it is susceptible to the sensitivity of the light sensor and its positioning/s in the solar field or due to wanted or environmental shading of the sensor. A key disadvantage is the inability of the control system to recover or find the direction of the sun during long periods of cloud cover.

Open loop tracking systems consist of a controller and a drive. During operation, the algorithm programmed in the control system, based for example on a clock, signals the actuator to change the collector position. Open-loop tracking systems can be further subdivided into pure time-controlled tracking systems or altitude/azimuth tracking systems. Timed tracking systems use algorithms based only on time, so that the collector is controlled by incremental movements. Altitude/azimuth trackers use algorithms based on astronomical data as a function of location and time.

Hybrid tracking has been developed to overcome the disadvantages of closed-loop and open-loop tracking. The basic strategy of hybrid tracking is to use an algorithm (open-loop) to track the sun and reduce misalignment by sensors (closed-loop). A typical hybrid loop tracking consists of the algorithm calculating the position of the sun and activating the drive, and later correcting the misalignment by the light sensor if it occurs.

Increased computer capabilities such as wireless measurement signal transmission and the use of data buses have led to the widespread adoption of hybrid control systems, which not only include the position of the collector in relation

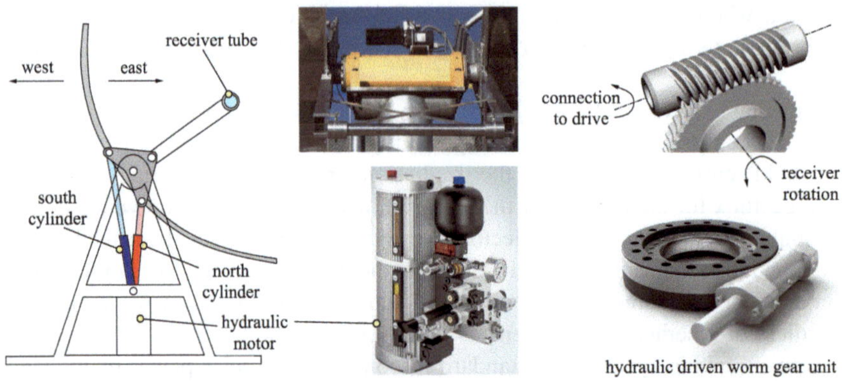

Fig. 7.28 Hydraulic drive operated receiver tracking principle (left), hydraulic motor gear box unit (top, middle); functional principle of backlash-free actuators based on a screw drive with a constructive design (right)

to the sun, but also model-based representations of the individual receiver circuit strings and the reference to the respective overall power plant operation strategy. Such systems not only take into account the plant itself and its different operating modes, such as start-up or shutdown, standby, control mode, but automatic switch to safety positions in case of extreme weather situations e.g. provided by external information provided by weather forecast systems.

Tracking, especially with parabolic receivers, places extreme demands on the drives and gearboxes, not only in terms of precision but also because of the enormous masses that must be operated without backlash under highly variable environmental conditions over long periods of time with as little maintenance as possible. Both hydraulic drives, which the figure outlines, and either hydraulic or worm gear units driven directly by electric motors have become established on the market. Figure 7.28 illustrates some options of tracking systems and drives being installed in solar fields.

7.2.2 Linear Fresnel Collectors

The focusing properties of large parabolic reflectors can be simulated by small reflector elements distributed over some suitable (ground or roof) plane surface. This allows large concentrator systems to be built up from small elements, avoiding the large structures and cleaning accessibility problems associated with very large reflectors. A single-axis linear Fresnel collector (LFC) system is composed of many long row reflectors that together focus sunlight on elevated linear tower receivers with one or more absorber tubes running parallel to the reflector rotational tracking axis. This allows a large size array to be constructed inexpensively using very similar or identical long focal length (and therefore almost flat) glass mirror elements.

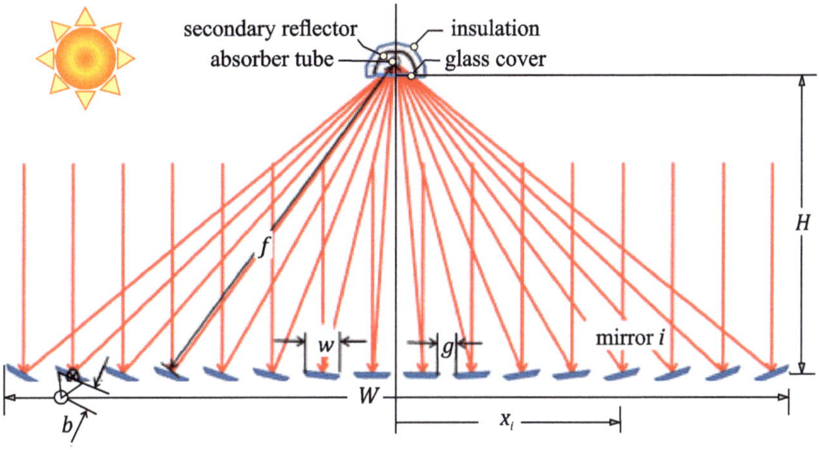

Fig. 7.29 Geometrical transversal cross-section of a primary reflector array with relevant geometric parameters (w = mirror width, g = gap between neighbouring mirrors, b = distance between rotational axis and mirror axis, f = focal length)

Today's usual single-axis tracking LFCs differ from parabolic troughs in that the reflectors are composed of many long row segments which focus collectively on an elevated long linear tower receiver running parallel to the reflector rotational axis (Fig. 7.29). Unlike for parabolic troughs, the LFC receiver is fixed in space, and the reflectors rotate to maintain focus on the fixed receiver. The reflector rows all track through the same number of degrees during the day as the sun moves but are inclined at different angles at any one time because they have different positions relative to the tower target. This allows a large size basic unit made up of identical relatively small reflectors of long focal length, which in turn leads to the ability to use almost flat, low-cost glass mirror elements.

The LFC approach to Concentrating Solar Power (CSP) today reaches a commercial maturity approaching that of trough systems. Specific commercial projects focused on different approaches have driven the development of LFC, showing the versatility of the concept. In a recent development, the first 50 MW_{el} commercial molten salt power plant using a linear concentrator is not based on the trough principle but on the LFC concept. Similarly, LFC collector designs exists for the application of solar process heat in industry. Different specifications lead to smaller collector dimensions, often lower temperature demands and thus more cost-effective material choices. The transportability of smaller collectors is key for the success in this market.

For both applications the fundamental advantages of LFC are essential:

- reflector production form low-cost flat mirrors,
- stationary long receivers,
- low wind loads due to smaller elements,

- high compactness of aperture leads to optimum area use,
- highly flexible optical design (high concentration) and easy adaptability to different heat transfer fluids.

The disadvantage of LFR compared to Parabolic Troughs is certainly the stronger sensitivity to low sun positions, especially in high latitude applications, and hence a lower optical efficiency which must be compensated by a larger solar aperture.

In contrast to a mirror of a parabolic trough, which consists of fixed segments or a single piece, the primary mirror of Fresnel collectors consists of n mostly identical primary mirror strips mounted on a steel structure fixed to the ground via a support system. In order to achieve a reduction of the solar image on the receiver, i.e. a large concentration ratio C, the primary mirrors are often marginally parabolically curved. Since the focal lengths f of the mirrors are very large in relation to their width W_{pr}, elastic curvature is possible. Depending on the focal length and mirror width, the maximum deflection at the center of the mirror is <1 mm. Several frames are connected to form a module, and the modules form a long row. Via motor drives the mirror strips (primary mirrors) are manipulated to reflect solar radiation onto the receiver depending on the position of the sun. Figure 7.30a shows the main geometrical parameters of an LFR collector primary mirror field in relation to the receiver.

The simplest example of a primary mirror strip curvature is a parabola. The parabolic contour is given by $y = ax^2$. From this approach results a focal length f of the primary mirror of $f = 1/(4a)$. A focal point (or a focal line) results with parallel incident light only for parabolic reflectors and there also only if the irradiation is parallel to the parabolic main axis, as sketched in Fig. 7.30b (top). If

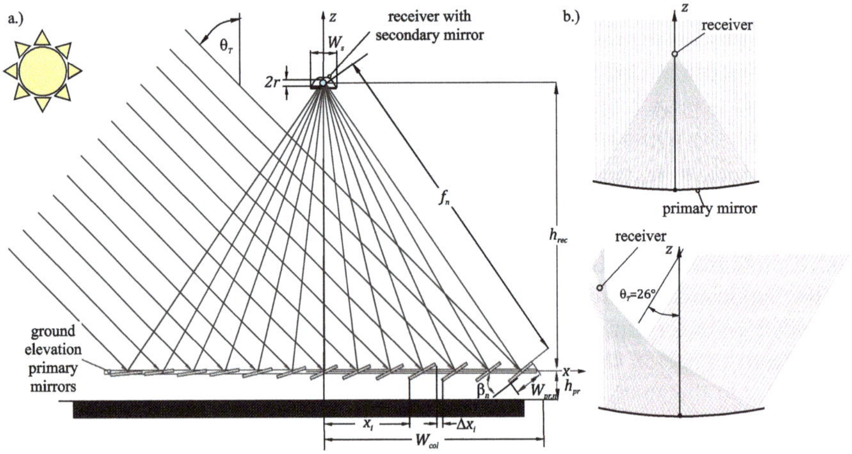

Fig. 7.30 **a** Geometry parameters and coordinate system defining the relation between primary mirrors and receiver of a LFR collector; **b** astigmatism due to deviation of incident radiation from the major axis of the parabola. Parallel irradiation $\theta_T = 0°$ (top), irradiation at $\theta_T = 26°$ (bottom)

the irradiation deviates from this, the focal point widens to a caustic. This widening leads to an aberration and is called astigmatism. Figure 7.30b (bottom) illustrates this for a transverse incidence angle of $\theta_T = 26°$. In a Fresnel collector, only the primary mirror, which is in direct alignment with the sun and the receiver, has approximately the condition for a focal line. However, exactly this mirror is shaded by the receiver. All other mirrors show a more or less strong astigmatism. The larger the ratio of primary mirror width $W_{pr,i}$ to target aperture and the smaller the ratio of primary mirror width $W_{pr,i}$ to focus length f, the stronger the spillage due to astigmatism.

The receiver unit of a linear Fresnel collector (LFR) contains a spectral solar selective absorber, often a secondary reflector, which e.g. is composed of compound parabolic concentrator (CPC), and sometimes a thermal insulation behind it. The receiver, which is permanently mounted at a height of several meters, should be protected from wind and dust by an anti-reflection coated glass pane or a cylindrical glass tube. The LFR collector design offers a vast variety design freedom. Subsequently, the model description of the integral and local LFR collector performance refers only to the standard design of the receiver composed of single tube design, which is located in the center of a housing consisting of a CPC mirror and covered at the bottom by a glass plate, which is illustrated in Fig. 7.34a.

7.2.2.1 Optics of LFC

Primary Mirror Field

The primary mirror or reflector array consists of multiple rows of slightly curved mirrors, rotating with half the angular speed of the sun in the transversal projection path over the rows. The shape ideally is a parabolic segment with a focal length determined by the distance to the receiver. The basic geometry illustrated Fig. 7.29 has variables; width of the mirror element (w_i), gap g_i between the mirror strips, the focal length (f_i) of each mirror to the receiver and the height (h_i) of a a segment to the absorber. which all may be functions of distance (x_i) of the mirror row i from the receiver tower.

There are several optical loss mechanisms related to geometry. Even for perfect reflectors, cosine losses occur due to off-normal incidence on the reflectors most of the time. Blocking considers the partial interception of reflected rays by mirror rows closer to the receiver tower. In addition, mirror rows may partially shade rows behind them as seen from the sun (shading). Additionally, any real mirror has a small but finite absorption and—if dust or a rough surface must be considered—scattering extinction both reduce the direct radiation reflected towards the receiver. Blocking and shading is dominant for low solar altitudes. Blocking and shading certainly can be minimized for a primary field if the outer mirror rows are positioned with a higher altitude above ground. The disadvantage might be that the receiver height must be increased to some extent.

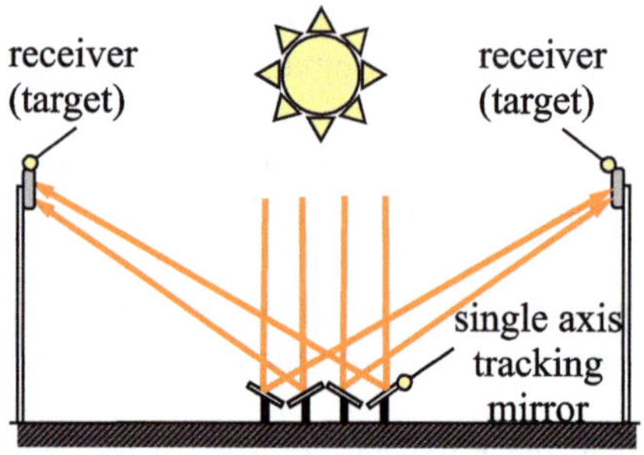

Fig. 7.31 Ray path of a compact linear Fresnel reflector (CFLR) showing interleaving of mirrors minimizing shading between mirrors based on Mills and Morrison (2000)

We may distinguish four principal classes of LFC primary mirror designs, see Abbas et al. (2018):

- Constant mirror width and constant gap (CW-CG),
- Constant mirror width and variable gap (CW-VG),
- Variable mirror width and constant gap (VW-CG),
- Variable mirror width and variable gap (VW-VG).

The advantage of constant mirror width is that the production line does not need to deal with different mirror sizes. Thus, it is a question whether variable mirror width options will be favored by a commercial development. Providing a variable gap to minimize blocking and shading losses however could be solved by mounting options at the expense of additional assembly effort and more careful design. Probably this design will result in a larger field area for the same reflector area.

A further design option is that the focal lengths of individual mirror rows can be constant or variable. The focal lengths should be adapted to the distance of the mirror rows to the receiver aperture. In practice it is too costly to produce different focal lengths for every row, so practical designs typically utilize only few (1–2) different focal lengths. This aspect is related to the question of constant width. A compromise could be to produce a small number of mirror groups, each with a different curvature and width. In addition to losses due to optical design there are other factors which degrade the quality of the solar image on the receiver. Rough, dusty, or scratched mirror surfaces as well as shape or contour inaccuracies widen the reflected beam. Also tracking inaccuracies should be avoided.

Optical design tries to optimize the size and efficiency of the reflector array (to increase concentration) with respect to economics. The further away a reflector row is from a receiver tower, the lower is its time-averaged optical efficiency. In financial terms every additional investment of adding a row has less and less addition to overall output. This is one reason for the compact linear Fresnel reflector approach, called CLFR and illustrated in Fig. 7.31, where the mirror rows can be reoriented, according to sun position, to reflect to a receiver in better position.

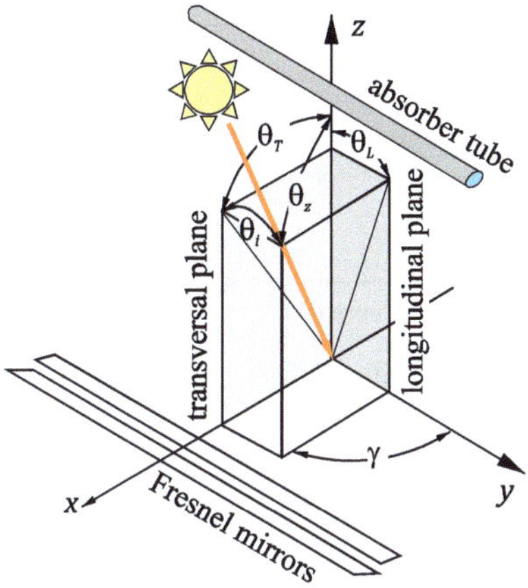

Fig. 7.32 Incidence angles for a linear Fresnel collector (LFC) according to Mertins (2006)

Secondary Optics

Larger concentration can be easily achieved for a linear Fresnel collector (LFC) by adding one more mirror row to the primary reflector field. However, to guarantee a good optical efficiency of the concentrator system a secondary concentrator often is used to collect the rays coming from the reflector field and redirect them to the absorber tube. Typically, the receiver aperture is larger than the absorbing area by a factor between 1.5 and 2. It is important that radiation impinging on the receiver aperture is directed towards the receiver tube. Non-imaging optics gives us several options to design such secondary concentrators. The principles proposed range from truncated compound parabolic concentrators (CPC) to elliptic or hyperbolic concentrators, which are discussed in the Sect. 7.2.2.2 of dealing with LFC receivers. As usually there is slightly more than one reflection on average in a secondary concentrator, the reflectivity should be chosen very high to maximize efficiency. Moreover, any absorption by the reflector heats up the secondary concentrator due to the high intensity. This must be considered carefully in the choice of the secondary reflector material. If temperatures become too high in the cavity, deformation and degradation will be the result.

Incidence Angle Modifier

A combination of primary concentrator with a receiver and possibly a secondary concentrator results in an angle-dependent optical efficiency of the collector. With two incidence angles (θ_i) the situation may be characterized as depicted in Fig. 7.32.

Let us define the angles transversal, the longitudinal and the solar longitudinal incidence angles θ_T, θ_L and θ_{LS} as

$$\theta_T = \tan^{-1}(|\sin\gamma| \cdot \tan\theta_z); \qquad (7.25)$$

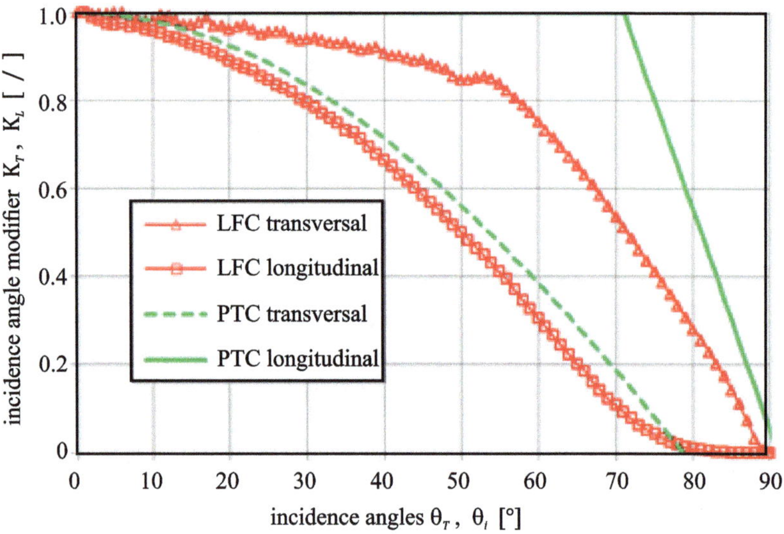

Fig. 7.33 Simulated incidence angle modifiers IAM (relative to DNI) for the Linear Fresnel Collector (LFC) for transversal (K_T) and longitudinal (K_L) directions, in comparison to the Parabolic Trough Collector (PTC) (here transversal K_T describes the row-to-row shading) from Morin et al. (2012)

$$\theta_L = \tan^{-1}(\cos\gamma \cdot \tan\theta_z); \tag{7.26}$$

$$\theta_{LS} = \sin^{-1}(\cos\gamma \cdot \sin\theta_z). \tag{7.27}$$

The incidence angle modifier function (IAM) describes the optical efficiency for off-normal incidence angles normalized by optical efficiency at normal incidence. It is usually written as the product of a transversal and a longitudinal function:

$$IAM = K(\theta_{LS}; \theta_T) = K_T(\theta_{LS} = 0; \theta_T) \cdot K_L(\theta_{LS}; \theta_T = 0). \tag{7.28}$$

The Fig. 7.33 illustrates the transversal and longitudinal IAM (K_T, K_L) of a LFC collector in comparison with a parabolic trough collector (PTC) as a function of the incidence angle ($\theta_{i,T}$).

In a comparison of design options theoretical maximization of optical efficiency should not be the target. This approach certainly leads to designs with high concentration and preserving etendue, however, it must be kept in mind that at the end manufacturing, construction and operating cost will also determine the final economic performance. Complex solutions with variable component geometries, with changing altitude of mirror rows, with complicated construction manuals will produce extra costs and sources of errors. The decision which approaches

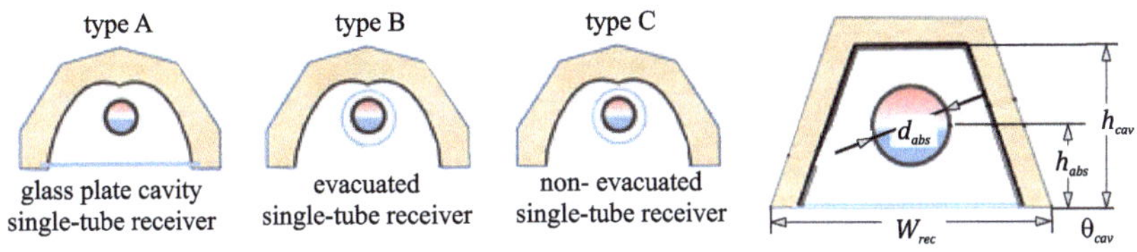

Fig. 7.34 Configurations for single-tube receivers with secondary reflector according to Montes et al. (2017) and Tsekouras et al. (2018)

to use and which to abandon due to cost certainly depends very much on experience and production technology of an individual commercial entity. Therefore, a purely academic design recommendation may fail to meet the expectations. The optimized design is always a compromise between theoretical and commercial considerations.

7.2.2.2 LFC Receivers and Thermal Performance

Thermal efficiency and heat losses of the receivers are critical to collector performance. But there is not a single optimal design, as depending on the design of the primary reflector field, its size, shape, and construction can be quite different. From a fundamental viewpoint we know two basic receiver options: Single tube receivers and multiple tube cavity receivers.

While single tube receivers usually have a secondary reflector to improve the optical concentration of the LFC and to avoid high intercept losses, multi-tube cavity receivers are without secondary optics. As already stated, the average number of reflections in a secondary is slightly above 1, and about 3–4% of the energy is lost in this process. It is a critical issue that the surface might degrade due to soiling, corrosion, and high temperatures. These risks must be carefully compared with the possible advantage of high concentration. Cavity receivers without secondary utilizing one or several tubular absorbers on the other hand have the potential disadvantages of higher thermal losses, larger optical spillage, and limited primary reflector field width.

Single Tube Receivers

Three different basic configuration for single-tube receivers with secondary reflector have been investigated by Montes et al. (2017) or Tsekouras et al. (2018) and are illustrated in Fig. 7.34. The implications are quite different for all designs. Whereas in (b) and (c) the reflectors are open to environmental impact from humidity, sand etc., the evacuated tube receiver protects the absorber effectively. In (a) both absorber and reflector may be protected against sand, but there is a risk for overheating of the reflector material. A selective absorber coating on the tube is open to oxidation in air. Traditionally secondary CPC-reflectors with acceptance angle defined by field width are considered. More sophisticated approaches have been proposed, therefore, using secondary concentrators of different shape based on non-imaging optics.

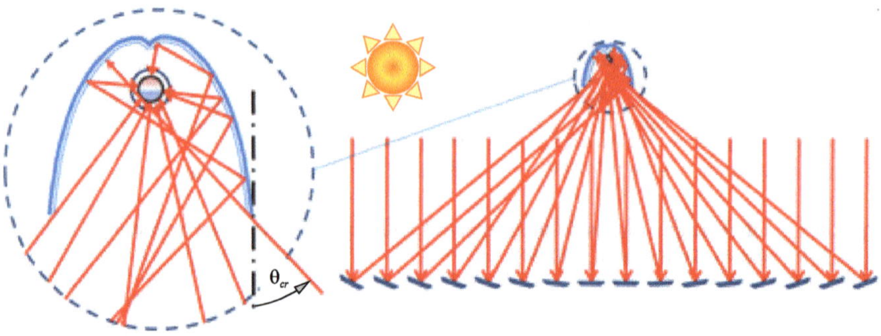

Fig. 7.35 CPC receiver with LFR

Design Options
The geometry used in most commercial approaches is a truncated CPC. A CPC reflector ideally redirects all rays from a certain field of view (within the critical acceptance angle θ_{cr}), which should cover the primary reflector field) towards the receiver surface as shown in Fig. 7.35. Truncation reduces the length of the CPC without much sacrifice in concentration (Rabl & Winston, 1976). The advantage of this design is a relatively shallow geometry using little reflector material. An ideal CPC with a cylindrical absorber would have a cusp nearly touching the absorber tube. To avoid that, the cusp often is flattened out, allowing some rays within the theoretical acceptance range to escape the concentrator. Experimental methods have been developed to assess the quality of the secondary concentrator early in development (Platzer, Heimsath, Hildebrandt, & Georg, 2008).

A similar concept from non-imaging optics using the Simultaneous Multiple Surface method yields a taller secondary concentrator but resulting in higher is called the LFR XX- SMS concept, which is depicted in Fig. 7.36a, see Canavarro et al. (2014), Guerreiro et al. (2015). This advanced optical design obviously has consequences on other issues like construction of the tower structure (due to higher wind loads) or concentrator material costs. It must be shown that the additional optical concentration leading eventually to better thermal performance at low irradiation, justifies the effort. According to (Guerreiro et al., 2015) a concentration of 73.7 can be reached using 16 mirrors 1.06 m wide within a field 20.1 m wide. The tower must be 9.8 m high. The optical efficiency is close to 70%.

A further secondary concentrator type has been designed by Chaves and Collares-Pereira also based on non-imaging optics, see Chaves and Collares-Pereira (2010). However, in this case the optics is designed for a modified primary mirror field where the rotational axes of the mirrors are not in one altitude level. The primary field mirrors of one segment between receiver towers, point to different receivers alternatingly, similarly to the CFLR concept. In this way the primary mirror field can be extremely compact minimizing the gaps. The receiver is a V-shaped absorber with two mirror wings. These concentrators are called Tailored Edge Ray Concentrators (TERC), see Fig. 7.36b. From a theoretical point this is

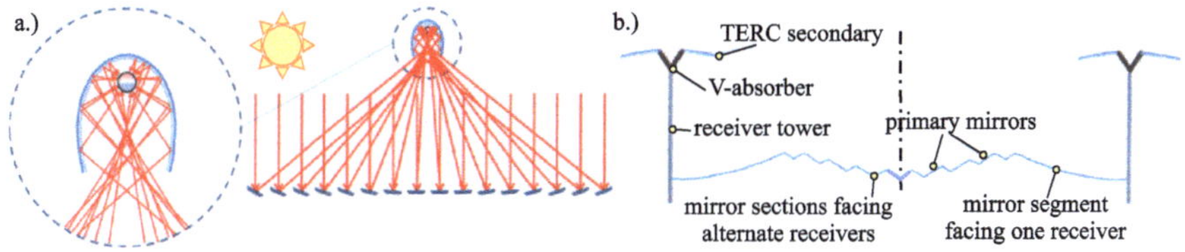

Fig. 7.36 Ray path in a LFR XX SMS collector concept (**a**) and geometrical set-up of TERC-concentrator (**b**) for several receivers according to (Chaves & Collares-Pereira, 2010)

an ideal optical concentrator with highest possible optical efficiency for infinitesimally small mirrors. Practically viewpoints lead to finite size mirrors and mirrors in one plane which brings the efficiency down.

Other secondary concentrator design principles have been put forward. The Adaptive Design Concentrator (ADC) is not defined by a set of equations but by a repetitive procedure to determine the next segment of the reflector from a previously defined one, cf. (Zhu, 2017).The resulting secondary shape looks very similar to a truncated CPC. The Segmented Parabolic Secondary Concentrator (SPSC) shape is composed of a parabolic trough broken at a given point to change its tilt. The result is a kind of reflecting cavity with 3 segments of curved walls. A recent comparative optical study concludes that the optical performance of ADC is very similar to that of a CPC, whereas the SPSC has a clearly lower performance which could be accepted if cost would be lower, see (Abbas et al., 2018).

Commercial Receivers

In commercial LFC the CPC geometry has been used similar to that shown in Fig. 7.37. It is of *Type a* described in Fig. 7.34. This receiver type is based on absorbers which should be selective and stable in air. The radiative-convective heat losses in the space between secondary walls and the aperture glazing are complex and can be modelled with CFD methods. However, for a faster assessment of heat loss dependence on temperature, incidence angles and irradiation levels a simplified method based on heat balance of different segments has been developed and used successfully (Heimsath, Cuevas, Hofer, Nitz, & Platzer, 2014a, 2014b, 2014c). The insulation level of the secondary to the ambient is certainly important as the hot air tends to stay in the top part of the cavity. Also, the air tightness level of the aperture window influences the wind dependence of heat losses. Air tightness stops not only forced convection inside the cavity but can also reduce dust deposition on the absorber and mirror surfaces. This may also be the reason that no LFC receiver seems to be of *Type c* (Fig. 7.34). That solution would be like the one of simple parabolic troughs for industrial use, where the absorber tubes are contained in a glass tube, but not evacuated. Mirror cleaning of the secondary mirror is tedious, especially for large tower heights. Therefore, the glass window for the receiver seems the better solution.

Using a vacuum tube receiver to reduce heat losses further is a viable alternative to the approach shown above. Then the geometry of *Type b* is chosen. It is a big

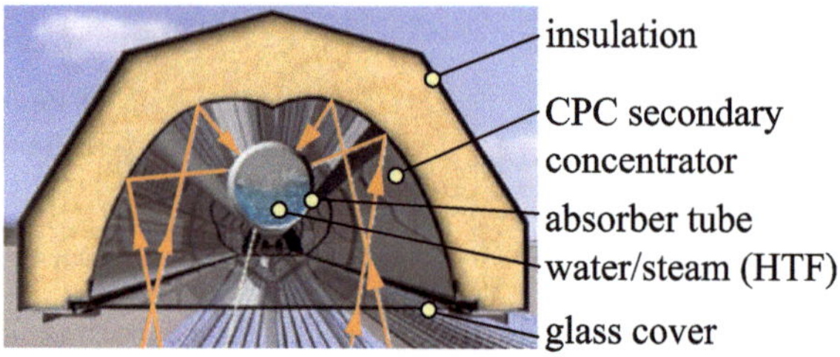

Fig. 7.37 Cross-sectional sketch of the NOVA-1 Novatec/Frenell receiver with secondary reflector (courtesy Frenell GmbH)

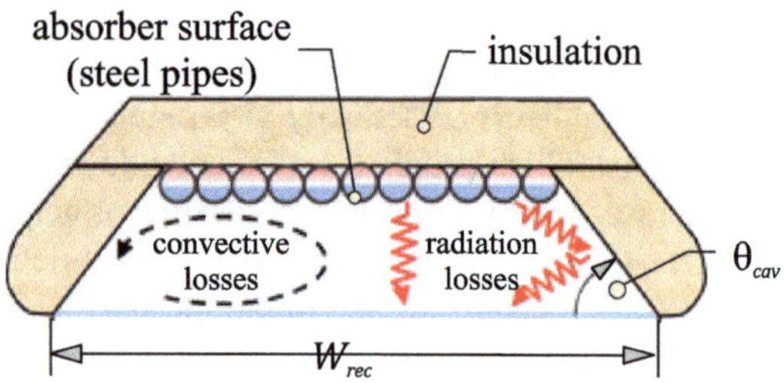

Fig. 7.38 Schematic cross-sectional sketch of a multi-tube cavity receiver

advantage that mirror surfaces are not heated anymore by convective and radiative losses from the absorber tube. Instead, they will be cooled to some effect by wind, which is important as the absorbed concentrated solar intensity still might heat up the reflector. As there is no need for thermal insulation, the receiver structure may be lighter and less voluminous, this reducing the shading of primary mirrors. Commercial companies here use standard size evacuated receiver tubes as used for Parabolic Troughs. The SuperNOVA collector of Frenell is also able to reach temperatures up to 550 °C when molten salt is used as a heat transfer fluid, see Morin, Mertins, Kirchberger, and Selig (2011).

Multi-tube Receivers

Multi-tube cavity receivers have been investigated theoretically by many authors in the past, looking into radiative and convective heat losses depending on the geometry with new studies published in recent years, see e.g. Di Canio, Tretyl, Jur, and Watson (1977–1979). A review on the newer studies is given in Bellos (2019).The designs cover usually 3–8 receiver tubes in a cavity and the parameters shown in Fig. 7.38 are varied in the different proposals. Also, here a glass or plastic cover as shown in Fig. 7.38 may protect the absorber tubes from dust and help to reduce thermal losses.

For the practical development, this receiver type has been chosen by the company Ausra which later became Areva Solar, see Conlon, Johnson, and Hanson (2011). These commercial cavity receivers in the early years used 8–16 tubes. Flow considerations, heat transfer capacity and availability of suitable tubes probably influenced this choice. It is important to separate the tubes in the receiver sufficiently to allow differential thermal expansion between the tubes which have different temperatures. This was analyzed first in a paper by Pye, Morrison, and Behnia (2003) looking into thermal convective and radiation losses (Fig. 7.38).

In later versions of the receiver the number of tubes was reduced to 8–10, with larger diameters of the tubes because larger primary fields needed sufficient flow. Areva Solar has also tested a collector technology line called SSG4 (Conlon, Johnson, & Hanson, 2011) with a superheated steam receiver pressurized to 9.2 MPa, unlike the previous saturated steam units.

Comparison of Heat Loss and Collector Efficiency

Only a few heat loss models of commercial collectors have been published and data are available. Although heat loss is not desirable, one must be cautious—higher heat losses do not necessarily imply lower efficiency, because the concentration also is very important. The main part of the collector energy balance is the solar gain where geometrical concentration and optical efficiency are multiplied by the direct solar irradiance DNI. The thermal losses are proportional to absorber area and influenced by receiver characteristics such as emissivity, evacuation, or wind protection by cover glass. For high solar fluxes the heat losses are often nearly negligible—for low solar fluxes they may play a greater role, i.e. when DNI is small or when incidence angle modifier or shading from other objects reduce the solar flux on the receiver.

Figure 7.39 presents heat loss data from company data as a function of the difference between mean absorber temperature and ambient temperature; data taken from Morin (2005), Bernhard et al. (2009), Industrial Solar GmbH (2022), Novatec Solar (2012). Depending on the solar irradiation and resulting temperature gradients in the absorber tubes, the fluid temperature during solar operation is lower than the absorber surface temperature. The heat resistance of the absorber tube and the convective heat transfer into the fluid cause a temperature drop. During solar gains the difference is typically around 10–30 °K. The F'-factor of the receiver is smaller than unity. For one-phase heat transfer fluid and high mass flow the mean fluid temperature may be estimated as the arithmetic mean between inlet and outlet temperature. For two-phase steam generation during the passage through the tubes the main part of the steam is wet, i.e. evaporation occurs at the saturation temperature. Therefore, apart from the small section for preheating and superheating before and after the evaporation section the fluid temperature is constant at the saturated steam temperature.

For the company Frenell published data exist for two collectors: The NOVA-1 collector is intended to produce saturated steam, while the SuperNOVA collector using the same concentration but having a special vacuum tube receiver for direct steam generation similar to the standard receivers used for parabolic troughs (see test data e.g. Burkholder & Kutscher, 2009). These receivers have thicker metal walls to contain the high pressure superheated steam. Similarly, the Industrial Solar LF-11 has an evacuated tube receiver but uses lower concentration. The SuperNOVA collector clearly has the lowest loss per aperture area (Fig. 7.39). This is not only due to the evacuated absorber (as the example of Industrial Solar LF-11

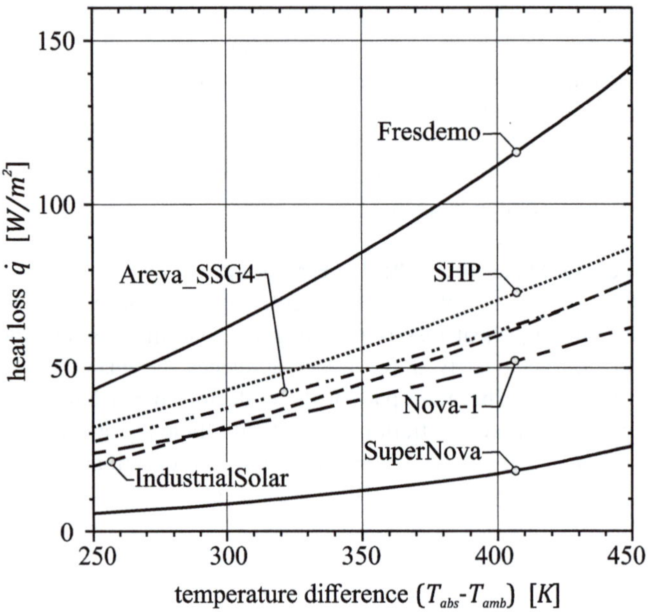

Fig. 7.39 Heat loss per aperture/reflector area for different commercial LFC developments

Table 7.8 Thermodynamical geometric (C_g) and intensity based (C_g^*) concentration factors of some commercial collectors

Collector type	Concentration C_g (thermodynamic)	Concentration C_g^* (intensity based)
SPG Fresdemo	34.1	107.1
SHP V1.0	17.6	55.2
Areva SSG4	20.2	63.6
ISG LF-11	25.0	78.6
Nova-1	54.6	171.4
SuperNova	54.6	171.4

shows), but also caused by the high geometrical concentration C_g. Thermodynamically the absorber tube circumference is relevant for the concentration, however typically in CSP industry the mirror aperture area is related to tube diameter which gives a factor π higher factor C_g^*. The reason given is that mainly the projected area (diameter $\times$ length of the tube) is illuminated by the focused sun. It may be seen as a kind of intensity-based concentration. In Table 7.8, therefore, both concentration factors are compared and stated for commercial collectors with heat losses shown in Fig. 7.39. The value of C_g^* may be compared to the concentration 78 of the Eurotrough parabolic trough collector.

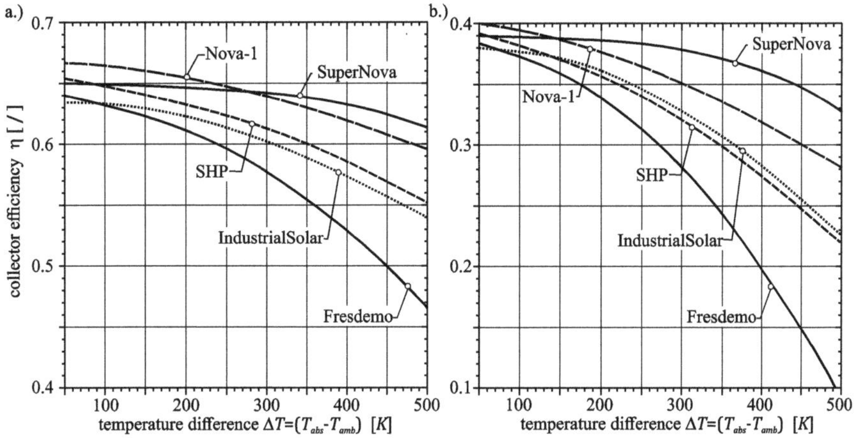

Fig. 7.40 Collector efficiency for commercial LFC as a function of the temperature difference between mean absorber and ambient temperature ΔT. **a** $DNI = 950$ W/m^2 and sun in the zenith ($IAM = 1$); **b** $DNI = 600W/m^2$ and low sun ($IAM = 0.6$)

When we consider and compare overall collector efficiency curves or collector output power, both the concentration and the optical efficiency is of utmost importance. The heat losses are less important especially for high DNI-values, as illustrated in Fig. 7.40a. The comparison with a case of lower DNI and less favourable incidence angle is shown in Fig. 7.40b. Due to off-angle incidence (sun not in the zenith resulting in an IAM of 60%) and lower DNI the solar gain is reduced, and all curves show lower efficiency. However, the curves of the different collectors show a wider spread compared to Fig. 7.40a. For the SSG4 collector no published data exist to estimate the optical efficiency, therefore the collector efficiency curve has been left out here.

Optical errors have a considerable effect on the thermal efficiency of a LFC. This means there is a large potential to improve first designs which only considered a part of the factors. In Fig. 7.41 the different optical loss effects for normal incidence are quantified. The case study shows optical loss effects for normal incidence ($\theta_T = 0°$; $\theta_i = 0°$). The first part shows losses due to the primary mirror field geometry (cosine, shading, blocking, η_{opt}) which are in more detailed outlined in Chap. 10, then receiver losses (acceptance of secondary geometry and receiver shading), and losses due to material properties (reflection, transmission, absorption). The light blue columns show a plausible range for the optical efficiency of a collector for lower quality components and high-quality components.

The geometrical losses of a standard LFC can be influenced by simple design optimizations using the parameters mirror width, gaps, receiver height and number of mirror lines. They are already very small. The receiver design still has some optimization potential, and better materials can contribute to a higher optical efficiency. For off-normal incidence the geometrical design is much more important,

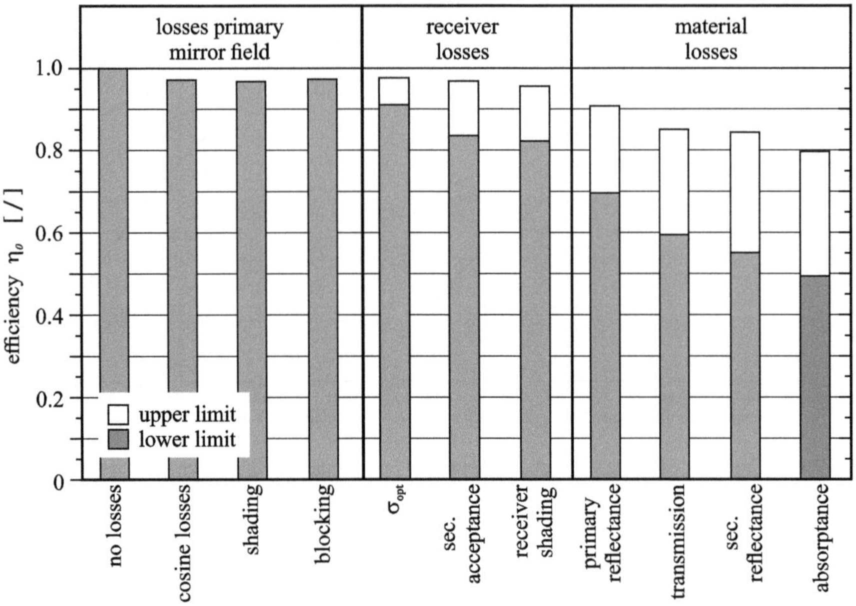

Fig. 7.41 Optical loss effects—overview on geometrical and material al losses from Heimsath et al. (2014a, 2014b, 2014c)

and the avoidance of mirrors blocking or shading each other is a key element of design.

The different optical losses of a linear Fresnel collector can be quantified for typical cases. There is a large range of the achieved optical efficiency depending on the design effort and material selection. Figure 7.41 shows the different effects indicating an upper and lower limit of the resulting optical efficiency η_0 by cumulating the effects of basic geometry, mirror shape and optical properties of the materials selected.

7.2.2.3 Evaluation Local LFR Collector Performance—Convolution Methods

Due to the complex geometrical Fresnel collector structure, the local optical modelling of an LFR collector is mostly executed numerically. In principle two methods can be applied:

- Monte Carlo based ray tracing methods (MCRT), similar as for PTC collectors.
- Convolution methods.

Both methods MCRT and convolution methods require a source description, i.e. a modelling of the sun disk accounting for the non-uniformity of the flux distribution on the solar disk and the widespread solar half-angle of 4.66 mrad (0.267°).

Almost all simulation assume that the rays pass through air with 100% transmittance, and the reflectance of the reflectors (primary and secondary mirrors is constant; the same is true for transmittance and refractive index of the receiver glass cover. The difference between both modelling approaches is that in contrast to the convolution method the MCRT solutions still contains the direction of the integral solar flux and by implementing an angle-dependent absorptance the product can be evaluated exactly and thus the solar conversion in the receiver tube is more accurately resolved. However, on the other hand, the MCRT methods need to simulate the ray paths of an enormously large number of photons to achieve a reasonable convergence of the solution. A convolution method provides a unique function as solution, while the MCRT provides rather results of a numerical experiment including uncertainties in each data point, which can be reduced by making many experiments. This is paid by higher computational effort especially when using a fine spatial discretization of all mirror elements to attain a high-quality solution. It can be shown that both approaches agree well and provide a code-to-code validation. The absolute typical computation time for MCRT exceeds that of the convolution more than an order of magnitude (Mertins, 2006), however, it can be faster, since it allows for parallelization using multi-processor computing.

A simpler modelling of just shading and blocking modelling can be provided much faster and are similar as for a PTC, however, requiring in contrast to PTC also consideration of transversal angle dependence (θ_T) in LFR collector. Details may be taken from Sharma et al. (2015), Canavarro et al. (2014) or Qiu et al. (2015).

For their simplified method (Sharma et al., 2015) have derived numerous analytical expressions to illustrate the correlations between the various geometric parameters and annual shading losses, annual blocking losses, and annual energy losses. These correlations are significant because they can be used to quickly and accurately evaluate the effects of a variety of factors on optical performance. Analyses of LFR collectors with different mirror shifts and widths show only marginal changes in annual efficiency. For example, the numerical calculations show that increasing the receiver height by 30% may increase the amount of energy by 8% and increasing the mirror length by 30% can increase the amount of energy by 24% for short collectors. Conversely, increasing the mirror width and decreasing the receiver height leads to an increase in shading.

Simulated Optical Efficiency of LFR Collectors

LFR concentrator, generally have inferior performance due to optical losses in the mirror field compared to a PTC. These include blocking of reflected radiation and shading of adjacent mirrors from incoming radiation, similar as in parabolic troughs, within the module itself, which are not present in a PTC. The reason for the reduced optical efficiency of an LFR collector arises from its geometrical set-up consisting of parallel arranged primary mirror strips with a co-planar tracking axis. By such a configuration the collector cannot conserve the solar beam etendue even if the Fresnel reflector (primary mirror) consists of an infinite number of infinitesimally small mirrors forming a quasi-straight line, as schematically shown

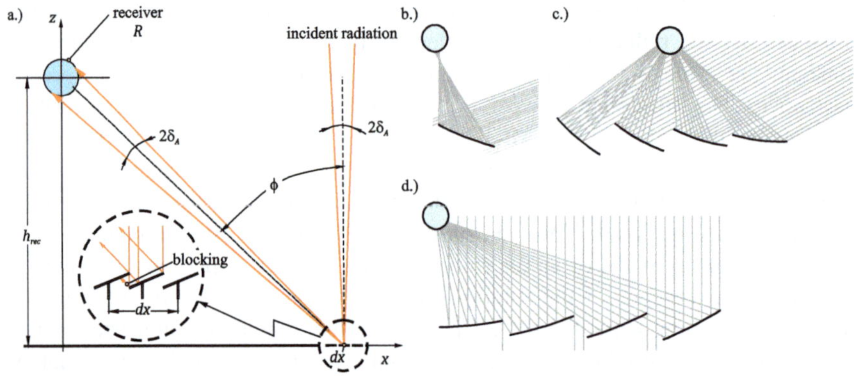

Fig. 7.42 **a** Etendue mismatch of incoming and emerging beam in a linear Fresnel collector array; **b** visualization of ray paths of cosine loss effect; **c** shadowing and **d** blocking/through losses

in Fig. 7.42a. In order to explain this essential feature, let us consider a small segment of the mirror on the x-axis deflecting a beam onto a circular receiver R. Therein, the infinitesimal incident etendue dU_i illuminating dx is given by:

$$dU_i = 2dx \sin \delta_A. \tag{7.29}$$

But the infinitesimal etendue dU_o emitted from dx within a cone $2\delta_A$ as well as inclined by an angle ϕ with respect to normal incidence to dx and pointing to the center of the receiver is given by:

$$dU_o = 2dx \sin \delta_A \cos \phi, \tag{7.30}$$

where $\cos \phi$ is the projection factor of dx in direction of the receiver. β itself depends on the position of the mirror in $x-$ direction and is given by

$$\beta(x) = \tan^{-1}(x/h_{rec}), \tag{7.31}$$

in which h_{rec} denotes the receiver height above the mirror plane. Thus, the beam width (etendue) of the incident light dU_i is greater than the etendue of the light emitted by dx dU_o, so the etendue is not preserved. The incident light that does not reach the receiver is lost due to shadowing or blocking between the heliostats or passes through the heliostats without being captured (through losses). To visualize shading, blocking or through losses they a shown in Fig. 7.42c–d, compared to cosine losses (see Fig. 7.42a) also present in PTC collectors.

In the case of cosine losses, which are shown schematically in Fig. 7.42b, only the primary mirror surface projected perpendicular to the incident solar radiation reflects. Therefore, the proportion of solar radiation related to the mirror surface corresponds to the cosine of the mirror normal and the direction of incidence. Depending on the position of the sun, the cosine losses are lowest for the primary mirror closest to the direct connection from the position of the sun and the receiver. This mirror never experiences a perpendicular incidence because it is shaded by the receiver itself. Particularly with larger transversal angles θ_T, shadowing of neighbouring primary mirrors occurs, as can be seen in Fig. 7.42c, whereby the shadowed part remains ineffective because it does not reach the receiver. It can be seen from Fig. 7.42c that the transverse angle at which shadowing begins depends mainly on the geometric arrangement of the primary mirror field and the receiver. A larger distance between the primary mirrors (called gap or shift) shifts the onset of shading to lower sun angles.

For small transversal angles especially for primary mirrors far out in x-direction, a part of their mirror surface may reflect radiation against the back of the neighbouring mirror (compare Fig. 7.42d). Increasing the receiver height h_{rec} reduces the tilt angle β_n, which means that mutual occlusion can be almost completely avoided. If the primary mirrors are pulled further apart, the masking losses also decrease and then have hardly any influence on the optical efficiency. However, this measure leads to a lower area utilization of the solar field, which is described by the fill factor FF of the solar field. The fill factor is defined as the quotient of the effective total primary mirror area A_{pr} and the solar field area A_{SF} occupation of the collector.

$$FF = A_{pr}/A_{SF}. \tag{7.32}$$

Primary Mirrors

If we assume primary reflectors of equal width, the curvature and inclination of each primary reflector must be adjusted so that the rays intersect the absorber at the focal point after reflection from all primary reflectors. The primary reflectors can have linear, parabolic, or circular profile shapes with a large radius of curvature. Furthermore, the primary mirrors can have the same focal length f or a variable radius of curvature among themselves. For a simplified description of the calculation process, we assume,

- that the primary mirrors perfectly follow the apparent motion of the sun,
- the solar radiation is vertically incident on the primary mirror, and
- the solar flux is constant within a cone with a half vertex angle $\delta_S \approx 4.65$ mrad.

For the sake of simplicity, we will consider reflectors with the same width. The angle of inclination β_n of each reflector is set so that the incident rays are reflected onto the absorber at the focal point. To prevent shadowing of a reflector by an adjacent reflector, a constant distance (called shift or gap) between two adjacent mirrors distance is provided. The $n-th$ reflector in the LFR array is thus characterised by three parameters, namely the lateral position (x_n), inclination (β_n) and shift (Δx_n). The angle that the $n-th$ primary mirror makes with the horizontal plane is the tilt angle (β_n) and is calculated according to:

$$\beta_n = \frac{1}{2}\tan^{-1}\left(\frac{x_n + \left(W_{pr}/2\right)\cos\beta_{n-1}}{f_n - \left(W_{pr}/2\right)\sin\beta_{n-1}}\right). \tag{7.33}$$

The gap (shift) between the adjacent mirrors shifts (Δx_n) is provided to enhance the performance of the LFR system by reducing the shade created by adjacent mirrors, particularly at large incidence angles and can is given by

$$\Delta x_n = W_{pr}\sin\beta_{n-1}\tan(2\beta_n + \varepsilon_0), \tag{7.34}$$

in which ε_0 solar subtense angle in mrad The lateral position of the $n-th$ mirror assuming collector symmetry can be acquired from the center of the mirror field through the relation:

$$x_n = x_{n-1} + W_{pr}\cos\beta_{n-1} + \Delta x_n \tag{7.35}$$

for $n > 1$. Equations (7.33)-(7.35) are solved iteratively to obtain the values of x_n, β_n and Δx_n, starting from the initial guess of $\beta_n = 0$, Δx_n, $x_0 = -W_{pr}/2$, $x_1 = W_{pr}/2$ and $n = 1, 2, ..., m$, with 'm' the total number of primary mirrors. The curvature radius of the primary reflectors depends on the focal length (f_n) and tilt angle (β_n) of the individual reflectors. The following Eqs. (7.36)-(7.38) are then used to determine the radius of curvature of circular/parabolic primary reflectors. Optical angle (ϕ_n) is defined as the angle between the optical axis and the line from the focus to the primary reflector being

$$\phi_n = \tan^{-1}\left(x_n/h_{rec}\right) \tag{7.36}$$

from which the focal length of the $n-th$ primary mirror is obtained by

$$f_n = \sqrt{x_n^2 + h_{rec}^2} \tag{7.37}$$

Thereby, the radius of the $n-th$ parabolic shaped ($r_{pr,n,p}$) or circular shaped (($r_{pr,n,c}$)) primary mirror is pre-scribed by

$$r_{pr,n,p} = f_n/2 \text{ or } r_{pr,n,c} = 2f_n/\cos\beta_n \tag{7.38}$$

Since the radius of curvature of the primary mirror is much larger than the receiver aperture, the profile of the primary reflector has little effect on the flux distribution arriving at the receiver, regardless of whether the primary mirror shape is circular or parabolic. The aim of the primary reflector optimization is to adjust the focal length and the radius of primary mirror curvature so that the energy absorption within the receiver cavity is maximized. Simulations show that the flux distributions to be achieved in the receiver cavity for different primary reflector profiles differ only marginally as long as there is a focus (Balaji et al., 2016).

Secondary Reflector and Receiver

With the aid of secondary reflectors, for a given combination of primary reflectors and absorbers, the target aperture can be enlarged, which allows the total heat flow on the absorber tube to be increased considerably. The receiver receives radiation from the primary mirror field from a large angular range, often more than ±50°. Due to this relatively large acceptance angle, the additional concentration achieved by the secondary reflector is rather small, so that the actual task of the secondary reflector is not to concentrate the radiation, but to redirect it in order to direct rays reaching the entrance aperture of the secondary receiver onto the absorber. Figure 7.43a shows the calculated beam path when using a planar primary mirror on a secondary receiver with a CPC profile, and Fig. 7.43b a primary mirror with a long focal length. If the divergence of the solar radiation (sunshape) is superimposed on the primary mirror in the form of an optical error (e.g. $\sigma_{opt} = 3.5$ mrad), it is clear from Fig. 7.43c that with the use of the CPC reflector all incident radiation continues to reach the absorber. The increased target aperture allows more primary mirrors to be installed relative to a target aperture, further reducing not only cost but also specific heat losses through convection. The latter is based on the fact that the secondary reflector system, which contains the absorber, is integrated into a thermally insulated structure that also serves as protection against wind. The use of a secondary reflector also facilitates the sun tracking process, the line focus of the primary reflector is not a point but a target area. One effect of using a secondary reflector is a more uniform circumferential distribution of solar irradiance on the absorber tube and thus a reduction in thermal loads and especially gradients, which is less than with parabolic troughs, which is illustrated in Fig. 7.43d for a computed solar flux distribution of on the perimeter of an LFR absorber tube. In contrast to the PTC (compare Fig. 7.13a), the absorber tube of the LFR experiences significant solar irradiation over almost the entire circumferential length. The secondary reflector geometry depends on the acceptance angle ($2\theta_{a,s}$) of the mirror shape, whereby the acceptance angle is defined as the angle over which the flux may deviate from the normal to the aperture, but still just reaches the absorber surface after the back reflection.

Figure 7.44 shows the angular dependence (θ_T) of the different loss mechanisms. The cosine losses here are defined differently from the PTC case: for the PTC the incidence cosine on the aperture plane is considered, whereas for the LFC often the incidence cosines on the individual mirror elements are taken into account. Therefore, an integrated cosine loss exists already for normal incidence $\theta_T = 0°$. The solid curve shows the cumulated cosine loss without considering the shading and masking losses on the geometric optical efficiency η_{opt} in the transverse plane. Since the mirror normal is the angle bisector of the irradiation direction and the mirror-receiver direction, the cosine losses do not approach zero even at flat sun angles.

If the sun's position deviates from the transverse plane by angle θ_i, part of the solar radiation reflected from the primary mirror field to the receiver is reflected beyond the ends of the receiver and can no longer be utilized. The usable portion can be estimated with the ratio of receiver height h_{rec} and collector length L_{col},

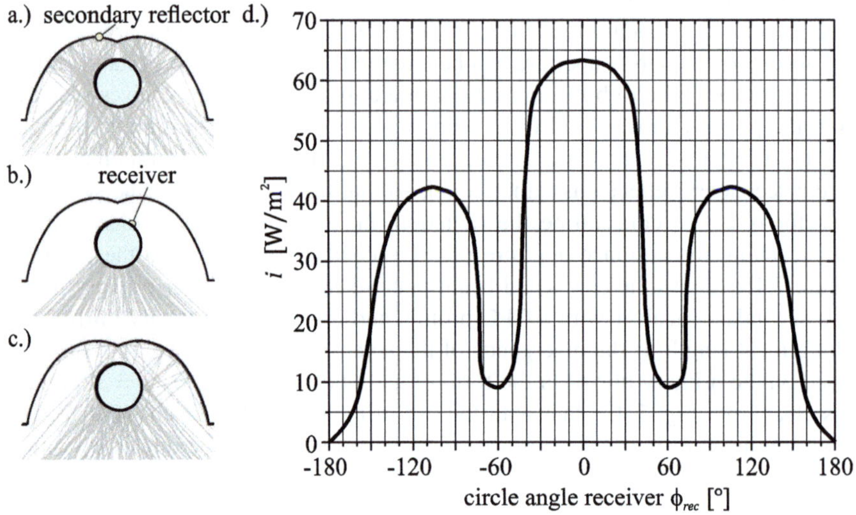

Fig. 7.43 MCRT calculated beam path in a secondary reflector (Mertins, 2006). **a** flat primary mirror ($f_{pr} \rightarrow \infty$), **b** primary mirror with large focal length f_{pr} ($f_{pr} = 17$ m); **c** primary mirror of large focal length ($f_{pr} = 17$ m) and optical error ($\sigma_{opt} = 3.5$ mrad) with additional divergence of solar radiation (sunshape); **d** computed solar flux distribution as a function of the circumferential angle ϕ_{rec} ($W_{pr} = 1.07$ m, $L_{rec} = 12$ m, $2r = 70$ mm, $h_{rec} = 8$ m, $\sigma_{opt} = 4.65$ mrad) modified from Balaji et al. (2016)

Fig. 7.44 Influence of cosine, shading and masking losses on the geometric efficiency η_{opt} as a function of the transverse angle θ_T for different collector geometries ($h_{rec} = 9$ m, $\Delta x = 0.1$ m)

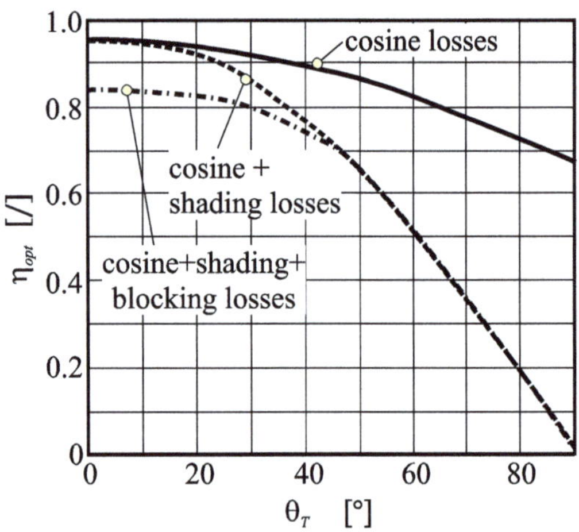

collector width W_{col} and the angle of incidence θ_{LS} (IEC6 2862-5-2, 2022):

$$\eta_{end} = 1 - \tan(\theta_{LS}) \cdot \frac{h_{eff}}{L_{col}};$$

$$h_{eff} = \sqrt{\left(\frac{W_{col}}{4}\right)^2 + h_{rec}^2}. \tag{7.39}$$

This effect is called end loss, and it is calculated for mirror rows and receiver of equal length and same position. Consequently, the shorter the absorber is in relation to the receiver height, the higher the end-of-row losses are. For a north–south orientation, row-end losses are greater at locations of higher northern latitude Φ. To avoid this, the receiver could be extended beyond the primary mirror field. Effort and benefit must be compared here. The receiver itself casts a shadow on the primary mirror field that migrates with time, so these shadowed areas do not contribute to the receiver gain. The shadow of the receiver runs moves across the primary mirror field as a function of the transverse angles (θ_T). This effect is called receiver shadowing. The portion obscured by the structure in this plane is of the order of about 2%, which is the quotient of the support span and the width of the supports. For each pass, the portion, called the reduction factor, is reduced by a flat 2%.

Geometrically, a linear Fresnel collector is defined by a multitude of geometric parameters, including:

- number of mirror rows N,
- distance between the mirrors Δx_n,
- mirror shape primary and secondary reflector (flat, parabolic, circular)
- mirror height,
- mirror width $W_{pr,n}$,
- height of the receiver assembly h_{rec},
- size of the receiver assembly,
- geometry of the receiver assembly,
- length of the collector assembly/line length (L_{rec}, L).

All the above geometric parameters can vary from one collector to another and it extremely difficult to optimize a linear Fresnel collector with respect to most or all the relevant parameters, especially in conjunction with the optical properties of the collector. In most cases, one starts from a basic empirical pattern of a linear Fresnel collector and optimizes the few selected parameters that are most important for overall optical and/or economic performance.

In this context, the question often arises as to which parameter is best suited as a comparative variable for evaluating the different collector configurations. At the collector level, the evaluation of the individual configurations is based on annual average values for a defined location. In the literature, different parameters can be found, such as the $\eta_{\text{opt, a}}$ annual optical efficiency ($\eta_{opt,a}$) and the$\eta_{\text{en,ann}}$ annual energetic efficiency ($\eta_{en,ann}$), but also $\eta_{\text{ex,ann}}$ annual exergetic efficiency ($\eta_{ex,ann}$), which are defined as follows.

$$\eta_{opt,ann} = \frac{\int_0^{365}\left[\eta_0 \cdot K(\theta_T, \theta_L) \cdot G_{b,n}\right]dt}{\int_0^{365} G_{b,n} \cdot \cos\theta dt}; \tag{7.40}$$

$$\eta_{en,ann} = \frac{\int_0^{365}\left[\eta_0 \cdot K(\theta_T, \theta_L) \cdot G_{b,n} - \frac{U_L}{C_g} \cdot \left(T_{avg} - T_{amb}\right)\right]^+ dt}{\int_0^{365} G_{b,n} \cdot \cos\theta dt}; \tag{7.41}$$

$$\eta_{ex,ann} = \frac{\int_0^{365}\left[\eta_0 \cdot K(\theta_T, \theta_L) \cdot G_{b,n} - \frac{U_L}{C_g} \cdot \left(T_{avg} - T_{amb}\right)\right]^+ \cdot \left(1 - \frac{T_{amb}}{T_{out}}\right) dt}{\int_0^{365} G_{b,n} \cdot \cos\theta dt}; \tag{7.42}$$

in which $G_{b,n}$ is the direct normal solar irradiance incident on the collector aperture with an incidence angle θ, $K(\theta_T, \theta_L)$ is the incidence angle modifier function for direct normal irradiance (including the incidence cosine), A_a the collector aperture area,T_{avg} is the average fluid temperature in the collector, and T_{out} the outlet fluid temperature of the collector loop. The annual energy efficiency considers only the positive values of the balance between collector gains and collector losses over the year, which implies that for lower irradiance values the collector must be switched off. The average fluid temperature T_{avg} must be estimated from the operational details similar to the outlet temperature T_{out}. When mass flow through the collector is varied to hold the outlet temperature constant, the values are more or less identical with the set point temperatures, however for fixed mass flow, the estimation is more difficult and might be calculated iteratively for each time step.

Indeed, the formulations of Eqs. (7.40) to (7.42) are simplified and have to be understood in their meaning. The optical efficiency $\eta_{opt,ann}$ provides a measure of how much solar radiation is provided to the receiver over a year, including the effects of off-normal incidence. In this view it can be conceived as a measure of the optical quality of the collector-assembly for a specific location in contrast to the optical efficiency at normal incidence $\eta_{opt,0}$, which is independent of latitude and climate. The annual energy efficiency gives a measure of the total thermal energy provided by the collector for a specified temperature range. And the exergetic annual efficiency includes as metric the Carnot efficiency calculated with the outlet temperature and hence gives a measure of possible work (or electric generation) to be provided.

The result of every local collector calculation is the determination of the optical efficiency, which includes the largest losses. Here, a distinction must be made between the optical efficiency at perpendicular incidence and the annual optical efficiency. Figure 7.45a shows the distribution of the different shares of the loss mechanisms in the optical efficiency $\eta_{opt,0}$ at normal incidence ($\theta_i = 0°$) and integrated over the year $\eta_{opt,ann}$ (Fig. 7.45b). The figure refers to a specific optimized Fresnel collector absorber, but the proportional distribution of other LFR collectors is similar in magnitude.

In the optical efficiency $\eta_{opt,0}$ at normal incidence, the material-specific properties such as absorption, reflection and transmission are of decisive influence. Large values provide information about the quality of the construction and the materials used.

The optical annual efficiency $\eta_{opt,ann}$, which is much more important for estimating annual yields, is mainly influenced by the geometry-related losses such as

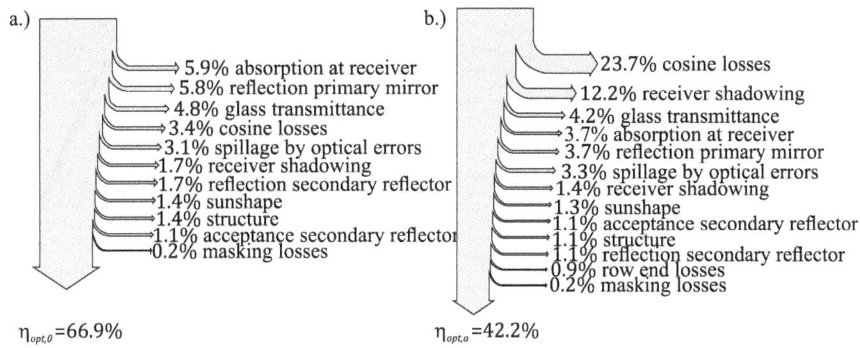

Fig. 7.45 Sankey diagrams for the optical loss fractions at normal incidence $\eta_{opt,0}$ (**a**) and annual yield $\eta_{opt,ann}$ (**b**) of an optimised Fresnel collector (Mertins, 2006)

cosine and shading losses. In typical optimisation for power plant planning, losses due to spillage and sunshape of about 4.5% are assumed. This value can be further reduced during fine optimisation of the collector geometry. However, it should be noted that a focus on reducing these losses is immediately detrimental to the levelized cost of electricity ($LCoE$) or heat, because this would require reducing the absorber or receiver height as well as the number of mirrors to reduce the length of the reflected rays and thus the spillage. A lower receiver height in turn increases the cosine losses due to the larger angle of incidence of the primary mirrors. A lower number of mirrors increases the specific heat losses and the specific costs of the collector.

Sensitivity Analysis of the Optical Losses in a Fresnel Collector
Typically, the first step of LFR collector design optimization refers to the distance between two adjacent rows of mirror strips, which requires a calculation of the intercept for different incidence angles in order to analyze the interactions between the rows in terms of cosine losses, blocking and shading, especially at high angles of incidence. The highest value of the intercept factor is usually reached when the distance between the mirror strips is zero, at which the Fresnel collector behaves similarly to a parabolic trough collector although it exhibits lower intercepts. This is not a practical solution, as it poses considerable challenges to both the mirror strips and their maintenance, installation, and cleaning, but also the installation of the tracking systems. The Fig. 7.46a illustrates the intercept γ of a LFR collector at a given site in north–south orientation at one time as a function of the receiver height h_{rec} for different mirror strip shifts Δx. Exceeding a mirror gap of $\Delta x = 0.1$ m yields a significant loss in the intercept. It can be also noticed in the figure that for receiver heights h_{rec} below 8–10 m the intercept drops to unacceptably low values. Small gap sizes Δx correspond to large fill factors FF (see Eq. (7.32)) being used in other optimization studies as initial step.

The second step is to analyze the time-of-day response of a collector. Since an LFR collector is single-axis tracking cosine losses play the greatest role integrally

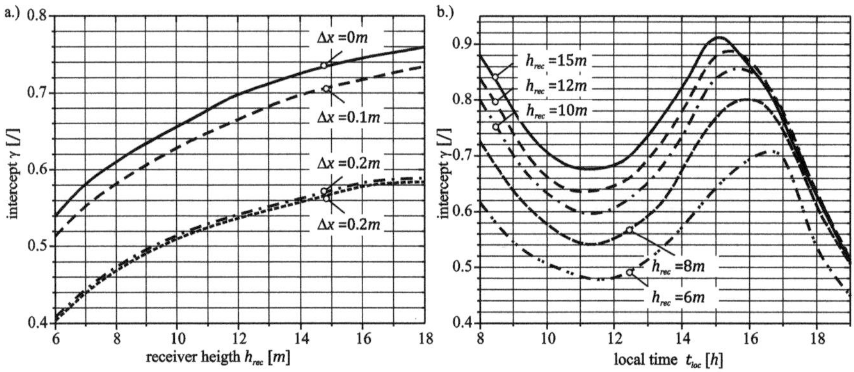

Fig. 7.46 Simulated intercept factors γ for a LFR-collector ($\Phi = 36.33°$, $\lambda = 10°21'$, $W = 1.25$ m) in north–south orientation; **a** gap width Δx for different receiver heigths h_{rec} at local time 10 am; **b** time-dependent intercept for different receiver heights h_{rec}; modified from Eddhibi et al. (2017)

over the year; these in turn depend on the orientation of the collector, the location, and the local time. In contrast to parabolic troughs, LFRs show a more complex time response of the diurnal cycle, as illustrated in Fig. 7.46b. The simulations show two peaks of the intercept factor throughout the day (from 8 am to 7 pm); the first in the morning and the second at the site shown at 3 pm. Such calculations are particularly relevant for control and regulation at the power plant level, as the thermal output of the solar field is usually selected to be greater than that of the power plant unit (solar multiple $SM > 1$) and thus feeds a thermal storage or buffer. The graphic also shows that the intercept depends essentially on the receiver height above the primary mirrors. Large intercept factors require receiver heights in the order of 7–10 m, a further increase of the receiver height is not sensible, since no significant additional gain in the intercept factor is achieved, on the contrary, the associated technical effort leads to a disproportionate increase in costs. As can also be seen from Fig. 7.46b, the intercept factor reaches its maximum when the angle of incidence is high and decreases at noon when the angle of incidence is close to zero, which is due to the orientation of the LFR collector considered ($N - S$). When LFR is oriented north–south, it follows the sun from east to west. It reaches the maximum value of the intercept factor twice, the first time in the morning and the second time in the afternoon, which corresponds to the lowest cosine losses. However, the lowest value of the intercept factor is reached at noon, resulting in the highest cosine losses. Compared to the east/west configuration, the north/south configuration has a more constant annual output; in summer it collects a larger amount of energy than the other configurations. Although it has a problem in winter, as it collects less collector yield than the East/West orientation (especially at higher geographic latitudes).

The next step is to optimize the height of the receiver above the primary mirror field together with the number of mirror rows. A time around noon is chosen

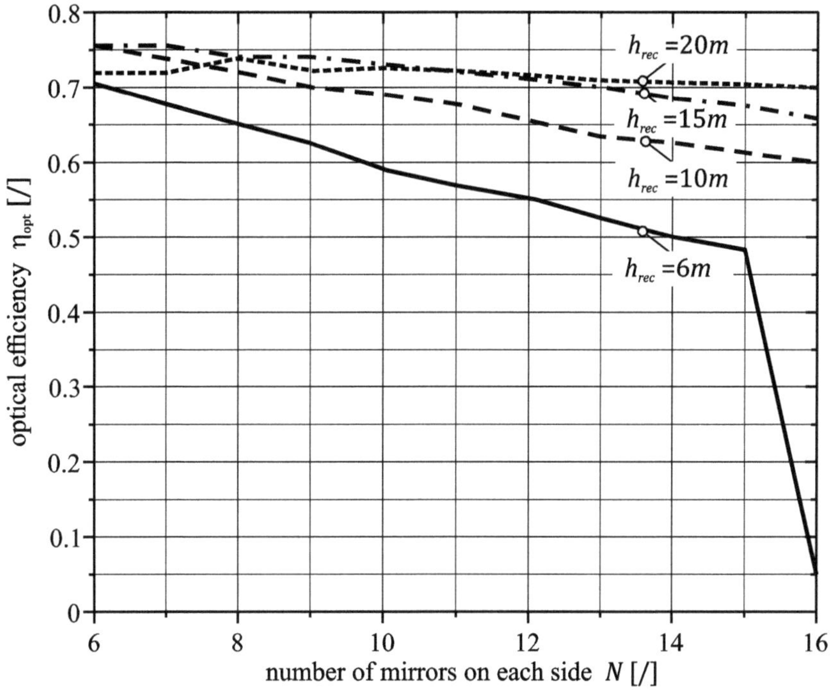

Fig. 7.47 Optical efficiency η_{opt} for a LFR-collector ($\Phi = 36.33°$, $\lambda = 10°21'$, $W = 1.25$ n) a one time instance in north–south orientation as a function number of mirrors on each side N and different receiver heights h_{rec} modified Eddhibi et al. (2017)

where the angle of incidence is close to zero to minimize blocking and shadowing effects and the line size is the optical efficiency. Figure 7.47 shows that the optical efficiency decreases monotonically as the number of mirror rows increases for a certain receiver height and a defined gap width. Therefore, the amount of blocking and shadowing produced by the rows far from the receiver has a greater impact than the amount of reflected solar radiation in the center of the primary field. The optical efficiency increases in direct proportion to the height of the receiver from the primary mirror field, as the blocking and shadowing effect is less the higher the receiver is placed. The spillage effect is defined as solar beams that miss hitting the receiver due to design specifications or errors. This effect increases for an LFR-collector that consists of a large reflective area and low receiver height as indicated in Fig. 7.47. The choice of optimal receiver height requires the consideration of blocking, shadowing, and scattering losses which all targeted to be reduced to a minimum. Thus, most commercial designs use about 9–12 mirors on each side with a receiver height form 7–10 m, allowing for a robust power up-and downscaling of the collector with a low sensitivity and a manageable tracking effort.

The diverse geometry parameters of an LFR collector also allow studies on the influence of the curvature of the primary mirrors, the shape of the secondary reflector, the number and size of the absorber tubes in the secondary receiver, on which there is a rich technical literature. Almost all concepts on the market have minimally cylindrically curved primary mirrors, which have a similar performance at typical receiver heights as parabolically curved ones but are significantly cheaper to manufacture (Abbas, Montes, Piera, & Martínez-Val, 2012). For secondary reflector shapes, truncated CPC designs have gained acceptance. As far as the number of absorber tubes is concerned, the standard is a centrally arranged single tube with diameters of 0.06–0.09 m, depending on the heat transfer medium and the pressure grading. Larger absorber tube diameters allow a reduction of the solar irradiance hitting the circumference, but at the same time the heat losses increase disproportionately (Morin et al., 2012).

Optical Errors Affecting LFR Collector Performance

In principle, the same geometric errors can occur in an LFR collector as in a parabolic trough (cf. 7.2.1.3.1), such as specularity error, displacement error due to shape or receiver deviation or load error. The segmented design and the considerably smaller component sizes of a LFR allow for considerably better manufacturing accuracies, provided all quality assurance aspects are met, so that their impact on the optical efficiency is smaller than for a PTC. On the other hand, tracking errors and slope errors are much more important for a LFR than for a PTC due to the larger focal distance. Compared to a PTC the standard deviation of the optical error in a LFR modifies with respect to the tracking error, which is doubled, to:

$$\sigma_{opt}^2 = \sum_i^n (a_i\sigma_i)^2 = \sigma_{spec}^2 + \left(2\sigma_{slope}\right)^2 + \sigma_{dis}^2 + (2 \cdot \sigma_{trac})^2. \tag{7.43}$$

The effects of the tracking error (σ_{trac}) on the optical efficiency and the local circumferential absorber concentration ratio at a tilt error $\sigma_{slope} = 2.5$ mrad are shown in Fig. 7.48a, b. Figure 7.48a shows that the optical efficiency $\eta_{i,opt}$ of the LFR decreases rapidly with increasing σ_{trac}, because the larger σ_{trac}, the more solar radiation reaches the secondary receiver aperture. Figure 7.48b depicts that the distribution of the local concentration ratio is preserved in its fundamental shape as long as $\sigma_{trac} < 2.5$ mrad; above this it varies more randomly if σ_{trac} is too large, e.g. $\sigma_{trac} = 3.5$ mrad. It should also be noted that local hot spots with high concentration ratio σ_{trac}, e.g. $\sigma_{trac} = 1.5$ mrad, may occur. In order to achieve high efficiency and to avoid a too high local solar flux, a high tracking accuracy of the order of $\sigma_{trac} = 0.5$ mrad is required, which can be realised by using a stepper motor.

Figure 7.48c, d illustrate the effects of slope σ_{slope} error at normal incidence. Again, the optical efficiency $\eta_{i,opt}$ decreases as σ_{slope} increases. More drastic is the decrease of the local concentration ratio (see Fig. 7.48d), which in the example decreases by more than one third when σ_{slope} increases from 0 to 2.5 mrad. The solar flux absorbed on the upper part of the absorber is only of the order of about 35% of the total flux for $\sigma_{slope} = 2.5$ mrad, and this allows the system to use the

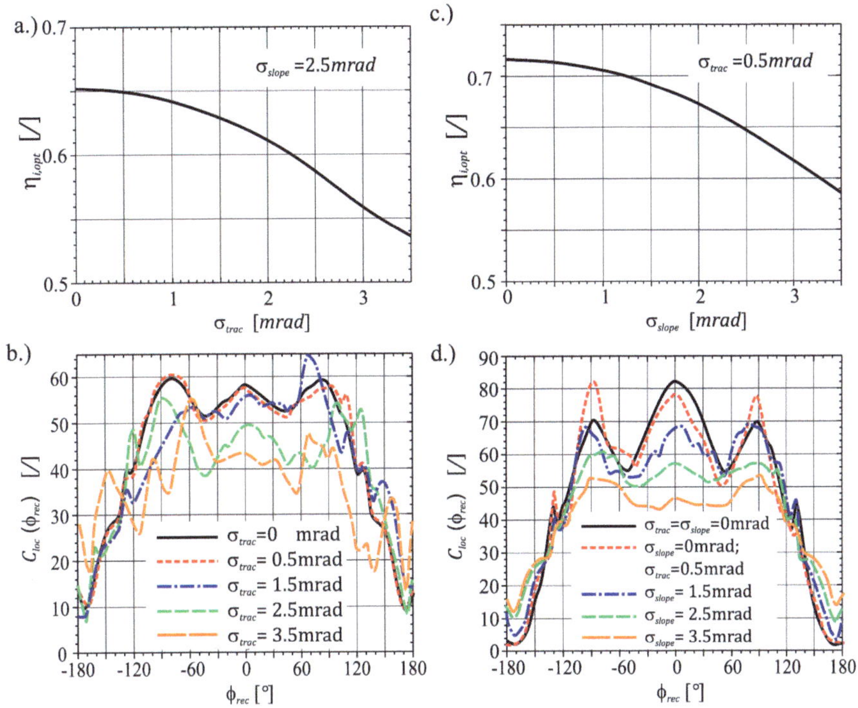

Fig. 7.48 Effects of tracking (σ_{trac}) and slope (σ_{slope}) errors in a LFR-collector at normal incidence ($W_{pr,i} = 0.6$ m, $h_{rec} = 8$ m, $\Delta x = 0.85$, CPC half acceptance angle $\theta_A = 56°$) modified from Qiu et al. (2015). Optical efficiency $\eta_{i,opt}$ (**a**) and local concentration ratio (**b**) as function of circumferrential receiver anlge ϕ_{rec} for different tracking errors σ_{trac} in (mrad at a slope error of $\sigma_{slope} = 2.5$ mrad; Optical efficiency $\eta_{i,opt}$ (**c**) and local concentration ratio (**d**) as function of of circumferrential receiver anlge ϕ_{rec} for different slope errors σ_{slope} in (mrad) at constant tracking error $\sigma_{trac} = 0.5$ mrad

entire circumferential surface of the absorber. On the one hand, this means that a more uniform radiation flux can be achieved by sacrificing a small efficiency with a reasonable slope error, e.g. $\sigma_{slope} = 2.5$ mrad, that can be achieved with existing technologies. On the other hand, the system cost can be reduced by using mirrors with slope error in this range.

7.2.2.4 Technical Designs of Linear Fresnel Systems

Mechanical Set-Up of a Linear Fresnel Collector

The design of linear Fresnel collectors consists of rows of mirrors arranged in parallel, which in turn consist of segmented primary reflectors placed near the ground at a height of about 0.5–2 m. Due to this design, the area-specific weight is very low; optimized systems have <30 kg/m^2 and thus have a much lower value than parabolic trough collectors. The design results in very low wind loads and thus changes in optical performance due to torsion are small. Moreover, as well as the

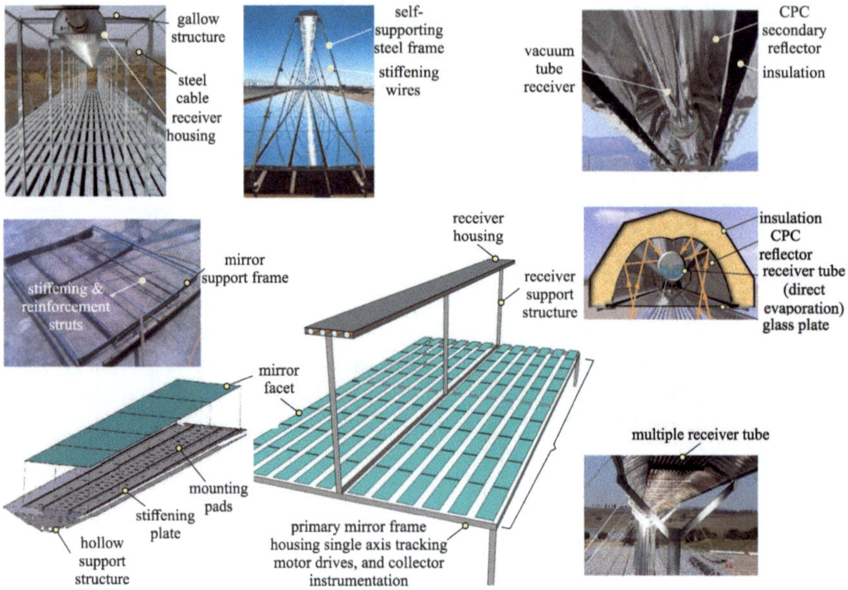

Fig. 7.49 Set-up of a linear Fresnel collector; top left gallow structure (Bernhard et al., 2009); self-support structure receiver structure (top middle); mirror facet support structure types: frame type (left middle), plate type (left bottom). Right: Vacuum type receiver in open CPC (top) former AREVA), single tube receiver (direct evaporation, FRENELL, middle); open multi-tube cavity receiver (Abbas et al., 2016), bottom)

good weight distribution places low demands on the foundations. For a parabolic trough the concrete foundations constitute about two thirds of the total weight of about 118 kg/m^2 which is not needed for a light-weight linear Fresnel construction (Viebahn, Kronshage, Trieb, & Lechon, 2008). The segmented structure does not require heavy machinery for installation. The light-weight support system has as well as modular design allows for relatively low collector costs.

The mirror rows can be erected with close spacing and allow a high land use factor with about 63% of the plant area used as aperture whereas for the Andasol PTC plant in Spain the aperture is only 26% of the total plant area. This saves valuable ground space especially in more densely populated regions, or for smaller process heat installations, the integration on the horizontal roof is more effective. The primary mirrors can be cleaned automatically, for example with small robots and almost without water consumption, and individual mirror rows are easily accessible; both aspects allow relatively low operation and maintenance costs.

The receiver of the LFR collector is fixed on a stand structure mostly made of steel and does not move, which in turn does not require flexible high-pressure and high-temperature pipe compensators as e.g. in parabolic trough systems. It is located several meters above the ground. A typical picture of the mechanical set-up of LFR collector is shown in Fig. 7.49 center.

For the elevation of the receiver support, there are concepts with a steel gallows structure fixed to the ground by steel cables like in FRESDEMO (Fig. 7.49 top left) or cantilevered steel frame structures. There are also several concepts for primary strip supports, all based on a steel frame. The uniaxially movable mirror facets are mounted on the frame. On this frame structure not only, the tracking motors/gearboxes are installed but also the instrumentation and controls for each collector. For the facet carriers on which the primary mirrors are located, there are concepts based on a frame structure with support struts on which the mirrors are glued or screwed. Other concepts are based on a rigid aluminum or steel body in which support pads are integrated to fix the mirror facets.

The receivers are usually fixed and suspended from the steel structure at a height of several meters. This allows good accessibility and automatic cleaning. For the receiver elements, secondary receivers based either on a CPC structure (Fig. 7.49 right top and center) or on a W-trough have prevailed, which in turn are located in a thermally insulated steel housing to keep heat losses low along the often several hundred meter long piping lines. The receiver tubes themselves vary considerably depending on the choice of heat transfer medium and the design of the selective receiver tube. Some concepts are based on receiver tubes that rely on vacuum tubes, as in the parabolic trough. This allows small convective heat losses and eliminates the need for a glass plate cover on the bottom of the secondary reflector (Fig. 7.49, top right), but results in a costly cleaning of the same. Another concept, especially for direct water evaporating receivers, relies on a single tube placed in the center of a CPC secondary reflector, which is covered at the bottom by a planar glass plate. This allows easy cleaning but higher convective heat losses. Another design consists of several parallel receiver tubes placed in a trapezoidal cavity, also open at the bottom, located on the top of the secondary receiver.

7.2.2.5 Applications and Outlook

The typical fields of application of linear Fresnel collectors cover a considerably wider spectrum than that of power plant applications as compared to parabolic trough systems. Existing LFR collectors are operational in the following application areas such as

- solar power plants with and without storage,
- solar process heat such as textile processing (bleaching, dyeing, ...) or food processing (drying, distillation, evaporation),
- solar-assisted heating/comfort cooling,
- sea water desalination and wastewater treatment,

In general, the PTC shows higher optical efficiency than the LFR because the segmented primary reflectors result in optical losses due to blocking and shadowing, and the aperture is horizontal and not tilted towards the sun as for the PTC. The row end optical losses of the LFR cannot be neglected if the collector length

(L_{col}) is small, cf. Equation (7.39). Because there is a longitudinal and a transversal $IAM(\theta_T, \theta_L)$ in the LFR, which are both smaller than one (in contrast to the PTC), for a single collector there is intrinsically a smaller optical efficiency than for the PTC, especially noticeable at higher geographical latitudes. On the other hand, the row-shading effect of PTC rows also has a strong impact on the optical efficiency for low sun altitude in the mornings and evenings.

Due to the robust, simple, and cost-effective structure and the lower requirements for maintenance and operation, the smaller area requirements, and the horizontal installation of the primary field an LFR collector is often quite attractive for applications in industry where space on roofs and ground is limited. Two optical aspects seem to be less in favor of a LFR than a PTC collector.

- lower overall optical efficiency at high transversal incidence angles,
- larger variation of optical efficiency during the day as a consequence.

The first aspect may be compensated by lower collector cost per area, the second aspect plays no role when thermal storage can be used to manage the mismatch between hourly generation and demand. When no vacuum receivers are used, certainly the length-specific heat losses are higher than for the standard PTC collectors, however, that is also true for PTC collectors without vacuum receivers used in industrial applications. Moreover, higher possible concentration can compensate this.

Although the maturity of LFC technology in several aspects: optics, flexibility, and aptness to heat transfer media, thermal efficiency and system integration has been shown already, there are still not enough demonstration plants worldwide. LFC must market its specific advantages and the still large potential for technical and economic improvement. On the other hand, there is still an appreciable optimization potential.

To play on the advantages of linearly concentrating systems, several issues must be emphasized:

- Modularity and repetitiveness of reflectors and receivers.
- Short distance between concentrator and receiver.
- Piping of fields.
- Compactness of solar fields.
- Scalability of power plant size.
- Cleaning solutions for soiled mirrors.
- Possibility for maintenance during operation.
- Free choice of collector orientation.
- Minimization of system costs.
- Adaptability to variable land flatness, slope, and form.

The potential of LFC is considerable when advanced optical designs are realized. As already the use of molten salt technology with temperatures up to 550 °C have been demonstrated, and the combination with direct two-tank storage solutions is straightforward, the different daily generation profile with a peak at noon compared with the typical "Hat-curve" of parabolic troughs does not seem to matter anymore. Whether the linear Fresnel collector or the point-focus "Fresnel" collector, the solar tower using heliostat fields have advantages, is therefore not clear for this temperature range, but possibly for even higher temperatures. For the classical "solar salt" consisting of a mixture of 40%KNO_3/60%$NaNO_3$ at 550 °C, other local conditions and of course economics will have the highest impact on selection.

A cost-effective improvement of an LFR collector is often associated with the design of the mirror-receiver configuration. In principle there are four measures to increase LFR collector efficiency:

- increasing absorber thermal performance,
- design of the primary reflector shape and arrangement,
- design of the secondary reflector, and finally,
- reduction of optical losses.

The individual measures will be briefly outlined below, details of which can be found in the literature listed.

Absorber Thermal Performance

Absorber optimization has two main goals; increasing thermal efficiency and homogenizing the circumferential radiation flux distribution.

The heat transfer side can be improved by optimizing the heat transfer fluid, for example by optimizing heat transfer using finned structures or by mixing additives into the heat transfer fluid, which show only marginal improvement potential (Bellos, 2019). Much more effective, however, is the reduction above all convective heat losses at the receiver, which are significantly greater than the heat radiation losses. Studies show that wind protection and the use of vacuum tubes increase the annual thermal efficiency by up to 10% (Cagnoli et al., 2018).

The non-uniform heat flux on the receiver allows for higher optimization potential and the literature is full of a variety of new structures designed to achieve better heat flux uniformity. A summary can be found in Bellos (Bellos, 2019). The most effective way seems to be to guide the target lines of the primary mirrors so that a part of the heat flux reaches the upper part of the absorber. With this measure, the optical efficiency is not affected. In the same direction, the use of segmented parabolic secondary mirrors (SPC) with the use of variable different target lines. With both measures, both a higher optical efficiency and a larger circumferential receiver flux distribution can be achieved, which is significantly better than the conventional TC (Trapezoidal Concentrator) and the CPC (Compound Parabolic Concentrator).

Primary Reflector Design

The primary reflector is a key component of the LFR system because it redirects the solar rays to the absorber or secondary reflector, and therefore its structure determines the final optical efficiency. Innovative primary reflector designs aim at minimizing shadowing and blocking effects, such as the Compact Linear Fresnel Reflector (CLFR) concept, where the solar rays are directed to two linear receivers, without compromising the ground use efficiency as illustrated in Fig. 7.31. However, the elimination of shading and blocking effects of the CLFR lead to losses in the concentration ratio. The use of the third dimension in positioning the mirrors, i.e. having primary mirror axis at different heights, may also reduce shading and blocking, allowing larger concentration factors.

Secondary Reflector and Receiver Design
The use of a secondary reflector in LFR systems is common practice to improve optical efficiency, and many different shapes have been proposed in the literature as presented in this chapter and Chap. 3. The different shapes show not only different optical performance, but also manufacturing cost. This is often more important than the pure chase for higher efficiency. Very critical for a good cavity receiver is the stability of the reflector and the low absorptivity for the concentrated radiation. Cleanliness and cleaning options of the optical system are important to consider in the design.

Optical End Loss Reduction
The minimization of end losses in LFR systems is possible. (Zhu et al., 2017) proposed the scalable linear Fresnel reflector (SLFR). The structure can constantly adjust to the longitudinal incidence angle of the sun, avoiding end losses, as outlined in Fig. 7.50a. The optical efficiency of a standard LFR can be raised up to 9%. However, a key advantage of the LFR collector, the low cost simple single-axis fixed mirrors are given up in this approach. Another advanced concept envisions a two-axis tracking system in which the primary reflector can slide along V-shaped rails to compensate for end losses. This design, shown in Fig. 7.50b, increases the optical efficiency compared to the single-axis tracking LFR collector by 8% to 50% depending on the normalized collector efficiency, especially at short collector lengths, moving receivers were also analysed to reduce the end losses. The effects of lengthening the receiver and moving the receiver, as shown in Fig. 7.50c, achieved an annual yield increase of about 50% when the length of the receiver is doubled, but the thermal losses also grow. Therefore, this type of measures has great importance Fresnel collectors with short length dimensions.

7.3 Point Focusing Systems

There are two main lines of development for point-focusing receivers. These are on the one hand so-called solar towers, where a heliostat field focuses the solar irradiation onto a central receiver installed close to the top in a tower. Each of the individual biaxial tracking mirrors is therefore similar to a Fresnel mirror, which redirects the solar radiation to the receiver with different focal lengths. The basic beam path of a solar tower concept is shown in Fig. 7.51a. With the help

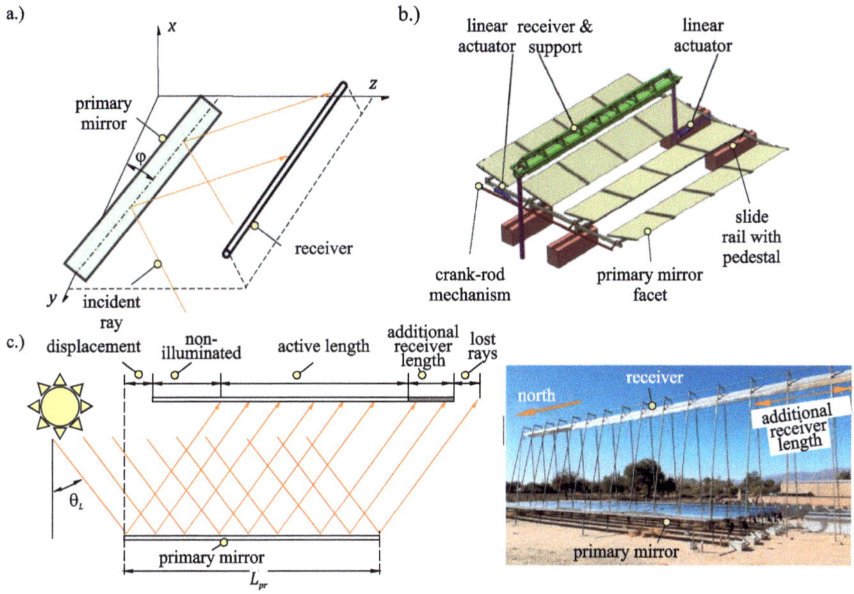

Fig. 7.50 Design measures to reduce end losses of LFR collectors; **a** scalable linear Fresnel reflector concept by Zhu et al. (2017); **b** two-axis tracking LFR system as proposed by Yang et al. (2018); **c** receiver extension along longitudinal direction as proposed and realized (modified from Bellos, 2019)

of large heliostat arrays, solar towers with more than 100 MW nominal readings have already been put into operation.

The second line of development are so called parabolic dishes, where a parabolic monolithic or segmented mirror focuses to a focal point, where the receiver is coupled to a heat engine (Stirling engine or micro gas turbine). The entire setup composed of mirror and receiver plus heat engine unit must track the position of the sun biaxially, as illustrated in Fig. 7.51b. As it is designed as full functional unit the dish systems perform power scaling by replicating the base unit. An intrinsic disadvantage is that the storage of thermally generated excess power is technically only possible to a limited extent. Therefore, dish systems are mostly used where either high temperature or electricity is to be generated independently of the grid.

7.3.1 Dish Collectors

Dish concentrating collectors use paraboloidal mirrors that track the sun and focus solar radiation into a receiver where it is absorbed and transferred to a heat engine/generator or alternatively into a heat transfer fluid that is transported to a ground-based plant. Dish concentrators due to their principle have the highest

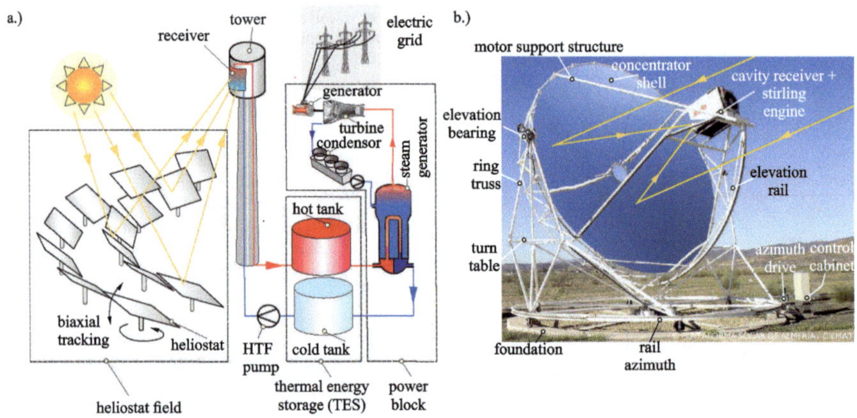

Fig. 7.51 Components and radiation solar beam path of point-focusing solar thermal systems; **a** solar tower concept **b** solar dish concept (EuroDISH © Ciemat, La Plataforma Solar de Almería (PSA))

optical efficiencies, the highest concentration ratios, and the highest overall conversion efficiencies of all the concentrating collectors. Most development with dish concentrators involved the use of receiver integrated Stirling engines for direct production of electricity. However also other concepts are possible.

The modular approach was seen in the beginning as an advantage of Dish-Stirling systems, however, the cost reduction of the directly competing photovoltaic (PV) systems, has been far more successful than expected. Technological progress, large market volume, and strong competition let PV panel cost fall to fall below 400US$/kW in many markets and therefore outperformed the CSP technologies for electric power generation without storage. Thermal energy storage (TES) is possible in principle for dish engine systems but suffers from small size leading to cost drawbacks and considerable technical challenges. The attempts to develop storage solutions for dish have not brought a breakthrough so far. Dish engine systems without storage directly compete with PV and have clearly lost this competition. Consequently, most of the companies active with dish systems have failed commercially or moved to other areas. However, as the efficiency advantages of dishes are considerable, they might have a potential in direct solar chemical processing at very high temperatures, which can only be achieved by dish technology.

7.3.1.1 Concentrator Optics

The ideal case of a concentrator focusing parallel rays onto a focal point in 3 dimensions is a rotational paraboloid or parts of that. If the focal point is on the center axis of the full parabola, then the rotationally symmetric reflector forms a rotational paraboloid with a height h, as illustrated in Fig. 7.52. Mathematically it is described as a parabola of focal length f in three-dimensional space. The light rays incident parallel to the y-axis are focused by the parabola contour to the focal point F of the parabola. Provided that the lines FP and PQ are of equal

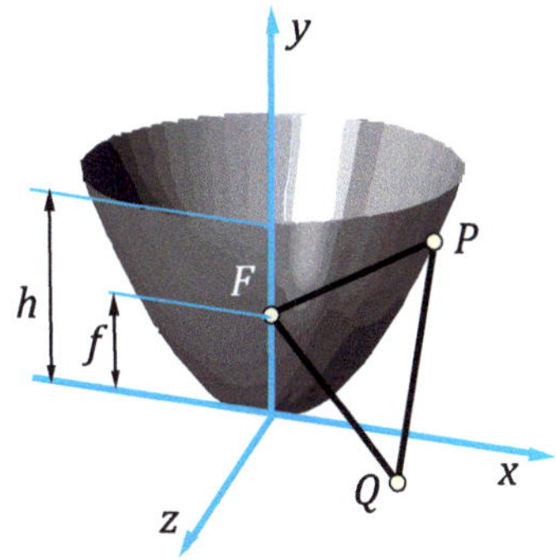

Fig. 7.52 Sketch of a rotational paraboloid of height h

length, the tangent is always correct to focus all vertical beams into the focal point F. Similar to the two-dimensional case of the trough as the focal length f of the parabola increases, a flatter paraboloid becomes necessary to focus the light in a point-like manner. The advantage of the flatter reflector area is less material requirement for the same aperture area (i.e. constant concentration ratio). The disadvantage is that when the distance to the focal point increases, possible mirror or tracking errors have a larger effect.

Using this optical element collectors using a symmetric rotational reflecting mirror ideally with the cross section of a parabola are called a parabolic dish collector (PDC). In the PDC all rays incident parallel to the parabolic axis are reflected by the mirror to a focal point where an absorbing element is located. Divergent irradiation within a cone of half-angle δ leads to a widening up of the focal point to a finite size focal spot when the absorber plane is going through the focal point. Depending on the angle formed with the axis in the incident plane the rays would intersect at a smaller or larger distance than the focal distance. Thus, the permissible acceptance angle θ_a also depends on the distance between focal point and the reflector. In the case of the parabolic contour, the ray length is maximum for the rays coming from the rim.

It should be noted that also partial sections from a paraboloid can be used as a concentrator. If the axis of the paraboloid is not included in the aperture, we call the reflector an off-axis paraboloid. In such a case the advantage is that the receiver can be positioned more flexibly with respect to the mirrors (see later in Fig. 7.53). It is also possible to move the reflecting area in such a way that the sun always is focused on a fixed spot in space. Then we talk about a fix-focus concentrator. This so-called Scheffler dish is presented in Sect. 7.3.1.1.2.

Parabolic Dish Optics

A parabolic dish requires a solar tracking system following the sun position in 3 dimensions all day long. The axis of the paraboloid must point towards the source, the sun. The movement of this dish collector thus has two degrees of freedom and is a two-dimensional tracking in angular space. The incidence angle for a dish is the angle between paraboloidal axis and solar beam direction. Due to the

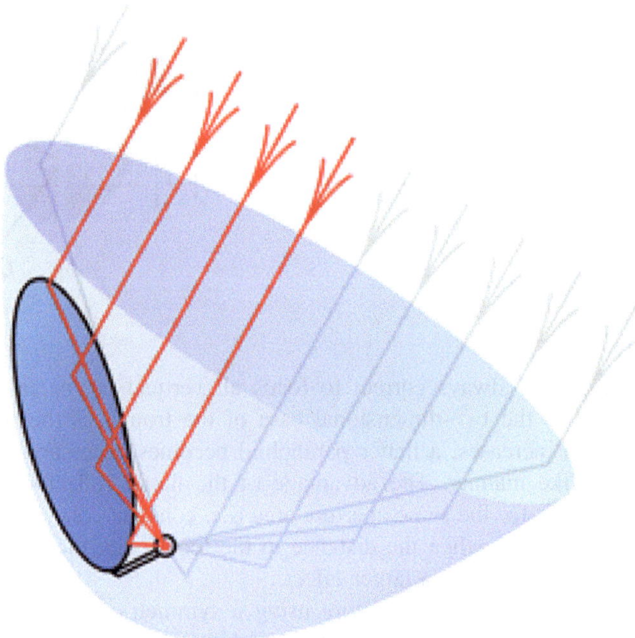

Fig. 7.53 Schematical view of an off-axis paraboloidal concentrator

two-dimensional tracking no cosine effect reduces the incident solar flux on the aperture.

The most important PDC parameters are geometric concentration ratio C_g, acceptance angle (θ_a), rim angle (ψ_r), and peak optical efficiency. The geometric concentration ratio, C_g, is the ratio between the collector aperture area and the total absorber area. For a dish collector we distinguish between flat receivers and spherical receivers. Depending on that we have different concentration ratios the geometric concentration ratio for a flat receiver is given by Eq. (7.8). Here, the shading of the reflector by the receiver itself is neglected as it depends on the coating of the receiver: If the absorber is only on one side (viewing the reflector) the area of the receiver should be not considered as aperture—then $C_{g,flat}$ has to be subtracted from Eq. (7.44); if the receiver is absorbing on both sides, the full width of the dish is considered as aperture and Eq. (7.44) is correct:

$$C_{g,flat} = \frac{\frac{\pi}{4} \cdot W^2}{\frac{\pi}{4} \cdot d^2} = \frac{W^2}{d^2} = \left(\frac{\sin 2\psi_r}{2 \cdot \sin \theta_a}\right)^2, \tag{7.44}$$

and it is maximal for a rim angle of $\psi_r = 45°$.

The geometric concentration ratio for a spherical receiver is given by Eq. (7.45)–here nothing should be subtracted as we assume absorption over the whole sphere:

$$C_{g,sphere} = \frac{\frac{\pi}{4} \cdot W^2}{\pi \cdot d^2} = \frac{W^2}{4 \cdot d^2} = \left(\frac{\sin \psi_r}{2 \cdot \sin \theta_a} \right)^2, \tag{7.45}$$

and it is maximal for a rim angle of $\psi_r = 90°$. In Eqs. (7.44) and (7.45) d describes the diameter of the receiver area (disc or sphere), W is the parabola width, θ_r the rim angle of the dish and θ_s the solar divergence half-angle. The acceptance half-angle, θ_a, is the maximum angle that can be formed by a ray with the paraboloid axis that it intercepts still the receiver area. To capture all direct beam solar radiation the minimum acceptance half-angle should be at least the solar divergence half-angle $\theta_s = 0.27°$ which contains the solar disk as seen from the Earth. It must be chosen by the designer which part of the circumsolar radiation shall be captured as well. The small acceptance angles do demand very accurate and expensive solar-tracking systems, but high concentration values can be achieved.

$$C_{g,sphere} = \frac{\frac{\pi}{4} \cdot W^2}{\pi \cdot d^2} = \frac{W^2}{4 \cdot d^2} = \left(\frac{\sin \psi_r}{2 \cdot \sin \theta_a} \right)^2. \tag{7.46}$$

The rim angle, ψ_r, which is directly related to the concentrator arc length, can be calculated as a function of the parabola focal distance (f) and aperture width (W) by

$$\tan \psi_r = \frac{8 \cdot f \cdot W}{W^2 - 16 f^2} \tag{7.47}$$

For parabolic dishes often, flat receivers are used therefore usual rim angles are in the range of 45°. Smaller rim angles are not advisable because they reduce the aperture surface and lead to comparatively large focal lengths.

Optical losses are very important in a PDC due to the extremely high concentration of the total solar flux incident on the PTC aperture plane. Optical losses are associated with the following four parameters:

Reflectance $\rho < 1$ of the collector reflective surface. Typical solar spectral reflectance values of the best clean silver-based low iron glass mirrors are around 0.93–0.94. Reflective films have usually a specular reflectance $\rho < 0.9$.

Intercept factor γ. This is the fraction of the direct solar radiation reflected by the reflecting dish that finally reaches the active surface of the receiver. The intercept factor for 0° incidence angle is caused by optical imperfections (micro-roughness, errors in the parabolic shape, mechanical deformation of the support structure) and some shadowing by the receiver support or structure.

Absorptance of the receiver coating α. This parameter quantifies the amount of energy absorbed by the spherical or flat receiver area divided by radiation reaching it. This parameter is typically 0.96 for black receivers. The temperatures reached can be extremely high exceeding 1000 °C. There are no selective absorbers stable at these temperatures. On the other hand, the high concentration C_g in the range on

1500–4000 leads to good thermal efficiencies. In fact, often small cavity receivers with small open aperture windows are incorporated in designs to reduce convective and radiative losses.

Multiplication of these parameters (intercept factor, mirror reflectance and receiver absorptance) for an incidence angle $\theta_i = 0°$ gives the peak optical efficiency of the PDC $\eta_{opt,0}$:

$$\eta_{opt,0} = \rho \cdot \alpha \cdot \gamma. \tag{7.48}$$

In total the $\eta_{opt,0}$ should be in the range of 0.9 for clean, good-quality PDCs. The effect of non-normal incidence angles for parabolic dishes does not apply, because the aperture should always track the sun position.

Scheffler Dish Optics

The Scheffler reflector, named after the inventor Wolfgang Scheffler, is a solar concentrator design which maintains a fixed focus while only having a single axis tracking mechanism. It is applied to produce steam for solar cooking, industrial processes an electricity generation. Ideally the reflector is an off-axis segment of a paraboloid mirror, see Fig. 7.53. The fix focus characteristics of the Scheffler dish is achieved by adjusting seasonally the curvature of the reflector in dependence of the solar declination and tracking the sun throughout the day about an axis parallel to the earth's axis of rotation. This design makes the construction and operation of high temperature solar concentrators accessible to developing nations. Because of the fix focus feature, it is possible for example to cook food in a pot at the concentration point inside a building, hence persons can be close to the focus point without being at risk to be harmed.

For a detailed ray tracing simulation of a single Scheffler Reflector as well as an array of reflectors a flexible algorithm has been developed by van Rooyen, Heimsath, Nitz, and Platzer (2011). The results are visualized in a 3D map for varying solar declination and solar hour angle. The occurring shading of the array of Scheffler Reflectors is also studied. The modelled reflective surface of a Scheffler Reflector comprises of rectangular planar mirror tiles of variable size. The algorithm adjusts the curvature of the paraboloidal segment depending on the solar declination (Fig. 7.54).

The dish typically is produced with a cross bar structure consisting of square steel or aluminum bars and angle sections to form a shape of parabola with metal channels bolted above the square bars for fixing mirrors. Small rectangular flat mirrors of solar grade mirrors or aluminum mirror sheets are then fixed on the structure to approximate the paraboloidal shape. The shape of the flexible support structure is adapted seasonally by telescopic clamps or strings. The daily tracking can be done with a small motor drive and gears to rotate the complete structure at 15° per hour with centrer of gravity in the rotational axis. A design guideline for the construction of small dishes can be obtained from Solare Brücke (2022). Expanding the technology to larger dishes was a prerequisite for commercial projects. In India several companies produce steam generation systems mainly

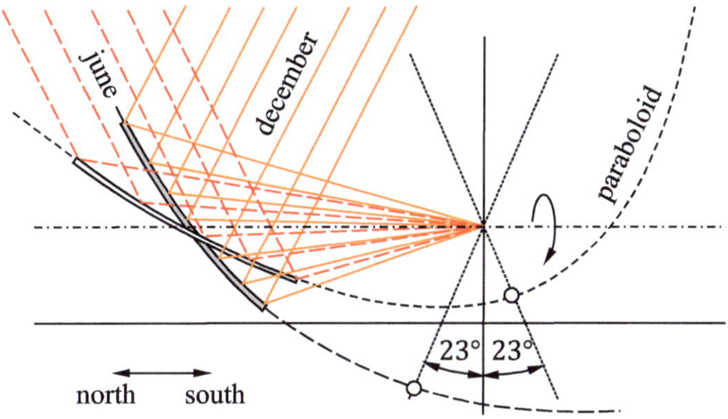

Fig. 7.54 Declination of the solar orbit for which the curvature of the dish segment is seasonally adapted. In this way optical axis of the paraboloid is tilted ±23° according to the declination

for cooking, and a 1 MW solar power plant "India One" has been constructed using a field of 770 dishes each 60 m^2 aperture and an iron cavity steam generator at the focal point (Brahma Kumaris, 2011).

7.3.1.2 Practical Considerations and Experiences Parabolic Dish Collectors

A dish system consists of a paraboloidal shaped concentrator, tracking system, a receiver, optionally an engine with a generator. All dish systems are based on the same three basic assemblies, as shown in Fig. 7.55:

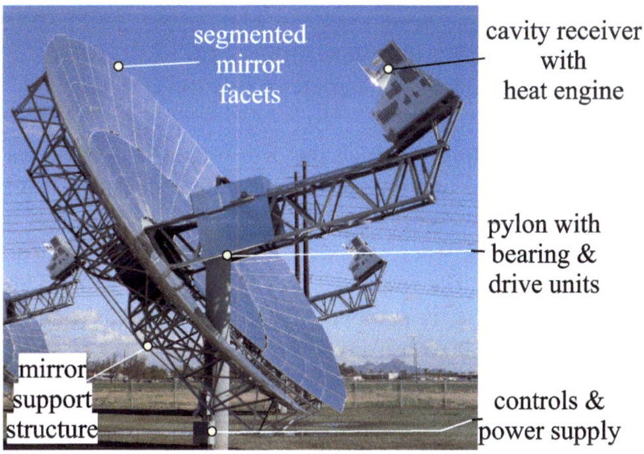

Fig. 7.55 25 kW Disk-Stirling base unit of Maricopa Solar Project CSP Project modified from United Sun Systems International Ltd. (2012)

- solar dish concentrator and support structure (reflector unit),
- receiver installed in the focal point (with/without heat engine) and
- solar tracking system (mechanical structure with bearings and drives).

and a system control unit. The concentrator tracks the sun biaxially in such a way that the optical axis of the concentrator always points to the sun. A reflective mirror surface on the paraboloidal concentrator, either metallized glass or plastic or anodized aluminum, reflects incident sunlight to a small region called the focus. The ideal shape of the reflecting surface of a solar concentrator is a paraboloid. Due to the three-dimensional structure, there is a multitude of different engineering solution options regarding the positioning of bearings and tracking systems, the connection of the receiver with the reflector system, which results from the sizes of the respective system.

Facetted reflectors (i.e., consisting of individual spherical or parabolically shaped segments) and full-surface reflectors are encountered in different designs. For facetted concentrators, several mirror segments are mounted on a supporting structure. The segments are attached and oriented individually. The segments may either consist of glass mirrors or substrates covered with reflecting foil or thin-glass mirrors. For full-surface concentrators, the entire concentrator surface, e.g. in the metallic membrane technology, is shaped parabolically by a forming process.

Practical considerations of wind load and weight lead to typical aperture sizes of 1–25 m diameter. The receiver captures the radiation as high-temperature thermal energy in a fluid that is either the working fluid for a receiver-mounted engine or is used to transport the energy to ground-based processes. A typical engine size is 3–25 kW_{el}. In the overwhelming number of cases Stirling engines were used, but also micro-turbines using the Brayton cycle were tested. Maintenance of the engines in the past has been a major cost driver. Maximum size of dishes up to now is 500 m^2 (developed by ANU—Australian National University). The bigger systems utilized rather steam receivers, high-temperature heat transfer fluid or compressed air receivers for the micro-turbines (Brayton cycle).

Tracking may be accomplished generally in two ways:

- Azimuth-elevation tracking in which the dish rotates in a plane parallel to the earth's surface (azimuth) and around an axis perpendicular to it (elevation). A rotation about both axes is necessary throughout the day.
- Polar-equatorial tracking where the dish rotates about an axis parallel to the Earth's axis of rotation. The rotational speed is constant with 15°/h. The other axis, the declination axis, is perpendicular to the polar axis and needs to be adjusted only gradually <0.5 per day.

Many pylon-based dish designs use slewing drives for azimuth and linear drives for the elevation movement. Both azimuth and elevation drives can profit from the large market for tracking PV and heliostats where similar solutions are used. One difference to PV applications is however that high stiffness, low backlash, and highly accurate positioning is necessary. As slewing drives are a considerable cost factor, new and advanced solutions are needed. Turntable-based dish mounting and tracking (e.g. EuroDish) is characterized by small and cost-effective gears but needs many structural parts to be assembled. They also need large foundations which must be equipped with rails for big systems. In the design of the tracking system always the compromise of accuracy and cost must be found.

High receiver surface temperatures of 600–800 °C or even 1000 °C and more result in considerable thermal losses by convection, infrared radiation, and heat conduction. Thermal radiation losses increase with the power of 4 in temperature (in Kelvin) and are dominant at high temperatures. This can give an unacceptably low efficiency if simple open receivers were used. Thus, efficient receivers for dish systems are cavity receivers with a small opening (aperture) through which concentrated sunlight enters.

As with the other approaches to concentrator design, the concept of using mirrored dishes to focus the sun has been around since 200 BC. One of the earliest technical implementations of a dish system was by Augustin Mouchot, who built dish-driven engine systems as early as 1864 and displayed one at the WorldExhibition in 1878 in Paris. The dish technology of today was revitalized in the early 1980s following the oil price shock in the 1970s. Many dish prototypes were designed and built by researchers and commercial companies, using different materials like stretched silvered plastic membranes, bonded thin and segmented thick glass mirrors, and even from tiny flat mirrors bonded to a paraboloid support material.

Receivers were often direct insolation receivers for Stirling engines, but also heat pipes and spherical steam receivers. Commercial developments stimulated larger technology progress for Dish-Stirling in the decade until 2010, with companies like Stirling Energy Systems (SES), Schlaich Bergermann & Partner (spb), Infinia, HelioFocus, Solar System Pty, South West Solar Technologies mainly in US, Australia and Israel. However, the rise of PV brought a rapid end to commercial project developments.

Nevertheless, new developments have appeared again, this time tackling also the question of storage integration or direct chemical reactors in the focus. An overview of these activities is given by Schiel and Keck (2012).

The achievable concentration ratios C, are higher than those of central receiver towers; geometric concentration ratios of peak $C = 1.3 \cdot 10^4$ have already been achieved, see Hafez et al. (2017). This is due, among other things, to the fact that the focal length is much smaller than for towers, and spillage, atmospheric attenuation, as well as surface and tracking errors are smaller. However, the power densities realized so far correspond only to those of solar towers. The theoretically achievable temperatures due to the high concentration ratios are far above those of available receiver materials, temperatures of more than 800 °C have been reached in experiments. This results in significantly higher thermal–mechanical energy conversion efficiencies. Dish/Stirling systems achieve more than 30% peak solar-to-grid, with an average solar-to-electric efficiency in the range of 16–25% under typical conditions. Power conversion options include Stirling engines

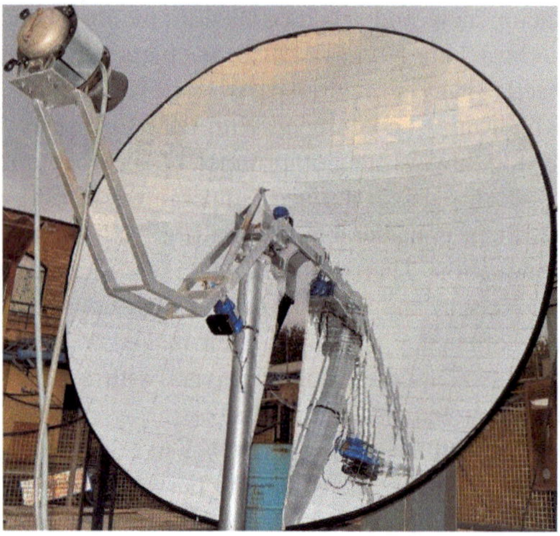

Fig. 7.56 1 kW free-piston Stirling engine installed in the Kerman pilot plant, using helium as fluid at maximum operating conditions of 800 K and 1 MPa from Gholamalizadeh and Chung (2017)

or microturbines, which require more compressor work than Rankine cycle conversion systems, but the waste heat from condensation is eliminated and water consumption is also minimal. Subsequently the individual assembly units are discussed.

Structural Concentrator Design

From previous design considerations, we know that the reflecting mirrors must be positioned on an approximately parabolic structure, and the thermal output results from the projected concentrator area. Typical solar concentrators are about 3–10 m in diameter, although large concentrators up to 17 m in diameter have also been fabricated for Stirling engines for power generation. Typical focal lengths range from 4 to 12 m. Almost all support structures and columns at these dimensions are made of steel for stiffness reasons. There are several structural designs for the reflector. The simplest type of structure is a closed reflector surface, as in the Eurodish collector (compare Fig. 7.51b), the other option is that of segmented facets mounted on a common support structure, compare Fig. 7.55. Closed reflector structures require shells exhibiting a continuous paraboloidal surface. They either consist of one large piece or are composed of precisely manufactured segments connected to each other. If they are made of one large piece, often a membrane type is chosen, see e.g. in from Schiel et al. (2004). Stretched membrane heliostats have been built with reflecting areas of more than 200 m^2 and up to 17 m in diameter. One disadvantage of a large component is the transport of the shell, so many other concepts show a simplified structure in which the tensioned membrane is replaced by sandwich panel mirrors supported at the circumference by a ring beam. The panels can either be reinforced plastic layers with a foam core and stiffened by means of a radial rib along the centerline, or consist of thin glass mirrors bonded to a plate, also stiffened radially to form the reflective surface, as shown in Fig. 7.56.

Fig. 7.57 Fresnel paraboloid collector with cavity receiver for process heat installed at a Hotel site in India modified from Brossard (2010)

The most frequent applied is a segment assembly, in which the reflecting surface is constructed of a space frame/glass mirror type.

In contrast, in a multi-facet design, each facet has its own curvature and alignment characteristics according to its position within the reflector assembly, and each facet can be aligned individually. The entire system, of course, requires an optical axis. This can be achieved by assembling the facets in the form of a large paraboloid. A multi-facet design has the advantage of not having to construct a large paraboloid surface in one step, which is time-consuming and expensive. In addition, facet constructions allow the use of very small spherical curvatures, since each individual facet has a small rim angle focal ratio. Assembling facets with "marginal" errors allows a good approximation to a paraboloid. The production of small facets is also advantageous because small components can be manufactured with very high precision and, in addition, fine adjustment to the focal point can be made during on-site installation.

A special case of multi-surface construction is a central Fresnel collector, where many single mirror facets are arranged to form a Fresnel lens, similar to a lighthouse. Figure 7.57 illustrates such a design of a reflector for solar process heat supply installed heliostat at the Maurya Sheraton Hotel in New Delhi (India).

Table 7.9 Range of reflectivity (ρ) and emissivity (ε) of some selected reflector options realized in solar dish systems modified from Hafez et al. (2017)

Reflector composite	Operation temperature T (°C)	ρ (/)	ε (/)
Polymer film/non-metal	125	0.98	0.02
Aluminum Acrylic	95	0.98	0.02
Silver Acrylic	95–100	0.92–0.97	0.02–0.08
Aluminum	100–500	0.86–0.97	0.03–0.14
Silver/copper with polymer	95	0.97	0.03
Polished stainless steel	80	0.50	0.50
Cermet	90	0.95	0.05

As with the solar towers, both azimuth-elevation and polar tracking of the mirror assembly have been realized for the dish systems, although on large systems the former has a preference that it can be rolled.

There are also a number of concepts of mirror facet shapes that have found their way into experimental implementation, ranging from flat mirrors, spherical mirrors, parabolic mirror segments, to cylinder mirrors, conical constructions to elliptical mirrors; an overview about realized mirror shapes and solar tracking types can be found in Hafez et al. (2017).

Reflector Material

The optical power incident on the receiver aperture ($\dot{q}_{rec}$) surface is given by the product of the concentrator reflectivity (ρ), the concentrator geometric concentration ratio (C), and the direct normal irradiance (DNI) in (W/m^2)

$$\dot{q}_{rec} = \rho \cdot C \cdot DNI \tag{7.49}$$

As can be seen from Eq. (7.49), in addition to the shape of the mirror, the reflectivity of the mirror material is of particular importance for achieving high power densities per unit area on the receiver. In principle, any material with good reflectivity is suitable, but in dish systems, mirrors with silver ($\rho = 0.95$) or aluminum ($\rho = 0.9$) as reflector material are in the foreground of applications due to their high reflectivity, especially when the receiver system is coupled with a heat engine.

However, in addition to applications for electricity generation, which require the highest possible receiver temperatures, dish systems are also being used as a cost-effective alternative for the decentralized provision of process heat, and so there are several demonstrators in which reflective polymer films, silver/aluminum with polymer films, polished stainless steel and cermet have been used as reflector material. It should be noted that reduced reflectivity automatically results in heating of the reflector, which should be considered in the appropriate combination of materials chosen. Table 7.9 shows the reflectivity or emissivity of various materials used in dish reflectors.

Also, the optical collector losses in dish systems result from geometric errors, the limited reflectivity of the mirror material and the radiation distribution of the sun. Similar as with solar towers, the geometric errors can be divided into several subtypes. These are composed of the relative optical errors of beam quality (σ_{bq}), astigmatism (σ_{ast}) and tracking (σ_{trac}) and the sun shape error σ_{sun} (typically assumed to $\sigma_{sun} \approx 2.5$ mrad). Theσ_{tot} total optical error (σ_{tot}) is then given by

$$\sigma_{tot} = \sqrt{S_0^2 \cdot \left(\sigma_{sun}^2 + \sigma_{bq}^2 + \sigma_{ast}^2 + \left(2 \cdot \sigma_{trac}^2\right)\right)}, \tag{7.50}$$

where S_0 is the actual distance from the dish to the focus point. In contrast to solar towers the atmospheric attenuation plays hardly any role due to the small distance between reflector and receiver, which is only a few meters.

In the beam quality factor (σ_{bq}), deviations of the mirror curvature from the ideal shape and imperfections of the reflecting surface due to waviness and roughness from manufacturing and assembly are taken into account. These are called slope errors (σ_{slope}), compare Sect. 7.2.2.3.2. Statistically, they are assumed to be uniformly distributed and described by a circular normal distribution. The slope error σ_{slope} is defined with respect to the surface normal vector; therefore, its effect is doubled in the reflected beam.

$$\sigma_{bq}^2 = \left(2 \cdot \sigma_{slope}\right)^2. \tag{7.51}$$

Receiver and Heat Engine

Conventional dish receivers for power generation are based on the use of gaseous heat transfer fluids with Stirling machines that use either hydrogen, helium or nitrogen, or Joule-Brayton based miniature gas turbines that use air as a heat carrier. The receiver is responsible for absorbing the reflected and concentrated solar radiation and transferring the heat to a heat-carrier fluid. A detailed description of concepts used in the design of receivers for Dish/Stirling Engine systems is found in Stine and Diver (1994).

A very advanced concept is the AMTEC receiver where sodium is vaporized at low pressure in the receiver and the vapor is passed through a solid electrolyte through an AMTEC assembly, compare Fig. 7.71, thereby generating electricity directly, whereby the waste heat (>600 °C) is subsequently converted further in a thermal engine. These concepts require dedicated design measures for which details may be taken from Wua, Xiao, Cao, and Li (2010) or Tanaka (2010).

The receivers are usually designed as a cavity receiver with a small opening through which the concentrated sunlight enters. The geometry of the receiver opening serves to limit the radiation emission and convection losses, with the optimum depth being one to 1.5 times the diameter of the aperture. Various methods can be used to transmit the absorbed radiation to the fluids. It is also possible to use secondary concentrators at the receiver opening to increase the intercept without increasing the size of the receiver aperture, or to reduce the opening diameter for a specific intercept. These are then trumpet-shaped in relation to the surface and

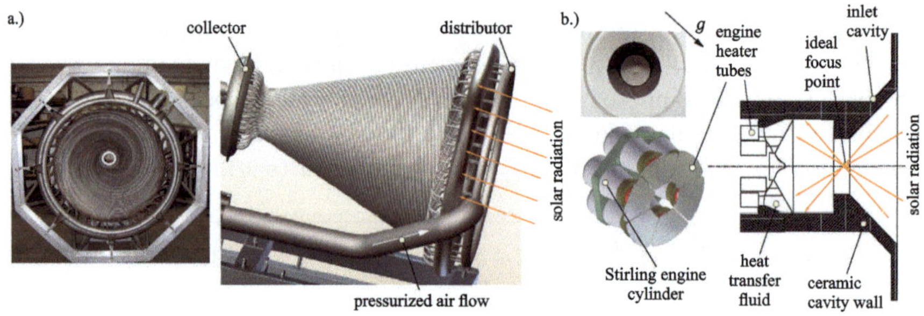

Fig. 7.58 **a** Front and side view of a direct illuminated pressurized air receiver of a gas turbine from BraytonEnergy LLC (2022), **b** cross-sectional cut to the principle of a indirect illumination cavity receiver type (left top-front view, left bottom connection to Stirling engine)

transmit the reflections from a large area through the cavity receiver opening similar to those used in solar tower concepts. The result is an increase in the intercept without the need to increase the receiver opening. A secondary concentrator generally improves the performance of a parabolic receiver and may reduce the negative impact of one or all components of the optical fault, but a secondary concentrator increases the cost of the collector, since the reflection coefficient for dish systems must be very high due to the high concentration ratio and therefore requires a well-designed cooling system. In the construction of the receivers themselves there are two different concepts:

- In the case of directly illuminated tube receivers, small parallel arranged tubes are used, through which the working fluid enters directly into the area of the concentrated solar radiation. The tube surfaces form the absorption surface. To increase the residency time in the solar radiation zone, especially for gas-cooled receivers, these are then pitched, as depicted in Fig. 7.58a for a pressurized air receiver of a gas turbine.
- Other receivers use an indirect illumination concept with a focus at the narrowest point of the cavity, cf. Fig. 7.58b. At the lower end, on a cone, is the absorber behind which the heat transfer fluid is arranged. There are two sub-species; one option transfers the heat directly to the plates of a Stirling engine, or a heat-pipe concept is used, in which a liquid metal absorber surface evaporates and condenses at the heat plates of the Stirling engine (reflux receiver).

7.3.2 Solar Towers—Central Receiver Systems

A concentrating solar tower system (STS), also called a central receiver system (CRS), essentially consists of three core components.

- A solar tower with a receiver installed at a certain height. If the conversion of the thermal receiver power does not take place in the tower itself, it contains the lines for the coolant supply and removal from the receiver.
- The receiver itself, in which the optical conversion of the solar radiation into heat takes place and which is dissipated with the help of the coolant. The concept and design of the receiver essentially determine the arrangement of the heliostat field and the height of the tower.
- The mirror field, which is decoupled from the coolant system, consists of heliostats that track the position of the sun on two axes, in which the solar radiation is focused on the receiver with the help of target point strategies based on measurement data from sensors in the solar field.

All three components must be precisely coordinated to enable a cost-optimal, competitive solar tower system and are discussed in detail in the following sections. The core component, which determines the arrangement of the heliostats around the tower, is the receiver mounted on the tower, which can be up to 250 m in height, and there are several designs. At present, it is clear that no receiver technology is economically superior to any other. Heliostat design, receiver type, material choices and control issues are intimately coupled, and will be discussed in detail in the following sections. But, first a short overview shall be given to demonstrate the complexity of the overall design.

Like the parabolic dish a solar tower is working with two-dimensional concentration on a focal point, not a line. The solar tower with the heliostat field is the corresponding equivalent of the linear Fresnel collector. Also, here the ideal paraboloidal shape is broken up according to the Fresnel principle into individual facets. These mirror facets are movable in two angular directions with their support structure and called heliostats.

7.3.2.1 Receiver Configuration Options

The receiver usually is the starting point in the design consideration. The heliostat components, control and field arrangement are very much depending on the chosen receiver which in turn is connected to the choice of heat transfer fluid used in the collector. The concentrating collector as defined here therefore consist of one or multiple receivers mounted on towers and connected to a heat transfer fluid loop transporting the energy to a power block or to a storage system, plus a suitable heliostat field for the receiver(s). The two fundamental design options of receivers are

- panel receivers mounted around a tower viewing the whole landscape around the tower (360°–receiver, Fig. 7.59a), a sub class is formed by external star (Fig. 7.59b) or rectangular receivers,
- cavity receivers with a limited view factor, to which simple cavity receivers as shown in Fig. 7.59c, but also face down receivers (Fig. 7.59d) and beam-down receivers belong (Fig. 7.59e).

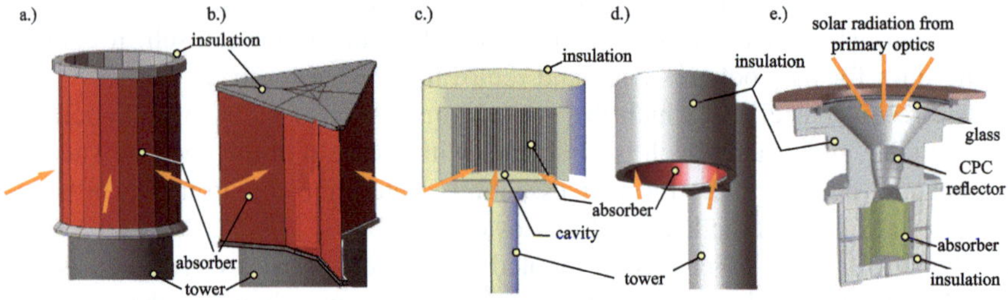

Fig. 7.59 Different receiver concepts of solar towers; external receivers with 360° solar field using external tubular receivers (**a**) or star receiver (**b**) and cavity receiver types sinple cavity (**c**), face down receiver type (**d**) and beam down receiver with secondary optics (**e**)

A further option is the use of a secondary optics which redirects the rays from the heliostat field to the ground. Such a concept referred to as beam down concept, allows the receiver to stand on the ground, open at the top, and eliminates the need for coolant routing to the turret, but at the expense of secondary optics with optical losses. Figure 7.59 schematically illustrates the different receiver type options, which will be discussed in more detail in Sect. 7.3.2.3.

The receivers with a full 360° view factor utilize a heliostat field distributed around the tower, which is more or less approximating a circular arrangement, shown in Fig. 7.60a. Depending on latitude and other factors like weather characteristics, intended load curve, available space etc. the circular form might be distorted and also the tower might not be in the center of gravity of the field. For higher latitude a larger fraction of the heliostats might be favorable on the side of the tower opposite the dominant sun position (in the Northern hemisphere on the North of the tower). Cavity receivers with a limited view factor use and receive only reflected radiation from heliostats within this view angles, which is depicted in Fig. 7.60b. Therefore, the heliostat field is asymmetric positioned on one side of the tower.

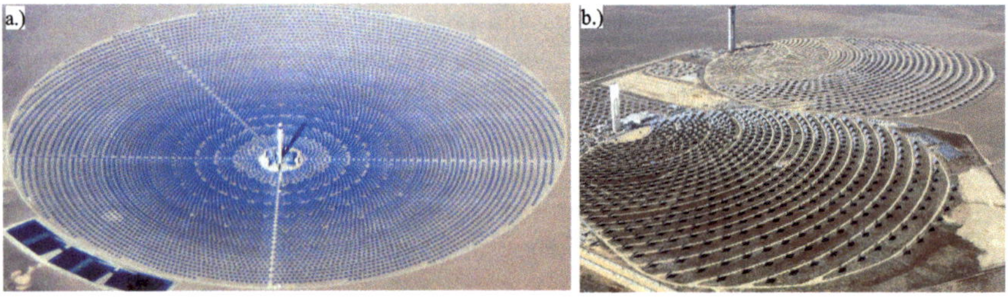

Fig. 7.60 Circular heliostat field with 360° receiver (left) from Crescent Dunes and (**a**) asymmetric heliostat field with cavity receiver at PS10/20 (right) (**b**) modified from (commons.wikimedia.org)

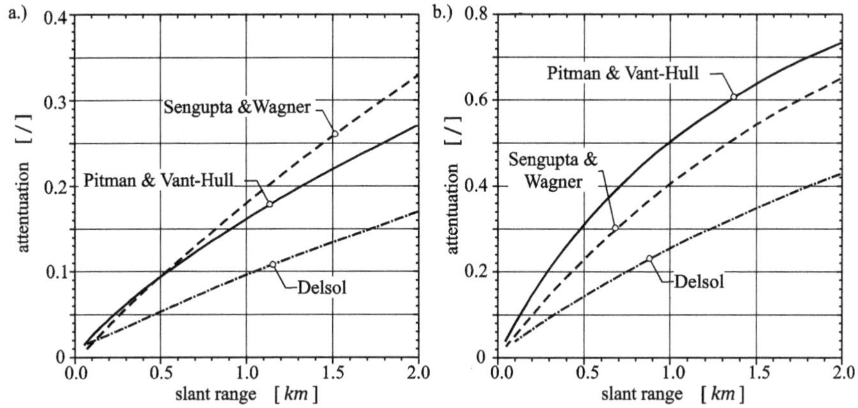

Fig. 7.61 Calculated atmospheric attenuation from the heliostat to the receiver over the slant range distance for different models. **a** clear day, **b** hazy day from Cardemil et al. (2014) for a 100 m high tower, Bom Jesus da Lapa/Bahia Brazil ($\Phi = 13.25$ *S*) *DNI* level (990 W/m^2) and water vapor concentration ($v = 11.2$ g/m^3)

The distance of the individual heliostats from the receiver varies very much. Heliostats very close to the tower have a short overall distance, but the incidence angle for the receiver is larger than for heliostats further away. For a cavity receiver therefore, it might be advantageous to tilt it towards the ground, which improves the optical acceptance from near-by heliostats. Depending on the haze or dust in the atmosphere the beam between heliostats at the outer rim of the field (say 1–2 km away) can be attenuated appreciably. An example of attenuation depending on visibility is shown in Fig. 7.61.

Consequently, the heliostats too close and too far away contribute less to the overall thermal power in the receiver. It is important that the coverage of heliostats on the ground is optimized to reduce required area. However, several factors must be considered and respected:

- Heliostat from a row in front of the considered heliostat should not block reflected beam radiation—the heliostat needs a free view to the receiver at least most of the time.
- Heliostats in the direction of the sun should not shade the considered heliostat—a condition which cannot be fully fulfilled especially for low sun altitudes.
- In no circumstance should heliostats be able to mechanically touch each other (in some state of rotation).
- Heliostats must be cleaned and maintained—therefore some maintenance roads have to be planned inside the fields e.g. for the vehicles for spray-washing.

A very common approach to avoid most of blocking and shading is the staggered arrangement of heliostats: The heliostat behind a row of heliostats is always put

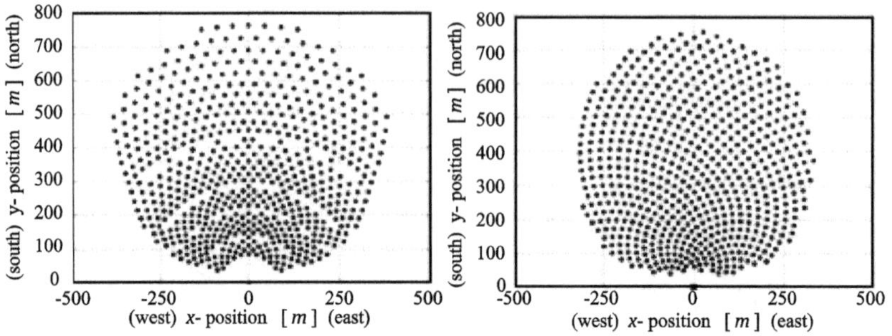

Fig. 7.62 Standard and optimized heliostat field layout, for concepts of optimization see e.g. Schöttll et al. (2019) or details in Chap. 10

into the gap between two heliostats in front. It is the same arrangement as singers in a choir have when the want to see the conductor!

The consequence is that in a row at a larger radius the distances between heliostats become larger and larger. This is often compensated by having this staggered arrangement not over the whole field but only in certain areas where positioning starts anew in each sector. In Fig. 7.62 a standard field layout and an optimized on is shown. The difference in annual optical efficiency is less than 0.6% but this may make a difference in economy.

The selection of the heliostats for a specific receiver and a specific position in a heliostat field has already several challenges:

- selection of heliostat size and shape (from flat to facetted to curved) suitable to receiver aperture,
- selection of tracking axes and control,
- heliostat support structure and stability,
- positioning of the heliostats in a field (minimization of mutual blocking and shading) depending on tower position and height, and,
- concentration factor using the optimization of heliostat area (or number).

The individual heliostats themselves have very different sizes. The reflective area of a heliostat ranges from 1 to 200 m^2. The ideas behind that are very well summarized in the review by Blackmon (2020) which also provides an overview over existing heliostat designs. There are three categories of costs for heliostats:

- Cost proportional to area (reflector material),
- size dependent costs (drive units, pedestals, motors etc.),
- fixed costs for each heliostat (sensors, encoders, limit switches etc.).

Qualitatively, it is obvious that many small heliostats need a lot of work for calibration they use many controllers, sensors, drives and motors. Also, one needs

Fig. 7.63 **a** eSolar heliostats (2 m^2) in a fixed array at the Sierra SunTower installation from Ricklin et al. (2014); **b** Segmented large scaled mirrors (118 m^2) from Crecent Dunes solar thermal facility with heliostats modified from National Renewable Energy Laboratory-NREL (2017)

more support structures, and it is important many steps in installation, maintenance and operation are automatized. Large heliostats on the contrary reduce these efforts, but they must be stable against large wind loads, the structures become quite massive, and the shape of the reflectors must be adapted whereas for very small heliostats even flat mirrors are adequate. So, simplified it is a consideration of few high-quality heliostats against a large number of mass-produced heliostats. It depends on materials and components selected, on the receiver size, on the installation and calibration routines, on the efficiency needed due to the size of the heliostat fields which approach small versus large heliostats is more cost-effective. However, using one technical solution shows also that a size optimization led to appreciable cost reductions of 30–40%. It seems possible that with reasonable assumptions heliostats might be produced well below 100USD/m^2. The Fig. 7.63 give an impression on small and large scaled heliostats. The heliostats are discussed in more detail in Sect. 7.3.2.5.

Traditionally, the heliostat field is being placed in a flat area, where only minimal ground height variations are existing. However, in principle heliostats can also be placed in a three-dimensional hilly landscape. Placing the heliostat on a slope facing the dominant sun position can even be used to improve the optical efficiency.

The optical design requires a complex determination of the optical path via a heliostat to the receiver to achieve an efficient, highly concentrating optical system. At the same time, the concentrating solar radiation hitting the receiver during optical conversion generates temperatures that reach the limits of the currently available structural and functional materials.

To ensure efficient and, above all, sufficiently quantitative predictive control of the heliostats and thus power generation in the receiver at all times, both the control and regulation of the thermal and the widely spatially distributed optical system (heliostats) with the sensors must be optimal and be coordinated at any time.

This requires many optimizations that are difficult to access analytically, which is why the development of solar towers in the past was essentially limited to experimental demonstrators. Only the massive increase in the computing power available, supplemented by innovative, high-precision production processes, advances in structural and functional material development, wireless signal transmission and refined modeling approaches make it possible today to predict and design efficient solar towers and their operational behavior. This computational design architecture has made the development of large solar towers so attractive that the growth rate in the construction of new solar tower systems, often referred to as solar power towers (SPT), has exceeded that of linear concentrator systems over the past decade.

For solar tower collectors many different receiver types and corresponding heat transfer media already have been developed and used in prototypes. The most common commercial heat transfer media are direct steam, where the steam is produced directly in the receiver tubes at a high pressure with temperature up to 480 °C, and molten salt (so-called Solar Salt, typically a 40 : 60 wt% mixture of potassium and sodium nitrate) where temperature up to 565 °C are possible. Fluids are usually circulated from and to the receiver consisting of tubes coated black. The standard here is a non-selective highly temperature-stable paint. To reach higher temperatures, the materials for the receiver and the heat transfer media must be changed. Other salt mixtures (nitrates, carbonates or even chlorides) are being investigated. The problem of carbonates and chlorides is the higher corrosivity of these hot salts. Liquid metals like NaK are also interesting candidates. It seems to be important that the fluid material is chemically stable, also against oxidation because of the fire hazard of leaks, and can be bought at reasonable cost. It is key that materials cannot harm the environment when leaking. A completely different approach is to use air as a heat carrier. The advantage of air is that it is non-corrosive, poses no environmental hazard and it has virtually no cost. It is stable even for temperatures up to 1500 °C. The disadvantage is the low heat capacity or density. The latter one can be somewhat improved when compressed air is used. But still large air channels and in operation a high electrical consumption due to ventilators are needed. In Chap. 11 the interlinks and interdependency between coolant choice and power plant concept are analyzed in more depth.

7.3.2.2 Receiver Design

There are different options described in literature to categorize different receiver types. A uniform way of classifying receiver concepts, however, is challenging, because the receiver design with the associated selection of functional materials and heat transfer fluid (HTF) routing within the receiver is irrevocably linked to the size and aperture angle of the receiver aperture and the heat dissipation capability of the selected heat transfer fluid, and thus determines the size and shape of the heliostat array. For example, from an optical point of view, it is easy to build a heliostat field that produces a peak concentration of more than 2 MW/m^2 on a receiver, but only a few receivers are able to dissipate more than 1–2 MW/m^2, such as liquid metals or molten salts, even with optimal coolant configurations, due to their heat transfer properties and capabilities.

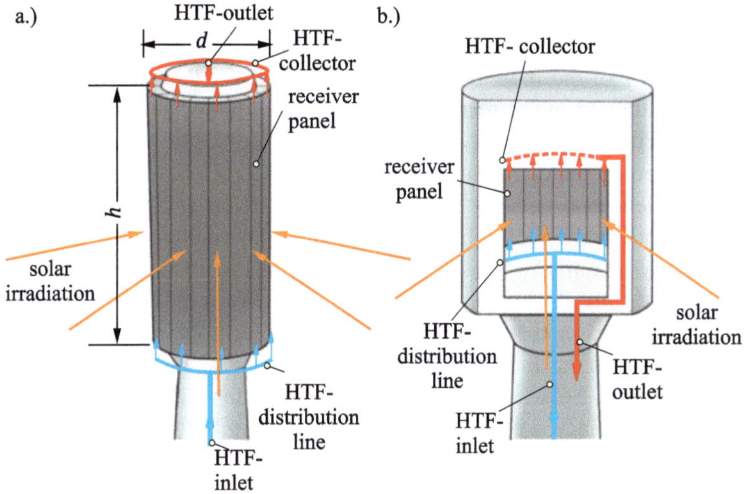

Fig. 7.64 Schematic diagram of an external **a** and a cavity **b** receiver concept used in solar tower installations

Principal Design Options

A basic distinction of receiver types is the question whether it is an external receiver, as outlined in Fig. 7.64a, in which the concentrated solar radiation is captured on the circumference of the tower, or a cavity receiver, in which the receiver is located in a cavity within the tower, as shown in Fig. 7.64b. In the case of the latter, there are further options: secondary concentrator optics can enlarge the aperture for a smaller cavity opening, beam-down or face-down cavities as shown in Fig. 7.59c–e are used as well. The orientation of the cavity can be optimized in terms of view factor and thermal losses.

It must be emphasized that the receiver design decisively depends on the heat transfer fluid (or coolant) used in the receiver, which determines both the maximum possible fluid temperature and the usable temperature difference between inlet and outlet. A holistic receiver development strategy provides a multi-dimensional optimization problem comprising the interaction of

- constraints on functional material properties for optical conversion,
- permissible heat flux density on the receiver design,
- coolant flow configuration,
- radiation and convection heat losses of the receiver,
- heliostat field design.

Therefore, before individual receiver designs are presented, various aspects in the development of receivers are discussed in a general way. This then also explains why there are many efficient solutions for central receiver types.

Constraints on Functional Material Properties for Optical Conversion
The design of the receiver directly affects via the attainable temperature the thermal efficiency of the individual receiver concept. On the other hand, the temperature level attained in the receiver poses constraints not only to the functional materials (absorptivity, emissivity, reflectivity) to achieve a high optical conversion but also to structural materials in terms of corrosion, erosion, and thermal stresses. Stress is experienced especially in transients of start-up and shut down. Central receiver concepts strive for high coolant outlet temperatures, therefore selective absorbers with low emissivity (ε) to limit radiation losses at temperatures above 500 °C are a technical challenge. The critical wavelength for the step-change from low reflectivity in the solar range to high reflectivity in the thermal range is tricky because of the increasing overlap of solar and thermal spectrum with higher temperatures. Solar absorption is more important than emissivity. The aim of the receiver is to increase the net energy captured in the receiver. For example, for a solar flux density of 1 MW/m^2 a reduction in the receiver absorptivity α by 0.02 translates in a decrease of heat gain of 20 kW/m^2. To compensate that with a reduction in radiative heat loss for an assumed surface temperature of 800 °K a reduced emissivity ε from 1.0 to 0.12 is required. This is hardly possible. A typical compromise for the critical wavelength should be between 1.5 and 2 μm. Other challenges are the operating conditions. Especially in atmospheric conditions with high oxygen content, corrosion can hardly be avoided. Receiver pipes undergoing a structural phase transition below the absorber temperature increase thermal stresses on the coatings.

Impact on Heliostat Field Design
A limited receiver view factor, e.g. of a billboard or cavity receiver rather than a cylindrical external receiver, may enhance the power density in MW/m^2, While a cylinder has a 360° field of view to the environment, a billboard type is limited to a 180° field of view, often even less, although a downward sloping front aperture can bring a larger ground area into the field of view. Using a polar aligned billboard design with a 120° field of view to the heliostats, high performance can be achieved around noon, as most heliostats have an angle of incidence of sunlight close to 0° ("cosine effect" is marginal). However, if the azimuth of the sun changes (at other times of the day), the optical power on the target aperture decreases rapidly. Conversely, with an external cylindrical receiver, only a moderate power is obtained around noon, but when the azimuth of the sun changes from its noon value, which can be more than 100° in summer, it results in only small changes in optical efficiency. This is because the nearly circular nature of the surrounding heliostat field, approximately maintains the average value of the cosine over the entire field. The effects of determining the receiver concept therefore directly involve a restriction in the design and shape of the heliostat array. The latitude of the tower location as well as the target power has an impact on the decision, whether a billboard, a cavity or an external cylindrical 360°-receiver is being used.

If, however, the selected working fluid allows high receiver outlet temperatures or low heat transfer coefficients (e.g. gases), a cavity concept is preferred, since the thermal losses with cavity concepts are considerably lower than those of an external receiver directly facing the environment.

One of the most important advantages of central receivers is the relatively short pipe lengths from the receiver to storage and demand site. Decentralized tower concepts, which are currently also considered, partially eliminate this advantage.

Effect of the Permissible Heat Flux Density on the Receiver Design

Most commercial receivers only allow surface power densities in the order of up to 1.0 MW/m^2 (e.g. for molten salt) with a certain safety margin from a heat transfer point of view, so that larger receivers are used than the minimum size of the optically achievable focal spot area. The ratio of height to diameter of an external receiver, often referred to as an aspect ratio, should be larger than unity, so that heliostats in the near range can be effective, since they see optically only a substantially shortened receiver. Since the target a certain outlet temperature, a minimum pipe length is required for a given flux density distribution. This usually leads to an aspect ratio between 1 and 2. The permissible solar flux is then achieved by defining several heliostat target planes (two to six) on the receiver, to each of which specific parts of the heliostat field are assigned. An intelligent strategy that estimates the radius of the beam and shifts the image of the heliostats associated with the outer target planes so that their beam remains just on the receiver results in a very low spillage, while the peak flux is significantly reduced (by a factor of 2–4, depending on the optimization strategy). A second requirement may be that the ratio between the north and south side of the flux density be limited to ensure a solar flux which is as uniform as possible on the receiver surfaces to minimize circumferential flux gradients and thus thermal stresses.

Coolant Flow Configuration

Depending on heat transfer fluids there are both open and closed receiver concepts.

- Open concepts are possible with the heat transfer fluid air. Ambient air is heated by sucking it through a fixed structure where the heat is generated by solar absorption. The other option is direct optical conversion on particles which are carried by an air flow are freely falling in the gravitational field. Open concepts have the advantage that there are no heat transfer losses due to heat conduction via walls or other structures. Conversely, they are susceptible to convective/radiative heat losses through wind.
- In closed receiver concepts, the optical conversion is realized pipe receiver panels. The generated heat flux is transported conductively through the pipe walls, and then transferred by forced convection to a fluid. Closed concepts inevitably imply a pressure-resistant pipe network as well as corresponding conveying machines (e.g. pumps, compressors, ….). In contrast to open concepts, closed arrangements allow both large coolant mass flow densities and a significantly larger heat transfer density to be achieved than in open arrangements. This

enables very compact receiver dimensions, which also mean low heat losses to the ambient.

There are also several experimentally validated concepts with respect to flow guidance. The flow guidance within receivers plays an important role, as it influences both the heat loss and the thermomechanical stress distribution, resulting from the temperature distribution, especially during transients. All receiver concepts presented in the remainder of this chapter aim at a homogeneous temperature distribution to keep the thermo-mechanical stress levels low. To achieve the highest possible coolant outlet temperature from the receiver over the longest possible period of the day, several options are possible for flow control. The simplest solution is the simple upward flow through a tube or a plate and the adjustment of the flow in such a way that the required coolant outlet temperature is reached at the upper end of the tube, as illustrated in Fig. 7.65a. Often, however, the vertical extension of the receiver is not sufficient to achieve the desired temperature lift along its length. An increase in the residence time then can be achieved with a meandering flow path, as shown in Fig. 7.65b by splitting a tube array into segments (top) through which flow passes in succession or a plate design with an integrated meander flow channel (bottom). If even longer residence times are required, the concept of a quasi-volumetric receiver can be used, in which cooling channels are positioned normal to the solar irradiation and flow is alternated several times to achieve the desired target outlet temperature. The concept of an open volumetric air receiver and a quasi-volumetric closed receiver is shown in Fig. 7.65c and d.

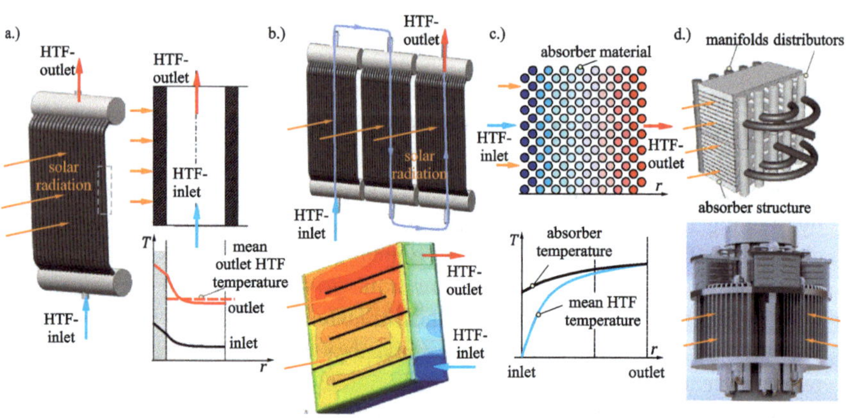

Fig. 7.65 **a** Billboard receiver and its typical HTF temperature evolution along a single tube (right); **b** Meander flow receiver to enlarge HTF residence type (top) subsequent bundle array type (bottom) monolithic billboard type receiver (for sodium/salt as HTF); **c** volumetric open air receiver; **d** closed gas/direct steam generator volumetric receiver with indirect cooling modified from Kalbhor (2014)

The basic form of the flow channels is either tube bundles or plate elements. Both designs allow a modular construction and are therefore suitable for both external and cavity receiver designs. The general structure of the tubular receiver consists of parallel tubes arranged in rows, whose outer surfaces are heated by the concentrated solar rays. A tubular receiver is simple, easy to build, inexpensive and easy to monitor and maintain. The tubular design allows for parallel but also staggered tube arrangements. Another key advantage is that heat transfer relationships for tubes are reliably known and available with high accuracy for both single-phase and two-phase flows at many Prandtl numbers (i.e. quasi for arbitrary fluids) regardless of their orientation, which substantially facilitates the predictive performance determination of the receiver.

Monolithic plate concepts are becoming increasingly important, as they allow a high degree of receiver coverage and thus have large heat flux densities and at the same time smaller temperature gradients than tube concepts. Due to their construction, they have a lower compressive strength than tube concepts; but additive manufacturing processes, especially in low-pressure receiver concepts with liquid salts or liquid metals as heat carriers, allow fluid-adapted high surface utilization rates at extremely high surface power densities.

Radiation and Heat Losses of the Receiver

In the case of the cavity receivers, the radiation and convection heat losses are lower in comparison with external receiver designs with a surrounding heliostat field. They therefore usually show a greater annual optical efficiency but require higher towers and greater tracking accuracy in order to "see" all heliostats in a northern (or southern) heliostat field in comparison to a surrounding field with the same given power. However, also the hot surfaces of a cavity receiver emit thermal radiation via the aperture. The internal volume of the cavity is filled with air, heated by the internal temperatures at the absorbing surfaces. with the gradient to the low temperatures at the aperture entrance can create turbulent buoyant flows that can carry large convective heat fluxes to leave the cavity. Depending on the aperture opening and the wind conditions prevailing at the aperture, the area specific heat dissipation from the cavity can even exceed that of external receiver concepts.

Cavity receivers are particularly suitable for systems with very high temperatures, where a small ratio between aperture and absorber area is advantageous and high fluxes can be absorbed. However, a cavity must both support and enclose the heat-absorbing surface. It is clearly larger than the support structure required for an external receiver with the same absorbed flux. It must also be well thermally insulated to reduce heat conduction losses through the large surface area. Cavity receivers therefore require large, heavy, and expensive enclosures, such that low-cost steel frame constructions as towers are usually ruled out. In addition, cavity receivers require a higher tower due to the limited cone angle.

7.3.2.3 Receiver Types

In central receivers, liquids (e.g. water, molten salt, liquid metals), gases (e.g. air,$s-CO_2$, helium) and solid media (e.g. ceramic particles) are used as heat transfer and/or storage media. Almost all fluid receivers use plate segments (billboard, staggered, monolithic) or tube bundle arrangements which have a (sometimes) selective coating on the side receiving the solar radiation and are cooled in the core by the heat transfer fluid (HTF). The tube fields can be contained in a cavity receiver or be arranged in a cylindrical or cubic configuration.

Due to the variety of commercially realized concepts but also in prototypes already proven receiver concepts, here the receiver types are divided according to the HTF type in

- gas-cooled receiver,
- liquid cooled receiver and
- particle receiver,

for which only the functional principle is explained as well as typical temperatures, pressures and thermal efficiencies are addressed in their order of magnitude. This also includes a presentation of the advantages but also challenges, otherwise reference is made to the secondary literature listed in the mentioned sources. An overview can be found in Ho (2017) or Coventry et al. (2015).

Gas-Cooled Receivers

Gas-cooled receivers allow intrinsically high receiver temperatures without any phase change of the fluid and, when using inert environmentally friendly gases, low costs for the heat transfer fluids and marginal environmental requirements for the construction of a plant. Conversely, due to their low thermal conductivity and low heat capacity, gases require high flow rates to realize a sufficiently good heat transfer so that the receiver material as well as the pipelines remain within material sustainable limits. The low thermal inertia requires good control and regulation logic to largely eliminate flow instabilities. Thus, the demands on the structural material increase and so do its costs. Conversely, the use of pressure-charged closed systems allows direct conversion into electrical energy with the aid of the Joule-Brayton process at high thermodynamic efficiencies but requires the use of expensive high-temperature and high-pressure piping as well as greater parasitic losses due to high specific work by operation of the compressors. With the gases themselves, thermal energy storage of a possible thermal surplus thermal power from the solar field is only possible with indirect thermal energy storage concepts such as brick Cowper or rock bed storage. Three sub-families of gas-cooled receivers are known in the literature:

- volumetric air receivers,
- closed (pressure-charged) receivers,
- gas-particle mixture receivers.

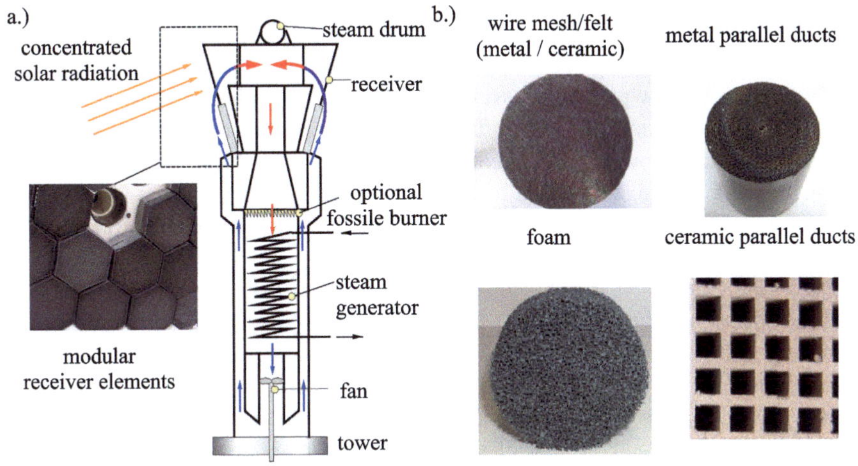

Fig. 7.66 **a** Schematic design of a solar tower with open air receiver. **b** Different designs of the absorber structures of volumetric open-air receiver

Volumetric Air Receivers

Open air receivers are a simple and safe design of a receiver system in terms of plant engineering, since even a possible failure of the fan does not immediately lead to destruction of the receiver unit. Figure 7.66a shows the schematic design of an open-air receiver system in which a steam generator is integrated into the tower, which can be supported by a fossil burner if necessary.

The receiver unit consists of a volumetric absorber that receives the solar radiation and is usually modular. The air is drawn through channels or honeycomb blocks that are irradiated by the concentrated sunlight. The air heats up as it passes through the irradiated structure, and ideally the light should penetrate deep into the receiver so that the warmest part of the receiver is inside, away from the aperture, so that the heat radiation losses to the surroundings remain low. However, most studies have shown that the opening of the volumetric receiver is the hottest part, resulting in significant radiation losses to the environment. The receiver structure consists of either a metallic or ceramic material, which is constructed either as a knitted fabric, felt, foam or ordered in parallel channel structures, as Fig. 7.66b shows. Here, the concentrated radiation penetrates the absorber material and generates heat, which has led to the term volumetric receiver. With the help of a fan, ambient air is drawn in from outside through the absorber material and the air flow absorbs the heat stored in the absorber. With open air receivers, the choice of material determines the maximum turbine inlet temperatures. With Inconel alloys up to 950 °C and with ceramic silicon carbide up to 1000 °C air temperatures have already been realised in experiments (Avila-Marin, 2011), and power densities of more than 800 kW/m^2 at receiver efficiencies of about 70% have also been demonstrated. Air as a heat transfer medium has many advantages here; it is non-toxic, not excessively corrosive, available everywhere in any quantity, and is also non-combustible and easy to handle.

One disadvantage of the open-air receiver concept is the low power density due to the fixation on low process pressure. This requires high air flow rates, which in turn only allow for smaller temperature differences. Impressive advantages should not be overlooked, however. With a modular design of the absorber unit, individual segments can be changed without lowering the pressure in the system. This allows high availability. The optimization potential of this operating concept essentially comprises two points.

- The reduction of heat losses on the front side. Analytical considerations as well as experiments show that a two-zone receiver, whose sun-facing side consists of a transparent honeycomb structure to which a ceramic structure is connected downstream, allows receiver efficiencies of more than 90%. The principle of operation corresponds to that of a transparent thermal insulation, see Pitz-Paal et al. (1991). The quartz honeycomb directs solar radiation into the receiver, so it reaches the ceramic absorber part of the receiver almost unhindered. The heat radiation losses from the ceramic receiver parts are partially absorbed in the quartz glass part and can be recovered as usable heat, since the quartz glass channels are also cooled by the gas flow. This results in low temperatures in the quartz glass structure, especially at the receiver inlet.
- The second way is to increase the air recirculation rate (the ratio between the air mass flow recirculated and the total air mass flow through the receiver). Heat recovery of the heat still present in the air after a heat exchange to the power conversion system increases the thermal efficiency of air-cooled solar thermal receivers. Typical optimal air recirculation rates are around 70%.

Other deficits that can be reduced but not eliminated by intelligent design is the soiling of the receiver structure by sand, and the heterogeneous heating of the receiver. Large temperature gradients cause strong changes in the temperature-dependent density and viscosity of the air, which lead to flow instabilities. Absorber materials with low porosity seem to eliminate this to a large extent.

Closed (Pressure-Charged) Receivers

In the case of gas-cooled pressure-charged receivers, the selective absorber, which is subject to high thermal loads, is cooled with the aid of a forced convective fluid flow. The pressure charging principle intrinsically requires a closed gas circuit in which the gas is compressed to pressures of up to several MPa with the aid of a compressor and is conveyed through corresponding cooling openings in the optically selective part of the receiver in which the optical conversion of the concentrated solar radiation takes place. As a result, heat is removed from the receiver unit either directly to the power conversion system or to a secondary circuit by means of a heat exchanger. There are several different concepts here.

Due to the higher pressure in the gas line network, higher power densities can be dissipated in the receiver and the achievable heat transfer coefficients can be higher than those of an open system by a factor of 100 ($h > 10^4$ W/(m^{2}K)) and even reach values of liquid-operated receivers. The closed (pressurized) gas receiver faces the solar field and all receivers tested successfully to date are of the cavity type, in which the concentrated solar radiation generated by the heliostat field is guided into the cavity via a secondary receiver. In the case of the absorber cavity, either the absorption can take place directly on the cooling channel, which can be either conical, cylindrical, or hemispherical. Some concepts increase the heat transfer surface, for example, by integrating silicon carbide foams, which additionally also increases the residence time of the heat transfer fluid in the receiver. Conversely, this measure shown in Fig. 7.67a means an increase in the pressure loss. Other flow guides with direct conversion on the cooling channels either place on a pipe bundle (Fig. 7.67b) or on the circumferential flow guide (Fig. 7.67c). A disadvantage of the direct conversion at the cooling channel is that the pressure-carrying tubes are exposed to extreme temperatures which correspondingly presuppose the use of high-temperature-certified, pressure-resistant materials. At the same time, the high temperatures lead to continuous growth of an oxide layer on the selective absorber material, which has a low thermal conductivity. In indirect, pressure-charged receiver concepts, this limitation is circumvented by the use of a quartz glass window, and the conversion can take place either on porous (Fig. 7.67d) or on cushion-shaped absorbers (Fig. 7.67e), which are surrounded by gas on both sides by means of cooling slots and thus do not require any strength functions. Figure 7.67f shows a further decoupling of the absorption with a simultaneously large heat transfer surface, in which a porcupine structure assumes the absorption function. Only a few demonstration projects rely on a purely solar application, the power spectrum of which is limited to the order of magnitude of a few MW, while the concepts aimed at a commercial application, some of which are listed in Table 7.10, are aimed at solar-fossil hybrids in order to substantially reduce the CO_2 footprint of fossil power plants in a sustainable manner.

One of the key challenges of a pressurized gas-cooled receiver is its low thermal inertia, which intrinsically leads to high lateral and transverse temperature gradients associated with correspondingly high thermo-mechanical stresses, and the optical loss and cost of the transparent window opening. Therefore, the operation of gas-cooled receivers requires special attention to control and regulation to uniformly achieve the desired receiver outlet temperatures necessary to operate a gas turbine or downstream power conversion system at high efficiency. Special technical precautions must also be taken to ensure receiver integrity during irradiation power peaks or fluctuations, for example, in the event of misalignment of heliostats or compressor pressure or flow rate fluctuations. Technical options include shutters or fast-acting adjustment mechanisms of the secondary receivers, which take corrective action by means of sensors installed in the solar field and the pipe network of the coolant.

Pressure-charged receivers appear to be less suitable for large power plant capacities in purely solar operation also in the future, since in addition to the limited possibilities for thermal energy storage, the design of large-scale receivers in particular requires technically complex constructions. As a thermal energy storage

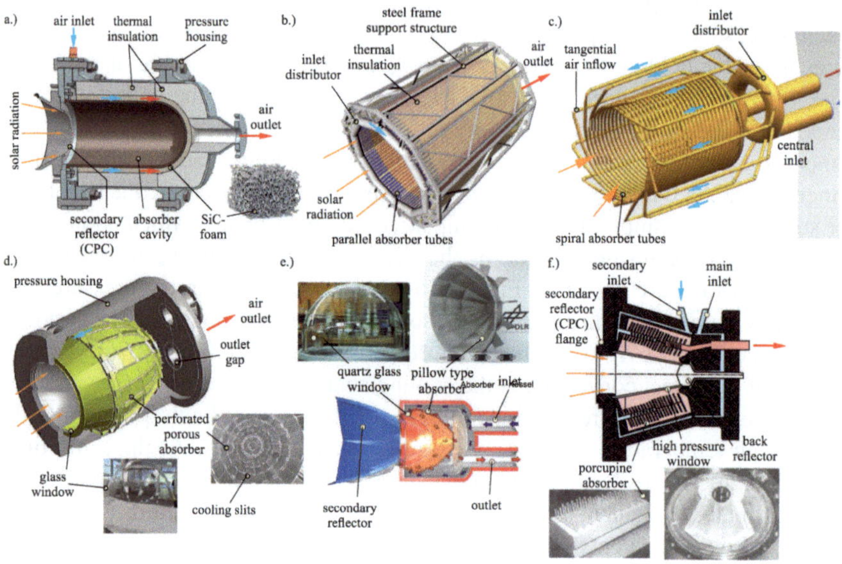

Fig. 7.67 Technical versions of pressurized air receiver; **a** Cylinder receiver with supporting foam heat transfer surface (Poživil et al., 2014); **b** Cylindrical tube bundle receiver according to Korzynietz et al. (2016); **c** tangent cooled cylinder receiver according to Wang et al. (2015); **d** indirect barrel receiver with porous absorber structure modified from del Río et al. (2015); **e** indirect hemispherical receiver with rear-cooled cushion absorber (Deutsches Zentrum für Luft- und Raumfahrt (DLR), 2008); **f** indirect receiver with porcupine-like absorber plates modified from Karni et al. (1997)

system for pressure-charged receivers, the only direct storage option is a particle bed, in which gas solids are heated and, when heat is required, this heat is fed back to the gas compare (Stein & Buck, 2017), which is used in particular during start-up to quickly reach the operating temperature. However, this requires a second compressor.

Another consideration is the choice of structural materials, because in pressurized receivers, a large part of the piping network must withstand temperatures of more than 550 °C, so the use of stainless steel is largely ruled out. In the demonstrators realized so far with temperatures up to and above 1000 °C, nickel alloys (Inconel), aluminum silicates or silicon carbide (SiSiC) are used (Avila-Marin, 2011).

A challenge for all receivers operated with air is the oxidation of metallic structural materials and as an alternative to air as a heat transfer medium, nitrogen and especially helium are suitable for closed Joule-Brayton circuits, mainly due to their inertness; for corresponding concepts see Forsberg et al. (2007), Dunham and Iverson (2014) and the literature listed therein. Although the choice of the cooling medium due to its thermo-physical properties is essential in the design of the cooling channel geometries and dimensions of the receiver, unless material limits are exceeded, the choice of the HTF is more of an argument of downstream electrical

Table 7.10 Characteristic data of some pressure charged gas cooled receiver design experimentally studied according to the design illustrated in figures

Technology orientation	Low pressure PSI	SOLUGAS	SOLGATE	PEGASE	DIAPR
Max. heat flux density (MW/m^2)	4.36	0.7–1	0.4	1	–4
Design type	Cavity	Cavity	Cavity	Cavity	Cavity
Inner cavity shape	Cylindrical with foam	Cylindrical	Conical	Cylindrical	Conical
Absorbing surface	Plane surface	Annular tubes bundle	Screen behind glass	Spiral cylindrical tube	Porcupine absorber behind glass
Peak surface temperature (°C)	900	1000	1140	1100	1200
Fluid	Air (He)	Air	Air	Air	Air
Max. fluid temperature (°C)	553 (619)	800	950	980	1000
Pressure (MPa)	0.4 (0.6)	0.9	0.65	0.4	2
Thermal efficiency (/)	0.77 (0.78)	Up to 0.81	0.79	~0.7	0.8
Conversion type	Solar only	Solar-fossil	Solar-fossil	Solar-fossil	Solar only

power conversion, so this aspect will be discussed in more detail in Chap. 11 in the context of the Joule-Brayton cycle for the different gases. In this context, the prejudice that power densities of more than 1 MW/m^2 cannot be reliably cooled with the aid of a gas-cooled wall should also be dispelled. Helium-cooled divertors have shown in pulse tests that, with appropriate design, power densities of more than 1 MW/m^2 can be removed even with large gradients and in fast transients, see You et al. (2018). Nevertheless, a dedicated mode of operation is required, which is difficult to achieve in the free field of a solar tower.

Small Particle Gas Receivers

This is a subgroup of pressure-charged receivers in which microparticles are used that are present as a suspension in the carrier gas. The solar radiation concentrated over the transparent front screen leads to a volumetric absorption at the particles and thus to a uniform heating in the fluid volume in contrast to the purely gaseous variants, which are characterized by a large temperature gradient of the wall and bulk of the fluid. These types of receivers have so far been tested more in an academic environment, as there are several challenges to overcome. For uniform mixing, it must be ensured that the fluidized gas-particle bed is homogeneously mixed at all flow rates so that there is a constant particle concentration and thus a constant volumetric optical conversion. At the same time, the design must be such that no recirculation zones can occur where particles agglomerate and change the flow configuration and thus the heat dissipation. In addition, a suitable window configuration must be found in such a way that the window transparency is maintained during the lifetime even under changing pressures.

Liquid Cooled Receiver

Liquid cooled receivers are probably the most common type of high temperature receivers in solar towers. Liquid heat transfer media have the advantage over gases that they have a high density, allowing comparatively large mass flows to be realized for cooling at low velocities. Thereby the pressure losses and the temperature rise per unit length in a receiver tube element is kept low, which reduces pumping power and minimizes the thermo-mechanical stresses in the structure. Another property is their incompressibility, which shrinks the pumping work required to circulate the heat transfer fluid. Thus, the mean flow velocities required to dissipate the receiver heat is the range of a few meters per second only, whereas with gases mean flow velocities of more than 50 m/s are not uncommon. This leads not only to lower corrosion and erosion rates on the structural material walls, but also to liquid-cooled receivers being more compact (i.e. smaller) than gas-cooled receivers at the same total thermal power. Depending on the heat transfer fluid, there are different designs.

Water/Steam Receivers

Water and direct steam central receivers are probably the longest known receiver design, in which the concentrating solar radiation is converted in the receiver and the heat is transferred to the water. In principle, external receivers can be used as receivers, as for example in the Ivanpah Solar Electric Energy Generating System using billboard type panel receivers, see Fig. 7.68a, but also cavity designs of the receivers have been realized, as in the demonstrator PS10 in Seville/Spain or Khi Solar One in South Africa, see Fig. 7.68b, and c. Direct evaporation systems have the advantage of avoiding exergy losses and costs associated with heat exchangers between the receiver and the power unit. However, to achieve high collector efficiencies, and thus high thermodynamic efficiencies in a downstream power conversion system, high pressures are required. While the first demonstrators such as PS10 operated at a pressure of 4.5 MPa using a saturated steam process, modern plants operate at pressures and temperatures comparable to fossil-fired power plants, Ivanpah and Khi for example at 12–16 MPa and 530–560 °C receiver outlet temperature. At such high temperatures, a direct transition from water to superheated steam within one vertical flow path in the receiver is no longer possible, so the flow path is divided into three segments, one of single-phase heating of the water, one of evaporation, and a final one of superheating of the steam. This is done by taking advantage of the fact that the radiance is not uniformly distributed over the receiver surface and designing the flow path within the receiver so that preheating and superheating are positioned in regions of lower irradiance.

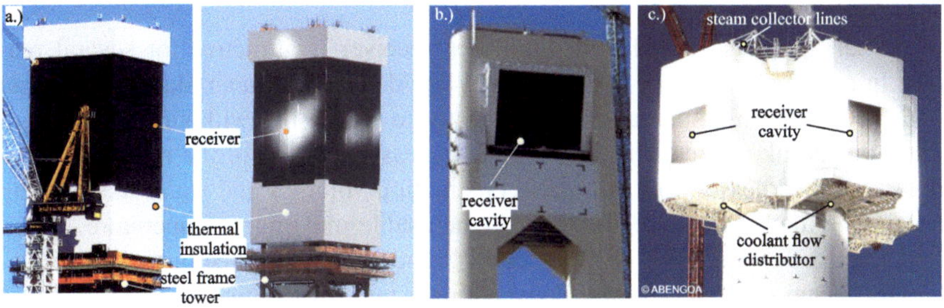

Fig. 7.68 **a** External direct superheated steam generating receiver of Ivanpah, CA Solar Electric Energy Generating System (left) no focussing and partial illumination (right); **b** single cavity saturated steam receiver of PS10 Seville Spain; **c** multiple direct superheated steam generating receiver of Khi Solar One SA (image Abengoa Solar)

In the early development stages of direct evaporation receivers, saturated steam systems were used, such as in PS10/PS 20. Due to the relatively low pressure (4.5 MPa) and thus low temperatures (receiver outlet temperatures approx. 300 °C), the efficiencies of the steam turbine are rather low, but the technical operating risk is low as well. The low pressure not only has the advantage that the tube walls for pressurization in the receiver tube can be kept relatively thin compared to high-pressure cooled systems and thus the heat transfer is greater, but it also offers the possibility of short-term storage of excess thermal power, for example in a steam accumulator or a steam drum.

In the case of high-pressure systems whose steam parameters exceed 10 MPa, the receiver design is more complex; not only must the flow path be divided into three parts because the heat transfer coefficients for water, saturated steam and superheated steam are different, but if reheating is used, an additional further flow path must be integrated. Direct thermal energy storage of excess thermal energy in a steam accumulator is not feasible, since it requires such large-scale pressure-resistant containers that they cannot be built with currently available materials. However, the high superheated steam temperatures allow the heat to be transferred by means of a heat exchanger to an indirect thermal energy storage facility, such as a liquid salt storage facility.

The advantages of using water are obvious. It is a technology known from power plant technology, so that not only the availability of the components—with the exception of the receiver—but also maintenance chains and quality assurance procedures are established and thus the capital risk is low. At the same time, the technical challenges during normal operation are high, e.g. high temperature gradients with variable solar irradiation and the associated thermal stresses can only be controlled by active flow control. The low flexibility in thermal storage (often none), means low capacity factors for the power plant and thus the availability of the power plant. Nevertheless, one of the largest tower power plants in the world, the Ivanpah power plant with 377 MW_{el}, is based on a water-cooled receiver technology.

Liquid Salt Receiver

Liquid salt as a heat transfer medium in a receiver is attractive for several reasons. In addition to the low specific price, liquid salts have a similar high thermal conductivity and a comparable heat capacity to water, but without undergoing a phase change like water. This allows a low-pressure design of the receivers using small wall thicknesses with compact construction volumes and water-like cooling structures. Another positive argument is the usable maximum temperature, which allows temperatures in the order of 600 °C and thus fits perfectly to a Clausius-Rankine steam cycle of a power conversion system. In addition, the use of liquid salts allows the use of direct thermal storage concepts, in which the heat transfer fluid of the receiver and the thermal storage fluid are identical. Compared to indirect concepts, this eliminates the need for components such as heat exchangers and pumps. A disadvantage of all salts is that they have a relatively high melting point, so that suitable measures must be taken to prevent "freezing" under all possible operating conditions. Measures range from draining strategies, recirculation concepts to auxiliary electrical heating systems. Further challenges with salts are the corrosion and dissociation behavior (compare Sect. 5.3.2.4), which require monitoring.

The first project of a molten salt receiver was deployed in Gemasolar/Spain, where on a 140 m tower in an external cylindrical receiver salt is heated from 290° to 565 °C to drive a 20 MW_{el} steam turbine. The excess thermal energy of the receiver is fed into a direct two-tank thermal energy storage system, which has a maximum storage capacity of 15 h of the downstream electrical power block. The Crescent Dunes power plant, which is more than a factor of 5 larger in electrical power output when commissioned, has a 195 m high tower on which an external receiver is mounted that is also cylindrical and about 30 m tall, as shown in Fig. 7.69a. The thermal energy extracted from the receiver is fed into a direct two-tank thermal energy storage system with a capacity of 10 h. The tower, the storage facility, and the receiver are very compact. The compactness of the tower, storage, and power block is illustrated in Fig. 7.69b.

Almost all liquid salt cooled projects realized so far rely on the external receiver, although in the case of the tower, steel frame structures for smaller power units do represent a cost-effective alternative to reinforced concrete structures of large power classes, as shown in Fig. 7.69c.

Ho and Iverson (2014) have also analyzed in detail alternative salt mixtures such as fluoride, chloride, and carbonate salts, which exhibit a higher temperature stability than the most common nitrate-based salt mixtures. But they conclude that while collector efficiencies of up to 90% are achievable, each of the salt classes considered has particular challenges. Fluoride salts are stable up to 1000 °C even under high heat flux, so that salt dissociation plays a minor role, but the melting points of the alternative salt mixtures are considerably higher than the nitrate-based salts. Chlorides have a lower melting point but their corrosion behavior towards based structural materials such as steel increases according to an Arrhenius function with increasing temperatures. Carbonate salts are cheap but can dissociate at high temperatures to form CO_2, so they require stabilizers and thus coolant

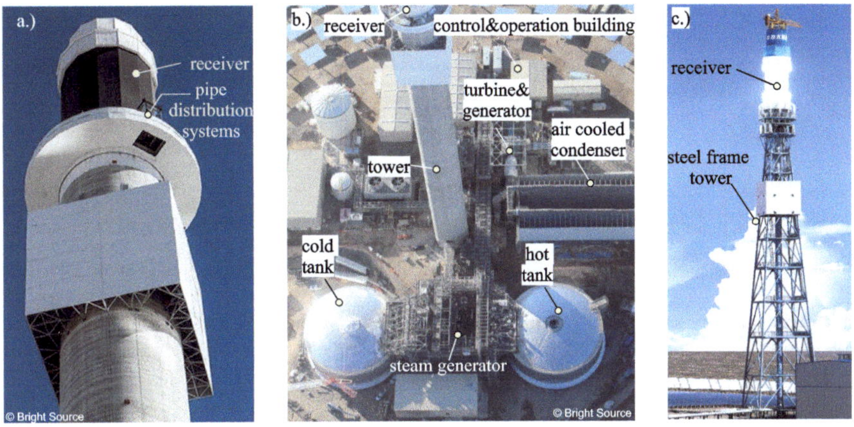

Fig. 7.69 Molten salt tower receiver installed. Crescent Dunes power station external cylindrical receiver (**a**) with the storage and power block parts at the foot of the tower (**b**), fotos modified from (BrightSource Energy Inc., 2023); external molten salt receiver of Délingha, China 10 MW tower on a steel frame tower (image (CosinSolar, 2023))

chemistry control. The trend in research here is more and more towards binary and ternary mixtures, which on the one hand exhibit lower melting point temperatures and simultaneously to increase the dissociation temperature limit to increase the usable temperature spread. Another focus is to improve the aging behavior and reduce corrosion, more details can be found in Vignarooban et al. (2015) and the literature cited therein.

Liquid Metal Receiver

The motivation for the use of liquid metal cooled receivers comes from the limitation of previously tested salts to temperatures around 600 °C. Liquid metals, such as sodium, lithium but also metals such as tin or lead or lead alloys allow here a single-phase low-pressure operation up to temperatures of 850 °C in case of sodium or over 1000 °C for other liquid metals, whereby the limitation is essential determined by the constraints of the tube structure material. The main advantage of liquid metals over salts is their outstandingly good thermal conductivity, especially in the case of sodium, which allows significantly larger heat fluxes compared to other HTF's. This not only results in smaller thermo-mechanical stresses due to the smaller temperature differences between front and back in the receiver structure, but also allows significantly smaller receiver areas with the same total power, which conversely requires greater optical accuracy of the heliostat field.

The advantages of liquid metals and especially of sodium as the heat transfer fluid compared to molten salts are particularly evident at high power densities in the optimum geometry configuration. The more compact design of the receiver (more than 45% smaller surface area) leads to a substantial reduction in convection and radiation losses, whereby the influence of the reduced surface area in particular is far more significant than the losses due to the possible significantly higher

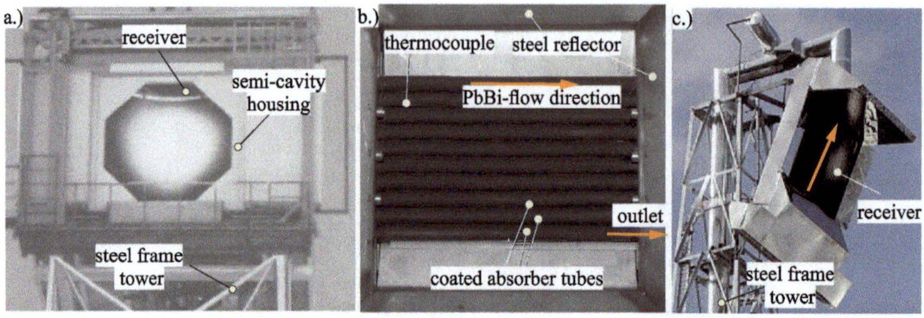

Fig. 7.70 **a** Sulzer sodium half-cavity receiver installed in Almeria/Spain (Kesselring & Selvage, 1986), **b** PbBi-eutectic cooled reflector receiver in Karlsruhe/Germany (Flesch, 2021), **c** Sodium cooled external plate receiver Jemalong/Australia (Coventry et al., 2015)

absorber temperatures, see Boerema et al. (2012). The potential for an optimized sodium system compared to molten salts is estimated to be of the order of 20%.

Although the potential of liquid metal cooled receivers was identified early on, field testing has so far only been carried out in the kW range in demonstration facilities, with the initial focus on sodium, see Fig. 7.70a; but liquid lead–bismuth has also been tested in receivers. Heat flux densities of more than one MW/m^2 could be dissipated with both types of receivers. All Fig. 7.70a–c show that both horizontal and vertical up flow configurations were analyzed, as well as cavity, half-cavity, but also shallow external receiver designs.

Despite the numerous advantages, several fluid-specific challenges to a designer have prevented market penetration by now. Common to all liquid metals is that their price is significantly higher than that of solar salts and their heat capacity is lower compared to salts, making direct thermal storage less economically attractive, which can only be compensated to a limited extent by the higher usable temperature difference.

There are also some arguments that stand in the way of widespread use. Metals such as tin and lead, as well as their alloys, are largely chemically inert in air, but have a high density that is not desirable on tall towers. In addition, their chemical compatibility with structural materials, especially steels, is difficult and they require dedicated oxygen control systems to both prevent dissolution of the steel alloy components within the HTF and simultaneously prevent oxidation of the coolant. Such oxidation problems or incompatibilities with metallic structural materials are alien to alkali metals such as lithium, sodium, or the eutectic sodium–potassium alloy, so that they have been successfully used in nuclear systems for decades. They can be cleaned unproblematically in-situ by means of cold traps. Moreover, corrosion and erosion play only a marginal role along the entire temperature domain. Their challenge lies in the design of a safe receiver system that excludes contact with oxygen or water and water vapor. Alkali metals, e.g. sodium are self-igniting in air just above the melting point if the temperature exceeds about 140 °C. The reaction of sodium with water, for example, releases a large enthalpy of 285.5 KJ/mol. Leakages of sodium-bearing tubes were also the reason for the termination of the sodium-cooled receiver project in Almeria. Nevertheless, the use of sodium has increased again in recent years in the context of successful tests on prototype solar towers.

Alkali metals, unlike all other heat transfer fluids, open up not only the use in thermodynamic cycles but also a direct conversion into electricity. As fluids with a high specific electrical conductivity, the option of direct thermal-electrical conversion exists, on the one hand, with the help of the AMTEC (Alkali Metal Thermal Electric Converters) process or using a magnetohydrodynamic (MHD) generator.

Thermo-electric converters as AMTEC transform thermal energy into electrical energy. Therein, absorbed heat is transported with a pressure difference across a solid electrolyte (in case of sodium it is a beta"-alumina solid electrolyte or BASE) to generate an electro-chemical potential. This is a direct transformation of electro-chemical energy to electricity, which allows to describe the AMTEC process through the same electro-chemistry principles as in a galvanic cell. At the interface of the electrolyte, the sodium atom separates from the electron. Driven by the pressure difference, the sodium atom diffuses through the electrolyte to the cathode and recombines there, while the electron is fed into the circuit at the anode, as illustrated in Fig. 7.71. Due to the stripping reaction and recombination, the sodium loses heat so that the temperature at the anode is lower than at the cathode side. Details of the AMTEC process can be found in Díez de los Ríos Ramos (2018). AMTEC has a theoretical efficiency close to that of the Carnot cycle about ~45% can be theoretically obtained even for liquid sodium at 800 °C, however, in practise about half of the value is achieved. The conversion efficiency even increases for vaporized sodium in a two-phase sodium receiver and has been proposed as topping cycle solar power tower (Hering et al., 2012), since the heat level of 600 °C at the cathode is suitable to operate a Rankine based steam turbine.

The other option is given by a liquid metal magnetohydrodynamic generator (LMMHD) producing electricity by accelerating the electrically conducting liquid through a nozzle with a velocity u_0 perpendicular an applied magnetic field B, according to $u_0 x B$. To accelerate the liquid metal flow, there are several possibilities. The simplest way is to use the heat of the receiver to drive a buoyant flow, however, attaining only small velocity and a poor balance while, boiling in a two-phase liquid metal receiver yields higher velocities and temperatures, which can be further enhanced by gas injection. However, the latter requires two-phase flow liquid metal receivers. All these concepts have been analyzed mostly only conceptually, and experimentally only on a small scale (Coventry et al., 2015).

Particle Receiver

Particle-based receivers use small solid particles that are heated by concentrated sunlight. The particles can be heated to temperatures above 1000 °C, and used to generate process heat or electricity. A major advantage of particle receivers is that the particles can be irradiated directly, eliminating the heat transfer resistance and flow limitations associated with indirect heating through tube or duct walls. However, indirect receiver concepts also exist in which concentrating solar radiation,

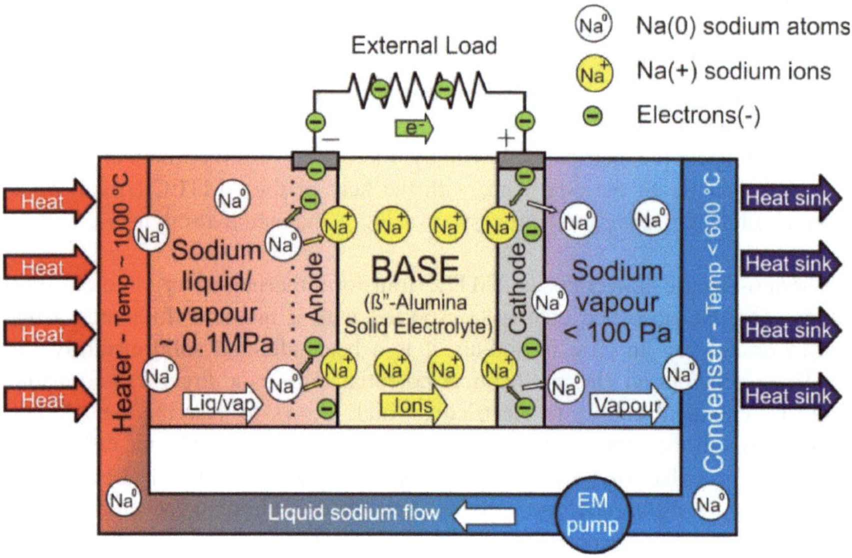

Fig. 7.71 Schematic diagram of the operation of an AMTEC generator for direct thermo-electric conversion

like gas-cooled receivers, are either heated directly through a glass wall or heat up as a fluidized bed against a wall.

A major advantage of direct particle receivers without any walls is that the heat is absorbed directly in the particles, so that solar flux limitations and as a result thermal stresses or relative thermal elongation of the absorber structure as well as pressure increases due to volumetric expansion of the heat carrier are eliminated. However, particle flows are not completely free of temperature limitations, effects particle agglomeration as well as clumping on all transport routes must be prevented or limited to a large extent. There are a variety of different designs of particle-based receivers, including free-falling or rotating cylindrical flow, tubular receivers with or without fluidization, concepts with multi-path recirculation, respectively, and downward-facing configurations; an overview is given in Ho (2017) and Ho and Iverson (2014).

As mentioned, particle receivers can be categorized as either direct or indirect particle heating receivers, where in direct particle receivers, particles are directly irradiated as they pass through the receiver, while in indirect particle receivers, either tubes or other enclosed volumes are used to transport and heat the particles. The direct particle receiver class scopes the following concepts:

- open gravity driven particle receivers,
- guided (or obstructed) falling particle receivers,
- rotating centrifugal receivers and
- fluidized particle receivers,

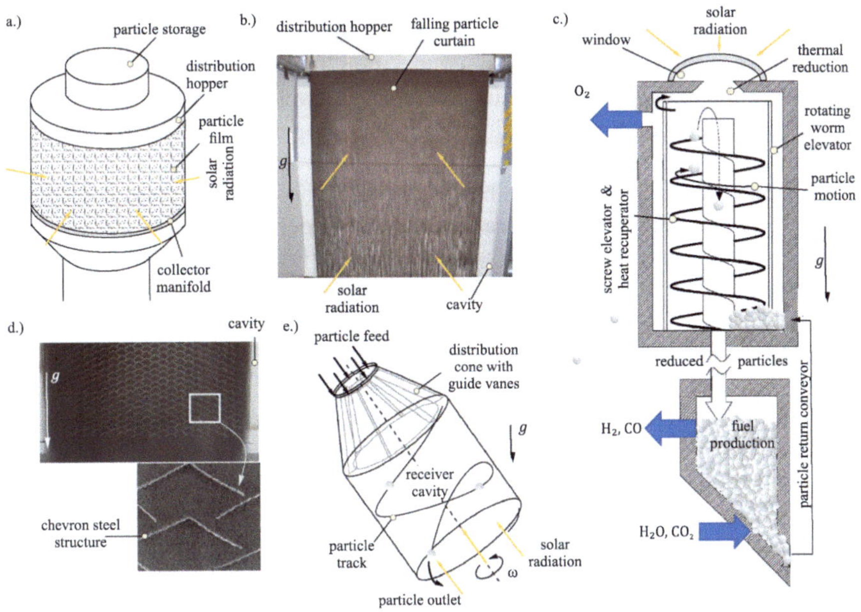

Fig. 7.72 Various direct particle receiver design concepts: **a** external receiver for a surround heliostat field; **b** Gravity driven falling particle curtain cavity receiver (**b**) and guided falling particle receiver with array of staggered chevron-shaped structure (**c**) both modified from Ho (2016); **d** thermo-chemical particle receiver from Ermanoski, Siegel, and Stechel (2013); rotating centrifugal receiver principle (**e**)

while to the indirect particle receiver relate to:

- gravity-driven particle flow through enclosures (plates/tubes),
- fluidized particle flow through tubes.

Direct Particle Receivers

- *Open-gravity driven particle receivers*

In an open gravity-driven particle receiver, particles fall in the form of a particle curtain along a wall in an external receiver type (Fig. 7.72a) or inside a cavity as a free curtain as depicted in Fig. 7.72b, where they are irradiated by incident concentrated sunlight. The curtain is created by a hopper above the receiver, which is fed with the particles by a conveyor.

To control the mass flow through the receiver to achieve the desired particle outlet temperature for a given radiant flux density, sliders or adjustable orifices are used. Although larger particle mass fluxes lead to an increase in the volume fraction of solid material in the curtain and thus a higher absorption rate of sunlight, which allows a higher thermal efficiency of the receiver to be achieved, the additional shading and blocking reduces the outlet temperature of the particles. One option to increase the outlet temperature is to increase the residence time of the particles in the zone of focused solar radiation, by multiple recirculation of the particles in the cavity, with the temperature successively increasing in each recirculation step. However, this requires additional temperature-resistant conveying equipment.

In open cavities, particles can be lost through the aperture, especially if the wind blows parallel to the aperture and the cavity depth is shallow. One option to prevent this is to circulate air or provide an air curtain in front of the particle curtain to reduce particle discharge from the receiver and thus convective losses.

- *Guided (or obstructed) falling particle receivers*

Flow-guided particle receivers aim from the start to extend the residence time of particles in concentrated sunlight by slowing the velocity of particle flow through the introduction of obstacles or porous structures, while ensuring direct absorption of concentrated solar energy.

Obstacles can be, for example, staggered mesh grids such as chevron-shaped grid structures made of stainless steel, as shown in Fig. 7.72d, in which Ho (2016) experimentally demonstrated more than 300 °C per meter of illuminated particle path temperature rise at 1 MW/m^2 irradiation. In another design, a spiral ramp is used, where particle flow is provided by a mechanically excited vibration in addition to gravity and has been successfully tested in experiments. Principally, this design requires beam-down optics, with large particle flows posing a technical challenge in this concept. In another concept study with beam-down optics, shown in Fig. 7.72c, particles are conveyed upward to the aperture by a screw elevator, where they are heated by the concentrated radiation and then fall down the hollow tube of the screw. This particle-based receiver design has been developed as part of a thermo-chemical reactor to reduce particles, which are then oxidized to produce either hydrogen or carbon monoxide.

- *Rotating centrifugal receivers*

The basic principle of operation is that on the top of the receiver, via a funnel, particles are fed into a cone provided with vanes into a rotating receiver which is open at the bottom, and into which concentrated sunlight enters from below, as shown in Fig. 7.72e. The centrifugal force of the rotating container causes the particles to spiral along the walls of the container, while being irradiated all the way by the concentrated sunlight, and finally leaving the container at the bottom, where they are trapped in a collector. Experiments with a downward tilted receiver and an irradiance of 670 kW/m^2 have determined particle exit temperatures of 900 °C and receiver efficiencies of about 75% (±4%). Challenges include maintaining a constant and sufficient mass throughput of particles at larger scales, the need for parasitic energy, and the reliability of a large rotating receiver system, especially in the presence of circumferential velocity variations.

The challenge of all more directly guided receiver designs is that they are open to ambient and thus associated to a particle loss. In addition, local overheating of structures located directly in the irradiation poses a lifetime problem, as does their erosion during operation. Other design-dependent technical challenges to be solved include non-uniform irradiance distributions on the particle curtain, variable mass flow rates, wind effects through the aperture, reliability of particle handling equipment.

Indirect Particle Receivers

- *Gravity-driven particle flow through enclosures (plates/tubes)*

Indirectly heated particle receivers consist of transparent structures (tubes or flat channels) where the particles flow downwards under gravity within the housing. The structures are irradiated by concentrated solar radiation, and heat absorption by the particles occurs both via the radiation and by heat transfer with the wall. Experiments here show that heat transfer in tubes is limited where wall contact of the particles is lost. Challenges include ensuring sufficient mass flow, ensuring homogeneous radiation flux within the cavity.

- *Fluidized particle receivers*

Fluidized bed receivers have been experimentally analyzed at Sandia, where in a vertical transparent silica tube, particles are fluidized with compressed air from below and solar irradiation occurs from above. The particle transport to achieve the required mass flow and at the same time fluidize the particle flow in the receiver tube is one of the challenges to be overcome to avoid local hot spots where the allowable material temperatures of the structures are exceeded. In addition, a suitable window for the pressurized receiver as well as the development of a solid–gas suspension system that ensures a homogeneous particle concentration and temperature inside the vessel is required.

The advantages of all indirect receiver designs are the absence of particle loss through the aperture opening and thus lower heat losses than with an open cavity receiver. However, the development of the additional interface is still a technical problem that has not yet been solved if similarly, high temperatures as in open receiver concepts are to be achieved.

7.3.2.4 High Temperature Structural and Functional Materials

Since the temperatures in the receiver take on the highest values in the whole system, a receiver design is intrinsically at the limits of the available structural materials as well as the functional materials.

The consideration of both aspects is essential for a sustainable receiver concept, which must ensure high availability under all time scales and at the same time a long lifetime. Failure mechanisms, manufacturing procedures as well as in-service operation, maintenance or replacement must be considered.

Structural Materials

The choice of structural materials is coupled to that of the heat transfer fluid which determines the maximum range of the operating temperatures.

In principle, several classes of materials are available where material temperatures above 500 °C are feasible.

- Fe—based alloys,
- Ni—based alloys,
- Co—based alloys,
- Alloys of refractory metals (W, Mo, Nb or Ta),
- Alloys based on intermetallic phases,
- Precious metal alloys (Pt),
- Ceramic Matrix Composites (*CMC*), such as Si_f–SiC composites.

The creep rupture strength which's determination is very complex, is the basis for the design of high-temperature components and is specified with the quantity $R_{m/time/temperature}$. There are two high temperature ranges. One with a time-independent material behavior, i.e. operation in the hot yield stress range, and one with a time-dependent material behavior, the so-called creep rupture range. The limit is the temperature at which the intersection of the yield strength at 0.2% elongation at a given temperature and the creep rupture strength R_m for 10^5 hours at the same temperature intersect.

High-temperature corrosion describes the reaction of a material with its environment at elevated temperature. In the absence of an electrolyte four aspects affects affect material performance.

- Oxidation. Here, a distinction must be made between external oxidation, which describes the formation of an oxide coating at the interface of the material and the HTF, and internal oxidation.
- Carburization, which reflects the internal carbic formation.
- Nitriding with internal or external nitride formation and
- sulfidation.

Regarding corrosion, its alternating effect with mechanical stress must also be analyzed. Two aspects play a role here; stress corrosion cracking, which leads to inter- and trans crystalline cracks due to chemical attack under stress, and fatigue corrosion cracking, which involves a combined effect of cyclic stress and aggressive media.

Figure 7.73a illustrates schematically the limits of creep rupture strength R_m at 10^4 h of some potentially applicable high-temperature material classes as a function of temperature. According to this figure, martensitic steels can be used up to about 600 °C with adequate strength, while the use of austenitic steels pushes the limit upwards by a maximum of 100 °C. Significantly higher temperatures are only attainable with nickel-based alloys, which are used in particular in gas turbine construction. For all metallic materials, however, the main failure mechanisms at

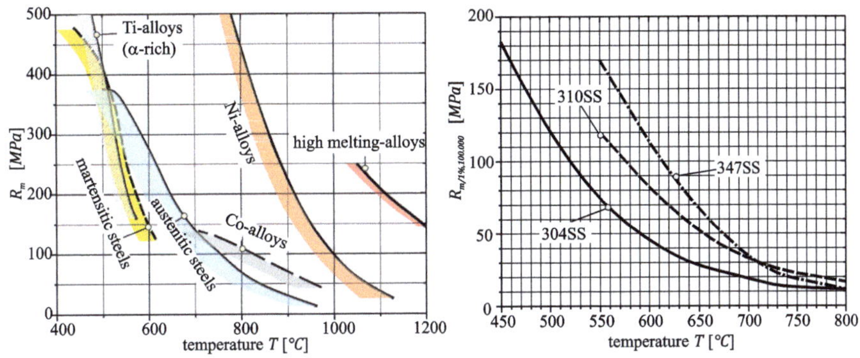

Fig. 7.73 **a** Tensile strength limits R_m in (MPa) of some structural material classes as a function of temperature; **b** Creep strength of some austenitic stainless steels at 1% creep over 10^5 h as a function of temperature modified from Davis (2000)

high temperatures, i.e. for $T > 0.4 \cdot T_m$ with T_m the melting temperature of the metal/metal alloy, are creep and high-temperature corrosion.

Before describing the properties of the individual material classes in rudimentary form, it is useful to first introduce a few definitions.

Steels

All iron-based materials use measures that impede the movement of dislocations in the material and slow diffusion to increase creep strength and high-temperature corrosion resistance. These include particle hardening, foreign phases, or the introduction of foreign particles, for example oxide dispersion hardened steels (ODS).

For temperatures up to 600–620 °C, martensitic steels (9–12% chromium content) can be used as a high-temperature material characterized by good creep rupture strength (creep resistance), high thermal conductivity reducing thermal stresses and a good scale resistance. With suitable adjustment of the water chemistry, martensite can also be used due to the formation of a corrosion-resistant magnetite layer; this also applies to liquid metals and lead alloys. Only in the case of air receivers, oxidation at temperatures above 600 °C can lead to the formation of a continuously growing iron oxide layer which tends to flake off.

The austenitic steels are high-temperature resistant and, depending on the alloy composition, can be used up to 700 °C, but with low strength there. They are characterized by good creep resistance due to partial carbide formation either by MC ($M = Ti, Nb$) or $M_{23}C_6$ ($M =$ Cr, Fe, Mo). Due to the high chromium content (Cr > 16.5%) they exhibit a very good high temperature corrosion resistance, so that except for liquid lead alloys or tin as heat carrier a long life is guaranteed. A disadvantage of all austenite is their low specific thermal conductivity, which can lead to high thermal stresses, especially during transients, and a high sensitivity to thermal fatigue.

When selecting a material for receivers, a high number of thermal cycles must be considered, whereby the number of thermal cycles during the entire service life is greater than that of conventional thermal power plants. The lower design limit is at least $2 \cdot 10^4$ cycles for an estimated service life of 30 years, which is why 10^5 cycles are usually assumed for design purposes. This thermal cycle number applies not only to the receiver but to all components of the heat transfer network but occurs more severely in the receiver. In addition, the receiver is not isothermally loaded, so several temperature ranges and resulting thermal stresses must be considered. Important aspects include consideration of loss of ductility and tensile strength, creep and carburization rates, fatigue crack growth rates, and oxidation resistance. Figure 7.73b shows the loss of material strength at 1% creep of various autentic steels as a function of operating temperature. High thermal stresses are also the trigger for crack growth, which is observed with increasing temperatures in all metallic materials.

Nickel and Nickel-Based Alloys

Nickel and nickel-based alloys have a face-centered cubic (fcc) lattice as the basic configuration and do not exhibit allotropic modifications; they show good corrosion resistance when chromium is alloyed in. Creep strength enhancement can be achieved by alloying measures, with nickel offering a myriad of alloying element options, such as Fe, Cr, Co, Mo, W, Nb, Ta, Al, Ti, Si, C, B, Zr, Hf All nickel alloys exhibit low thermal conductivity and thus high susceptibility to thermal stress; at the same time, the alloy also shows high coefficients of thermal expansion, and thus the tendency to thermal fatigue.

In recent years, nickel-based superalloys have been developed, which are ultimately two-phase microstructures consisting of a disordered γ-matrix and an ordered γ'-phase. With the aid of targeted process control during production, different solidification states can be set, thereby reducing the number of grain boundaries, and increasing the proportion of γ'-phase. For example, single-crystalline solidified nickel superalloys for turbine blades have no grain boundaries and are extremely creep-resistant up to approx. 1100 °C. The oxidation behavior in air is good at high temperatures due to the formation of an oxidic protective layer but must be checked in individual cases. Combustion experiments show that the protective layer formation is reduced in the presence of CO_2. Another disadvantage is the limited weldability of nickel alloys. Further, Ni and its alloys exhibit a good solubility to all high mass number metals, making the choice of a lead or tin alloy as a cooling medium difficult and requiring the development of appropriate protective coatings and in-situ HTF coolant chemistry control and conditioning equipment.

The variation of crack growth rates (da/dN) for a fixed value of stress intensity factor range (ΔK) is shown in Fig. 7.74 for different Inconel alloys (600, 625, 718 and $X - 750$) and stainless steels. The observed variations could be explained by the temperature dependence of the elastic modulus and the yield strength. The measurements show that the crack growth rate increases dramatically with increasing temperature, from room temperature to about 900 °C by a factor of 25. Since

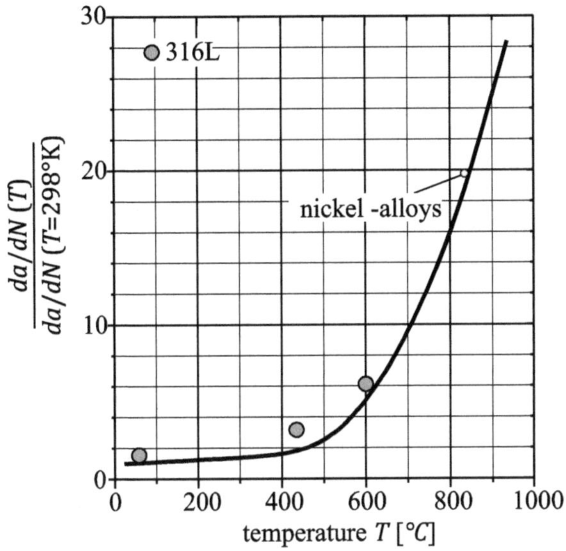

Fig. 7.74 Relative change of fatigue-crack-growth rates (da/dN) compared to room temperature of nickel alloys from Viswanathan (1993) and 316 L as a function of temperature T (°C) for a stress intensity factor range ($\Delta K = 30$ MPa$\sqrt{\text{m}}$)

the experimental data were obtained for different ΔK differences and temperature ranges, the different material classes cannot be compared directly. But, when the values are analyzed for a constant stress intensity factor range (ΔK), as shown in Fig. 7.74, a clear trend of increasing crack growth rate with increasing temperature can be seen. The figure shows that the growth rates at temperatures up to about 50% of the melting point (0.5 T_m = 550–600 °C) show little influence versus temperature; above this threshold, crack growth increases rapidly with increasing temperatures. The crack growth rates for all materials at temperatures up to 600 °C relative to the rates at room temperature can be estimated with a maximum correlation factor of 5 (2 for ferritic steels).

Taken together, both Figs. 7.73 and 7.74 show that a holistic receiver design must consider not only surface heat flux provided by the concentrated solar radiation but also thermomechanical and thermocycling effects in all expected temperature ranges to make a reliable lifetime prediction.

Cobalt-Based Alloys

These alloys are suitable for use up to 800 °C and are characterized by good weldability compared with nickel alloys. They exhibit good creep resistance based on solid solution hardening with W, Mo, Ta, Nb and particle hardening with carbides. At the same time, they show a lower temperature dependence of the creep resistance than the nickel alloys. Although nickel alloys already have a high price compared to steels, cobalt alloys are generally even more expensive. Other disadvantages are their lower strength compared to nickel alloys, since no hard phase content is possible. Also, lattice transformation of Co can occur during overheating, resulting in a decrease in creep resistance. Therefore, the fcc matrix of cobalt alloys is usually stabilized with Fe and Ni, further details on nickel, cobalt and their alloys may be taken from Davis (2000).

High Melting Alloys

Well-known from the construction of conventional power plants are niobium and molybdenum alloys, as well as tungsten in the field of nuclear fusion. There is a tendency for the yield strength $R_{m/time/temperature}$ to shift from niobium via molybdenum to tungsten to ever higher values well above 1200 °C for W. The challenge with these alloys, apart from the high specific price and the relatively small selection of possible joining processes with conventional materials, is above all the scaling resistance under gas atmospheres containing oxygen or nitrogen, where special protective coatings have already been developed for some applications. At temperatures above 1000 °C, hydrogen embrittlement also plays a role, negatively affecting this mostly bcc-oriented metal matrix alloys in terms of strength and fracture behavior.

Ceramic Matrix Composites (CMC)

In many technical high-temperature applications, such as in flame tubes, turbines or furnaces, the commonly used nickel-based superalloys are being replaced by ceramic matrix composites (*CMC*), whereby characteristic properties comparable to metals are achieved through the use of long fibers and the embedding of the ceramic. The most important high-temperature fiber-reinforced ceramics currently available on an industrial scale are C/C, C/SiC, SiC/SiC and Al_2O_3/Al_2O_3, which differ from the ceramic base material in the following respects:

- higher elongation at break of up to one percent,
- significantly higher crack resistance,
- extreme thermal shock resistance,
- better dynamic load capacity,
- anisotropic properties oriented to the fiber orientations.

Fiber-reinforced ceramics are usually manufactured in three steps:

- Placement and fixation of the fibers in the desired component shape,
- placement of the ceramic matrix material between the fibers, with five basic options; deposition of ceramics from a hot gas mixture, generation by pyrolysis of a polymer containing Si and C atoms, generation by chemical reaction or by sintering at low temperatures (1000°–1200 °C) and production by electrophoresis with ceramic powder.
- Finishing and, if required, further post-treatment steps such as the application of coatings.

Accordingly, the secret of fiber-reinforced ceramics with high crack resistance is to ensure during manufacture that bridging across the crack is possible. This presupposes that the matrix can slide along the fibers, i.e. that it is only weakly mechanically connected to the fibers and that the connection between fibers and matrix allows sliding. When used as a structural material, attention should be paid to the form in which the composite ceramic is present, i.e., either in two-dimensional form as stacked fabric plies or as crossed plies of one-dimensionally aligned fibers before. This type of *CMC* exhibits anisotropic behavior, as crack propagation between these layers is not hindered by any fiber bridging, making potential delamination a weak point of the material. Quasi-isotropic material behavior can be specifically achieved by introducing three-dimensional fiber structures.

SiC/SiC-CMC*s* reinforced with continuous fibers of silicon carbide are of particular interest for applications in high-temperature solar thermal receivers because they exhibit high temperature stability in oxidizing atmospheres due to a high melting point and the formation of a protective silicon dioxide layer, lower density (~2.6 g/cm^3), higher specific strength, and lower creep rate at high temperatures compared to nickel-based superalloys. The fiber reinforcement leads to completely different failure mechanisms compared to monoclinic, more brittle ceramics or metallic materials. Silicon carbide as a matrix has a high thermal conductivity, but this decreases with increasing temperature; conversely, the fiber has a low thermal conductivity. As a composite material, Sic–SiC reaches values on the order of steels ($\lambda \approx$ 12–22 W/(mΔK)) and has been successfully in performance tests for solar applications, see Ortona et al. (2015).

Although SiC/SiC-CMCs have excellent temperature stability, the presence of water vapor can lead to volatilization of the silica by formation of silicon hydroxide. This requires a coating to be applied to the CMC, and these environmental barrier coatings (*EBC*) are produced by thermal spraying, usually silicate-based. The coatings are multi-layered and consist of a silicon bond, one or more intermediate layers, and a silicate top layer with low silica activity. This low activity results in less formation of silicon hydroxide when exposed to water vapor, and thus improved corrosion resistance. However, these silicate-based coatings are susceptible to cracking, which leads to the formation of silica deposits on the silicon binder layer, eventually resulting in severe corrosion or spalling of the coating. Newer robust processes rely on pack cementation to produce oxides other than silica.

Selective Absorber Materials for Tower Receivers

Next-generation central tower receivers are expected to attain heat transfer temperatures exceeding 650 °C to achieve higher efficiency of the downstream power conversion system and thus better economics. Higher temperatures and long lifetimes, which imply high cycling, are particularly challenging for the optically solar selective absorber (SSA) receiver material, which must maintain high absorptivity ($\alpha \to 1$ for $\lambda \leq 2.5$ μm) at these high temperatures to maximize absorbed energy while maintaining low emissivity ($\varepsilon \to 0$) to minimize heat radiation losses. In contrast to linear focusing collectors, the volumetric heat release in high temperature receivers is not at least one order of magnitude greater, but also the temperature. Thus, ensuring the functionality of SSA, both in its structural design and composition, represents a unique development challenge, even if the physical and material-specific concepts for achieving spectral selectivity do not differ from those elaborated in Chap. 3.

In all windowless receiver concepts, such as open-air receivers but also liquid cooled receivers or the particle receivers, the selective efficiency (η_{sel}) is therefore the key parameter. It is defined as the energy retained by the absorber, considering both absorption and emission, in relation to the energy incident on the surface. Other requirements for the absorber material include:

- high thermal conductivity,
- high temperature chemical and structural stability through entire operation range,
- low thermal expansion,
- low creep during thermal cycling and long durability (>25 years),
- high resistance to oxidation,
- compliance with local regulations on use of materials for health, safety, or environmental reasons,
- inexpensive and easy to apply to the absorber tube,
- good adhesion of the absorber to the heat transferring surface.

In addition, external absorbers, i.e. those not protected by a cover, should ideally be dirt and abrasion resistant to minimize cleaning, repair, and maintenance costs. Furthermore, the oxidation resistance to air and the stability of its structure during its lifetime are of particular importance to ensure the availability of the overall system. In the case of window-based receiver concepts, such as pressure-charged gas-cooled receivers, the high-temperature behavior and the optical properties of the glass also play a special role, while some requirements for the receiver material, such as the oxidation behavior, among others, become somewhat less important.

First High Temperature Absorber Materials

For the parabolic trough collector technology, often cermets ($M - Al_2O_3$, $M = Mo, W, Ti, Ta$ et al.) are used as solar selective absorber (SSA) coatings, but although there the durability in a vacuum receiver is proven, the application in atmosphere in tower receivers with enhance oxidation and ageing accelerated by increasing temperatures poses a significant challenge.

The absorber substrates themselves are made of different materials with different melting points, either aluminum (600 °C), different steels (about 1500 °C) or ceramics (1000–3000 °C). The choice of material for the absorber tubes depends first on the temperature range for which the receiver system is designed—possibly well above 1000 °C for tower systems, but if solar radiation is absent, the temperature can drop below zero degrees overnight. Untreated metals such as aluminum,

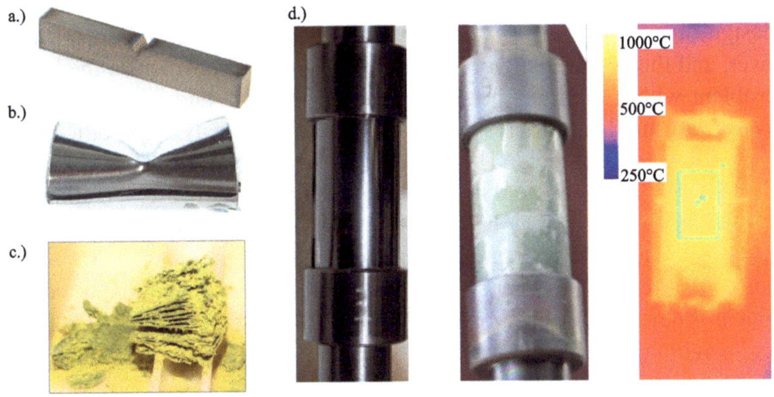

Fig. 7.75 Development chain of an intrinsic $W-Ti$ laminate **a** Notched impact specimen of the laminate from Reiser et al. (2015), **b** plastic deformation of a $W-Ti$ laminate tube; **c** delamination of the tube under oxygen; **d** W-laminate tube for the Almeria test before (left) and after the high-temperature tests (middle) and IR image during operation (right)

copper, steel but also tungsten is ruled out because they can be highly reflective. With the help of coatings, however, this can be changed. Coatings which withstand high temperatures in atmosphere, however, also are quite rare. The emissivity for the high temperature, say at 600 °C is usually large even with an optimized optical design of the selective absorber and its critical wavelength. The different selective absorber concepts and examples of high-temperature selective coatings are presented in Sect. 3.3.4.

The main product on the market for solar towers is a proprietary non-selective coating called Pyromark 2500. However, absorbers coated with Pyromark must be repainted on an annual basis, which reduces collector availability. The manufacturer, Tempil, states that Pyromark 2500 resists temperatures up to 1093 °C and can be applied on steel, aluminum, alloys, and ceramics. Pyromark 2500 has a very high solar absorptance ($\alpha = 0.96$), but unfortunately it also has a very high thermal emittance ($\varepsilon \approx 0.9$) at the temperatures exceeding 600 °C.

Modern manufacturing and fabrication methods as well as the low requirement for low emissivity imply that intrinsic absorbing absorbers should not be excluded from general considerations. Figure 7.75 shows such a development chain for tungsten. By means of an alternating diffusion welded titanium and tungsten foils (Fig. 7.75a), a ductile tube was produced which also behaves ductile under plastic loading as illustrated in Fig. 7.75b. However, it is not stable to oxidation in air and delaminates, as depicted in Fig. 7.75c. Applying a titanium carbide coating, an oxidation-resistant high-temperature-stable selective absorber tube is obtained, as shown in Fig. 7.75d, which can easily withstand temperatures above 1000 °C with high absorption.

7.3.2.5 Heliostats

The design of a solar tower system, i.e. the size and position of the receiver on the tower, and the determination of the number, type and position of the heliostats is a problem with a high number of degrees of freedom and can only be solved with computational help, the results of which allow to analyze the efficiency and the yield of the heliostat field. This is always a two-step process.

First, a large heliostat field is generated as input and after positioning the selected heliostats (which may be of several types for near and far field positions), for this initial field the annual energy that each heliostat provides to the receiver is evaluated. To calculate this energy, yield an optical model is required to determine the optical efficiency not only of each individual mirror, but also its interaction with the neighboring mirrors, which scopes numerous loss effects occurring in the heliostat field, namely:

- cosine loss, i.e., the reduction of effective mirror size due to off-axis reflection,
- shadowing of specular surfaces by neighboring heliostats,
- blocking of reflected light by adjacent heliostats,
- attenuation of reflected light between heliostat and receiver due to absorption and scattering processes in the atmosphere,
- reflected light reaching the receiver plane but not entering the aperture ("spillage") due to incomplete heliostat alignment and/or an oversized reflection image.

If the annual yield of the initial heliostat field is known as well as its annual integral optical efficiency; however, one is still far away from the optimal heliostat arrangement for the tower and the question arises, how can an optimal arrangement look like? This question is dealt with in Chap. 10 on the high temperature solar fields. In the following we would like to present the heliostat as a component with a large degree of freedom concerning design and material selection.

Heliostat Design Options

The cost of the heliostat array accounts for about more than one-third of the total plant cost of a solar tower power plant, so a variety of efforts have been made in recent years to optimize this aspect, the most recent reports are Coventry et al. (2016) or Pfahl et al. (2017). Simultaneously, an individual heliostat must meet several criteria such as the accuracy of the device and the local conditions at the heliostat field. The list of these criteria has a major impact on design and cost. They include the required optical quality, any external loads to be sustained, the required performance of the tracking system, protective measures for the heliostat and receiver, the lifetime of individual components, and the overall heliostat system.

Before we turn to the design criteria, it is first useful to identify the individual components of a heliostat. An individual heliostat as depicted in Fig. 7.76 essentially consists of five assemblies, which have different functionalities:

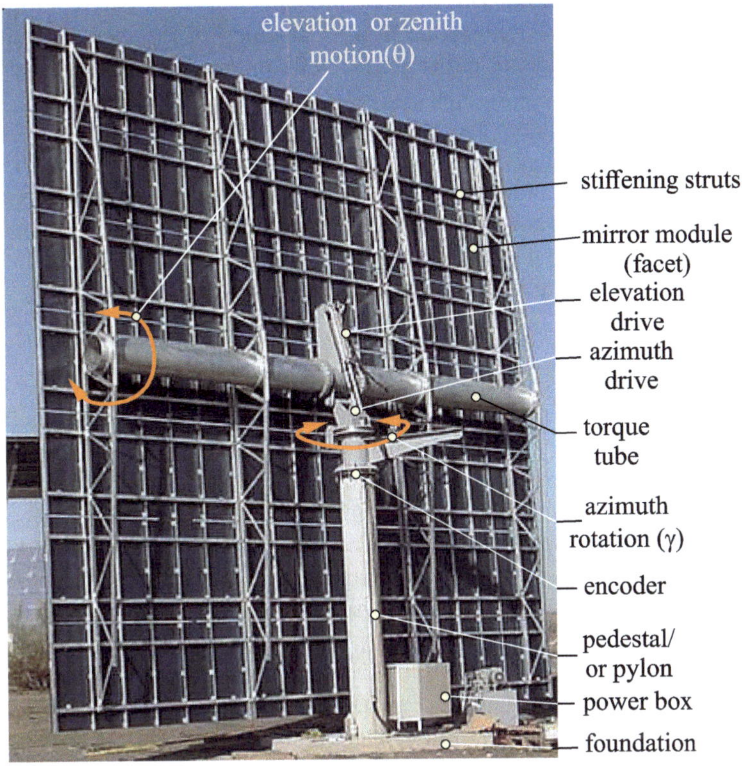

Fig. 7.76 Typical T-type heliostat with pedestal and frame assembly (modified foto from © Plataforma Solar de Almería)

- (canted) mirrors using a highly reflecting surface to project the sunlight to the receiver termed as *reflecting module*.
- structure or mirror fixation, which is realized by means of an adjustable support structure serving degree of freedom,
- pylons and *foundation* for connection to the ground,
- *mirror canting, measurement and control devices* to focus the mirror (canting) and determination of position of the mirror surface and its control,
- *drive mechanisms* for biaxial tracking of the mirror surface.

Heliostats may have different shapes and are not all rectangular. The company SBP Sonne for example designed a heliostat, the Stellio type, with pentagon shape (see https://www.sbp.de/projekt/stellio-heliostat/). This has the advantage that heliostats can be positioned closer to each other, as the risk of touching with edges is reduced.

Heliostats with membrane mirrors are a second option showing the variability of design. The concentrator consists of two membranes that are connected by a compression ring. A vacuum pump creates a slight negative pressure in the space between the two membranes. This negative pressure leads to a curved deformation of the concentrator. This enables the desired bundling of the incident radiation. The membranes are made of either plastic foils or metal foils, the latter being more durable.

The advantage of a membrane heliostat compared to a faceted heliostat is that good optical quality is achieved with low production and assembly costs. The time-consuming alignment (canting) of the individual elements is no longer necessary, as it is only a matter of a concentrator surface that has to be aligned. Disadvantages of the membrane heliostat are the higher costs compared to the faceted heliostat and a larger self-consumption of electricity caused by the vacuum pump.

Heliostat Sizing

The cost-optimal size of a heliostat is a question that cannot be answered easily. Current projects show that the focus is on two size classes: medium-sized heliostats with mirrors up to 40 m^2 and very large mirrors in the size class of approx. 120 m^2. The design is always the result of an optimization process, which is dependent on costs of individual components, on the approach to manage complexity versus replicability. The main advantage of large heliostats is that fewer components, foundations and drives are required per square meter of mirror area, but at the same time the static and dynamic wind loads increase significantly as well as the requirements for high power and especially high torque motors with high precision (Blackmon, 2020).

Conversely, the main advantage of smaller heliostats is that the wind loads are smaller and the superstructures lower, and simultaneously the requirements for drives and foundations (weight per m^2) are reduced. The reason for the lower area-specific weight is that an increase in the chord length of the mirror panel quadratically increases the mirror area and, therefore, all other dimensions of the heliostat also increase according to the chord length so that the loads caused by the wind peaks remain the same. However, with the increase in size, the volume and mass of the heliostat grow to the third power of the chord length. Higher weight means higher costs, so smaller heliostats seem to be more advantageous due to their lower specific weight, if additional costs are not eaten up by more effort in setup, adjustment, and maintenance during operation. Communication and control algorithms in the required IT are much more extensive than for large heliostats. With more advanced computing and cheaper hardware there could be a tendency towards smaller heliostats in future, but as of today there is no clear breakthrough. And as mentioned earlier, the selection of heliostats is dependent on plant size and receiver type, so that different approaches might be successful in different projects.

Reflecting Module/Mirror Facets

The reflective module of a heliostat usually consists of one or more mirrors (also called facets) arranged in a usually spherical pattern (canting) to concentrate the reflected light onto the receiver surface. In addition, each mirror is often mounted slightly concave to help concentrate the reflected beam to a focus. Flat mirrors are also occasionally used to simplify the design. Subdivision into facets is not only for technical reasons in manufacturing, but also facilitates handling during construction and allows better control of curvature for larger heliostats. Typical facet sizes are between 1 and 10 m^2. The module must perform accurately under a wide range of operating conditions such as varying wind speeds and ambient temperatures should have the following characteristics:

- high optical performance (e.g., reflectance/transmittance, specularity, geometric configuration).
- low specific weight,
- long lifetime ($\approx 25y$) and low maintenance costs.

Typical heliostat mirror facets are based on so-called glass/metal structures. This is a sandwich construction in which a second surface mirror (i.e., the reflective material is covered by a transparent material placed between the sun and the reflective surface). This double layer is bonded to the support structure using an adhesive and sealed at the surrounding facet edges. The silver reflective surface is thus protected from one side by the glass but must also be protected from the back. In monolithic mirrors, a copper layer is used with a protective layer for the copper. Mirrors made of laminated glass use glass on both the front and back sides, so the reflective film is sandwiched between two layers of glass. The reflective layer is made of silver or aluminum, with silver being advantageous over aluminum due to its high reflectivity ($\rho \approx 0.95$–0.97) over a wide range of wavelengths because its reflectivity ($\rho \approx 0.88$–0.92) is lower and alumnium has an undesirable absorption band at about 800 nm.

The transparent cover is made almost exclusively of low-iron glass because it is nearly inert, abrasion resistant, and inexpensive and available in large quantities, and has high strength and simultaneous low absorption across the wavelength spectrum. Current commercially fabricated reflectors exhibit a reflection coefficient of $\rho \approx$ ~0.93–0.94 without excessive corrosion or UV degradation. The development goals are, on the one hand, to increase the reflectance to values of $\rho \approx 0.95$–0.96 and to use low-maintenance anti-soiling coatings to reduce operating costs by cleaing the mirrors. Another disadvantage of the glass-metal composite is the relatively high area specific weight, about 7.5 kg/m^2 for a common 3 mm thick mirror, which must be supported by the mechanical structure and positioned by the drive mechanism with appropriate moments.

Mirror Support Structure

The support structure of a heliostat consists of two large assemblies:

- Base segment, which anchors the assembly to the floor.
- Frame structure, which holds the mirror facets rigidly in position.

Most often, the basic construction of medium or large heliostats is a so-called T-type, where the torque tube and the pylon together form a "*T*"; both are connected by the drive system for the azimuth and elevation movement. The base provides the necessary ground clearance for the reflection module. The frame assembly supports the arrangement of the mirror modules or facets. The modules are joined to the assembly using adhesives or other mechanical fasteners. Each mirror module typically has a slight concave curvature and is also tilted (aligned) with respect to the plane of the support structure to better focus reflected sunlight on the receiver to improve performance. The torque tube is connected to the drive system, while the mirror facets are attached to the cross beams. Trusses are preferred, especially for larger heliostats, because their depth can be varied to achieve the required stiffness with little weight loss. In some designs, the beams connected to the mirrors are replaced by perforated pressed plates. The azimuth/elevation gear is installed at the connection between the base and the torque tube. The aim of the design is to give the heliostat structure the necessary stiffness and to minimize deflections due to wind loads. For large heliostats, this means a heavy torque tube, which increases the specific weight, material, and cost of the heliostat structure. One of the great advantages of the T-type design (where there are open or closed designs), with multiple facets is the ability to set the heliostat downward (compare Fig. 7.76), where setting this safety position prevents damage to the reflective module in hailstorms or similar conditions. However, this comes at the cost of slightly reduced optical efficiency due to the smaller effective area.

Glass and steel have different thermal expansion ratios. Therefore, the mirrors are often connected to the support structure by pins that are flexible in both directions of the mirror plane. In the example shown in Fig. 7.77a, steel strips are used that are flexible in one direction and connected to the cross-members by riveted joints, allowing rotation in the other direction. Only the central pins (above the torque tube) hold the gravity load of the mirrors in the upright orientation and are attached to the cross members by two riveted joints. The pins are bonded to the mirrors by pads that increase the area of force application and rotate to avoid high stresses in the glass. The height of the pads perpendicular to the mirror plane determines the shape of the mirror panels. The number of pads is determined by the wind loads and the required accuracy of the parabolic curvature of the mirror field. For smaller heliostats, wind and other loads are significantly lower, allowing lighter frame structures that either require only a very small pedestal, or are mounted on a floor structure (compare Fig. 7.77b) shared by several adjacent heliostats to avoid a costly foundation. Small mirrors also do not require a true frame structure (especially a torque tube), so simple structures are sufficient to provide stiffness to the mirror. For small facets with sufficient adhesive thickness, the flexibility of the adhesive is able to compensate for the differences in thermal expansion between steel and glass, and no pins are required.

Foundation

The function of the foundations is static protection, which naturally depends on the size and weight of the structure. It is important to prevent rotation of the pylon. For large heliostats the bases are mounted on a reinforced concrete pier or concreted directly into the ground, lighter heliostats allow mounting on an above ground ballast frame or ballast foundations consisting of a concrete ground anchor

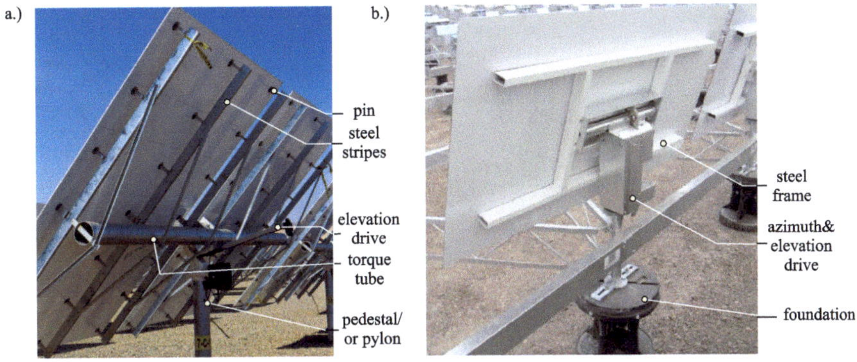

Fig. 7.77 **a** T-Type Heliostat with connection of glass mirrors and steel support structure via flexible pins (for compensation of different thermal expansion of steel and glass) and pads modified from Pfahl et al. (2017); **b** Low cost support structure for small heliostats (Téllez, Villasante, & Burisch, 2014)

weighted with natural site material, more details can be found in Coventry et al. (2016).

Mirror Canting, Measurement and Control

The heliostat's task is to reflect sunlight onto a predetermined target on the receiver. This requires a procedure that involves several aspects.

- alignment of the individual mirror facets of a heliostat to the receiver aperture.
- calibration measurement of the flux profile on the target.
- control logic to adjust the heliostats.

The term canting refers to the alignment of the individual mirror facets of a heliostat on the carrier structure with respect to the receiver and has as its goal to maximize the thermal yield on the receiver. For small heliostats consisting of a single facet, the curvature of the facet and the applied canting are identical, resulting in a continuously canting surface.

Many studies have shown that the radius of curvature of the heliostat should be different along the tangential and sagittal planes to reduce aberrations in the reflected image. Therefore, the energy gained from a particular heliostat field can be maximized using a targeted curvature technique. The applied curvature technique also depends on the tracking method of the heliostat. All heliostats have two axes of motion, most of which provide azimuth and elevation rotation as shown in Fig. 7.78a.

Regarding azimuth-elevation tracking usually a first vertical axis (azimuth) and a second horizontal axis are used, although a first horizontal axis is also possible. The methodology for this tracking type chosen for this process is, however, not unique, but can be classified into two groups: on-axis and off-axis. In case of on-axis canting the facets are aligned to give a perfect image, when the heliostat

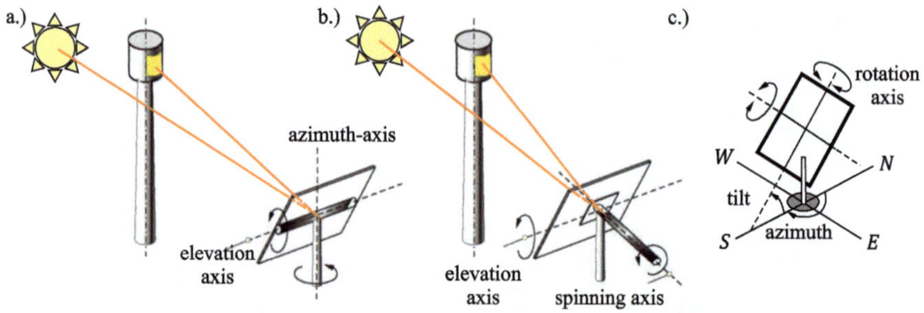

Fig. 7.78 Schematic sketch for heliostat tracking options **a** azimuth-elevation tracking; **b** target aligned tracking; **c** polar-mounted elevation tracking

center, the target and the sun are collinear, i.e. are on a line. For off-axis canting, the facets are arranged to optimize the convergence of all images on the solar receiver for a particular time and day of the year.

An alternative tracking option is the polar-mounted elevation tracking as shown in Fig. 7.78c. Here the first axis is fixed in north–south orientation parallel to the earth's axis (polar mounting). The rotation around this axis tracks the daily solar path. Due to the changing declination of the solar orbit the tilt angle must be adjusted every day so that the reflected rays hit the receiver. For this heliostat tracking the corners of the rectangular mirror stay parallel to the north–south direction. This implies that heliostats can be placed closer to each other than for the azimuth-elevation tracking. A more compact assembly is possible, which is beneficial for the near field.

A third alternative option is the target-aligned tracking, in which the target-aligned heliostats allow rotation around the fixed axis connecting the center of the heliostat to the target, so that the second axis of the surface is vertical to the plane defined by the sun, target and mirror center. A second rotation tilts the mirror perpendicular to this plane so that the reflected image is within the receiver target area (Fig. 7.78b). In principle, this procedure reduces the astigmatism of rectangular shaped heliostats but generates additional effort, since each individual heliostat must be aligned to the target.

The procedures for setting a correct heliostat canting can be divided into two broad categories mechanical and optical options, an overview can be found in Yellowhair and Ho (2010) among others.

- Mechanical methods (e.g., using gauge blocks or inclinometers) involve calculating the required facet canting angles in advance and manually measuring and adjusting each facet to achieve the desired angles. These methods are labor intensive, tedious, and prone to several errors, primarily due to structural gravity sag and local slope errors at the measurement point.
- Optical methods (such as photogrammetry, fringe reflection, parallel laser beam projection, scanning laser projection, or reflection of a known image) typically

rely on applying a correction to the mirror surfaces so that an undistorted reflection of a known image is obtained. Optical methods are not only able to correct for facet tilt, but also measure focal length errors and generally provide higher accuracies. The application of the canting process requires precision and labor and is therefore a factor that contributes to the overall cost of the heliostat during the manufacturing and installation phases.

The heliostat has the task of reflecting sunlight onto a pre-determined target. Therefore, the heliostat controller must know the orientation of the mirror to determine deviations and precisely control the actuators to minimize optical deviations. The conventional approach is calibration measurements and open-loop control. Modern control systems rely on closed-loop systems that use sensor signals for feedback. The pre-requisite for an open loop is very low statistical error and stable, predictable systematic behavior. Further, they assume that the heliostat model represents all real-world imperfections. Closed-loop systems are limited only by the backlash of the actuators, the measurement accuracy of the feedback sensor, and the motor control. Difficult influences such as deformation due to gravity, wind, and temperature can be compensated for as long as the initial sensor-heliostat calibration remains intact. Closed loop tracking accuracy is usually better, which reduces the spillage to achieve the desired overall flux distribution on the receiver.

Drive Mechanisms

The actuators control the position of the heliostat frame assembly by rotating or inclining the structure to project solar radiation onto the fixed target. The drive unit consists essentially of two assemblies, the power source, and the transmission unit. The requirements for the heliostat drive unit demand only low speed but high torque. For optical efficiency, however, high transmission precision and low backlash are required, since tracking errors in the range of mrad can already lead to a significant power loss of more than 10% of the total power. In addition, the above-mentioned requirements for the drives should be maintained during the typical service life of about 25 years, with the lowest possible maintenance expected, which means that all seals, wear, and abrasion of the drives should be designed to last as long as possible. These stringent requirements mean that the drive and gearbox are a cost factor of the same order of magnitude as the mirrors (Coventry et al., 2016). The mechanical transmission of forces and movements from the energy source (motor or hydraulic pump) to the rotary axes of the heliostat can be achieved using different technical concepts. Important for the selected transmission solution are the transmission ratio, the transmission accuracy, and the stiffness, whereby the efficiency of the transmission is of secondary importance. There are a number of different drive solutions also used in industry, which meet the requirements for required power, speed and accuracy, so that there is no one-size-fits-all solution; mostly a drive concept consisting of an electric motor combined with a mechanical transmission gear is used in commercial heliostat systems, including worm and spur gears, screw and nut, chain, harmonic drives, whose characteristics are briefly outlined below and illustrated in Fig. 7.79.

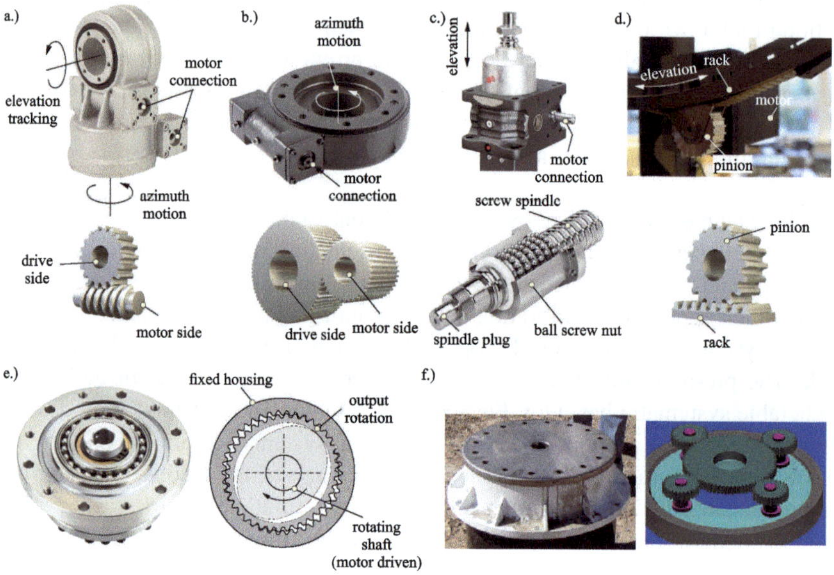

Fig. 7.79 Different types of gearboxes for tracking heliostats and their operating principles; **a** Worm gearboxes, **b** spur gear/planetary gearboxes, **c** screw-nut gearboxes, **d** chain/rack-pinion gearboxes, **e** harmonic gear drive, and **f** planocentric gearboxes adapted and modified from Coventry et al. (2016)

- Worm gears allow large gear ratios in each stage (10–30), with unfortunately low efficiencies. Backlash can be reduced by preloading the gears but increases wear and thus aging. They are wide used as drive mechanisms of Heliostats in central receiver plants.
- Spur gears provide smooth mechanical transmission at moderate gear ratios (2–5) and therefore require a large number of stages. Construction ions such as planetary gears offer larger gear ratios in each stage. The efficiencies are high, but the irreversibility of the gear cannot be guaranteed.
- Screw and nut gear drive are used when linear motion is part of the transmission chain, screw and nut solutions can be considered to achieve large gear ratios. When using ball screws, the smoothness and efficiency are considerable, but the cost is higher, and the system is reversible. Friction screw and nut solutions, on the other hand, allow irreversible step designs.
- Chain-pinion or rack-pinion solutions offer the possibility of low-cost transmission. In principle, they are based on contact forces. In case of the chain option, the chain is not firmly connected to the assembly to be moved, and thus the achievable stiffness is lower, this also applies to accuracy. However, high transmission ratios can be achieved at high efficiencies.
- Harmonic drive gears, also known as "expansion shaft gears," rely on contact and deformation to provide high gear ratios of up to 100, high accuracy, high

efficiency, and backlash-free transmission of rotary motion. They are often used in high-precision solutions, such as in machine tools or robotics, but are costly.

- Planocentric drives (Winsmith drives) offer exceptional gear reduction (over 30000 : 1) in an extremely compact package; unfortunately, heliostats seem to be the only application for this concept, and they cannot benefit from industrial system applications.

7.4 Comparison of Optical Performance of Parabolic Trough, Linear Fresnel, and Solar Towers

It is not possible to conclude from the optical performance of the individual line- or point-focusing systems alone, which of the concepts is better or worse, because for a complete solar thermal system its optical and thermal behavior, storage possibilities, the temporal and temperature suitability for the respective application includes many more parameters to be considered. For example, the different orientation of a parabolic trough or a linear Fresnel might produce different solar heat generation profiles during a day, but when storage is included to match demand and generation, the exact profile is not important at all. An optical and thermal performance can be determined for each of the presented mirror-receiver concepts. Using the optical performance, the consequences in receiver absorbed power and heat transfer to the fluid can be derived as well as thermal losses. The required quality of the optical components (mirrors, absorbers, glass) for a certain power demand can be determined, not necessarily representing a cost optimized. The overall optimum, however, is dependent on additional factors such as thermal energy storage capacity or the storage concept, spatial limitations, and many other parameters more.

Nevertheless, it is useful to compare the individual technologies in terms of their time-dependent optical performance. In the following, a linear Fresnel collector is compared with a parabolic trough and a central receiver tower qualitatively and quantitatively since these concepts offer a power (up-)scaling at least in principle. The comparison considers equal primary reflector/heliostat characteristics, as far as possible equal receiver characteristics, which are summarized in Table 7.11, and is based on the article of Kincaid et al. (2018). However, here it has to be noted that the characteristic properties may vary for different products of the same collector type and the intention here is to illustrate the characteristic differences of the technologies.

Figure 7.80 illustrate the optical efficiency (η_{opt}) as a function of transverse (θ_T) and longitudinal (θ_L) incidence angle at the same location for all three technologies, determined using a Monte Carlo ray tracing calculation, where the longitudinal angle of the tower is defined as the angle with respect to the east–west axis. The PT design exhibits the highest optical performance except at large longitudinal incidence angles, where the central receiver takes larger values. The

Table 7.11 Optical properties of a linear Fresnel (LF), a parabolic trough (PT) and a central receiver (CR) system for comparison of optical performance

Component	Linear Fresnel	Parabolic Trough	Central Receiver
Primary reflector			
Slope error σ_{slope} (mrad)	3	3	1.53
Specularity σ_{spec} (mrad)	1.5	1.5	1.5
Tracking σ_{trac} (mrad)	1	1	1
Optical error σ_{opt}(mrad)	6.5	6.3	3.95
Reflectivity ρ(/)	0.94	0.94	0.94
Cleanliness η_{cl}(/)	0.97	0.97	0.97
Concentration ratio C(/)	37	26	1089
Receiver			
Absorptance α(/)	0.96	0.96	0.94
Transmittance τ(/)	0.97	0.97	–
Cleanliness η_{cl}(/)	0.98	0.98	–
Bellow shading f_{bel}(/)	0.96	0.96	–
End loss f_{end}(/)	0.98	0.98	–
Secondary reflector			
Reflectivity ρ(/)	0.94	–	–
Slope error σ_{slope} (mrad)	2	–	–
Specularity σ_{spec} (mrad)	1.5	–	–

parabolic trough technology generally has a high optical efficiency, which is insensitive to changes in transverse incidence angle until shadowing effects from the adjacent rows become significant. The Fresnel collector has a similar high optical efficiency as the central receiver at low angles of incidence, but the CR technology shows higher performance at higher angles of incidence.

Figure 7.81 depict the stacked optical losses due to various error sources as a function of the transverse and longitudinal incidence angles of the three technologies. To get a picture of the general characteristic differences three technologies, we divide the optical losses into three categories. The cosine losses of the collector denote the effect in which the aperture is not perpendicular to the incident solar rays, while the intercept loss without cosine losses subsumes shadowing, blocking and spillage losses. The optical losses due to the component properties are represented by the reflectivity, the absorptivity of the receiver, or the transmittance of the glass envelope shielding the absorber (if present).

For uniaxial, i.e. line-focused collectors that track in the transverse plane, the cosine loss is less than for point-focused CR. Conversely, in the longitudinal direction, the cosine loss is greater for both line-focused collectors than for the CR. As the transverse incidence angle increases, the shadowing and blocking losses grow rapidly for the LF collector. In contrast, the cosine effect is responsible for most

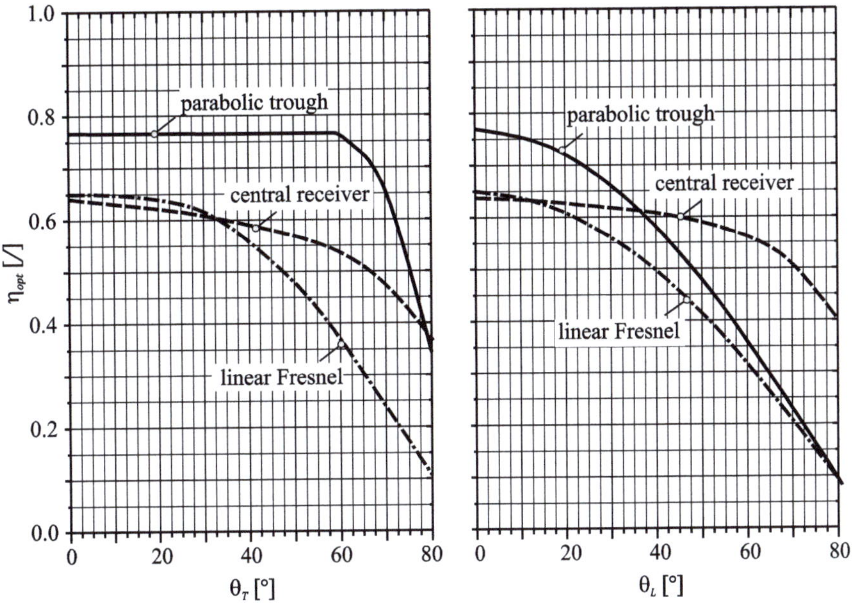

Fig. 7.80 Optical efficiency (η_{opt}) as a function of transverse (θ_T, left) and longitudinal (θ_L, right) incidence angles for parabolic trough (PT), central receiver (CR), and linear Fresnel (LF) collectors

of the optical losses in the longitudinal direction for both the LF and PT collectors since the reflectors cannot track in this direction. If the row spacings in the PT field are sufficiently large, a shadowing effect only occurs at transverse incidence angles of 70° and above, but then grows rapidly. The CR collector behaves fundamentally different due to its biaxial tracking and shows in both directions a comparable dependence of the optical loss as a function of transverse and longitudinal incidence angle (which is also due to the circular heliostat field in this example). Even at normal incidence the cosine loss is significant at the CR, because the inclination angle of a heliostat increases with the distance from the tower. Since a heliostat field increases in size for a given tower height, its cosine loss also increases even at normal incidence. The same holds for line-focusing systems as the aperture size increases; however, the contribution to optical losses is small compared to CR collectors due to the difference in scale.

Annual Efficiency
Annual optical efficiency ($\eta_{opt,a}$) is defined as the average efficiency weighted by DNI at each hour of a year:

$$\eta_{opt,a} = \frac{\sum_{t=0}^{8760h} DNI \cdot \eta_{opt,0} \cdot IAM(\theta_T, \theta_L)}{\sum_{t=0}^{8760h} DNI} \tag{7.52}$$

in which $\eta_{opt,0}$ denotes the the optical efficiency at zero degree incidence.

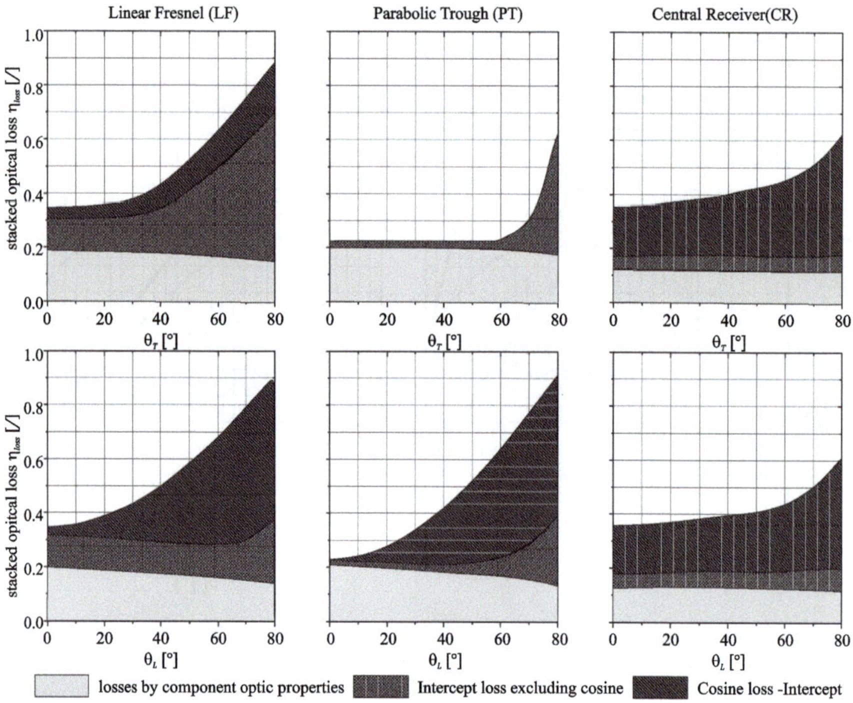

Fig. 7.81 Comparison of calculated stacked optical losses as a function of transverse (θ_T, top row) and longitudinal incidence angle (θ_L,bottom) for linear Fresnel, parabolic trough, and central revceiver at the same site, modified from Kincaid et al. (2018)

The three collector systems behave completely differently in summer and winter as shown in Fig. 7.82. In summer, the PT collector shows the highest and most balanced efficiency throughout the day. The LF system shows a comparable performance to the central receiver but drops opposite it in the morning and evening. In winter, on the other hand, the central receiver is superior to linear technologies, while in particular the linear Fresnel system suffers from a relatively low efficiency at high angles of incidence. The temporarily higher optical efficiency of the PT system compared to the central receiver in the morning and afternoon results from its lower sensitivity to transverse angles of incidence. During the day, the transverse angle of incidence approaches zero at midday, but the longitudinal angle of incidence is higher in the sky than in the morning hours due to the low sunlight. This leads to a lower optical efficiency for the PT receiver at noon.

The power generated on three consecutive days by the different collector systems per square meter of mirror area in summer and winter are shown in Fig. 7.83.

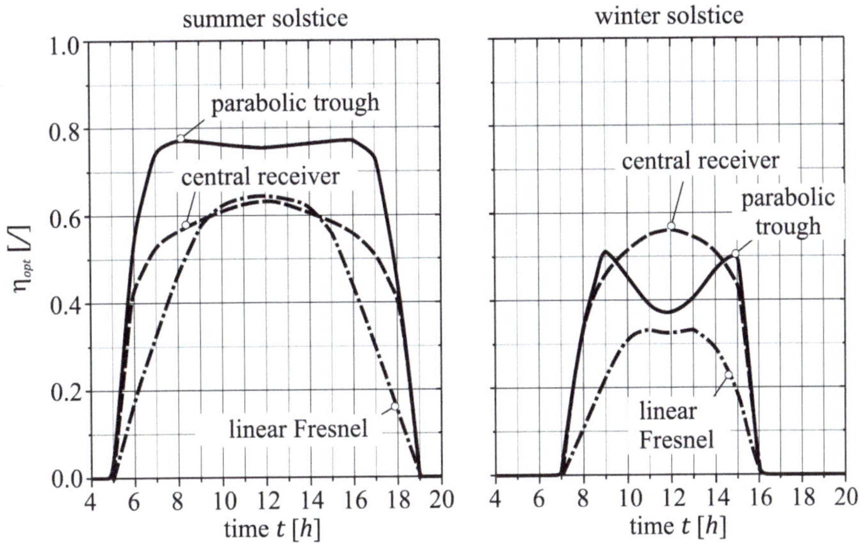

Fig. 7.82 Simulated hourly optical efficiency η_{opt} in summer (**a**) and winter (**b**) by linear Fresnel, parabolic trough, and central receiver technology, site: China Lake, California $\Phi = 35°39$ °N

The PT collector shows the highest power extraction in the summer week while the central tower receiver is superior in the winter.

The calculated monthly average optical efficiency of the respective technologies normalized by the hourly DNI is shown in Fig. 7.84a, assuming a minimum operating stow angle of 10°. The parabolic trough achieves the highest efficiency during most of the year, while the solar tower shows the most balanced values as an annual average.

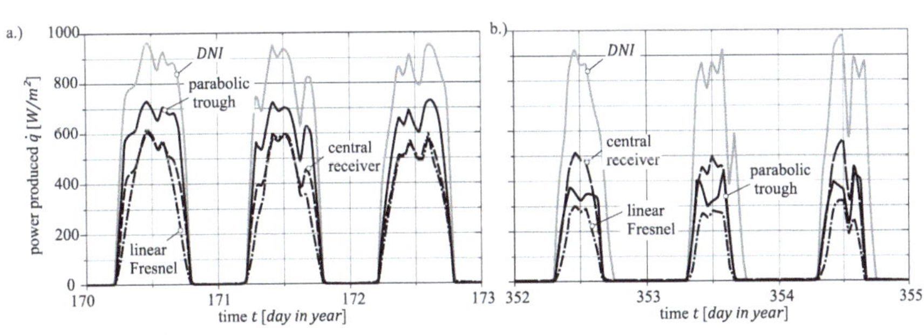

Fig. 7.83 Computed specific produced power (per m^2 mirror aperture) in summer (**a**) and winter (**b**) by linear Fresnel, parabolic trough, and central revceiver technology during three days, site: China Lake, California $\Phi = 35°39$ °N

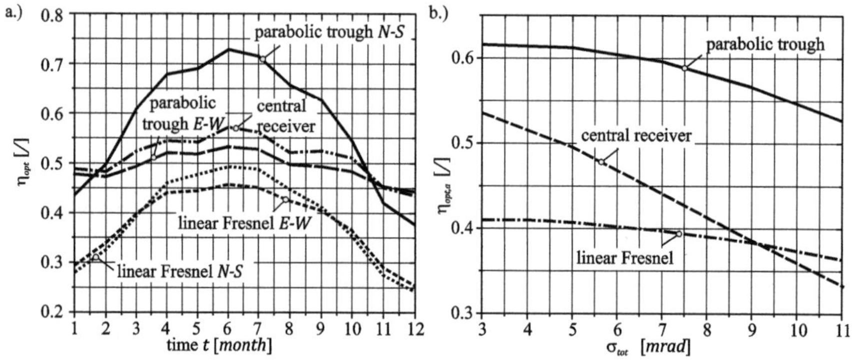

Fig. 7.84 Sensitivity of optical efficiency (η_{opt}) as a function of collector orientation over the annual cycle (**a**) and as a function of total optical error (σ_{tot}), **b** of parabolic trough, linear Fresnel, and solar tower; site: China Lake, California $\Phi = 35°39$ °N

Line-focusing collectors show different characteristics depending on their orientation north–south (N–S) or east–west (E–W) as illustrated in Fig. 7.84a. For the parabolic trough, the east–west orientation leads to a decrease in annual efficiency from 60 to 49%, but a greater uniformity in the annual mean, which may be desirable in the application. The sensitivity of the linear Fresnel system with respect to the orientation is rather low, while for central receivers, provided they have a circumferential field, the orientation does not play a role.

It has to be emphasized that the efficiency consideration is not sufficient to determine optimum installation cases. A lower efficiency might be compensated by a larger aperture area, if the area-specific costs are smaller than for the competitive solution. A conclusion whether a CRS, A PTC or LFC system is best for a certain application cannot be drawn from the consideration of the annual and diurnal optical performance.

Sensitivity of Optical Performance with Respect to Total Optical Collector Error
The optical performance of a CSP system is significantly affected by the total optical error (σ_{tot}) of the beam path, so small values imply high technical cost both in term of investment but also maintenance. Figure 7.84b illustrates the annual optical efficiency ($\eta_{opt,a}$) as a function of the total collector optical error (σ_{tot}). All technologies show a monotonic decrease with increasing total optical error, with the sensitivity being small for linear Fresnel systems and largest for the central tower, so that for $\sigma_{tot} > 9$ mrad the optical efficiency of the tower falls below that of the linear Fresnel collector. The parabolic trough is more sensitive to optical efficiency than a linear Fresnel collector, but even at large optical errors it still shows efficiencies of the order of an optimal central receiver system.

References

Abbas, A., Montes, M., Piera, M., & Martínez-Val, J. (2012). Solar radiation concentration features in Linear Fresnel Reflector arrays. *Energy Conversion and Management, 54*, 133–144. https://doi.org/10.1016/j.enconman.2011.10.010

Abbas, R., Montes, M., Rovira, A., & Martinez-Val, J. (2016). Parabolic trough collector or linear Fresnel collector? A comparison of optical features including thermal quality based on commercial solutions. *Solar Energy, 124*, 198–215. https://doi.org/10.1016/j.solener.2015.11.039

Abbas, R., Sebastián, A., Montes, M., & Valdés, M. (2018). Optical features of linear Fresnel collectors with different secondary reflector technologies. *Applied Energy, 232*, 386–397. https://doi.org/10.1016/j.apenergy.2018.09.224

ALANOD GmbH & Co. KG. (2021). *Efficiency-solar-surface.* Product catalogue, Egerstr. 12· 58256 Ennepetal. Germany. Retrieved December 10, 2022, from www.alanod-solar.com

ASTM G173-03. (2020). *Standard tables for reference solar spectral irradiances: direct normal and hemispherical on 37° tilted surface*, West Conshohocken, PA, 19428-2959. ASTM Headquarters. https://doi.org/10.1520/G0173-03R20

Avila-Marin, A. L. (2011). Volumetric receivers in solar thermal power plants with central receiver system technology: A review. *Solar Energy, 85*, 891–910. https://doi.org/10.1016/j.solener.2011.02.002

Balaji, S., Reddy, K., & Sundararajan, T. (2016). Optical modelling and performance analysis of a solar LFR receiver system with parabolic and involute secondary reflectors. *Applied Energy, 179*, 1138–1151. https://doi.org/10.1016/j.apenergy.2016.07.082

Bendt, P., Rabl, A., Gaul, H. W., & Reed, K. A. (1979). Optical analysis and optimization of line focus collectors. In: *NASA STI/Recon Technical Report N 80.*

Bellos, E. (2019). Progress in the design and the applications of linear Fresnel reflectors—A critical review. *Thermal Science and Engineering Progress, 10*, 112–137. https://doi.org/10.1016/j.tsep.2019.01.014

Bernhard, R., de LaLaing, J., Kistner, R., Eck, M., Eickhoff, M., Feldhoff, J., … Morin, G. (2009). Linear fresnel collector demonstration at the PSA—Operation and investigation. In *Proceedings of 15th SolarPACES Conference*, Berlin, 15–18 September 2009.

Blackmon, J. B. (2020). Heliostat size optimization for central receivers. In K. Lovegrove, & W. Stein (Eds.), *Concentrating solar power technology* (2nd ed., pp. 585–632). Elsevier Woodhead Publishing.

Boerema, N., Morrison, G., Taylor, R., & Rosengarten, G. (2012). Liquid sodium versus Hitec as a heat transfer fluid in solar thermal central receiver systems. *Solar Energy, 86*, 2293–2305. https://doi.org/10.1016/j.solener.2012.05.001

Brahma Kumaris. (2011). *India One.* Retrieved from https://solar.brahmakumaris.com/links/

BraytonEnergy LLC. (2022). SolarCAT Solar Compressed Air Turbine. 75B Lafayette Road, Hampton, NH 03842. Retrieved December 21, 2022, from https://www.braytonenergy.net/our-projects/solarcat/

BrightSource Energy Inc. (2023). 1999 Harrison Street Suite 2150, Oakland, CA 94612, California, USA. Retrieved February 27, 2023, from https://www.brightsourceenergy.com/breakthrough-technologies

Brossard, D. (2010, May 16). *Heliostat—Maurya Sheraton Hotel—New Delhi* (Flickr.com, Ed.) Retrieved May 3, 2023, from https://www.flickr.com/photos/string_bass_dave/4609871733/in/photostream/

Burkholder, F., & Kutscher, C. F. (2008). Heat-loss testing of Solel's UVAC3 parabolic trough receiver: Technical report. In *Heat-loss testing of Solel's UVAC3 parabolic trough receiver: Technical Report*. National Renewable Energy Laboratory.

Burkholder, F., & Kutscher, C. (2009). *Heat loss testing of Schott's 2008 PTR70 Parabolic Trough Receiver.* National Renewable Energy Lab. (NREL), Golden, CO, USA. https://doi.org/10.2172/1369635

Cagnoli, M., Mazzei, D., Procopio, M., Russo, V., Savoldi, L., & Zanino, R. (2018). Analysis of the performance of linear Fresnel collectors: Encapsulated vs. evacuated tubes. *Solar Energy, 164*, 119–138. https://doi.org/10.1016/j.solener.2018.02.037

Canavarro, D., Chaves, J., & Collares-Pereira, M. (2014). Simultaneous multiple surface method for Linear Fresnel concentrators with tubular receiver. *Solar Energy, 110*, 105–116. https://doi.org/10.1016/j.solener.2014.09.002

Cao, F., McEnaney, K., Chen, G., & Ren, Z. (2014). A review of cermet-based spectrally selective solar absorbers. *Energy & Environmental Science, 7*(5), 1615–1627.

Cardemil, J., Starke, A., Scariot, V., Grams, I., & Colle, S. (2014). Evaluating solar radiation attenuation models to assess the effects of climate and geographical location on the heliostat field efficiency in Brazil. *Energy Procedia, 49*, 1288–1297. https://doi.org/10.1016/j.egypro.2014.03.138

Chaves, J., & Collares-Pereira, M. (2010). Etendue-matched two-stage concentrators with multiple receivers. *Solar Energy, 84*(2), 196–207. https://doi.org/10.1016/j.solener.2009.10.022

Conlon, W. M., Johnson, P., & Hanson, R. (2011, July 12–14). Superheated steam from CLFR solar steam generators. In *Proceedings of ASME 2011 Power Conference* (Vol. 1, pp. 301–307). https://doi.org/10.1115/POWER2011-55174

CosinSolar. (2023). SUPCON SOLAR Delingha 10MW Tower CSP Plant. No. 307, Liuhe Road, Binjianga, District, Hangzhou, Zhejiang, China. Retrieved Febraury 27, 2023, from http://www.supconsolar.com/en/index/index.html

Coventry, J., Andraka, C., Pye, J., Blanco, M., & Fisher, J. (2015). A review of sodium receiver technologies for central receiver solar power plants. *Solar Energy, 122*, 749–762. https://doi.org/10.1016/j.solener.2015.09.023

Coventry, J., Campbell, J., Xue, Y., Hall, C., Kim, J.-S., Pye, J., … Beath, A. (2016). *Heliostat cost down scoping study*. Final Report, ASTRI—Australian Solar Thermal Research Initiative.

Davis, J. R. (2000). *Nickel, Cobalt, and their alloys-ASM specialty handbook* (Vol. 1), Materials Park, OH 44073, USA. ASM International. Retrieved from www.asminternational.org

del Río, A., Korzynietz, R., Brioso, J., Gallas, M., Ordóñez, I., Queroa, M., & Diaz, C. (2015). Soltrec–Pressurized volumetric solar air receiver technology. *Energy Procedia, 69*, 360–368. https://doi.org/10.1016/j.egypro.2015.03.042

Deutsches Zentrum für Luft- und Raumfahrt (DLR). (2008). Solarstrahlungsempfänger (Solarreceiver) auf dem Turmkraftwerk in Almería. Institut für Technische Thermodynamik, Solarforschung.

Di Canio, D., Tretyl, W., Jur, F., & Watson, C. (1977–1979). *Line-focus solar thermal central receiver research study*. Report DOE/ET/20426-1, FMC Corporation, Santa Clara, CA.

Díez de los Ríos Ramos, N. (2018). *Technology developments for the Alkali Metal Thermal to Electric Converter (AMTEC) Test Facility*. Dissertation, Faculty of Mechanical Engineering, Karlsruher Institut für Technologie (KIT), Karlsruhe Germany.

Dunham, M., & Iverson, B. (2014). High-efficiency thermodynamic power cycles for concentrated solar power systems. *Renewable and Sustainable Energy Reviews, 30*, 758–770. https://doi.org/10.1016/j.rser.2013.11.010

Eddhibi, F., Amara, M., Balghouthi, M., Qoaider, L., & Guizani, A. (2017). Analytic optical design of a Linear Fresnel solar collector with variable parameters. *Journal of Materials and Environmental Sciences, 11*(8), 4068–4084.

Ermanoski, I., Siegel, N., & Stechel, E. B. (2013). A new reactor concept for efficient solar-thermochemical fuel production. *Journal of Solar Energy Engineering, 135*(3), 031002 (10 p.). https://doi.org/10.1115/1.4023356

Flesch, J. (2021). *LBE-cooled tube receiver performance—Design aspects and high-flux operation in a solar furnace*. Dissertation, Karlsruhe Institute of Tchnology (KIT), Faculty Chemical Egnigneering and Process Technology, Karlsruhe.

Forsberg, C., Peterson, P. F., & Zhao, H. (2007). High-temperature liquid-fluoride-salt closed-brayton-cycle solar power towers. *Journal of Solar Energy Engineering, 129*(2), 141–146. https://doi.org/10.1115/1.2710245

García-Cortés, S., Bello-García, A., & Ordóñez, C. (2012). Estimating intercept factor of a parabolic solar trough collector with new supporting structure using off-the-shelf photogrammetric equipment. *Applied Energy, 92*, 815–821. https://doi.org/10.1016/j.apenergy.2011.08.032

Gholamalizadeh, E., & Chung, J. D. (2017). Exergy analysis of a pilot parabolic solar dish-stirling system. *Entropy, 19*(10), 509 (12 p.). https://doi.org/10.3390/e19100509

Guerreiro, L., Canavarro, D., & Collares Pereira, M. (2015). Efficiency improvement and potential LCOE reduction with an LFR-XX SMS plant with storage. *Energy Procedia, 69*, 868–878. https://doi.org/10.1016/j.egypro.2015.03.112

Hafez, A., Soliman, A., El-Metwally, K., & Ismail, I. (2017). Design analysis factors and specifications of solar dish technologies for different systems and applications. *Renewable and Sustainable Energy Reviews, 67*, 1019–1036. https://doi.org/10.1016/j.rser.2016.09.077

Heimsath, A., & Nitz, P. (2019). The effect of soiling on the reflectance of solar reflector materials—Model for prediction of incidence angle dependent reflectance and attenuation due to dust deposition. *Solar Energy Materials and Solar Cells, 195*, 258–268. https://doi.org/10.1016/j.solmat.2019.03.015

Heimsath, A., Bern, G., van Rooyen, D., & Nitz, P. (2014a). Quantifying optical loss factors of small linear concentrating collectors for process heat application. *Energy Procedia, 48*, 77–86. https://doi.org/10.1016/j.egypro.2014.02.010

Heimsath, A., Cuevas, F., Hofer, A., Nitz, P., & Platzer, W. J. (2014b). Linear Fresnel collector receiver. *Energy Procedia, 49*, 386–397.

Heimsath, A., Cuevas, F., Hofer, A., Nitz, P., & Platzer, W. J. (2014c). Linear Fresnel collector receiver: Heat loss and temperatures. *Energy Procedia, 49*, 386–397. https://doi.org/10.1016/j.egypro.2014.03.042

Hering, W., Stieglitz, R., & Wetzel, T. (2012). Application of liquid metals for solar energy systems. *EPJ Web of Conferences, 33*, 7. https://doi.org/10.1051/epjconf/20123303003

Hildebrandt, C. (2009). *Hochtemperaturstabile Absorberschichten für linear konzentrierende solarthermische Kraftwerke.* Dissertation, Universität Stuttgart, Institut für Thermodynamik und Wärmetechnik, Stuttgart.

Ho, C. K. (2016). A review of high-temperature particle receivers for concentrating solar power. *Applied Thermal Engineering, 109*, 958–969. https://doi.org/10.1016/j.applthermaleng.2016.04.103

Ho, C. K. (2017). Advances in central receivers for concentrating solar applications. *Solar Energy, 152*, 38–56. https://doi.org/10.1016/j.solener.2017.03.048

Ho, C., & Iverson, B. D. (2014). Review of high-temperature central receiver designs for concentrating solar power. *Renewable and Sustainable Energy Reviews, 29*, 835–846. https://doi.org/10.1016/j.rser.2013.08.099

IEC 62862-5-2. (2022). *Solar thermal electric plants Part 5-2. Systems and components. General requirements and test methods for large-size linear Fresnel collectors.*

Industrial Solar GmbH. (2022). Fresnel Collector LF-11 Datasheet. Basler Str. 115, D-79115 Freiburg, Germany. RetrievedMarch 29, 2022, from https://www.industrial-solar.de/en/technologies/fresnel-collector/

IRENA -International Renewable Energy Agency. (2013). *Renewable Power Generation Costs in 2012: An overview.* Abu Dhabi, C67 Office Building, Khalidiyah (32nd) Street, United Arab Emirates. Retrieved from www.irena.org

ISO 9845-1. (1992). *Solar energy—Reference solar spectral irradiance at the ground at different receiving conditions—Part 1: Direct normal and hemispherical solar irradiance for air mass 1,5.* ICS:27.160 Solar energy engineering.

Janotte, N., Feckler, G., Kötter, J., Decker, S., Herrmann, U., Schmitz, M., & Lüpfert, E. (2014). Dynamic performance evaluation of the HelioTrough® collector demonstration loop-towards a new benchmark in parabolic trough qualification. *Energy Procedia, 49*, 109–117. https://doi.org/10.1016/j.egypro.2014.03.012

Kalbhor, A. (2014). Modified solar central receiver in concentrated solar power systems. *International Journal of Engineering and Technical Research (IJETR), 12*(2).

Karni, J., Kribus, A., Doron, P., Rubin, R., Fiterman, A., & Sagie, D. (1997). The DIAPR: A high-pressure, high-temperature solar receiver. *Journal of Solar Energy Engineering—Transactions of the ASME, 119*, 74–78. https://doi.org/10.1115/1.2871853

Kesselring, P., & Selvage, C. (1986). *The IEA/SSPS solar thermal power plants—Facts and figures—Volume 1: Central Receiver System (CRS).* International Energy Agency (IEA).

Kincaid, N., Mungas, G., Kramer, N., Wagner, M., & Zhu, G. (2018). An optical performance comparison of three concentrating solar power collector designs in linear Fresnel, parabolic trough, and central receiver. *Applied Energy, 231*, 1109–1121. https://doi.org/10.1016/j.apenergy.2018.09.153

Korzynietz, R., Brioso, J., del Río, A., Quero, M., Gallas, M., Uhlig, R., … Teraji, D. (2016). Solugas—Comprehensive analysis of the solar hybrid Brayton plant. *Solar Energy, 135*, 578–589. https://doi.org/10.1016/j.solener.2016.06.020

Kurup, P., & Turchi, C. S. (2015). *Parabolic trough collector cost update for the system advisor model (SAM).* 15013 Denver West Parkway,Golden, CO 80401. National Renewable Energy Laboratory (NREL). Retrieved from https://www.nrel.gov/docs/fy16osti/65228.pdf

Lee, C.-Y., Chou, P.-C., Chiang, C.-M., & Lin, C.-F. (2009). Sun tracking systems: A review. *Sensors, 9*, 3875–3890. https://doi.org/10.3390/s90503875

Lüpfert, E., Zarza-Moya, E., Geyer, M., Nava, P., Langenk, J., Schiel, W., … Mandelberg, E. (2003). Eurotrough collector qualification complete—Performance test results from PSA. In *Proceedings of ISES Solar World Congress* (p. 12), June 14–19, Göteborg, Sweden

Mertins, M. (2006). *Technische und wirtschaftliche Analyse von horizontalen Fresnel-Kollektoren.* Dissertation, Universität Karlsruhe, Faculty for Mechanical Engineering, Karlsruhe.

Mills, D. R., & Morrison, G. (2000). Compact Linear Fresnel reflector solar thermal powerplants. *Solar Energy, 68*(3), 263–283. https://doi.org/10.1016/S0038-092X(99)00068-7

Moga, D., Sita, I. V., Stroia, N., Dobra, P., Moga, R., & Petreus, D. (2015). Sensing and control strategies in tracking solar systems. In *Proceedings of International Conference on Control Systems and Science* (pp. 989–995), 23–25 May 2015. https://doi.org/10.1109/CSCS.2015.143

Montes, M. J., Abbas, R., Muñoz, M., Muñoz-Antón, J., & Martínez-Val, J. (2017). Advances in the linear Fresnel single-tube receivers: Hybrid loops with non-evacuated and evacuated receivers. *Energy Conversion and Management, 149*, 318–333. https://doi.org/10.1016/j.enconman.2017.07.031

Morin, G. (2005). *Weiterführende Forschung und Entwicklung zum Fresnel-Konzept sowie Vorbereitung eines Demonstrationsvorhabens.* Report MAO1-GM-0512-E02, Fraunhofer ISE, Freiburg.

Morin, G., Mertins, M., Kirchberger, J., & Selig, M. (2011). SuperNOVA–construction, control & performance of steam superheating linear Fresnel collector. In *Proceedings of 17th SolarPACES International Symposium.* Granada, Spain.

Morin, G., Dersch, J., Platzer, W., Eck, W., & Häberle, A. (2012). Comparison of Linear Fresnel and parabolic trough collector plants. *Solar Energy, 86*, 1–12. https://doi.org/10.1016/j.solener.2011.06.020

Nagarajan, S., & Barshilia, H. (2010). *Review of sputter deposited mid- to high- temperature solar selective coatings for flat plate/evacuated tube collectors and solar thermal power generation applications.* Bangalore 560017, India. National Space Laboratories.

National Renewable Energy Laboratory-NREL. (2017, August 4). Retrieved May 3, 2023, from Mirror Field Crescent Dunes Solar Thermal Facility. https://www.flickr.com/photos/nrel/35965146980/in/photostream/

Novatec Solar. (2012). Turnkey Solar Boiler. Karlsruhe, Germany. Retrieved December 12, 2021, from https://www.energy-xprt.com/products/solar-boiler-251465

Ortona, A., Yoon, D., Fend, T., & Smirnova, O. (2015). Tubular Si-infiltrated SiCf/SiC composites for solar receiver application—Part 2: Thermal performance analysis and prediction. *Solar Energy Materials & Solar Cells, 140*, 382–387. https://doi.org/10.1016/j.solmat.2015.04.034

Pfahl, A., Coventry, J., Röger, M., Wolfertstetter, F., Vásquez-Arango, J.-F., Gross, F., … Liedke, P. (2017). Progress in heliostat development. *Solar Energy, 152*, 3–37. https://doi.org/10.1016/j.solener.2017.03.029

Pitz-Paal, R., Morhenne, J., & Fiebig, M. (1991). A new concept of a selective solar receiver for high temperature applications. *Solar Energy Materials, 24*, 293–306. https://doi.org/10.1016/0165-1633(91)90070-2

Platzer, W., Heimsath, A., Hildebrandt, C., & Georg, A. (2008). Quality control of concentrating collector components for the optimization of performance. In *Proceedings of 14th SolarPACES International Symposium*, March 4–7, 2008, Las Vegas, USA.

Pottler, K., Ulmera, S., Lüpfert, E., Landmann, M., Röger, M., & Prahl, C. (2014). Ensuring performance by geometric quality control and specifications for parabolic trough solar fields. *Energy Procedia, 49*, 2170–2179. https://doi.org/10.1016/j.egypro.2014.03.230

Poživil, P., Agab, V., Zagorskiy, A., & Steinfeld, A. (2014). A pressurized air receiver for solar-driven gas turbines. *Energy Procedia, 49*, 498–503. https://doi.org/10.1016/j.egypro.2014.03.053

Pye, J., Morrison, G., & Behnia, M. (2003). Modelling of cavity receiver heat transfer for the Compact Linear Fresnel reflector. In *Proceedings of ISES Solar World Congress-Solar Energy for a Sustainable Future* (p. 10), 26–29 November 2003. Gothenburg, Sweden.

Qiu, Y., He, Y.-L., Cheng, Z., & Wang, K. (2015). Study on optical and thermal performance of a linear Fresnel solar reflector using molten salt as HTF with MCRT and FVM methods. *Applied Energy, 146*, 162–173. https://doi.org/10.1016/j.apenergy.2015.01.135

Rabl, A. (1985). *Active solar collectors and their applications* (1st ed.). Oxford University Press.

Rabl, A., & Winston, R. (1976). Ideal concentrators for finite sources and restricted exit angles. *Applied Optics, 15*(11), 2880–2883. https://doi.org/10.1364/AO.15.002880

Reiser, J., Franke, P., Weingärtner, T., Hoffmann, J., Hoffmann, A., & Rieth, M. (2015). Tungsten laminates made of ultrafine-grained (UFG) tungsten foil—Ageing of tungsten–titanium (W–Ti) laminates. *International Journal of Refractory Metals and Hard Materials, 51*, 264–274. https://doi.org/10.1016/j.ijrmhm.2015.04.032

Ricklin, P., Slack, M., Rogers, D., & Huibregtse, R. (2014). Commercial readiness of eSolar next generation heliostat. *Energy Procedia, 49*, 201–208. https://doi.org/10.1016/j.egypro.2014.03.022

Riffelmann, K., Richert, T., Nava, P., & Schweitzer, A. (2014). Ultimate Trough®—A significant step towards cost-competitive CSP. *Energy Procedia, 49*, 1831–1839. https://doi.org/10.1016/j.egypro.2014.03.194

Schiel, W., & Keck., T. (2012). Parabolic dish concetrating solar power systems. In K. Lovegrove, & W. Stein (Eds.), *Concentrating solar power technology* (pp. 284–321). 80 High Street, Sawston, Cambridge CB22 3HJ, UK. Woodhead Publishing.

Schiel, W., Hunt, B., Keck, T., Schweitzer, A., Weinrebe, G., & Partner, S. B. (2004, May 25–27). Solar thermal power plants for central or distributed electricity generation. In *Proceedings of Power-Gen Europe* (p. 22).

SCHOTT Solar CSP GmbH. (2013). Advanced qualification methods for parabolic trough receivers. Riyadh, November 20th 2013, Kingdom Saudi Arabia. Retrieved December 20, 2022, from https://www.dlr.de/sf/Portaldata/73/Resources/dokumente/veranstaltungen/saudisch-d_ws_riad_11-2013/SCHOTT_Solar_Saudi-German_CSP-WS_Riad_20112013.pdf

SCHOTT Technical Glass Solutions GmbH. (2018). *BOROFLOAT® 33 & Functional Coatings-Product information.* Technical Information sheet, Otto-Schott-Strasse 13,07745 Jena,Germany. Retrieved December 18, 2022, from www.schott.com/borofloat/coatings

Schöttll, P., Rohani, S., Leonardi, E., Pisani, L., Les, I., Mutuberria, M., & Nitz, P. (2019). Solar field heliostat selection based on polygon optimization and boundaries. *AIP Conference Proceedings, 2126*, 8. https://doi.org/10.1063/1.5117565

Sharma, V., Nayak, J. K., & Kedare, S. (2015). Comparison of line focusing solar concentrator fields considering shading and blocking. *Solar Energy, 122*, 924–939. https://doi.org/10.1016/j.solener.2015.10.011

Siemens Concentrated Solar Power, Ltd. (2011). Siemens UVAC 2010. http://www.energy.siemens.com/hq/pool/hq/power-generation/renewables/solar-power-solutions/concentrated-solar-power/downloads/E50001-W320-A104-V2-4A00_DA_UVACreceiver_US_ohne%20Israel.pdf. Siemens AG, Energy Sector, Freyeslebenstrasse 1, 91058 Erlangen, Germany.

Solare Brücke. (2022, September 20). Retrieved from Scheffler reflector: http://www.solare-brucke.org/index.php/en/die-scheffler-reflektoren

Stein, H., & Buck, R. (2017). Advanced power cycles for concentrated solar power. *Solar Energy, 152*, 91–105. https://doi.org/10.1016/j.solener.2017.04.054

Stine, W., & Diver, R. (1994). *A compendium of solar dish/stirling technology.* Sandia National Laboratories.

Tagle-Salazar, P., & Nigam, K. R.-S. (2020). Parabolic trough solar collectors: A general overview of technology, industrial applications,energy market, modeling, and standards (D. Gruyter, Ed.) *Green Processing and Synthesis, 9*, 595–649. https://doi.org/10.1515/gps-2020-0059

Tanaka, K. (2010). Concept design of solar thermal receiver using alkali metal thermal to electric converter (AMTEC). *Current Applied Physics, 10*, S254–S256. https://doi.org/10.1016/j.cap.2009.07.031

Téllez, F., Villasante, C., & Burisch, M. (2014). *State of the art in heliostats and definition of specifications-survey for a low cost heliostat development.* Deliverable 12.1, EU Project STAGE STE.

Tsekouras, P., Tzivanidis, C., & Antonopoulos, K. (2018). Optical and thermal investigation of a linear Fresnel collector with trapezoidal cavity receiver. *Applied Thermal Engineering, 135*, 379–388. https://doi.org/10.1016/j.applthermaleng.2018.02.082

United Sun Systems International Ltd. (2012, December 31). Maricopa dish-stirling plant. Retrieved March 29, 2023, from https://commons.wikimedia.org/wiki/File:Maricopa_Dish-Stirling_plant_01.jpg

van Rooyen, D. W., Heimsath, A., Nitz, P., & Platzer, W. (2011). An optical model of the fix-focus scheffler reflector. In *Proceedings of 30th ISES Solar World Congress* (pp. 4527–4537), Kassel, 28 August–2 September 2011. https://doi.org/10.18086/swc.2011.28.30

Viebahn, P., Kronshage, S., Trieb, F., & Lechon, Y. (2008). *Final report on technical data, costs, and life cycle inventories of solar thermal power plants.* Deliverable 12.2 Project 502687 NEEDS EU 6th Framework Programme.

Vignarooban, K., Xu, X., Arvay, A., Hsu, K., & Kannan, A. (2015). Heat transfer fluids for concentrating solar power systems—A review. *Applied Energy, 146*, 383–396. https://doi.org/10.1016/j.apenergy.2015.01.125

Viswanathan, R. (1993). *Damage mechanisms and life assessment of high-temperature components* (2nd ed.). Metals Park, Ohio 44073. ASM International.

Wang, K., Wu, H., Wang, D., Wang, Y., Tong, Z., Lin, F., & Olabi, A. (2015). Experimental study on a coiled tube solar receiver under variable solar radiation condition. *Solar Energy, 122*, 1080–1090. https://doi.org/10.1016/j.solener.2015.10.004

Wua, S.-Y., Xiao, L., Cao, Y., & Li, Y. (2010). A parabolic dish/AMTEC solar thermal power system and its performance evaluation. *Applied Energy, 87*, 452–462. https://doi.org/10.1016/j.apenergy.2009.08.041

Yang, M., Zhu, Y., & Taylor, R. (2018). End losses minimization of linear Fresnel reflectors with a simple, two-axis mechanical tracking system. *Energy Conversion and Management, 161*, 284–293. https://doi.org/10.1016/j.enconman.2018.01.082

Yellowhair, J., & Ho, C. (2010). Heliostat canting and focusing methods: An overview and comparison. *Energy Sustainability*, 609–615. https://doi.org/10.1115/ES2010-90356

You, J. H., Visca, E., Barett, T., Böswirth, B., Creszensi, F., Dompatil, F., … Vorpahl, C. (2018). European divertor target concepts for DEMO: Design rationales and high heat flux performance. *Nuclear Materials and Energy, 16*, 1–11. https://doi.org/10.1016/j.nme.2018.05.012

Zarza Moya, E. (2020). Parabolic-trough concentrating solar power systems. In K. S. Lovegrove (Ed.), *Concentrating solar power technology* (2nd ed., p. 832). Woodhead Publishing. https://doi.org/10.1016/C2018-0-04978-6

Zhu, G. (2017). New adaptive method to optimize the secondary reflector of linear Fresnel collectors. *Solar Energy, 144*(1), 117–126. https://doi.org/10.1016/j.solener.2017.01.005

Zhu, Y., Shi, J., Li, Y., L., W., Huang, Q., & Xu, G. (2017). Design and thermal performances of a scalable linear Fresnel reflector solar system. *Energy Conversion and Management, 146*, 174–181. https://doi.org/10.1016/j.enconman.2017.05.031

Zou, B., Yang, H., Yao, Y., & Jiang, J. (2017). A detailed study on the effects of sunshape and incident angle on the optical performance of parabolic trough solar collectors. *Applied Thermal Engineering, 126*, 81–91. https://doi.org/10.1016/j.applthermaleng.2017.07.149

MIX
Papier aus verantwortungsvollen Quellen
Paper from responsible sources
FSC® C105338

If you have any concerns about our products,
you can contact us on
ProductSafety@springernature.com

In case Publisher is established outside the EU,
the EU authorized representative is:
Springer Nature Customer Service Center GmbH
Europaplatz 3, 69115 Heidelberg, Germany

Printed by Libri Plureos GmbH
in Hamburg, Germany